7th International Conference on Concrete Repair (ICCR 2019)

Concrete Solutions

MATEC Web of Conferences Volume 289 (2019)

Cluj-Napoca, Romania
30 September - 2 October 2019

Editors:

Michael G. Grantham
Calin Mircea

ISBN: 978-1-5108-9257-6

Printed from e-media with permission by:

Curran Associates, Inc.
57 Morehouse Lane
Red Hook, NY 12571

Some format issues inherent in the e-media version may also appear in this print version.

For additional information, please contact EDP Sciences – Web of Conferences
at the address below.

EDP Sciences – Web of Conferences
17, Avenue du Hoggar
Parc d'Activité de Courtabœuf
BP 112
F-91944 Les Ulis Cedex A
France

Phone: +33 (0) 1 69 18 75 75
Fax: +33 (0) 1 69 28 84 91

contact-edps@webofconferences.org

Additional copies of this publication are available from:

Curran Associates, Inc.
57 Morehouse Lane
Red Hook, NY 12571 USA
Phone: 845-758-0400
Fax: 845-758-2633
Email: curran@proceedings.com
Web: www.proceedings.com

TABLE OF CONTENTS

OPTIMIZING APPROACH ON FIBRE ENGINEERED CEMENTITIOUS MATERIALS WITH SELF-HEALING CAPACITY (SH-FECM) BY THE USE OF SLURRY LIME (SL) ADDITION 1
Baeră Cornelia, Szilagyi Henriette, Matei Claudiu, Hegyi Andreea, Lăzărescu Adrian, Mircea Anamaria Cătălina

PRODUCTION OF CONCRETE COMPATIBLE BIOGRANULES FOR SELF-HEALING CONCRETE APPLICATIONS 9
Sonmez Merve, Ersan Yusuf Cagatay

DURABILITY OF SELF-HEALING CONCRETE 17
De Belie Nele, Van Belleghem Bjorn, Erþan Yusuf Caðatay, Van Tittelboom Kim

AN EXPERIMENTAL AND NUMERICAL STUDY ON VASCULAR SELF-HEALING CEMENTITIOUS MATERIALS 25
Jefferson Anthony, Selvarajoo Tharmesh, Freeman Brubeck, Davies Robert

EFFECT OF OXYGEN PERMEABILITY ON CORROSION PROPERTIES IN MORTAR SPECIMENS MIXED WITH BACILLUS SUBTILIS 29
Sugitani Saaya, Kawaai Keiyu, Ujike Isao

EFFECT OF CRACK PATTERN ON THE SELF-HEALING CAPABILITY IN TRADITIONAL, HPC AND UHPFRC CONCRETES MEASURED BY WATER AND CHLORIDE PERMEABILITY 35
Negrini Alberto, Roig-Flores Marta, Mezquida-Alcaraz Eduardo J., Ferrara Liberato, Serna Pedro

EFFECT OF MICRO-CLIMATE VARIATIONS ON CARBONATION RATE OF CONCRETE IN THE INLAND ENVIRONMENT 43
Aminu Alhassan Yunusa, Apeh Sunday

ABRASION RESISTANCE OF HIGH EARLY-STRENGTH CONCRETE FOR RAPID REPAIR 49
Ghafoori Nader, Maler Matthew O., Najimi Meysam, Hasnat Ariful

PROPERTIES OF HIGH EARLY-STRENGTH TYPE V CEMENT CONCRETE FOR RAPID REPAIR 53
Ghafoori Nader, Maler Matthew O., Najimi Meysam, Hasnat Ariful

PREDICTION OF THE SWELLING CAPACITY OF SALT-INSENSITIVE SUPERABSORBENT POLYMERS IN CONCRETE BY RHEOLOGICAL MEASUREMENTS 57
Mazanec Oliver, Assmann Alessander, Eissmann Diana

THE ROLE OF SHRINKAGE REDUCING ADMIXTURES AND SUPPLEMENTARY CEMENTITIOUS MATERIALS IN VOLUME STABILITY AND STRENGTH DEVELOPMENT 63
Papayianni Ioanna, Kesikidou Fotini, Alt Philip Henes

MIX-DESIGN AND CONTROL OF EXPOSURE CLASS OF DURABLE CONCRETE 69
Paulini Peter

EXPERIMENTAL STUDY OF THE INFLUENCE OF CRACK WIDTH DUE TO CORROSION BEHAVIOUR USING DIFFERENT MEASUREMENT METHODS AND DIFFERENT EXPOSURES 77
Schneider Martin, Gardener Georg

HOW CONCRETE REMOVAL TECHNIQUES AFFECT THE BONDING BETWEEN CONCRETE AND REPAIR MORTAR 84
Yazdi Mohammad Ali, Dejager Elien, Debraekeleer Mats, Gruyaert Elke, Van Tittelboom Kim, De Belie Nele

THE EFFECTS OF ANODE DISTANCE AND CORROSION ACTIVITY ON CURRENT DISTRIBUTION FOR ICCP SYSTEMS 89
Bhuiyan Shamir, Law David, Ward Liam, Saliba Justin

IMPRESSED CURRENT CATHODIC PROTECTION WITH EMBEDDED ANODES – FIELD TRIAL ON THE VIADUC DE SAINT-CLOUD IN FRANCE 93
Bouteiller Véronique, Damien Guillaume, Boinski Frédéric, Schneider Julien, Michaux Christophe

NEW DEVELOPMENTS FOR CORROSION PROTECTION OF CONCRETE STRUCTURES IN AUSTRALIA 100
Cheaitani Atef, Cheytani Samir

ZINC LAYER ANODE APPLIED ON REINFORCED CONCRETE STRUCTURES 107
Hallopeau Xavier, Moucadeau Eugène, Giorgini Roberto, Lesieutre Olivier, Meillier Arnaud, Annede-Villeau Camille

RESIDUAL PROTECTION ON A WHARF STRUCTURE FOLLOWING APPLICATION OF IMPRESSED CURRENT CATHODIC PROTECTION 113
Bhuiyan Shamir, Law David, Nicholls Peter

CATHODIC PROTECTION OF REINFORCEMENT IN CONCRETE – EXPERIENCE AND DEVELOPMENT OVER 30 YEARS ... 120
Polder Rob, Peelen Willy

APPLICATION OF CFRP AS ANODE IN CATHODIC PROTECTION FOR STEEL REINFORCED CONCRETE – A REVIEW ... 126
Zhang Emma Qingnan, Tang Luping

INSIGHTS FROM CORROSION CURRENT MEASUREMENTS ON CORROSION MECHANISMS IN REINFORCED CONCRETE AND ON THE EVALUATION OF OTHER CORROSION DATA ... 130
Schneck Ulrich

INSTANT PERFORMANCE VERIFICATION AFTER ELECTROCHEMICAL CHLORIDE EXTRACTION BY ENHANCED CORROSION TESTING ... 138
Schneck Ulrich

GALVANIC CHLORIDE EXTRACTION BY AN EMBEDDED ZINC ANODE: ION DISTRIBUTION MAPPED BY LASER INDUCED BREAKDOWN SPECTROSCOPY (LIBS) ... 148
Schwarz W., Wilsch Gerd, Pichlhöfer A., Ebell G., Völker T.

A FRESH LOOK AT DEPOLARISATION CRITERIA FOR CATHODIC PROTECTION OF STEEL REINFORCEMENT IN CONCRETE ... 153
Rathod Nikita, Slater Peter, Sergi George, Seveviratne Gamini, Simpson David

MONITORING AND TESTING CORROSION AND RESPONSIVE BEHAVIOUR IN GALVANIC SYSTEMS. ... 161
Stone Christian, Glass Gareth, Roberts Adrian, Davison Nigel

CASE STUDIES OF HYBRID AND GALVANIC SYSTEMS ON BRIDGE STRUCTURES ... 168
Stone Christian, Donadio Michel, Christodoulou Christian, Davison Nigel

RESULTS ON THE LONG-TERM PERFORMANCE AND EFFECTIVENESS OF A GALVANIC CP SYSTEM APPLIED ON A LIGHTWEIGHT CONCRETE BRIDGE WITH PRESTRESSED STEEL AS DETERMINED BY DEPOLARISATION VALUES AND CURRENT DENSITIES ... 177
van den Hondel A.J. (Hans) MSc

BEHAVIOUR OF COLUMN CONSTRUCTED WITH FRP TUBES FILLED WITH CONCRETE ... 186
Donchev Ted, Srebrenkoska Vineta, Petkova Diana, Albkeirat Mohammad, Akhundzada Hikmatullah

IMPLICATIONS UPON USING EXTERIOR POST-TENSIONING TENDONS TO RETROFIT A LOW STRENGTH CONCRETE WAFFLE SLAB ... 191
Faur Andrei, Jacint Virág

HYBRID FIBRE-REINFORCED GEOPOLYMER (HFRG) COMPOSITES AS AN EMERGING MATERIAL IN RETROFITTING AGING AND SEISMICALLY-DEFICIENT CONCRETE AND MASONRY STRUCTURES ... 202
Guades Ernesto, Stang Henrik, Fisher Gregor, Schmidt Jacob

EXPERIMENTAL TESTING OF MEDIUM SCALE GFRP REINFORCED CONCRETE FRAMES ... 210
Hopartean George, Donchev Ted, Petkova Diana, Georgopoulos Costas, Limbachiya Mukesh, Parnada Noel

EXPERIMENTAL STUDY REGARDING THE INFLUENCE OF FIBRE TO MATRIX COMPATIBILITY ON GENERAL PERFORMANCE OF FIBRE ENGINEERED CEMENTITIOUS MATERIALS (FECM) ... 215
Mircea Anamaria, Mircea Călin, Szilágyi Henriette, Baeră Cornelia, Hegyi Andreea

INFLUENCE OF THE SURFACE MODIFICATION BY SANDING OF CARBON TEXTILE REINFORCEMENTS ON THE BOND AND LOAD-BEARING BEHAVIOR OF TEXTILE REINFORCED CONCRETE ... 222
Morales Cruz Cynthia, Raupach Michael

CORROSION PROTECTION OF STEEL REINFORCEMENT BY USING SURFACE APPLIED CORROSION INHIBITORS ... 232
Seyhan Engin Cueneyt, Goodwin Fred, Huang I-Wen (Evan)

REINFORCED CONCRETE CORROSION PREVENTION/REDUCTION BY HYDROPHOBIC IMPREGNATION ... 238
Donadio Michel, Bakalli Mirdash, Dan Zeno

OPPORTUNITIES REGARDING THE POTENTIAL USE OF THE SELF-CLEANING CONCEPT WITHIN URBAN CONTEMPORARY ARCHITECTURE IN ROMANIA ... 245
Grebeni°an Elvira, Szilagyi Henriette, Hegyi Andreea, Mircea Călin, Baeră Cornelia

SELF-CLEANING CONCRETE FOR LANDSCAPING APPLICATIONS ... 251
Mircea Diana-Maria

COMPARATIVE STUDY ON SHOTCRETE REPAIR SOLUTIONS OF BURIED STEEL PIPELINES ... 255
Mircea Călin GR, Blaga Sergiu, Dăbăcan Ovidiu

ESTIMATION OF IN-SITU CONCRETE STRENGTH USING DRILLING RESISTANCE 261
Karatosun Serkan, Asan Muhammet, Gunes Oguz
NEW WARNING SENSORS TO DETECT CORROSION RISK IN REINFORCED CONCRETE 267
Khadra Mahdi, Marie-Victoire Elisabeth, Bouichou Myriam, Crémona Christian, Vildaer Stéphanie
COMPARISON OF 2D AND 3D ULTRASONIC PULSE ECHO IMAGING TECHNIQUES FOR STRUCTURAL ASSESSMENT. 274
Misak Lubos, Corbett David, Grantham Michael
EFFECTIVENESS OF TWO FIELD METHODS OF SATURATING NEAR SURFACE CONCRETE ON THE WATER PERMEABILITY OF IN SITU CONCRETE 279
Yang Kai, Basheer Muhammed, Nanukuttan Sreejith, Bai Yun, Long Adrian
PERFORMANCE OF PATCH REPAIRS ON HISTORIC CONCRETE STRUCTURES: A PRELIMINARY ASSESSMENT 285
Arato Gonçalves Ana Paula, Macdonald Susan, Marie-Victoire Élisabeth, Bouichou Myriam, Wood Chris
A NOVEL APPROACH TO IN-DEPTH FAÇADE ASSESSMENTS: IMPROVED CORROSION TEST METHODS FOR EMBEDDED STEEL FRAMING IN HISTORIC MASONRY CLAD BUILDINGS 289
Crevello Gina, Matteini Irene, Noyce Paul
CHALLENGES OF PRESERVING MODERNIST CONCRETE 296
Harrer Ann, Gaudette Paul
INFLUENCE OF CARBONATION ON THE RESISTANCE OF CONCRETE STRUCTURES TO CHLORIDE PENETRATION AND CORROSION 301
AL-Ameeri Abbas S., Imran Rafiq M., Tsioulou Ourania
PREDICTING RESIDUAL SERVICE LIFE OF CONCRETE INFRASTRUCTURE: A CONSIDERABLY CONTROVERSIAL SUBJECT 310
Gulikers Joost
CONDITION ASSESSMENT AND SERVICE LIFE PREDICTION OF A MARINE CONCRETE STRUCTURE: AN INTERESTING CASE STUDY 318
Gulikers Joost
ASR: PRACTICAL INVESTIGATIVE TECHNIQUES AND FIELD MONITORING SYSTEMS USED TO ASSESS ASR FOR SERVICE LIFE MODELING. 324
Matteini Irene, Noyce Paul, Crevello Gina
EXPERIMENTAL STUDY OF THE INFLUENCE OF CRACK WIDTH DUE TO CORROSION BEHAVIOUR USING DIFFERENT MEASUREMENT METHODS AND DIFFERENT EXPOSURES 333
Schneider Martin, Gardener Georg
COMPARISON BETWEEN THE PREDICTED PERFORMANCE CURVE AND THE MARKOV CHAIN MODELS FOR STRUCTURAL PERFORMANCE OF INFRASTRUCTURE COMPONENTS 340
Semaan Nabil, Dib Youssef
THE INFLUENCE OF SAPS ON CHLORIDE INGRESS IN CRACKED CONCRETE 347
Van Mullem Tim, Caspeele Robby, De Belie Nele
PRETENSIONED BFRP REINFORCED CONCRETE BEAMS: FLEXURAL BEHAVIOUR AND ESTIMATION OF INITIAL PRESTRESS LOSSES 355
Pavlović Ana, Donchev Ted, Petkova Diana, Limbachiya Mukesh, Almuhaisen Refad
MODELLING THE BEHAVIOUR OF CONCRETE SHEAR WALLS WITH BFRP REINFORCEMENT 360
Rahman Hamid, Donchev Ted, Petkova Diana
DURABILITY OF CONCRETES PREPARED WITH CRYSTALLINE ADMIXTURES 367
Sideris Kosmas K., Chatzopoulos A., Tassos Ch., Manita P.
FIELD INVESTIGATION OF SLAB DETERIORATION DUE TO CARBONATION IN A CONCRETE BUILDING 372
Araya Eleni B., Ekolu Stephen, Solomon Fitsum
STRUCTURAL CONDITION ASSESSMENT OF A REINFORCED CONCRETE DRIVEWAY 376
Berkowski Piotr, Dmochowski Grzegorz
CONCRETE REPAIR CHALLENGES IN THE MIDDLE EAST 385
Brueckner Rene
STRUCTURAL FAILURE OF OLD REINFORCED CONCRETE RIVER WHARF 392
Dmochowski Grzegorz, Berkowski Piotr
LIFE CYCLE ASSESSMENT OF REFURBISHMENT SCENARIOS TO CONFINED TANK STRUCTURES AT A WASTE WATER TREATMENT PLANT 400
Donadio Michel, Heider Catrina, Dan Zeno

HYBRID SANDWICH PANELS FOR BUILDING USES: FOCUS ON GLASS FIBRE REINFORCED POLYMER AND MINERAL MATRIX INTERFACE. 405
Djama Khaled, Michel Laurent, Ferrier Emmanuel, Gabor Aron
CORROSION AT LOW MOISTURE CONTENT IN BOTH CARBONATED AND CHLORIDE POLLUTED CONCRETE – VILLA E-1027, A CASE STUDY 414
Marie-Victoire Elisabeth, Bouichou Myriam
INVESTIGATION OF THE PRESTRESSED BEAMS AND GLASS-WINDOW PANELS AT A SHOPPING CENTRE 421
Mircea Andreea-Terezia
FLEXURE BEHAVIOUR OF REINFORCED HIGH STRENGTH CONCRETE ELEMENTS AFFECTED BY CORROSION 425
Negrutiu Camelia, Sosa Ioan, Heghes Bogdan
STUDY ON ESTIMATION OF CREEP BEHAVIOUR OF CONCRETE AT EARLY AGE CONSIDERING TEMPERATURE EFFECT 431
Ohashi Kayo, Arai Jun-ichi, Mizobuchi Toshiaki
THE INFLUENCE OF DEFECTS AT THE STEEL/CONCRETE INTERFACE FOR PITTING CORROSION INITIATION STUDIED THROUGH X-RAY COMPUTED TOMOGRAPHY AND IMAGE ANALYSIS 437
Rossi Emanuele, Nijland Timo, Çopuroðlu Oðuzhan, Polder Rob, Šavija Branko
RESEARCH AND PRACTICES OF LARGE COMPOSITE EXTERNAL WALL PANELS FOR ENERGY SAVING PREFABRICATED BUILDINGS 444
Shi Yunxing, Zhang Yangang, Ni Kun, Liu Wei, Luo Ye
MECHANICAL PROPERTIES OF ALKALI ACTIVATED GEOPOLYMER PASTE USING DIFFERENT ROMANIAN FLY ASH SOURCES – EXPERIMENTAL RESULTS 449
Lăzărescu Adrian, Mircea Călin, Szilagyi Henriette, Baeră Cornelia
DEVELOPMENT OF SUSTAINABLE CONCRETE REPAIR MATERIALS VIA MICROENCAPSULATED AGENTS 456
Litina Chrysoula, Al-Tabbaa Abir
Author Index

CONCRETE SOLUTIONS 2019
7th INTERNATIONAL CONFERENCE ON CONCRETE REPAIR
Cluj-Napoca, Romania
30 of September – 2 of October, 2019

URBAN INCD INCERC

aci American Concrete Institute

rilem

CONCRETE SOLUTIONS SINCE 2003

2003 SAINT-MALO (FRANCE)

2006 SAINT-MALO (FRANCE)

2009 PADUA (ITALY)

2011 DRESDEN (GERMANY)

2014 BELFAST (NORTHERN IRELAND)

2016 THESSALONIKI (GREECE)

2019 CLUJ-NAPOCA (ROMANIA)

THE TECHNICAL UNIVERSITY of CLUJ-NAPOCA

The Concrete Society
Your concrete community

Institute of Concrete Technology

The Proceedings of all conferences are available either from Concrete Solutions or the Publisher

Please enquire if interested.

Sika®

BASF
We create chemistry
MASTER® BUILDERS SOLUTIONS

CORE SPONSORS

corrPRE
Special Anodes Manufacturing

MC
BE SURE. BUILD SURE.

PRINCIPAL SPONSORS

TECNOSERVICE DEDICATED TO QUALITY EQUIPMENT S.R.L

EXHIBITOR

Introduction

The Seventh International Conference on Concrete Repair, (ICCR2019) was held in Cluj Napoca, Romania from the 30 September to the 2 October 2019. The Conference was the latest in a sequence of ICCR International Conferences (St. Malo, 2003 and 2006; Padova, 2009; Dresden 2011; Belfast 2014; Thessaloniki 2016). The organization was a collaborative venture between GR Technologie Ltd (Concrete Solutions) and the Technical University of Cluj Napoca in Romania

The estimated cost to repair reinforced concrete structures has been said to be some €175 per square meter of exposed surface. In the United States alone, the annual direct cost of corrosion in highway bridges alone is roughly €7.3 billion, including maintenance, repair, replacement, and the cost of capital. Indirect costs, including traffic delays and lost productivity, may run 10 times that number.

But corrosion is more than an economic issue. In June 1983, a 100-foot section dropped out of the Mianus River Bridge in Greenwich, Connecticut, U.S.A., killing three motorists and critically injuring three others. The steel pins that joined sections of the bridge had decayed. In May 2000, in Concord, North Carolina, U.S.A., more than 100 people were injured when steel strands corroded in a pre-stressed concrete pedestrian bridge and the structure collapsed onto the highway below. In 2006, The De La Concorde Overpass near Montreal collapsed. The collapsed section crushed two vehicles under it, killing five people and seriously injuring six others who went over the edge while travelling on the overpass. More recently in Italy in August 2018, a partial collapse of the Ponte Morandi Bridge in Genova was blamed on corrosion in the cable stays and killed 43 people who were crossing the bridge at the time. The disaster caused a major political controversy about the poor state of infrastructure in Italy and raised wider questions about the condition of bridges across Europe.

So the problems of dealing with maintenance and repair of structures are not just financial, they affect people's lives too!

Concrete Solutions has historically developed a name as a conference that brings together researchers from the Academic world with Engineers and Scientists who are practitioners in the field of concrete repair. These vary from Owners and Specifiers, to Engineers and Technicians, but include Architects and those engaged in preserving historic structures. The conference has received excellent support by researchers and practitioners from around the world, with authors being drawn from numerous research and industrial organisations from 22 countries for this 7th Conference. These Proceedings contain papers presented at the conference, classified into a total of 11 themes:

Self-healing Concrete
Patch Repair
Electrochemical Repair
Strengthening Materials and techniques/Repair with Composites
Surface Protection Methods and Materials
NDT and Diagnosis of Problems
Repair and Preservation of Heritage Structures
Service Life Modelling
Whole Life Costing
Case Studies
Sustainable Repair

Only original contributions were considered for inclusion in the conference proceedings and all papers submitted were subjected to a full process of peer review. The review of manuscripts was undertaken by members of the International Scientific Advisory Board and other identified leading experts, acting independently on one or more assigned manuscripts. This invaluable assistance, which has greatly enhanced the quality of the Proceedings, is gratefully acknowledged. The full conference papers are published by MATEC Web of Conferences, a provider of open-access proceedings in Materials Science, Engineering and Chemistry. This was done in order to give the papers better acessibility in the international research and industry communities. This hard copy of the proceedings contains abstracts of all accepted papers, in order to provide an overview on the topics covered and to guide conference participants in selecting technical sessions that are of most interest to them.

Concrete Solutions and the Technical University of Cluj-Napoca are pleased to acknowledge the support and sponsorship of the following organisations:-

CORE SPONSORS

Sika Europe

BASF Construction Systems

PRINCIPAL SPONSORS

Corr Pre Special Anodes Manufacturing

MC-Bauchemie SRL

EXHIBITORS

TecnoService Equipment Srl

INTERNATIONAL SUPPORTERS

The American Concrete Institute (ACI)

RILEM

Concrete Society UK

The Institute of Concrete Technology

INCD URBAN INCERC Romania

Finally, the editors wish to give special thanks to the authors for their efforts at producing and delivering some excellent and thought provoking papers of a very high standard. We are sure that these Proceedings will be used by many working in this critical field, and that the papers and the discussions at the Conference will form a suitable base for discussion and provide suggestions for future development and research.

Michael Grantham
Călin Mircea
Editors

Dr. Veronique Bouteiller, IFSTTAR, France
Dr John Broomfield - Corrosion Consultant - UK
Mr. David Corbett - Proceq SA, Switzerland
Mr. Richard Day, The Concrete Society, UK
Prof. Frank Dehn - MFPA Leipzig, Leipzig Institute for Materials Research and Testing
Dr Ted Donchev - Kingston University,
Mr John Drewett - Concrete Repairs Ltd., UK
Ass. Prof. Raafat El-Hacha, University of Calgary, Canada
Prof. Dr. Vyatcheslav R. Falikman, Structural Concrete Association, NIIZHB, Russia
Prof. Nader Ghafoori, University of Nevada, USA.
Dr. Chris Goodier - Loughborough University, UK.
Prof. Michael Grantham - Concrete Solutions (GR Technologie Ltd.) and Sandberg LLP
Ass. Prof. Oguz Gunes, Istanbul Technical University, Turkey
Prof. Peter Hewlett, University of Dundee, David Ball Group and the Institute of Concrete Technology
Mr Steve Hold, Independent Consultant, Cardiff, UK
Prof. Dr. Carlos Jurado Cabañes, Polytechnic University of Madrid and Ingecal Ingenieros S.L. Spain.
Ass. Prof Dr. Iqbal Khan, King Saud University, Saudi Arabia
Dr. David Law, RMIT University, Australia
Prof. Dr. Minoru Kunieda, Nagoya University, Japan
Prof. Christophe Lanos - IUT Rennes, France
Dr. Bryan Magee, University of Ulster
Prof. Carmelo Majorana - University of Padova, Italy
Dr Elisabeth Marie Victoire - Laboratoire de Recherche des Monuments Historique, France
Prof. Dr. Ing. Viktor Mechtcherine - Technische Universitat Dresden
Prof. Calin GR Mircea, Technical University of Cluj-Napoca, Romania (Conference Co-Chair)
Prof. Ioanna Papayianni, Aristotle University of Thessaloniki
Dr. Ioan Pepenar, ICECON Research Institute for Construction Equipment and Technology, Romania
Prof. Radhakrishna Pillai, Indian Institute of Technology, Madras, India
Prof. Dr. Rob Polder, TNO Built Environment and Geosciences, Netherlands.
Prof. Peter Robery – Robery Forensics, and Leeds University UK
Dr. Ulrich Schneck -Citec GmbH- Dresden, Germany
Dr Ioannis Sfikas, Mott McDonald, UK
Ass. Prof. Dr. Kosmas Sideris - Democritus University of Thrace, Greece
Dr. Henriette Szilagyi, URBAN-INCERC, Cluj Napoca Branch, Romania
Dr. Jonathan Wood, Structural Studies and Design, UK

edp sciences

Web of Conferences

Statement of Peer review

In submitting conference proceedings to *Web of Conferences*, the editors of the proceedings certify to the Publisher that

1. They adhere to its **Policy on Publishing Integrity** in order to safeguard good scientific practice in publishing.
2. All articles have been subjected to peer review administered by the proceedings editors.
3. Reviews have been conducted by expert referees, who have been requested to provide unbiased and constructive comments aimed, whenever possible, at improving the work.
4. Proceedings editors have taken all reasonable steps to ensure the quality of the materials they publish and their decision to accept or reject a paper for publication has been based only on the merits of the work and the relevance to the journal.

Title, date and place of the conference

Concrete Solutions 2019, 7th International Conference on Concrete Repair
30 Sept - 2 Oct 2019, Cluj Napoca, Romania

Proceedings editor(s):

Michael G Grantham
Calin Mircea

Date and editor's signature

11.08.2019

Optimizing approach on Fibre Engineered Cementitious Materials with Self-Healing capacity (SH-FECM) by the use of slurry lime (SL) addition

Cornelia Baeră[1*], *Henriette* Szilagyi[2], *Claudiu* Matei[3], *Andreea* Hegyi[2], *Adrian* Lăzărescu[2,4] *Anamaria Cătălina* Mircea[2,4]

[1]NIRD URBAN-INCERC, Timişoara Branch, 300223 Traian Lalescu 2, Timişoara, Romania
[2]NIRD URBAN-INCERC, Cluj-Napoca Branch, 400524 Calea Floreşti 117, Cluj-Napoca, Romania
[3]NIRD URBAN-INCERC, Bucureşti Branch, 021652 Pantelimon 266, Bucureşti, Romania
[4]Technical University of Cluj-Napoca, Civil Engineering Faculty, 400020 Daicoviciu 15, Cluj-Napoca, Romania

Abstract. SH-FECM (Fibre Engineered Cementitious Materials with Self-Healing capacity), developed at NIRD "URBAN-INCERC" Cluj-Napoca Branch in the last five years, are consistently based on Engineered Cementitious Composites (ECCs) concept, elaborated in the early nineties at Michigan University (USA). They all represent a cement-based typology of dispersed reinforcement composites able to develop high deformability by the means of multiple cracking pattern under loading, leading to a cumulative set of valuable material features: metal like behaviour when subjected to loads, brittle failure prevention, increased self-healing potential via the compositional design, etc. The process of development and constant improvement of the SH-FECMs represents a long term theoretical and experimental approach, aiming to establish the optimum raw materials (mostly locally available) compatibility within the cementitious matrix so that the composites would present superior performance under comparative evaluation. This paper presents the first results, evaluated as positive for both, fresh and hardened state materials, regarding the inclusion of Slurry Lime (SL) addition as replacement of the initial lime powder addition (L) in the material composition. The long-term effects are on ongoing investigation, but the initial results are clearly promising, starting from a better fresh state aspect and evaluating for faster setting time and improved early age mechanical behaviour. The beneficial effects are also in terms of economic and ecological aspects, considering that the used lime slurry (SL) addition represents an actual waste resulted from a local, natural stone processing factory. Its use as direct addition in the SH-FECMs mixtures could represent an efficient recycling and waste prevention action, with long term beneficial potential, in terms of Circular Economy principles.

1 Introduction

Fibre Engineered Cementitious Materials with Self-Healing capacity (SH-FECM) represent strain hardening, fibre reinforced composites, able to develop width & shape controlled microcracks under increasing load. This kind of cracking pattern represents the foundation of co-dependent physical and mechanical intrinsic features of the material: a) high deformability, metal like behaviour, preventing the sudden, brittle failure of classic plain concrete of cementitious materials; b) improved self-healing potential, as a direct consequence of small cracks and of dispersed reinforcement, (which not only blocks crack widening but also represents a skeleton where precipitation products, emerging from concrete structure in direct relation to the environmental conditions, can settle and facilitate the small crack closing [1, 2, 3].

The SH-FECM were developed at Incerc Cluj by initial adaptation of the Engineering Cementitious Composites (ECCs) concept of Victor Li [1] to the specificity of local conditions and raw materials [4, 5]. The ECCs and consequently, the SH-FECMs belong to the High Performance Fibre-Reinforced Cementitious Composites (HPFRCC) family, combining innovative features, theoretically integrated in the matrix functionality by the means of micromechanics tools [1, 2, 6, 7, 8, 9, 10, 11, 12, 13] in order to amplify some intrinsic material properties. Consequently, the strain hardening behaviour under loading, which ensures development of successive, opening controlled microcracks under loading, ensures by default the optimisation of material self-healing capacity: small cracks are easier to fill and seal with the healing products, generated due to specific self-sealing properties of the cement based materials [1, 2].

The Self-Healing concept in the area of cementitious materials represents a research topic that recently gained substantial attention from academia, research groups and also the construction industry, due to its consistent potential towards durable infrastructure development and maintenance. Preventive solutions and design of smart materials which can better ensure the infrastructure

* Corresponding author: cornelia.baera@incd.ro

safety and durability requirements, with reduced costs in terms of rehabilitation and repair along their life spam, represents new alternatives that recently delivered encouraging results [3].
The design, production and further improvement of the SH-FECMs at Incerc Cluj by using locally available raw materials represented one of the major research directions of the last years [15, 16]. The high fly ash (FA) content of the SH-FECM binding system also represents an attractive possibility of valorising the huge fly ash quantity, by-product / waste of the energy industry in Romania. Statistical data showed that in 2017 for instance, one Romanian power plant reported a production of approx. 650,000 tonnes of fly ash, 50,000 tonnes of slag and 50,000 tonnes of gypsum Unfortunately, just a small FA amount, less than 25%, approx. 162,000 t/year is used in the construction industry (as addition for concrete and mortars, cement, soil stabilisation – mainly for the roads, infrastructure, etc.). The rest is simply converted into large quantities of industrial waste, stored in waste dumps and experiencing fast degradation of their potential as binding/filler raw material (due to complete exposure to outdoor climatic conditions) and simultaneously generating ecological discomfort [17].

Identifying new, innovative, high performance materials which could integrate large amounts of mineral additions (by-products/wastes, inert, latent or pozzolanic hydraulic potential as part of the binding system) represents one of the major goals of Incerc Cluj research in recent years [4, 5, 17]. This research topic is strongly founded and encouraged by Romania's need develop the principles of Sustainability and Circular Economy, as part of its adherence to the European Union.

2 Experimental procedure – materials and methods

2.1 General approach

The current study evaluates the possibilities of optimising of SH-FECMs by using slurry lime addition and consequently, its influence and compatibility with the cement (C) and fly ash (FA) matrix. The mechanical and self-healing capacity evaluation of the material is also studied, with respect to the previous, stable mix containing lime powder (L), used as an inert filler for optimizing the fresh state, bleeding prevention, and matrix packing optimisation [4, 5].

2.2 Raw materials

The component materials for the SH-FECM mixes are mostly locally available: the binders (cement (C) and fly ash (FA)), the aggregate (fine silica sand (FSS)), and additions. The fibres used for the dispersed reinforcing of the composites are provided by the Japanese company Kuraray and they seem to generally provide excellent results in worldwide ECC development approaches [1, 2].

SH-FECM promotes the ECC *typical high fly ash (FA) content binding system*; FA even exceeds the cement (C) content and the specific ratio FA/C = 1.2. ECC typically represents high FA composites, with FA/C values ranging from 1.0 to even 1.5 or even more [1, 2, 4, 16], in order to provide proper matrix features for ensuring the strain hardening behaviour under loading and also the late self-healing potential, due to the long term FA pozzolanic activity [3].

Ordinary Portland Cement (OPC) was used (Fig. 1, a) and class F Fly-Ash (FA) (Fig. 1, b) produced by a local power station and characterised by the XRF analysis presented in Table 1.

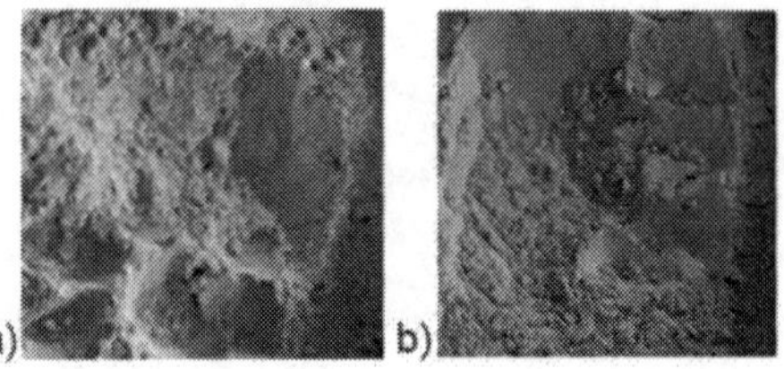

Fig. 1. The binding system: a) Ordinary Portland Cement, (C) b) Class F Fly Ash, (FA)

Table 1. XRF analysis data for the fly ash samples.

FA Oxides, %					
SiO_2	**Al_2O_3**	**Fe_2O_3**	**CaO**	**MgO**	**SO_3**
53.61	26.16	7.58	2.42	1.49	0.26
Na_2O	**P_2O_5**	**TiO_2**	**Mn_2O_3**	**L.O.I.**	-
0.59	0.12	1.04	0.08	3.57	-

The Aggregate represents Bega Minerale SA (Cluj, Romania) silica sand. The initial 0.5 mm maximum grain size sand (SS), (Fig. 2, a) was later replaced by a finer sand (FSS), 0.3 mm maximum grain size (Fig. 2, b), in the process of matrix packing optimisation and consequently for an increased strain hardening behaviour under loading, in accordance to previous recommendations [2].

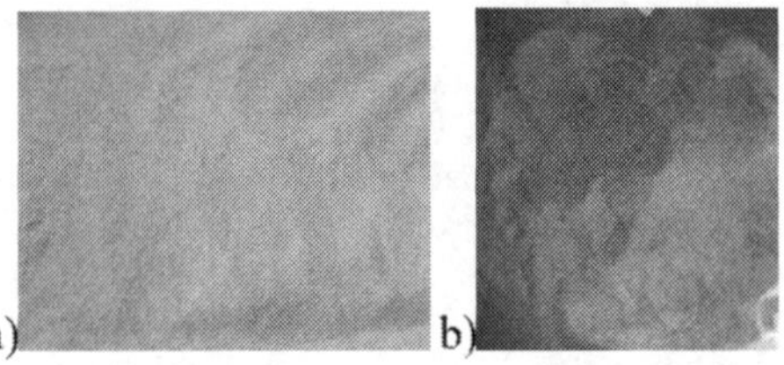

Fig. 2. SH-FECM Aggregate: a) Silica Sand (SS); b) Fine Silica Sand (FSS)

The polymeric fibres are the Kuraray (Japan) Polyvinyl Alcohol (PVA) fibres (Fig. 3), lubricated (by 1.2% mineral oil) and generally confirming excellent performance for ECC development. The manufacurer's data sheet mentions the following Kuralon K-II fibre

RECS15x8 fibre characteristics: diameter 39µm, cut length 8.0±1.0 mm, tensile strength (tenacity) 12.0±2.5 cN/dtex (1560±325 MPa), Elongation 6.5±1.5%, Modulus - 320 cN/dtex. The high tensile strength and the elasticity properties of the fibres are proved to be crucial for the composite performance. Lower tensile strength polymeric fibres (Polypropylene fibres), used for a comparative evaluation, failed to achieve close results, especially in long term testing [4].

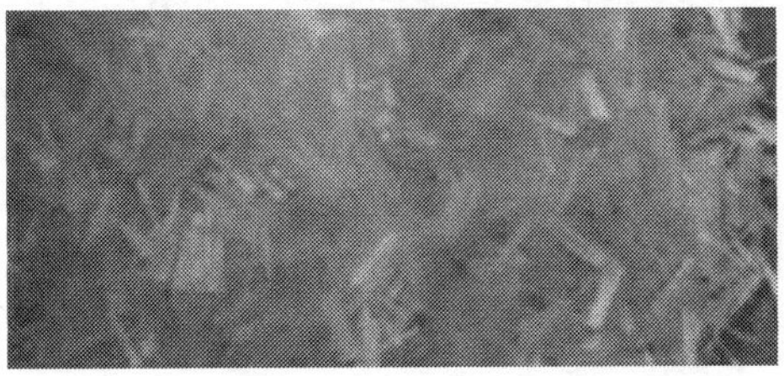

Fig. 3. PVA Kuralon K-II fibre (RECS15x8) (Kuraray, Japan)

The 2% (by volume) fibre content in the mix was proved to be the optimum for the dispersed matrix reinforcing [1, 2, 4]: those with a lower fibre content generally failed in ensuring the desired cracking pattern under loading; a higher fibre content does not involve additional benefits to the overall composite behaviour, but at the same time, it generates increased costs and also poor workability during mixing, with higher admixture addition.

The *secondary addition* consisting of limestone (filler or slurry) was included in the mix design for better control of the fresh state, namely preventing the bleeding effect when adding the fibre, balling of the fibre and also for ensuring the homogeneity of the mix and its creamy, desired texture [1]. The hardened state characteristics are also improved: besides better matrix packaging induced by increasing the fine particle content, the Self-Healing potential of the material is stimulated [3]. Initially, *limestone filler (L)*, available on the regular building materials market, was used (Fig 3, a). The last experimental procedures included the novelty of substituting the filler (L) by *limestone slurry* (SL) (Fig. 3, b).

a)

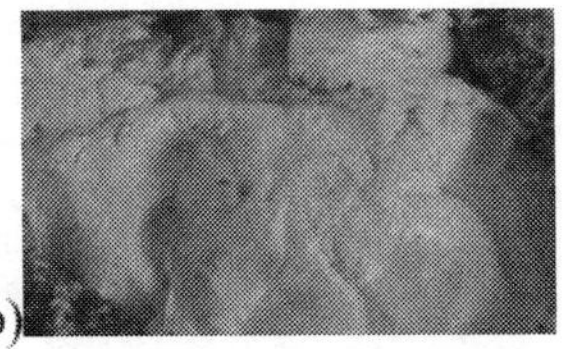
b)

Fig. 3. SH-FECM Secondary additions: a) Limestone filler (L); b) Limestone slurry (SL), waste, with natural humidity

The limestone slurry (SL) represents a waste generated by the local producer of natural stone finishing products, which showed increased interest in finding an ecological and cheap solution for valorising it and consequently, elimination of the limestone slurry waste dumps (Fig. 4). A major direction of the research developed at Incerc Cluj-Napoca was to identify, evaluate and improve the potential of local mineral additions (by products or waste), as partial or complete raw material replacement cement-based composites for the construction industry, and thus obtaining reliable composites in terms of applicability field, physical, mechanical and durability performance. The limestone slurry (SL) represented one of the targeted wastes of the study, considered as secondary addition, besides fly ash (FA), considered as primary addition and substantial cement replacement within the matrix. The SL initial evaluations generated very promising results, especially when combined with other additions and especially with the fly ash (FA) [18].

Fig. 4. Limestone slurry (SL) waste dumps

The low water to binder ratio (W/B), crucial for mechanical performance, was kept low by means of the *polycarboxylate superplasticizer* Glenium 51 (HRWR), (BASF), traditionally used for FECM development [2, 4, 5, 16].

2.3. Mix design and mixing procedure

Table 2 presents the mix proportions of the SH-FECM mixes and the specific ratios with respect to the cement (C) content (C = 1.0). Two SH-PVA-FECM mixes are considered for the current evaluation: 1) the SS-L-PVA-FA/C (1.2), developed by using the coarser silica sand (SS) and the limestone filler addition (L); 2) the FSS-SL-PVA-FA/C (1.2), developed by using the fine silica sand (FSS), with maximum grain size 0.3 mm and the Limestone slurry addition (SL).

Table 2. Mix proportions of SH-FECM.

Components	FECM Mixes	
	SS-L-PVA-FA/C (1.2)	FSS-SL-PVA-FA/C (1.2)
C*	1.0	1.0
FA	1.2	1.2
B	2.2	2.2
SS/FSS	0.8	0.8
L/SL	0.18	0.18
W	0.65	0.65
HRWR	0.021	0.021
FIBRE	2.0%	2.0%
W/C	0.67	0.67
W/B	0.30	0.30
Liq./B	0.31	0.31

**C = 1.0;*

Supplementary Symbols: B = C + FA (Binding system); SS – Silica Sand (max. grain size < 0.5 mm); FSS = Fine

Silica Sand (max grain size <0.3 mm); L = limestone filler; SL = Slurry Lime; W – Water; HRWR-Superplasticizer; Liq. = W+ HRWR;

The supplementary powder addition (L) for the SS-L-PVA-FA/C (1.2) was established to 100 Kg/m^3 [4]. The slurry lime (Fig. 3 and 4) represents a lime paste with variable but easily determinable water content. The L replacement was performed by simply adding the equivalent of SL paste. The SL paste quantity to be added in the mix was calculated by initial evaluation of natural water content (by drying it until constant mass was reached and reporting it to the initial value) and considering as reference the L filler content (dry powder). In addition, the total water (W) in the mix was also adjusted considering the liquid input provided by the paste. The mixing sequences [5], established in accordance to the EN 196-1 [19] methodology by using the typical laboratory mixture suitable for low volume batches, are correspondingly adapted: the SL paste was added immediately after the dry mixing of the binding materials and aggregate, in order to ensure a homogeneous dispersion. The lime paste dispersed homogeneously and provided a nice, cohesive texture to the entire mix.

3 Experimental procedures: results and discussions

3.1 Fresh state aspect

Both mixtures developed the homogeneous, creamy mix (Fig. 4), associated with the ECCs fresh state aspect [1, 2]. The SS-L-PVA-FA/C (1.2) mix already experienced an improvement regarding the fresh state workability (Fig. 4, a) and bleeding control, by using the limestone filler addition (L), when compared to the L free mixes. A clear improvement was noticed when replacing the L to SL addition (Fig. 4, b), as the material becomes more cohesive and homogeneous; these feature are preserved even after the fibre addition which happens in the last sequences of the mixture technology, usually considered the most vulnerable steps considering the risks of balling, bleeding and segregation.

a)

b)
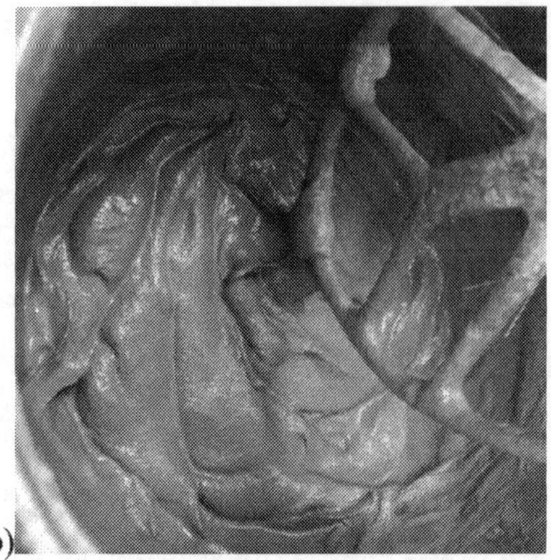

Fig. 4. SH-FECM fresh state aspect: a) SS-L-PVA-FA/C (1.2) mix; b) FSS-SL-PVA-FA/C (1.2) mix

Placing into the moulds is performed easily, especially for the FSS-SL-PVA-FA/C (1.2), as the composites developed the necessary self-compacting behaviour which make vibration unnecessary.

3.2 Mechanical properties

The flexural and compressive strengths as relevant, basic mechanical characteristics for mix evaluation, were performed by using the EN 196-1 [19] and EN 1015-11 [20] methods. The relevant testing ages were considered at 28 days and also at 90 days: the late age testing is considered to bring important information for the high content fly ash composites, taking into account the late pozzolanic reaction of fly ash. Prismatic specimens, 40 x 40 x 160 mm^3, removed from the moulds 24h after casting, were cured by water immersion at T (21 ± 2)°C until testing.

Table 3 presents the flexural strengths determined by 3PB test performed at 28 and 90 days and the strength increase recorded at the final testing, with respect to the representative 28 day values. The compression tests, performed on the half prisms resulted after 3PB tests, had a similar approach: 28 and 90 days compressive strength and the corresponding strength gain form initial testing to the final one. The obtained values are presented in Table 4. The graphical representation of the strengths are presented in Figure 5 (Flexural test) and 6 (Compression test).

Table 3. Mechanical strengths: 3PB Flexural Resistance.

SH-FECM Mixes	3PB Flexural resistance (MPa)		
	28 d	90 d	Strength gain (%)
SS-L-PVA-FA/C (1.2)	17.0	28.2	65.9
FSS-SL-PVA-FA/C (1.2)	23.8	26.4	10.8

https://doi.org/10.1051/matecconf/201928901001

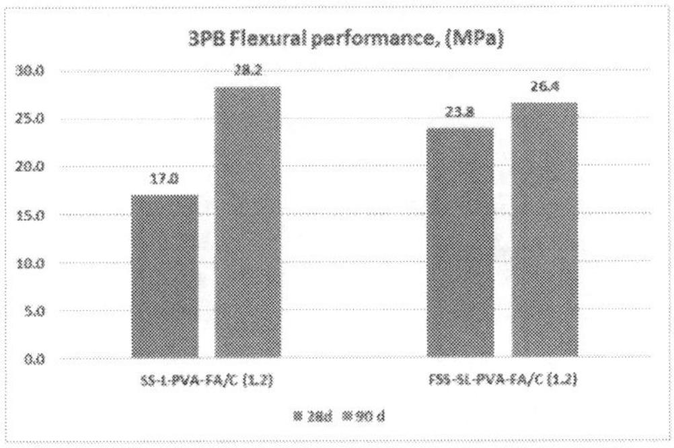

Fig. 5. SH-FECM, Graphical 3PB Flexural Resistance at 28 and 90 days

The analysed SH-FECM mixes show comparable results regarding the flexural performance (3PB). The slurry limestone (SL) mix seems to develop faster strengths at 28 days, but longer term evaluation (90 days) proves a clear gap recovery of the limestone filler (L) mix. The compressive results are even more balanced, proving the efficiency of replacing the limestone filler with the lime paste. Both mixes show impressive tensile performance and failure without loss of integrity.

Table 4. Mechanical strengths: Compression

SH-FECM Mix	Compression resistance (MPa)		
	28 d	90 d	Strength gain (%)
SS-L-PVA-FA/C (1.2)	52.5	72.8	47.0
FSS-SL-PVA-FA/C (1.2)	53.2	72.8	38.5

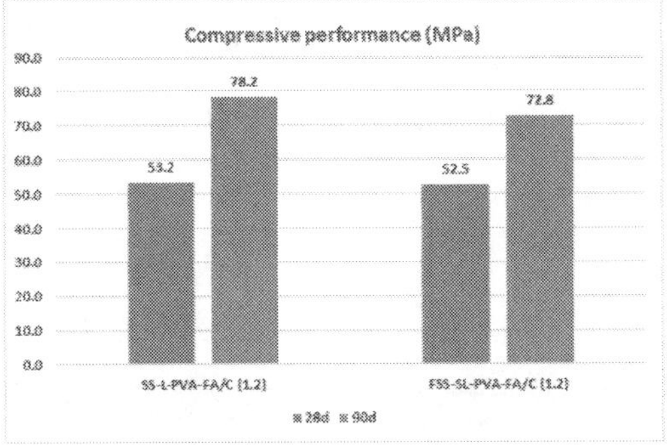

Fig. 6. SH-FECM Compressive Strength

3.3 Microcracking pattern (MC) at 28 days

The composite ability of developing the successive steady state multiplecracking network (MC) under loading generates the deformability potential and the strain hardening behaviour, characteristic for the ECC materials. The flexural 4-point bending test (4PB) at 28 days performed on coupon specimens, 200x40x8 mm provides comparative values of the two mixes. The deflection of the materials under loading is presented in Fig. 7.

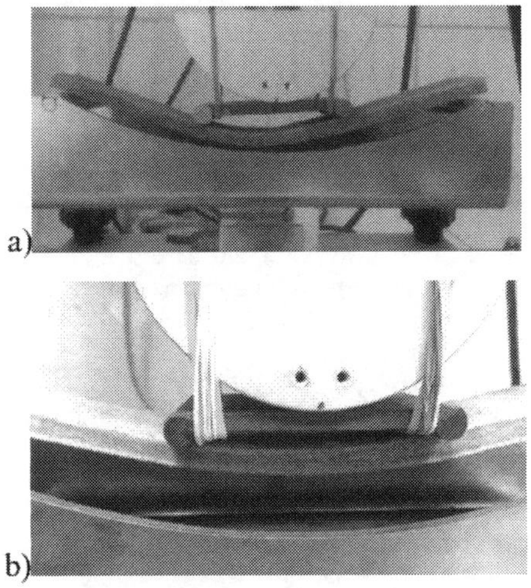

Fig. 7 4PB for SH – FECM: SS-L-PVA-FA/C (1.2) specimen; b) FSS-SL-PVA-FA/C (1.2) specimen

The testing device used for both mixes is the adapted 4PB displacement control UNIFRAME MINI. The network of the multiple cracks is characterised by the general aspect, the number of cracks developed until failure is considered and also by their openings (minimum, mean and maximum values); the crack typology (with or without branches) is also monitored.

The microscopic visual analyses were performed by using a stereomicroscope Leica S8 APO, equipped with digital camera Leica DMC2900. The LAS Analysis Bundle softwear was used for analysis of the data acquisition

The analytical results are summarised in Table 5.

Table 5. Microcracking potential (MC): 4PB testing at characteristic ages of 28 days

SH-FECM Mix	MC interval			
	Crack number (n)	Crack opening		
		w_{max} (μm)	w_{med} (μm)	w_{min} (μm)
SS-L-PVA-FA/C (1.2)	18-22	192	29	10
FSS-SL-PVA-FA/C (1.2)	11-13	48	29	10

It was noticed that MC capacity SS-L-PVA-FA/C (1.2) mix overcomes the FSS-SL-PVA-FA/C (1.2)'s, in terms of crack numbers. Still, both mixes show nice 4PB cracking pattern and controlled microcracking developing, reaching the expected strain hardening typology.

3.4 The Self-Healing capacity

The Self-Healing (SH) capacity of the cementitious composites SH-FECM, is a relatively new concept associated to durability of material and also to its ability to contribute to a sustainable development. SH means a reduced resource and energy consumption for its production and also for the preservation of its functionality, in optimum conditions and with minimum repair and maintenance input [3].

The self –healing approach implies two distinct but also interfering directions: a) Self-Closing (SC) of the cracks, meaning the material intrinsic capacity of sealing, complete or just partial, the cracks that occur within its structure due to state of stress; b) the Self-Repairing (SR) potential, namely the partial or complete, physical and mechanical performance regain of the microstructural material damage, related to the initial, virgin state performance.

The SS-L-PVA-FA/C (1.2) and FSS-SL-PVA-FA/C (1.2) mixes are analysed considering crack closing (SC) and also repairing of the microstructure (SR). The evaluation of SH performance is performed classically, by inducing a controlled state of damage, applying the healing curing and then the final, comparative analyses of the healed specimens (SH) with respect to the reference sample, (virgin state, VS).

The healing curing generally implies exposing the damaged samples to a pre-defined number of alternating cycles with variable relative humidity (RH) and temperature (T).

The experimental procedure for SH-FECM SH evaluation consisted of 28 wet and dry, alternating cycles (Wet: 12 h immersion in tap water, T (20 ± 2) °C; Dry: 12 h exposure to air, at T: (21 ± 3)°C and RH: (50 ± 5)%.

The current study is mainly focused on Self–Repairing (SR) evaluation. The Self–Repairing (SR) efficiency is performed by using the 3PB flexural test: prismatic specimens were subjected to flexural stress until the pre-cracked state is visually induced (SH samples); the Damage coefficient (D%) is quantified with respect to the flexural strength determined in identical conditions (similar specimens and similar testing, namely 3PB until failure is induced, (R samples) (Eq. 1).

After healing exposure, namely the 28 wet and dry alternating cycles, the SH samples are re-tested again to failure and the results are reported compared to the Virgin state samples (VS), representing identical prismatic specimens, cured by immersion in tap water, T (20 ± 2) °C till the testing of the healed samples. The results are expressed quantitatively, via Recovery coefficient (R%), (Eq. 2). At the same time, a qualitative evaluation is performed, via visual analysis.

The Damage coefficient (D%):

$D\% = 100 \times (R_{ti,28d\,SH} / R_{ti,28d\,R})$, (%) (Eq. 1)

Where:

$R_{ti,28d\,R}$ - 3PB flexural strength determined on R samples, at the age of 28 days;

$R_{ti,28d\,SH}$ – pre-cracked 3PB flexural strength determined on SH samples at the age of 28 days;

The Recovery Coefficient (R%):

$R\% = 100 \times (R_{ti,56d\,SH}) / R_{ti,56d\,VS}$ (Eq. 2)

Where:

$R_{ti,56d\,SH}$ – 3PB flexural strength determined on SH samples at the age of 56d (retesting after SH curing);

$R_{ti,56d\,VS}$ – 3PB flexural strength determined on VS samples at the age of 56d.

Table 5 presents quantitative evaluation of repairing capacity of SH-FECM mixes via 3PB testing, at 28 days, and 56 days respectively, after 28 wet and dry curing cycles.

The qualitative analysis is performed by visual analysis of the cracking pattern, during the initial testing, after the healing curing and again, after re-testing the healed specimens. Visual analyses, microscopically performed, confirmed complete crack closing (100% SC efficiency) on all faces of the specimens, for both of the composites, which indicates important information regarding their durability performance. Crack sealing (SC) which is visually evaluated on the surface on the crack, does not imply crack filling with CSH or $CaCO_3$ precipitation products deep inside the crack, ensuring a regain of mechanical properties as well, but even a superficial crack closing can ensure protection against corrosion or other aggressive agents.

For both composites, SS-L-PVA-FA/C (1.2) and SL-PVA-FA/C (1.2), more than 50% of the evaluated cracking patterns produced during re-retesting of the healed specimens, showed new crack(s) developing under re-loading, different from the initial one(s); the initial crack(s) remain(s) sealed during the re-testing, thus offering a reliable proof of the SH curing efficiency: the composites are able to initiate a high strength sealing products within the cracks, comparable to the virgin state material (Figure 8).

Table 6. Quantitative evaluation of SR capacity of SH-FECM mixes by using the Flexural 3PB testing at 28 days

FECM Mix	$R_{ti,}$ 28d R (MPa)	$R_{ti,}$ 28d SH (MPa)	D% (%)	$R_{ti,}$ 56d SH (MPa)	$R_{ti,}$ 56d VS (MPa)	R% (%)
SS-L-PVA-FA/C (1.2)	17.0	10.3	***60.6***	23.1	24.4	***94.7***
FSS-SL-PVA-FA/C (1.2)	23.8	9.5	***39,7***	23.4	24.2	***96.5***

Figure 9 emphasises the crack network evolution, microscopically evaluated, along the Self-Repairing experimental procedure: the initial crack initiation during the pre-cracking stage; crack sealing after curing of the samples; retesting up to failure of the SH specimens and identification of the cracking pattern, which proves to be different than initial one in more than 50% of the evaluated specimens.

https://doi.org/10.1051/matecconf/201928901001

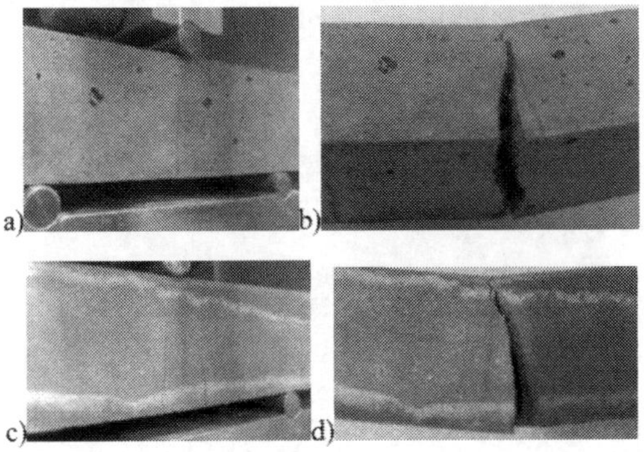

Fig. 8 Self-Repairing (cracking pattern) of SH – FECM: a) 3PB pre-cracked, healed crack; b) retesting: new failure crack for SS-L-PVA-FA/C (1.2) specimen; c) 3PB pre-cracked, healed crack; d) retesting: new failure crack for FSS-SL-PVA-FA/C (1.2) specimen

Both composites show similar, increased SH potential, both related to crack closing (Fig. 8 and Fig. 9) but also in terms of mechanical regeneration capacity of the material.

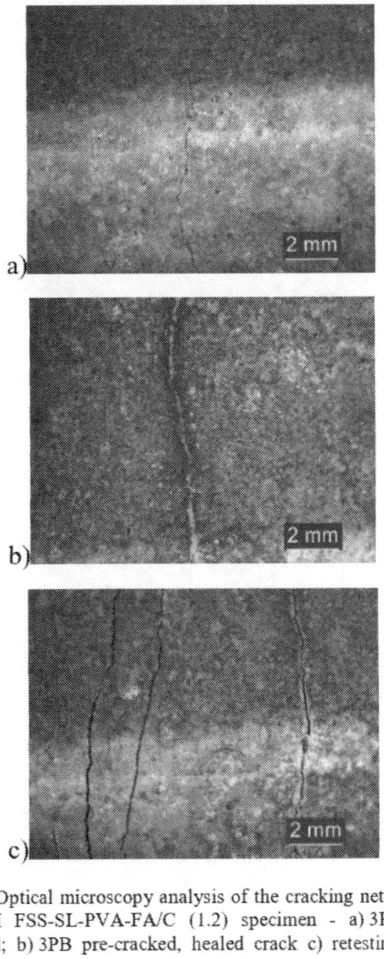

Fig. 9 Optical microscopy analysis of the cracking network of the SH FSS-SL-PVA-FA/C (1.2) specimen - a) 3PB pre-cracked; b) 3PB pre-cracked, healed crack c) retesting: new failure cracks

4 Summary and conclusions

The experimental study regarding the mix design optimization approach for the SH-PVA-FECM, by replacing the limestone filler (L) with slurry limestone paste (SL), a waste generated by local industry, generally proves positive results. The fresh state aspect of the mixes shows clear improvement of the SL mixture, more cohesively than the initial L mixture. At the same time, the fresh state features of the FSS-SL-PVA-FA/C (1.2) indicate future possible reduction of liquid content. This approach is an on-going study for material optimisation.

In terms of mechanical performance, the materials indicate a similar range of values which encourage the use of the SL waste as addition to the high performance composites. The Self-healing capacity is also comparable, supporting the viability of the replacement.

Another important conclusion of the study is the good match within the SH-FECM matrix of the class F fly ash (FA) and the Limestone slurry (SL), which develops good overall performance, both regarding the fresh state and also the hardened state of the material. This could open the path of a high ecologically impacted valorisation of both additions: the fly ash and also the slurry limestone, with increased beneficial, environmental effect.

Acknowledgements

The current experimental program was carried out in the framework of the PhD. Research Program of the first author at the Technical University of Cluj-Napoca. All the experimental work has been carried out in the laboratory of NIRD. URBAN-INCERC, Cluj-Napoca Branch, Romania and was supported by the Research Programmes for Smart Specialization, Sustainable Territorial Development, Environment Preservation and Resilience of Building Heritage – "CONCRET", PN 18 35 04 03: "Research for valorising the inert, hydraulic late or pozzolanic mineral additions in innovative cementitious materials for resilient structures, in the context of implementing the "Circular Economy" concepts in Romania and PN 19 33 04 02: "Sustainable solutions to ensure health and safety of the population in the open innovation context and environmental protection", financed by the Romanian Government.

References

1. V.C. Li, *Engineered Cementitious Composites (ECC) Material, Structural, and Durability Performanc,* in Nawy E. (Editor), Concrete Construction Engineering Handbook, Chapter 24, CRC Press (2008), pp. 1-78
2. D. Snoeck, *Self-healing and microstructure of cementitious materials with microfibres and superabsorbent polymer,* PhD. Thesis, Ghent University, Ghent, Belgium (2015)
3. N. De Belie, E. Gruyaert, A. Al-Tabbaa, P. Antonaci, C. Baera, D. Bajare, C. Litina, *A review of*

self-healing concrete for damage management of structures, Advanced Materials Interfaces **5** (2018) 17

4. C. Baeră, H. Szilagyi, A. Lăzărescu, *Developing Fibre Engineered Cementitious Materials with Self-Healing abilities (SH-FECM) by using polyvinyl alcohol fibres and supplementary powder addition*, Wit Derkowski et al. (Eds.), Proceedings of the fib Symposium 2019 held in Kraków, Poland (2019), pp. 387-294
5. C. Baeră, H. Szilágyi, C. Mircea, A. Hegyi, D. Snoeck, *Fibre engineered cementitious materials (FECM) optimisation by the use of hybrid polypropelene fibres and supplementary powder addition*, Proceedings of the 17th International Multidisciplinary Scientific GeoConference & EXPO – SGEM 2017, Albena, Bulgaria (2017), pp. 169-176
6. V.C. Li, C.K.Y. Leung, *Steady state and multiple cracking of short random fiber composites*, ASCE Engineering Mechanics **188** (1992) 11, pp. 2246-2264
7. V.C. Li, C. Wu., S. Wang, A. Ogawa, T. Saito, *Interface Tailoring for Strain-Hardening Polyvinyl Alcohol - Engineered Cementitious Composites (PVA-ECC)*, ACI Materials Journal **99** (2002) 5, pp. 463-472
8. V.C. Li, H. Stang, H. Krenchel, *Micromechanics of Crack Bridging in Fibre-Reinforced Concrete*, Materials and Structures **26** (1993) 8, pp. 486-494
9. V.C. Li, S. Wang, C. Wu, *Tensile strain-hardening behavior of polyvinyl alcohol engineered cementitious composites (PVA-ECC)*, ACI Materials Journal **98** (1997) 6, pp. 483-492
10. V. Mechtcherine, F. de Andrade Silva, M. Butler, D. Zhu, B. Mobasher, S.-L. Gao, E. Mäder, *Behaviour of strain-hardening cement-based composites under high strain rates*, Journal of Advanced Concrete Technology **9** (2011) 1, pp. 51-62
11. E.-H. Yang, *Designing added functions in Engineered Cementitious Composites*, University of Michigan, Ann Arbor (2008), pg. 293.
12. E.-H. Yang, Y. Yang, V.C. Li, *Use of High Volumes of Fly Ash to Improve ECC Mechanical Properties and Material Greenness*, ACI Materials Journal **104** (2007) 6, pp. 303-311
13. V.C. Li, E. Herbert, *Robust Self-Healing Concrete for Sustainable Infrastructure*, Journal of Advanced Concrete Technology **10** (2012) 6, pp. 207-218
14. D. Snoeck, N. De Belie, *Mechanical and self-healing properties of cementitious composites reinforced with flax and cottonised flax, and compared with polyvinyl alcohol fibres*, Biosystems Engineering **111** (2012) 4, pp. 325-335
15. H. Szilagyi, C. Baeră, A. Hegyi, A. Lăzărescu, *Romanian resources of waste and industrial by-products as additions for cementitious mixtures*, Proceedings of the 18th International Multidisciplinary Scientific GeoConference SGEM 2018, 30 June – 9 July, Albena, Bulgaria (2018), pp. 325-332
16. C. Baeră, *Self-Healing capacity of Advanced Engineered Cementitious Composites*, PhD. Thesis, Technical University of Cluj-Napoca, Cluj-Napoca, Romania (2016)
17. A. Lăzărescu, C. Baeră, H. Szilagyi, *A Review on the Development of Fly Ash-Based Geopolymer Materials Using Romanian Local Raw Materials*, Proceedings of the EUROCOALASH 2019, Dundee, Scotland, UK (2019), pp. 145-156
18. H. Szilagyi, C. Baeră, A.C. Mircea, A. Lăzărescu, *Dezvoltarea betonului verde prin integrarea compozițională a adaosurilor minerale locale (subproduse sau deșeuri)*, Paper proceedings of the research conference on constructions, economy of constructions, architecture, urbanism and territorial development, INCD URBAN-INCERC, Bucharest, Romania **15** (2019), pp. 147-154
19. ASRO, *SR EN 196-1:2016 Methods of testing cement. Determination of strength*, Romanian Standards Association, Bucharest, Romania (2016)
20. ASRO, *SR EN 1015-11:2002/A1:2007 - Methods of test for mortar for masonry. Determination of flexural and compressive strength of hardened mortar,* Romanian Standards Association, Bucharest, Romania (2007)

Production of concrete compatible biogranules for self-healing concrete applications

Merve Sonmez[1], and *Yusuf Cagatay* Ersan[1,2,*]

[1]Abdullah Gul University, Department of Civil Engineering, TR-38080, Kayseri, Turkey

[2]Hacettepe University, Department of Environmental Engineering, TR-06800, Ankara, Turkey

Abstract. Recently, cost-efficient nitrate reducing biogranules were suggested as an alternative to axenic microbial cultures for development of microbial self-healing concrete. In a marine environment, biogranule containing microbial self-healing concrete showed simultaneous self-healing of cracks and immunisation against rebar corrosion. Yet, information about the production strategy of these biogranules and their compatibility with a mortar matrix is limited. This study presents the production of biogranules and their compatibility with mortar specimens when incorporated at dosages between 0.36% to 4.30% w/w cement (0.25% to 3% of bacteria w/w cement). In-house produced biogranules composed of 70% bacteria and 30% of minerals w/w of biogranule were used for the compatibility tests. In test mortars, calcium formate (CF) and calcium nitrate (CN) were used as regular nutrient admixtures, and nutrient content was set identical in every batch. Up to 2.9% incorporation, biogranules had no significant influence on the fresh properties of mortar. More than 2.9% incorporation caused poor workability and a 26% decrease in 3-Day compressive strength of biomortar specimens. Overall, the biogranules produced are compatible with a cementitious matrix up to 2.9% w/w cement, and even up to 3.6% if early age strength is not essential, which makes biogranules one of the most compatible microbial healing agents among the suggested agents in the literature.

1 Introduction

The formation of microcracks is a prevalent issue in reinforced concrete structures, which requires continuous and delicate monitoring for detection, and labour intensive maintenance. Unless treated properly, microcracks can pave the way for reinforcement corrosion and compromise durability of the structures. In the last decade, significant effort was put on development of self-sensing and self-healing concrete in order to minimize external maintenance practices. Exploiting microbes and microbial induced calcium carbonate precipitation (MICP) can be mentioned among the popular self-healing concrete development strategies. Metabolic by-products (i.e. CO_2) produced by heterotrophic microbes has appeared to be useful for creating an autonomous repair system in concrete, as microbially produced CO_2 turns into CO_3^{2-} (carbonate) in the alkaline environment of concrete and precipitates as $CaCO_3$ (calcite) inside the cracks.

Various microbes have already been tested for the proof of the principle and several useful strains were identified for the development of microbial self-healing concrete (Table 1). Therefore, it is safe to claim that the proof and design stages of microbial self-healing concrete are almost over. Currently, the field requires optimization and improvement of the developed microbial self-healing concrete to enable cost-efficient upscaling of the technology. Therefore, economic feasibility, practicality of the proposed self-healing strategy and the delicacy of the microbial healing agent should be evaluated carefully. Moreover, the constraints for the dosages of typical components of microbial self-healing concrete (nutrients, protective carriers and microbial healing agents) should be determined.

One of the suggested microbial cultures for development of microbial self-healing concrete is granulated non-axenic culture, so called activated compact denitrifying core (ACDC) [1]. One of the advantages of granulated microbial cultures are their layered structure obtained with systematic placement of bacteria based on the required redox potentials of their metabolic activities [2]. ACDC biogranules were defined to have a core community composed of highly active nitrate reducing microbiome that were surrounded by aerobic bacteria, extracellular polymeric substances and calcium based minerals [1]. The layers surrounding the core community ensure the self-protection capability of ACDC biogranules and avoid the need for protective carriers such as porous aggregates or microcapsules [3], which make it advantageous over axenic cultures in terms of practicality, shelf-life and cost-efficiency. ACDC biogranules specifically granulated for concrete application were determined to be compatible with a cementitious matrix and are useful for crack healing in concrete at a bacteria dosage of 0.50% w/w cement [1,4]. Moreover, at the same dosage, ACDC granules simultaneously inhibit reinforcement bar corrosion, by

*Corresponding author: yusuf.ersan@agu.edu.tr

producing the anodic corrosion inhibitor NO_2^-, and heal a 300-µm-wide crack in 28 days in marine-similar conditions [5]. The former provides a bio-based alternative to the use of chemical corrosion inhibitors and enhances the sustainability of the cementitious materials. The added corrosion inhibition function may also accelerate the transition of the technology from concept to market.

Table 1 Bacteria successfully applied and suggested for development of microbial self-healing concrete

Study	Bacterial strain - pathway	Protective carrier
Wiktor and Jonkers [6]	*Bacillus alkalinitrilicus* – aerobic oxidation of organic carbon	Light weight aggregates
Wang et al. [7,8]	*Bacillus sphaericus* – urea hydrolysis	Hydrogels Microcapsules
Silva et al.[9,10]	*Bacillus sphaericus* – urea hydrolysis	Diatomaceous earth
	Cyclic EnRiched Ureolytic Powder (CERUP)	Self-protected culture
Ersan et al. [1,11]	*Diaphorobacter nitroreducens* – anoxic oxidation of organic carbon (nitrate reduction metabolism)	Granular activated carbon
	Activated compact denitrifying core (ACDC) anoxic oxidation of organic carbon (nitrate reduction metabolism)	Self-protected granular culture
Palin et al. [12]	*Bacillus halmapalus* – aerobic oxidation of organic carbon	Calcium alginate beads
Tziviloglou et al. [13]	*Bacillus cohnii* – aerobic oxidation of organic carbon	Expanded clay
Khaliq and Ehsan [14]	*Bacillus subtilis* – aerobic oxidation of organic carbon	Light weight aggregates
Davies et al. [15]	*Bacillus pseudofirmus* – aerobic oxidation of organic carbon	Expanded perlite

Owing to the aforementioned advantages, ACDC biogranules are worth investigating further in detail for up-scaling of microbial self-healing concrete technology. Concrete admixtures should be evaluated for their compatibility with the cementitious matrix at different dosages and such information about the compatibility of ACDC biogranules remains unknown. Therefore, this study aims to define the dosage constraints for the use of ACDC for development of microbial self-healing concrete by evaluating the fresh and hardened properties of various cementitious composites containing 0.36% to 4.30% of ACDC biogranules w/w cement (0.25% to 3% of bacteria w/w cement).

2 Materials and methods

2.1 Cultivation of self-protected biogranules

Biogranules with an activated denitrifying core were used as the microbial agent in tested biomortar mixtures. Fresh biogranules were produced in a cylindrical sequencing batch reactor (effective h= 60 cm, Ø= 12.4 cm and 50% volume exchange ratio). Old dry biogranules, which had been cultivated in a previous study [1,4] were ground to a size of less than 0.212 mm and used as seed for cultivation of the fresh biogranules in this study. The bioreactor was operated with four batch-cycles/day, and each cycle consisted of three periods. The period sequence was set as 180 minutes of anoxic period followed by 175-180 minutes of aerobic period and finally 0-5 minutes of settling period. Anoxic period was obtained by 120 minutes simultaneous fill/draw period with an upflow influent flow rate of 0.5 mL/s and an extra 60 minutes of sole anoxic period. The aerobic period was obtained by aeration with an upflow air velocity of 0.8 cm/s.

The reactor was fed with an alkaline minimal nutrient medium (Table 2) to enable cultivation of microorganisms, resilient enough to survive the alkaline and micronutrient deficient concrete environment. The pH of the feed solution was set to 10.2 by using 10 M NaOH solution.

Table 2. Composition of the minimal nutrient medium prior to pH adjustment to 10.2

Compound	Concentration (mg/L)
$NaHCOO$	3239
$NaNO_3$	331
$Ca(NO_3)_2$	267
$MgSO_4.7H_2O$	89
KH_2PO_4	13

At the end of 2 months of operation, biogranules occupied 90% of the overall bioreactor content and thus we started to harvest biogranules for the compatibility tests. Granule harvesting was repeated biweekly for the following 5 months. Harvested biogranules were dried in a ventilated electric oven at 60°C for 48 hours. Dry granules were further allocated into three portions based on their particle sizes (0.45 to 2 mm) and stored in closed containers at room temperature. The allocated portions had sizes of 0.45 to 0.85 mm, 0.85 to 1mm and 1 to 2 mm. Dry granules having particle sizes either more than 2 mm or less than 0.45 mm were returned to the bioreactor.

The bacteria content of the biogranules was determined by using the ratio of volatile suspended solids (VSS) to total suspended solids (TSS). Bacteria and biogranule contents presented in this study represent the

VSS and TSS, respectively. VSS and TSS analyses were conducted according to the standard methods [16].

2.2 Chemical admixtures serving as nutrients for microbial induced $CaCO_3$ precipitation

As mentioned above, the investigated biogranules use anoxic oxidation of carbon via nitrate reduction metabolism for both energy generation and growth. Therefore, to induce $CaCO_3$ precipitation, the biogranules require an organic carbon source as an electron donor and a nitrate source as an electron acceptor. Commercially available concrete admixtures, namely, calcium formate (CF) and calcium nitrate (CN), were used in mortar mixtures as sources of electron donor and electron acceptor, respectively. Throughout the experiments, the nutrient contents of different biomortars were kept identical to clearly distinguish the compatibility of biogranules with the mortar matrix.

2.3. Testing compatibility of ACDC biogranules with mortar

2.3.1 Preparation of the mortar mixes

The biomortar specimens containing identical amounts of nutrients and various amount of bacteria were prepared according to the standard procedure described in EN 196-1. DIN EN 196-1 standard sand (1350 g), CEM I 42.5R cement (450 g) and, tap water (225 g) were used as structural ingredients with a weight ratio of 3:1:0.5. The CF and CN contents of the biomortars were 5.00% (22.50 g) and 2.00% (9.00 g) w/w cement (Table 3). These admixtures' percentages were previously optimized by considering the effect of their weight percentages on the bioavailability of nutrients as well as the fresh and hardened properties of mortar [17].

Table 3. Composition of the tested mortar and biomortar mixtures

Mortar mixtures[1]	ACDC[2] Content (g)	CF (g)	CN (g)
Reference	-	-	-
Control	-	22.50	9.00
Bio-0.25%	1.61	22.50	9.00
Bio-0.50%	3.21	22.50	9.00
Bio-0.75%	4.82	22.50	9.00
Bio-1.00%	6.43	22.50	9.00
Bio-1.50%	9.64	22.50	9.00
Bio-2.00%	12.86	22.50	9.00
Bio-2.50%	16.07	22.50	9.00
Bio-3.00%	19.28	22.50	9.00

[1] Sand:cement:water contents 1350 g: 450 g:225 g were fixed in all mixes and the percentages represent wt bacteria/wt cement in biomortars.

[2]ACDC: Biogranules (70% bacteria + 30% minerals) with an activated compact denitrifying core. CF: Calcium formate. CN: Calcium nitrate.

Together with a plain mortar mixture, hereafter referred to as reference, which didn't include microbial or chemical admixtures, a control mortar mixture that included chemical admixtures (5% CF and 2% CN), and a series of biomortars with a bacteria content varying in the range of 0.25% to 3.00% were investigated (Table 3).

The particle size distribution of the incorporated dry biogranules, which varied between 0.45 to 2 mm, were set identical in each biomortar mix. Accordingly, in each biomortar mix, 50 % of the total biogranule content were between 1.00 to 2.00 mm in size, 35% of the total biogranule content were between 0.85 to 1 mm in size and 15% of the total biogranule content were between 0.45 to 0.85 mm in size (Table 4).

Table 4 Fixed size distribution of the dry biogranules prior to biomortar incorporation

Mortar mix	Size distribution of the incorporated dry biogranules (g)		
	0.45-0.85 mm	0.85-1.00 mm	1.00-2.00 mm-
Bio-0.25%	0.24	0.56	0.81
Bio-0.50%	0.48	1.12	1.61
Bio-0.75%	0.72	1.69	2.41
Bio-1.00%	0.97	2.25	3.21
Bio-1.50%	1.45	3.37	4.82
Bio-2.00%	1.93	4.50	6.43
Bio-2.50%	2.41	5.62	8.04
Bio-3.00%	2.89	6.75	9.64

2.3.2 Evaluation of the fresh mortar properties

Throughout the experiments, fresh mortar properties were determined via setting test and flow table test. In order to evaluate the strength development of different biomortars, compressive strength tests were conducted on Day 3, Day 7 and Day 28.

Setting tests were performed based on the ASTMC807 standard by using automatic vicat equipment (Matest E044N Vicatronic, Italy). Based on ASTM C191-04a and NBN EN-1008 standards, 60 minutes and 90 minutes were set as the acceptable lower limit for initial setting and final setting times, respectively. In ASTM C191-04a, for a single operator within the laboratory, the standard deviation of setting time of hydraulic cement is given as 12 minutes for initial setting time and 20 minutes for the final setting time throughout the range of 49-202 minutes and 185-312 minutes, respectively. Since no standard value about the standard deviation of setting times of cement mortar is available, same standard deviations were used to identify significant differences between different biomortars tested in this study.

Flow table tests were conducted according to the NBN EN 1015-3 standard. Considering the NBN EN 1015-3 standard, the maximum acceptable deviation between subjected specimen flow and the reference flow was determined as 10% for significance. Obtained data from these tests were used to determine the consistency and workability of different biomortars.

2.3.3 Evaluation of the hardened mortar properties

Compressive strength tests were conducted to evaluate the hardened properties of different mixtures and determine the strength development of different biomortars. Tests were conducted according to the ASTM C109 standard. Cubic specimens of 50x50x50 mm size were poured by using cubic moulds and moulded specimens were stored in sealed bags for 24 hours. After demoulding, cubic specimens were stored again in sealed bags at room temperature until the test day. Specimens were tested via compressive strength testing machine with a maximum load capacity of 300 kN (UTC-5700, Turkey) on Day 3, Day 7 and Day 28.

3 Results and discussion

3.1. Properties of the obtained biogranules

Fresh new biogranules with activated compact denitrifying core were successfully produced by using old dry biogranules (Figure 1).

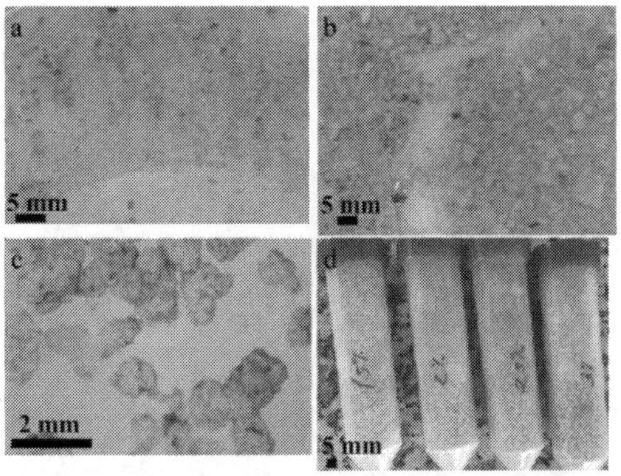

Figure 1 Successful production of biogranules with activated compact denitrifying core (ACDC). Old dry ACDC used as seed for production of fresh ACDC (a); fresh ACDC on a tray prior to drying (b); dry fresh ACDC of 1-2 mm in size under microscope (c); dry ACDC ready for mortar incorporation (d).

Biogranules produced in this study showed similarities to the ones produced in previous studies. Solid matter analyses (VSS:TSS ratio) revealed that, by weight, 70% and 30% of the biogranules were bacteria and inorganic matter, respectively, which was similar to the composition of previously produced biogranules [1,4]. Therefore, it can be claimed that the production of ACDC biogranules with a similar organic and inorganic matter content is possible even if the operational parameters slightly vary. Yet, the exact microbial composition of the fresh biogranules was unknown, which prevents their further comparison with the previously produced ACDC biogranules.

The size distribution of the harvested granules, after drying at 60°C, are given in Figure 2. Since, granules having particle sizes between 0.45 and 2 mm were used for the tests, approximately 34% of the harvested granules were not useful (Figure 2). The least proportion of the dry granules had particle sizes larger than 2.00 mm (Figure 2). Approximately 30% of the dry granules were between 1.00 to 2.00 mm in size, which corresponds to almost half of the testable dry granules (Figure 2).

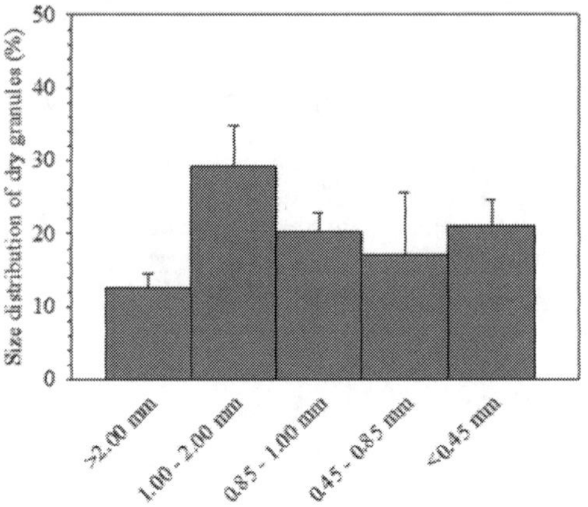

Figure 2 Particle size distribution of the dry fresh ACDC biogranules. Error bars represent the standard deviation of the mean of different harvests (n= 10)

3.2. Maximum tolerable biogranule dosage in fresh mortar

The effect of biogranules on fresh cement mortar was investigated by comparing setting and flow behaviour of different biomortar mixes with reference and control mortars. Obtained setting times and flow data for the tested fresh mortar mixes are given in Figure 3 and Table 5, respectively.

Both initial and final setting times of control mortar and all of the tested biomortars were significantly different from the reference mortar according to the defined standard deviation (12 minutes for initial setting and 20 minutes for final setting) in ASTM C191-04a. Initial and final setting values for control mixture were 55 ± 12 minutes and 172 ± 20 minutes, respectively. These findings were in good agreement with the previous findings where initial and final setting times of CF5.00-CN2.00 mixture were found as 65 ± 12 minutes and 146 ± 20 minutes [17].

Incorporation of bacteria, in the form of ACDC, with a dosage up to 2.50 % w/w cement did not cause any significant change in the setting times of the fresh cement mortar when compared to the control mixture. When the dosage was increased to 3.00%, a significant decrease occurred in the initial setting time. The recorded initial setting time for the corresponding biomortar mix was 34 ± 12 minutes, which was far from the acceptable initial setting limit (60 minutes) defined in ASTM C191-04a.

In all biomortar mixtures, ACDC biogranules had no influence on the final setting time when compared to the control mortar. These results were in parallel with the previous observation.

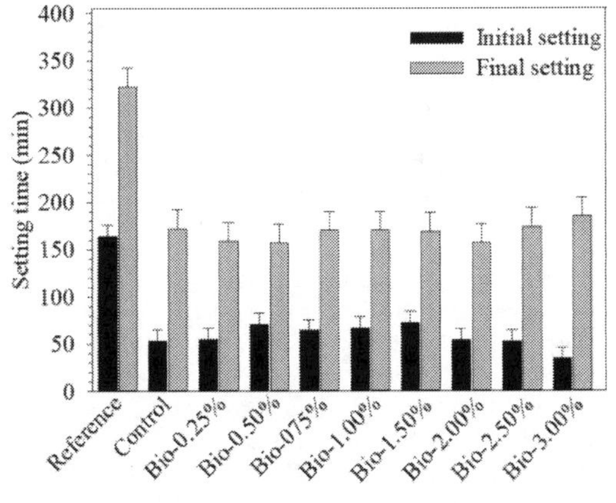

Figure 3 Variation of initial and final setting times in biomortars. Error bars represent standard deviation values of ±12 min for the initial setting time and ±20 min for the final setting time according to ASTM C191-04a.

Ersan et al. [18] incorporated bacteria, in the form of ACDC biogranules, with a dosage of 0.50% w/w cement into a mortar mixture that contains 2.00% calcium formate and 3.00% calcium nitrate. It was reported that ACDC has no significant influence on either of the setting times of the mixture when compared to the control mortar containing calcium formate and calcium nitrate. When the influences on setting times are considered, findings in this study enable us to extend the tolerable ACDC dosage in fresh biomortars up to 3.60% w/w cement (2.50% of bacteria w/w cement).

Flow table results were evaluated in two different ways and presented in Table 5. At first, flow values of different mixtures were compared with the flow value of reference mortar. Secondly, the comparison was made considering the flow value of control mortar. Both comparisons revealed that when bacteria were added in the form of ACDC, addition of bacteria up to 2.00% w/w cement had no influence on workability of the mortar. However, addition of bacteria (as ACDC) more than 2.50% w/w cement negatively affected the rheology of mortars, decreasing the workability and thus resulting in poor compaction (Figure 4).

Either comparisons based on reference or control flow values led to contradictory conclusions. On the one hand, Bio-2.50% and Bio-3.00% mortars had flow values that corresponded to 96% and 91% of the reference flow (Table 5). Based on this comparison, both mortars were in the tolerable zone in terms of workability criteria given in NBN EN 1015-3. On the other hand, when compared to the control flow, flow values of Bio-2.50% and Bio-3.00% mortars were 93% and 89%, respectively, which led to elimination of Bio-3.00% due to its poor workability. When the flow table test results were combined with the compaction quality observations of different biomortars, it was found out that among the two approaches, comparisons with respect to control mixture were more meaningful in terms of determining the impact of biogranule dosage on workability, particularly for bacteria dosages higher than 2.00% w/w cement. Even though the flow values of Bio-2.50% and Bio-3.00% mortars were in the acceptable zone, good compaction of such mortars was not possible (Figure 4).

Table 5 Flow table results of different mixtures with respect to reference and control mortars[1]

Mix names	Flow (mm)	Flow (% vs R)	Flow (% vs C)
Reference	118	100	98
Control	121	103	100
Bio-0.25%	116	98	96
Bio-0.50%	124	105	103
Bio-0.75%	125	106	103
Bio-1.00%	123	104	102
Bio-1.50%	123	104	102
Bio-2.00%	121	103	100
Bio-2.50%	113	96	93
Bio-3.00%	108	91	89

[1]R: Plain mortar as reference, C: Control mixture contains 5% of calcium formate and 2% of calcium nitrate w/w cement. The percentages in biomortars represent wt bacteria/wt cement.

Figure 4 Compaction problems observed in Bio-2.50% and Bio-3.00% after demoulding.

Considering both workability and setting time results of different biomortars, the maximum tolerable biogranule dosage in bimortars could be determined as 3.60% w/w cement (2.50% of bacteria w/w cement, in the form of ACDC). Biogranule dosages higher than 3.60% w/w cement are not recommended since they negatively affect the initial setting time and the compactability.

When organic matter is in contact with mortar matrix, set retardation occurs [19]. Microbial cells are composed of organic matter, and when cells die, the organic matter is released into the mortar matrix, which causes a delay in the setting time [18]. For instance, addition of an unprotected denitrifying microbial culture, *Diaphorobacter nitroreducens*, with a dosage of 0.5% w/w cement into the mortar mix increased both initial and final setting time of mortar by 40 minutes [18]. The setting results were significant in terms of the self-protection capability of biogranules. Findings reveal that the addition of bacteria in the form of biogranules avoids possible set retardation effect of bacteria, especially at high dosages.

The three stages that the loss of microbial healing agent is highly possible are during mixing, during the setting period and during the shrinkage of pores due to further hydration [20,21]. Yet, in this study, no set retardation was observed up to a bacteria dosage of 3.00% w/w cement, which confirmed that the bacteria inside the ACDC were not in contact with the mortar matrix whilst the mortar was setting. Therefore, it can be claimed that the mineral coating surrounding the biogranules was effective in protecting the bacteria during mixing and setting periods. The results were in accordance with the previous findings [3], which reported the survival of self-protected ACDC after mortar incorporation.

3.2. Compatibility of ACDC biogranules with hardened concrete

Compressive strength tests were conducted on Day 3, Day 7 and Day 28 and the strength development of different mixtures were evaluated accordingly. There was no significant difference between the early age strengths of the reference specimen (σ=29±4 MPa) and the control specimen (σ=30±2 MPa). This observation was consistent with our previous findings [17]. Furthermore, increasing the ACDC content up to 2.86% w/w cement (Bio-2.00%) did not cause any significant change in the 3-Day compressive strength of the mortar (Figure 5). However, further increase in the biogranule content (i.e. Bio-2.50% and Bio-3.00%) caused a decrease of 26% in 3-Day compressive strength values. Nonetheless, at the end of 7 days, all specimens had similar compressive strength values, which were around 40 MPa, and at the end of 28 days, the control mortar and biomortars even surpassed the reference mortar in terms of compressive strength. The 28-Day compressive strength of the reference mortar was recorded as 43±3 MPa (Figure 5). The compressive strength values of control mortar and biomortars were similar to each other and varied around 56 MPa, which corresponds to a 30% increase in compressive strength. The latter was consistent with our previous findings that addition of calcium formate and calcium nitrate with a dosage of 5% w/w and 2% w/w, respectively, led to 20% increase in compressive strength with respect to the 28 days compressive strength of reference mortar [17]. Therefore, the observed strength increase could be attributed totally to the positive influence of chemical admixtures (i.e. calcium formate and calcium nitrate) on strength development. Therefore, it was safe to claim that biogranules produced in this study were compatible with the mortar matrix and did not influence the hardened properties of the mortar significantly. The negative influence of ACDC biogranules was observed only at high incorporation dosages (i.e. Bio-2.50% and Bio-3.00% where ACDC content were 3.60% and 4.30%, respectively) and only on early age strength development. Yet, it is not clear whether the poor early age strength development was due to a high amount of biogranules or due to poor compaction of the corresponding specimens (Figure 4). Excepting Bio-2.50% and Bio-3.00% mortars, all the other biomortars showed decent strength development at early age.

Many researchers claim that direct addition of bacteria improves early age strength development due to MICP [22–24]. In this study, even though the nutrients and biogranules were added directly into the mixture, no significant influence was observed in early age strength development. There might be two reasons for this observation. Firstly, ACDC biogranules have a $CaCO_3$-$Ca_3(PO_4)_2$ coating which minimizes interaction of the core bacteria with the nutrient containing mortar matrix. Therefore, resuscitation of ACDC and MICP were unlikely whilst undergoing early age strength development. Secondly, the biogranules were produced at pH around 10 and further dried. In order to resuscitate the biogranules a pH value of around 10 is required and the pH of a mortar mix normally varies around 13.5, which might be the reason why no MICP occurred during the early age strength development.

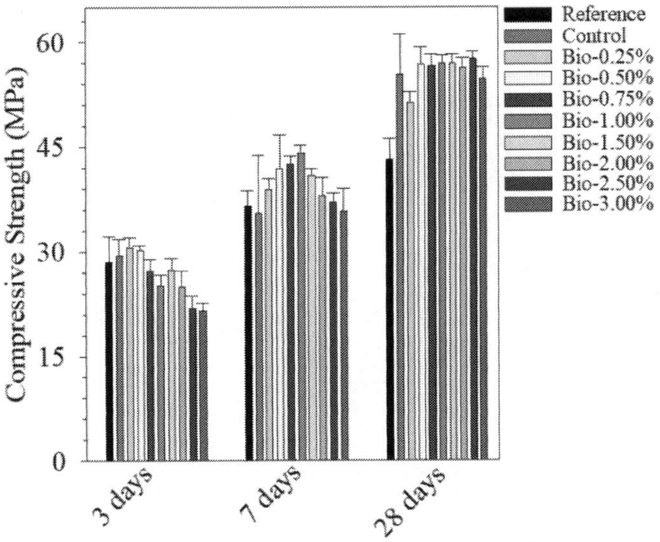

Figure 5 Compressive strength development in concrete with different amount of biogranules. Error bars represent the standard deviation of the mean (n=3) Statistical significance was determined based on one-way ANOVA test (p=0.05).

4 Conclusions

Biogranules with an activated compact denitrifying core (ACDC) can be produced from old dry ACDC granules and the properties of produced fresh granules are similar to the old ones.

A $CaCO_3$-$Ca_3(PO_4)_2$ coating surrounding the ACDC biogranules minimizes the interaction between the mortar matrix and the core bacteria, and contributes to the biogranules' compatibility with mortar.

ACDC biogranules are compatible with fresh mortar and do not interfere with their setting and workability if their dosage in the mortar mixture is around 3.60% w/w cement (i.e. 2.50% of bacteria in the form of ACDC w/w cement) or less. At higher dosages, the initial setting time drops to 34 minutes, which is intolerable for site applications. Furthermore, workability and compaction quality decreases at higher dosages.

ACDC biogranules have no influence on strength development and they are compatible with the hardened mortar. Yet, if their dosage in the mortar mixture is around 3.60% w/w cement (i.e. 2.50% of bacteria in the form of ACDC w/w cement) or higher, then early age strength development is negatively affected and the 3-Day compressive strength of biomortar specimens decreases by 26%.

Considering the influence of ACDC biogranules on both fresh and hardened properties of cement mortar, it can be stated that the tolerable biogranule content in biomortars is around 3.60% w/w cement (i.e. 2.50% of bacteria in the form of ACDC w/w cement). The best performance is achieved when the ACDC content is less than 3.60% w/w cement.

Acknowledgments

The research leading to these results received funding from the Scientific and Technological Research Council of Turkey (TUBITAK) under grant agreement no 118M768 (Determination of the optimum biogranule content in biogranule containing self-healing bioconcrete).

References

[1] Y.Ç. Erşan, E. Gruyaert, G. Louis, C. Lors, N. De Belie, N. Boon, Front Microbiol. **6**, 1228 (2015).

[2] D. Gao, L. Liu, H. Liang, W.-M. Wu, Crit Rev Biotechnol. **31**, 137–152 (2011).

[3] Y.Ç. Erşan, H. Verbruggen, I. De Graeve, W. Verstraete, N. De Belie, N. Boon, Cem Concr Res. **83**, 19–30 (2016).

[4] Y.C. Ersan, D. Palin, S.B. Yengec Tasdemir, K. Tasdemir, H.M. Jonkers, N. Boon, N. De Belie, Front Built Environ. **4**, 70 (2018).

[5] Y.Ç. Erşan, K. Van Tittelboom, N. Boon, N. De Belie, Sci Rep. **8**, 14092 (2018).

[6] V. Wiktor, H.M. Jonkers, Cem Concr Compos. **33**, 763–770 (2011).

[7] J. Wang, D. Snoeck, S. Van Vlierberghe, W. Verstraete, N. De Belie, Constr Build Mater. **68**, 110–119 (2014).

[8] J. Wang, H. Soens, W. Verstraete, N. De Belie, Cem Concr Res. **56**, 139–152 (2014).

[9] F.B. Silva, *Up-scaling the production of bacteria for self-healing concrete application*, PhD Thesis,

Ghent University, (2015).

[10] F. Silva, N. De Belie, N. Boon, W. Verstraete, Constr Build Mater. **93**, 1034–1041 (2015).

[11] Y.Ç. Erşan, E. Hernandez-Sanabria, N. De Belie, N. Boon, Cem Concr Compos. **70**, 159–170 (2016).

[12] D. Palin, V. Wiktor, H.M. Jonkers, Biomimetics. **2**, 13 (2017).

[13] E. Tziviloglou, V. Wiktor, H.M. Jonkers, E. Schlangen, Constr Build Mater. **122**, 118–125 (2016).

[14] W. Khaliq, M.B. Ehsan, Constr Build Mater. **102**, 349–357 (2016).

[15] M. Alazhari, T. Sharma, A. Heath, R. Cooper, K. Paine, Constr Build Mater. **160**, 610–619 (2018). doi:10.1016/j.conbuildmat.2017.11.086.

[16] APHA, AWWA, WEF, *Standard Methods for Examination of water and wastewater*, 22nd Edition, (2012).

[17] Y.Ç. Erşan, Y. Akın, *Optimizing nutrient content of microbial self-healing concrete*, Proceedings of the Sixth International Symposium on Life-Cycle Civil Engineering (IALCCE 2018), 28-31.10.18, Ghent, Belgium, (2019).

[18] Y.Ç. Erşan, F.B. Silva, N. Boon, W. Verstraete, N. De Belie, Constr Build Mater. **88**, 196–203 (2015).

[19] B. Khan, B. Baradan, Sci Vis. **8**, 71–78 (2002).

[20] H.M. Jonkers, A. Thijssen, G. Muyzer, O. Copuroglu, E. Schlangen, Ecol Eng. **36**, 230–235 (2010).

[21] Z.B. Bundur, A. Amiri, Y.C. Ersan, N. Boon, N. De Belie, Cem Concr Res. **98**, 44–49 (2017).

[22] S. Ghosh, M. Biswas, B.D. Chattopadhyay, S. Mandal, Cem Concr Compos. **31**, 93–98 (2009).

[23] N. Chahal, R. Siddique, A. Rajor, Constr Build Mater. **37**, 645–651 (2012).

[24] Z. Basaran-Bundur, M.J. Kirisits, R.D. Ferron, Cem Concr Res. **67**, 237–245 (2015).

https://doi.org/10.1051/matecconf/201928901003

Durability of self-healing concrete

Nele De Belie[1,*], *Bjorn* Van Belleghem[1], *Yusuf Çağatay* Erşan[1,2,3], *Kim* Van Tittelboom[1]

[1]Magnel Laboratory for Concrete Research, Department of Structural Engineering, Faculty of Engineering and Architecture, Ghent University, Technologiepark Zwijnaarde 60, B-9052 Ghent, Belgium
[2]Centre for Microbial Ecology and Technology (CMET), Faculty of Bioscience Engineering, Ghent University, Coupure Links 653, B-9000 Ghent, Belgium
[3]Department of Environmental Engineering, Hacettepe University, Beytepe Campus, TR-06800, Ankara, Turkey

Abstract. Application of self-healing concrete reduces the need for expensive maintenance and repair actions. However, the durability of self-healing concrete has only been scarcely investigated. Here, recent results are presented regarding the resistance of self-healing concrete to chloride ingress. For self-healing concrete with macro-encapsulated polyurethane, chloride profiles and electron probe microanalysis indicated that this mechanism was efficient to reduce the chloride penetration into the crack and from the crack into the concrete matrix [1]. Furthermore, the corrosion behaviour of reinforced concrete specimens subjected to cyclic exposure with a NaCl solution was studied [2]. The electrochemical measurements indicated that autonomous crack healing could significantly reduce the corrosion in the propagation stage. No visual damage could be detected on the rebars after 44 weeks of exposure. On the contrary, cracked specimens without integrated self-healing mechanism, reached a state of active corrosion after 10 weeks of exposure and after 26 weeks clear pitting damage was observed on the rebars. While self-healing by encapsulated polyurethane is complete after one day, bacteria-based products take several weeks to heal a 300 µm crack. Bacterial granules containing denitrifying cultures released nitrite as an intermediate metabolic product which protected the reinforcement during the crack healing process [3].

1 Introduction

The appearance of small cracks (< 300 µm in width) in concrete is almost unavoidable, not necessarily causing a direct risk of collapse for the structure, but surely impairing its functionality, accelerating its degradation, and diminishing its service life and sustainability. The loss of performance results in the need for increased investments on maintenance and/or intensive repair/strengthening works. While concrete contains inherent self-healing properties, this autogenous healing mechanism is only efficient for small cracks. The phenomenon itself has been well studied, but the maximum healable crack sizes mentioned range between 10 to 100, sometimes 150 µm [4]. Autogenous healing is difficult to predict or rely on and only occurs in the presence of water. Therefore, concrete has been adapted to stimulate autogenous healing and in addition, mechanisms for autonomous healing have been invented. Even stimulated autogenous mechanisms (e.g. by introduction of superabsorbent polymers into the concrete) are generally limited to healing crack widths of about 100-150 µm, taking several weeks or even months to heal cracks completely, and heavily rely on the environmental conditions (mainly presence of water). In contrast, most autonomous self-healing mechanisms can heal cracks of 300 µm, even sometimes up to more than 1 mm, and usually act faster (complete healing obtained in a time span of 1 day up to 3-4 weeks depending on the system). Autonomous healing agents include micro- and macro-encapsulated polymers or minerals, and bacteria based systems (encapsulated or not). For encapsulation, a whole range of shell materials have been investigated and tailored for use in cementitious matrices [4]. Researchers have made progress regarding changeable properties (flexible in fresh concrete to survive concrete mixing, while brittle in hardened concrete to release the contents at crack occurence) and improved bond with the cementitious matrix. The suggested optimum dosages usually range between 0.5% to 10% by weight of cement, to obtain sufficient healing in combination with a limited effect on mechanical properties of concrete. The long-term stability of encapsulated polymers remains a point of concern (especially due to the unavoidable permeability of the capsule shells). This is less problematic when encapsulating bacterial spores or mineral additives, since their "reactivity" with moisture penetrating through the shell is much lower.

A comprehensive review on self-healing concrete, aiming for damage management of structures, was recently compiled in a combined effort by COST CA15202 "SARCOS" members [4]. This review reveals the key challenge that the self-healing additions up-to-date are produced at lab scale and self-healing efficiency is only shown at paste/mortar level. The rare existing demonstrators up-scaled in concrete structures often

* Corresponding author: nele.debelie@ugent.be

show insufficient [5, 6] or not yet proven [7, 8] self-healing efficiency. Further, they lead to new questions, such as effect of variability in the design parameters and longevity of the embedded system [9]. One of the reasons for reduced efficiency after upscaling to concrete is the significant dilution of the additives when maintaining the dosage relative to cement weight; however, keeping the same dosage in proportion to the total volume, results mostly in an unacceptable strength decrease and high healing agent costs. Another challenge mentioned by the cited review paper is that the durability of self-healing concrete elements when exposed to, for instance, chlorides has only been scarcely investigated. No long-term durability results of self-healing concrete are available, and even results for accelerated durability testing in lab conditions are scarce. Mostly, durability is assessed indirectly through parameters such as gas and water permeability, surface resistivity and capillary absorption. Although presently more than 200 studies reported on improved durability aspects of concrete due to bacteria-mediated limestone formation, virtually all are based on short-term laboratory studies with indirect indicators. It is suggested that further research should focus on durability of the healed structures e.g. resistance to chloride diffusion and carbonation, corrosion, freeze/thaw, salt crystallization, etc. [4].

2 Literature on durability of self-healing concrete

Limited studies are available on durability of self-healing concrete. Mostly, durability is linked to indirect indicators like water absorption or water permeability. The few articles that treat actual durability mostly focus on chloride ingress and a few on actual reinforcement corrosion.

2.1 Chloride ingress in lab-scale specimens

In our previous studies [10, 11], capsules containing polyurethane (PU) prepolymers were embedded in the concrete to release their content when cracks appear. The resistance to chloride penetration of autonomously healed mortar was evaluated experimentally by using the NT Build 443 diffusion test. At the age of 28 days, after saturation in a 4 g/l $Ca(OH)_2$ solution for 7 days, the specimens were placed in a closed container with a 165 g/l NaCl solution. In cracked mortar, the chloride diffusion coefficients in the zone immediately around the crack significantly increased compared to uncracked mortar. There was a limited effect of the crack width for cracks of 100 μm to 300 μm), in which case the calculated service life decreased from 103 to 23 years (about 80%). Autonomous crack healing had a beneficial influence on the resistance against chloride diffusion. However, for about one third of the cracks the healing mechanism failed, probably due to shifting of the capsules, excessive capillary forces in the capsules, etc. Nevertheless, the service life of autonomously healed structures by means of encapsulated polyurethane increased on average by about 100% compared to cracked structures, e.g. for self-healed cracks of 200 μm a service life of 44 years was calculated based on the mean apparent diffusion coefficient. However, there was a lot of variability depending on the proper functioning of the healing mechanism. If the healing mechanism works efficiently, the crack is sealed well and almost no chlorides will penetrate into the crack, resulting in a predicted service life similar to sound structures.

Van Belleghem et al. [12] applied a similar system with macro-encapsulated PU, and after an accelerated diffusion test, the chloride concentration in the cracked zone was reduced by 75% or more as a result of the self-healing action. As a result, the service life of steel reinforced self-healing concrete slabs in marine environments could amount to 60–94 years as opposed to only seven years for ordinary (cracked) concrete. Subsequent life cycle assessment calculations indicated important environmental benefits (56%–75%) for the ten CML-IA (Center of Environmental Science of Leiden University–Impact Assessment) baseline impact indicators, which are mainly induced by the achievable service life extension.

Van den Heede et al. [13] proposed a modified chloride migration test based on the one prescribed in NT Build 492 to check self-healing concrete for marine environments. They used samples with an artificially induced crack (width: 0.3 mm, depth: 25 mm) in the bottom cast surface of concrete cylinders and have screened four macro-encapsulated polymer-based healing agents that way. A highly repeatable self-healing efficiency value of 100% could only be obtained for a commercial low-viscosity PU precursor.

The findings of Darquennes et al. [14] clearly indicated that autogenous healing at early age limits the chloride penetration and improves the structural durability, particularly for mixtures containing blast-furnace slag. Moreover, it was shown that the chloride migration test is an adequate method to monitor the self-healing process. Finally, self-healed specimens could be simply regarded as specimens characterized by a smaller crack.

Sahmaran et al. [15] investigated the chloride transport properties of engineered cementitious composites (ECC) under combined mechanical and environmental loads. Immersion and salt ponding tests were conducted to determine chloride ion transport properties. Under high imposed bending deformation, the preloaded ECC beam specimens revealed microcracks less than 50 μm and an effective diffusion coefficient significantly lower than that of the similarly preloaded reinforced mortar beam because of the tight crack width control. Through the formation of microcracks, a significant amount of self-healing was observed within the ECC cracks subjected to NaCl solution exposure.

Li and Li [16] investigated engineered cementitious composite's (ECC) durability in terms of maintaining its unique tensile characteristics under combined mechanical loading and aggressive chloride conditions. ECC specimens were preloaded to 0.5, 1.0, and 1.5% tensile strain levels; immersed in chloride solution for 30, 60, and 90 days; and reloaded until failure. This study revealed that the reloaded specimens retained multiple microcracking behaviour and tensile strain capacity greater than 2.5%, while the average crack width increased from 50 µm to 100 µm. Self-healing in ECC under chloride exposure was evident in terms of recovery of initial material stiffness and tensile strain capacity. Under severe marine environmental conditions, ECC remained durable.

Liu et al. [17] studied Engineered Cementitious Composites (ECC) with self-controlled tight microcracks and self-healing capacities under aggressive sulfate and chloride conditions. Resonant frequency (RF) and mechanical properties were determined for ECC specimens that were preloaded to 1% strain and exposed to sulfate and sulfate-chloride solutions. The performance of ECC was found not to be adversely affected by the aggressive solutions. Instead, self-healing of the microcracks was observed leading to partial recovery of the mechanical properties. It was also found that ECC tends to heal faster and more completely in sulfate solutions than in water.

Abro et al. [18, 19] used the rapid chloride ion migration test under the steady-state regime to assess the efficiency of self-healing of cracked mortar containing crystalline and expansive admixtures. The target crack widths were achieved by splitting cylindrical specimens into two pieces and reassembling them by a steel band with silicon tapes of required thickness attached between them on both edges. They suggested to choose a high potential (36 V) for the steady-state migration test to minimize the test duration (to 24-36 h) because otherwise the ongoing self-healing would affect the measurements. The self-healing capacity with respect to resistance to chloride penetration could be evaluated with this test, and an index for the recovery of resistance to chloride penetration due to self-healing could be defined by using the rate of chloride ions that pass through the crack. The diffusion coefficients obtained increased almost linearly in proportion to the crack width. For larger crack widths, they found that the mixture incorporating calcium sulfoaluminate as an expansive admixture and bentonite as a swelling agent, together with a crystalline admixture including di-sodium carbonate and organic calcium ions, showed better self-healing performance than the mixture incorporating only the crystalline admixture. Also, depending on the types of self-healing techniques, it was noted that there is a limit in the crack width that can be healed.

Van Mullem et al. [20] compared the chloride ingress in cracked mortar with and without superabsorbent polymers (SAP). The specimens were saturated in a chloride solution during 1 or 5 weeks after which the chloride ingress could be visualised using silver nitrate. The specimens, which were healed prior to chloride immersion, had a significantly lower chloride ingress. The SAPs were able to delay the chloride ingress, as well as limit the influence of the crack on the chloride ingress.

Asensio et al. [21] carried out an experimental campaign, involving the study of 342 samples, half of which contained a crystalline admixture. An enhanced recovery of strength and an improved crack sealing ability were observed for samples containing the healing agent.

Azarsa et al. [22] presented results on the influence of crystalline admixtures on the strength, self-healing, and durability characteristics of concrete, using the rapid chloride permeability (RCP), surface/bulk electrical resistivity and water permeability test, self-healing test, compressive strength test and salt ponding test. The water permeability coefficient decreased by three times whereas the self-healing ratio increased by a higher rate by adding crystalline admixtures to the concrete mix.

Liu et al. [23] studied the mechanism and self-healing kinetics of cement paste with chelating agent in marine environment. They found that cracks with an initial width of 400 µm could heal completely within 2 days. Brucite and calcite were the main minerals of the reaction products of self-healing in cement paste with chelating agent submersed in seawater, and their percentage in the reaction products of self-healing was about 60% and 30%, respectively.

Perez et al. [24] investigated the durability of self-healing micro-concrete by freeze-thaw cycles and salt spray test. The ultra-high-strength reinforced micro-concrete contained an autonomous healing mechanism based on the reaction of an epoxy compound enclosed within silica microcapsules and amine functionalised silica nanoparticles distributed within the cementitious matrix. The results of capillary water absorption tests, complemented by electron microscopy analysis, confirmed the enhanced durability of the autonomously healed material in both the freeze-thaw and the salt spray tests, as compared to the reference micro-concrete.

Palin et al. [25, 26] studied beads consisting of calcium alginate encapsulated bacterial spores and mineral precursor compounds for application in low-temperature marine environments. The material displayed an excellent crack-healing capacity, reducing the permeability of 0.4 mm-wide cracks by 95%, and 0.6 mm-wide cracks by 93% following 56 days of submersion in artificial seawater at 8 °C. Healing of the cracks was attributed to autogenous precipitation, autonomous bead swelling, magnesium-based mineral precipitation, and bacteria-induced calcium-based mineral precipitation in and on the surface of the bacteria-based beads.

2.2 Chloride ingress in real-scale specimens

Araujo et al. [27] carried out real-scale tests on concrete beams of 2.5 m x 0.40 m x 0.20 m reinforced with four ribbed steel bars (Ø=10 mm). Poly(methyl methacrylate) and glass capsules were used to encapsulate the healing agent. The beams were positioned slightly tilted with the cracks in the upper surface and a 3 wt% NaCl solution was flowing over the beams for 1 day per week, during 6 consecutive weeks. The tests revealed that the resistance of cracked concrete against chloride could be increased when incorporating self-healing properties. Most sampled cracks were completely healed, with chloride ingress being similar to uncracked concrete.

2.3 Corrosion monitoring of embedded steel reinforcement

Van Tittelboom et al. [28] carried out an accelerated corrosion test on mortar samples with embedded reinforcement bars, which showed that the presence of cracks accelerates the onset and propagation of corrosion at the interface between the rebar and the cementitious matrix. They highly increased the corrosion rate by using a high chloride concentration and by applying a potential difference of 15 V over the length of the rebar. They found that both manual crack healing and autonomous healing with polyurethane delayed the onset and propagation of corrosion. The difference in corrosion behaviour was proven by visual evaluation of the rebar surface and by X-ray radiographic analysis.

3 Chloride ingress in cracked self-healing concrete with macro-encapsulated polyurethane

Our recent research focused on an accurate measurement of (perpendicular-to-crack) chloride ingress in case of autonomous crack healing by encapsulated polyurethane [1]. Chloride profiles were obtained by grinding followed by potentiometric titration and by Electron Probe Micro Analysis (EPMA).

Autonomous crack healing was achieved by embedding glass capsules (length 35 mm, internal diameter 3.00 mm, wall thickness 0.175 mm) filled with a polyurethane (PU) precursor in the concrete matrix. At the moment of crack appearance, the polyurethane precursor is released from the capsules and polymerizes within 24 to 48 hours when it comes in contact with moisture in the concrete matrix. Artificial standardized cracks were created in cylindrical concrete specimens (Ø = 100 mm; h = 50 mm) by introducing thin brass plates with a thickness of 300 μm into the fresh concrete, which were removed after 28 d of curing at 20°C and a relative humidity (RH) of > 95%. The removal of the plates in the specimens with the autonomous healing mechanism caused breakage of the capsules which were mounted through holes in the brass plates. After 48 h curing of the PU, the specimens were coated with an epoxy coating on all sides except the test surface. Three traditional and three self-healing specimens were subjected to an accelerated chloride diffusion test according to NT Build 443 (immersion in a 165 g/l NaCl solution at 20°C) for a total period of 605 days. Then, the specimens were split along the direction of the crack. One half was used for profile grinding after which the total chloride content of the powders was determined through an acid-soluble extraction in a nitric acid solution followed by a titration against silver nitrate. The other half was split again and a piece of 50 mm x 40 mm was polished, embedded in an epoxy resin and analysed by EPMA (pixel size of 50 μm). After a calibration procedure [30], the chloride contents can be obtained from the EPMA analysis.

In the traditional concrete samples, a very high chloride concentration was noticed at the crack wall (1.39 m% per concrete mass near the crack mouth and 1.1 m% deeper down in the crack). Deeper down in the crack, the influence of the crack caused increased chloride contents until 14 to 18 mm away from the crack wall. Self-healing specimens showed a much smaller increase in chloride content near the crack wall, especially deeper down in the crack, where the chloride concentration at the crack wall was only 0.27 m%. Furthermore, an increased chloride concentration was only seen up to 4 mm away from the crack near the crack mouth, while in the deepest zone of the crack, the influence of the crack could be seen until a distance of 8 mm from the crack surface. Overall, at all positions, the measured chloride contents were much below those measured in the traditional concrete. These conclusions can also be clearly drawn from the chloride analysis done by EPMA. The healed specimens show much less chloride ingress along the crack than the traditional cracked specimens (Fig. 1). Therefore, it is clear that self-healing with macro-encapsulated polyurethane will reduce the chloride concentrations in the crack and reduce the perpendicular-to-crack chloride penetration.

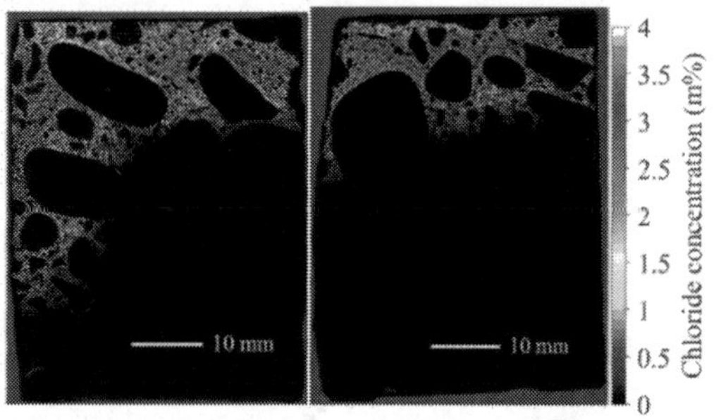

Fig. 1. Example of EPMA maps showing chloride penetration for cracked specimens of traditional concrete (left) and cracked specimens of self-healing concrete (right). The crack is at the left-hand side of the images. Redrafted after [1].

4 Corrosion of steel reinforcement in self-healing concrete with macro-encapsulated polyurethane

Subsequently, the corrosion behaviour of reinforced concrete specimens subjected to cyclic exposure with a NaCl solution was studied [2]. Steel reinforced concrete prisms with dimensions of 120 mm x 120 mm x 500 mm were made for corrosion monitoring, based on a design by Kessler et al. [29].

The steel reinforcement (BE500S) consisted of two parts which were electrically separated from each other: a centrally located steel bar anode (∅ = 10 mm, length 250 mm, concrete cover 25 mm) and two cathodic reinforcement cages made by four longitudinal rebars (∅ = 8 mm) and five stirrups (∅ = 6 mm), connected by an insulated copper wire. Both the anodic and cathodic part of the reinforcement were connected to the exterior by means of a copper wire. The self-healing specimens contained two layers of six capsules with low (200 mPas at 25°C) or high viscosity PU (6700 mPas at 25°C) in the centre of the concrete prisms, one layer above and one layer below the anodic rebar.

After 28 d curing at 20°C and > 95% RH, a crack of around 300 μm was created by means of a three-point bending test. Capsule breakage resulted in release of the PU precursor inside the crack and 48 h of storage at 20 °C and 60% RH allowed the PU to polymerize. The concrete specimens were then exposed to 44 cycles of consecutively 1 day with 33 g/l NaCl solution in the cracked area (and 1.15 g/l $Ca(OH)_2$ solution further away from the crack), followed by a 6 day dry period, as described in [2]. The macro-cell corrosion current between anode and cathode (I_{macro}) was measured approximately once every hour with a low resistance ammeter. Weekly electrochemical measurements were performed at the end of the wet period with a potentiostat (Gamry Interface 1000E) using an internal Ag/AgCl reference electrode that was embedded in the concrete near the location of the anodic rebar, or with an external saturated calomel reference electrode (SCE) placed in the NaCl solution above the middle of the prisms. The corrosion potential (E_{corr}) was measured while anode and cathode were connected. Subsequently the anode and cathode were disconnected and after depolarization (16 h), the open circuit potential (OCP) of anode and cathode was measured and the corresponding driving potential (ΔE) was calculated as the difference of these potentials. Thereafter, the linear anodic and cathodic polarization resistance, $R_{P,A}$ and $R_{P,C}$ respectively, were measured. Finally, electrochemical impedance spectroscopy (EIS) was performed to determine the development of the concrete resistance (R_E). When all measurements were performed, the anode and cathode were connected again. All electrochemical measurements were performed weekly during the first 12 weeks of the wet-dry cycles and thereafter every two to four weeks. Only the macro-cell corrosion current and corrosion potential will be discussed here (Figure 2), more information can be found in [30].

The uncracked samples showed a non-active state of corrosion along the whole exposure period. No macro-cell corrosion current was detected and the corrosion potential was nearly constant at values of -70 to +80 mV versus Ag/AgCl. The cracked samples showed clear evidence of corrosion initiation within the first three weeks of exposure to NaCl solution: there is a steep rise in the macrocell corrosion current and a drop of the corrosion potential ranging from 120 to 280 mV. Generally, the corrosion behaviour of the self-healing samples with high viscosity PU (PU_HV_CAPS) was very similar to the corrosion behaviour of the cracked samples. However, the self-healing samples with low viscosity PU (PU_LV_CAPS) behaved more closely to the uncracked samples. The macro-cell corrosion current was negligibly small and the anodic polarization resistance was comparable to the values found for the uncracked samples. The mean rate of volumetric loss of steel only amounted to 0.042 mm³/week, which was 14 times lower than what was found for the untreated cracked samples. The rebars also showed no or very limited signs of damage due to corrosion after visual inspection.

5 Corrosion of steel reinforcement in self-healing concrete with bacterial granules

Self-healing by encapsulated polyurethane can proceed quickly due to the fast mechanism of crack filling and polyurethane curing. On the other hand, self-healing through biominerals produced by bacteria takes more time.

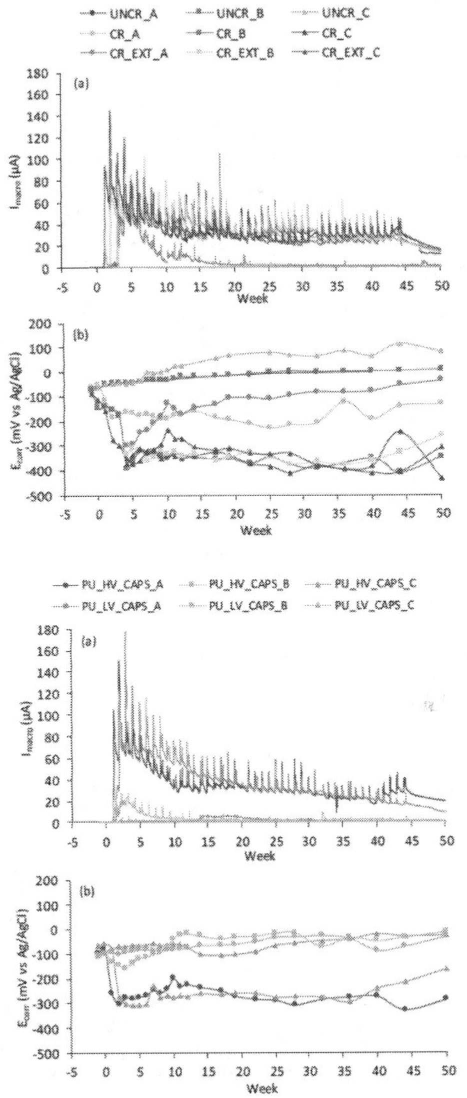

Fig. 2. The electrochemical corrosion parameters macro-cell corrosion current and corrosion potential for uncracked (UNCR) and cracked (CR) samples (top) and for self-healing samples with high (HV) and low (LV) viscosity PU (bottom). Redrafted after [30].

Complete healing of a 300 μm crack will generally take several weeks. Therefore, it was investigated whether the latter approach can avoid corrosion of cracked concrete in a chloride containing environment. Self-protected bacterial granules containing either ureolytic or denitrifying cultures were integrated in mortar specimens containing a steel rebar. Denitrification can be advantageous when the necessary nutrients are considered. Yeast extract, urea, calcium lactate and calcium glutamate used for urea hydrolysis, are organic compounds which may negatively affect concrete properties. In the case of denitrification, commercial concrete admixtures such as calcium formate and calcium nitrate can serve as nutrient source for NO_3^- reducing bacteria without any negative effect. Self-healing strategies using ureolytic bacteria lack the preventive action to avoid exposure of the steel surface to corrosive substances during the healing period. However, recent proof-of-concept studies show that it is possible to achieve nitrite production and hence corrosion inhibition by using NO_3- reducing bacteria [31].

Microbial NO_3^- reduction was investigated by using a non-axenic self-protected NO_3^- reducing culture called ACDC, produced as described in [3]. In corrosion experiments, as a control culture, a self-protected non-axenic ureolytic culture was used (CERUP, produced by Avecom NV, Belgium), with an effect on mortar properties similar to ACDC, while it can induce crack sealing through urea hydrolysis. Admixtures of $Ca(HCOO)_2$ and $Ca(NO_3)_2$ were used in the mortar with the ACDC, and $CO(NH_2)_2$ was used as a nutrient for the CERUP. A series of mortar specimens (40 mm × 40 mm × 160 mm) was prepared according to EN 196–1, with an embedded central smooth steel rebar (Ø = 8 mm). The microbial test mortar contained ACDC (0.5% wt/wt cement), $Ca(NO_3)_2$ (3% wt/wt cement) and $Ca(HCOO)_2$ (2% wt/wt cement) in addition to the components present in plain mortar. The positive control mortar contained NO_2^- (1.6% wt/wt cement) as a corrosion inhibitor in the form of $NaNO_2$ (2.4% wt/wt cement) and $Ca(HCOO)_2$ (2% wt/wt cement). Abiotic control mortar contained the nutrients of the microbe-based test batch, $Ca(NO_3)_2$ (3% wt/wt cement) and $Ca(HCOO)_2$ (2% wt/wt cement). The microbial self-healing control mortar contained CERUP (0.5% wt/wt cement) and $CO(NH_2)_2$ (5% wt/wt cement). The prepared mortar specimens were cured at a temperature of 20°C and a RH of >90% for 28 days. Artificial cracks were formed after 24 h curing according to the procedure described previously [32], by using 300 µm-thick brass plates with a semi-circular notch (Ø = 8 mm) against which the smooth rebar was positioned. At the end of the 28-days curing period, cracked and uncracked pieces (38 mm × 40 mm × 15 mm) were sawn from the prisms. The widths of the artificially created cracks were measured with a stereomicroscope after creation and at the end of the 28-days and 120-days experimental periods. Corrosion testing was done by placing conductive, insulated copper wires in contact with one end of the steel rebar to create an electrical connection with the data logging system. The cracked surfaces of the specimens (40 mm × 15 mm) were immersed in 0.5 M Cl^- solution and kept in contact with the solution for 28 and 120 days. Throughout the experiments, the dissolved oxygen concentration in the bulk solutions was 7.3 ± 0.2 mg/L and the pH was 9.1 ± 0.1. The electrochemical corrosion potential was monitored by recording the open circuit potential (OCP) versus a reference saturated calomel electrode (SCE) for 120 days. All data (versus reference SCE) were converted to values versus the standard hydrogen electrode (SHE), on the basis of the given equilibrium potentials. The obtained data were compared with the limit OCP value (–250 mV vs SHE). At the end of the experiments, the steel rebars were taken out of the tested specimens, cleaned and weighed.

The first 28 days of the Cl^- exposure experiment can be defined as the pre-sealing period and half of the specimens (3 of each series) were used to determine the corrosion behaviour of the rebars at that time. The other half of the mortar specimens were monitored further in Cl^- solution, until day 120 (combination of pre-sealing and post-sealing exposure to salt solution).

Among the cracked mortar specimens, only bacteria-based mortars showed complete healing of the cracks in a short period of time. Mortar containing ACDC granules showed complete crack closure in 28 days similar to the self-healing control mortar containing ureolytic bacteria. The healing material was stable in both mortars until the end of the experiment.

The OCP results revealed that rebar corrosion initiated in the cracked plain mortar after 16 days exposure to Cl^- solution (significant drop below –250 mV) whereas it took 44 days for the rebar in the abiotic control batch (Fig. 3). In both cases, after corrosion initiation, the OCP values dropped sharply below the critical value (–250 mV). The rebars in the self-healing control started to corrode on day 28. Although the crack was completely sealed in 28 days, the rebar protection performance of the self-healing control did not become any better than that of the other negative controls (i.e. abiotic control and plain mortar) during the post-sealing period between 28 and 120 days. The rebar embedded in the mortar specimens containing ACDC always showed OCP values higher than –250 mV throughout the 120-days experimental period, similar to those embedded in the positive-control mortars containing NO_2^-. In both types, no sharp decrease occurred in the OCP value throughout the test. In addition, the mass losses measured from the steel reinforcement showed that ACDC containing mortar could reach the rebar protection performance of uncracked mortar in both pre- and post-sealing periods, similar to the positive control. For the rebars embedded in ACDC containing mortars, the mass loss corresponded to 50% of the loss recorded for negative control specimens. When corrosion inhibition occurred, the rebars lost less than 2% of the total material (average rebar mass 5.5±0.2 g, n=30) at the end of 120 days exposure to salt solution. More information can be found in [3].

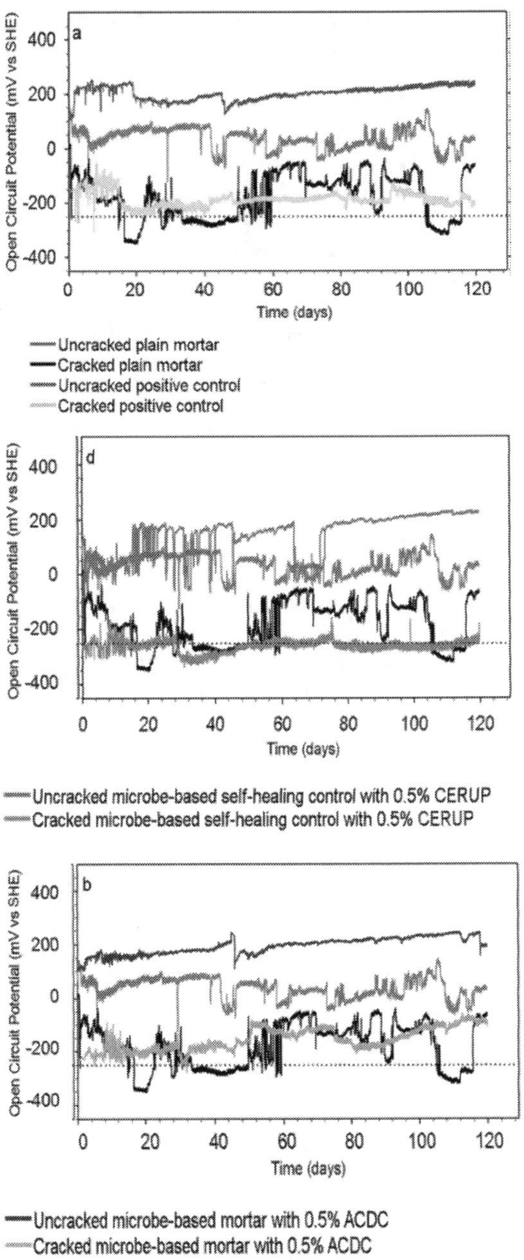

Fig. 3. OCP versus SHE for uncracked and cracked normal mortar (all graphs), uncracked and cracked positive control containing chemical nitrite as corrosion inhibitor (top), uncracked and cracked self-healing mortar with ureolytic CERUP (middle), uncracked and cracked self-healing mortar with denitrifying ACDC (bottom). Redrafted after [3].

6 Conclusions

Self-healing concrete has been the topic of increasing research attention over the last decade. Several methodologies to realize self-healing have been developed, but their effect on concrete durability and service life has only been scarcely investigated. Mostly, durability is assessed indirectly through parameters such as water absorption or permeability of cracked and healed concrete. Nevertheless, more recently several researchers also started investigating chloride ingress and corrosion of embedded reinforcement steel. In the current article, this literature is summarized. Furthermore, some highlights of our own recent findings are presented, regarding durability of self-healing concrete with macroencapsulated polyurethane on the one hand, and with granulated denitrifying bacteria on the other hand.

Through analysis of specimens that had been submersed in chloride solution, by titration on drilled powders and EPMA, it was shown that self-healing with macro-encapsulated polyurethane will reduce the chloride concentrations in the crack and reduce the perpendicular-to-crack chloride penetration. Furthermore, electrochemical measurements on reinforced prisms indicated that the corrosion behaviour of the self-healing samples with low viscosity PU behaved very similar to the uncracked samples. The macro-cell corrosion current was negligible and the anodic polarization resistance was comparable to the values found for the uncracked samples. The volumetric loss of steel only amounted to 0.042 mm^3/week, which was 14 times lower than what was found for the untreated cracked samples.

Self-healing implies that a crack formed in a concrete structure will activate a healing mechanism, leading to crack filling, regain of liquid tightness and/or mechanical properties. However, complete healing takes time, ranging from several hours (for encapsulated polymers) to several weeks (for encapsulated mineral or crystalline additions and bacteria). In this time span, corrosion of reinforcement steel can be initiated. In a simple corrosion test, it was shown that in the case of bacteria-based healing; only the denitrifying cultures were able to protect the reinforcement as efficiently as if the specimens contained a chemical nitrite based inhibitor. This can be ascribed to the release of nitrite as an intermediate metabolic product by the bacterial granules with a denitrifying core. In cracked mortar, rebar corrosion initiated after 16 days exposure to 0.5 M Cl^- solution (drop of open circuit potential below -250 mV). In the case of bacteria based self-healing using a ureolytic culture, corrosion continued as well, even though complete crack healing was seen after 4 weeks.

The proposed self-healing mechanisms will in future have to be tested for their efficiency in larger concrete elements and more realistic environmental conditions.

Acknowledgements

Part of the described research was performed in the framework of the ISHECO project (Impact of Self-Healing Engineered Materials on steel COrrosion of reinforced concrete) under the program SHE (Engineered Self-Healing materials). The research was funded by SIM (Strategic Initiative Materials in Flanders) and VLAIO (Flanders Innovation and

Entrepreneurship). The financial support from the foundation for this study is gratefully acknowledged.

References

1. B. Van Belleghem, Y. Villagrán Zaccardi, P. Van Den Heede, K. Van Tittelboom, N. De Belie, Con Build Mat (to be published)
2. B. Van Belleghem, S. Kessler, P. Van Den Heede, K. Van Tittelboom, N. De Belie, Cem Con Res **113**, 130-139 (2018)
3. Y.C. Erşan, K. Van Tittelboom, N. Boon, N. De Belie, Nature - Sci Rep **8**, 14092 (2018)
4. N. De Belie et al., Adv Mat Int, 1800074 (2018)
5. K. Paine et al., *RILEM pro123*, 387-394 (2018)
6. M. Araujo et al., Cem Con Comp **89**, 2602018 (2018)
7. N. De Belie N et al., *SLD*, Delft, 429-436 (2018)
8. R. Davies et al., Front Mat, 00051 (2018)
9. A. Kanellopoulos et al. Smart Mater Struct **26**, 45025 (2017)
10. M. Maes, N. De Belie, N. *Proc Concrete Solutions 6th Int. Conf. On Concrete Repair*, CRC Press, Thessaloniki, 409-415 (2016)
11. M. Maes, K. Van Tittelboom, N. De Belie. Con build mat **71**, 528-537 (2014).
12. B. Van Belleghem, P. Van den Heede, K. Van Tittelboom, N. De Belie. Materials **10**, 5 (2017)
13. P. Van den Heede, B. Van Belleghem, M.A. Araujo, J. Feiteira, N. De Belie, N. Key Eng Mat **761**, 152-158 (2018)
14. A. Darquennes, K. Olivier, F. Benboudjema, R. Gagne. Con build mat, **113**, 1017-1028 (2016).
15. M. Sahmaran, M. Li, V.C. Li. ACI mat J **6**, 604-611 (2007)
16. M. Li, V.C. Li, ACI mat J **108**, 333-340 (2011)
17. H.Z. Liu, Q. Zhang, C.S. Gu, H.Z. Su, V. Li. Con Build Mat **153**, 984-956 (2017).
18. F.R. Abro, A.S. Buller, K.-M. Lee, S.Y. Jang. Materials **12**, 1865 (2019)
19. F.R. Abro, K.S. Min, S.Y. Jang, K.-M. Lee, A.S. Buller. *Proc. ICSHM*, Yokohama (2019)
20. T. Van Mullem, R. Caspeele, N. De Belie. *Proc. Concrete Solutions 7th Int. Conf. on Concrete Repair*, Cluj Napoca. MATEC Web of Conferences (2019)
21. E.C. Asensio, S. Rigamonti, E. Gastaldo, L. Ferrara. *Proc. ICSHM*, Yokohama (2019)
22. P. Azarsa, R. Gupta, A. Biparva. Cem Con Comp **99**, 17-31 (2019).
23. H. Liu, X. Wang, X. Wu, H. Huang. Proc. ICSHM, Yokohama (2019)
24. G. Perez, J.L. Garcia Calvo, P. Carballosa, et al. Mag con res **23**, 1231-1242 (2017)
25. D. Palin, V. Wiktor, H.M. Jonkers. Biomimetics **2** (3), 13 (2017)
26. D. Palin, V. Wiktor, H.M. Jonkers. Smart mat struct **25** (8), 084008 (2018)
27. M. Araújo, S. Chatrabhuti, S. Gurdebeke, N. Alderete, K. Van Tittelboom, J.-M. Raquez, V. Cnudde, S. Van Vlierberghe, N. De Belie, E. Gruyaert, Cem Con Comp **89**, 260-271 (2018)
28. K. Van Tittelboom, B. Van Belleghem, M.N. Boone, L. Van Hoorebeke, N. De Belie. Adv Mat Int, 1701021 (2017)
29. S. Kessler, F. Hiemer, C. Gehlen, Beton- und Stahlbetonbau. **112**, 198–206 (2017)
30. B. Van Belleghem, PhD, Ghent University (2018).
31. Y.C. Erşan, H. Verbruggen, I. De Graeve, W. Verstraete, N. De Belie, N. Boon, Cem Con Res **83**, 19-30 (2016)
32. B. Van Belleghem et al. Con Build Mat **110**, 154–162 (2016)

MATEC Web of Conferences **289**, 01004 (2019) https://doi.org/10.1051/matecconf/201928901004
Concrete Solutions 2019

An experimental and numerical study on vascular self-healing cementitious materials

Anthony Jefferson [1]*, *Tharmesh* Selvarajoo[1], *Brubeck* Freeman[1], *Robert* Davies[1],

[1]Cardiff University, School of Engineering, UK

Abstract. This paper gives an overview of a combined experimental-numerical study on vascular self-healing (SH) systems for cementitious composite materials. The work aimed to bridge the gap between numerical and experimental investigations for this type of SH system and to provide a set of data for developing, calibrating and validating a finite element model for these materials. The study investigated both healing-agent transport and mechanical damage-healing processes, including healing-agent curing. The experimental programme included mechanical tests on notched concrete beams and compact direct-tension specimens with inbuilt vascular healing systems, as well as tests to measure the transport properties of healing-agent within discrete concrete cracks and through the concrete matrix. The new coupled model employs elements with embedded strong discontinuities to simulate cracks and mechanical healing behaviour. A damage-healing constitutive model is described that simulates multiple damage-healing 'events'. This mechanical model is coupled to discrete and continuum flow models that simulate healing-agent transport. The transport model accounts for pressurised and capillary flow, as well as curing-dependent flow properties. The main focus of this contribution is to show how these parallel programmes of work were combined so that the experimental observations guided the numerical developments and modelling questions were answered using experimental findings.

1 Introduction

Interest in self-healing concrete has grown considerably in recent years due to the potential of these materials to greatly improve the durability properties of concrete structures [1]. Much of the work undertaken on these materials has been experimental in nature but there has also been some numerical research aimed at producing computational models for these materials [2]. Whilst considerable progress has been made on models that simulate certain aspects of self-healing behaviour, there remain some key behavioural characteristics that current models do not represent well, which includes the simulation of simultaneous and multiple damage-healing 'events'. A further problem with much previous numerical work is that the models have been developed, calibrated and validated using little, or sometimes no, experimental work [2]. To remedy this situation, a combined experimental-numerical study has been undertaken at Cardiff University on a particular vascular self-healing system. The present paper gives an outline description of this work. A full description will be provided in a series of forthcoming publications.

2 Experimental Programme

The experimental programme of work focussed on the behaviour of a vascular self-healing system, which has healing-agent supply channels embedded in cementitious structural elements. The healing agent used was cyanoacrylate (PC20).

The various test series undertaken are summarised below.

(i) Notched fracture beam tests with a range of a) fixed healing periods, b) loading rates and c) healing-agent supply pressures (See Figs. 1 and 8).

Fig. 1. Notched concrete beam tests with embedded SH

(ii) Uniaxial tension tests, which involved forming a discrete crack in a doubly notched specimen and subsequently supplying healing agent to the crack. The healing agent was then allowed to cure. In this experimental series, multiple tests were conducted with different fixed crack-opening displacements and different curing periods.

(iii) Dynamic flow tests, which measured the variation of the capillary contact angle with flow velocity for the selected healing agent (See Fig. 2).

* Corresponding author: author@e-mail.org

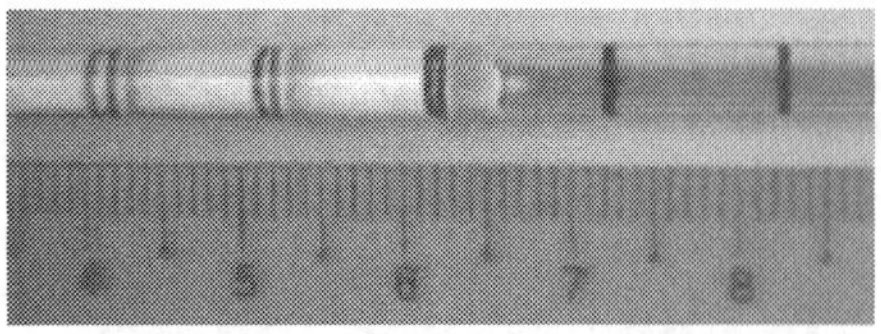

Fig. 2. Healing agent dynamic flow test

(iv) Curing front tests, in which the curing rate of healing agent was measured in a shallow 'bath' of agent adjacent to a concrete substrate (See Fig. 3).

Fig. 3. Cyanoacrylate curing with a concrete substrate

(v) Sorption tests that involved measuring the flow of healing agent into a concrete sample through a natural cracked surface (See Fig. 4).

Fig. 4. Concrete specimen with a natural crack face in a bath of cyanoacrylate

An illustrative graph from a test on a 255x75x75mm concrete beam (see Fig. 8), which had a central 5mm notch and embedded channels for supplying healing agent, is given in Fig. 5. In the test shown, the healing agent was pressurised to 0.1 bar above atmospheric pressure. The mechanical load was applied such that the Crack Mouth Opening Displacement (CMOD), measured with a clip gauge, increased at a rate of 0.0002mm/s.

It is evident from this test that the damage and healing processes were concurrent after the crack had formed. The results also include a control test in which no healing agent was supplied to the channels.

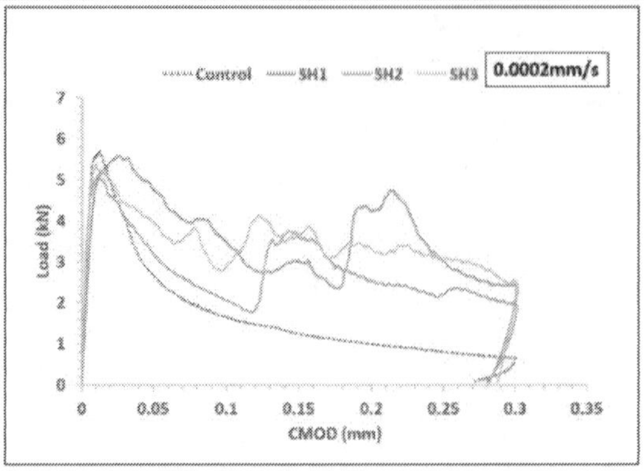

Fig. 5. Illustrative results from a damage-healing test.

3 Numerical model

The coupled numerical model includes components for simulating the flow of healing agent in discrete cracks, the transport of healing agent through a micro-cracked (and uncracked) concrete matrix, healing agent curing, cracking and healing. The key components of the model are now summarised.

3.1 Capillary flow model

Capillary flow in discrete cracks is simulated using the model proposed by Gardner et al [3], which is based on a modified Lucas-Washburn equation. The method has been previously validated against experimental data for a range of healing agents. The flow rate at the meniscus is given by equation (1);

$$\dot{z} = \left(p_{c0}(1-\beta_s) - \rho g z \sin(\varphi)\right) \Big/ \left(\frac{2\beta_m}{b(z)} + \eta\right) \tag{1}$$

where $\eta = A(z)\int_0^z 1 \Big/ \left(\frac{b(x)\beta_w}{2} + \frac{k(x)}{\mu}\right) A(x)\,dx$

and in which the superior dot denotes the time derivative, z is the capillary rise height; $b(x)$ is the channel (crack) opening; β_w, β_s and β_m are the wall, stick-slip and meniscus correction parameters respectively; φ is the capillary inclination angle; ρ is the density of liquid; g Is the gravitational acceleration; μ is the dynamic viscosity; t is the capillary rise time (s); the effective permeability term $k = b^2/12$, and α is the inclination of the capillary wall.

An example set of validation data for the flow of healing agents in natural cracks in concrete specimens is given in Fig. 6 [3].

 https://doi.org/10.1051/matecconf/201928901004

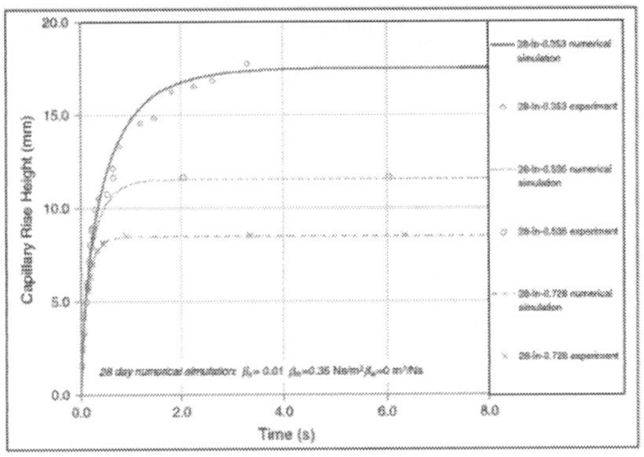

Fig. 6. Capillary rise height data at 28 days for inclined planar crack configuration.

3.2 Damage-healing model

The damage-healing part of the model is applied to a 'crack-plane', which is defined as the mid-surface of a narrow band of material that contains a macro-crack or an array of micro-cracks. The model relates the crack-plane traction vector ($\tilde{\boldsymbol{\tau}}$) to the relative displacement vector ($\tilde{\mathbf{u}}$), for which the damage-healing constitutive relationship is as follows:

$$\tilde{\boldsymbol{\tau}} = (1-\omega)\tilde{\mathbf{k}}^e\breve{\mathbf{u}} \; + \; h\,\tilde{\mathbf{k}}^h\left(\breve{\mathbf{u}}-\breve{\mathbf{u}}_h\right) \qquad (2)$$

in which $\omega \in [0,1]$ is the damage parameter, $h \in [0,\omega]$ is the proportion of healed material and $\breve{\mathbf{u}}_h$ is the permanent healing relative-displacement that ensures that the healing material forms in a stress-free state. $\tilde{\mathbf{k}}^e$ and $\tilde{\mathbf{k}}^h$ are elastic constitutive matrices for the original and healed material respectively. The inelastic component of $\tilde{\mathbf{u}}$ is computed from equation (3).

$$\breve{\mathbf{u}} = \tilde{\mathbf{u}} - \left(\tilde{\mathbf{k}}^e\right)^{-1}\tilde{\boldsymbol{\tau}} \qquad (3)$$

The crack opening at any position is assumed to be equal to the normal component of $\breve{\mathbf{u}}$, which is denoted by the scalar u_f.

The cohesive zone model has been applied to a 4-noded element with embedded strong discontinuity within a research finite element program developed at Cardiff University.

3.3 Healing agent curing

The curing of cyanoacrylate can be simulated using the following curing function (ϕ);

$$\phi(t,t_0) = 1 - e^{\frac{-\langle t-t_0\rangle}{\tau_h}} \qquad (4)$$

in which t and t_0 represent the current and initial time respectively, and τ_h is a healing time parameters. <> is a McCauley bracket, defined as $\langle x\rangle=0$ if $x\leq 0$ *and* $=x$ if $x>0$.

A comparison between equation (4) and some experimental data measured by the Cambridge Polymer group (2004) for the curing of a droplet of cyanoacrylate on glass is shown in Fig. 7.

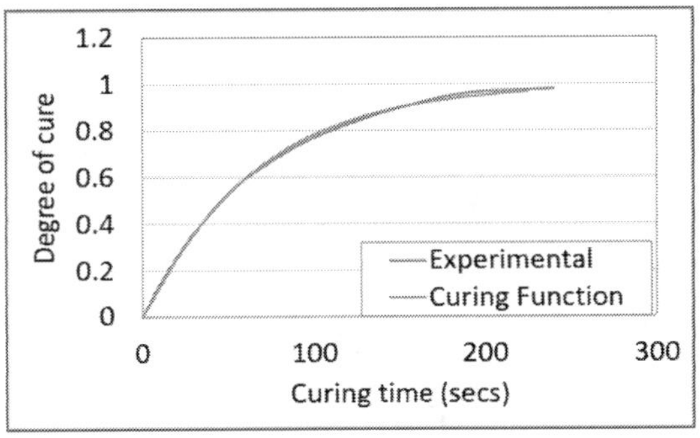

Fig. 7. Comparison between curing function (4) and experimental data from reference Cambridge Polymer Group (2004)

The volume of cured material (v_{cured}) is obtained from the following convolution integral;

$$v_{cured} = \int_{t_0}^{t} \phi(s,t_0)\frac{dv}{ds}\,ds \qquad (5)$$

During a time increment, new healing material may arrive, healed material may damage and damaged material may re-heal. To account for these factors the relative area of curing and cured material (a) at a particular point is given by;

$$a = a + \Delta a_c - \Delta a_{redam} + \Delta a_{rec} \qquad (6)$$

in which Δa_c is the incremental area of virgin filled crack, Δa_{redam} is the incremental area of re-damaged material and Δa_{rec} is the incremental area of re-filled cracks;

The total relative proportion of healed material is then computed from equation (7).

$$h = h\cdot e^{-\Delta t/\tau_h} + a\cdot(1-e^{-\Delta t/\tau_h}) \qquad (7)$$

This expression takes advantage of the semigroup property of the exponential function (equation 4) [4].

A much fuller description of the model, including full details of the damage-healing model, a description of the element with strong discontinuity, the diffuse surface curing model, the method used to track the increments of material arriving and curing, the matrix flow component and the computational details of the coupled non-linear solution procedure will be provided in a forthcoming journal paper on the model.

4. Example

The coupled model has been used to consider the set of experiments described in Section 2(i) of this paper. A particular example is presented here for illustration. Many more examples are presented in the forthcoming journal paper.

The test setup is illustrated in Fig. 8. The beam was loaded such that the crack mouth opening displacement (CMOD) increased at a rate of 0.001mm/s. The healing agent was supplied throughout the duration of the test. In addition, a control test was undertaken in which no healing agent was supplied.

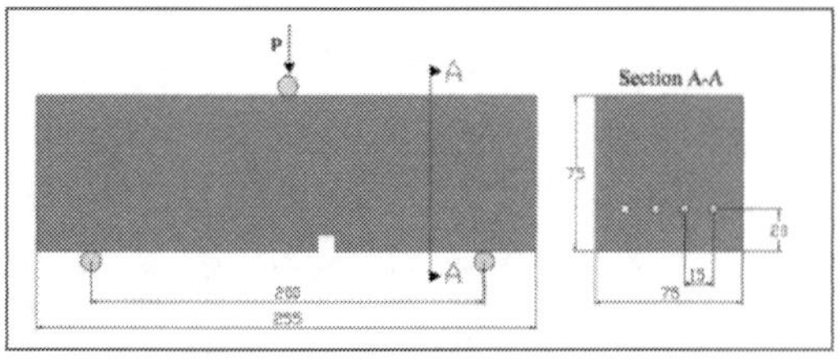

Fig. 8. Experimental arrangement

The component of the deformed mesh around the central crack is shown in Fig. 9. This shows the extent of healing at a time of 97 seconds from the start of the test.

A comparison between the model predictions and experimental data is provided in Fig. 10, in which Th and Th3 denote experimental results from healing tests. It may be seen that the coupled model provides a reasonable representation of the experimental behaviour.

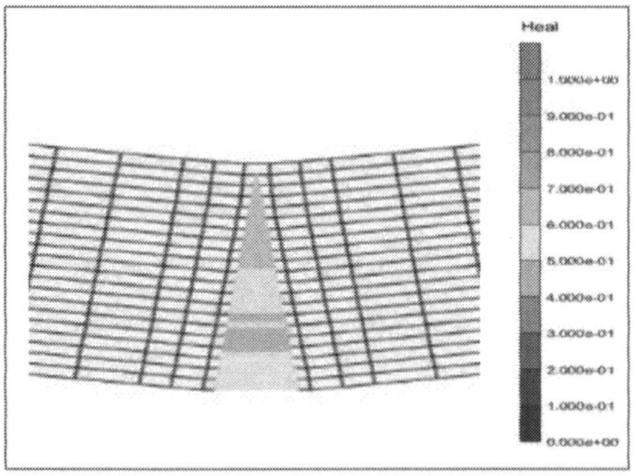

Fig. 9. Finite element mesh in deformed configuration with degree of healing

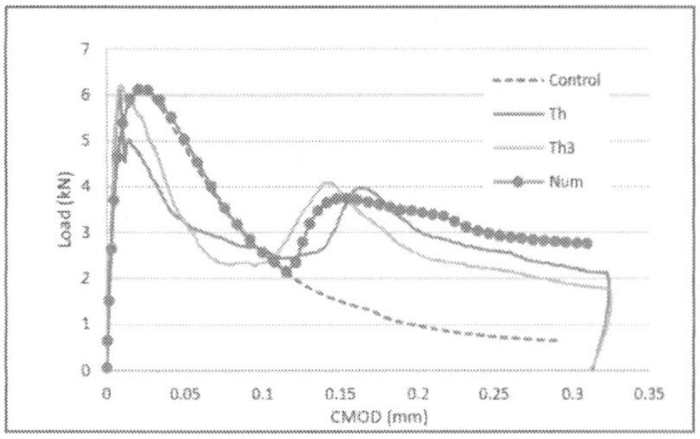

Fig. 10. Comparison between experimental and numerical results.

5 Concluding remarks

The data gathered in the experimental programme characterised the transport and curing properties of a healing agent in cracked cementitious structural elements. The tests also provided detailed information on damage-healing behaviour and shows how these processes interact when they occur simultaneously.

The coupled finite element model is able to represent the flow behaviour of a healing agent in a discrete crack and in a micro-cracked continuum. The model is also able to reproduce complex damage-healing behaviour with good accuracy.

References

1. N. De Belie at al. Review of Self-healing Concrete for Damage Management of Structures. Advanced Materials Interfaces. DOI: 10.1002/201800074
2. A. Jefferson, E. Javierre, B. Freeman, A. Zaoui, E.A.B. Koenders, and L. Ferrara.. Research Progress on Numerical Models for Self-Healing Cementitious Materials. Advanced materials interfaces, DOI: 10.1002/admi.201701378, (2018).
3. D.R. Gardner, A.D. Jefferson, A. Hoffman, R.J. Lark. Simulation of the capillary flow of an autonomic healing agent in discrete cracks in cementitious materials, Cement and Concrete Research , 58, (2014).
4. J. Mergheim, P. Steinmann. Phenomenological Modelling of Self-Healing Polymers Based on Integrated Healing Agents. Computational Mechanics 52 (3), 681–92, (2013).

Acknowledgement

The authors gratefully acknowledge the financial support from the UK Engineering and Physical Sciences Research Council (EPSRC) research grant EP/P02081X/1, Resilient Materials for Life (http://RM4L.com/).

Effect of oxygen permeability on corrosion properties in mortar specimens mixed with bacillus subtilis

Saaya Sugitani[1], *Keiyu* Kawaai[1,*], and *Isao* Ujike[1]

[1]Ehime University, Civil and Environmental Engineering, 3, Bunkyocho, Matsuyama, Ehime, Japan

Abstract. The effect of the degree of water saturation on oxygen permeability in mortar specimens mixed with an aerobic microorganism of bacillus subtilis (natto) under various water saturation conditions is explored in this study. In the experimental program an immersion test was conducted using distilled water and salt water (10%) under room conditions controlled at a constant temperature of 20°C. The oxygen permeability on the surface of steel bars embedded at different saturation conditions in mortar specimens were examined by electrochemical measurements e.g. half-cell potential measurements, cathodic polarization techniques. Based on the results obtained, the rate of oxygen permeability of the mortar specimens mixed with the bacillus subtilis (natto) was clearly reduced from the age of 28 days to 91 days. The humid environment is advantageous to the metabolic activity of bacillus subtilis (natto) which is likely to occur in saturated conditions compared to those in relatively dry environments. This led to more negative half-cell potentials in the mortar specimens mixed with the bacillus subtilis (natto) especially under the saturated conditions. The variations of half-cell potentials could be attributed to variations of cathodic polarization properties which is illustrated by decreased oxygen permeability in the specimens.

1 Introduction

Corrosion of steel bars in concrete proceeds in the presence of water and dissolved oxygen via electro-chemical reactions. When a corrosion cell is formed on the steel bars in concrete owing to chloride attack or carbonation, iron metal is ionized at the anodic region and is oxidized to Fe^{2+} while oxygen is concomitantly reduced at the cathodic regions. The cathodic reactions require dissolved oxygen present in pore solution, thus suggesting that the oxygen permeability is a significant factor contributing to propagation of corrosion.

Ingress of oxygen into the depth of steel bars embedded in concrete is likely to occur through the processes of diffusion of oxygen gas in the dried pore structure, dissolution of oxygen gas into the pore solution, and diffusion of dissolved oxygen as shown in Fig.1. Kawaai et al., (2018) showed that the rate of oxygen permeability can be decreased owing to reduced availability of dissolved oxygen through the metabolic processes of bacillus subtilis (natto) in mortar specimens cast with a partial replacement of mixing water with culture solution. This led to higher corrosion resistance even if bending cracking occurs in the mortar specimens. The results suggest that there is a good possibility that microcell and macrocell corrosion current density are decreased, which could be attributed to the enhancement of resistance against the cathodic reactions. This study examines the effect of the degree of water saturation on the oxygen permeability in mortar specimens mixed with bacillus subtilis (natto). In particular, the difference of metabolic activity which can be affected by the

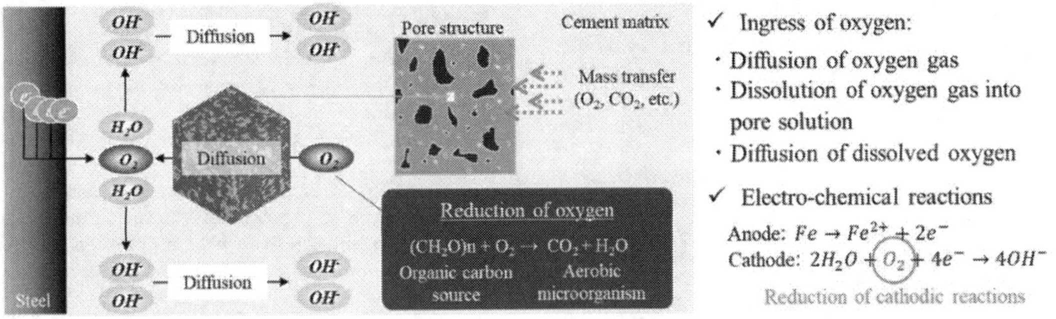

Fig. 1. Schematic of oxygen permeability and cathodic reactions in corrosion processes

* Corresponding author: kkawaai@cee.ehime-u.ac.jp

Table 1. Chemical composition of culture media.

Chemicals	g/L
NH_4Cl	5
NH_4NO_3	1
Na_2SO_4	2
$MgSO_4 \cdot 7H_2O$	0.1
$CaCl_2$	1000
K_2HPO_4	3
KH_2PO_4	1
$C_6H_{12}O_6$	0.5
H_2O	1000
$C_{10}H_{16}N_2O_3S$	0.5
Trace element	1mL

Table 2. Chemical composition of trace element.

Chemicals	mg/mL
$Na_2B_4O_7 \cdot 10H_2O$	88
$CuSO_4 \cdot 5H_2O$	393
$Fe_2(SO_4)_3 \cdot 6H_2O$	910
$MnCl_2 \cdot 4H_2O$	72
$(NH_4)_6Mo_7O_{24} \cdot 4H_2O$	37
$ZnSO_4 \cdot 7H_2O$	8807
H_2O	1000

Table 3. Mix proportion.

Cases	Unit Weight kg/m^3				Unit Weight g/m^3	
	W	C	S	N1	AE-WRA	AE
OPC	278	505	1363	0	1668	556
N25	209	505	1363	70	1668	556

Note: N1 (bacillus subtilis) is mixed after cultivated for 24 hours. Mixing water (tap water) is replaced with a replacement ratio of 0.25.

availability of water and dissolved oxygen is investigated using segmented steel bars embedded in mortar specimens under various wet/dry environments.

2 Experimental programs

This section provides detailed information about experimental programs carried out in this study. In particular, materials and testing methods associated with evaluation of oxygen permeability via electro-chemical measurements are described.

2.1 Cultivation of bacillus subtilis (natto)

The rod-shaped (1 μm width and 2-3 μm length) bacillus subtilis (natto) is resistant to unfavourable environmental conditions including salinity and extreme pH through the formation of an endospore at times of nutritional stress until conditions become favourable. The application would be classified into two categories. An approach is that the bacillus subtilis (natto) is mixed in mixing water before concrete specimens are cast. If they could actively metabolize organic carbon sources provided, the dissolved oxygen concentration is reduced through the aerobic processes. The other approach is that a mixture containing the bacillus subtilis (natto) is sprayed into cracks in concrete. The application is very similar to the liquid-based repair techniques (Ujike et al., 2014; Wiktor and Jonkers, 2015). As mentioned in the introduction, it is clear that the bacillus subtilis (natto) could be compatible with the alkaline environment in pores or cracks saturated with higher pH solution.

The bacillus subtilis (natto) used in this study was cultivated using culture media which is typically used for minimal medium for E. coli. and the details of chemical composition are summarised in Tables 1 and 2. The mixing process should be started with the chemicals listed in the top of the table in order. They were diluted 100 times in mass using distilled water. For the cultivation procedure, nutrients and food such as an organic carbon source which was glucose ($C_6H_{12}O_6$) in this study should be provided and aeration apparatus was also utilized in the container.

2.2 Mortar specimens

Fig. 2 shows the schematic of prism mortar specimens used in this study. The specimens, of cross section 50 x 50 mm, length of 150 mm were cast with mortar mixtures with and without bacillus subtilis (natto). Segmented steel bars (20 mm each) comprising 3 steel elements which are connected by epoxy resin were embedded in the specimens. The cover depth was 20 mm and duplicate specimens were made to confirm the consistency of the results. It should be noted that the specimens were sealed by epoxy resin except for an exposed surface. Table 3 shows the mix proportions of the mortar mixtures used in this study. The cement used was high early strength Portland cement (3.14 g/cm^3) in accordance with JIS R 5210. The fine aggregate was crushed sand obtained from sandstone with specific

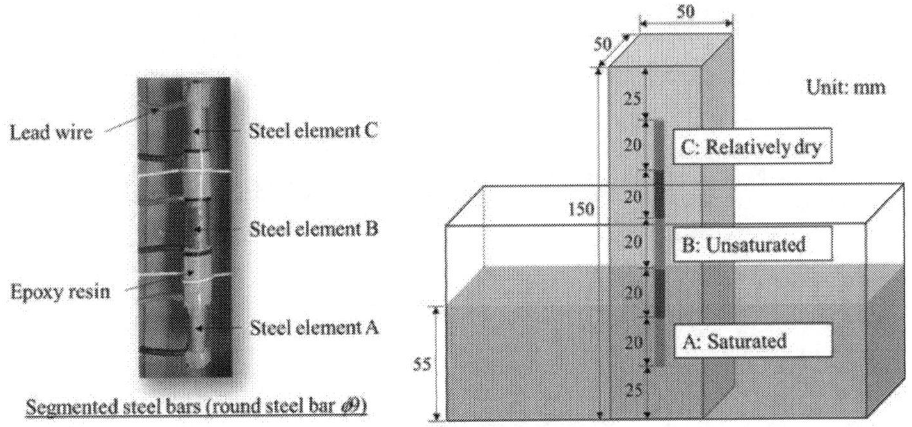

Fig. 2. Schematic of mortar specimen and segmented steel bars

gravity, water absorption, and fineness modulus of 2.61 g/cm^3, 1.06%, and 2.92 respectively. The OPC mixture was a control mixture and N25 was a mixture with culture solution of bacillus subtilis (natto) with a replacement ratio of 0.25. The specimens were cured in a water tank until the age of 28 days immediately before the immersion test was started.

2.3 Electro-chemical measurements

Electro-chemical measurements were carried out via half-cell potential measurements, AC impedance method, and cathodic polarization measurements. The half-cell potential was measured in accordance with ACTM-C 876 from the concrete surface using a Ag/AgCl reference electrode saturated with KCl. The potentials based on a $Cu/CuSO_4$ electrode were estimated by the readings. The set-up for the oxygen permeability measurement is similar to polarization resistance measurements. The electro-chemical measurement by a potentiodynamic test comprised a working electrode (embedded steel bars), counter electrode (stainless steel plate), and a reference electrode (Ag/AgCl). When DC current was impressed by the rate of 1 mV/s, the current density was measured by a potentiostat. It should be noted that no IR correction was made on the results in this study. The limiting current density was assumed when the potential difference reached 860 mV and oxygen permeability in concrete specimens was calculated based on Eq. (1) (Nagataki et al., 1996).

$$\frac{dQ}{dt} = -\frac{i_{lim}}{nF} \tag{1}$$

where dQ/dt: the rate of oxygen permeability (mol/cm^2/sec); i_{lim}: limiting current density (A/cm^2); F: Faradays constant (96,500 coulombs/mol); and n: the

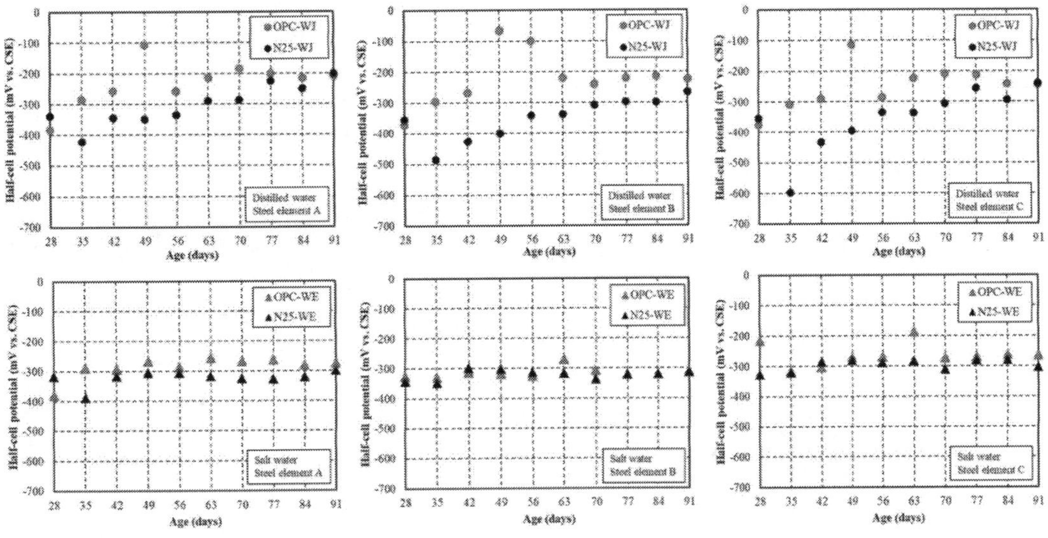

Fig. 3. Results of half-cell potential in the cases of steel elements A, B, and C in the OPC and N25 specimens

number of electrons exchanged equal to 4.

3 Results and discussion

This section reports on the results of immersion tests carried out on the mortar specimens. Variations of half-cell potential and oxygen permeability measured by cathodic polarization techniques under various saturation degree are central to this study.

3.1 Results of half-cell potential and corrosion current density

Fig. 3 shows results of half-cell potentials measured on the steel elements A, B and C after being exposed to the immersion tests. As can be seen, the half-cell potentials of steel elements A, B, and C in the N25-WJ were observed to be lower compared to those observed in the OPC-WJ. They were clearly lower than -350 mV vs. CSE which is a threshold value indicating that possibility of corrosion is greater than 90%. The difference with regards to potentials between OPC-WJ and N25-WJ mixtures was significant until the age of 91 days in this study. On the other hand, the potentials measured in the OPC-WE and N25-WE mixtures were comparable, which ranged from -200 to -400 mV vs. CSE. The typical ranges of potentials of normal steel in concrete are reported by RILEM technical commetee (2003). The potentials measured in water saturated concrete without oxygen generally show more negative values compared to those of chloride contaminated concrete. The effect of the addition of the bacillus subtilis (natto) under the saturated condition (steel element A) is subsequently discussed.

The potential difference between OPC-WJ and N25-WJ was calculated until the age of 182 days as shown in Fig. 4. It should be noted that the half-cell potential greater than -200 mV vs. CSE in the OPC-WJ is considered to be consistently -200 mV taking into consideration the variations of potential measurements. When the values calculated based on the potential

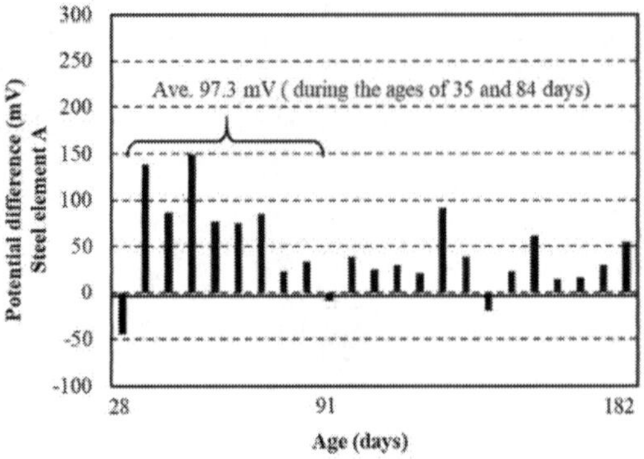

Fig. 4. Potential difference (OPC-WJ and N25WJ)

difference are positive, this indicates that the potential of N25-WJ is relatively lower. The results showed that the potential difference was about 97.3 mV on average during the ages of 35 and 84 days. In contrast, the corrosion current density measured on the N25-WJ specimens were marginal and there is a slight possibility that corrosion of steel bars in the specimens occurred. Thus, the possibility of corrosion based on the results of half-cell potentials are not consistent with those of corrosion current density measured by the polarization resistance.

In principle, the half-cell potential of steel bars is lowered on condition that limiting current density is decreased owing to the reduction of oxygen concentrations which react in the cathodic polarization processes. In addition to the immersion tests using mortar specimens, polarization tests were carried out on the steel elements immersed in aqueous solution in this study. The set-up is similar to that used for the mortar specimens as shown in Fig. 5. The aqueous solution was prepared using distilled water and a culture solution of bacillus subtilis (natto). This was intentional to investigate the effect of dissolved oxygen concentrations on the cathodic reactions.

Fig. 6 shows the results of cathodic polarization

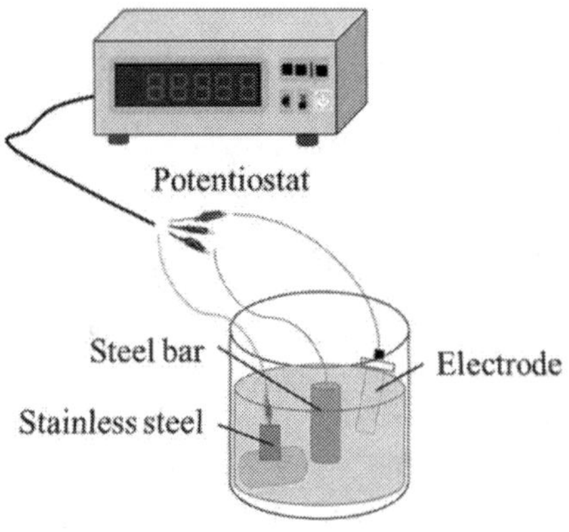

Fig. 5. Polarization measurements on round steel bar

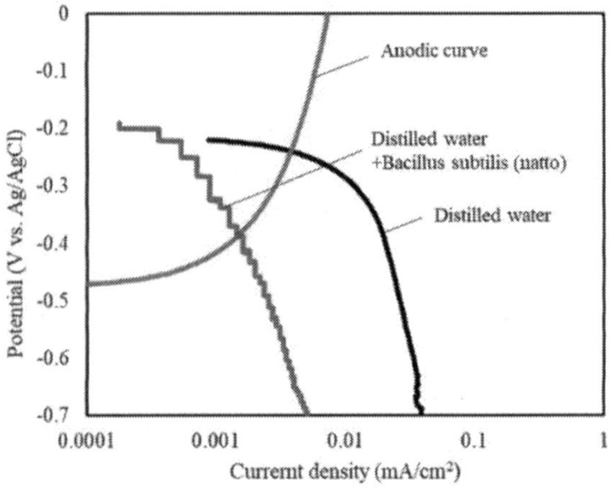

Fig. 6. Cathodic/Anodic polarization curves in aqueous solution

curves in the cases of steel bars immersed in distilled water and the culture solution respectively. For illustration, the anodic polarization curve when corrosion occurs is also depicted in the figure. As can be seen, the cathodic polarization curve tends to shift on the left side in the case of the culture solution of bacillus subtilis (natto). This is attributed to the smaller limiting current density caused by the reduced availability of oxygen on the cathodic reactions in the culture solution of bacillus subtilis (natto). This led to lower half-cell potential at the cross point of two polarization curves in principle. Thus. The results suggest that dissolved oxygen concentration affects the half-cell potential measurements which is illustrated by the more negative potentials observed until the age of 91 days as mentioned previously. Thus, the lower potentials were found to show the evidence of decreased oxygen concentrations associated with the aerobic metabolic process by the bacillus subtilis (natto) mixed in the mortar specimens during the periods.

3.2 Results of oxygen permeability

Fig. 7 shows the results of rate of oxygen permeability measured at the ages of 91 and 182 days. It should be noted that the results of N50 mixtures were also included in order to show the variations during the ages. In the case of steel element A, the results of the rate of oxygen permeability are plotted in the vicinity of the line (y=x). Although the permeation front of water could reach to the depth of steel bars embedded in the mortar specimens, the metabolic activity is likely to take place under wet conditions. Thus, the rate of oxygen permeability is generally lower compared to those observed in other steel elements. In addition, the variations during the time intervals were smaller under the saturated conditions. In contrast, as can be seen in the figures, the rate of oxygen permeability is generally higher in the cases of steel element B and C, which could be attributed to the dry environment in which moisture is evaporated from the pores and the activity of bacillus subtilis could be limited in part. This led to larger variations observed in the rate of oxygen permeability in those locations. Based on the results and discussion mentioned above, the effect of the presence of bacillus subtilis (natto) is most pronounced under the saturated conditions.

Fig. 8 and Fig. 9 show the rate of oxygen permeability measured on the steel element A at the ages of 28 and 91 days. In the case of N25-WJ, the rate of oxygen permeability is decreased by 28.6% and 48.9% at the ages of 28 and 91 days respectively. Based on the results obtained, the rate of oxygen permeability is clearly decreased from the age of 28 days to 91 days, which is consistent with the results reported by Kawaai et al., (2018). The results suggest that the consumption of dissolved oxygen associated with the metabolic processes of bacillus subtilis (natto) could contribute to the reduction of rate of oxygen permeability on the cathodic reactions. In addition, the lowered half-cell potentials observed especially until the age of 91 days could be attributed to the reduction of dissolved oxygen

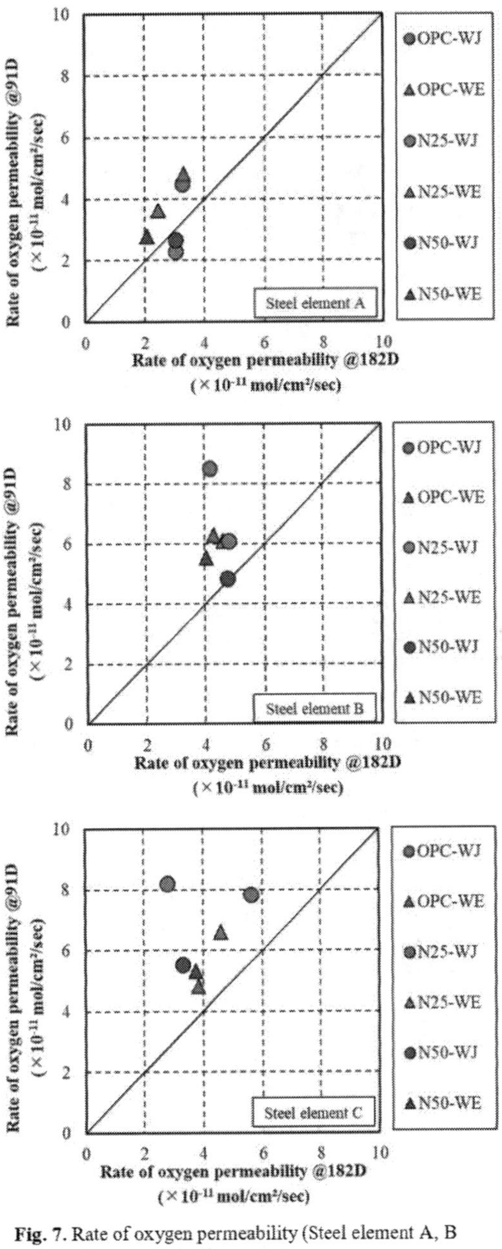

Fig. 7. Rate of oxygen permeability (Steel element A, B and C@91 and 182 days)

concentrations which can be illustrated by the changes of cathodic polarization curves. This needs further research especially for the interpretation of possibility of corrosion based on the half-cell potentials e.g. ASTM C 876.

4 Summary

The rate of oxygen permeability is clearly decreased from the age of 28 days to 91 days in the presence of bacillus subtilis (natto) under the saturated conditions. In addition, there is a good possibility that the half-cell

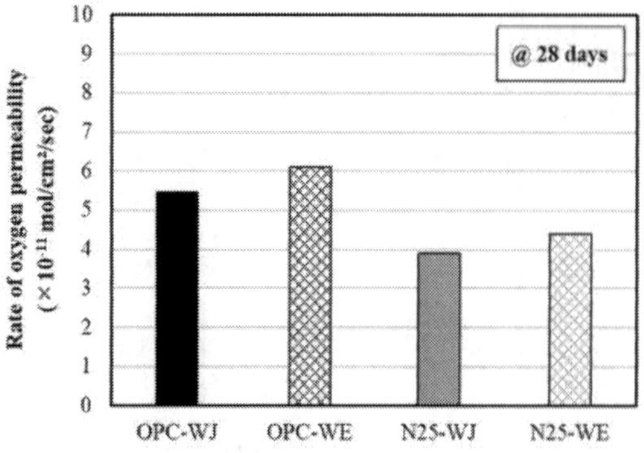

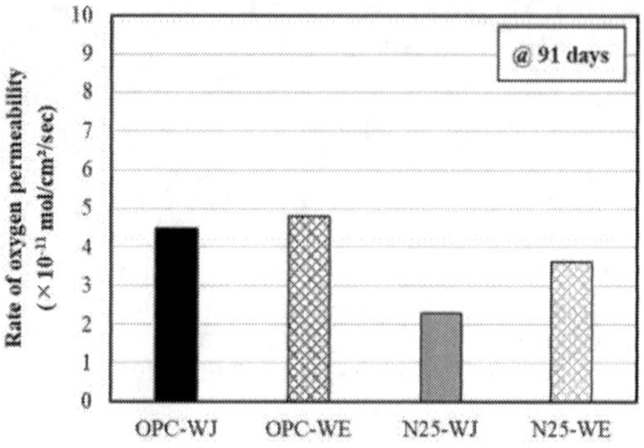

potential is lowered owing to reduced availability of dissolved oxygen concentrations in the presence of bacillus subtilis (natto), which can be illustrated by the changes of cathodic polarization curves. The lower potentials were found to show the evidence of decreased oxygen concentrations associated with the aerobic metabolic process by the bacillus subtilis (natto) mixed in the mortar specimens during the periods. The results suggest that the possibility of corrosion should be assessed taking into consideration other electro-chemical measurements e.g. polarization resistance not only based on the half-cell potentials.

Financial support by Grant-in-Aid for Young Scientists (A), Grant No. 17H04933 and ueda memorial foundation is gratefully acknowledged.

References

1. ASTM C876-91 R99, "Standard test method for half-cell potentials of uncoated reinforcing steel in concrete", ASTM International, West Conshohocken, PA. (1991)
2. I. Ujike, F. Kubo, K. Kawaai, and S. Okazaki, "Influencing Factors Affecting Microbial Metabolic Processes of Bio Materials Used for Leakage Repairs", Concrete Solutions 5th International Conference on Concrete Repair – Grantham et al. (Eds), CRC Press, pp. 127-133 (2014)
3. JIS R 5210, Portland cement, (2009)
4. K. Kawaai, I. Ujike, S. Fujioka, T. Nishida, and A. Saito, "Cathodic Polarization Properties of Steel Bars in Concrete Mixed with Aerobic Microorganism", Final Conference of RILEM TC 253-MCI, pp.259-266 (2018)
5. RILEM technical committee, "Recommendation of RILEM TC 154-EMC: 'Electrochemical T3echniques for Measuring Metallic Corrosion'", Materials and Structures, Vol.36, pp. 461-471 (2003)
6. S. Nagataki., N. Otsuki., A. Moriwake, and S. Miyazato, "The experimental study on Corrosion Mechanism of Reinforced Concrete at Local Repair Part", Journal of Materials, Concrete Structures and Pavements", JSCE, 544(32), pp. 109-119 (1996)
7. V. Wiktor, and H.M., Jonkers, "Field Performance of Bacteria-Based Repair System: Pilot Study in a Parking Garage", Case Studies in Construction Materials, Vol. 2, pp. 11-17 (2015)

Effect of crack pattern on the self-healing capability in traditional, HPC and UHPFRC concretes measured by water and chloride permeability

Alberto Negrini[1,2], *Marta* Roig-Flores[1,*], *Eduardo J.* Mezquida-Alcaraz[1], *Liberato* Ferrara[2], *Pedro* Serna[1]

[1] Universitat Politècnica de València, Instituto de Ciencia y Tecnología del Hormigón, Valencia, Spain
[2] Politecnico di Milano, Milan, Italy

Abstract. Concrete has a natural self-healing capability to seal small cracks, named autogenous healing, which is mainly produced by continuing hydration and carbonation. This capability is very limited and is activated only when in direct contact with water. High Performance Fibre-Reinforced Concrete and Engineered Cementitious Composites have been reported to heal cracks for low damage levels, due to their crack pattern with multiple cracks and high cement contents. While their superior self-healing behaviour compared to traditional concrete types is frequently assumed, this study aims to have a direct comparison to move a step forward in durability quantification. Reinforced concrete beams made of traditional, high-performance and ultra-high-performance fibre-reinforced concretes were prepared, sized 150×100×750 mm^3. These beams were pre-cracked in flexion up to fixed strain levels in the tensioned zone to allow the analysis of the effect of the different cracking patterns on the self-healing capability. Afterwards, water permeability tests were performed before and after healing under water immersion. A modification of the water permeability test was also explored using chlorides to evaluate the potential protection of this healing in chloride-rich environments. The results show the superior durability and self-healing performance of UHPFRC elements.

1 Introduction

Concrete has a natural self-healing capability to seal small cracks, named autogenous healing, which is mainly produced by continuing hydration and carbonation [1]. At early age, continuing hydration of unhydrated cement particles is presumably the main cause for autogenous healing of cracks, while for older elements, carbonation would be the main mechanism [2]. Previous studies of the literature reported that Ultra-High-Performance Fibre-Reinforced Concrete (UHPFRC) and Engineered Cementitious Composites (ECC) are able to heal cracks for low damage levels, due to their cracking pattern with multiple micro-cracks and high cement content.

Regarding the self-healing properties of high performance concrete, in the mid-90s, Jacobsen et al. investigated autogenous healing of high-performance concrete with water/binder ratio of 0.40 and silica fume contents between 0 and 5% by the weight of cement. They achieved clear improvement in terms of resonance frequency values but not in terms of strength [3, 4]. Healing products were observed bridging cracks after the self-healing process and were identified as C-S-H [3]. Portlandite and ettringite were observed locally, as well. They analysed the samples with electron microscopy and only observed complete filling of the cracks under 5 µm. Granger et al. [5] analysed the recovery of stiffness and peak load of UHPC for specimens pre-cracked to have 10 µm of residual crack width, achieving stiffness recovery for specimens healed in water for 10 weeks. The stiffness of the filling products was close to those of primary C–S–H. The slow recovery was thought as caused by the development of the mechanical properties of the new crystals or of the weakness zone between new crystals and primary C–S–H.

The geometry of cracks, namely, crack width, length, depth, and cracking pattern (branched crack or accumulated crack) may determine their autogenous healing degree. Edvardsen [6, 7] reported that crack width is the parameter that most affects the healing process, since the narrower the cracks, the more efficient the autogenous healing. Therefore, by limiting and controlling the crack width, the potential for autogenous healing can be substantially improved. The use of fibre-reinforced concrete to improve autogenous healing has focused mainly on Strain-Hardening Cementitious Composites (SHCC) or Engineered Cementitious Composites (ECC). These materials have a high volume of fibres, which permit the post-cracking stress re-distribution and a ductile behaviour, thus featuring multiple cracking with very small cracks (<0.05-0.07 mm). Yang et al. [8, 9] showed that ECC healed cracks could be of sufficient strength that new cracks are produced during the reloading stage, and the self-healed ECC material remained ductile. Their studies showed that, for small damage, specimens healed in cycles of

* Corresponding author: marroifl@upv.es

water immersion and drying, recovered resonant frequency and stiffness. Ferrara et al. [10] evaluated autogenous healing in High Performance Fibre-Reinforced Cementitious Composites (HPFRCCs) exposed to different conditioning environments, comparing different levels of crack openings. Their results showed that HPFRCCs are able to seal the cracks, and to recover their pristine level of mechanical performance, which was favoured by the presence of water, even in form of high air humidity or in the case of wet and dry cycles.

Several studies have been performed to study self-healing in marine environments. One of those studies, performed by Darquennes et al. [11], studied the effect of self-healing of early age cracks (<28 days), with cracks around 0.4 mm, in mixes with blast-furnace slag, to limit chloride penetration and how chloride resistance could be improved by self-healing. The authors demonstrated that self-healing decreased the cumulative chloride concentration at the end of the migration test as well as the chloride concentration rate at steady state.

However, there is a lack of information on how autogenous healing of cracks could protect the interior of the concrete matrix, and its reinforcement, from the entrance of aggressive agents.

This research compares self-healing in a traditional concrete with a high-performance concrete and two types of ultra-high-performance concrete reinforced with steel fibres, pre-cracked to have comparable damage levels. A modification of the water permeability test performed in previous research [11, 12] was explored using a water column with chlorides, to evaluate the protection produced by healing underwater without chlorides, if the samples are exposed afterwards to aggressive chloride-rich environments.

The objectives of this study are:

- To provide a methodology to evaluate self-healing in Ultra-High-Performance Fibre-Reinforced Concrete based on the study of permeability in a cracked state;
- To characterize concrete durability by means of water permeability and chloride penetration tests;
- To analyse the cracking pattern, durability and self-healing performance of Ultra-High-Performance Fibre-Reinforced Concretes and traditional concrete types.

2 Materials and methodology

The concretes analysed in this study were Conventional Concrete (C20/25, named C1), High-Performance Concrete (C60/75, named C2) and the two Ultra-High-Performance Fibre-Reinforced Concretes (named C3 and C4). The experimental phase process consisted of casting the different concrete mixes, pre-cracking reinforced beams and to study the options for permeability tests for the evaluation of self-healing. Each concrete batch cast had a volume of 70 litres, to produce: 4 steel reinforced beams 100×150×750 mm^3, 2 prisms 100×100×500 mm^3 and 4 cubes 100×100×100 mm^3.

2.1. Materials and mix design

The mixes were designed to represent two types of traditional concretes without fibres, with different water/cement (*w/c*), water/binder (*w/b*) ratios and cement contents to compare the effects of the composition, and two different UHPFRCs, with different fibre contents, to produce different deformation and crack responses. Table 1 shows the composition of all these mixes.

The cement used was CEM I 42,5 R-SR5 from Lafarge, and the undensified microsilica from Elkem. Different aggregates were used according to the type of concrete produced. For traditional and high-performance concretes (C1 and C2) gravel of the categories 7/12 and 4/7 (d_{min}/D_{max}) and natural sand were used in the mix. In the traditional concrete C1 mix, limestone filler was also added to compensate for the lack of fines of this sand. In UHPFRC fine silica sand (0/0.6 mm) and medium silica sand (0/1.2 mm) were used. In UHPFRC mixes, silica flour from Quarzfin U-S 500 was used for the finer fractions. Sika ViscoCrete-20 HE superplasticizer was used to achieve the high workability of these self-compacting mixes. In addition, in the UHPFRCs mixes, high tensile strength micro steel fibres 0.22×13 mm were used, while in traditional and high-performance concrete, no fibres were added. These fibres provided the mix with high ductility and multi-cracking behaviour, increasing compressive strength. In addition, the fibres contributed to keeping crack width fixed after the pre-cracking and to the stability of the samples during the permeability tests.

Table 1. Mix design of the four concrete types.

kg / m^3	C1	C2	C3	C4
Cement I 42.5 R-SR5	350	450	800	800
Silica Fume		50	175	175
Water	192	160	160	160
w/c	*0.591*	*0.356*	*0.200*	*0.200*
w/b	*0.591*	*0.320*	*0.164*	*0.164*
Aggregate 7/12	600	200		
Aggregate 4/7	300	600		
Natural sand	950	950		
Limestone filler	60			
Siliceous sand - medium			565	
Siliceous sand - fine			302	1092
Siliceous sand - flour			225	
Short fibres			160	130
Plasticizer Sika 20 HE		3.5	30	30

2.2. Concrete characterization

Control tests were performed on cubes and prisms with the purpose of evaluating the difference between mixes (C1, C2, C3 and C4). For each concrete batch, four 100×100×100 mm^3 cubic specimens were cast to determine the compressive strength at 28 days, according to EN 12390-3, applying a load rate of 8 kN/s and 0.8 MPa/s. The average results of compressive strength obtained are displayed in Table 2.

For mixes C3 and C4, two 100×100×500 mm^3 prismatic specimens were cast to study their tensile

Table 2. Control tests results: compressive strength and tensile behaviour for UHPFRCs.

Concrete	Compr. strength	Flexural strength		Inverse Analysis – Tensile constitutive model					
	f_{cm}	σ_{loc}	δ_{loc}	f_t	f_{tu}	e_{tu}	E	w_d	w_o
Code	*MPa*	*MPa*	*mm*	*MPa*	*MPa*	$\cdot 10^{-3}$	*MPa*	*mm*	*mm*
C1	38.5	-	-	-	-	-	-	-	-
C2	82.3	-	-	-	-	-	-	-	-
C3	149.6	18.22	0.66	9.49	6.83	2.356	47800	1.60	2.40
C4	121.1	16.66	0.70	7.75	6.84	2.526	39100	1.92	2.89

properties using a simplified inverse analysis method from a four-point bending test on unnotched specimens. The method followed was proposed in [13], and defines the tensile constitutive behaviour as a function of six parameters: elastic modulus (E), cracking strength (f_t), ultimate cracking strength (f_{tu}) and its associated strain ($\varepsilon_{t,u}$), crack opening at the intersection of the line that defines the initial slope to the w axis (w_0) and the characteristic crack opening ($l_f/4$), being l_f the fibre length. Table 2 shows the constitutive behaviour in tensile obtained for C3 and C4 mixes. As expected, the C3 mix reported a tensile strength value of about 22% higher than the C4 due to its higher content of steel fibres.

2.3. Pre-cracking method

In this work, for the study of self-healing, reinforced concrete beams were used. The reinforcement introduced in the samples consisted of 2ø8 lower reinforcing bars and 6ø6 stirrups to avoid shear failure during the pre-cracking stage. These beams were pre-cracked afterwards to the desired cracking level at the age of 28 days. The dimensions, number and location of the reinforcing bars and stirrups are shown in Fig. 1, as well as the relative distances and points of support when performing the four-point bending test.

Before starting the test, two rows of small discs were glued to the side of each sample, setting up an upper row and a lower row, to be able to measure the strain variation at two different levels with a Demountable Mechanical Strain Gauge (DEMEC) before, during and after every loading-unloading cycle.

The procedure for controlling crack width was based on calculating the strain at six intervals separated by initial distance of 100 mm. First, the change in length between the points with the DEMEC was detected and then strain (ε) was calculated through the following formula:

$$\varepsilon = 1000 \cdot (\Delta L_0 - \Delta L_x) / l_0 \quad (1)$$

Where:
- ε is the strain level expressed in ‰
- ΔL_0 is the length between 2 DEMEC points at zero load, expressed in mm
- ΔL_x is the length between 2 DEMEC points at x load, expressed in mm
- l_0 is the reference length of 100 mm

This method allows control of the resulting strain at two different heights: at the lower row (points 1-5) and at the upper row (points 6-10). The location of these measuring points is displayed in Fig 2, where each couple of points (such as 6-8, 7-9 or 8-10) were located at a distance of 100 mm. The lower row (points 1-5) were located at the reinforcement level, while the upper row (points 6-10) were located at middle depth of the beam.

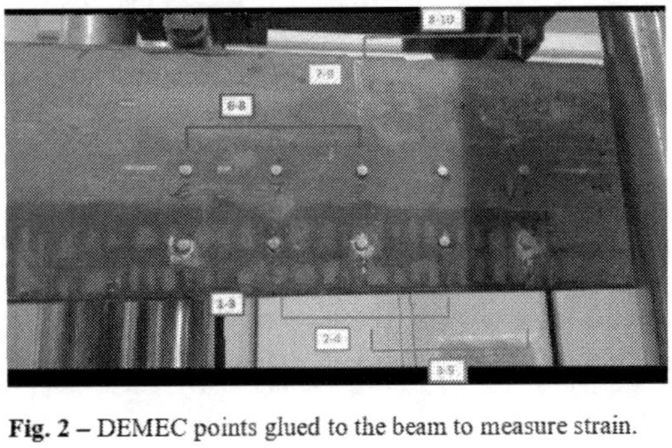

Fig. 2 – DEMEC points glued to the beam to measure strain.

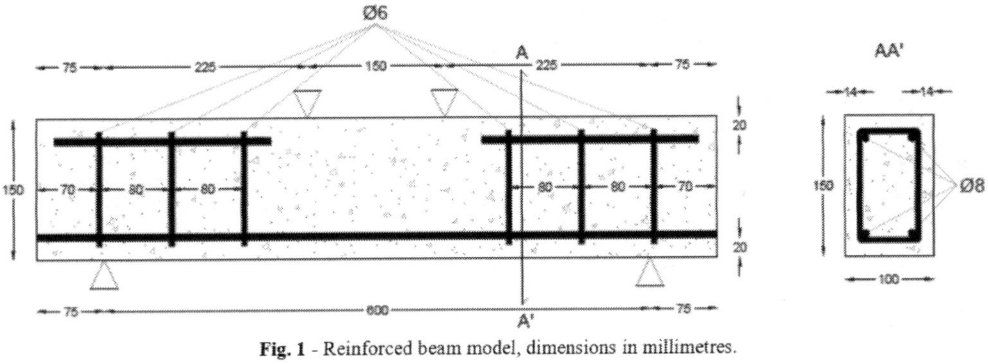

Fig. 1 - Reinforced beam model, dimensions in millimetres.

The cracking target was based on achieving a large crack, or rather, a residual average strain ε_{avg} between 1‰ and 2‰ after unloading at the lower DEMEC row (reinforcement level). Considering the different values of strain obtained in the intervals (1-3, 2-4, 3-5) between the DEMEC points due to the random appearance of the cracks and to give homogeneity to the results, the condition of large crack was considered reached when the averaged strain of the three intervals of the lower DEMEC row exceeded the value of 1.5‰.

During loading, the strain value of 2‰ was frequently reached and exceeded. This large crack condition represents a condition beyond the service limit state, since, it is the value in which rebars start to yield.

The procedure followed with the UHPFRCs (C3 and C4) during the controlled cracking consisted of gradually increasing the load from zero, stopping at certain loading values, where the change in length between the DEMEC points was measured and consequently, through a spreadsheet, strain (‰) was obtained in each interval (1-3, 2-4...). Strain was usually measured at zero tons (reference condition of zero strain), 6 tons, 8 tons and 10 tons. After reaching 10 tons, loading-unloading cycles were performed, increasing after 0.5 tonnes in each cycle until reaching the target strain ($\varepsilon_{avg} \geq 1.5$‰). These loading-unloading cycles were needed to reach the desired deformation values after unloading, since UHPFRCs mixes recovered a great part of their deformation. This procedure allowed the desired target to be reached with an average residual strain value close to 1.5‰ at the lower row and < 1.0‰ at the upper row.

For traditional and high-performance concretes (C1 and C2) the procedure followed was equivalent but reaching lower loads because of their lower flexural strength. A maximum load of 6 tons was reached, but due to the greater difficulty in controlling the crack opening with this type of concrete, very high residual strains were achieved, in the order of 3‰ - 4‰. In fact, under the same load, in traditional concrete, the strain value turns out to be very different in the intervals belonging to the same DEMEC line. On the contrary, all UHPFRC beams and pair of DEMEC points showed similar strains. This is produced by the cracking pattern of traditional and high-performance concretes without fibres, characterized by only few cracks. When these cracks appear, their location is unpredictable. Those locations where strain values were higher were where cracks appeared.

The strain values obtained with DEMEC points were complemented with measurements made with an optical microscope at either the upper or lower row levels to correlate strain rates and crack widths, a critical relation for the evaluation of the durability parameters and the entrance of aggressive agents.

2.4. Self-healing evaluation

After this pre-cracking stage, the crack width of all beams was evaluated and afterwards, the beams were sawed to obtain smaller specimens and perform permeability tests before and after healing.

2.4.1 Preparation of samples

After the pre-cracking stage, cracks were all located in the central section of the beam while the lateral portions of the beam remained uncracked. For this reason, beams were sawn to obtain four smaller specimens of the size of 150×150×50 mm^3 to perform the permeability tests for the evaluation of self-healing (Fig 3).

Since some of the tests performed to evaluate self-healing are destructive, The A and B samples were distributed as follows: A samples were tested before healing and B samples after healing. These A and B samples were twin samples coming from the same section of the beam and thus had the same deformation.

2.4.2 Adapting water permeability tests

Previous tests performed for traditional fibre-reinforced concretes [3, 4] using a high-pressure water permeability test showed the correlation between higher values of damage and higher values of water flow. Thus, this high-pressure water-permeability test was used as a first approach for this experimental campaign.

In this case, however, traditional concretes showed specific features during the preliminary tests. When performing high-pressure water permeability tests to C1 and C2 concretes, water pressure and the compression applied to fix the sample to the plates from the device, produced the breakage of some of the specimens. In addition, since the cracks passed through the complete section of the sample, they would need complete sealing of the lateral cracks to permit water to flow only through the interior of the sample.

Fig. 3 – Prismatic specimens obtained after sawing the reinforced beams.

For these reasons, to evaluate self-healing capability in traditional concrete, it was decided to perform a water permeability test in low-pressure conditions. This method consisted of using a PVC tube ($ø_e$ = 75 mm, length around 60 cm) glued to the specimens with a resin (in this case, Sikaflex 11 FC). This low-pressure water permeability test is based on calculating the time taken by a 50 cm column of water to empty, passing completely through the cracked specimen. If the cracks were too small to allow the column to empty within a few hours, it was decided to calculate the decrease in water-height of the water column in a fixed time (3 hours).

This low-pressure water permeability test has been performed only on the traditional and high-performance concrete samples (C1 and C2), and on single cracks, since its cracking pattern produces only localized cracks. Thanks to this test, it is possible to measure the degree of permeability of these cracked specimens. With this test, the same samples can be tested a second time after healing takes place and thus direct assessments on autogenous healing effects can be performed.

In the case of UHPFRC samples (C3 and C4), they were able to resist the high-pressure water permeability test; however, water was not passing through their cracks. This happened despite the high strain values obtained during the pre-cracking stage. In fact, all the UHPFRCs tested had no water passing through the specimens in 5 minutes of testing time at the pressure of 2 bars. Furthermore, no moisture was even detected at the base of the specimens at the end of the tests, proof that water was not able to cross the cracked section. Only in the case of one of the specimens, some drops of water were observed coming out, however, it was not possible to measure them due to their small weight, approx. from 0 to 10 grams (see Table 3).

Because of that, another modified water permeability test was proposed. This test is based on the use of sodium chloride and its reactivity to silver nitrate as an indicator to show where water was able to penetrate. This test is destructive and thus the same samples cannot be tested a second time after healing.

Table 3. Results of high-pressure water permeability test performed on UHPFRCs samples.

Concrete mix	Average crack width (mm)	Amount of water passing through the sample
C3	0.025	-
C3	0.025	-
C4	0.025	-
C4	0.050	0-10 grams

This proposed test consists of a version of the low-pressure water permeability test, using the same PVC tubes with an outer diameter of 75 mm. Every tube was filled with salt water with a concentration of 35 g NaCl/litre, and the salted water column was left for 3 days over the samples. The water level was controlled every working day, so it remained around 50 cm, and in case of visible evaporation, tap water was added to maintain the same pressure level. After three days, the tube was removed, and the specimen was cut transversely to the direction of the cracks with a circular saw for concrete (Fig. 4).

In order to measure the chloride penetration depth, firstly, specimens were left to dry and afterwards a silver nitrate ($AgNO_3$) solution with a concentration of 0.1 mol/l was sprayed on the both cracked surface where the cut was made. After the chemical reaction, the area colored in white is the area of interest, where silver ions reacted with the chloride ions, while the brown one is formed when the silver ions react with the hydroxyl compounds.

Fig. 4 – Specimen after performing chloride penetration test with the rests of resin and the transversal section that will be cut marked.

2.4.3 Self-healing conditions

To promote autogenous healing, specimens were left immersed in a tank full of deionized water for 28 days. This healing time was considered enough for producing most of the autogenous healing reactions, in part, due to the small size of the cracks.

The water was chosen to be deionized water from a cationic and anionic resin (proportion of 60% anionic of strong base and 40% cationic of strong acid) to avoid the presence of chlorides in the tap water that could condition the results of the chloride penetration tests after healing.

As previously explained, the twin specimens (B sides) were immersed in water to heal, since the procedure for detecting chloride penetration did not allow the same sample to be tested twice. After one month in water, it was expected that cracks in the specimens had closed or partially closed on the surface, thus being able to assess a recovery of the durability properties.

3 Results

This section shows the results obtained when analysing the cracking pattern and the autogenous healing capability of these four concrete types using the aforementioned permeability tests.

3.1. Cracking pattern

UHPFRCs showed higher mechanical performance during tests and a cracking pattern with multiple micro-cracks. After unloading they did not show visible localized cracks, whereas traditional concretes did. In fact, while on the one hand traditional concrete presented only few visible cracks, measurable through the use of a crack width meter, UHPFRC on the other hand featured a large number of widespread cracks, not visible to the naked eye. Then, the positions of the cracks were analysed with the use of the microscope and were marked on the samples.

Table 4 shows the differences in deformation and number of cracks obtained for the two beams tested per concrete type. In general, C4 specimens, compared to C3 specimens, at the same maximum load reached or even lower, always suffered greater deformations, due to their lower fibre content, but with a comparable value of average number of residual cracks and average crack width.

When comparing the appearance of the first visible crack under load between traditional concrete and UHPFRC, it has been observed that in the former the first crack appeared at about 2 tons whereas in the latter at about 8 tons. This cracking behaviour of UHPFRC is achieved thanks to its particular mix design consisting of small size of the constituents, high cement content and steel fibres capable of limiting crack opening. The amount of fibres in the mix turned out to be of great importance in the control of crack opening. In fact, the C4 mix, having a lower amount of fibres (130 kg/m^3 instead of 160 kg/m^3 of the C3 mix) showed lower flexural strength and lower performance in terms of control of the crack opening.

Fig 5 and Fig 6 show examples of pictures of the cracks obtained for a traditional concrete and a UHPFRC type, respectively, taken at the same magnification level. This difference in crack width will have significant effects on the transport properties, durability and their autogenous healing potential

Table 4. Crack analysis results after pre-cracking all the reinforced beams.

Type	Max. Load	DEMEC 1-5			DEMEC 6-10		
		ε_{avg}	N° cracks	ω_{avg}	ε_{avg}	N° cracks	ω_{avg}
	Ton	‰		*mm*	‰		*mm*
C1	6.2	3.70	2	0.200	1.66	2	0.100
	6.0	2.95	2	0.200	1.26	2	0.100
C2	6.6	4.33	3	0.150	2.56	3	0.100
	6.5	4.35	2	0.250	2.78	2	0.150
C3	12.0	1.40	12	0.025	0.81	12	0.015
	13.0	1.55	11	0.025	0.90	11	0.015
C4	11.0	2.26	20	0.025	1.21	11	0.015
	11.0	2.19	14	0.050	1.56	10	0.030

Fig. 5 – Crack of 0.1 mm in traditional concrete at 200x magnification

Fig. 6 – Crack of 0.01 mm in an UHPFRC at 200x magnification

3.2. Water permeability

Water permeability tests performed as described in the subsection "2.4.2 Adapting water permeability tests" showed that UHPFRCs samples, despite the large cracking level corresponding to an average crack width of 0.025 mm, did not allow the movement of water through the cracks. In addition, traditional and high-performance concrete types, without fibres, had stability problems with the high-pressure water permeability device. Thus, this section shows the results obtained with the low-pressure water permeability for concretes without fibres (C1 and C2) and chloride penetration water permeability test for UHPFRCs (C3 and C4).

3.2.1 Low-pressure water permeability

This test was used to evaluate autogenous healing in four samples of concrete without fibres. two of the samples were of conventional concrete (C1-1 and C1-2) and two of high-performance concrete (C2-1 and C2-2). From the results obtained for water level decrease and the time taken, water flow can be calculated. The results obtained with the four specimens tested with this test are presented in Table 5.

Before starting the tests, an additional crack measurement was made on the specimens, to verify if crack width was kept the same compared to the previous analysis, since the sawing process could have modified the crack width. It was noticed that cracks were wider in the inner section of the beam, namely, at the interface between the sample A and B, where the cut was made, whereas on the surface cracks were around 0.050 mm narrower. These results are also displayed in Table 5.

Table 5. Results of high-pressure water permeability test performed on UHPFRCs samples.

	Crack width		Water flow passing through the sample (gr/min)	
Specimen tested	**Surface 1 (mm)**	**Surface 2 (mm)**	**Before healing**	**After 28d healing**
C1-1	0.05 + 0.15	0.20	0.922	1.106
C1-2	0.05 + 0.15	0.20	0.092	0.000
C2-1	0.1 + 0.15	0.30	12.763	15.802
C2-2	0.1 + 0.15	0.30	11.061	8.480

Regarding the recovery of properties due to autogenous healing, neither traditional nor high-performance concretes showed noteworthy improvements after performing the low-pressure water permeability tests. The results obtained show that half of the samples experienced slight reductions in the water permeability test due to autogenous healing, but half of them experienced slight increases. Thus, these results indicate the high variability and low reliability of autogenous healing in traditional concrete types with cracks around 0.20 mm.

3.2.2 Water permeability through chloride penetration

This test was used to evaluate autogenous healing in four samples of UHPFRCs. two of the samples from C3 mix and two of them from C4 mix.

The results achieved indicate that, despite the small size of the cracks (around 0.025 mm), water containing chlorides was able to penetrate through the matrix in both, C3 and C4 mixes. Thus, in both cases, 3 days of exposure of salted-water, which simulates the exposure conditions of a marine environment, produced a considerable penetration of chlorides through the cement matrix.

Fig. 7 displays pictures showing the results obtained for C3 and C4 mixes before and after healing. In these pictures, chloride penetration is displayed as the white or lighter areas.

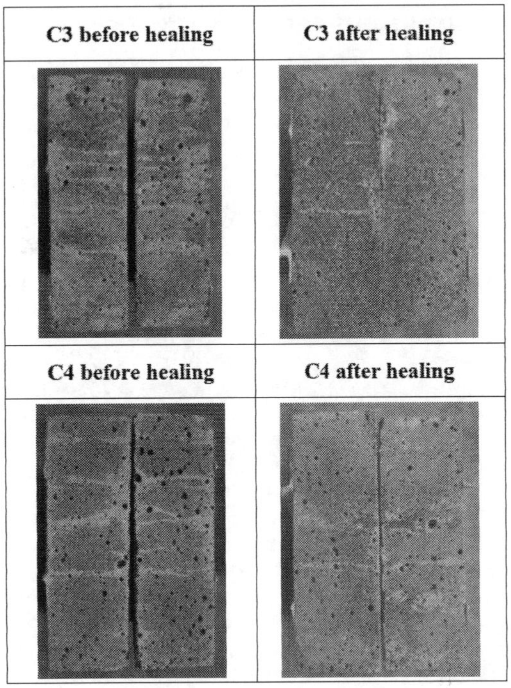

Fig. 7 – Comparison between chloride penetration in C3 and C4 before and after healing

Analysing the results obtained before healing, the C4 mixes showed a greater chloride penetration through the concrete matrix than the C3 ones, which is demonstrated by the bigger visible chloride area. This is due to the different composition of the mixes, mainly due to the difference in fibre contents (lower in C4). In addition, in the C3 mixes there is a much more widespread penetration of chloride, while in the mixes of C4 it is possible to note the clear presence of localized cracks.

Comparing the results obtained before and after healing for these two UHPFRCs mixes, it is clearly demonstrated that the precipitates closing the cracks of C3 mixes were able to stop salted-water penetration, and thus, provide a good protection against corrosion. The C4 mix provided a good improvement, but there were still a couple of cracks passing through the section of the sample, showing that they had lower autogenous healing capability and/or protection against chloride penetration.

Regarding the visual crack closure, it should be mentioned that most of the UHPFRCs narrower cracks (< 0.1 mm) were closed or partially closed, with a yellow precipitate inside the cracks due to the corrosion of the steel fibres. However, in the traditional and high-performance concretes, no visual crack closure was detected due to their greater cracks' width.

4 Conclusions and future work

This work aimed to propose water-permeability methodologies to evaluate self-healing of concrete samples that present different types of cracking patterns and sizes. The following related conclusions can be drawn:

The methodology proposed in this work consists of easy to implement steps in the laboratory, that allow the evaluation of self-healing in traditional and UHPFRCs. The phases of the methodology are: pre-cracking, where the ability to control the cracking procedure reaching the desired level of strain has been demonstrated; the phase of crack analysis, where by means of optical microscope together with software for the image analysis it is possible to detect and measure every crack; and finally the permeability tests, which have proved to be able to detect the variation in durability properties due to the crack healing.

- Concrete samples without fibres could be tested with the low-pressure water permeability test, by the evaluation of the water flow passing through the cracks before and after healing took place. In these samples, autogenous healing is not able to produce significant improvements of the durability.
- UHPFRC samples were able to totally prevent the entry of water liquids in the low and high-pressure water permeability tests. However, with the chloride penetration water permeability test, the results indicated the clear entrance of salted-water. In these samples, autogenous healing displayed a clear improvement in the water permeability test proposed, which was even greater in the mixes with higher contents of fibres, and thus, lower crack width for the same strain level.

Further studies are still being performed that will include additional cracking levels as well as self-healing promoters.

This project has received funding from the European Union's Horizon 2020 research and innovation programme under grant agreement No. 760824, project ReSHEALience.

References

1. De Belie, N., et al. 2018 "A Review of Self-Healing Concrete for Damage Management of Structures." Advanced Materials Interfaces (2018): 1800074.
2. K. Van Tittelboom and N. De Belie, "Self-Healing in Cementitious Materials—A Review," Materials, vol. 6, pp. 2182-2217, 2013.
3. Jacobsen, S., Granl, H.C., Sellevold, E.J. & Bakke, J.A., 1995. High Strength Concrete - Freeze/Thaw Testing and Cracking. Cement and Concrete Research, 25(8), pp.1775-80.
4. Jacobsen, S. & Sellevold, E.J., 1996. Self Healing of High Strength Concrete after Deterioration by Freeze/Thaw. Cement and Concrete Research, 26(1), pp.55-62.
5. Granger, S., Loukili, A., Pijaudier-Cabot, G. & Chanvillard, G., 2005. Mechanical characterization of the self-healing effect of cracks in Ultra High Performance Concrete (UHPC). In Proceedings Third International Conference on Construction Materials, Performance, Innovations and Structural Implications, ConMat, 5, pp.22-24.
6. Edvardsen, C.K., 1996. Wasserdurchlässigkeit und Selbstheilung von Trennrissen in Beton. Berlin: Beuth.
7. Edvardsen, C., 1999. Water Permeability and Autogenous Healing of Cracks in Concrete. ACI Materials Journal, 96(4), pp.448-54.
8. Yang, Y., Lepech, M.D., Yang, E.-H. & Li, V.C., 2009. Autogenous healing of engineered cementitious composites under wet–dry cycles. Cement and Concrete Research, 39, pp.382–90.
9. Yang, Y., Yang, E.-H. & Li, V.C., 2011. Autogenous healing of engineered cementitious composites at early age. Cement and Concrete Research, 41, pp.176-83.
10. Ferrara, L., Krelani, V., Moretti, F., Roig-Flores, M. and Serna-Ros, P., 2017. Effects of autogenous healing on the recovery of mechanical performance of High Performance Fibre Reinforced Cementitious Composites (HPFRCCs): part 1. Cement and Concrete Composites, 83, pp.76-100.
11. Darquennes, A., Olivier, K., Benboudjema, F. and Gagné, R., 2016. Self-healing at early-age, a way to improve the chloride resistance of blast-furnace slag cementitious materials. Construction and Building Materials, 113, pp.1017-1028.
12. Roig-Flores, M., Moscato, S., Serna, P. & Ferrara, L., 2015. Self-healing capability of concrete with crystalline admixtures in different environments. Construction and Building Materials, 86, pp.1-11.
13. Roig-Flores, M., Pirritano, F., Serna, P. & Ferrara, L., 2016. Effect of crystalline admixtures on the self-healing capability of early-age concrete studied by means of permeability and crack closing tests. Construction and Building Materials, 114, pp.447-57.
14. López, J.Á., Serna, P., Navarro-Gregori, J. and Coll, H., 2016. A simplified five-point inverse analysis method to determine the tensile properties of UHPFRC from unnotched four-point bending tests. Composites Part B: Engineering, 91, pp.189-204.

Effect of micro-climate variations on carbonation rate of concrete in the inland environment

Yunusa Aminu Alhassan[1], and *Sunday* Apeh[2]

[1,2]Department of Civil Engineering, The Federal Polytechnic Idah, Nigeria

Abstract. In an inland environment, carbonation is the primary cause of initiation for potential corrosion of steel in reinforced concrete. This problem has been exacerbated over recent years by increased urbanisation and vehicular traffic, resulting in higher atmospheric carbon dioxide contents – a problem typical of economically active cities throughout the world. It is important that designers of reinforced concrete structures respond to these variations through appropriate specifications to ensure that structures perform satisfactorily over their intended service lives. This paper is part of a study undertaken to assess the carbonation of concretes exposed to a range of micro-climate variations in inland environments, particularly with variations in carbon dioxide content, temperature and relative humidity conditions with the intent of developing a prediction model for the rate of carbonation. Concretes samples were prepared using three binder types representing variations of blends with FA, GGBS and four w/b ratios ranging from 0.4 to 0.75 and subjected to different degrees of initial water curing (3, 7, 28 days). These samples were placed in three exposure conditions: indoors in laboratory air, outdoors sheltered from rain and sun and outdoors fully exposed to the elements. The depths of carbonation of these samples were monitored over a period of 24 months in order to determine the rates of carbonation. Concrete samples in the outdoor sheltered sites presented the highest rate of carbonation. Although samples in this exposure site carbonated faster, the risk of reinforcement corrosion is likely to be low because the samples are protected from direct moisture effect. Keywords: Carbonation, Corrosion, Reinforced concrete, Micro climate, Inland environments

1. Introduction

The major cause of deterioration in reinforced concrete is the corrosion of the steel reinforcement. The steel is susceptible to corrosion in the presence of chloride ions or if it becomes depassivated when the alkalinity of the concrete at the location of steel is reduced by carbonation. Carbonation is a chemical reaction between carbon dioxide, in the presence of moisture, and calcium hydroxide. The ability of concrete carbonization resistance is a very important index of durability [1]. The durability of reinforced concrete structures is mainly embodied in concrete cover to reinforcement. Corrosion of reinforcement caused by concrete carbonization is complicated. If measures are not taken, the time relative to the whole service life from reinforcement corrosion to structural failure is short. As a result, the service life of reinforced concrete can be shown by the ability of concrete carbonization resistance [2, 3].

Many researchers have shown that the concrete carbonization reaction depends on a number of factors. One is the quality of concrete, such as the intercommunicating pore structure and $Ca(OH)_2$ content of concrete, the former determines the diffusion speed of CO_2 in the concrete [4, 5]. The latter is related to the ability of concrete absorbing CO_2. Another very important aspect is the environmental conditions, such as temperature, humidity, CO_2 concentration, etc. [6]. The diffusion ability of CO_2 in concrete is partly dependent on the quality of concrete, including pore structure and moisture state, etc. At the same time, the ability of absorbing CO_2 during the concrete carbonization process is mainly dependent on the type and content of cement.

Zhang and Jiang [7] considered the influence of the environmental relative humidity, the concentration of CO_2, concrete water–binder ratio and cement content on carbonization and put forward a carbonization prediction model. The model proposed by Niu et al [8] considered the environmental temperature, humidity, the concentration of CO_2 and the influence of the quality of concrete itself. Since the 1970s, the use of by-products like fly ash, ground granulated blast-furnace slag (GGBS) [9], silica fume (SF), rice husk ash, and metakaolin in cement and concrete production has gained rapid development [10, 11]. The incorporation of these mineral admixtures in concrete has been of great interest and is gradually being applied to practical projects, because it could not only reduce concrete production cost, conserve energy and resources, reduce environmental pollution impact [9, 12], but also improve resistance to the deterioration of aggressive chemicals and permeation of aggressive materials [13, 14].

Generally, it is agreed that carbonation does not occur in the same way in all mixes, nor does it occur in all circumstances. Different mixes will exhibit distinct carbonation and the same mix exposed to different

Corresponding author: alhassanay@gmail.com

environments will not show the same carbonation. Hence, this study's main objective is to try to understand the conditioning factors so as to explain concrete carbonation. The author's intent is to develop knowledge of the rate of carbonation in the inland environment at the microclimate level, that will enable the reliable prediction of potential corrosion and hence the service life of reinforced concrete structures.

2. Experimental Programme

2.1 Materials and mix proportion

Three types of common cements were used in concrete mixtures prepared under the investigation. Ordinary Portland cement (OPC) of class CEM I 42.5 with or without 30% fly ash (FA) or 50% ground granulated blast-furnace slag (GGBS) was used. Since the type and proportions of supplementary cementing materials (SCMs) play a significant role in carbonation behavior of cementitious systems, it is important that the class of cement type used in concrete is known. Since the SCMs were incorporated by blending with CEM I, an equivalent standard cement class has been identified for each of the blended cement types used in mixtures, as given in Table 1. It can be seen that nearly all standard cement types typically used in structural concretes are represented in the investigation. Fine aggregate (quarry sand) and coarse aggregate (uncrushed gravel) of 19 mm nominal size were used. The fine aggregate was of medium grading in accordance with BS 882:1992.

Table 1. Cement types used in concrete mixes

Cement type used	CEM I	30%FA	50%GGBS
Equivalent standard cement (EN 197-1, 2000)	CEM I	CEM II/B, CEM IV/A	CEM III/A, CEM IV/B

The concrete mixtures consisted of water-binder (w/b) ratios = 0.4, 0.5, 0.6, 0.75 and binder contents = 300, 350, 400, 450 kg/m^3 made using a 19 mm granite coarse aggregate and granite crusher sand of standard gradation [15]. A commercially available plasticizer, Chryso Plast 90 was added in concrete mixes of 0.4 and 0.5 w/b, to achieve slumps between 40 and 95 mm. All the other mixtures had slumps ranging from 80 to 140 mm, without use of the plasticizer. Table 2 gives details of the mixtures used in the investigation.

2.2 Sample preparation and testing

Concrete 100 mm cubes were cast using the mixtures in Table 2 for carbonation measurements. All the samples were cast and compacted in accordance with BS EN 12390-1 (2000). After casting, the samples were covered under damp hessian and polyethylene sheets for 24hour. The samples were demoulded the following day and then immediately kept in a mist room at 20 ± 2 °C and 98 ± 2% RH prior to exposure. In the investigation, the cube samples were exposed at outdoor sites to undergo carbonation under natural climate. The outdoor exposure environments were three different urban sites consisting of: the rooftop of a one storey building, samples stored at this location are considered not sheltered; the parking garage of a three storey building, samples exposed under this condition may be regarded as indoor; and beneath a foot bridge spanning over a busy highway, samples under this condition would be considered sheltered from sun and rain.

Prior to exposure of samples to the designated sites, cubes were removed from curing baths at ages of 3, 7, 28 days and air-dried under laboratory conditions for 14 days. All four sides running from the finishing surface through to its opposite end surface, were coated with two layers of epoxy coating, leaving the two adjacent opposite sides uncoated. A water-based epoxy obtained from StonCore Africa, was used. After sufficient drying of surface coatings over a 24 hour period, samples were taken to their respective exposure sites and placed at a spacing of at least 50 mm from each other such that the two non-coated surfaces were exposed to air. For each mixture, a set of four cubes, each set having been cured for 3, 7 or 28 days, was exposed at each of the three sites. At each exposure site, environmental factors were monitored using a 24-hour, automated data logger set up to record the micro-climatic conditions. The relative humidity (RH), CO_2 concentration and temperature were monitored for one year during the period under study. Data was recorded every minute and downloaded monthly into a computer hard-drive.

Carbonation measurements were carried out for concrete cubes of 100 mm biannually for 2 years of exposure in a varying micro-climatic condition. The samples were broken into two halves at six months interval. After splitting the concrete sample, the freshly split surface was cleaned and sprayed with a phenolphthalein pH indicator. The indicator used was a phenolphthalein 1% ethanol solution with 1 g phenolphthalein and 90 ml 95.0V/V% ethanol diluted in water to 100 ml. In the non-carbonated part of the sample, where the concrete was still highly alkaline, a purple-red colour was obtained. In the carbonated part of the sample where the alkalinity of concrete is reduced, no coloration occurred. The average depth of the colourless region was measured from three points, perpendicular to the two edges of the split face, both immediately after spraying the indicator and at 24 hour later.

Table 2. Concrete mixtures used in the investigation

Cement type	Mix No.	Binders (kg/m^3)			Aggregate (kg/m^3)		WC (kg/m^3)	w/b	SP (l)	Slump (mm)
		CEM I	FA	GGBS	CA	SA				
CEM I	PC-40	450	-	-	990	800	180	0.4	4.00	45
	PC-50	400	-	-	990	877	200	0.5	3.25	75
	PC-60	350	-	-	1015	869	210	0.6	-	100
	PC-75	300	-	-	1050	810	225	0.75	-	140
30FA	FA-40	315	135	-	990	800	180	0.4	1.75	55
	FA-50	280	120	-	990	877	200	0.5	1.70	90
	FA-60	245	105	-	1015	869	210	0.6	-	125
	FA-75	210	90	-	1050	810	225	0.75	-	135
50GGBS	SL-40	225	-	225	990	800	180	0.4	1.65	40
	SL-50	200	-	200	990	877	200	0.5	1.60	50
	SL-60	175	-	175	1015	869	210	0.6	-	66
	SL-75	150	-	150	950	900	225	0.75	-	105

CA - coarse aggregate, SA - fine aggregate, WC - water content, w/b - water/cementitious ratio, SP - plasticizers,

3. Results and Discussion

3.1 Micro-climate variation at exposure sites

Figures 1, 2 and 3 presents the average microclimatic conditions recorded at each exposure site through a 12-month period. From Figure 1, it can be observed that the variability in CO_2 concentration between the exposure sites is not significant but follow a defined pattern with the indoor site having higher CO_2 concentration. However, slightly higher CO_2 values are observed for the indoor exposure site in most months and this can be attributed to the enclosed nature and the lack of free flow of air. The indoor exposure site is a basement parking garage and would be influenced by CO_2 from the exhaust fumes of cars. The outdoor exposed condition, unsheltered from rain and sun presented the least CO_2 concentration. This can be ascribed to its height above ground level and distance from sources of CO_2 pollution, since land use influences the CO_2 concentration.

Generally, based on the CO_2 variations presented in the figure and all other factors being equal, concrete samples exposed to the indoor condition should have the highest carbonation rates while concretes exposed outdoor and unsheltered should present the least carbonation rates. This is based on the fact that the rate of diffusion of CO_2 into concrete increases with increase in CO_2 concentration between the internal and external environment of the concrete. The implication of the CO_2 variations for the durability performance and service life of concrete structures are that structural elements exposed to indoor conditions will be exposed to higher carbonation rates and should therefore be more carefully designed and constructed if its design service life in terms of the initiation limit state (ILS) is to be achieved.

From Figure 2, it can be seen that there exists a governing trend indicating a general decrease in the RH from May to September, followed by a steady decrease from October to February.

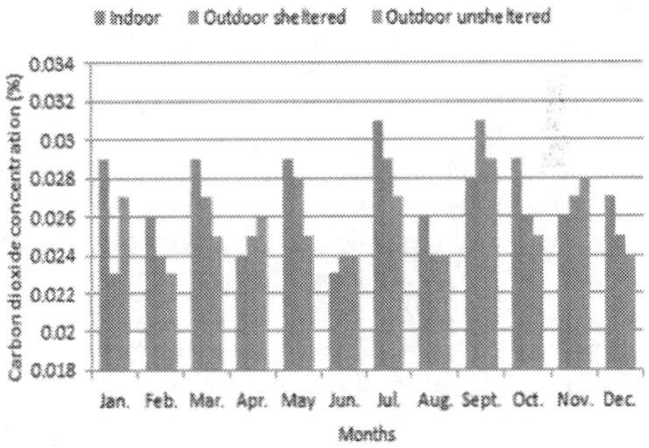

Fig. 1 Average monthly carbon dioxide variations for the exposure sites

The higher RH recorded in the raining period is related to precipitation associated with rainfall which is typically experienced from April to September.

Based on the RH results for the exposure sites and since the external RH influences the internal humidity of the concrete, it is evident that concrete samples exposed indoors had the lowest internal moisture content. Concretes exposed outdoors but sheltered from rain and sun, had moderate internal humidity judging from the exposure site RH values presented in Figure 2. While the moisture content for concrete samples exposed outdoor to rain and sun are high for most part of the year as the average monthly RH for this site is higher compared to the other sites.

Keeping other factors equal, the RH result presented shows that concrete cube samples exposed outdoor but sheltered from rain and sun will have the highest carbonation rates because this exposure site RH profile mostly lies in the zone for maximum carbonation rate [16, 17, 18]. The implication of the RH results for durability performance and service life of reinforced concrete elements is that structural elements exposed outdoor but sheltered will present the highest carbonation rate. Thus, such elements should be more carefully designed and constructed if its design service life in terms of the ILS is to be achieved. It is also evident that concrete elements exposed outdoor to rain and sun are at risk of corrosion due to the high RH as a results of the precipitation during raining months. Although the rate of carbonation may be very low for concrete elements exposed to the outdoor condition because of the low or high saturated pore structure which depend on the period of the year.

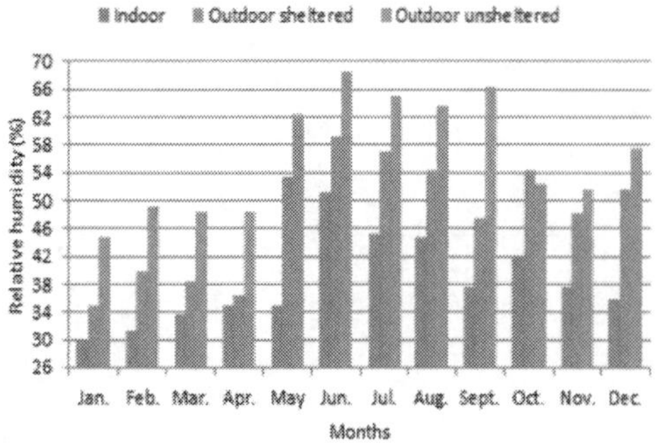

Fig. 2 Average monthly relative humidity variations for the exposure sites

Figure 3 shows the average monthly temperature variation within a span of one year for the three exposure sites. Generally, temperature decreases for all exposure sites from January to June and then increases from July to December. From the temperature measurements recorded for these sites, it is evident that the indoor exposure site experienced warmer conditions for most part of the year while it was cooler for the outdoor unsheltered. Although, the influence of temperature variations is not very significant for carbonation in concrete [19], results of the temperature variation shows that carbonation rates should be higher for concrete cube samples exposed indoors, while it may be similar for the other two exposure sites since their temperature profiles are similar.

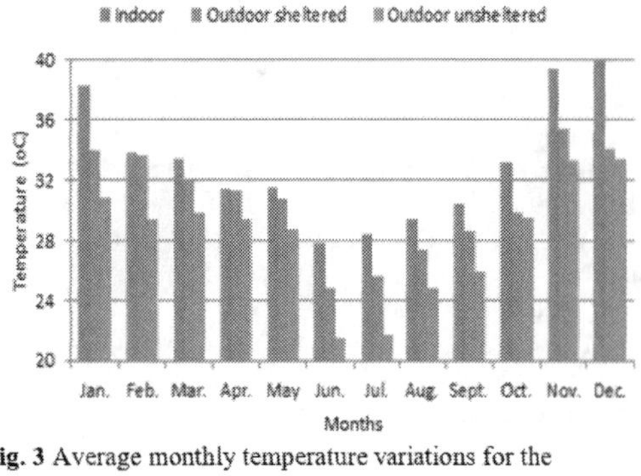

Fig. 3 Average monthly temperature variations for the exposure sites

3.2 Natural carbonation

Carbonation results obtained for the various concretes exposed to the natural inland environments are presented in Figure 4. As earlier mentioned, concrete samples were exposed to natural indoor and outdoor sites and monitored for carbonation at ages of 6, 12, 18 and 24 months. Although the present study investigated the effects of the variability in microclimatic conditions on the rate of concrete carbonation, the influence of SCMs and curing periods are evident from the presented results. Concrete samples incorporating SCMs shows an increase in carbonation regardless of the exposure site conditions. While a longer initial moist curing duration generally resulted in a lower carbonation rate for all concretes type investigated. This is attributed to the fact that as curing increases hydration in concrete, the pore space in concrete reduces, particularly in the near surface zone. Additionally, curing influences the chemical properties of the concrete, for instance in unblended concrete it increases the $Ca(OH)_2$ content of the concrete thereby improving its buffering effect against CO_2 [20] However, curing reduces the $Ca(OH)_2$ content in blended concretes but improves the concrete microstructure [3, 21]. For example, in plain concretes, the carbonation rate decreased significantly with curing age within the first seven days. In the case of concrete containing SCMs, the carbonation rate continued to decrease even after 7 days of curing again pointing to the later hydration effect of these SCMs.

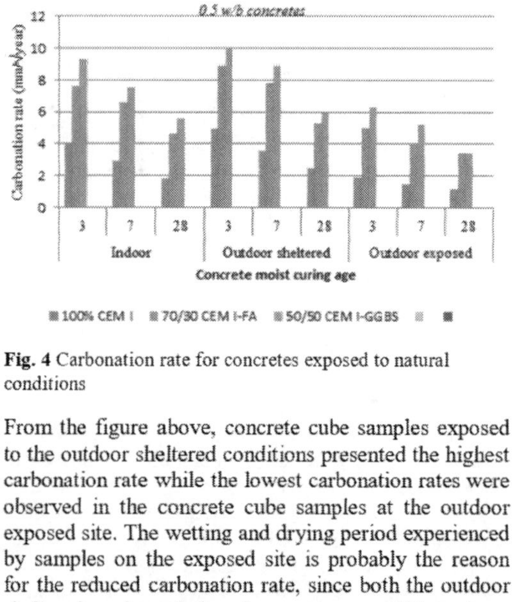

Fig. 4 Carbonation rate for concretes exposed to natural conditions

From the figure above, concrete cube samples exposed to the outdoor sheltered conditions presented the highest carbonation rate while the lowest carbonation rates were observed in the concrete cube samples at the outdoor exposed site. The wetting and drying period experienced by samples on the exposed site is probably the reason for the reduced carbonation rate, since both the outdoor sheltered and unsheltered exposure conditions have similar micro-climates in terms of CO_2 content, *RH* and temperature as noted in Figures 1, 2 and 3 respectively. In addition, concrete cube samples exposed outdoor to rain and sun have higher internal humidity during the summer months from precipitation effects, thus reducing CO_2 ingress during this period. Furthermore, these concretes had improved pore structures from rapid hydration effects as a result of its high internal moisture content during the summer months.

The micro-climatic factors known to influence the carbonation rate of concretes are the CO_2 concentration,

RH and temperature with the most important being RH. The RH surrounding the concrete influences the concretes internal humidity which then affects the ingress of CO_2. For instance, [22] noted an insignificant effect on concrete carbonation rate for samples with compressive strength less than or equal to 40 MPa when the CO_2 concentrations were varied between 7% and 18%. While for concretes with strength higher than 40 MPa, variation of a greater order of magnitude in CO_2 concentration may be required to show any significant effect on the carbonation rate. The rate of carbonation in concrete is also weakly sensitive to temperature within the range 20-40 °C [23]. A similar observation regarding the insensitivity of temperature to the rate of carbonation in concrete was also noted by Loo et al., (1994) in their work.

4. Conclusions

The main aim for this paper was to investigate the influence of micro-climate variations on the carbonation rate of concretes in inland environments. Through the carbonation study, several conclusions can be drawn.

a. Increasing the duration of moist curing and reducing the amount of addition of supplementary cementing materials has a significant effect on decreasing the rate of carbonation in concrete.
b. Concrete samples in the outdoor exposed sites presented a lower rate of carbonation compared to the sheltered sites but higher than for the indoor sites.
c. Concretes element exposed outdoor are at risk of reinforcement corrosion when compared to the other two sites. This is as a result of the higher relative humidity and the cyclic wetting during rain period.
d. The micro-climate condition, especially the relative humidity, dictates the rate of carbonation of concrete in the inland environment. Thus, concrete samples in the outdoor sheltered sites presented the highest rate of carbonation. Although samples in this exposure site carbonate faster, the risk of reinforcement corrosion is likely to be low because the samples are protected from direct moisture effect.

This work is based on the research supported in part by the Tertiary Education Trust Fund (TETFund) of Nigeria. The first author also wishes to acknowledge the inspiration provided by Prof Yunus Ballim of the School of Civil & Environmental Engineering, University of the Witwatersrand, Johannesburg, South Africa.

References

1. L. Bertolini, B. Elsener, P. Pedeferri, E. Redaelli, R.B. Polder. Corrosion of steel in concrete: prevention, diagnosis, repair: Wiley. com. (2013).
2. P.K. Mehta. Durability of concrete--fifty years of progress? ACI Special Publication, 126. (1991)
3. P. Mehta, P. Monteiro, P. Concrete: Structure, Properties, and Materials, Prentice-Hall, Englewood Cliffs, NJ, (1993).
4. L. Wu, G.X. Pu. Influencing factors of concrete carbonation and prediction model of carbonation depth. Technique of Seepage Control 8 (3), 10–12 (in Chinese with English abstract). (2002)
5. X.C. Zhou, X.C. Experimental study on the influence of mineral additive to the concrete air permeability. Shanxi Architecture 37 (10), 101–103 (in Chinese with English abstract). (2011)
6. Z.L. Meng, F. Zhu, H. Zhou, J.S. Qian. Carbonation and ways of preventing carbonation of high-volume fly ash concrete. Building Science Research of SiChuan27 (3), 50–54 (in Chinese with English abstract). (2001)
7. Y. Zhang, L.X. Jiang. A practical mathematical model of concrete carbonation depth based on the mechanism. Industrial Construction 28 (1), 16–19 (in Chinese with English abstract). (1998).
8. D.T Niu, Z.P. Dong, J.X. Pu. Random model of predicting the carbonated concrete depth. Industrial Construction 29 (9), 41–45 (in Chinese with English abstract). (1999).
9. A. Oner, S. Akyuz. 2007. An experimental study on optimum usage of GGBS for thecompressive strength of concrete. Cement and Concrete Composites 29, 505–514. (2007).
10. H.S. Shi, B.W. Xu, X.C. Zhou. Influence of mineral admixtures on compressive strength, gas permeability and carbonation of high performance concrete. Construction and Building Materials 23, 1980–1985. (2009)
11. R. Siddique, J. Klaus. Influence of metakaolin on the properties of mortar and concrete: a review. Applied Clay Science 43, 392–400. (2009)
12. E.H. Yang, Y.Z Yang, V.C. Li. Use of high volumes of fly ash to improve ECC mechanical properties and material greeniness. ACI Materials Journal 104 (6), 620–628. (2007).
13. D.M. Roy, P. Arjunan, M.R. Silsbee. Effect of silica fume, metakaolin, and low calcium fly ash on chemical resistance of concrete. Cement and Concrete Research31, 1809–1813. (2001).
14. H.W. Song, V. Saraswathy. 2006. Studies on the corrosion resistance of reinforced steel in concrete with ground granulated blast-furnace slag — an overview. Journal ofHazardous Materials B138, 226–233. (2006).
15. F.S. Fulton. Fulton's concrete technology, 9th Edition (Gill Owens, ed.), Cement and concrete Institute, Midrand, Johannesburg, RSA. (2009).
16. G. Verbeck. (1958). *Carbonation of hydrated Portland cement*. Cement and Concrete, 17-36. (1958).
17. M. Richardson. Fundamentals of durable concrete, modern concrete technology: Published by Spon Press, London. (2002)
18. B.M. Fernandez, S. Simons, C. Hills, P. Carey. A review of accelerated carbonation technology in the treatment of cement-based materials and sequestration of CO_2. Journal of Hazardous Materials, 112(3), 193-205. (2004)

19. Da Silva, F., Helene, P., Castro-Borges, P., & Liborio, J. (2009). Sources of variations when comparing concrete carbonation results. Journal of materials in civil engineering, 21(7), 333-342. (2009).
20. Ballim, Y. (1994). Curing and the durability of concrete. PhD Thesis, University of the Witwatersrand.
21. V.G. Papadakis. Effect of supplementary cementing materials on concrete resistance against carbonation and chloride ingress. Cement and Concrete Research, 30(2), 291-299. (2000)
22. Y. Loo, M. Chin, C. Tam, K. Ong. (1994). A carbonation prediction model for accelerated carbonation testing of concrete. Magazine of Concrete Research, 46(168), 191-200. (1994).
23. V.G. Papadakis, C.G. Vayenas, M.N. Fardis. Fundamental modelling and experimental investigation of concrete carbonation. American Concrete Institute Materials Journal. 88(4): 363-373. (1991).

MATEC Web of Conferences **289**, 02002 (2019) https://doi.org/10.1051/matecconf/201928902002
Concrete Solutions 2019

Abrasion resistance of high early-strength concrete for rapid repair

Nader Ghafoori[1], *Matthew O. Maler*[1], *Meysam Najimi*[2], *Ariful Hasnat*[1]

[1]Department of Civil and Environmental Engineering and Construction, University of Nevada, Las Vegas 4505 S Maryland Pkwy Street, Las Vegas, USA
[2]Department of Civil, Construction, and Environmental Engineering, Iowa State University, Ames, IA 50011, USA

Abstract. This paper examines the abrasion resistance of high early-strength concrete developed for rapid repair of highways and bridge decks. The cement types chosen for this study included ASTM Type III, ASTM Type V, and Calcium Sulfoaluminate (CSA) cements. A cement content of 386 kg/m^3 (650 lb/yd^3) was maintained for all studied concretes. Test samples were tested after 24 hours and 28 days of curing in order to evaluate compressive strength and depth of wear. Test results revealed that the opening time to attain minimum required compressive strength for CSA cement concrete was one hour, whereas the values for Type V and Type III cement concretes were 8.5 and 6 hours, respectively. After 24 hours curing, CSA cement concrete displayed the highest strength, but lowest resistance to wear. The 28-day cured CSA cement concrete produced the highest strength and resistance to abrasion, while Type III cement concrete showed a similar strength, but lower resistance to wear, when compared to those of the Type V cement concrete.

1 Introduction

With more than 60% of the US interstate highway system paved with concrete, surface wear due to traffic loading or environmental conditions is one of the main reasons that can reduce the quality of ride. As a result, major maintenance and repair works are required to maintain these huge chains of road network [4]. In recent years, significant advancement has been made in the development of fast repair techniques of reinforced concrete pavements. High Early-Age Strength (HES) concrete was traditionally regarded as a concrete that achieves a loading strength in a matter of days rather than weeks. However, in the last 10-15 years, this time has been reduced down to a matter of hours [1]. The required minimum compressive strength is determined on a case-by-case basis depending on a given project. In construction applications for structures and dwellings the minimum allowed by the International Building Code is 17.24 MPa (2500 psi) (IBC 2012), while in road and bridge construction the minimum requirement is typically 20.68 MPa (3000 psi) [2].

Concrete mechanical properties play an important role in the abrasion of concrete [5-8]. Siddique and Khatib [9] found that an increase in compressive strength and modulus of elasticity of concrete improves the overall abrasion resistance. Experimental work on the abrasion resistance of highway concrete pavements subjected to heavy traffic from studded tires has been carried out. Increasing the concrete strength from 50 to 100 MPa (7100 to 14,300 psi) reduced the depth of wear of concrete by roughly 50% [9].

In recent years, Calcium Sulfoaluminate Cement (CSA) has become widely used in concrete repair works due to its outstanding properties of volume stability and rapid strength gain [10]. In comparison to Type V cement, Type III and CSA cement are expensive. In addition, some of their long-term properties may be compromised in exchange for rapid strength development. Therefore, they tend to be used as emergency repairs and in high priority situations such as inner-city freeway exchanges and airport runways, or situations in which a significant cost is associated with the elapsed closure time. In this context, an experimental program was undertaken to compare the effect of different cement types on abrasion resistance and compressive strength development at different curing ages for rapid repair of concrete pavements and highways.

2 Experimental Procedures

2.1 Materials

Locally available fine and coarse aggregates were used. Their size distribution complied with the requirements of ASTM C136 and ASTM C33 for coarse and fine aggregates. ASTM Type III, Type V, and Calcium Sulfoaluminate (CSA) cements were used to produce standard high early-strength concretes. Table 1 provides the composition and Blaine Fineness of the cements used in this study. The CSA cement had the highest fineness when compared with that of the Type III and Type V cements. Polycarboxylate high-range-water-reducer (HRWR), Accelerating admixture, and air-entrainer

* Corresponding author: hasnat@unlv.nevada.edu

(AE) were also used to maintain the desired fresh and early-strength properties of concrete.

2.2 Mixture constituents and proportions

Three mixture proportions were selected to evaluate the compressive and wear properties of concrete. The mixture constituents and proportions are given in Table 2. The total cement content was kept constant at 386 kg/m^3. As listed in Table 2, three different water-to-cementitious ratios, 0.35, 0.275, and 0.40 were used for Type III, Type V, and CSA cement mixtures, respectively. The basis for selection of different w/c was to achieve the earliest opening time (the highest early-age strength development) while providing adequate fresh properties. The workability of the studied concretes was kept constant at 125±25 mm (5±1 in). To accelerate the hardening process, a higher amount of accelerating admixture was used for Type III and Type V cement concretes. A small amount of accelerating admixture was also used in the CSA cement concrete mixture. All studied concretes contained air-entraining admixtures (AE) to achieve a uniform air content of 4±1%.

Table 1. Cement composition and Blaine fineness.

Potential Compounds	ASTM Type III	ASTM Type V	CSA (Rapid set) *
C_3S (%)	50	58	Proprietary information
C_2S (%)	24	16	
C_3A (%)	11	4	
C_4AF (%)	4	12	
Blaine Fineness (m^2/kg)	496	420	>500

* C_3S and C_3A are in the form C_4A_3S

Table 2. Mixture proportions

Mixture id	w/cm	C* (kg/m^3)	FA* (kg/m^3)	CA* (kg/m^3)	WRA* (kg/m^3)	Acc* (kg/m^3)	Air entrainer (kg/m^3)
Type III-AE	0.35	386	898	1097	1.65	7.71	1.9
Type V-AE	0.275	386	933	1141	3.24	7.71	2.1
CSA-AE	0.40	386	865	1057	2.20	1.21	0.11

*Note : *AE = Air Entrained ; C = Total Cement ; FA= Fine Aggregate ; CA= Coarse Aggregate ; Acc= Accelerator*

2.3 Mixing, sampling, curing, and testing

A pan-style counter-current mixer, having a constant speed of 60 RPM, was used. After pouring freshly-mixed concrete into the mold, specimens were densified for a period of approximately 7 seconds using a vibrating table. Both one day and 28 days moist curing were utilized to cure concrete specimens. To evaluate compressive strengths, cylindrical specimens with dimensions of 102 mm x 203 mm (4 in. x 8 in.) were used and tested in accordance with ASTM C39. Resistance to wear of the studied high early-age strength concretes were evaluated in accordance with the ASTM C779, Procedure A, Ball Bearings. This test method was selected as it best simulates rolling traffic and the severe wearing found in industrial and commercial applications, as well as highway pavements exposed to traffic wear from vehicles having tires equipped with snow chains or studs. Two 152 mm x 152 mm x 152 mm (6 in. x 6 in. x 6 in.) cube-shaped samples for each curing type were used to obtain the wear depth. The testing machine used for this study is shown in Figure 1.

Fig. 1. Abrasion resistance testing machine.

3 Results and Discussion

3.1 Setting time

When Type V Portland cement was used, the initial setting time of HES concretes was at approximately 85 minutes, whereas the final setting time was 169 minutes. For Type III cement, the initial setting time was recorded at 70 minutes and the final setting time was 129 minutes. The reduction in the setting times in Type III, as compared to those of Type V Portland cement, is due to Type III cement having an increased C_3S content and higher Blain Fineness. It is well established that an increase in water-to-cement ratio leads to an increase in setting times. Since the studied concretes made with Type III Portland cement had a higher water-to-cement ratio, it would suggest that its C_3S and Blain Fineness had greater influence on setting time than water-to-cement ratio. High early strength concretes containing CSA cement were not evaluated for setting times since its manufacturer assigned 20 minutes for initial setting and 30 minutes for final setting.

3.2 Opening time

The time to achieve the required 21 MPa (3000 psi) compressive strength (The opening time) using Type III, Type V, and CSA cement concrete is given in Figure 2. It is observed that the opening time for Type III cement concrete was lower as compared to that of Type V cement concrete. This is due to the higher fineness and presence of a high amount of C_3S in Type III cement. Also, air-entrainer acted as a retarder to the concrete and increased the opening time for both Type III and type V cement concretes. As per manufacturer recommendation, the CSA cement concrete mixture was made with addition of retarder with a1 hour opening time. All three mixtures exceeded the minimum opening time compressive strength requirement of 21 MPa.

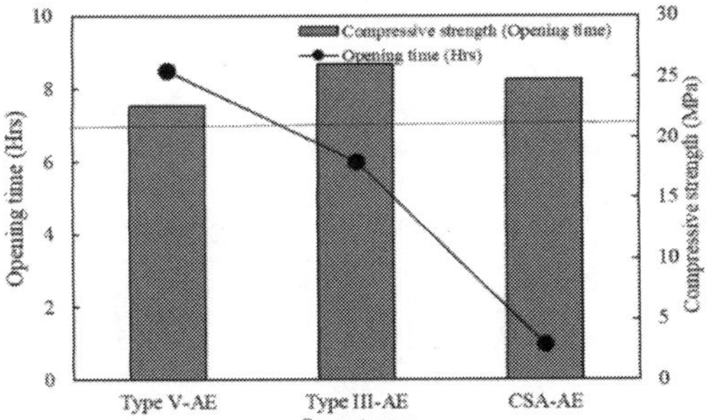

Fig. 2. Opening time and compressive strength of concretes made with Type III, Type V, and CSA cement.

3.3 Compressive strength

The influence of curing age and cement type on the compressive strength are shown in Figure 3. When Type V cement was used, the 28 day compressive strength of the concrete increased by 41%, as compared to the 24-hour compressive strength of the same mixture. For Type III cement concrete, the 28 day compressive strength of the concrete increased by 31%, as compared to the 24-hour compressive strength of the same mixture. For CSA cement mixtures, the 28-day compressive strength of the concrete increased by 20%, as compared to the 24-hour compressive strength of the same mixture. Concrete with Type V cement showed a higher compressive strength development at 28 days as compared to the 24 hours compressive strength. This is mainly due to presence of lower C_3S in Type V cement concrete as opposed to that of the Type III and CSA cement concrete mixtures. On the whole, in both 24 hours and 28 days tests, CSA cement showed the highest compressive strength amongst the three studied mixtures.

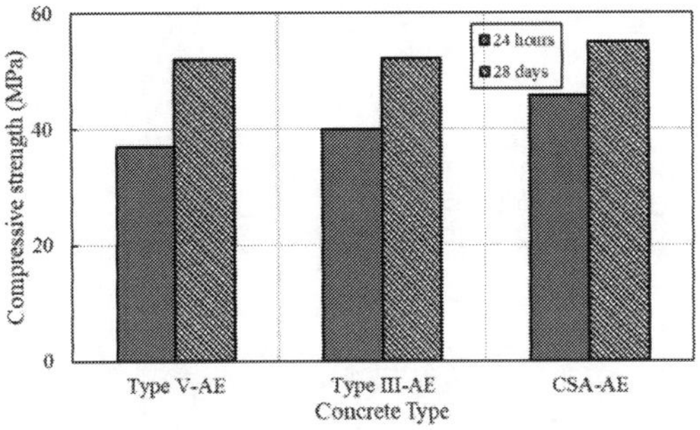

Fig. 3. Effect of curing age on compressive strength of concretes made with Type III, Type V, and CSA cement.

3.4 Abrasion resistance

The depth of wear of the concretes at different test ages is presented in Figure 4. The use of Type V cement in the mixture resulted in a 23% lower abrasion depth at 28 days than that at 24 hours. On the other hand, concrete containing Type III cement produced a 16% lower abrasion depth at 28 days than that of the 24-hour cured specimen. A significant improvement was observed in abrasion resistance for the CSA cement concrete mixture with 44% lower abrasion depth at 28 days testing in comparison with 24 hours testing. Figure 5 represents the abrasion depth and compressive strength of the studied concretes at 28 days curing age. Increases in the compressive strength reduced the depth of wear significantly. As can be seen, the compressive strength of the CSA cement concrete mixture also had the highest compressive strength and the lowest depth of abrasion amongst all three studied concrete mixtures. Table 3 shows the percentage increase in compressive strength and reduction in abrasion depth of the studied concretes when the curing age increased from 24 hours to 28 days.

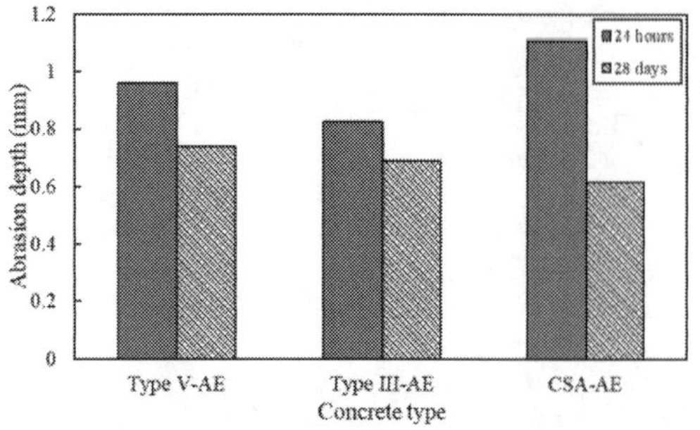

Fig. 4. Effect of curing age on abrasion depth for Type III, Type V, and CSA cement.

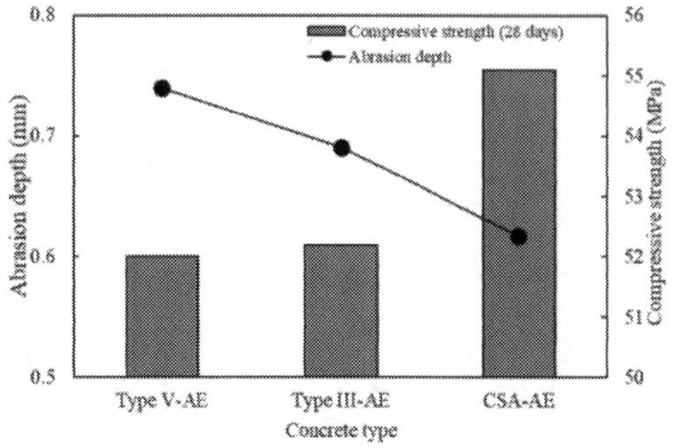

Fig. 5. Effect of curing age on failure time of Type III and Type V cement concrete.

Table 3. Influence of curing age on strength and resistance to abrasion

Mixture Type	% increase in Compressive strength	% decrease in depth of wear
Type V-AE	41	23
Type III-AE	31	16
CSA-AE	20	44

4 Conclusions

Based on the results of the investigation, the main conclusions are as follows:

- All studied concretes attained the minimum opening time compressive strength of 21 MPa within a short period of time. The resulting opening times were 8.5, 6, and 1 hours for Type V, Type III, and CSA cement concretes, respectively. CSA cement concrete showed the highest 24-hour and 28-day cured compressive strengths amongst the three studied concretes. However, 28 days curing reduced the strength gap amongst the three studied mixtures.
- At 24 hours curing, CSA cement concrete showed the highest depth of abrasion in comparison with that of the Type III and Type V cement concretes.
- Once curing age was extended to 28 days, CSA cement concrete displayed the lowest depth of wear and the highest compressive strength.

Acknowledgements

This study was made possible by the financial contributions of The US Federal Highway Administration through UTC Solaris Consortium. Thanks are also extended to a number of producers and manufacturers for providing materials used in this investigation.

References

1. T.R. Naik & B.W. Ramme: *High early strength concrete containing large quantities of fly ash.* In: ACI Materials Journal, **86** (1989) 2, PP. 111-116.
2. M.O. Maler: *High Early-Age Strength Concrete for Rapid Repair.* In: MSc Diss. University of Nevada, Las Vegas (2017).
3. P. Suraneni, J. Monical, E. Unal, Y. Farnam, & J. Weiss: *Calcium oxychloride formation potential in cementitious pastes exposed to blends of deicing salt.* In: ACI Materials Journal **114** (2017) 4, PP. 631-641.
4. K. Amini, S. Sadati, H. Ceylan, & P.C. Taylor: *Effects of mixture proportioning, curing, and finishing on concrete surface hardness.* In: ACI Materials Journal **116** (2019) 2, PP. 119-126.
5. T.R. Naik, S.S. Singh, & M.M. Hossain: *Abrasion resistance of concrete as influenced by inclusion of fly ash.* In: Cement and Concrete Research **24** (1194) 2, PP. 303-312.
6. A. Nanni: *Abrasion resistance of roller compacted concrete.* In: ACI Materials Journal **86** (1989) 6, PP. 559-565.
7. N. Ghafoori & H. Diawara: *Abrasion resistance of fine aggregate-replaced silica fume concrete.* In: ACI Materials Journal **96** (1999) 5, PP. 559-569.
8. N. Ghafoori & B.M. Sukandar: *Abrasion resistance of concrete block pavers.* In: ACI Materials Journal **92** (1995) 1, PP. 25-36.
9. O.E. Gjørv, T. Baerland, & H.R. Ronning: *Abrasion resistance of high-strength concrete pavements.* In: Concrete International **12** (1990) 1, PP. 45-48.
10. J.S. Yeung, M.C. Yam, & Y.L. Wong: *1-Year development trend of concrete compressive strength using Calcium Sulfoaluminate cement blended with OPC, PFA and GGBS.* In: Construction and Building Materials **198** (2019), PP. 527-536.

Properties of high early-strength Type V cement concrete for rapid repair

Nader Ghafoori[1], *Matthew O. Maler*[1], *Meysam Najimi*[2], *Ariful Hasnat*[1]

[1]Department of Civil and Environmental Engineering and Construction, University of Nevada, Las Vegas 4505 S Maryland Pkwy Street, Las Vegas, USA
[2]Department of Civil, Construction, and Environmental Engineering, Iowa State University, Ames, IA 50011, USA

Abstract. This study examines the suitability of ASTM Type V cement concrete for rapid repair applications. To this end, experimental results on transport and durability properties of high early-strength concretes using ASTM Type V cement were compared with those of a more traditional cement used for rapid repair, i.e. Type III cement. A cement content of 445 kg/m^3 (750 lb/yd^3) was maintained for all studied concretes. The experimental program included compressive strength, absorption, rapid chloride migration, corrosion resistance, and mass loss due to freezing and thawing regimes. The results of this study revealed that use of Type III and V cements were both effective for concrete rapid repair applications. The opening time to reach the minimum compressive strength of 21 MPa (3000 psi) was found dissimilar. Type III cement concrete showed better strength properties at early ages due to its high fineness. However, as curing age was extended to 24 hours and 28 days, Type V cement concrete produced higher strength results. Moreover, Type III cement concretes failed to display better performance in transport properties, corrosion, and frost resistance when compared to that of the studied Type V cement concretes.

1 Introduction

The infrastructure annual report published by the American Society of Civil Engineers (ASCE) depicted that the infrastructure of the United States received an overall D$^+$ grade where A stands for best and F is for the worst [1]. More than two trillion dollars have been needed to repair American infrastructure alone [2]. As a result, extensive research has been conducted on repair and retrofitting of existing structures [9-11]. For structures like buildings, it is easy to repair without making a colossal effort. However, for a reinforced concrete bridge or concrete highway repair, the scenario is quite different. Before repairing a deteriorated bridge or a highway pavement, significant planning and considerable time are required to make the transportation infrastructures serviceable. Recent studies have found that traffic congestion dominates the life-cycle social costs due to construction-related congestion, operation interruption, wasted time and fuel cost, vehicle damage as a result of poor road conditions, safety concerns, and environmental impact during maintenance and repair operations [3].

In this context, in recent years, several propriety cementitious materials and admixtures have been developed for rapid repair of concrete structures [4]-[6]. However, for all practical purposes, most of them do not deliver as promised. Additionally, these newly-developed materials are far more expensive than the parent material used for construction. To this end, viable alternatives are needed for rapid repair works. ASTM Type III cement has high fineness and high C_3S content that helps in early age strength development, making it to be a viable solution for rapid repair [7]. However, the concerns over its wide-spread availability and long-term strength and durability have given an impulse to consider other ASTM cement types for rapid repair [8].

2 Experimental Procedure

2.1 Materials

The fine and coarse aggregates used in this study were obtained from a local quarry in Nevada. The physical properties of the coarse and fine aggregates are presented in Table 1. ASTM C136 and ASTM C33 were used to determine size distribution of the aggregates. ASTM Type III and Type V cements were used in this study. Polycarboxylate high-range-water-reducing (HRWR), and air-entraining (AE) admixtures were used to maintain the desired fresh properties of concrete.

2.2 Mixture constituents and proportions

A total of four mixture proportions were selected to evaluate the mechanical and transport properties of concrete. The selected mixture proportions are given in Table 2. The total cement content was kept constant. To accelerate the hardening process, a higher amount of accelerating admixture was used. Among the four studied mixes, two of them contained air-entraining

* Corresponding author: hasnat@unlv.nevada.edu

Table 1. Aggregate properties.

Aggregate type	Bulk specific gravity (OD)	Bulk specific gravity (SSD)	Absorption (%)
Fine aggregate	2.76	2.78	0.81
Coarse aggregate	2.73	2.76	0.82

Table 2. Mixture proportions

Mixture id	w/cm	C* (kg/m³)	FA* (kg/m³)	CA* (kg/m³)	HRWR* (kg/m³)	Acc* (kg/m³)	Air entrainer (kg/m³)
Type III	0.35	445	848	1037	1.65	8.90	-
Type III-AE	0.35	445	848	1037	1.37	8.90	2.30
Type V	0.275	445	890	1088	2.91	8.90	-
Type V-AE	0.275	445	890	1088	2.80	8.90	2.50
*Note : *AE = Air Entrained ; C = Total Cement ; FA= Fine Aggregate ; CA= Coarse Aggregate ; Acc= Accelerator*							

admixtures (AE). A constant workability of 125±25 mm (5±1 in.) was maintained for all studied concretes.

2.3 Mixing, sampling, curing, and testing

A pan-style counter-current mixer with a capacity of 0.0283 m³ (1 ft³) and a constant speed of 60 RPM was used. Upon molding, specimens were densified for a period of approximately 7 seconds, longer as workability decreased, using a vibrating table operating at 6200 Hertz. Two different curing types; namely 24 hours curing and 28 days moist curing were utilized. Cylindrical specimens with dimensions of 102 mm (D) x 203 mm (L) (4in. x 8 in.) were used to evaluate compressive strengths of the studied concretes in accordance with ASTM C39. Three transport and durability properties were evaluated. These properties were: Absorption, Rapid Migration (RMT), and Accelerated corrosion Tests. The absorption properties of the studied high early-age strength concretes were evaluated in accordance with ASTM C642 using cylindrical specimens with a diameter of 102 mm (4 in.) and a height of 51 mm (2 in.). These properties included absorption after immersion, absorption after immersion and boiling, and volume of permeable pore space (Voids). Rapid Migration Tests (RMT) of the studied high early-age strength concretes studied were conducted in accordance with NT Build 492 (Figure 1a). The corrosion tests were conducted using 102 mm (4 in.) x 152 mm (6 in.) cylindrical specimens in accordance with FM S-522 (Figure 1b). To evaluate the durability properties of the studied high early-age strength concretes, the mass losses due to freeze/thaw with de-icing salt tests were conducted in accordance to ASTM C672.

3 Results and Discussion

3.1 Compressive strength

The required time to achieve 21 MPa compressive strength (The opening time) using Type III and Type V cement concrete is given in Figure 2. It is observed that, the opening time for Type III cement concrete was lower as compared to that of Type V cement concrete. This is due to the higher fineness and presence of a high amount of C_3S in Type III cement. Also, air-entraining admixture acted as a retarder to the concrete and increased the opening time for both Type III and type V cement concretes. The effect of curing age and cement type on the compressive strength is shown in Figure 3 Concrete with air-entrainment showed lower compressive strength compared to the non- air-entrained concrete. This is mainly due to presence of higher porosity in air-entrained concrete compared to the non air-entrained concretes. In contrast to the early age strength performance of type III cement concrete, its 24 hour and 28 day cured strengths fell below those Type V cement concrete.

(a) (b)

Figure 1. Test setup to evaluate the transport properties of concrete (a) Rapid migration test, (b) Accelerated corrosion test.

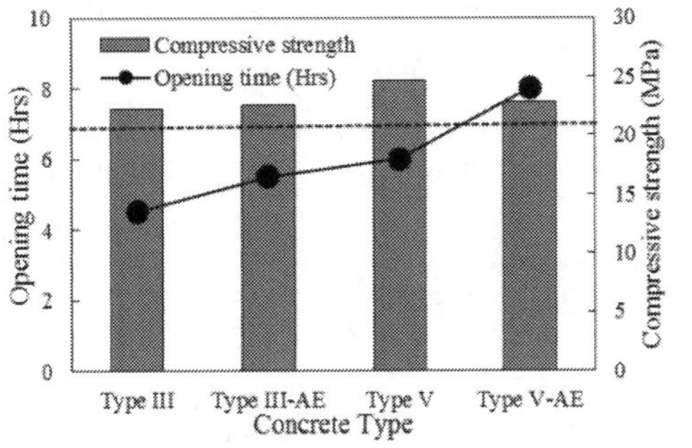

Fig. 2. Opening time-compressive strength of concretes made with Type III and Type V cement with and without air-entrainment (AE).

3.2 Absorption capacity

Absorption and permeable void percentage of concrete at different test ages is presented in Table 3. The absorption and void percentage found to be high when air-entraining admixture was added to the mixtures. Curing played an important role in reducing the absorption and percentage of permeable voids in the concrete. With 28 days curing, the percentage voids reduced significantly for both studied Type III and Type V cement concretes. The absorption and volume of voids of Type V cement concretes were considerably lower than those of Type III cement concretes.

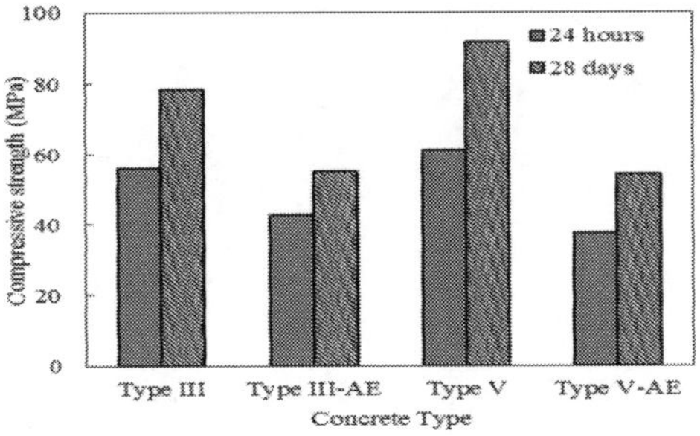

Fig. 3. Compressive strength of concretes made with Type III and Type V cement with and without air-entrainment (AE).

Table 3. Absorption and permeable voids of concrete at different curing ages.

Mixture ID	Curing age	Absorption after immersion (%)	Absorption after immersion and boiling (%)	Volume of permeable voids (%)
Type III	24 hours	4.56	4.69	11.20
	28 days	2.01	2.14	5.19
Type III-AE	24 hours	4.65	5.00	11.26
	28 days	2.76	3.05	6.89
Type V	24 hours	2.18	2.66	7.17
	28 days	0.89	1.32	3.33
Type V-AE	24 hours	3.42	4.02	8.98
	28 days	1.84	2.12	4.73

3.3 Rapid chloride migration test (RMT)

The chloride penetration depth through a rapid chloride migration test (RMT) is presented in Figure 4. The studied 24 hour cured Type III cement concrete performed slightly better than the companion Type V cement concrete for both non air-entrained and air-entrained mixtures. However, after 28 days curing, the chloride penetration for Type V cement concrete was less than that of the Type III cement concrete.

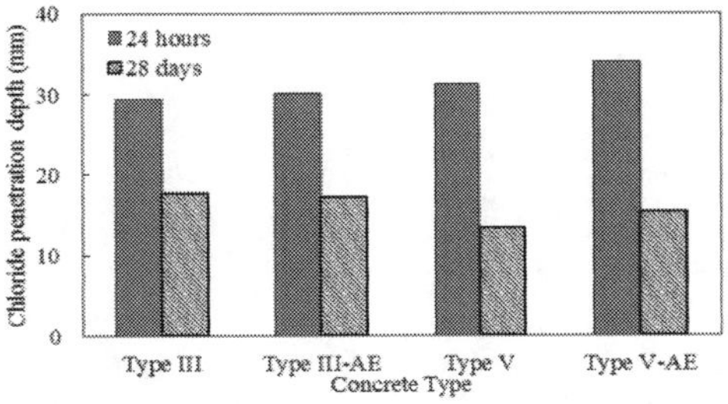

Fig. 4. Effect of curing age on chloride penetration depth for Type III and Type V cement.

3.4 Corrosion resistance

Resistance to corrosion was measured by the average time for the specimen to fail. Figure 5 presents the average failure time for Type III and Type V cement concretes at different curing ages. No variation was observed in average failure time for 24 hour cured concretes. Once curing time was extended to 28 days, the studied Type V cement concrete outperformed Type III cement concrete.

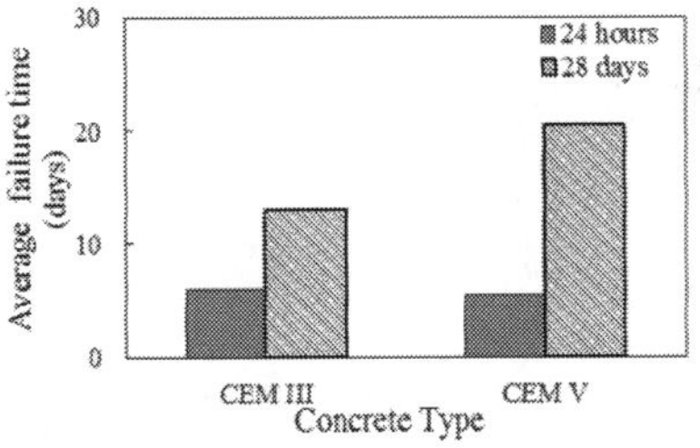

Fig. 5. Effect of curing age on failure time of Type III and Type V cement concrete.

3.5 Mass loss due to freezing and thawing

The percentage mass loss due to freezing and thawing is given in Figure 6. The studied Type III cement concretes displayed nearly 50% more mass loss than Type V cement concrete for the 28 days cured samples. In the case of 24 hours curing, the effect of cement type was not substantial.

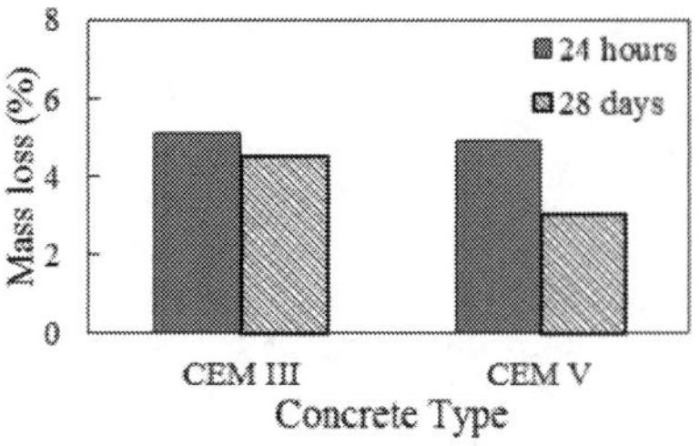

Fig. 6. Ultimate mass loss of concrete subjected to 25 freezing and thawing cycles.

4 Conclusions

Based on the results of the investigation, the main conclusions are as follows:

- Within a short period of time, all studied concrete attained the minimum compressive strength of 21 MPa which is required for rapid repair. On average, for non air-entrained concretes, Type III cement concrete met the opening strength requirement about 90 minutes before the Type V cement concrete did. For air-entrained Type V cement concrete, it took nearly 150 minutes more to reach 21 MPa than that of companion Type III cement concrete with air-entrainment.
- Age of curing had a significant effect on both compressive strength and transport properties of the studied concretes. As the curing age was extended to 24 hours and 28 days, Type V cement concrete showed higher strength and better resistance in varying degrees against chloride permeation, corrosion, and freezing and thawing.

Acknowledgements

This study was made possible by the financial contributions of the Nevada Department of Transportation and the US Federal Highway Administration through the UTC Solaris Consortium. Thanks are also extended to a number of producers and manufacturers for providing materials used in this investigation.

References

1. American Society of Civil Engineers. ASCE annual report Infrastructure report card (2017). Accessed February17, 2018 (https://www.infrastructurereportcard.org/)
2. P. H. Emmons & D. J. Sordyl: *The state of the concrete repair industry, and a vision for its future*. In: Concrete repair bulletin, **19** (2006) 4, PP.7-14.
3. G. A. Keoleian, A. Kendall, J.E. Dettling, V.M. Smith, R.F. Chandler, M.D. Lepech, & V.C. Li: *Life cycle modeling of concrete bridge design: Comparison of engineered cementitious composite link slabs and conventional steel expansion joints.* In: Journal of infrastructure systems **11**(2005) 1, PP. 51-60.
4. C.K. Leung & T. Pheeraphan: *Very high early strength of microwave cured concrete*. In: Cement and Concrete Research **25** (1995) 1, PP.136-146.
5. S.S. Seehra, S. Gupta, & S. Kumar: *Rapid setting magnesium phosphate cement for quick repair of concrete pavements—characterisation and durability aspects.* In: Cement and Concrete Research **23** (1993) 2, PP. 254-266.
6. S. Popovics, N. Rajendran, & M. Penko: *Rapid hardening cements for repair of concrete.* In: ACI Materials Journal **84** (1987) 1, PP. 64-73.
7. L. Mo & V. C. Li: *High-Early-Strength Engineered Cementitious Composites for Fast, Durable Concrete Repair-M aterial Properties*. In: ACI Materials Journal **108** (2011) 1, PP. 3-12.
8. M.O. Maler: *High Early-Age Strength Concrete for Rapid Repair*. In: MSc Diss. University of Nevada, Las Vegas (2017).
9. J. Silfwerbrand: *Improving concrete bond in repaired bridge decks*. In: Concrete international **12** (1990) 9, PP. 61-66.
10. K.R. Hindo: *In-place bond testing and surface preparation of concrete.* In: Concrete International, **12** (1990) 4, PP. 46-48.
11. M.A.A. Aldahdooh, N.M. Bunnori, M.A.M. Johari, A. Jamrah, & A. Alnuaimi: *Retrofitting of damaged reinforced concrete beams with a new green cementitious composites material*. In: Composite Structures **142** (2016), PP. 27-34.

Prediction of the swelling capacity of salt-insensitive superabsorbent polymers in concrete by rheological measurements

Oliver Mazanec[1,*], *Alessander* Assmann[2], and *Diana* Eissmann[3]

[1] Regional Business Segment Management Admixture Systems Europe, BASF Construction Solution GmbH, Trostberg, Germany
[2] Application Technology - Construction Additives, BASF Construction Solution GmbH, Trostberg, Germany
[3] Production, Chemist, BASF Construction Solution GmbH, Trostberg, Germany

Abstract. A study on the macroscopic viscosity of a flowable, fine-grained high-strength mortar was conducted in order to estimate the swelling capacity of two salt-insensitive superabsorbent polymers (SAPs). In doing so, rotational rheometer measurements of the mortar modified with dry pre-mixed SAP were performed at several water-to-cement ratios (W/C), i.e. in the range 0.32 to 0.40. The swelling capacity of each SAP was estimated by correlating the fitted macroscopic viscosity vs. solid particle fraction curve with the curve of the plain mortar without addition of SAP, produced at the same W/C. The swelling capacities were found to yield 28 g pore water per g dry SAP (g/g) and 35 g/g, respectively. In addition to this new approach, two more practical relevant test methods were evaluated with respect to swelling capacity: first, comparative mortar spread tests, which were based on the use of the same fresh mortars but with less amount of superplasticizer, and second, tea bag tests with plain SAP immersed in an artificial pore solution. Only minor differences were found for the estimated swelling capacities. Hence the conclusion may be drawn that all three test methods are suitable for determining the swelling capacity of SAP in a cementitious environment.

Key words: macroscopic viscosity, salt-insensitive superabsorbent polymers, swelling capacity, tea bag test.

1 Introduction

The main goal of using SAP as construction additive to concrete and cement-based dry mortars, is in the ability to gain control on water balance and functional pore design. So far, the application of SAP for the purpose of internal curing and durability enhancement of concrete has been the main focus [1]. As well as the amount of entrained water and pores, the spatial distribution of the already saturated cavities dictates the success of this technology [2]. Whereas size and shape of the dry particles are adjustable within production related limits [3], a test method for the reliable prediction of the swelling capacity of SAP in a cementitious environment is required. The swelling capacity is defined as the ratio of absorbed fluid to the mass of dry SAP and determines the final size and distribution of the already saturated SAP pores.

The absorption kinetics of SAP are mainly driven by osmosis. The ion concentration in the swelling medium and the molecular structure of the SAP itself, e.g. cross-linking density and anionicity, have the highest influence on the swelling capacity [4], [5]. Furthermore, the particle size of SAP appears to influence both the swelling capacity, and the speed of absorption [6]. In [6], Fick's second law of diffusion was applied to describe the swelling process.

Most investigations on the swelling capacity of SAPs refer to gravimetric, and optical investigations of bulk SAP material immersed in water, or an artificial pore solution, refer to [7] for a list. Characterization of swelling capacity under real conditions is much more difficult. In other words, real conditions mean changing ion concentration with hydration time and mechanical stress applied during mixing or caused by the net weight of the fresh mortar after being placed in the mould.

A commonly used approach is to estimate the swelling capacity by adjusting the amount of added water required, whilst obtaining the same slump flow of a reference mixture without SAP addition. A refinement of this approach was presented by Mönnig [8]. He compared the mortar spread as a function of additional mixing time. The test method is referred to as comparative mortar spread test which is described in Chap. 3.2.

Trtik et al. used neutron tomography to visualize the water uptake of SAP in fresh mortar [9]. However, only a very large sized SAP particle of ~1 mm in dry state could be observed. A comparable experiment for estimating the water uptake of SAP is high resolution

NMR relaxation, which was conducted by Nestle et al. [10].

Another method for estimating the water absorption of internal curing agents is suggested by Johansen et al. [11] and was applied to wooden pulp fibres. The absorption capacity was derived from the difference in W/C of a reference cement paste, and a cement paste with internal curing. Both show the same rate of heat evolution under isothermal conditions, despite the difference in total W/C. However, this approach conflicts with the fact that, an internally cured cement paste reaches a higher degree of hydration compared to a reference with lower W/C and, consequently, develops a higher heat of hydration.

The volume fraction of spherical SAP pores, derived from the respective area fraction of the voids by evaluating polished cross-sections of hardened mortar samples, can also be used to determine the swelling capacity of SAP. First results on this approach were published by Jensen [12].

In this paper, a new test method for the estimation of the swelling capacity of SAP is presented and compared to existing methods.

It is based on elaborative rheometer measurements and refers to the correlation of the fitted macroscopic viscosity vs. solid particle fraction curve of a flowable, fine-grained high strength mortar.

2 Experimental investigation

2.1 Materials

Two salt-insensitive SAP samples based on acrylamide were used for the different test series (SAP A and SAP B). The polymers contain acrylamide, anionic monomers and crosslinker. SAP B has a higher relative density of anionic groups than SAP A, both samples have an irregular particle shape.

The cement selected for the rheological investigations was Ordinary Portland Cement (CEM I 52.5N from Heidelberg Cement). A reference mortar was prepared from 900 g OPC and 1350 g CEN standard sand. The W/C was 0.35. The reference mortar mixture contained 54 vol.-% paste (cement and water) and 46 vol.-% sand. The solid volume fraction ϕ was 0.47.

To disperse the cement particles, the mixtures were prepared with a powdered polycarboxylate ether superplasticizer (PCE) from BASF Construction Solutions GmbH (Melflux AP 101 F), and its dosage rate is expressed in percentage by mass of cement.

2.2 Methods

Test methods for the characterization of superabsorbent polymers are standardized by EDANA, an association of the nonwoven industry. A standard test for the absorption capacity is the so-called tea bag method [13]. A known amount of SAP is placed in a sealed tea bag and put in a test solution. After the swelling time, the tea bag is removed, hung up to remove excess liquid and then weighed. The result is the absorption capacity [4]. In order to study the absorption kinetics as a function of time, the swelling time was varied from 2 to 240 minutes. As the test solution, a synthetic cement pore fluid with the following composition (mmol/l) was used in order to simulate test conditions of a real cementitious system: [Na+] = 98.5, [K+] = 181.6, [SO42-] = 86.2, [Ca2+] = 10.0 with pH = 13.1.

The comparative mortar spread tests were performed following the method defined in reference [8]. The mortars were prepared by mixing with a Rilem pan-type mixer. In doing so, the dry pre-mixed blend of 1350 g sand, 900 g cement, 0.45 g PCE and 2.0 g SAP A or 1.5 g SAP B, was added to the mixing bowl already containing the mixing water. In the case of plain reference mixtures without SAP, the amounts of mixing water were 311 g, 315 g and 319 g. In case of SAP modified mortars, 371 g of water were added. After 60 seconds of mixing, the bowl was scraped to ensure homogeneity and mixing was continued for another 120 seconds. The mortar spread was measured on a shock table (15 shocks), 4 minutes after first water contact and subsequently, in steps of 7 minutes, up to an age of 67 minutes. Between each spread measurement the mortar was mixed again for 2 minutes to break up agglomerates.

For the rheological investigations the mortar was mixed according to DIN EN 196, using a Rilem pan-type mixer. After mixing, the mortar was poured into a Hägermann cone which was placed on a glass plate. The superplasticizer dosage was chosen to yield a flow spread of 280±10 mm five minutes after water addition. To achieve the target flow spread, a superplasticizer dosage of 0.125% by weight of cement was required.

The rheological measurements for yield stress and viscosity were performed 10 min after water addition using a rotational rheometer (Rheotest RN 4) equipped with a vane geometry with a radius of 20 mm. The cup was filled with the mortar, and before commencing the measurements, each mortar was remixed for 10 s to break up agglomerates, enabling the subsequent observation of structure formation. Thereafter, an increasing shear rate ramp up to 100 s^{-1} was applied followed by a decreasing shear rate ramp. Only the decreasing ramps are shown in the following.

To determine the rheological properties of the mortars, the solid volume fraction ϕ of the plain mortar, without addition of SAP, was varied in steps from 0.40 to 0.52. The packing density remained unchanged since the relative proportion of the materials was kept constant.

3 Results and discussion

3.1 Absorption kinetics observed by standard tea bag method

Figure 1 illustrates the amount of synthetic cement pore fluid which is absorbed by 1 g SAP A and SAP B, as a function of the swelling time. The water

https://doi.org/10.1051/matecconf/201928902004

absorption immediately starts after the tea bag is put in the test solution. Approximately 90% of the total absorption capacity is reached after 10 minutes swelling time. No loss in the absorption capacity is observed over time, i.e. at least for two hours, indicating that both SAP are salt-insensitive. Salt-insensitive means that there is no break-down in water retention of SAP exposed to cement pore fluid over time. The final swelling capacities measured after 240 minutes are 28 g pore fluid per g dry SAP A (g/g), and 35 g/g for SAP B. SAP B shows higher absorption rates compared to SAP A, which is due to the higher anionic charge density of SAP B.

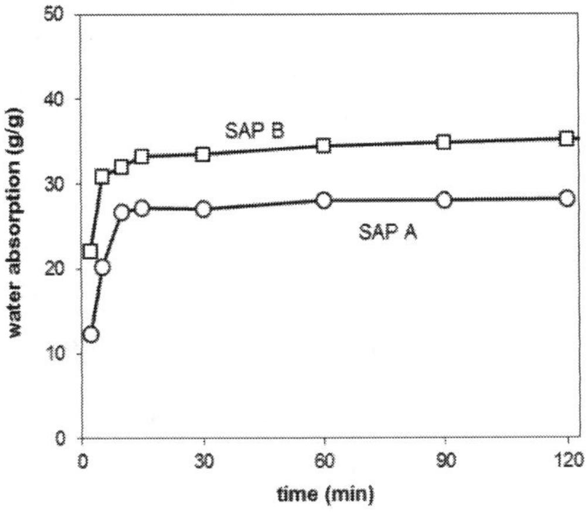

Fig. 1. Absorptivity of the SAP samples in synthetic cement pore fluid.

3.2 Swelling capacity of SAP estimated by comparative mortar spread tests

Figure 2 demonstrates the measured mortar spread as a function of the age of the mortar. The dotted lines represent the linear regressions of the test results. For the estimation of the swelling capacity, the reference line which is the closest to the line of the SAP modified mortar is the most suitable one. In case of SAP A modified mortar (left diagram) it is the reference mortar produced with 315 g water.

In the case of the SAP B modified mortar (bottom diagram) interpolation leads to a hypothetical reference mortar produced with approximately 314 g water. It is expected that the difference in the total water content, i.e. 56 g water and 57 g water, was absorbed by the added SAP, i.e. 2.0 g SAP A and 1.5 g SAP B, respectively. This consideration, which includes that water stored in SAP, does not contribute to spread formation, leading to the swelling capacity of SAP, i.e. 28 g/g for SAP A and approximately 38 g/g for SAP B.

In the case of SAP B, the comparative mortar spread tests led to a swelling capacity which is about 8% higher compared to the results of the tea bag method.

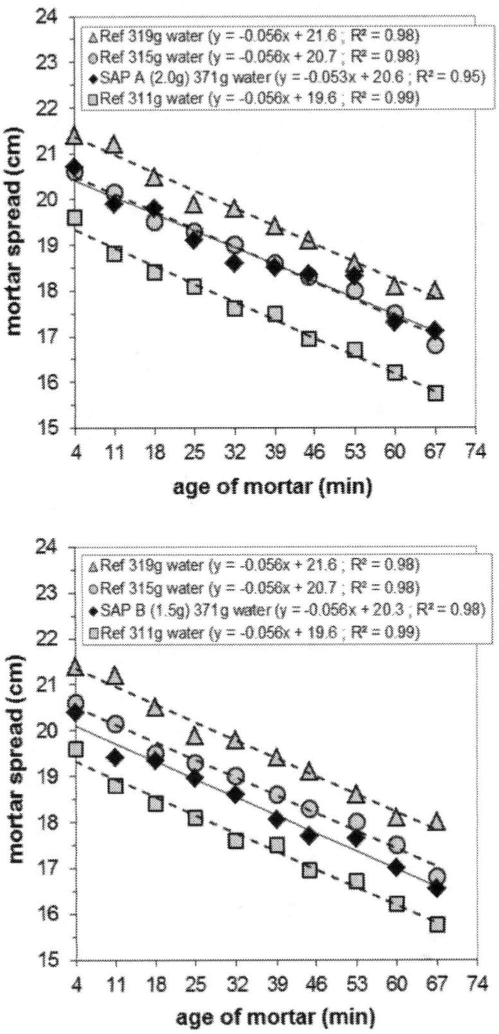

Fig. 2. Mortar spread of SAP A (top) and SAP B (bottom) modified mortar as a function of age compared to curves of plain reference mortars with variable water content. Linear regression of the results is represented by the dotted lines.

3.3 Swelling capacity of SAP estimated by apparent shear viscosity measurements

The rheological behaviour of a material is described by its yield stress and its viscosity. The apparent viscosity as a function of the shear rate at different superplasticizer dosages is shown in Figure 3. For the cement mortar a typical shear thinning behaviour (i.e. apparent viscosity is decreasing with shear rate) could be observed. It is known that colloidal attractive interactions between cement particles are the origin of the yield stress of cement mortars, and that the competition between colloidal interactions and viscous dissipation is the origin of the shear thinning behaviour at low shear rates [14]. By increasing the superplasticizer dosage, the shear thinning behaviour is less pronounced. On one hand, this can be attributed to the PCE that adsorbs onto the surface of the cement particles and, on the other hand, to steric/electrostatic repulsion effects; the PCE is able to decrease the attractive colloidal interactions. The yield stress of the system decreases along with its contribution to apparent viscosity at low shear rates.

The flow curves for the cement mortar show a plateau in viscosity at a shear rate of 30 s^{-1}. At this shear rate, viscous forces dominate the macroscopic behaviour of the system, i.e. the mortars behave like Newtonian fluids, which means that their viscosities tend towards a constant value [15]. In this case, the contribution of the viscous forces to the macroscopic stress is proportional to the shear rate. They are at the origin of the value of plastic viscosity μpl, which captures the magnitude of hydrodynamic dissipation.

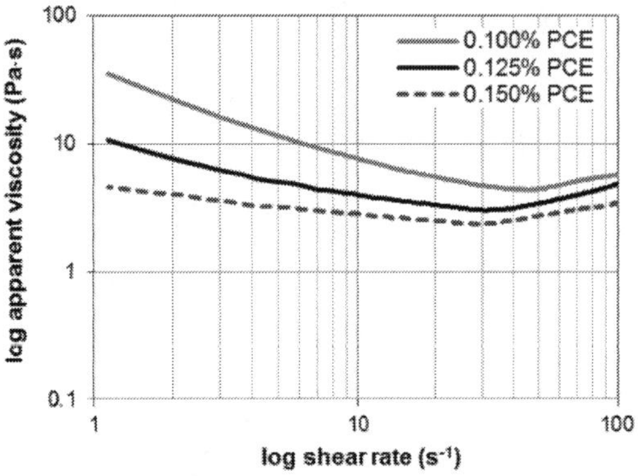

Fig. 3. Apparent viscosity vs. shear rate for different superplasticizer dosages.

In Figure 4 (top diagram) the apparent viscosity as a function of shear rate for various dosages of SAP A is shown. Due to the addition of SAP A the apparent viscosity increases, but a similar shear thinning behaviour to the plain mortar could be observed. The same behaviour was measured by decreasing the W/C ratio from 0.35 to 0.32 at constant PCE dosage of 0.125% cf. Figure 4 (bottom diagram).

The effect of SAP dosage on the apparent viscosity at 30 s^{-1} is shown in Figure 5 (top diagram).

SAP B which has the lowest crosslinking density, showed the greatest effect on the apparent viscosity. This correlates well with the results shown.

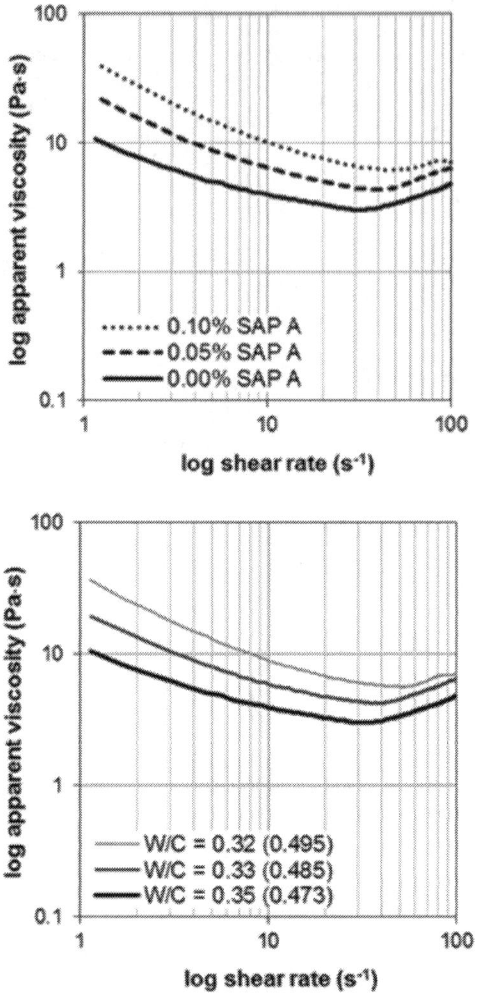

Fig. 4. Apparent viscosity as a function of shear rate of cement mortar at constant W/C of 0.35 and variable dosage of SAP A (top) and at different W/C ratios and solid volume fractions ϕ, respectively (bottom).

Due to the higher water absorption capacity of SAP B, the amount of free water in the cement mortar is more reduced compared to SAP A. The reduction of the free water results in an increase of the solid volume fraction.

https://doi.org/10.1051/matecconf/201928902004

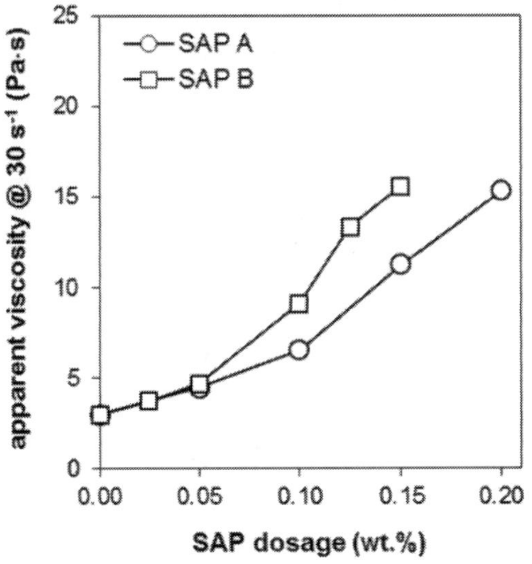

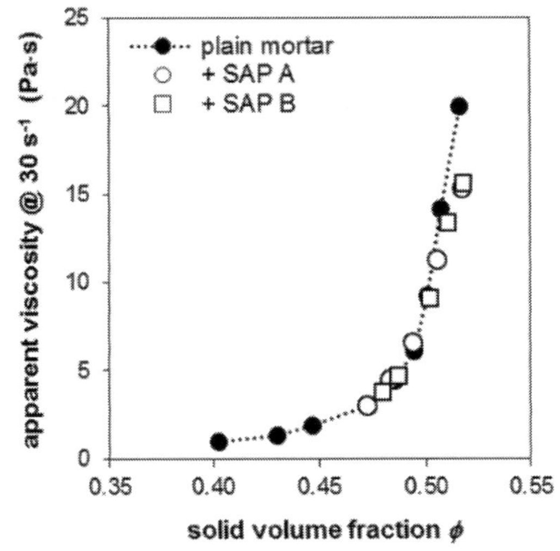

Fig. 5. Effect of SAP dosage on apparent viscosity for cement mortar with a constant solid volume fraction ϕ of 0.47 (top) and apparent viscosity vs. solid volume fraction for the plain mortar and mortars with different SAP dosages (bottom).

Recent studies suggest that the viscosity increases by increasing the solid volume fraction [14, 16]. To characterize the swelling and absorption capacity from the two SAPs, the viscosity has been plotted as a function of the solid volume fraction for the plain mortar, see dotted baseline in Fig. (bottom). It can be noted that the apparent viscosity of the plain mortar increased significantly by decreasing the water content of the mixture. In the next step, it was calculated according to the following equation, how much water is stored by the SAP to fit well with the baseline. It was found that SAP A absorbed 28 g/g and SAP B absorbed 38 g/g.

$$\phi = \frac{\phi_{solid}}{(\phi_{solid} + (\phi_{water} - \phi_{water\ in\ SAP}))}$$

here ϕ is the volume solid fraction, ϕsolid is the volume of solids in the mixture and ϕwater is the water content of the mixture.

With this approach it is possible to calculate an estimated swelling capacity of SAP in flowable cement mortars, as well as determining the necessary amount of extra water to compensate an increase in viscosity caused by addition of SAP.

4 Summary and conclusion

A study on macroscopic viscosity of a flowable, fine-grained high-strength mortar was conducted in order to provide a new approach for the estimation of the swelling capacity of SAP in cementitious mortar systems. This new approach is based on rotational rheometer measurements. The swelling capacity is obtained from the correlation of the fitted macroscopic viscosity vs. solid particle fraction curve of an SAP modified fresh mortar, with the curve of a plain reference mortar produced at different W/C. It can be concluded that the rheological properties of a flowable cement mortar modified with SAP are solely influenced by the water uptake of SAP, and not by any interparticle actions. From the comparison of the estimated swelling capacities of two SAP samples with results from standard tea bag tests, on the one hand and comparative mortar spread test on the other, the following conclusion may be drawn: All three test methods lead to quite similar results. In the case of sample SAP A all three test methods resulted in an estimated swelling capacity of 28 g/g. In case of sample SAP B the tea bag method led to 35 g/g. whereas the comparative mortar spread tests, and the rheometer measurements led to a value of 38 g/g. Since the deviation is less than 10%, and both the comparative mortar spread tests and the rheometer measurements are much more time consuming, the tea bag method (applied in synthetic cement pore fluid) is still recommended as a rapid test for the estimation of the swelling capacity of SAP.

References

[1] V. Mechtcherine, et al.: "Effect of internal curing by using superabsorbent polymers (SAP) on autogenous shrinkage and other properties of a high-performance fine-grained concrete: results of a RILEM round robin test", Materials and Structures 47 (3), 2014, pp. 541-562

[2] A. Assmann: "Physical properties of concrete modified with superabsorbent polymers", Stuttgart, University of Stuttgart, Department of Construction Materials, Ph.D. thesis, 2013

[3] F.L. Buchholtz & A.T. Graham (eds.): "Modern superabsorbent polymer technology", New York, Wiley-VCH, 1998

[4] S. Friedrich: "Superabsorbent polymers (SAP)", chapter 3 in [7]

[5] C. Schröfl, V. Mechtcherine, M. Gorges: "Relation between the molecular structure and the efficiency of superabsorbent polymers (SAP) as concrete admixtures to mitigate autogenous shrinkage", Cement and Concrete Research 42 (6), 2012, pp. 865-873

[6] L.P. Esteves: "Internal curing in cement-based materials", Aveiro, Universidade de Aveiro, Ph.D. thesis, 2009

[7] V. Mechtcherine, H. W. Reinhardt (eds.): "Application of superabsorbent polymers in concrete construction. State-of-the-art report of the RILEM TC 225-SAP", Springer, Heidelberg/Germany, 2012, ISBN 978-94-007-2732-8

[8] S. Mönnig: "Superabsorbing additions in concrete – applications, modelling and comparison of different internal water sources", Stuttgart, University of Stuttgart, Department of Construction Materials, Ph.D. thesis, 2009

[9] P. Trtik, et al.: Neutron tomography measurements of water release from superabsorbent polymers in cement paste, in: W. Brameshuber (ed.): "Additions improving properties of concrete", Bagneux, RILEM Publications, 2010 (RILEM Proceedings; 77), pp. 175-185

[10] N. Nestle, et al.: "Water balance & pore structure development in cementitious materials in internal curing with modified superabsorbent polymer studied by NMR", Microporous and Mesoporous Materials 125 (1-2), 2009, pp. 51-57

[11] N.A. Johansen, et al.: "New method for determination of absorption capacity of internal curing agents", Cement and Concrete Research 39 (1), 2009, pp. 65-68

[12] O.M. Jensen: "Autogenous phenomena in cement-based materials", Aalborg University, Department of Building Technology and Structural Engineering, 2005, ISBN 87-91606-00-4

[13] EDANA – Europoean Disposals and Nonwovens Association: "Recommended test method – free swell capacity", ERT 420.2-02, 2002

[14] N. Roussel, A. Lemaitre, R.J. Flatt, P. Coussot: "Steady state flow of cement suspensions: A micromechanical state of the art", Cement and Concrete Research 40 (1), 2010, pp. 77-84

[15] J. Hot, N. Roussel: "Influence of adsorbing polymers on the macroscopic viscosity of concentrated cement pastes", in: Proceedings of the Tenth International Conference on Superplasticizer and other Chemical Admixtures in Concrete, Prague, 2012

[16] L. Struble, G.K. Sun: "Viscosity of Portland cement pastes as a function of concentration", Advanced Cement Based Materials 2, 1995, pp. 62-69

The role of shrinkage reducing admixtures and supplementary cementitious materials in volume stability and strength development

Ioanna Papayianni[1,*], *Fotini* Kesikidou[1], and *Philip Henes* Alt[1]

[1]Laboratory of Building Materials, School of Civil Engineering, Aristotle University of Thessaloniki

Abstract. Shrinkage is one of the main reasons for mortar and concrete failures like curling, crack formation and de-bonding. It is a complex phenomenon due to many factors involved, such as the type and amount of cement, water to binder ratio, binder to aggregates ratio and the type and granulometry of the aggregates, relative humidity, air temperature and the temperature of concrete. To reduce this phenomenon, Shrinkage Reducing Admixtures (SRAs) have been studied over the last 30 years. On the other hand, investigation in the field of Supplementary Cementitious Materials (SCMs) has indicated that their use in concrete may improve its volume stability depending on their percentage and the type of the material. In this paper, the addition of a Shrinkage Reducing Admixture and Supplementary Cementitious Materials like ladle furnace slag, calcareous fly ash and limestone filler, were investigated. Their influence, separately and in combination, in volume stability and strength development of cement mortars was identified. Capillary absorption and open porosity were also determined.

1 Introduction

Shrinkage is one of the main reasons for mortar and concrete failures like curling, crack formation and de-bonding. The loss of moisture from fresh concrete results in a reduction in volume. If the shrinkage movement is opposed by some external or internal restraint, stresses will develop. When the stress exceeds the tensile capacity of the concrete, cracks are formed, causing an increase in surface permeability and promoting concrete deterioration and corrosion of embedded reinforcement (1) (2) (3) (4).

Shrinkage is a complex phenomenon due to the many factors involved, such as the type and amount of cement, water to binder ratio, binder to aggregates ratio and the type and granulometry of the aggregates, relative humidity, air temperature and the temperature of concrete. There are several types of shrinkage. Plastic shrinkage, which occurs when fresh concrete loses moisture, autogenous shrinkage due to chemical reactions during the hydration process, thermal shrinkage and finally drying shrinkage, which is caused by the movement and loss of water (3) (5) (6).

To reduce this phenomenon, Shrinkage Reducing Admixtures (SRAs) have been studied over the last 30 years. They are designed to decrease drying shrinkage by reducing the surface tension that is developed when moisture is lost from very small pores during the drying of concrete (7) (8). Research has shown around 50 to 60% shrinkage reduction at 28 days and around 40 to 50% at the age of 90 days, with the use of SRAs. However, loss of compressive strength up to 15% has also been reported (9) (10).

On the other hand, investigation in the field of Supplementary Cementitious Materials (SCMs) has indicated that their use in concrete may improve its volume stability depending on their percentage and the type of the material (11) (12). The addition of SCMs generally increases the pore refinement of the concrete and creates smaller pores in the matrix, causing the reduction of shrinkage (5) (13) (14) (15). Furthermore, the autogenous shrinkage is said to be reduced at an early age due to the slow hydration reaction of SCMs (16).

In this paper, the addition of a Shrinkage Reducing Admixture and Supplementary Cementitious Materials like ladle furnace slag, calcareous fly ash and limestone filler, were investigated. Their influence in volume stability and strength development of cement mortars was identified. Capillary absorption and open porosity were also determined.

2 Experimental

2.1 Mixing

Twelve mortar mixes were produced, six of them with the addition of the SRA and six without. European Standard, EN196-1, was followed for the mixing procedure. Siliceous sand (0-4) mm was used as aggregate in a 1:3 binder: aggregate ratio. Chemical analysis of the binders is presented in Table 1 and composition of the mixtures in Table 2.

* Corresponding author: papayian@civil.auth.gr

Table 1. Chemical composition (%w.t.), specific gravity and fineness of the binders.

Binder	CEM I 42.5	CEM IV 32.5	LFS	Fly ash*
SiO_2	19.6	17.07	25.23	29.12
Al_2O_3	2.4	7.27	3.31	15.24
FeO_3	8.11	2.7	3.14	6.59
CaO	66.84	54.67	53.94	40.72
MgO	3.91	2.34	5.26	4.73
Na_2O	0.57	0.48	0.22	0.8
K_2O	1.08	0.58	0.48	1.58
L.I.	1.91	14.26	7.18	0.22
Specific Gravity (g/cm^3)	3.1		2.65	2.28
Particle size volume (%)	10-15μm 6-7%	10-25μm 5-6%	60-70μm >5%	retained on 45 μm sieve 43%

*$CaOH_{free}$ = 10.8%

Table 2. Composition of the mortars

Mix	Binder		W/B	Workability (cm)	Remarks
	Cement	SCM			
I42.5	1	-	0.54	15	Cement I42.5
I42.5-S	1	-	0.54	16	1%SRA
S	0.8	0.2	0.58	15	20% replacement of cement with Ladle Furnace Slag (S)
S-S	0.8	0.2	0.58	17	1%SRA
SF	0.7	0.2/ 0.1	0.54	15	20% (S), 10% limestone filler (F)
SF-S	0.7	0.2/ 0.1	0.54	16	1% SRA
FA	0.7	0.3	0.64	15.7	30% Fly Ash (FA)
FA-S	0.7	0.3	0.64	16	1% SRA
FAS	0.5	0.3/ 0.2	0.56	16	30% (FA), 20% (S)
FAS-S	1.5	0.3/ 0.3	0.56	16.5	1% SRA
IV 32.5	1	-	0.54	15	Cement IV 32.5
IV 32.5-S	1	-	0.54	15	1% SRA

Water was added in order to achieve workability of 16±1cm, according to EN 1015-3. All mixes marking with –S indicate the addition of the SRA (1% by mass of the binder). Water to binder ratio was around 0.54 - 0.58, with the exception of the mixture with fly ash (0.64). The use of SRA seems to improve slightly the workability of the mortars. No superplasticizer was used in any of the mortars, in order to reduce the influence of other mechanisms in the hydration process. Prismatic specimens (40x40x160) mm were produced and tested at the ages of 7 and 28 days. After demoulding, the specimens were kept in a climate chamber with 20° C and RH>95%.

Table 3. Density of the mortars in the fresh and hardened state

Mixture	ρ_{fresh} (g/cm^3)	ρ_{hard} (g/cm^3) 7days
I42.5	2.24	2.13
I42.5S	2.26	2.18
S	2.3	2.19
S-S	2.18	2.16
SF	2.45	2.13
SF-S	2.24	2.11
IV 32.5	2.3	2.19
IV 32.5-S	2.23	2.16
FA	2.26	2.1
FA-S	2.21	2.1
FAS	2.24	2.09
FAS-S	2.23	2.08

From the measured density of the fresh mortar (Table 3), it seems that there is a weight loss from 0.4% to 8%, comparing the mixtures with and without the SRA. The smallest loss was observed for the mortar with fly ash and ladle furnace slag (FAS) and the highest for the combination of ladle furnace slag and limestone filler (SF). Different behaviour was shown by the mix with I42.5 cement, as it appears to gain weight when the SRA was added. Similar results were noted and for the hardened state density.

2.2 Mechanical characteristics

The mortars were tested under flexural and compressive strength (EN 1015-11) and dynamic modulus of elasticity was determined at the ages of 7 and 28 days. According to Table 4, it seems that mortar mixes with SRA exhibited a very slight difference in flexural strength at 7-d and 28-d age (from 3% to 10%) increase or decrease. In the case of the respective compressive strength (Table 5) a small drop in strength (from 6% to 18%) is observed in mixtures with SRA apart from mixtures containing 20% ladle furnace slag, which showed a decrease in strength up to 40%. Values taken for E_{dyn} (Table 6), measured by ultrasonic pulse velocity, generally follow those of compressive strength. However, they are not so representative because they are significantly influenced by the moisture content of mortar specimens.

It could be said that the mechanical characteristics are influenced to a low degree which is more obvious in compressive than in flexural strength. There is also a small drop in fresh density measurements except for CEM I42.5 mortar that showed a slight increase. Similar results were noted for the hardened mortars' density.

Table 4. Flexural strength at 7 and 28 days.

Mix	σ_K (MPa)	
	7 days	28 days
I42.5	3.43	4.00
I42.5-S	3.55	4.27
S	3.67	4.35
S-S	3.80	3.77
SF	3.17	4.72
SF-S	2.79	4.49
FA	3.14	4.65
FA-S	2.83	5.03
FAS	3.34	4.22
FAS-S	2.81	4.60
IV32.5	4.08	4.95
IV32.5-S	3.32	4.68

Table 5. Compressive strength at 7 and 28 days.

Mix	σ_Θ (MPa)	
	7 days	28 days
I42.5	37.71	45.98
I42.5-S	35.46	40.57
S	39.27	45.32
S-S	22.27	31.61
SF	25.01	29.07
SF-S	21.95	26.28
FA	22.10	31.20
FA-S	19.83	28.74
FAS	21.50	26.58
FAS-S	16.96	26.87
IV32.5	35.30	33.29
IV32.5-S	28.83	34.79

Table 6. Dynamic modulus of elasticity at 7 and 28 days.

Mix	E_{dyn} (GPa)	
	7 days	28 days
I42.5	32.31	35.86
I42.5-S	36.14	36.05
S	35.42	34.92
S-S	31.49	32.09
SF	28.89	34.62
SF-S	31.41	35.79
FA	26.82	34.30
FA-S	23.63	34.40
FAS	27.83	33.98
FAS-S	25.09	32.30
IV32.5	32.36	38.66
IV32.5-S	28.47	36.71

2.3 Physical characteristics

In order to test the ability of the materials in shrinkage conditions, volume stability was measured for 28 days. During this period, the specimens were kept in a climate chamber with 20° C and 60% of relative humidity. The measurements are shown in Figures 1-6. Capillary absorption, according to EN 1015-18, is shown in Figures 7-12. Finally, open porosity was measured at the age of 28 days according to the RILEM CPC11.3 method and is presented in Table 7.

Regarding volume stability measurements in Figure 1 to Figure 6, it could be said that the SRA addition has a clear positive effect in reducing shrinkage deformations in the case of mortars with CEM I42.5 and CEM IV32.5. In the mortars with the SCMs the effect is neutral (as shown in Figure 3) or a little different in comparison with the respective mortars without SRA. As shown in Figure 5, mortars with FA exhibited a significant expansion with the addition of SRA. It is worth mentioning that mortars with SCMs even that with blended type CEM IV32.5 showed lower deformations (better volume stability) in comparison to mortar with CEM I42.5, although the water/cementitious ratio was in these mortars higher as shown in Table 2.

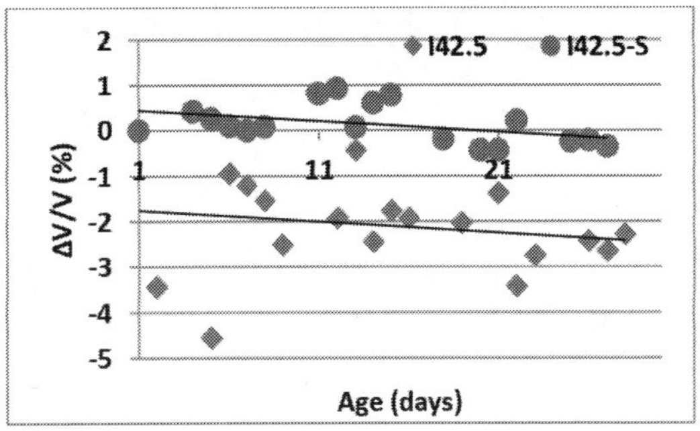

Fig. 1. Volume stability of I42.5 and I42.5-S mixtures

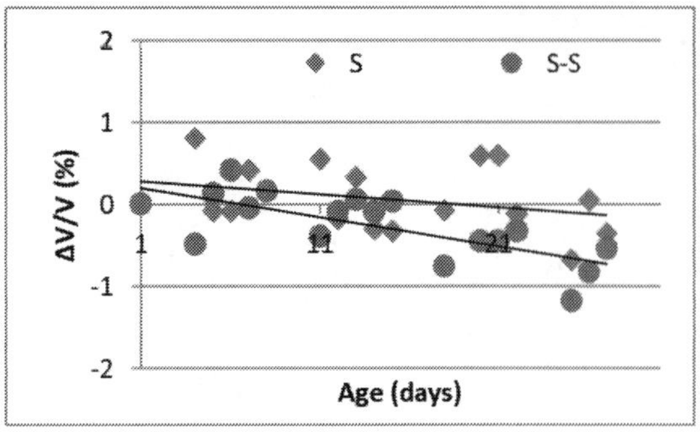

Fig.2. Volume stability of S and S-S mixtures

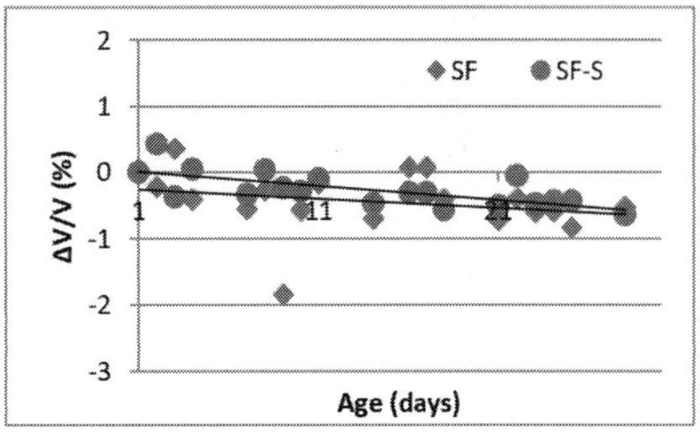

Fig.3. Volume stability of SF and SF-S mixtures

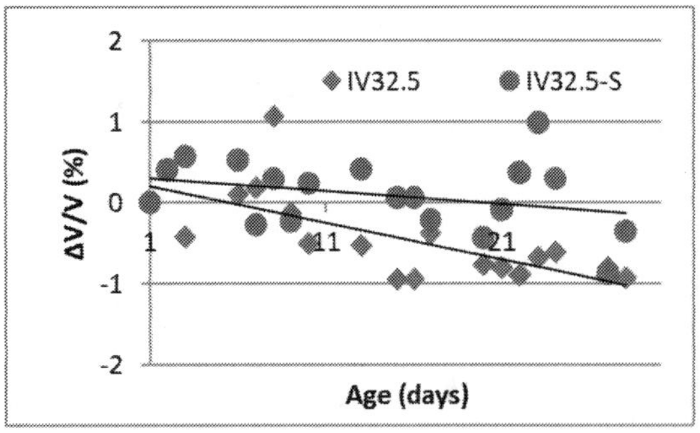

Fig.4. Volume stability of IV32.5 and IV32.5-S mixtures

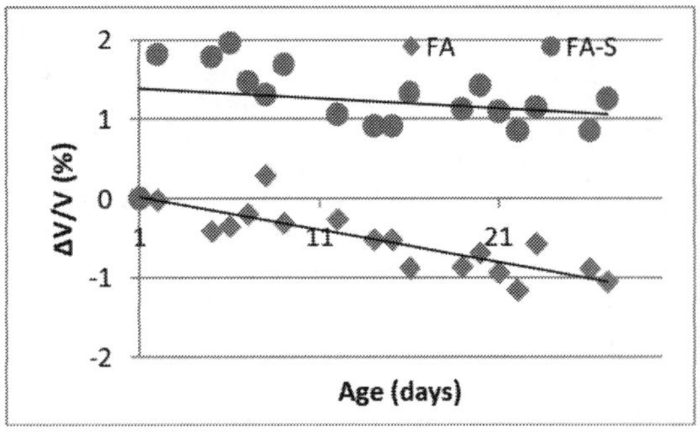

Fig.5. Volume stability of FA and FA-S mixtures

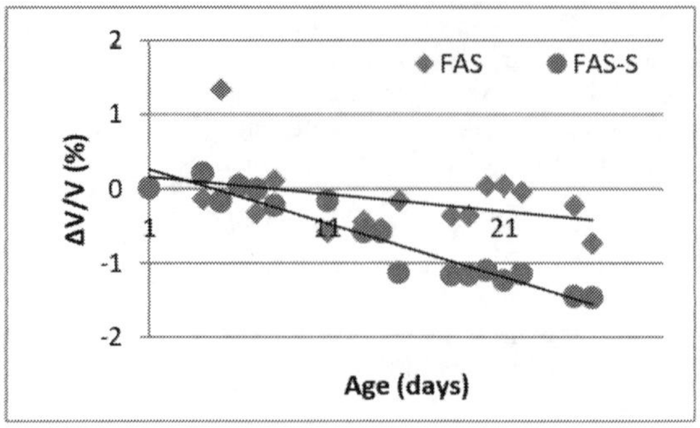

Fig.6. Volume stability of FAS and FAS-S mixtures

Concerning the porosity measurements of the mortar specimens, in most cases there is an increase of 28-d porosity, with the addition of SRA, apart from mortars with CEM I42.5 and those with SF (ladle slag + filler). Therefore, it could be said that porosity or other included voids open to surface are not reduced with addition of SRA.

Table 7. Open porosity at 28 days.

Mix	Porosity (%)
I42.5	6.12
I42.5S	5.19
S	4.88
S-S	7.5
SF	7.27
SF-S	6.65
FA	7.3
FA-S	6.63
FAS	6.82
FAS-S	7.25
IV32.5	5.51
IV32.5-S	6.15

Capillary absorption of mortars with SRA (Figures 7 to 12) showed that it is similar to the respective mortar without SRA or a little lower. In the case of S –S mortar there is a slight increase of absorption in mortar with SRA. It means that the continuity of pores somehow is not favored or is blocked with the addition of SRA.

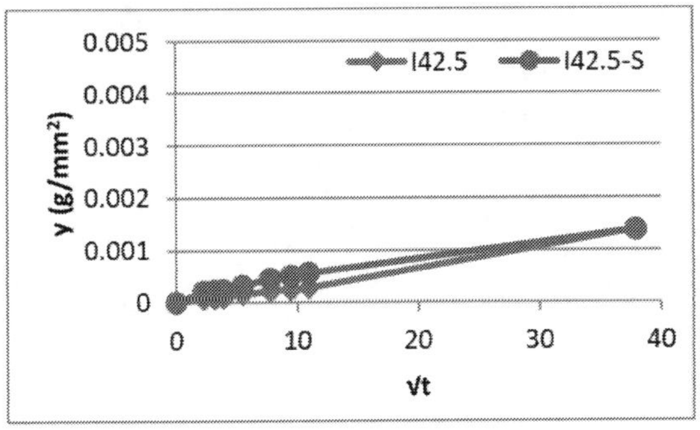

Fig.7. Capillary absorption of I42.5 and I42.5-S.

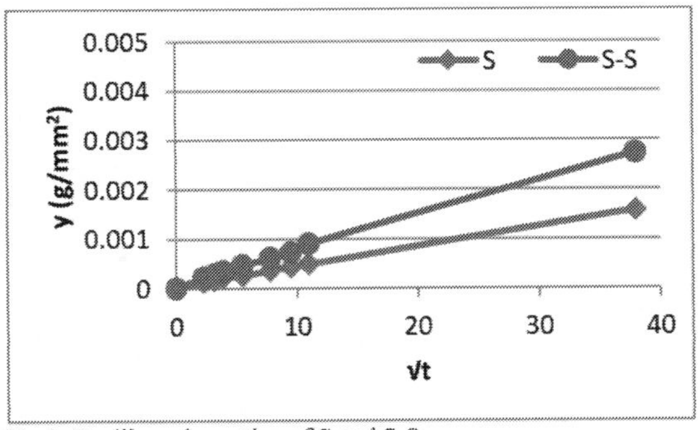

Fig.8. Capillary absorption of S and S-S.

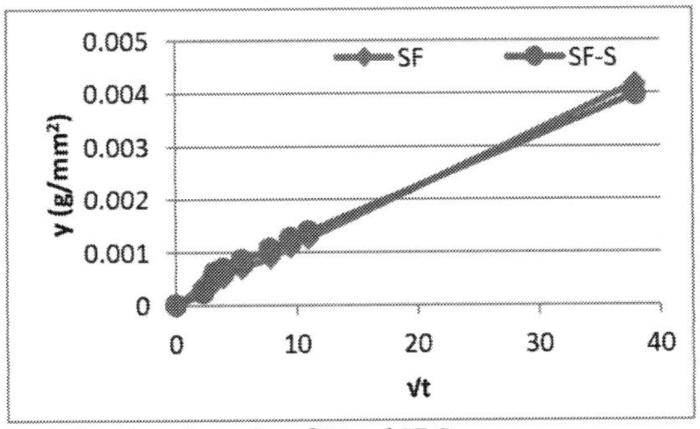

Fig.9. Capillary absorption of SF and SF-S.

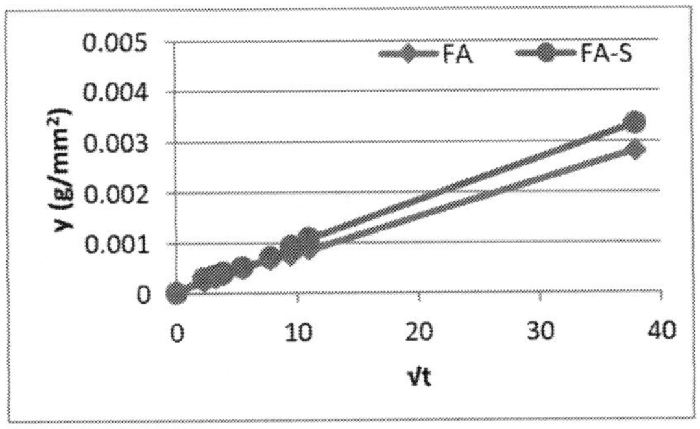

Fig.10. Capillary absorption of FA and FA-S.

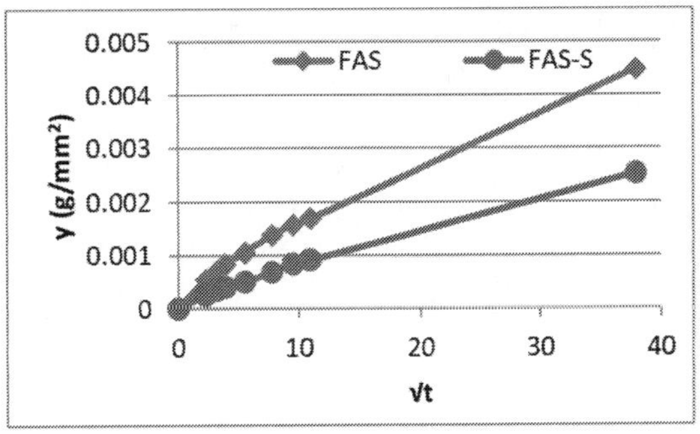

Fig.11. Capillary absorption of FAS and FAS-S.

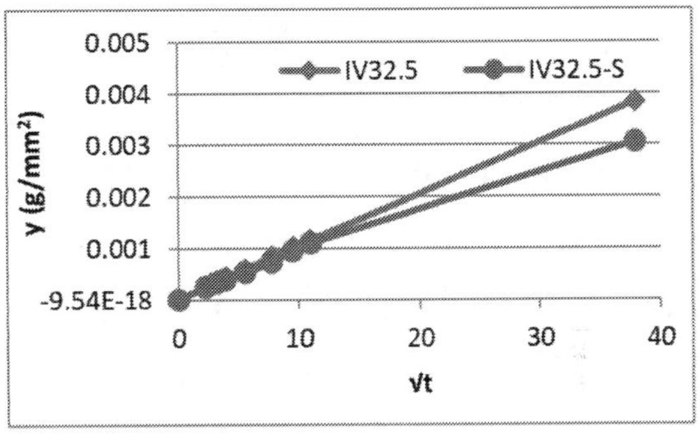

Fig.12. Capillary absorption of IV32.5 and IV32.5-S.

3. Conclusions

The effectiveness of SRA seems to be influenced by the binders of the mortars. The performance of it is clearly positive in mortars with CEM I42.5 , but in the case of use of SCM(s) the reduction in shrinkage deformations is dependable on the binding system and further investigation is needed to describe the SRA performance with different type SCM(s) in the system. It should be mentioned that the SCM(s) addition itself reduces shrinkage deformation.

A slight but clear decrease in compressive strength was presented by all mortars with SRA while flexural strength was unchanged. It could be also said that a slight mitigation of capillary elevation was observed with addition of SRA. It must be said that the test results concern the addition in all mortar mixtures of the same dosage of SRA (1% by mass of the binder). Investigation about the optimization of the required dosage to each binding system has not been made.

4. References

1. F. Tittarelli, C. Giosuè, S. Monosi. Combined Use of Shrinkage Reducing Admixture and CaO in Cement Based Materials. *Mat. Sc. and Eng.* **245**. (2017)
2. P.K.Mehta, P.J.M. Monteiro. *Concrete: Structure, Properties and Materials*. s.l. : The McGraw Hill Companies, (2006).

3. Holland, T. Using shrinkage reducing admixtures. Concrete Construction, Publication #C99C015 Copyright© 1999 https://www.concreteconstruction.net/_view-object?id=00000153-8bc0-dbf3-a177-9ff9f20a0000.
4. C.M. Neubauer, H.M. Jennings and E.J. Garboczit. Mapping drying shrinkage deformations in cement-based materials. *Cem. and Con. Res.***10**, 27,1605-1612, (1997)
5. D. W. Mokarem, R. E. Weyers, D. S. Lane. Development of a shrinkage performance specifications and prediction. *Cem. and Con. Res.* **35**, 918-925, (2005)
6. J. Weiss, P. Lura, F. Rajabipour and G. Sant. Shrinkage-Reducing Admixture: Effects on Durability of High-Strength Concrete. *ACI Mat. J.* July-August, **110**, 4, 365-374, (2013)
7. Y. Li, J. Li. Capillary tension theory for prediction of early autogenous shrinkage of self-consolidating concrete. *Con. and B. Mat.* **53**, 511–516, (2014)
8. Bentz, D. P. Influence of Shrinkage-Reducing Admixtures on Early-Age Properties of Cement Pastes. *J. of Adv. Con. Tech.* . October, **4**, 3, 1-7, (2006)
9. G. Sant, A. Eberhardt,D. Bentz, J. Weiss. Influence of Shrinkage-Reducing Admixtures on Moisture Absorption in Cementitious Materials at Early Ages. *J. of Mat. in Civil Eng.* March, **22**, 3, 277-286, (2010)
10. J. Weiss, P. Lura, F. Rajabipour, G. Sant. Performance of Shrinkage-Reducing Admixtures at Different Humidities and at Early Ages. *ACI Mat. J.* September-October, **105**, 5, 478-486, (2008)
11. E. Anastasiou, I. Papayianni. Use of calcareous fly ash in SCC. *Eurocoalash.* September, Thessaloniki, Greece (2012).
12. A. Liapis, E.Anastasiou, M. Papachristoforou, I. Papayanni. Feasibility study and criteria for EAF slag utilization in concrete products. *J. of Sust. Metal.* March **4**, 1, 68-76 (2018).
13. J. Li, Y. Yao. A study on creep and drying shrinkage of high performance concrete. *Cem. and Con. Res.*, **31**, 1203-1206, (2001).
14. Haque, M. N. Strength development and drying shrinkage of high strength concretes. *Cem. and Con. Comp.* **18**, 333-342, (1996).
15. M. Collepard, S. Biogini. Effect of water/cement ratio , pozzolanic additions and curing time on the chloride penetration. *Proc. Of Int. ERMCO Conf.* Stavanger., 606-615 (1989).
16. J. Liu, Z. Ou, J. Mob, Y. Wang, H. Wu. The effect of SCMs and SAP on the autogenous shrinkage and hydration process of RPC. *Con. and B. Mat.* November **155**, 239-249 (2017)

Mix-Design and Control of Exposure Class of Durable Concrete

Peter Paulini

Material Technology Unit, University of Innsbruck, Austria
Corresponding author: peter.paulini@aon.at

Abstract. Durability requirements for concrete are determined by climatic, chemical and mechanical effects, depending on the climatic location and intended use of structural parts. The European standard EN 206-1 defines different types of exposure classes and specifies requirements for concrete, such as type and minimum quantity of cement, maximum w/c ratio, minimum compressive strength or minimum entrained air pore content. Additional characteristics required for a higher workability often lead to contradictory situations in practice and cannot be satisfactorily resolved by varying the mix design of the concrete. Based on a real life scenario where a C35/45 XD3 concrete had been used in the construction of a multi-storey car park, which presented serious cracking after only a short service time, this paper proposes a number of methods designed to assess the quality and durability of concrete. The proof of exposure classes at the hardened building concrete is not regulated in existing national standards. Therefore, in many court disputes it is difficult to prove the quality of the hardened concrete because of lacking regulations. However, it is possible to verify the w/c-ratio using the capillary porosity. The basic relationships between cement content, w/c-ratio and capillary porosity are derived and discussed. Higher proportions of cement, which are necessary to achieve a better workability or higher strength, result in a reduced modulus of elasticity and increased shrinkage as well as a higher risk of cracking. The associated increase of capillary pore volume and of permeability reduces the service life of concrete. The determination of performance-based concrete properties such as permeability, chloride migration or electrical conductivity further complements the assessment of concrete durability. These properties can be used for estimating the resistance of the capillary pore structure to specific damage mechanisms.

1. Introduction

The notion of sustainability in modern economics totally changed the understanding of durability of materials in the construction sector. Typically, reinforced concrete (RC), the most prominent construction material, is a composite material consisting of cement paste as the binder, natural stone as the aggregate and steel as the reinforcement. The pore structure, which is of prime importance in this discussion, could be qualified as its fourth component. The pores can either be partly or fully saturated with water or, predominantly near the surface areas, be dry and filled with air. In most natural stones the proportion of pores is very small, below 1% in volume, and often not joined up in a capillary system. Therefore, only hardened cement paste (HCP) has a continuous capillary pore structure in which electrical, chemical and physical damage processes mainly can take place. As we know, monuments built in natural stone have survived centuries bearing witness to ancient civilisations. If we were to consider concrete as a binary composite material consisting of durable natural stone and a capillary porous matrix of CSH-gel in which the potential for any damage were contained, a law of durability could be drawn up by a simple pragmatic logic: "*reducing the proportion of binder to a minimum while maintaining workability and achieving the degree of strength required*". This requirement not only has a technical but also a strong ecological and economic aspect.

Furthermore, however, in RC structures, the reinforcing material must also be resistant to corrosion. Corrosion-inducing processes typically start at the surfaces of a structural element, e.g. if attacked by saline solutions. Steel is a meta-stable material which oxidizes in the presence of water and oxygen to form iron oxide. In the alkaline environment of a saturated concrete pore water where $12<pH<13.5$, normal oxygen corrosion is prevented by alkali passivation. Only in areas near to the surface areas of structural members portlandite $Ca(OH)_2$ can carbonate into insoluble $CaCO_3$ and the passivation is lost. This is a slow process requiring the presence of CO_2 and water, i.e. alternating dry and humid conditions. National standards state a required cover concrete in order to minimise corrosion resistance of reinforcement. The speed of progression of carbonation or chloride penetration is determined by the capillary pore structure

Table 1, Austrian concrete production grades acc. to ON B 4710-1 NAD 10 + 16 [18]

Type	Exposure Classes	w_0/b max.	entrained Air %-V	min. Binder kg/m³
B1	XC3	0.60		280
B2	XC3/XD2/XF1/XA1L	0.55		300
B3	XC3/XD2/XF3/XA1L	0.55	2.5 - 5.0	300
B4	XC4/XD2/XF1/XA1L	0.50		300
B5	XC4/XD2/XF2/XA1L	0.50	2.5 - 5.0	320
B6	XC4/XD2/XF3/XA2L/XA2T	0.45	2.5 - 5.0	360
B7	XC4/XD3/XF4/XA1L	0.45	4.0 - 8.0	340
B8	XC3/UB1	0.60		280
B9	XC3/UB2	0.60		280
B10	XC3/XD2/XF1/XA1L	0.55		300
B11	XC3/XD2/XF1/XA1	0.55		280
B12	XC3/XD2/XF1/XA1L	0.50		300

and defines the time periods of carrying out repairs on the structure.

Most national and international standards stipulate durability requirements by defining exposure classes. For each class, the type of chemical, climatic or mechanical attack is stated and subdivided into 3 to 4 levels of intensity. Exposure classes are based on a particular mix design and define requirements on the following parameters:

- cement/binder type with minimum content in some standards
- use of supplementary cementitious materials (fly ash, blast furnace slag, silica fume)
- maximum allowed w/c ratio, decreasing with higher level of attack
- minimum grade of concrete in some standards minimum concrete cover

Progressively lowering the w/c-ratio for higher levels of attack reduces the capillary pore volume in HCP. However, the specification of minimum cement content or minimum concrete strength counteracts this intended result and leads to the erroneous assumption that increasing the binder content would lead to higher durability. Reinforcement bars in RC structures are passivated at a minimum binder content of about 280-300 kg/m³ (EC2 NAD: C25/30 - C30/37). Higher contents of binder material are only required for certain classes of concrete strength or for improved workability and are opposed to increased durability. Some provisions in national annexes of EN 206 lead to situations where compliance with elevated exposure classes causes a reduction in durability, as shown in the example below.

The correct proportion of binder material in a concrete mix is the result of a process of optimization within a four-dimensional matrix of strength – workability – durability and price. In our further discussion we concentrate primarily on the one material parameter – pore volume – which is of utmost significance for all four dimensions. For optimal strength and durability, pore volume should be minimized, while sometimes in practice it is maximized for better workability.

2. Multi-storey Car Park – Court Case

Leoben city shopping mall (LCS) including a two-storey car park was built in 2006. After having been in service for only five years severe crack damage was discovered in the floor slabs of the parking decks. The weight bearing structure consists of 35 cm thick concrete slabs, point supported and also based on the outer walls of the building. The concrete slabs were treated with a 3 mm epoxy surface cover providing protection against de-icing humidity and carbonation. Surface water is drained through welded flat plastic drains connected to vertical guttering.

The engineer had opted for a C35/45 XD3 F3 type concrete for the RC structure. The type of concrete used in the deck slabs was equivalent to C 35/45 B7 according to ON B 4710-1 (Tab.1) [18]. This type of concrete fulfils requirements of exposure class XD3 and others. The conformity of concrete properties such as compressive strength (>53 MPa), air content (6.7%) and flow (F450 mm) was tested during building. The binder consisted of 410 kg/m³ cement CEM II-AS 42.5 and of 50 kg/m³ supplementary cementitious materials. Initial testing for this type of concrete was based on a batch water of 189 kg/m³. This concrete should contain a capillary pore volume of

$$V_{cp}=(189-0.24\cdot 450)/10=8.1\%\text{-Vol} \quad (0)$$

A court expertise confirmed that the prescribed quality of concrete C35/45 XD3 had been used and attributed the crack damage to insufficient steel reinforcement and to temperature-induced confining tensile stresses. Our

Table 2. Test data of five year old LCS concrete

Concrete Properties	Dimension	own tests		2nd Court expert
		near surface	core	
water saturated density	kg/m³	2268	2298	2202
capillary dry density	kg/m³	2020	2073	
mean total porosity	%-V	24.8	22.5	20.98
entrained air pores	%-V	--	--	6.6
capillary porosity	%-V	17.8	15.5	14.38
mean mix water	kg/m³	--	234	239
mean w/b ratio calculated	--	--	0.52	0.53
mean compressive strength	MPa	--	--	52.8
splitting tensile strength	MPa	--	3.09	--
surface pull tensile strength	MPa	--	--	3.1
Young modulus	GPa	--	--	23
longitudinal US velocity	m/sec	--	4485	--
rebound ratio microhardness	--	--	0.505	--
coefficient air permeability dry spec.	m²	3.33E-16	2.47E-16	--
mean shrinkage accelerated 50 °C	--	--	7.38E-04	--

own testing of core samples taken from the slabs caused doubts on the concrete quality XD3 and we concluded that the cracks were caused by shrinkage stress. This assessment was then confirmed in a second court opinion which also found a very low static Young module. Table 2 summarizes the concrete properties described in both expert reports. Quite obviously, the capillary pore volume of the tested concrete, obtained in both reports, was far above the expected volume of 8.1 %-Vol.

Technical arguments presented to a court should always be based on the state of technology. However, national standards rarely set out criteria for verifying the exposure class of a hardened concrete structure in the instance of damage. Proof of the w/b-ratio in a hardened concrete can best be performed determining the capillary pore content. Furthermore performance-based specifications enable us to establish a durability class ranking. This approach is presented in detail in the following chapters.

3. Capillary Pore Volume

The water consumption of cement clinker phases is known from their stoichiometric reaction equations. In the chemically bonded state water loses volume and leads to a chemical shrinkage. For individual clinker phases the water demand can be expressed relatively to mass (f_{wm}) or to volume (f_{wv}) [1]. The same is possible for the chemical shrinkage factor (f_s). With the proportions of Portland cement clinker these factors become (Eq. 1-3) [1]

$$f_{wv} = \frac{V_w}{V_c} = 0.7422 \cdot [C_3S] + 0.6874 \cdot [C_2S] + 1.2023 \cdot [C_3A] + \ldots + 1.3754 \cdot [C4AF] \quad (1)$$

$$f_{wm} = \frac{M_w}{M_c} = 0.2367 \cdot [C_3S] + 0.2092 \cdot [C_2S] + 0.4000 \cdot [C_3A] + \ldots + 0.3919 \cdot [C4AF] \quad (2)$$

$$f_s = \frac{V_{hy} - V_o}{V_o} = -0.2243 \cdot [C3S] - 0.1909 \cdot [C2S] + \cdots + 0.4455 \cdot [C3A] + 0.307 \cdot [C4AF] \quad (3)$$

Therein M_c and V_c denote mass and volume of cement and the values in brackets represent the mass proportions of clinker phases. The ratio of the two factors f_{wv} and f_{wm} is the density ratio of cement to water.

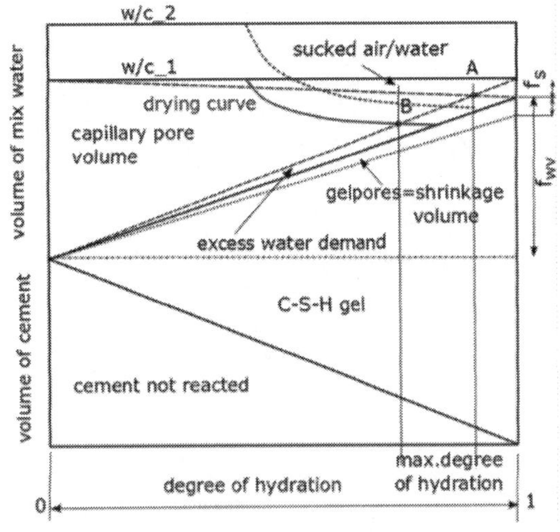

Fig.1 Linear hydration model of capillary volume in CSH-paste [1]

https://doi.org/10.1051/matecconf/201928902006

$$\frac{f_{wv}}{f_{wm}} = \frac{\rho_c}{\rho_w} \qquad (4)$$

For the various types of cement, the water demand factors change only slightly and are in the range of $0.80<f_{wv}<0.83$ and $0.24<f_{wm}<0.26$ respectively [1, 2]. In the literature, mass based water demand values of 0.23-0.25 are mentioned [3, 4]. In this case a value of 0.24 will be assumed for f_{wm}. Based on a linear hydration reaction model, the volumes in C-S-H-gel are shown in Figure 1. In practice, the above-mentioned factors for unknown stoichiometric reactions of supplementary cement additions vary considerably and should be measured.

Assuming a linear relationship with the degree of hydration α, the capillary pore volume V_{cp} in concrete becomes (Eq.5)

$$V_{cp} = \frac{C}{\rho_w} \cdot (wc - \alpha \cdot f_{wm}) = \frac{C}{1000} \cdot (wc - \alpha \cdot 0.24) \quad (5)$$

and the w/c-ratio for a given quantity of cement C can be expressed as in (Eq. 6)

$$wc = \frac{V_{cp} \cdot \rho_w}{C} + f_{wm} \cdot \alpha = \frac{V_{cp} \cdot 1000}{C} + 0.24 \cdot \alpha \qquad (6)$$

Under ambient air conditions the CSH-gel needs twice the chemical shrinkage volume to achieve full hydration (Fig.1). Therefore, the minimum required w/c-ratio to reach full hydration is appr. 0.24+2·0.065=0.37. A low w/c-ratio and climatic conditions where pore water evaporates can considerably reduce the maximum degree of hydration, especially in regions near to the concrete surfaces. The capillary suction reduces the available hydration water in near-surface regions of the structure causing higher pore volumes in those areas. Typically climatic, physical and chemical attacks are strongest at the member surface which is why durability in these areas is always lower than in core regions. Therefore any assessment of concrete quality must necessarily distinguish between near surface and core concrete.

Assuming sound aggregates the total volume V_t of the concrete specimen can be determined using Archimedes' principle according to national standards as difference between the vacuum saturated density ρ_s and the dry density ρ_d of 100°C dried concrete (Eq.7). Therein ρ_w denotes the water density. The apparent capillary pore volume V_{cp} is then obtained by subtracting the entrained air pore volume V_{eap} (0.01<d<1 mm) and the volume of compaction pores V_{co} (d>1 mm) (Eq.8). In order to calculate the ratio w_0/c of the fresh concrete at mixing time using Eq.(6), the apparent capillary volume, established by density tests on the hardened concrete has to be reduced by the chemical shrinkage volume V_{cs} (Eq.9). The degree of hydration α of the hardened concrete at testing time has to be chosen carefully. The remaining hydration potential of hardened concrete could also be tested on fine ground concrete powder using a buoyance weighing method [1].

$$V_t = \frac{\rho_s - \rho_d}{\rho_w} \qquad (7)$$

$$V_{cp} = V_t - V_{eap} - V_{co} \qquad (8)$$

$$V_{cs} = \frac{C \cdot \alpha \cdot f_s \cdot f_{wm}}{1000} \qquad (9)$$

In Figures 2a,b the capillary pore volume (Eq.5 V_{cp}) and wc-ratio (Eq.6) are shown for concrete with different binder p0roportions for full hydration. These values are also numerically tabulated in Table 3. Some requirements of BS 8500-1 for XD1 and XD3 for a 50 and 100 year service life are also included in Fig.2 a,b combined with the data obtained with the LCS concrete.

From Fig. 2a it is evident, that the requirements of a maximum w/c-ratio do not limit capillary pore volume. In Fig. 2b isolines of constant capillary porosity for durability potential classes as proposed by Baroghel-Bouny [9] and shown in Tab.3 are plotted so as to place types of concrete in this field (Fig. 2). Quite obviously the LCS concrete falls outside the w/c limit and within the range of a medium to low durable concrete. If binder proportions are increased, the w/c-ratio needs to be decreased in order to limit capillary volume. Table 3 also shows a classification of concrete durability by limiting the binder proportions in several stages.

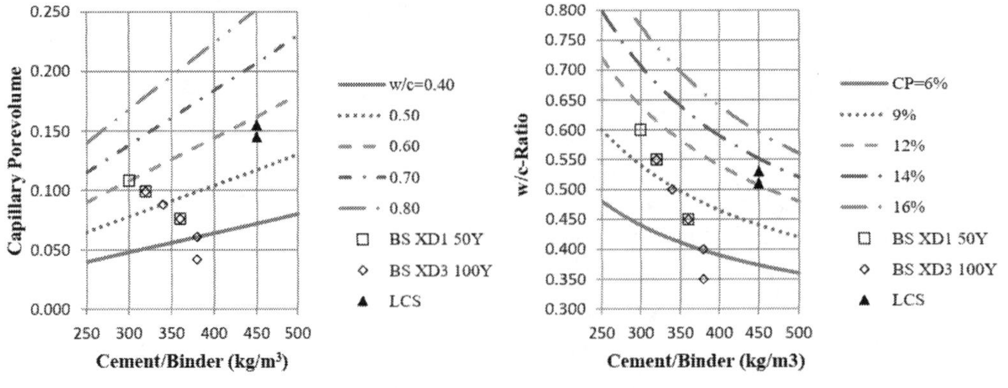

Fig 2 a,b Capillary pore volume and w/c ratio vs. binder of completely hydrated concrete and durability requirements of BS 8500-1 for XD1 and XD3 mixes vs. LCS concrete

Table 3. Apparent capillary pore volume (Eq.5) of completely hydrated concrete (f_{wm}=0.24) and durability classes with limited binder content

Binder in mix	w/b-ratio						
(kg/m³)	0.40	0.45	0.50	0.55	0.60	0.65	0.70
Durability class	very high		high		medium	low	very low
275	4.4%	5.8%	7.2%	8.5%	9.9%	11.3%	12.7%
300	4.8%	6.3%	7.8%	9.3%	10.8%	12.3%	13.8%
325	5.2%	6.8%	8.5%	10.1%	11.7%	13.3%	15.0%
350	5.6%	7.4%	9.1%	10.9%	12.6%	14.4%	16.1%
375	6.0%	7.9%	9.8%	11.6%	13.5%	15.4%	17.3%
400	6.4%	8.4%	10.4%	12.4%	14.4%	16.4%	18.4%
425	6.8%	8.9%	11.1%	13.2%	15.3%	17.4%	19.6%
450	7.2%	9.5%	11.7%	14.0%	16.2%	18.5%	20.7%
475	7.6%	10.0%	12.4%	14.7%	17.1%	19.5%	21.9%

4. Concrete Durability Indicators

Over the past two decades, more research has been done to increase the durability of concrete. Large infrastructures are often concrete structures designed for long service lives. Increased climatic, chemical and mechanical attack leads to higher demands on the quality of concrete. In addition to the description of durable concrete quality as defined in standards, performance-based test methods have been developed which are intended to enable both rapid classification of concrete grades and at the same time an assessment of their service life [5, 6]. Unfortunately, such methods have so far rarely been included in national standards because of the often still considerable scattering of measurement data.

Concrete durability is defined in relation to different types of damage resulting from specific transport or degradation mechanisms. Essentially, these relate to moisture and ion transport, electron conduction and gas transport. All these processes occur over longer time periods and become visible as structural damage only due to cumulative effects of repeated partial damage cycles. Test methods for assessing concrete damage are therefore time-consuming and have often been complemented by accelerated lab procedures. Such durability indicators enable us to correlate an accelerated damage process observed in the laboratory to one occurring in a real building [7-10]. A compilation and comparison of test methods for the durability of concrete can be found in [13, 14]. Baroghel-Bouny proposed very meaningful durability indicators for various test methods in [9, 10]. These data, complemented by the electrical conductivity data according to [11] and a proposal for maximum binder proportions are shown in Table 4.

Usually test methods for determining accelerated transport characteristics are performed in the laboratory and require a reference state. Capillary dry concrete is used for testing gas permeability whereas diffusion and electrical conductivity tests are carried out on totally saturated concrete [11]. For on-site tests constant reference conditions cannot be guaranteed and they are always performed on partially saturated concrete with undefined humidity conditions. Therefore, on site tests show large scattering and are hardly suitable for a long termed quality assessment of concrete. According to the

Table 4. Proposed Durability Indicators [9] and [11] expanded by maximum Binder proportion

Durability Potential	Dimension	very low	low	medium	high	very high
Porosity accesible to water	%-Volume	> 16	14 - 16	12 - 14	9 - 12	6 - 9
Effective chloride diffusion coefficient	10^{-12} m²/sec	> 8	2 - 8	1 - 2	0.1 - 1	< 0.1
Apparent chloride diffusion coefficient	10^{-12} m²/sec	> 50	10 - 50	5 - 10	1 - 5	< 1
Apparent gas permeability	10^{-18} m²	> 1000	300 - 1000	100 - 300	30 - 100	< 30
Intrinsic liquid water permeability	10^{-18} m³	> 10	1 - 10	0.1 - 1	0.01 - 0.1	< 0.01
Initial $Ca(OH)_2$ content	%-Mass	< 10	10 - 13	13 - 20	20 - 25	> 25
Surface resistivity AASHTO T 358	Ohm.m	< 120	120 - 210	210 - 370	370 - 2540	> 2540
maximum Binder proportion in mix	kg/m³	> 525	< 525	< 475	< 425	< 375

Swiss standard SIA 262/1 Annex E air permeability tests using the Torrent method may be performed with maximum concrete moisture of as much as 5.5% -M. This corresponds to a pore volume of more than 11%-V based on a dry density of approximately 2.0 t/m^3. Most concrete types with w/c-ratios of up to 0.55 would be fully saturated and would not allow any gas transport (Tab. 3). On the other hand, due to the suction effect of chemical shrinkage and evaporation, near-surface layers can only hydrate to lower degrees of hydration with higher pore volumes compared to core concrete. Nevertheless, on-site measurements of air permeability *kT* taken by the Torrent method produce values far below those obtained on dry samples. Air permeability *kT* may vary by a factor of 1000 in the saturation range $0<S<1$ [12, Fig. 7].

It follows from the above that performance characteristics for the capillary pore system can only be reliably established under reference conditions of capillary dry or totally saturated concrete. The damage mechanisms underlying most concrete exposure classes are dependent only on the transportability of the capillary system. If one plots the mean values of the conductivity coefficients listed in Table 4 against air permeability, the log-log representation shows a good linear correlation (Fig. 3). However, the marginal values for the "very high" durability potentials have been eliminated in the plot due to the greater scattering of these time-consuming tests.

Assuming that the same capillary system is equally effective for each transport mechanism, the exponents would have to be exactly one. Despite the fact that class limits of the transport coefficients listed in Tab.4 are only roughly defined and are distributed in a variation range, the exponents of their mean values in the power law are close to 1. The air permeability of LCS concrete in Fig. 3 falls in the range of "medium to low", which corresponds to the classification based on capillary porosity (Fig. 2b). Unfortunately, no chloride diffusion tests were performed during the damage assessment of LCS concrete, which would have led to a more accurate ranking for chloride resistance.

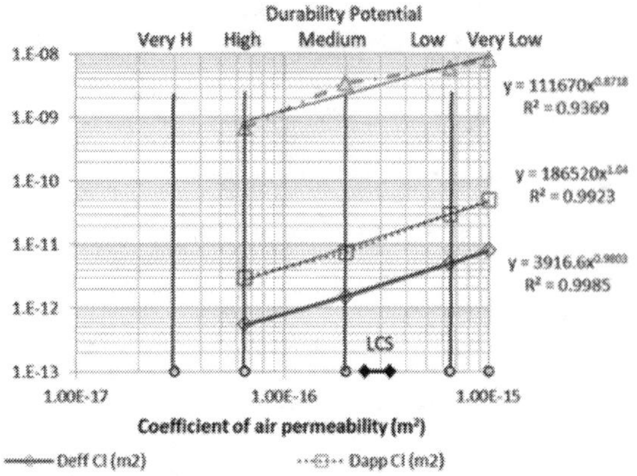

Fig.3. Mean transport coefficients of Tab. 4 vs. gas permeability

More sensitive and more reliable parameters for gas flow in porous materials have been described in [16] based on a power law. Transport laws based on power laws have been known for a long time and have been applied to the air permeability of concrete in [15]. A high correlation in concrete has also been found for coefficient of air permeability, carbonation depth and chloride ion diffusion in [17]. However, the type of cement and the usage of supplementary cementitious materials have a strong influence on the diffusion of chloride ions in concrete and may shift the relations given in Fig 3. Common understanding is that CEM I cements have a lower chloride binding capacity than mixed cements, especially using blast furnace slag. Our own tests have also shown low non steady state Cl migration coefficients D_{nssm} using condensed silica fume [20].

The actual damage to the LCS parking decks was caused by 0.3-1.2 mm wide separation cracks through the entire concrete slab. Saline melting water penetrated the concrete slab causing an accumulation of chloride in the crack area. Clearly the severe cracking was due to the binder-rich concrete mix resulting in high shrinkage. Furthermore, the surprisingly low Young modulus of the concrete contributed significantly to the high shrinkage deformations. Initially, the stiffness of the concrete was misjudged on the basis of ultrasonic (US) measurements taken from water-saturated samples and was ultimately established at 23 GPa by means of static tests on cylindrical cores. A more detailed analysis using US data on dry samples would also have yielded this result, as shown below.

The p-wave modulus M and the dynamic E-modulus E_{dyn} arise from Eq.(10) and (11)

$$M = {v_L}^2 \cdot \rho \tag{10}$$

$$E_{dyn} = M \cdot \frac{(1+\nu)\cdot(1-2\nu)}{1-\nu} \tag{11}$$

Ultrasonic waves propagate in concrete in the rigid composite material of HCP and rock aggregates. Water-filled capillary pores are too small and are not captured by the wavelengths at frequencies of 50-250 kHz. While the correct assumption of Poisson´s ratio is of paramount importance for the assessment of concrete stiffness (Eq.11). Establishing it on the basis of US shear waves is often difficult and inaccurate due to superimposed echoes. A simple and fast determination of the Poisson's ratio consists in using the rebound number Q of rebound hammers (Schmidt Hammer, Micro-hardness Tester). Assuming a reduced rebound number q_r Poisson´s ratio is calculated in Eq.(12)

$$\nu = \frac{1-q_r}{2} \tag{12}$$

The reduction of the rebound number is particularly necessary when testing materials with high stiffness. The calibration of Schmidt hammers on a test block leads to rebound numbers of Q=84%, which must be reduced by 50% (q_r = 0.42) for roughly the same moduli of elasticity of plunger and test block. By applying Eq.(12),

Poisson´s ratio of the calibration block becomes 0.29, falling well in the range for steel materials. If the stiffness differences of impactor and test material differ widely, the reduction has only a minor effect. Therefore, Micro-hardness testers with tungsten-carbide spheres as intender with E-moduli of 580-650 GPa are very suitable for concrete. For the LCS concrete, the mean rebound number of 0.505 (Tab. 2) results in a Poisson´s ratio of 0.248 without reduction, and 0.26 with a reduced rebound number (as used in Table 5). The static modulus of elasticity of a porous concrete is calculated here as the mean value of a lower bound and an upper bound limit according to Eq.(13). The determination of the LCS static E-modulus from US data is shown in Tab.5. A detailed description for NDT determination of Poisson´s ratio and relative modules will be given elsewhere

$$E_{st,m} = \frac{M}{4} \cdot (5 \cdot q_r - q_r^2) \qquad (13)$$

5. Discussion

The crack damage in the LCS parking decks can be attributed to two main causes:

- the required quality of a concrete C35/45 XD3 as supplied to the site, did not respect the maximum permissible w/b-ratio of 0.45. The w/b-ratio of the core concrete was about 0.53, which meant that an increased proportion of capillary porosity was present in the concrete
- based on Austrian standard ÖN B 4710-1 the concrete B7 with a w/b-ratio of less than 0.45 complies exposure class XD3, it also needs to fulfil the requirements of exposure class XF4, for which an additional fraction of entrained air-pores of 4-8%-V is needed. These air pores additions resulted in a binder-rich mix in a total pore content of 22-25%V for the LCS concrete.

The bundling of multiple exposure classes in Austrian standard ÖN B 4710-1 led to the situation that concrete plants in Austria normally only supply concrete classes listed in Tab. NAD 16 of this standard. Other specifications will only be delivered upon special order. The minimum cement content required by ÖN B 4710-1 for concrete grades complying with exposure classes XF are at least 20 kg/m³ higher than that of other comparable national standards. If, in addition, a higher concrete strength class is requested, the required proportion of cement increases even more strongly and reaches quite unusually high levels. In the damage scenario described above, the typical characteristic compressive strength in the structure (>42 MPa) was respected due to an equivalent binder content of 450 kg/m³ in the mix, but at the same time the concrete presented an extremely low Young modulus and high shrinkage. Therefore, it can be argued that the unfortunate bundling of exposure classes in the aforementioned standard played a significant role in causing the severe crack damage in the LCS parking decks.

Table 5. Ultrasonic data of LCS concrete and calculation of elastic modules acc. Eq.(11), (13)

US-Longitudinal					Poisson´s ratio=	**0.260**	
Specimen	Length	dry Density	US Time long	US Velocity long	p-wave Module	E dyn	E st,m
	(mm)	(kg/m3)	(my sec)	(m/s)	(GPa)	(GPa)	(GPa)
1.1m	121	1999	26.3	4600.8	42.31	34.58	24.01
1.2m	124	2147	28.3	4381.6	41.22	33.69	23.39
1.3m	134	2131	29.4	4557.8	44.27	36.18	25.12
1.4m	99	2004	22.9	4323.1	37.45	30.61	21.25
1.5m	122	2021	25.8	4728.7	45.19	36.93	25.65
2.2m	138	2037	30.3	4554.5	42.25	34.53	23.98
2.3m	134	2025	30.7	4364.8	38.58	31.53	21.89
2.4m	132	2155	30.3	4356.4	40.90	33.43	23.21
2.5m	134	2153	29.8	4496.6	43.53	35.58	24.70
Mean					**41.75**	**34.12**	**23.69**
StdDev					2.39	1.95	1.36

6. Conclusions

The methods to demonstrate compliance with exposure classes of concrete in cases of structural damage are generally not regulated in existing national standards and evidence in court cases can only be brought based on state-of-the-art procedures. One of these consists in demonstrating compliance with a given w/c ratio based on the verification of the capillary porosity present in the concrete. So far, performance-based transport criteria have only rarely been adopted by standards to substitute durability of concrete and can best be used as complementary arguments.

In this paper, we discuss the fundamental relationships between concrete mixing data and capillary porosity and present the results graphically. Furthermore, we demonstrate the correlation between transport coefficients of major damage mechanisms (Durability Indicators) in concrete. They can be linearized by approximation in a log-log representation and divided into resistance classes. In addition to commonly accepted indicators, we proposed to use the binder content of the concrete mixture as an additional constraint.

High capillary pore content does not only accelerate damage mechanisms by increasing transport coefficients but also reduces concrete stiffness. Such types of concrete are vulnerable to shrinkage deformations and have a higher crack risk. The paper also presents NDT methods for determining realistic values for the static Young´s modulus and Poisson´s ratio of concrete.

References

1. P. Paulini, Zement-Kalk-Gips **41** Jg.(10), 525-531, (1988) , https://bit.ly/2Xx11Bx
2. P. Paulini, N.V. Waubke, Bauingenieur, **65,** 373-379 (1990),
3. P.C Hewlett, F.M. Lea, *Lea's Chemistry of Cenemt*, Butterworth-Heinemann Ltd., (2004)
4. H.W.F. Taylor, *Cement Chemistry*, 2nd Ed., Thomas Telford Publishing, (1997)
5. H, Beushausen, L.F. Luco, *Performance-based specifications and control of concrete durability*, RILEM TC 230-PSC Report, Springer, (2016)
6. J. Kropp J., H.K. Hilsdorf, *Performance Criteria for Concrete Durability*, RILEM Report **12**, Taylor & Francis, (1995)
7. L. Basheer, J. Kropp, J. Cleland, Cement and Concrete Composites, **15**, 93-103, (2001)
8. T. Luping, L-O. Nilsson, P.A.M. Basheer, *Resistance of concrete to chloride ingress, Testing and Modelling*, Taylor & Francis, (2012)
9. V. Baroghel-Bouny, *Proc. 5th Int. Conf. Concrete under Severe Conditions Environment and Loading, CONSEC* 07, 241 – 248, (2007)
10. V. Baroghel-Bouny, K. Kinomura, M. Thiery, S. Moscardelli, Cement and Concrete Composites, **33**, 832-847, (2011)
11. AASHTO **T 358-17**, *Standard Method of Test for Surface Resistivity Indication of Concrete's Ability to Resist Chloride Ion Penetration*, (2017)
12. R. Torrent, F. Moro F., A. Jornet, RILEM Proceedings PRO, **89,** *Int. Workshop on performance-based specifications and control of concrete durability*, Zagreb 2014, 489-498, (2013 Ed)
13. D. Bjegovic et al., *Test methods for Concrete durability Indicators*, Chapter 4 in [5] (2016)
14. K. Yang, S. Nanukuttan, W.J. McCarter, A. Long, P.A.M. Basheer, Revista ALCONPAT, **8** (3) 246-263, (2018)
15. P. Paulini, F. Nasution, *Proc. 5th Int. Conf. Concrete under Severe Conditions Environment and Loading,* LCPC, Paris, 241-248, (2007)
16. P. Paulini, Transport in Porous Media, **128(**1), 303-318, (2019)
17. Katpady, D.N., Hazehara, H., Soeda, M. et al., Int J Concr. Struct Mater (2018)
18. Austrian Standard ON B 4710-1: 2007, Beton - Festlegung, Eigenschaften, Herstellung, Verwendung und Konformität - Teil 1
19. Swiss Standard SN 505 262/1 SIA 262/1:2013, Betonbau – Ergänzende Festlegungen
20. D. Moreau, *Comparative study of air permeability and chloride ingress into concrete using the Nordtest method NT Build 492*, Master Thesis, Univ. Innsbruck, (2017)

Experimental study of the influence of crack width due to corrosion behaviour using different measurement methods and different exposures

Martin Schneider[1]*, and *Georg* Gardener[2]

[1]Carinthis University of Applied Sciences, 9800 Spittal/Drau, Austria
[2]Carinthia University of Applied Sciences, 9800 Spittal/Drau, Austria

Abstract. Corrosion of reinforcing steel has a great influence in reducing the lifetime of concrete structures; Carbonation of the concrete pore solution causes surface corrosion on the steel and diffusion of chloride ions through the capillary system of the concrete cover causes pitting corrosion on the steel surface. Corrosion of metals is highly dependent on the environmental conditions. Exposure to chloride ions can be critical to the service life of reinforced concrete structures. The durability of reinforced concrete structures exposed to deicing salt or marine environments can be affected by impact of chloride ions. Detection methods for the rate of corrosion of non-destructive and destructive procedures were analysed. The potential mapping applied on the concrete surface was discussed as a standard method for corrosion detection and will be explained in detail including the application boundaries of the method. It is assumed that the corrosion behaviour of reinforcing steel depends on crack widths. To analyse that, 8 coated and 8 uncoated test samples with different concrete strength classes were used. The concrete objects were exposed to a 3% sodium chloride solution. The corrosion behaviour of reinforcing steel is analysed by using potential mapping with different reference electrodes (Ag/AgCl and Cu/CuSO4). The results show a significant correlation between crack size and protection system on the surface. The maximum crack width with a low indication of corrosion was found to be 0.1 mm.

1 Cracks in concrete

The special feature of the composite material reinforced concrete is that the concrete protects the embedded reinforcing steel from corrosion. If this protection of the reinforcement steel is lost, it results in an influence on the durability. This can be caused by excessively large crack widths. The permissible crack width is specified in technical regulations and standards. Consequently, cracks with limited crack width are no longer considered a defect or damage. It should be noted, however, that cracks with small widths promote the penetration of harmful substances such as CO2 and chloride. For the assessment of the corrosion of reinforcement steel, electrochemical examination procedures are available. A relatively inexpensive non-destructive method is the half-cell potential measurement [1].

1.1 Exposure classes

To ensure the necessary durability, concrete elements must be sufficiently resistant to chemical and physical influences from environment or use. These effects on the building material are divided into exposure classes. The following exposure classes record effects on the reinforcement in the concrete: XC (Carbonation), XD (Deicing), XS (Seawater) [1].

1.2 Causes of Cracks

Crack formation in concrete occurs when the locally present stress exceeds the local tensile strength of the material. It is irrelevant at first whether the crack causing stress is generated from an external static or dynamic load or whether this results from an obstruction of the deformation effort of the concrete – residual stress or constrained stress. Of essential importance is the fact that concrete is not completely homogeneous and isotropic with regard to its properties, in particular its tensile strength [2].

• In fresh concrete cracks occur as a result of rapid volume reductions at the near-surface concrete layer - caused by unfavourable temperatures, solar radiation, wind and low humidity [3].
• In building practice, cracks may occur because of loss of the hydration heat or by shrinkage, in particular by early shrinkage [3].

In order to ensure sufficient protection of the reinforcement and the necessary durability, the crack widths are limited in standards and guidance. The limit values of the crack widths is dependent on the exposure class and the design class and are determined according to ÖNORM 1992 1-1.

* Corresponding author: m.schneider@fh-kaernten.at

Table 1. Exposure classes [4].

Exposure class	**maximum crack width**
X0, XC1	0,4 mm (appearance)
XD1, XD2, XD3	0,3 mm

1.3 Reinforcement corrosion caused by cracks

The type of crack plays an important role when it comes to assessment. For example, bending cracks usually have a lower risk potential for a component than separating cracks. CO_2, chloride and water can penetrate the component through separating cracks. In addition, separation cracks often cause a change in the static system. This can lead to extensive load redistribution and further crack formation. Surface cracks caused by inadequate post-treatment do not have the same serious consequences as large cracks caused by a lack of reinforcement steel. Cracks that occur along the reinforcement are particularly critical because they can lead to composite failure [5].

Fig. 1. Typical damage pattern of pitting corrosion 5].

A special characteristic of reinforced concrete is the corrosion protection of the installed reinforcing steel by the concrete. Due to the high pH value of the concrete pore solution ranging from pH 12.5 to pH 14, a passive film is formed on the surface of the reinforcing steel, a few atom layers thick, to protect the steel underneath from further degradation. If the concrete loses its alkalinity or if a critical chloride content is exceeded on the steel surface, the corrosion protection of the reinforcing steel may be lost. If pollutants such as chlorides or CO_2 penetrate the concrete via a crack, rapid damage processes can occur to both the reinforcement and the concrete itself. This leads to depassivation and corrosion of the reinforcement steel [8]. The danger of chloride-induced pitting corrosion (local corrosion) lies in the combination of high corrosion rates of the steel with simultaneous invisibility of the damage process from the outside. This local corrosion leads to high corrosion rates, because a small corrosion scar results in a large cathode surface which accelerates the corrosion process. The steel removal rates associated with these high corrosion rates can reduce the steel cross-section of the reinforcement within a short time. At the same time, the damage processes in the interior of the component go on, so that the reinforcement steel can already be significantly damaged, without any appearance of damage on the concrete surface [6].

2 Corrosion due to chloride ingress

The corrosion protection of the reinforcement steel can be lost if the surrounding concrete loses its alkalinity or if a critical chloride content on the steel surface is exceeded. Small amounts of chloride in the concrete can be contained in the raw materials: cement, water or concrete aggregates. They are defined as natural chloride content of the concrete. Chloride ions diffuse into the concrete or are transported by capillary absorbed water and interact with the passive film of the steel when they reach its surface. If a critical concentration of free chloride ions on the steel is exceeded, the passive film of the steel fails and corrosion can begin. This phase of chloride penetration before the depassivation of the steel is called the initiation phase. With the beginning of corrosion, the propagation phase starts. The chronological sequence of the accumulated damage to the reinforced concrete structures exposed to chlorides is shown in Figure 2 [7].

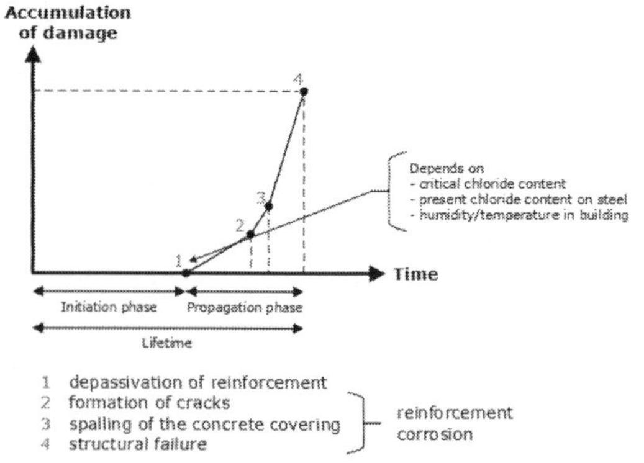

Fig. 2. Chronological sequence of accumulated damage to structures with chloride-induced corrosion [7].

2.1 Propagation Phase

A special type of corrosion is pitting corrosion. During this corrosion of steel in concrete, the positively charged iron ions (Fe^{2+}) go into solution, where the concrete pore fluid is the electrolyte. At the same time a reduction of the oxygen takes place. Under the formation of negatively charged hydroxyl ions (OH^-), the superfluous electrons at the steel/electrolyte interface react with water and oxygen, that is usually present and is dissolved in water in sufficient quantity. The formation of hydroxyl ions OH^- is called a cathodic sub-process. In reinforcing steel, but also in concrete (electrolyte), the charge balance is retained. Iron dissolution (formation of Fe^{2+}) is called an anodic sub-process. The processes described are simplified in Figure 3 below [8].

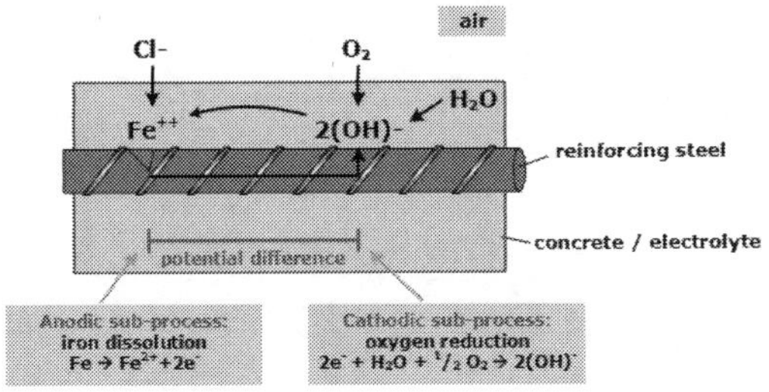

Fig. 3. Corrosion processes and elements involved [8].

Basically the corrosion process can be divided into two simultaneously running sub-processes. In the anodic sub-process, metal dissolution takes place. In this process the reinforcement steel loses two electrons to the environment and converts into Fe^{2+}.

$$Fe \rightarrow Fe^{2+} + 2\,e^{-} \quad (1)$$

Oxygen is reduced in the cathodic sub-process. This absorbs the electrons emitted in the anodic reaction and reacts with the pore water to form OH^{-}.

$$H_2O + O_2 + 2\,e^{-} \rightarrow 2\,(OH)^{-} \quad (2)$$

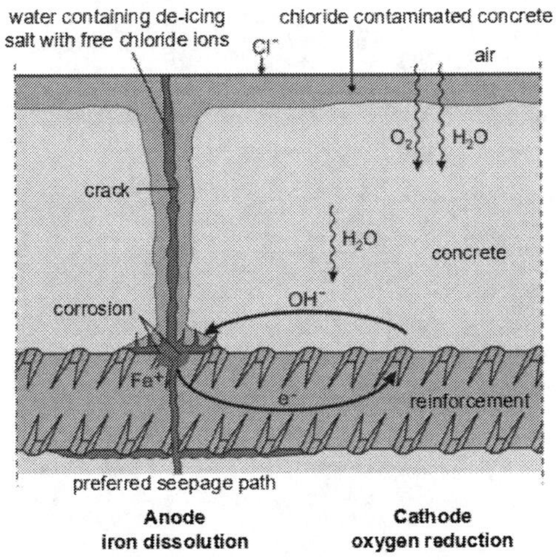

Fig. 4. Chloride-induced corrosion in the crack area [8].

2.2 Crack widths

According to ÖNORM B 1992 1-1, the crack width must be limited in order to ensure that the proper use of the structure, its appearance and durability are not affected. The standards show permissible characteristic crack widths (wk) in dependence on the exposure classes. For reinforced concrete elements exposed to exposure class XD, the crack width shall be limited to wk = 0.3 mm below the quasi-stable action combination. According to ÖNORM B 1992 1-1, if a crack width restriction is required, a minimum amount of reinforcement must be inserted in areas where tension is expected. The minimum reinforcement results from the balance of the tensile force before the incipient crack in the concrete and the tensile force after the incipient crack in the reinforcement. The actual load has no influence on the amount of steel, since the forces to be absorbed are purely a result of the tensile stresses in the concrete before the crack formation [4].

3 Experiment Setup

The aim of this laboratory experiment is to use the half-cell corrosion potential mapping to evaluate whether different crack widths have an effect on the corrosion behaviour of reinforcing steel. For this, 2 specimen series with two different exposure classes XD1 and XD3 as well as two different strength classes C30/37 and C35/45 were used. Half of the specimens were coated with epoxy resin. The cracks were artificially created by 3 point bending load. In order to determine the corrosion behaviour, the test specimens were coated with a sodium chloride solution and then observed during a period of one and a half months using half-cell corrosion potential measurement. Based on the resulting data, the behaviour of the reinforcement steel was evaluated.

3.1 Preparation of the test specimens

For the experiment, 8 coated and 8 uncoated test specimens with the dimensions 60x40x10 cm were produced and reinforced as shown in Figure 4.1 below.

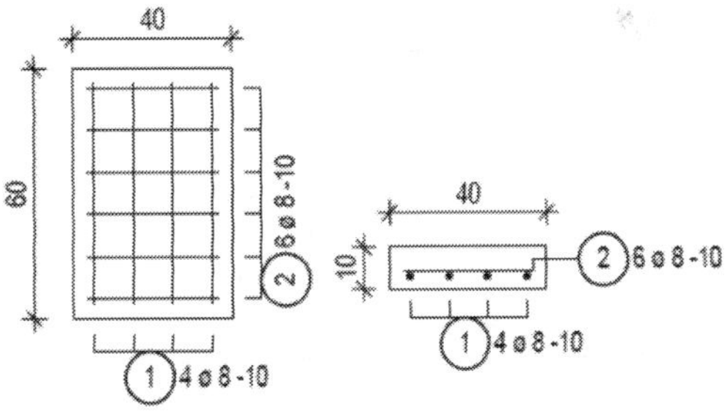

Fig. 5. Formwork and reinforcement plan of the test specimens.

The following concrete strength and exposure class was used for the production of the test specimens. Portland cement CEM I 42.5 N (270 kg) with an additive mixture consisting of slag sand, limestone and fly ash (40 kg) was used as binder.

Table 3. Concrete formulation.

class	exposure	w/c-ratio
C30/37	XD1	0.55
C35/45	XD3	0.45

Ribbed structural steel B500B was used as reinforcement for 14 test specimens and B550B for two others. The diameter was 8 mm and the bar-to-bar distance was 10cm. The concrete cover depth was 4cm. The coating used was a two-layer epoxy resin coating with a layer thickness of 0.5 mm.

3.2 Crack initiation

The cracks in the test specimen were created by a 3 point bending load. As shown in Figure 6, the concrete slabs were placed in a hydraulic testing machine and loaded in a controlled manner until the crack was created. The desired crack width was 0.1 - 0.3 mm. In order to prevent the crack from closing again, wooden wedges were attached to both ends of the crack.

Fig. 6. Crack initiation.

3.3 Salt exposure

All specimens were loaded by a central load according to the bending tensile test to initiate cracks with different crack widths. The test specimens were contaminated from 16.05.2017 to 22.06.2017. During this period, the concrete slabs were loaded with 3% sodium chloride solution. In the first 5 applications, the test specimen was loaded with sodium chloride for the rapid increase of the chloride. After that, the other treatments were only carried out with water.

3.3 Measurement setup

In the laboratory test, 16 plate-like test specimens with the dimensions 60 x 40 x 10 cm were used. The test specimens were stored under a roof for the entire duration of the measurement in an outdoor climate. To carry out the potential measurement, the reinforcement of the two test specimen groups was connected separately with a cable. As shown in Figure 8, the reference electrodes were attached to the top near the crack. The potential measurement was carried out with two different types of electrodes. On the one hand silver - silver chloride and on the other hand, copper - copper sulphate was used. This allows comparing several measurement results with each other.

Table 4. Cycle of application.

date	application
16.05.2017	100 ml NaCl solution, 3 % Cl
18.05.2017	100 ml NaCl solution, 3 % Cl
19.05.2017	100 ml NaCl solution, 3 % Cl
22.05.2017	100 ml NaCl solution, 3 % Cl
23.05.2017	100 ml NaCl solution, 3 % Cl
24.05.2017	100 ml water
29.05.2017	100 ml water
30.05.2017	100 ml water
31.05.2017	100 ml water
06.06.2017	100 ml water
13.06.2017	100 ml water
22.06.2017	100 ml water

Fig. 7. Experimental setup.

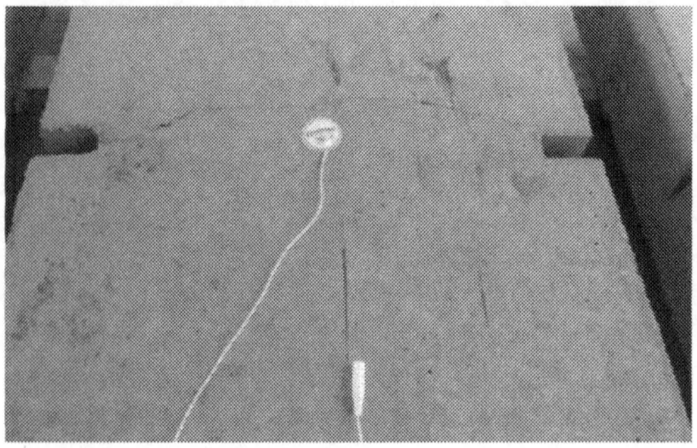

Fig. 8. Position of electrodes.

4 Results and evaluation

The electrochemical potential field measurement is a method for assessing the corrosion behaviour of the reinforcement of reinforced concrete components. By measuring the electrochemical potential difference, areas with a high corrosion probability can be detected. To determine the current potential at the steel/concrete interface of the reinforcement steel below the concrete surface, the potentials arriving at the surface are measured. The measurement is carried out by a high-impedance voltmeter connected to the reinforcement steel. If the electrode potential is determined at a standard electrode with the normal hydrogen electrode as reference, this is referred to as normal potential. The measured potential values can be interpreted as more negative potentials indicating a higher probability for corrosion processes. A difference is made here between two ranges. In the range above (more positive than) -200mV a low corrosion probability is given. In the range below (more negative than) -200mV a high corrosion probability applies [1].

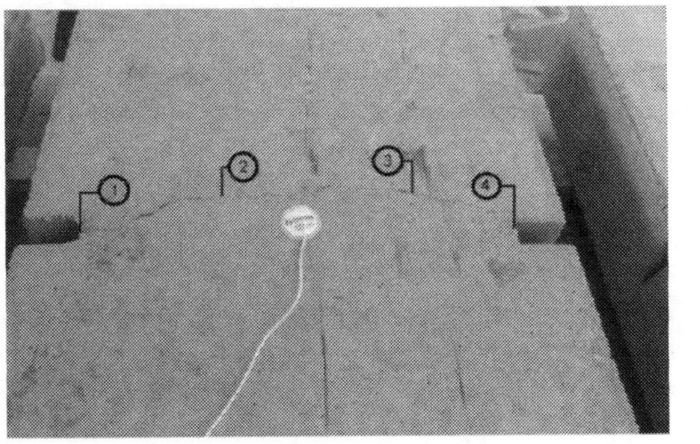

Fig. 9. Position of the measuring points

The measured crack widths are given in Table 5 below.

Table 5. Crack widths of uncoated and coated Specimens

Crack widths uncoated specimens [mm]					
	Measuring points				Mean value
Specimen No.	1	2	3	4	x
1	0,1	0,2	0,4	0,3	0,25
2	0,2	0,5	0,1	-	0,27
3	1	0,5	0,5	0,6	0,65
4	0,2	1	0,8	0,4	0,5
5	0,4	0,6	0,6	0,4	0,5
6	0,1	0,6	0,8	0,8	0,58
7	0,2	0,4	0,5	0,4	0,38
8	0,2	0,3	0,3	0,1	0,23

Crack widths coated specimens [mm]					
	Measuring points				Mean value
Specimen No.	1	2	3	4	x
1	-	0,3	0,4	-	0,35
2	0,6	1	1,2	0,6	0,85
3	0,4	0,5	0,4	0,2	0,38
4	0,5	1,4	1,2	0,6	0.93
5	1	1,4	1,4	0,7	1,13
6	0,1	0,2	0,3	-	0,20
7	0,3	0,4	0,4	0,2	0,33
8	0,1	0,2	0,2	0,1	0,15

4.1 Detection of cracks

In order to detect the crack widths as accurately as possible, these were measured at four points on the specimens, as shown in Figure 9. The results are summarized in the following diagrams.

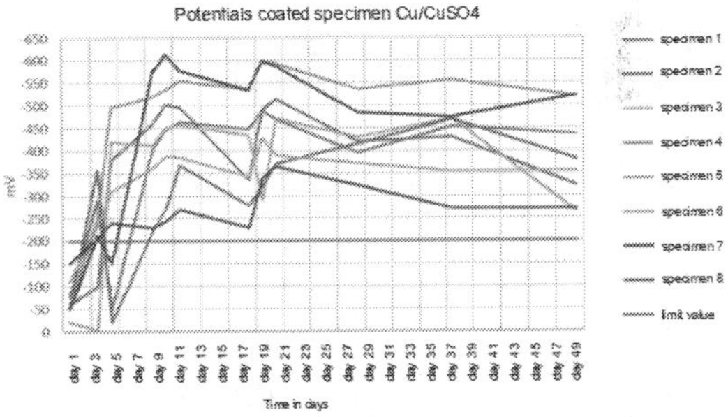

Figure 10. Potential mapping of coated samples Cu/CuSO₄.

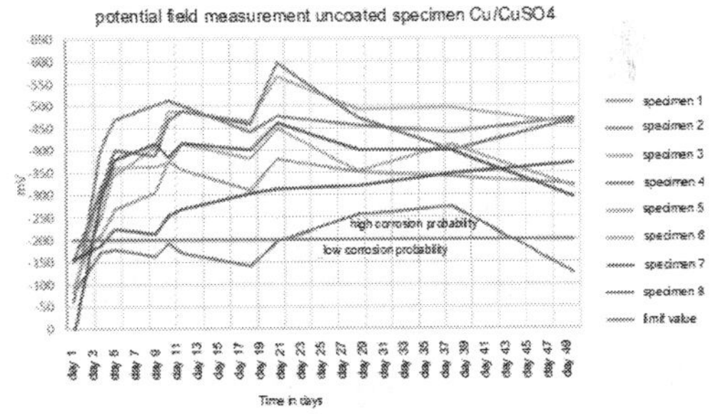

Figure 11. Potential mapping of uncoated samples Cu/CuSO₄.

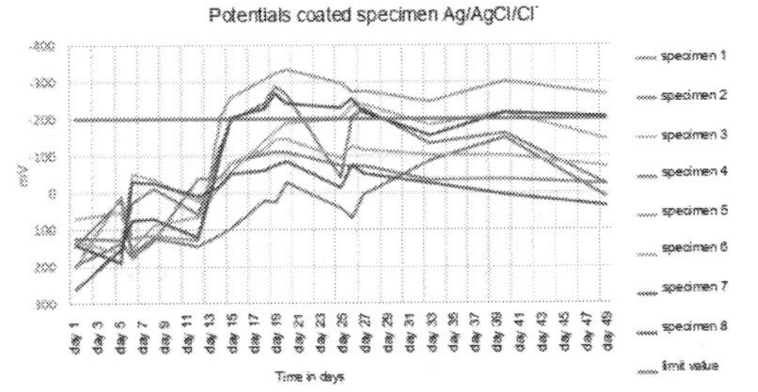

Figure 12. Potential mapping of coated samples Ag/AgCl/Cl.

https://doi.org/10.1051/matecconf/201928902007

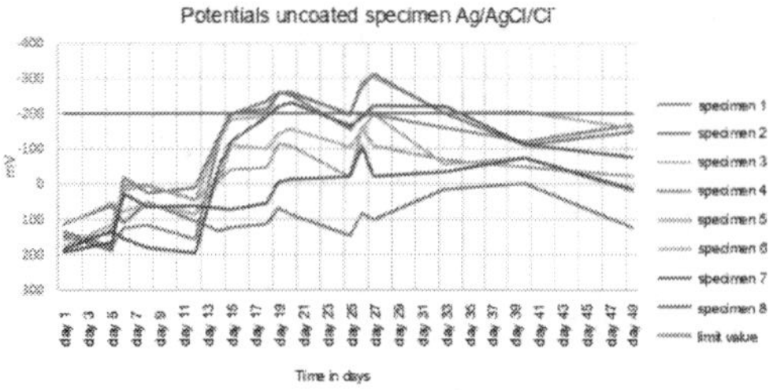

Figure 13. Potential mapping of uncoated samples Ag/AgCl/Cl.

4.3 Evaluation of the potential measurement with $Cu/CuSO_4$ electrode system

As can be seen in Figures 10 and 11, there are increased fluctuations in the corrosion potential during early application. These variations are due to the salt solution application, as the first five applications were carried out using sodium chloride. At the beginning of the measurements the coated specimens show greater variations than the uncoated specimens. However, the amplitudes of the corrosion graphs drop again right after the application. After further application, some peaks are still visible.

4.3.1 Coated Specimen

After the second application, 2 samples with a measured potential of -550 to -600 mV stand out: Test specimen 5 with a mean crack width of 1.13 mm and test specimen 8 with a mean crack width of 0.15 mm. After the fifth application, all 8 specimens were in a range below the -200mV limit. The fluctuation range of the measured potentials of the test specimens was -230mV after 28 days.

4.3.2 Uncoated Specimen

During the early applications using sodium chloride, the test specimens reacted quickly with few fluctuations. This effect is most clearly visible in sample 3, which also shows the largest average crack width of 0.65 mm. After 15 days, all samples exceeded the critical -200mV range. After 28 days the fluctuation range of the determined potentials was -300mV.

4.3.3 Comparison of the test series

In the comparison of the two test series, there should be a difference between the coated and uncoated series, as the moisture in the crack area is rather higher and cannot leak through the surface. However, no such effect could be observed. After being exposed to sodium chloride, 15 of the 16 samples of the two groups of test specimens reached the critical -200mV limit in which corrosion is very probable. After 28 days, the different crack widths did not show any significant similarity. No correlation was found between the crack widths and the corrosion potential of the individual specimens.

4.4 Evaluation of potential measurement with the Ag/AgCl/Cl electrode system

The measurement with the silver-silver chloride electrode system began 10 days before the first application of sodium chloride in order to check the functional capability of the electrode system by corresponding reference measurements. After the first application of sodium chloride a clear reaction was observed. As can be seen in Figures 12 and 13, the further applications are only weakly visible on the diagrams. This effect can be observed in both groups of the test specimens. This could be due to a contact problem of the reference electrodes. In comparison with the copper - copper sulfate electrode clear differences were determined within the potential measurement.

5 Summary

The aim of this study was to determine whether different crack widths may have an influence on the corrosion behaviour of reinforcing steel. Two different electrode systems were used: Silver - silver chloride - electrodes and copper - copper sulphate - electrodes. For the specimens. two different strength classes and two exposure classes were used. The measured potentials, Figs. 12 and 13, show that the silver -silver chloride – electrodes in this case were not suitable for this type of measurement because a contact problem of the electrodes led to irregular measurement results. Therefore, it was not possible to make a representative evaluation of the measurement results. For this reason, the results of the measured potentials of the copper - copper sulfate - electrodes were used. These measurement results show that no noticeable improvement can be achieved using an epoxy resin coating. Neither with the strength class nor with the exposure class could clear trends be observed. A correlation between the different crack widths and the corrosion rate could not be established. As a result, it is clear that different crack widths have no influence on the corrosion behaviour of the reinforcing steel.

References

1. DGZfP (Deutsche Gesellschaft für Zerstörungsfreie Prüfung), *Merkblatt B 03, Elektrochemische Potentialmessung zur Detektion von Bewehrungsstahlkorrossion*, Berlin (2014)
2. S.H. Müller, U. Nolting, M. Haist, *Symposium Beherrschung von Rissen in Beton*, Universität Karlsruhe (2010)
3. D. Bosold, A. Grünewald, *Zement-Merkblatt Betontechnik B18 "Risse in Beton"* (2014)
4. ÖNORM B 1992-1-1, Austrian Standard Institute (2018)
5. H.S. Müller, U. Nolting, M. Haist, *Symposium Dauerhafter Beton, Grundlagen, Planung und*

https://doi.org/10.1051/matecconf/201928902007

Ausführung bei Frost- und Taumittel-Beanspruchung, Universität Karlsruhe (2009)

6. M. Raupach, M. Kosalla, L. Wolff, L. Meyer, *Einfluss von Rissen auf die Korrosionsgefahr der Bewehrung in Parkdecks und deren Instandsetzung – Neue Erkenntnisse aus der Forschung und Praxis*, Technische Akademie Esslingen (2014)
7. M. Beck, *Zur Entwicklung der Eigenkorrosion von Stahl in Beton*, RWTH (2010)
8. C. Dauberschmidt, *Bauwerksdiagonose bei chloridbelastetem Beton- immer ein Entsorgungsfall?*, DBV-Tagung (2010)

How concrete removal techniques affect the bonding between concrete and repair mortar

Mohammad Ali Yazdi[1], *Elien* Dejager[1], *Mats* Debraekeleer[2], *Elke* Gruyaert[2], *Kim* Van Tittelboom[1], *Nele* De Belie[1]

[1] Magnel Laboratory for Concrete Research, Department of Structural Engineering, Faculty of Engineering and Architecture, Ghent University, Tech Lane Ghent Science Park, Campus A, Technologiepark Zwijnaarde 60, 9052 Ghent, Belgium
[2] Technology Cluster Construction, Structural Mechanics and Building Materials, Department of Civil Engineering, KU Leuven, Gebroeders De Smetstraat 1, 9000 Ghent, Belgium

Abstract. Many bonding failures between the substrate and repair mortar are attributed to too aggressive removal of the contaminated concrete and incorrect use of repair materials. To this end, in this study the effect of concrete removal techniques on substrates of different compositions and, as a result, on the bonding with repair mortar has been investigated. Substrate surfaces of different concrete compositions, micro-concrete (MC) and crushed stone concrete (CC), were treated by using three commonly used techniques: grit blasting (GB), jack-hammering (JH), and hydrodemolition/water-jetting (WJ). Automated Laser Measurements (ALM), sand patch tests and surface tensile strength measurements were applied to characterize the substrate surface. According to the results, the co-lateral effects of removal techniques such as WJ are dependent on the concrete mix composition. Moreover, WJ- and JH- treated samples achieved a lower surface tensile strength and bond strength compared to GB-treated slabs.

1 Introduction

Bond strength between concrete substrates and repair materials is influenced by parameters such as the saturation degree [1] and roughness [2] of the substrate surface, the applied removal technique [3], etc.

Based on the removal depth, concrete removal techniques are categorized into different classifications. For example, techniques such as grit blasting (GB) are used to remove or to clean the surface. Concrete removal techniques like jack hammering (JH) or high-pressure water jetting (WJ) are usually used when a higher removal depth is the target [4]. However, these techniques, i.e. JH and WJ are usually referred to as "aggressive". The former is claimed to create microcracks in the superficial layer of the substrate, and as a result, to weaken the bond strength [5,6,7]. Research results on the latter sometimes contradict each other. Some research studies have reported the presence of microcracks caused by WJ [6,8], while others did not [9,10].

In this paper, GB, JH and WJ were applied and compared. The surface of the substrates were first characterized by Automated Laser Measurements (ALM), volumetric sand patch tests and surface tensile strength measurements. The bond strength between the treated substrates and the repair material was later quantified by performing pull-off tests.

2 Materials and methods

2.1 Sample preparation

Two concrete mixtures of micro-concrete (MC) and crushed stone concrete (CC) were prepared. For each mixture, 18 slabs with dimensions of 300mm x 300mm x 100mm were made. Table 1 shows the mix composition of the concrete. The aggregate size is also shown in this table, with CC having the largest D_{max}. The effect of aggregate size on the contribution of removal techniques and the bond strength is later discussed. Superplasticizer was added to the system until the slump reached between 100 and 150 mm.

Table 1. The concrete substrate mix design and characteristics.

Materials	Micro-concrete (MC)	Crushed stone concrete (CC)
CEM I 52.5 N (kg/m^3)	470	340
Water (kg/m^3)	188	153
Sand (kg/m^3)	750	604
Gravel 2/8 (kg/m^3)	946	
Crushed limestone 2/6.3 (kg/m^3)		336
Crushed limestone 6.3/20 (kg/m^3)		967
Superplasticizer Glenium 27	x	
Superplasticizer Viscocrete 1020x		x
W/C	0.4	0.45
Compressive strength (N/mm^2)	35.3	37.0

At the age of 3 months, the slabs were exposed to grit blasting (GB), jack hammering (JH) or water jetting (WJ) removal techniques: GB with a pressure of 0.7 MPa, JH using Air Hammer model 29062 with 3000 strokes per min and WJ with a high pressure of 130 MPa were conducted. Afterwards, MasterEmaco S 5400 repair mortar was used to cover the surface of the treated substrates with a layer of ±5mm thickness to analyse the bond strength.

2.2 Roughness

2.2.2 Automated Laser Measurement (ALM) method

The roughness of the substrate surface was evaluated using Automated Laser Measurements (ALM). The height of the surface peaks and valleys (Z) was measured with a high precision laser beam (sensor ILD 1800-50 and interface optoNCDT 1800, Micro-Epsilon Mess-technik GmbH, Ortenburg, Germany; resolution = 5 microns) mounted on an in-house developed measurement table equipped with two stepping motors that control the movement in the X and Y direction [11], see fig.1.

Fig 1. The ALM table.

For each concrete substrate, the surface profile was obtained in the X-direction for 8 different Y-positions, and vice versa, see fig. 2. The traversing length was 250mm. Six samples per concrete composition and per pre-treatment technique were used for the ALM, however, for water-jetted slabs only one sample was used owing to the expense of the technique.

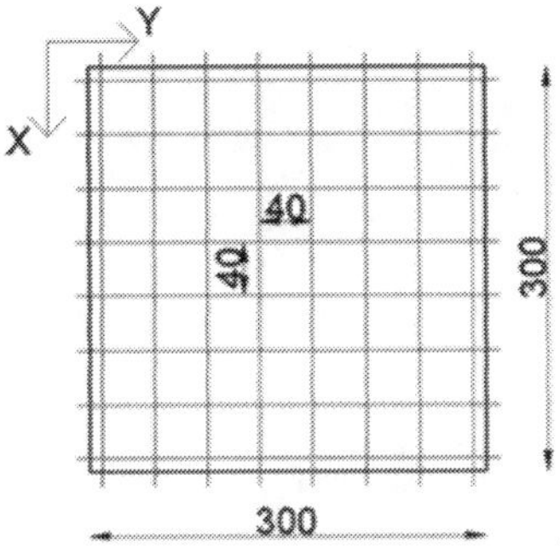

Fig 2. ALM direction.

In this study, the R_a and the R_q values were calculated according to the standard BS 1134 (2010) and were considered to evaluate the roughness of the substrate surfaces. The R_a centre-line roughness value (mm) is the arithmetic mean of the absolute ordinate values Z_i within the sampling length. And, the R_q root-mean-square roughness value (mm) is the root mean square of the ordinate values Z_i within the sampling length. Equations 2 and 3 mathematically express R_a and R_q, respectively.

$$R_a \approx \frac{1}{n}\sum_{i=1}^{n}|Z_i| \quad (2)$$

$$R_q \approx \sqrt{\frac{1}{n}\sum_{i=1}^{n} {Z_i}^2} \quad (3)$$

where n is the number of measurements, Z is the amplitude and i is the measurement point number.

Fig. 3 shows an example of a roughness profile for a MC sample with and without taking the skew into account.

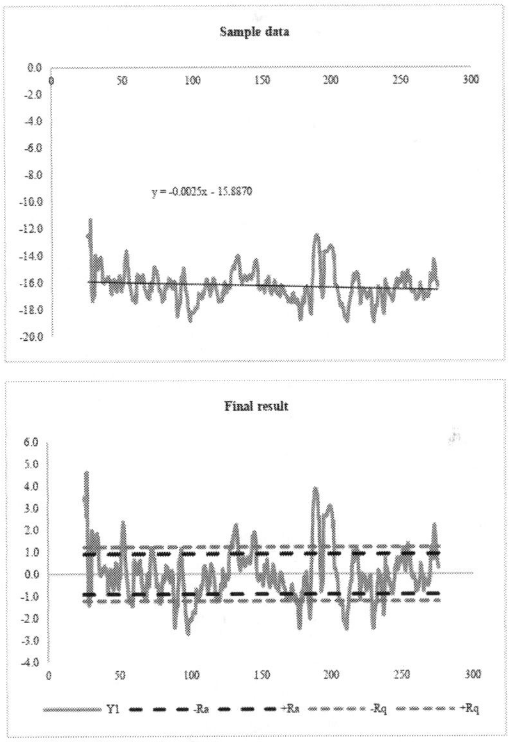

Fig 3. Roughness profile with (bottom) and without (top) skew adjustment.

2.2.1 volumetric sand patch test

The volumetric sand patch test method was performed according to EN 1766:2017. A heap was formed by placing 10 ml dried silica sand with a grain size of 0.05 mm to 0.1 mm in the middle of the pre-treated surface of the substrates. Afterwards, a disk was used to uniformly spread the sand into a circular shape, see fig 4. The diameter of the circle was measured three times (at angles of 0°, 60°, 120°) when no more sand could be moved. The roughness index R_i (mm) was finally calculated, as follows:

$$R_i = \frac{4V}{\pi d^2} 1000 \qquad (1)$$

where V is the volume of sand used in the test (ml) and d is the mean sand patch diameter (mm).

Fig 4. Sand patch test.

2.3 Adhesion strength

Three pull-off tests were performed according to NBN EN 1542 on each substrate before and after the repair mortar was applied. The former was to measure the surface tensile strength, and the latter was performed 28-days after the repair mortar was applied in order to evaluate the bond strength between the substrate and the repair material. In order to perform the test, three cores with a diameter of 50mm were drilled 15mm deep into the substrate surface, and then a dolly was glued to each core using epoxy resin. After 24h the pull-off test was conducted using the Proceq DY-2 device shown in fig. 5. Both strengths were finally calculated using the following equation (4):

$$f_{hs} = \frac{4F_h}{\pi D^2} \qquad (4)$$

where f_{hs} is the final strength (N/mm^2), F_h the failure load (N) and D the mean diameter (mm).

Fig 5. The pull-off device.

3 Results and discussion

3.1 Roughness

Fig. 6. shows the ALM results of GB-, WJ- and JH-treated substrates. According to BS 1134:2010, for machining processes that produce lays, the maximum value achieved from the results of measurements in different directions would be considered as a roughness parameter. Therefore, in this study, the value given in fig.6 for WJ-treated samples, is the highest roughness obtained in either X or Y. However, for other pre-treated samples the average of X and Y direction is plotted in fig.6.

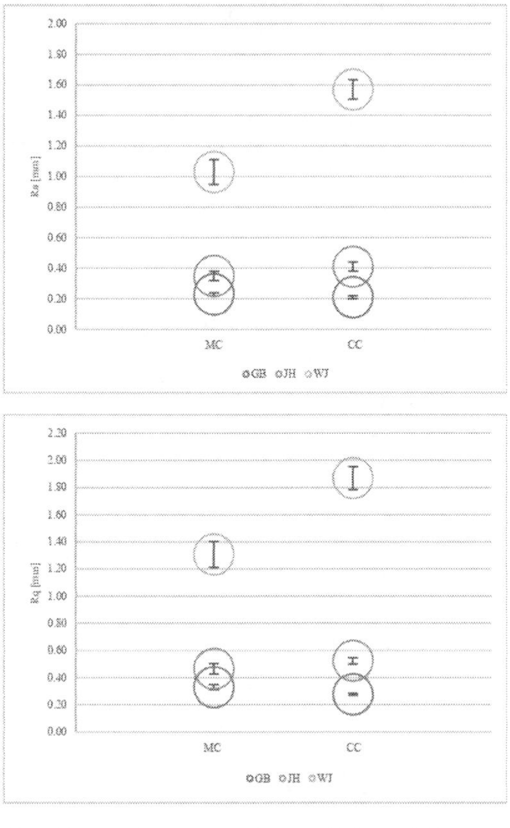

Fig 6. Degree of roughness, R_a(top) and R_q(bottom). Error bars represent the standard error.

As can be observed from fig.6, WJ results in a much rougher surface compared to GB and JH. When concrete compositions are compared, it is observed that the roughness is higher in WJ-treated CC mixtures compared to MC mixtures, showing the dependence of WJ on the concrete composition. The observation of a higher degree of roughness in CC compared to MC might be because of the difference in aggregate size of these mixtures. In fact, CC has larger aggregates than MC (see table 1). However, the contribution of other techniques such as GB and JH are shown not to be dependent on the mix composition.

Fig. 7 shows the sand patch test results of MC and CC mixtures. As can be seen in the figure, the degree of roughness was the highest for WJ-treated MC and CC mixtures compared to the other treated substrates. This finding corresponds with the outcome of the ALM. However, the mix composition effect on the roughness of the substrates is not expressed in the sand patch test results, showing that this test method might not be precise enough. This was also reported in [12]. With regard to GB- and JH-treated samples, a good overlap can be seen on the results achieved by the sand patch test, regardless of the mix composition.

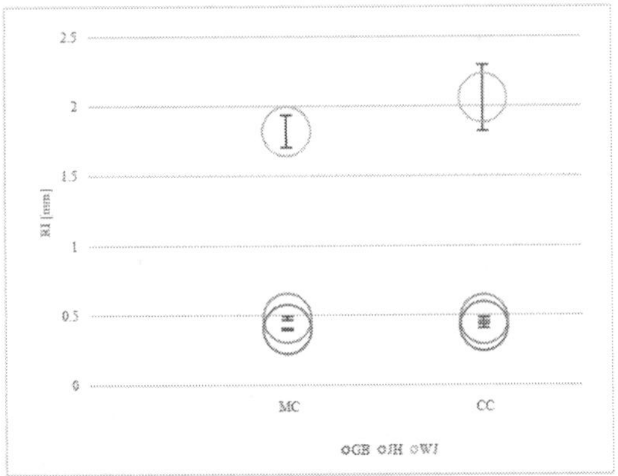

Fig 7. Sand patch test results. Error bars represent the standard error.

The Belgian Construction Certification Association has defined three main texture classifications, as specified within PTV-BB-563 (2007), depending on the roughness of the substrate surface, see Table 2.

Table 2. The texture classification of substrate surface.

Texture	Roughness index
Smooth	less than 0.2 mm
Sandblasted	Between 0.2 and 0.5 mm
Rough	Greater than 0.5 mm

Both ALM and sand patch test results show that GB- and JH- treated surfaces are placed in "sandblasted" classification. However, WJ-treated samples with roughness index of over 0.5 mm is classified as "rough", according to table 2.

3.2 Adhesion strength

The surface tensile strength and bond strength of pre-treated samples are shown in fig. 8. As can be seen, GB, regardless of the mix composition, achieved the highest surface tensile strength, followed by JH and WJ. The bond strength results shown in fig.8 follow the same trend as that of the surface tensile strength, except for WJ-treated MC samples, where a higher bond strength is seen compared to that of JH-treated MC slabs. In general, GB-treated samples have the highest bond strength and surface tensile strength compared to JH- and WJ-treated slabs. This could be because of the difference in the degree of roughness of these pre-treated mixtures. Moreover, the fact that the latter techniques, i.e. JH and WJ are much more aggressive than GB and thus belong to a different classification should not be forgotten. GB is mainly used for cleaning or removing the substrate surface, but JH and WJ techniques have the potential to remove the cover or even matrix concrete, depending on the removal energy. Therefore, the potential detrimental effects caused by JH and WJ on the concrete substrates, and as a result, on the strength, can be more pronounced.

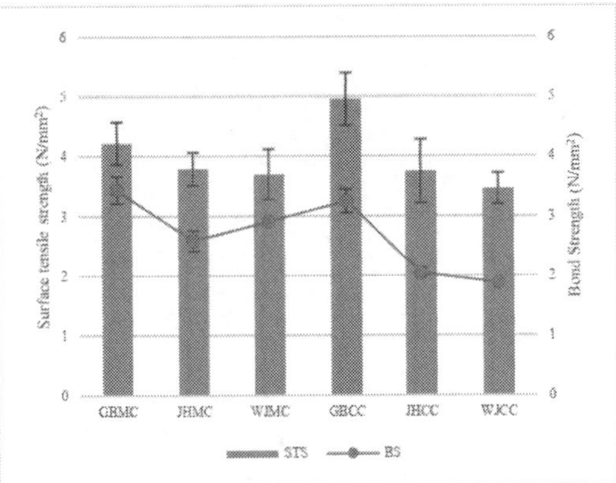

Fig 8. Surface tensile strength and bond strength results (STS: surface tensile strength and BS: bond strength). Error bars represent the standard error.

The presence of microcracks has been reported as one of the main reasons causing the reduction in bond strength of samples treated by JH [5,6,7]. Some microcracks caused by JH were also observed in this study, see fig.9.

Fig 9. An example of detected cracks caused by JH.

Fig. 10 shows an example of surface tensile strength failures in the concrete substrates of JH-treated CC and MC samples. Failures in the superficial layer of jack hammered substrates (see fig.10) have been widely reported in literature [6,13] and been attributed to the presence of microcracks at this layer.

As can be seen, failures were predominantly dependent on the mix composition. For example, for JH-treated MC samples almost no failures take place in the superficial layer, while for JH-treated CC samples 2 out of 3 failures are in a zone very close to the surface. Moreover, some irregularities were observed in the failures of CC samples (shown in fig.10), which can be because of the aggregate size and shape in these mixtures.

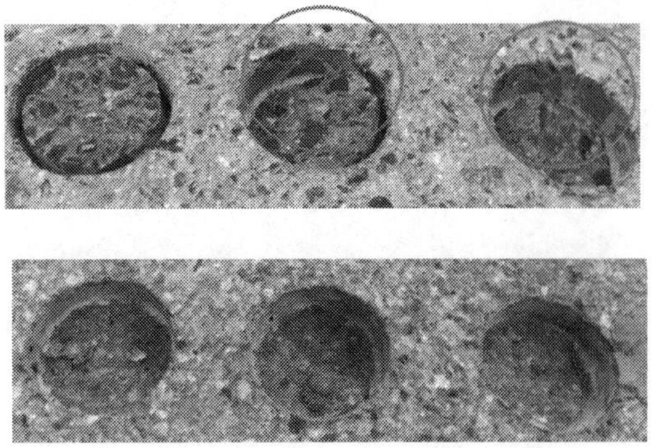

Fig 10. Surface tensile strength failures in the substrates of jack hammered CC (top) and MC (bottom).

4 Conclusion

The results of this study show a much higher degree of roughness in WJ-treated samples compared to GB- and JH-treated slabs. This was seen in both the ALM and the sand patch test results. However, the sand patch test method seems not to be a proper method for roughness measurement of substrates with a very rough surface. On the contrary, the ALM method gave more reasonable results when the roughness of different mix compositions (MC and CC) was compared.

WJ was shown to influence the mixtures based on their compositional features such as aggregate size. Both JH and WJ achieved a lower bond strength as well as surface tensile strength compared to GB. The reduction in JH-treated samples could be due to the microcracks created during the hammering process. However, the fact that GB belongs to a different classification, and therefore, possesses different applications and removal energy compared to JH and WJ should be taken into account for future research studies in this field.

References

1. L. Courard, B. Bissonnette, A. Garbacz, A.M. Vaysburd, and K.F. von Fay, *Guidelines for concrete surface preparation: 10 years research and experience*. in *MATEC Web of Conferences*. EDP Sciences, 2018.
2. F. Perez, M. Morency, B. Bissonnette, and L. Courard, *Correlation between the roughness of the substrate surface and the debonding risk*. in *2nd International Conference on Concrete Repair, Rehabilitation and Retrofitting, ICCRRR-2*. Cape Town, South Africa: CRC Press, 2008.
3. S. Matthews, *CONREPNET: Performance-based approach to the remediation of reinforced concrete structures: Achieving durable repaired concrete structures*. J. Build. Apprais., 2007. **3**(1): p. 6-20.
4. M.C. Vorster, J.P. Merrigan, R.W. Lewis, and R.E. Weyers, *Techniques for concrete removal and bar cleaning on bridge rehabilitation projects*. National Academy of Sciences, Washington, DC, 1992.
5. J. Silfwerbrand and H. Beushausen, *Bonded concrete overlays–bond strength issues*. in Proceedings ICCRRR 2005 International Conference on Concrete Repair, Rehabilitation and Retrofitting, Cape Town, South Africa, Taylor & Francis Group, London. 2006.
6. L. Courard, T. Piotrowski, and A. Garbacz, *Near-to-surface properties affecting bond strength in concrete repair*. Cem. Concr. Comp., 2014. **46**: p. 73-80.
7. C. Talbot, M. Pigeon, D. Beaupré, and D. Morgan, *Influence of surface preparation on long-term bonding of shotcrete*. Mater. J., 1995. **91**(6): p. 560-566.
8. L. Courard, B. Bissonnette, and N. Belair, *Effect of surface preparation techniques on the cohesion of superficial concrete: comparison of jack-hammering and water jetting*. in *Concrete Repair, Rehabilitation and Retrofitting*. Cape Town, South Africa: Taylor & Francis Group, 2006.
9. A. Garbacz, L. Courard, and B. Bissonnette, *A surface engineering approach applicable to concrete repair engineering*. Bulletin of the Polish Academy of Sciences: Technical Sciences, 2013. **61**(1): p. 73-84.
10. B. Bissonnette, L. Courard, H. Beushausen, D. Fowler, M. Trevino, and A. Vaysburd, *Recommendations for the repair, the lining or the strengthening of concrete slabs or pavements with bonded cement-based material overlays*. Mater. Struct., 2013. **46**(3): p. 481-494.
11. N. De Belie, J. Monteny, A. Beeldens, E. Vincke, D. Van Gemert, and W. Verstraete, *Experimental research and prediction of the effect of chemical and biogenic sulfuric acid on different types of commercially produced concrete sewer pipes*. Cement and Concrete Research, 2004. **34**(12): p. 2223-2236.
12. P.M.D. Santos and E.N.B.S. Julio, *comparison of methods for texture assessment of concrete surface*. ACI Mater. J., 2010. **107**(5): p.433-440.
13. A. Garbacz, M. Górka, and L. Courard, *Effect of concrete surface treatment on adhesion in repair systems*. Mag. Concrete. Res., 2005. **57**(1): p. 49-60.

The effects of anode distance and corrosion activity on current distribution for ICCP systems

Shamir Bhuiyan*, *David* Law, *Liam* Ward, and *Justin* Saliba

School of Engineering, RMIT University, Melbourne, VIC, 3001, Australia

Abstract. One of the most effective techniques to mitigate corrosion of steel reinforcement in concrete is impressed current cathodic protection (ICCP). This technique has been widely used for several decades; however, there is limited research systematically studying the various factors affecting the distribution of cathodic current over the area of steel reinforcement. Understanding how current is distributed in an ICCP system is crucial for its design to ensure that all areas of steel are sufficiently protected, and overprotection of certain areas is avoided. In this study, the effect of distance of the anode from the steel and level of corrosion on current distribution is investigated by applying a potential sweep and measuring the steel potential and current for specimens with three layers of steel at different depths. The findings from this study revealed an inverse relationship between cathodic polarisation with steel distance from the anode and that higher levels of corrosion can cause a more non-uniform current distribution favouring the closest bar to the anode.

1 Introduction

One of the key considerations in designing an impressed current cathodic protection system (ICCP) is the aspect of current distribution due to its non-uniform nature in most practical scenarios. In the context of reinforced concrete structures, it is critical to ensure that the total area of steel bars receives an adequate level of current for protection whilst avoiding overprotection of any areas. Current distribution is associated with throwing power which is the ability to polarize the steel at different depths, which in turn depends on a number of factors, namely magnitude of the supplied current, concrete resistance, corrosion rate and reinforcement geometry/arrangement [1-3].

Concrete resistance is a function of the material property (ie resistivity) and the physical dimensions of the reinforced concrete area. In the context of this study, the former is kept constant while the latter is varied by changing the anode-steel distance. Following Ohm's law, a path of greater resistance will result in a lower current for a given driving voltage, resulting in lower polarization. Consequently, the amount of current received is dependent on the position of the bars in relation to the anode. In a study conducted by Bertolini et al [2], wherein a slab with double-layered reinforcement was cast, it was found that the steel layer closest to the anode received between 70-90% of the total current, although this included bars which were positioned directly above one another. A similar range of results were reported by Xu et al [3] and Hassanein et al [1] for their experiments which also involved multi-layered steel reinforcement.

With regards to corrosion rate, this can be correlated directly with the resistance to polarization, with a higher corrosion rate implying a need for a greater amount of current to overcome it. Computer and theoretical models [2,4] have predicted that current requirements can increase several times over in order to provide the same level of polarization for a corroding bar compared to a passive bar. For concrete with actively corroding steel, the current distribution tends to be more uneven and favours bars closer to the anode, provided corrosion rates do not vary in the structure [1,3]. In cases where an active-passive couple exists, the current distribution is further complicated [5].

The objective of this study is to demonstrate the effects of anode distance and resultant corrosion rates on the level of polarization of the bars to develop an understanding of how current is distributed within three layers of reinforcement by applying a potentiodynamic scan. This paper reports the initial (0 and 5 months data) results to assess the effect of bar position of current distribution. This specifically addresses the period during the initiation of corrosion.

2 Methodology

In order to study the effect of distance of anode from the steel reinforcement on the level of cathodic polarisation, a simple setup was designed, wherein 3 steel reinforcement bars and a platinized titanium ribbon (anode) were cast in a concrete block measuring 350x100x100mm, with a design configuration as shown in figure 1a. The anode was placed above the bars in order to study the effect of distance alone. For reference purposes, the closest steel bar to the anode was named bar 1 and the second closest and furthest, bar 2 and bar 3 respectively. The horizontal spacing between the anode

* Corresponding author: Shamir.bhuiyan@rmit.edu.au

https://doi.org/10.1051/matecconf/201928903001

and bars 1, 2 and 3 was 25mm, 125mm and 225mm respectively whereas the vertical spacing between the anode and centreline of the bars was 50mm. The actual reinforced concrete test setup prior to and after casting is shown in figures 1b and 1c respectively.

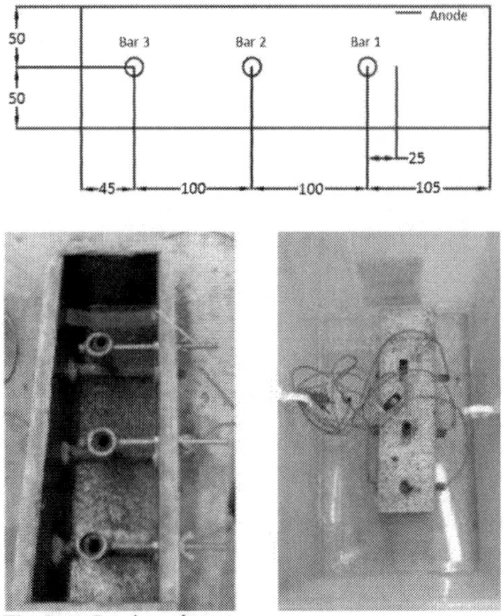

Fig. 1a-c. Specimen layout

The bars were 13mm diameter mild steel and spanned the width of the mould, as did the anode ribbon having a width of 20mm. $Cu/CuSO_4$ electrodes were utilized as reference electrodes and were positioned directly above each bar. The housings of the electrodes were made from PVC tubing which were plugged at the ends with porous wooden stoppers. Before each measurement, the tubes were filled with $CuSO_4$ solution and Cu wires were inserted into the solution. The concrete mix design, table 1, was a standard 40 MPa mix, using General Purpose (GP) cement. To study the effects of corrosion level on current distribution, 5% NaCl (3.5% Cl ion) by weight of cement was added to the wet mix to initiate corrosion of the bars which would allow the corrosion levels to vary over time. The specimen was periodically cycled with water (once per week) and stored in a sealed container to maintain a high-humidity environment to promote corrosion.

Table 1. Concrete mix design

Constituents	Quantity (Kg/m^3)
(GP) Cement	390
Water (w/c: 0.65)	254
Fine Aggregate	585
Coarse Aggregate (7mm)	585
Coarse Aggregate (10mm)	585

After casting the concrete, an initial set of readings was taken within 48 hours representing the initial corrosion initiation period, when passivation and corrosion initiation where occurring. A second set of readings was taken after a 5 month period by which point in time active corrosion was established. Potentiodynamic cathodic polarization scans were conducted in order to determine the current distribution in the system. Scans were conducted at a rate of 100mV/min from the rest potential to -800mV (corresponding to bar1) using a potentiostat (ACM Field Machine). This corresponded to a period of between 3-5 minutes depending upon the rest potential. The anode was connected as the auxiliary electrode, the steel bars as the working electrodes and the $Cu/CuSO_4$ electrode (bar1) as the reference electrode for operating the potentiostat. During the scans, the potentials and currents for each bar were recorded using a data-logger (DataTaker DT80). In order to measure the currents received by each bar, the bars were connected to 100Ω resistors in series and the voltage measurements across the resistors were logged. A multiplier of 0.01 was applied to convert voltage readings to amps. The wiring diagram is shown in figure 2.

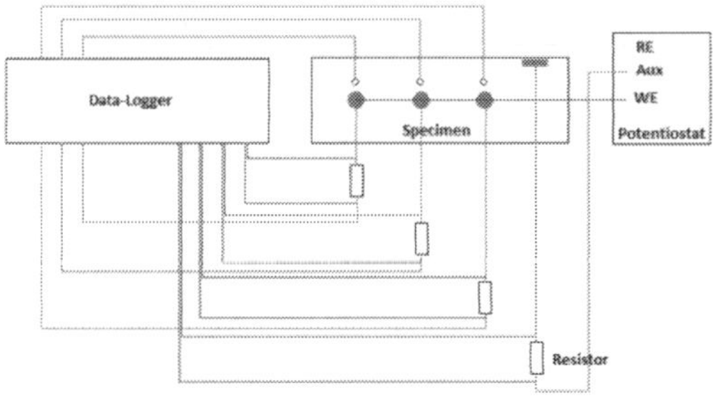

Fig. 2. Wiring diagram of experimental setup

3 Results

3.1. Data at 48 h (0 month)

Variations in bar current with anode current (total applied current measured across the anode) are shown in figure 3. Results were plotted up to 0.5 mA anode current, which corresponded to a current density of approximately 120 mA/m^2. Figure 4 shows the variation in potential for the three bars with anode current. As expected, bar 1 being closest to the anode received the greatest portion of the current (figure 3) and was thus polarized the most (figure 4). Bars 2 and 3 received considerably lower current, and thus the extent of polarisation was reduced accordingly. Additionally, bar 3 appeared to receive a higher current than bars 1 and 2 until 0.08 mA and 0.37 mA anode current respectively, despite it being further away from the anode.

It should be noted that there was an offset of the bar current for bars 2 and 3 at the beginning of the scan which could be due to a small macro-cell current flowing between them through the system. This is supported by the difference in the rest potential of bar 3 as compared to bars 1 and 2 (figure 4). The presence of this potential difference could account for the current flow observed.

Further analysis of the anode – bar current plots in figure 3 reveal the rate of change of bar current with respect to the anode current decreases with the anode-steel distance (ie highest in bar 1 and lowest in bar 3). These observations in slope variation suggest that although bar 3 appears to initially receive greater current than bars 1 and 2, it actually receives less of the cathodic current throughout the scan. These findings suggest that as the overall anode current is increased, a greater proportion of this current is received by the nearest bar and smaller proportions by the bars further away. This is important when considering overall design of cathodic protection systems in terms of receiving sufficient protective current by bars further away from the anode.

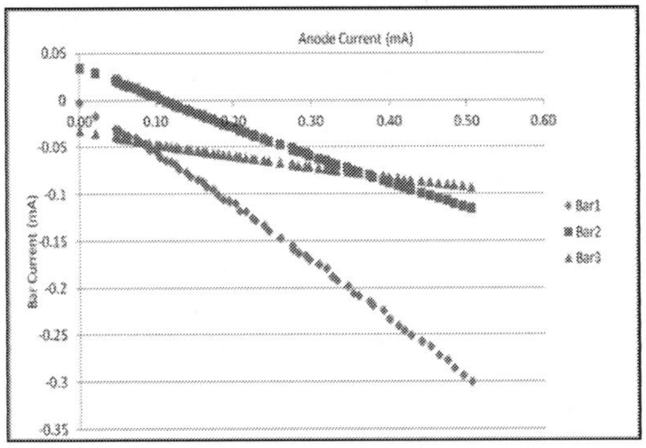

Fig. 3. Bar current vs anode current for 0 month

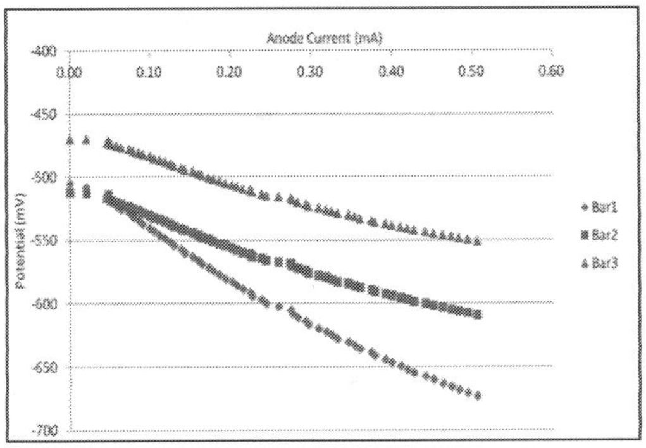

Fig. 4. Potential vs anode current for 0 month

3.2. Data after 5 months

Variations in bar current with anode current after 5 months is shown in figure 5. Similar trends were observed as for 0 month data, except that a greater proportion of current was received by the nearest bar, particularly at the higher anode currents (-0.3 mA for bar 1 for 0.5 mA anode current at 0 months, compared with –0.35 mA after 5 months). Figure 6 shows the variation in potential for the three bars with anode current after 5 months. Here, due to increased corrosion activity, the rest potentials of all bars had shifted in the negative direction, ranging between -550 and -600 mV. The rest potentials of all the bars were also closer together than the 0 month data, suggesting a more uniform corrosion state of the bars had been established after 5 months. The data indicates that the macrocell current was still evident, however to a lesser degree, as seen from their trends in figure 5. Bar 3 no longer showed a higher initial bar current than bar 1 and the point up till which bar 3 displayed a higher bar current than bar 2 had also receded to 0.18mA (anode current).

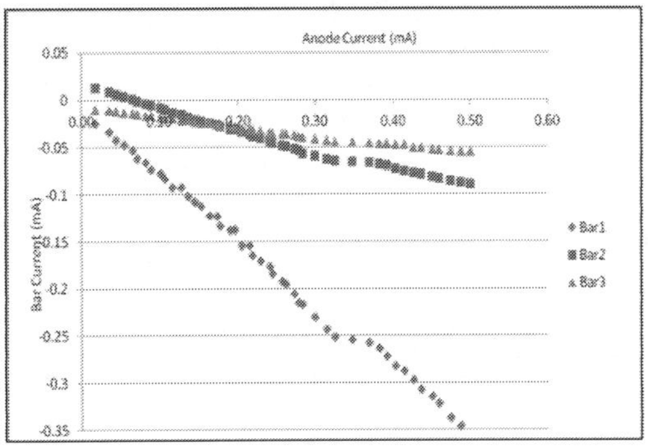

Fig. 5. Bar current vs anode current for 5 months

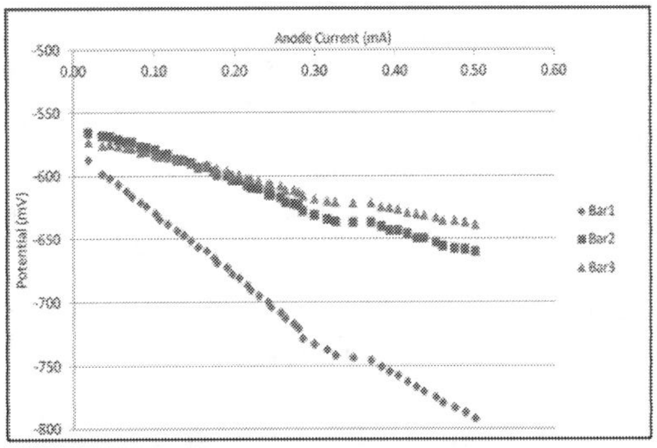

Fig. 6. Potential vs anode current for 5 months

3.3. Discussion

In order to compare and analyse the data, the 0 month and 5 months polarisation data (anode versus bar currents) have been combined and plotted in a bar chart, as shown in figure 7. Polarization over-potential, calculated as the difference between the open circuit potential and polarized potential, versus anode current data is plotted in a bar chart, figure 8. It is clearly observed that bar 1 received the greatest portion of the cathodic current for both data sets. For the 0 month scan, bar 1 received an average of approximately 55% of total current received across all the bars whereas for the 5 months scan, this percentage increased to 70%. This is in agreement with previous research [1-4]. The variation in current distribution between 0 month and 5 months is attributed to the establishment of corrosion of the bars. At 0 months, the bars were subject to both a passive state due to the hydration reaction of the cement and active state, due to corrosion initiation by the cast-in chlorides. At this point, corrosion rates tend to be sporadic [6] and it can be observed that bar 3 at 0 month had a different rest potential compared to bars 1 and 2, thus indicating a difference in corrosion rate After 5 months exposure, a similar degree of corrosion had been established on all bars, as indicated by the more negative, but similar potentials observed.

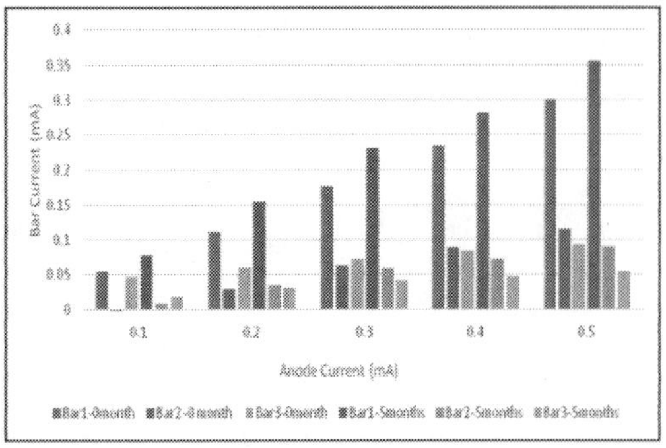

Fig. 7. Comparison of individual absolute bar currents at fixed anode currents

Figure 8 shows polarization trends were consistent, with bar 1 being polarized the most and bar 3 the least. The polarization trends generally followed the same trends as the bar current, i.e. increasing overpotential with increasing anode current. This is expected given that they are directly related. However, for 0.1-0.3mA anode current data (figure 7) it can be observed that bar 2 appeared to receive lower current than bar 3 although the corresponding data for the polarization in figure 8 shows that bar 2 was consistently polarized to a greater degree. It is hypothesized that the current results were offset due to the formation of an initial macrocell and that the actual current received at the bars was consistent with polarization trends throughout the scan. This is further supported by the gradient variations as observed in figures 3 and 5. In terms of the bar distances from the anode in this study representing varying layers of steel reinforcement, the increase in current being drawn by bar 1 at 5 months compared to 0 month indicates that the current was preferentially drawn to the top steel layer - more so when corrosion rates are more uniform and higher at 5 months compared to the initial data where corrosion rates are lower and the corrosion rates vary between the bars.

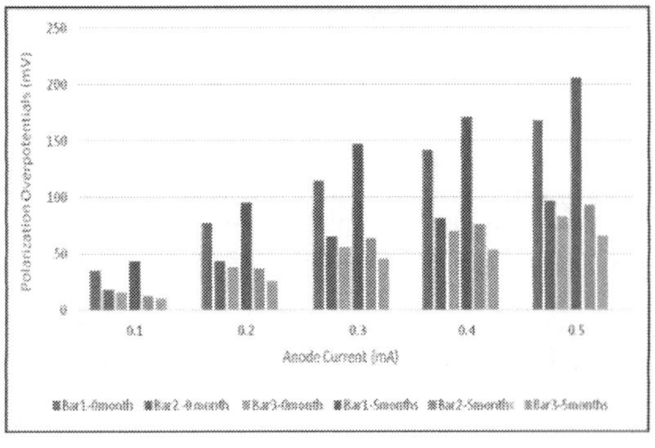

Fig. 8. Comparison of polarization overpotentials at fixed anode currents

Furthermore, the resistivity in the concrete would be expected increase between the initial data and the 5 month data due to hydration of the cement. This could itself lead to a change in the balance between the ohmic resistance and the polarisation resistance and hence the current distribution to the bars.
At 5 months the current distributions across the three bars are in agreement with predicted results in literature [2,4]. However, the results do indicate that the specific distribution of the current between the bars is influenced by the condition of the bars and the corrosion that is occurring on each bar relative to the other layers of steel. Thus, there may be an effect of dissimilar corrosion rates on the current distribution and hence the protection afforded to the steel. This may be important in the design of ICCP systems and further tests are required to validate the effect and assess the impact on the protection afforded to the steel.

4 Conclusions

Experimental results showed that bar1, which was closest to the anode, received between 55-70% of the total current, which is similar to findings stated in the literature for long term experiments. The effect of anode-steel distance was demonstrated, wherein a greater distance resulted in a greater path of resistance for the current and thus a lower degree of polarization for a given anode current. The 0 month data showed that bar 1, closest to the anode, received approximately 55% of the total current, which increased to 70% by 5 months. This is attributed to the increase in corrosion rates of the bars and the resitivity in the concrete.

References

1. A.M. Hassanein, G.K. Glass, N.R. Buenfeld, "Protection current distribution in reinforced concrete cathodic protection systems", Cement & Concrete Composites **24,** 159–167 (2002)
2. L. Bertolini et al, "Cathodic protection of new and old reinforced concrete structures", Corrosion Science **35,** 1633-1639 (1993)
3. J. Xu, Wu Yao, "Current distribution in reinforced concrete cathodic protection system with conductive mortar overlay anode", Construction and Building Materials **23,** 2220–2226 (2009)
4. G.K. Glass, A.M. Hassanein, "Surprisingly Effective Cathodic Protection", Journal of Corrosion Science and Engineering **4,** paper7 (2006)
5. A.A. Sagüés, S.C. Kranc, "On the determination of polarisation diagrams of reinforcing steel in concrete", Corrosion Science 48(8), 624–33 (1992)
6. S. Bhuiyan, D.W. Law, P Nicholls, C. Christodoulou, "Investigation of residual protection of steel following application of protective current", Construction and Building Materials **162,** 503-511 (2018)

https://doi.org/10.1051/matecconf/201928903002

Impressed current cathodic protection with embedded anodes – field trial on the Viaduc de Saint-Cloud in France

Véronique Bouteiller[1,*], *Guillaume* Damien[2], *Frédéric* Boinski[3], *Julien* Schneider[3], and *Christophe* Michaux[4]

[1]Université Paris-Est, EMGCU – MAST – IFSTTAR, 77447 Marne-la-Vallée, France
[2]DRIEA, IF/DiRIF/SIMEER/DIOA, France
[3]Cerema, Direction territoriale Ile-de-France, LEM, France
[4]In-Situ SA, Luxembourg

Abstract. This paper deals with the electrochemical repair of a road bridge, Viaduc de Saint-Cloud which was constructed in the 1970's in France. In 2011, a corrosion analysis was performed together with a durability study. It was demonstrated that the corrosion was induced by chloride ions coming from de-icing salts leaking from the deck. The induced corrosion was increased by the low cover depths resulting in a chloride penetration beyond the reinforcement depth. In order to evaluate the effectiveness and the reliability of Impressed Current Cathodic Protection (ICCP) with embedded anodes, a field trial was performed in 2012 on three span segments including one segment over pier (P2). The results of the ICCP design, the implementation and the monitoring are presented. It was found that two months were necessary before the cathodic protection would reach an equilibrium. Performance criteria as described in ISO 12696 dealing with cathodic protection passed on two segments but failed on the segment over pier. On the latter, where the congestion of reinforcement and prestressed tendons was higher, the anodic system needed some improvement. Finally, the results of the field trial allowed proposing a technical and an economic repair solution.

1 PRESENTATION OF THE BRIDGE

Viaduc de Saint-Cloud was constructed in the 1970s in France (Fig. 1) [1]. It belongs to the A13 highway just west from downtown Paris and allows more than 120000 vehicles per day to cross the River Seine between Saint-Cloud and Boulogne. The span length is 1100 meters. The viaduct is constituted by three-cell box girders (Fig. 2).

Fig. 1. Viaduc de Saint-Cloud (H. Delahousse, Ifsttar).

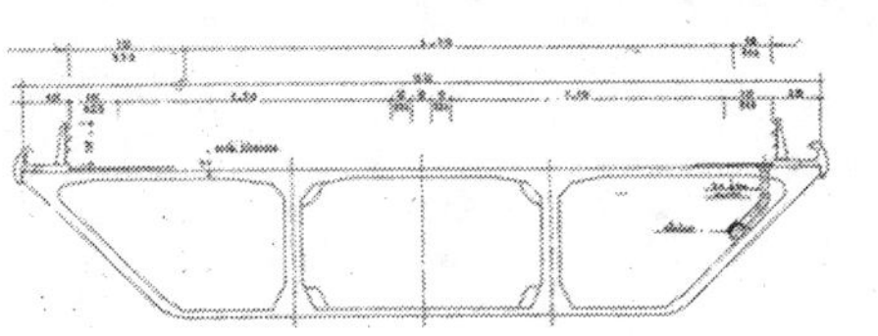

Fig. 2. Geometry of the segment: three-cell box girders.

2 ASSESSMENT

The bridge assessment was based on visual inspection, concrete durability and reinforcement corrosion investigations.

The results of a visual inspection and delamination survey (rust stains, concrete cracks or spalling damages) were helpful to define different areas from severe to negligible corroded zones. These results are gathered in Fig. 3. The field trial was done on pier P2 which was in a medium to severe state of corrosion and which has easier access (Fig. 4).

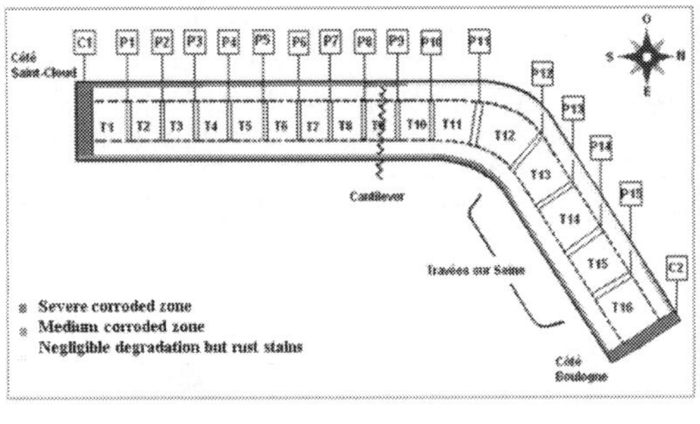

Fig. 3. Visual inspection results of the bridge.

* Corresponding author: veronique.bouteiller@ifsttar.fr

Fig. 4. Field trial on the three segments on pier P2 (from left to right: one segment over pier P2-0 and two span segments P2-1 and P2-2).

Fig. 5 presents the locations of the chloride cores in the segments and the chloride profiles versus concrete depth of the durability investigation [2]. Results show that chlorides have penetrated into the concrete deeper than the reinforcement which was at 10 to 20mm. Chlorides came from de-icing salts that were spread on the bridge in winter and that penetrated the concrete cover leaking through the waterproofing protection and flowing along the segments. It was concluded that the major cause of concrete deterioration was chloride induced corrosion of the steel. Concrete carbonation depth was less than 10mm.

Fig.6 represents the half-cell potential mappings for each segment [3]. From corrosion diagnosis, half-cell potential measurements were in the range 0 to -500mV. The results were consistent with a high probability of corrosion as demonstrated by the gradients in Fig. 6.

3 FIELD TRIAL

3.1 Rehabilitation technique

Classical "patch repair" had failed in 50% of cases after 10 years according to the European ConRepNet Project [4]. The two main causes of failure are debonding between original concrete and new concrete and the incipient anode effect (mainly in the case of chloride contaminated structures) [4, 5]. In the case of the Viaduc de Saint-Cloud this classical repair technique, including removal of contaminated concrete and application of new concrete, was not compatible with the heavy traffic and the structural load capacity of the bridge.

Cathodic protection is the only technology that can directly stop corrosion in reinforced concrete structures if designed, installed, and operated correctly [5-9]. Cathodic protection systems are capable of providing a significantly larger extension in service life depending on the service life of the anode material and the maintenance of the system. In order to address the corrosion problem of the Viaduc de Saint-Cloud, Cathodic Protection rehabilitation treatment was therefore proposed in accordance with NF EN ISO 12696 standard [10].

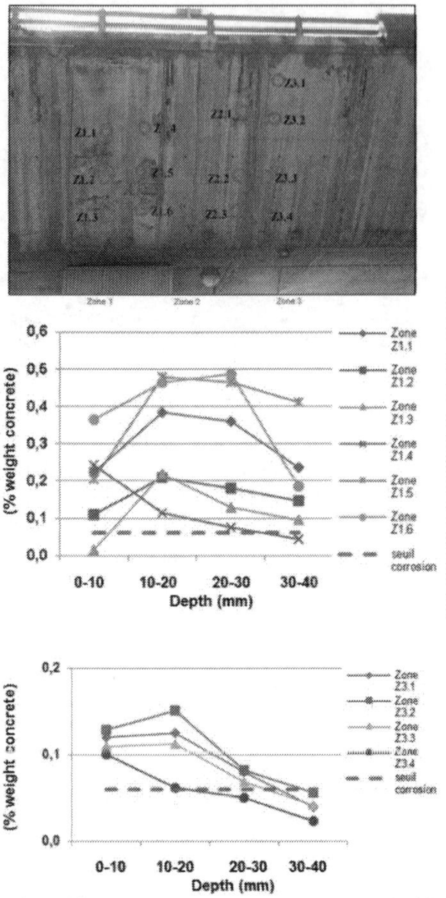

Fig. 5. Locations of concrete cores (top) and free chloride profiles versus depth on segment P2-0 over pier P2 (medium) and on segment P2-2 (bottom).

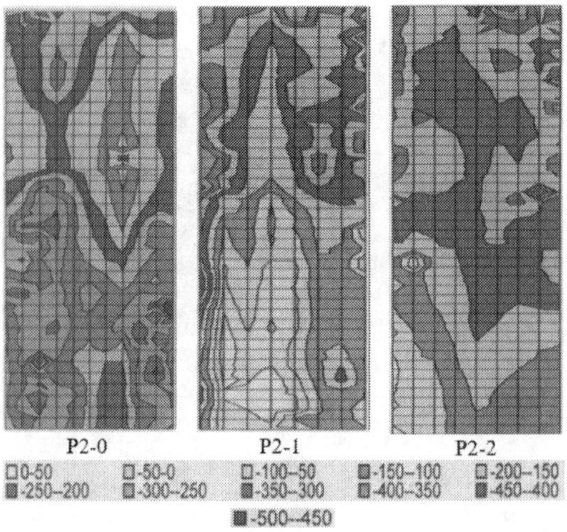

Fig. 6. Half-cell potential mappings for the three segments (mV, Cu/CuSO4)

Within the Cathodic Protection (CP) solutions, the impressed current (IC) technique was chosen rather than using sacrificial anodes (SA) because of the high content of chlorides present in the concrete - even deeper than the reinforcement. ICCP systems arrest steel reinforcement corrosion activity by supplying electrical current from an external source to overcome the ongoing corrosion current in the structure. A number of different anode system configurations are possible but in this case embedded anodes were chosen. They were drilled into the concrete from the interior of the girder offering three main advantages: (i) no traffic interruption, (ii) works could be done by daytime and (iii) no added load to the structure (as would have been the case with a titanium mesh and cementitious overlay).

It is to be noted that ICCP with embedded anodes in France is not a well-known technique and this trial aimed at improving our skills pointing out advantages and limits. The objectives for this research were to:
- Select and design the ICCP system
- Evaluate the performance of the ICCP with embedded anodes for reinforcement and presstressed tendons
- Monitor and control the system
- Establish recommendations for full-scale application

3.2 ICCP with embedded anodes – Design, monitoring and installation

The ICCP system with embedded anodes of the investigated field trial area on the three segments around P2 was designed considering several parameters:
- The surface area of the reinforcement to be protected was calculated for each segment,
- The prestressed tendons that could be in electrical continuity with the reinforcement were also calculated,
- The anodes (CPI) durAnodes (200mm long) or durAfat (50mm long) were made of MMO activated titanium; and were drilled into the concrete (the concrete thickness was around 200mm), and filled with the special compatible CP grout as developed by the anode supplier; locations (heights and depths) depended on the zones to be protected (one or two layers of reinforcement and space) as illustrated with a red cross in Fig. 7.
- The power source (and rectifier) and the targeted current density,
- The reference electrodes were essentially ERE20 (Mn/MnO_2) (Force Technology) or decreased probes (coupons) durAmmo (CPI). Eight reference electrodes per segment were chosen to enable appropriate monitoring and control of the system. They were embedded at different heights and depths in concrete in order to monitor the behaviour of the reinforcement (one or two layers) and of the post-tensioning tendons. Locations of the reference electrodes are identified with a blue cross in Fig. 7.
- The electronic devices (CPI), comprised a central system unit and three durAcenter slave devices (one for each segment).

Fig. 7 illustrates the zoning arrangement for each segment. It is to be noted that all the prestressed tendons of the three segments were gathered in zone 4 and were monitored with durAcenter 1.

A permanently installed monitoring system allowed us to assess the performance of the cathodic protection based on eight reference electrodes (Mn/MnO_2 type ERE20) per segment (Fig. 7). Remote monitoring was performed using a computer equipped with a modem.

Some pictures of the installation of the ICCP with embedded anodes in the Viaduc de Saint-Cloud are given in Fig. 8. During installation, polarity checks and continuity checks were tested.

3.4 Results

This section describes the findings obtained from the monthly monitoring of the segments over a period of 12 months and discusses in detail the findings.

No half-cell potential mapping was performed after the ICCP system installation as the external segment was repaired with a concrete overlay with a high resistance (to repair spalling and delamination or for aesthetic reasons).

With the monitoring system, output voltages and currents (Fig. 9) and instantaneous OFF steel/concrete potential (Fig. 10) were measured. The performance of the cathodic protection was evaluated based on the second criteria of the NF EN ISO standard 12696 [10]: "a potential decay over a maximum of 24 h of at least 100 mV from Instantaneous OFF". Tab.1 presents the monthly calculated values of the potential decay.

From all these results (Fig. 9, 10 and Table. 1), it can be observed that during the first two months, the performance criteria was not achieved for all the reference electrodes (Tab 1, reference electrodes 1, 3, 4, 5 and 6 for segment P2-0; references 5, 6 and 8 for segment P2-1) and therefore adjustments of the voltage (and thus current) were made as can be observed on Fig.9; Fig. 9 also shows that the currents were higher in spring and summer and lower in autumn and winter.

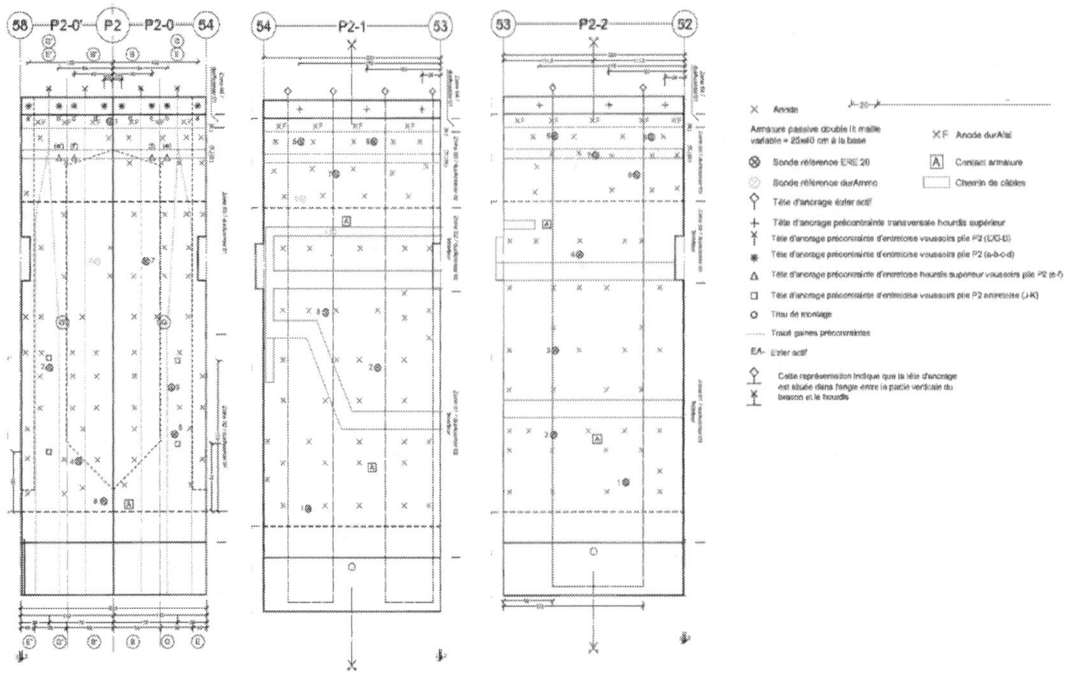

durAcenter 01 durAcenter 02 durAcenter 03

Fig. 7. Zoning arrangements for the three segments P2-0 (zones 2, 3 and 4), P2-1 (zones 1, 2 and 3) and P2-2 (zones 1, 2 and 3). Size of one segment is 2.2m wide (2m for segment over pier), 5.5m high and 0.2m thick.

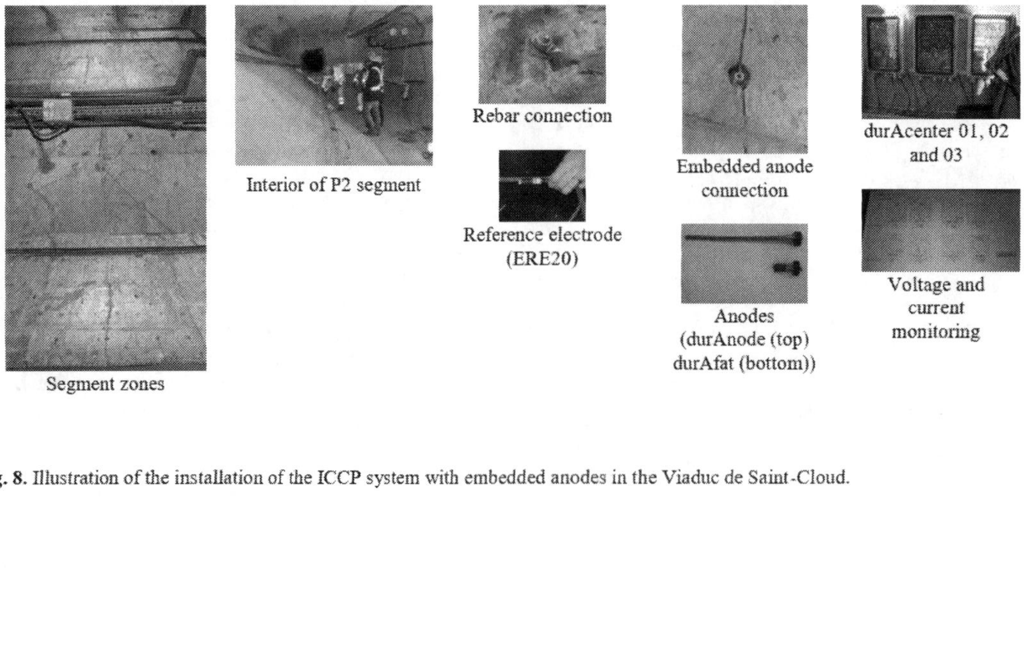

Segment zones; Interior of P2 segment; Rebar connection; Reference electrode (ERE20); Embedded anode connection; Anodes (durAnode (top) durAfat (bottom)); durAcenter 01, 02 and 03; Voltage and current monitoring

Fig. 8. Illustration of the installation of the ICCP system with embedded anodes in the Viaduc de Saint-Cloud.

https://doi.org/10.1051/matecconf/201928903002

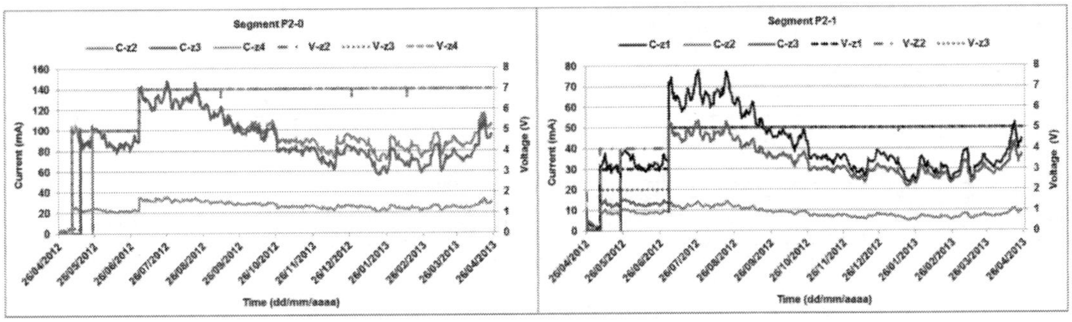

Fig. 9. Current and voltage measurements during the twelve months trial, segment P2-0 (left) and segment P2-1 (right).

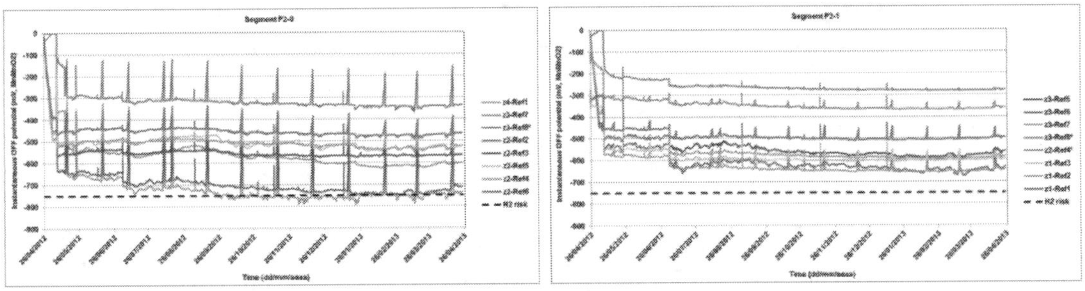

Fig. 10. Instantaneous OFF potential of the eight reference electrodes during the twelve month trial; segment P2-0 (left) and segment P2-1 (right) (the peaks correspond to the monthly verification of depolarisation ISO 12696).

Fig 10 illustrates instantaneous OFF potential of the eight references per segment depending on their locations (segment P2-0: 5, 2 and 1 references respectively in zones 2, 3 and 4; segment P2-1: 3, 1 and 3 references respectively in zones 1, 2 and 3).

Zone 4 of Fig 10 left includes the prestressed tendons of the three segments. These latter results also allowed us to verify that there was no risk of hydrogen embrittlement: the green curve relative to the reference located close to the prestressed tendons (zone 4) was not close to the hydrogen risk represented by the dotted line.

After the two months energizing period, the performance was 100% achieved for the three zones of segment P2-2 and segment P2-1 (except for reference electrode 8 but which was not a manganese reference) (Tab. 1). The conclusion was different for the segment over pier P2 (durAcenter 01 in Tab.1). In zone 4 (Ref 1) and zone 2 (Ref 5 only) the performance criteria failed. This result was explained by the presence of the whole prestressed tendons from the three segments in zone 4 and by the presence of concentrated longitudinal and transversal prestressed tendons in zone 2. In order to improve the cathodic protection where larger density of steel was present due to presstressed tendons, a larger anodic surface area was proposed (embedded anodes + ribbon or mesh).

Table 2 summarizes the averaged results obtained after the energizing period for the three segments. The design calculation was made based on an assumption of 15mA/m^2 of steel surface. The operating currents were in the range 2.1 to 8.8 mA/m^2 of steel surface. The current per anode per zone was in the range 1.2 to 3.2 mA and was in good agreement with the purchased anodes (maximum current of 4.4 mA or 3.0 mA respectively for durAnode and durAfat (CPI Denmark)). A life expectancy of more than 25 years was expected.

Table 1. Monthly potential decay values during the twelve months trial for segments P2-0, P2-1 and P2-2 (green colour indicates a pass value whereas orange or red colours a failed value).

Year	2012								2013			
Month	5	6	7	8	9	10	11	12	1	2	3	4
durAcenter 01	Segment P2-0 (zone 4 including the prestressed tendons of the three segments)											
z4-Ref1	33	27	34	39	37	35	33	29	28	28	27	28
z3-Ref7	344	352	399	381	380	384	362	352	356	312	308	323
z3-Ref8*	170	174	192	189	187	174	172	171	174	177	167	186
z2-Ref2	204	224	259	277	277	280	304	265	287	269	302	285
z2-Ref3	81	86	102	106	118	105	106	102	107	102	100	87
z2-Ref5	25	24	-6	31	37	27	26	12	16	16	16	21
z2-Ref4	72	78	88	102	112	109	112	83	96	65	75	85
z2-Ref6	82	85	91	103	111	107	106	95	97	89	83	86
durAcenter 02	Segment P2-1											
z3-Ref5	53	44	114	141	146	138	147	133	139	135	135	133
z3-Ref6	95	66	194	221	209	200	202	181	189	191	203	213
z3-Ref7	120	113	194	218	213	190	195	175	181	173	172	186
z3-Ref8*	60	58	89	110	102	81	81	69	70	64	64	76
z2-Ref4*	139	137	154	176	175	152	152	134	138	127	128	138
z1-Ref3	199	206	251	259	258	244	239	228	225	213	215	227
z1-Ref2	265	269	318	323	321	308	301	296	289	290	288	306
z1-Ref1	162	167	211	214	218	211	208	198	202	193	195	199
durAcenter 03	Segment P2-2											
z3-Ref8	187	214	270	289	302	286	290	282	283	278	282	280
z3-Ref7	121	127	151	164	168	163	165	164	165	163	164	158
z3-Ref6	82	97	121	144	152	144	153	144	152	147	148	139
z3-Ref5	78	78	102	116	119	118	124	123	131	133	127	124
z2-Ref4	288	292	318	330	334	313	315	296	302	293	295	292
z1-Ref3	310	314	375	375	376	360	353	343	342	335	334	334
z1-Ref2	337	346	419	414	416	406	400	395	388	383	382	389
z1-Ref1	122	126	157	159	162	153	149	144	145	146	142	145
*=durAmmo												

Table 2. Average results for the 10 months period after energizing.

	Segment	Protected zone as designed						Reference electrodes			Average values for results obtained from July 2012 to April 2013				Performance criteria
	Name	Zone number	Location (height of segment)	Concrete surface (m²)	Steel density (m² steel /m² concrete)	Anode type	Number of anode/zone	Ref n°	Number of reference electrode per zone	Depth (cm)	Voltage (V)	Total current/zone (mA)	Current density per zone (mA/m² steel)	Current per anode/zone (mA)	% month (pass)
durAcenter 01	P2-0/P2-1/P2-2	4	prestressed tendons	1.82	1.75	durAfat	15	1	1	10	7	28	8.8	1.9	0
	P2-0	3	medium	5.14	3.29	durAnode	32	7 8*	2	20 13	7	93	5.5	2.9	100 100
	P2-0	2	bottom	5.14	2.41	durAnode	31	2 3 5 4 6	5	10 10 10 10 10	7	100	8.1	3.2	100 90 0 40-50 40-60
	-	1	-	-	-	-	-	-	-	-	-	-	-	-	-
durAcenter 02	*durAcenter 01	4	prestressed tendons	1.82	-	durAfat	(4)	-	-	-	-	-	-	-	-
	P2-1	3	top	1.70	2.27	durAnode	14	5 6 7 8*	4	10 10 10 8	5	35	9.1	2.5	100 100 100 20
	P2-1	2	mednan	1.98	1.51	durAnode	5	4*	1	20	5	9	3.0	1.8	100
	P2-1	1	bottom	6.60	1.00	durAnode	31	3 2 1	3	10 20 10	5	44	6.7	1.4	100 100 100
durAcenter 03	*durAcenter 01	04	prestressed tendons	-	-	durAfat	(5)	-	-	-	-	-	-	-	-
	P2-2	03	top	1.70	2.23	durAnode	14	8 7 6 5	4	10 10 10 10	6	30	7.9	2.1	100 100 100 100
	P2-2	02	mednan	1.98	1.47	durAnode	5	4	1	20	5	6	2.1	1.2	100
	P2-2	01	bottom	6.60	0.97	durAnode	29	3 2 1	3	20 10 20	5	39	6.1	1.3	100 100 100

* DurAmmo

4 CONCLUSIONS

Cathodic protection systems are capable of providing a significantly larger extension in service life compared with other corrosion mitigation systems.

In France, to address the chloride induced corrosion of the Viaduc de Saint-Cloud where access and traffic management are challenging, Impressed Current Cathodic Protection with embedded anodes has been tested. In this ICCP system a small current is passed from a permanent anode to the reinforcing steel. It involves the permanent installation of a low voltage, controlled electrical system which passes a direct current to the steel so that the steel is made into a cathode, thus preventing the steel from corroding. The ICCP system was mainly installed from the interior of the bridge, thus allowing significant cost savings without interrupting the traffic which was another advantage. This electrochemical treatment provided protection to the steel that was monitored monthly and controlled over a period of twelve months.

Based on the results of this field trial, the following conclusions were drawn:

- The designed ICCP system was achieving a satisfactory protection for segments P2-1 and P2-2.
- For the segment over pier P2, the ICCP was partly unsuccessful as designed and needed some improvements mainly in the area involving the prestressed tendons,
- For prestressed tendons, the risk of hydrogen embrittlement was not observed. In the areas whith large reinforcement density (rebars and prestressed) the field trial has demonstrated the need to increase the value of the protective current. This could be achieved with the application of a "surface" anode (titanium mesh).

This research provided some technical information to establish recommendations for full-scale application of ICCP with embedded anodes in the Viaduc de Saint-Cloud which counts 518 segments, 16 segments over piers and 16 special segments. It also helped us to acquire technical skills to implement this electrochemical technique and provided financial information for the total cost of repair of the bridge. Finally, it is important to point out that the cost of the field trial only represented 2% of the full-scale repair cost and was therefore worth doing.

It should also be emphasized that people involved in this kind of electrochemical rehabilitation treatment need to acquire sufficient skills to understand how the system works. CP repair therefore needs trained personnel in house or formally certified persons [11].

For future research, the full scale bridge repair should be studied in order to estimate the long term durability of such a treatment and its maintenance.

5 ACKNOWLEDGEMENTS

The authors are grateful to the following persons involved in the field trial: P. Peyrac, A. Jacquel, K. Gremillon, T. Sauvageon et J. Desille from DiRIF, T. Kretz, B. Godart et L. Gaillet from Ifsttar, F. Landrin, M. Landrin et JL Saussol from Cerema, M. Grill from In-Situ SA, L. Serra, P. Ballerand et L. Fontaine from Spie Batignolles, S. Lorang et S. Butaud from Iraco, and J. Crouzillac from BAC. The authors are also grateful to the persons of the technical committee of Viaduc de Saint-Cloud for technical and financial discussion and particularly to E. Humbert who lead this committee.

This field trial was funded by the DiRiF (Direction inter-départementale des Routes d'île de France).

References

1. https://structurae.info/ouvrages/viaduc-de-saint-cloud (in French)
2. AFPC-AFREM, Durabilité des bétons : Mode opératoire recommandé pour l'extraction et le dosage des chlorures libres et totaux dans le béton, LMDC Toulouse, (1997)
3. B. Elsener, RILEM TC 154-EMC:Electrochemical Techniques for Measuring Metallic Corrosion - Recommendations - Half-cell potential measurements - Potential mapping on reinforced concrete structures. Materials and Structures, **36**: p. 461-471, (2003)
4. GP. Tilly, J. Jacobs, CONREPNET - Concrete repairs - Performance in service and current practice. United Kingdom, IHS BRE Press, ISBN 978-1-86081-974-2, (2007)
5. L. Etcheverry, D. W. Fowler, H. G. Wheat and D. P. Whitney, Evaluation of cathodic protection systems for marine bridge substructures, Texas Department of Transportation, Research report 7-2945, (1998)
6. L. Bertolini, F. Bolzoni, A. Cigada, T. Pastore, P. Pedeferri. Cathodic protection of new and old reinforced concrete structures, Corrosion Science, **35**(5–8):1633-1639, (1993)
7. L. Lazzari, P. Pedeferri, Cathodic protection, Polipress ed. Milano, ISBN 88-7398-020-1, (2006)
8. AA. Sohanghpurwala, Cathodic protection for life extension of existing reinforced concrete bridge elements, National Cooperative Highway Research Program Synthesis 398, editor. Washington DC, Transportation Registration Board, ISBN 978-0-309-09839-7, (2009)
9. Wilson K, Jawed M, Ngala V. The selection and use of cathodic protection systems for the repair of reinforced concrete structures. Construction and Building Materials, **39**:19-25, (2013).
10. NF EN ISO 12696 standard, Cathodic protection of steel in concrete, AFNOR, (2017)
11. NF EN ISO 15257 standard, Cathodic protection — Competence levels of cathodic protection persons — Basis for certification scheme, (2017)

New developments for corrosion protection of concrete structures in Australia

Atef Cheaitani[1], *and Samir Cheytani*[2]

[1]Remedial Technology Pty Ltd, Managing Director, Sydney, Australia
[2]Remedial Technology Pty Ltd, Corrosion Engineer, Sydney, Australia

Abstract. The corrosion protection, repair and maintenance of reinforced concrete structures located along the coast of Australia has been a challenging task for engineers and asset owners over the past thirty years. Most of these structures are bridges and wharves which are situated in marine environments and are subject to tidal exposure. These environmental conditions create challenges in planning repair work, especially if electrochemical systems such as cathodic protection are installed for corrosion protection of these assets. While impressed current cathodic protection (ICCP) is a proven technology which can provide long-term corrosion prevention solutions for marine structures, this technology has been viewed by many asset owners as overly complex and expensive. There is the perception that ICCP systems require costly permanent monitoring programs and have high maintenance costs. These perceptions have been supported by the frequent failure of power supply units, and in some cases, widespread defects associated with ICCP systems such as grout acidification in concrete elements situated in tidal zones. The perception about the complexity of ICCP technology has led to the selection of less effective galvanic-based systems for the protection of assets, and systems which require no monitoring and have low maintenance requirements. This paper will present information on the new developments which have emerged from long-term experience and research work related to the simplification in the design, installation and improved monitoring of ICCP systems.

1. Introduction

Impressed current cathodic protection (ICCP) for reinforced concrete structures has been employed on a large number of concrete structures in Australia over the past thirty years, and up until recently, ICCP has been the technology of choice by asset owners for the corrosion protection of chloride contaminated structures.

In spite of the proven long-term benefits of ICCP technology, in recent years some assets owners have been inclined to select galvanic-based anode systems for the corrosion protection of their assets. One of the primary reasons has been the perception that galvanic based anode systems offer the same level of corrosion protection to structures as ICCP systems, while eliminating the requirements for ongoing monitoring and maintenance work.

This paper will present the theory of ICCP and galvanic anode systems, highlight the advantages and disadvantages of each approach and present various information about recent new developments and improvements related to the simplification in the design, installation and monitoring of ICCP systems.

2. Corrosion Protection Systems

2.1 What is cathodic protection?

When steel corrodes in concrete, the electrochemical process is comparable to that of a battery. In a battery, electrons are generated because two dissimilar metals are exposed to an acidic solution (paste or gel in conventional batteries) which corrodes one metal and creates a harmless reaction in the other. This corrosion reaction at the 'anode' generates electrons that are consumed by the 'cathode'.

When steel reinforcement begins to corrode in concrete, a small area becomes the positive pole (anode) and another much larger area becomes the negative pole (cathode). The corrosion current flows out of the steel at the anode (the corroding part), passes through the concrete and to another part of the steel where there is no corrosion occurring (the cathode). This current flow is called the corrosion circuit and the steel dissolved at the anode forms iron oxide.

In a practical battery, the electrical connection between positive and negative poles can be disconnected. The circuit is then broken and the dissolution of metal stops.

* Corresponding author: atef.cheaitani@remedialtechnology.com.au

In concrete, the corrosion circuit is buried in the structure and the electrical current running through the concrete cannot easily be disconnected. Cathodic protection technology is based on stopping the current from running through the concrete by providing a new current from an external source via an external anode in contact with the concrete. The flow of electrons between the new anode and the reinforcing steel changes the previously positive poles (anodes) into current receivers. Thus, all of the reinforcement becomes the negative pole or cathode, and hence the name 'cathodic protection'. For concrete structures, there are two types of cathodic protection systems; impressed current cathodic protection (ICCP) and galvanic anode systems.

2.1.1 ICCP systems

Impressed Current Cathodic Protection (ICCP) technology is a well-established technology for the corrosion protection of concrete structures. All aspects related to the design, installation, monitoring and protection criteria for ICCP systems are documented in global standards such as the NACE Standard SP 2290-2007 (1), the International Standard ISO 12696:2012 (2) and the Australian Standard AS 2832.5-2008 (3).

With large numbers of ICCP systems installed in Australia over the past thirty years, there have been some issues associated with ICCP system design, installation and durability. These issues are mostly related to; complexity and poor reliability of power supply units and control systems; poor design, resulting in durability issues associated with CP system components such as junction boxes; inadequacy of grout properly encapsulating the anodes and shortfalls in establishing simple maintenance and monitoring programs for these systems.

These issues are often related to incompetent design, a lack of understanding by CP designers of the criteria for selecting control and monitoring systems, construction quality issues due to poor workmanship, and a lack of priority for the long-term durability requirements of system components especially when constructed in harsh and aggressive marine environments (4). In spite of these issues, it is generally accepted that a well designed and monitored ICCP system can deliver optimum corrosion protection to a concrete structure and meet the protection criteria in accordance with the applicable standards.

2.1.2 Galvanic systems

Galvanic anode systems are currently an area of substantial growth. This technology is becoming increasingly attractive because of its simplicity and low monitoring and maintenance requirements. The anode which is normally made from a metal such as zinc, is connected to the reinforcing steel and the potential difference between the zinc and the steel causes a small protection current to flow from the zinc to the steel.

Galvanic anodes in concrete are usually supplied with a proprietary backfill which provides space for the products of anodic dissolution. Most of the recent innovation and research in galvanic anode technology has been associated with the backfill material.

Hybrid anode systems differ from the purely galvanic systems. The hybrid treatment (5) consists of applying a temporary impressed current followed by permanent galvanic protection. The principle of this system is that during the initial impressed current phase, active pits are realkalised and this arrests active corrosion and returns the reinforcing steel to a passive state. Following the application of impressed current for a predetermined period of time, the passivity of the steel is then maintained by the sacrificial anode system.

For many years, galvanic anode systems have been installed in conjunction with concrete patch repairs to reduce the occurrence of the incipient anode effect. Normally in these applications, no permanent monitoring system is incorporated and there is no reliable data about the level of the long-term effectiveness or benefit of these anodes. The installation of galvanic anodes has been viewed as an additional low-cost measure to prolong the life of a patch repair.

Recently, galvanic based anode systems including hybrid anode systems have been installed for the corrosion protection of various structures. Based on some initial review of performance data of these systems, it appears that in structures with low corrosion activity and low chloride contamination, these systems offer some level of protection for a period of time after system commissioning.

Galvanic anode system after 8 months of operation				
Ref	Natural potential	IO (mV)	System OFF 24 h (mV)	24 h Decay (mV)
1	-248	-381	-326	55
2	-263	-383	-327	56
3	-274	-484	-404	80
4	-274	-478	-443	35
5	-242	-370	-336	34
6	-177	-245	-198	47

Fig. 1. Performance data from a galvanic anode system installed on a bridge structure after 8 months of system operation.

For galvanic systems, the moisture in concrete significantly affects the current and leads to seasonal changes reflecting wetter and drier conditions. Current output is lower during dry conditions. However, for structures with a high level of chloride contamination, high corrosion activity, and in some cases relatively high resistivity, the level of corrosion protection is limited and the protection criteria in accordance to the applicable standards cannot be achieved in most cases.

3 Areas of improvement for ICCP systems

The key areas of improvement for ICCP systems include the development of simple system designs, robust/heavy duty components, and systems which are simple to monitor and maintain over their design life. In order to achieve these objectives, based on extensive experience with the installation and monitoring of a large number of ICCP systems, there are various key areas which have been identified related to design, product selection, installation and monitoring and maintenance.

ICCP system after 8 months of operation					
Zone	Ref	Natural Potential (mV)	Instant OFF Potential (mV)	24 hr OFF (mV)	24 hr Decay (mV)
1	11	-355	-420	-293	127
	12	-283	-465	-234	231
	13	-262	-502	-224	278
	14	-324	-554	-274	280
	15	-263	-492	-224	268
	16	-273	-558	-241	317
2	21	-327	-549	-302	247
	22	-261	-509	-229	280
	23	-284	-327	-231	96
	24	-268	-505	-237	268
	25	-276	-532	-235	297
	26	-261	-365	-227	138

Fig. 2. Performance data from an ICCP system installed on a bridge structure after 8 months of system operation.

3.1 Design

There are two main aspects which may impact on the simplification of the design and cost reduction of an ICCP system. These areas are system zoning and design current density.

3.1.1 System zoning

In theory, designing ICCP systems with smaller zones (circuits) provides a higher level of control of the cathodic protection current in various parts of the structure. For large installations, the use of smaller circuits also requires more cabling, more reference electrodes and a larger number of power supply units for the system.

Recently, the authors of this paper were involved in the refurbishment of a relatively large ICCP system installed on Wharves 4 & 5 at the Port of Brisbane (6). As part of the refurbishment work, 13 control units incorporating 172 separate power supply units were moved from underneath the wharf to locations above the wharf. The original system of this ICCP installation included 172 circuits. The new control system included 48 circuits. The substantial reduction in the number of zones (circuits) was achieved by combining smaller circuits with the same exposure conditions and CP current requirements.

Following system re-commissioning, a comparison of CP performance data before and after the reduction in circuits indicates that larger cathodic protection zones offer the same level of corrosion protection and adequate current distribution to the embedded steel. The reduction in the number of circuits had a major impact on reducing the quantity of cables, conduits, power supply units and consequently the cost of ICCP system installation. Providing the embedded reference electrodes are placed as recommended by the applicable standards, larger circuits can be used for ICCP systems without compromising corrosion protection and current distribution.

Combined circuits 2016					
Zone	Section	Ref	IO	24h OFF	24h Decay
			2016	2016	2016
7	4	71A	-160	-26	135
	4	71C	-153	-39	115
	4	72A	-346	-197	149
	4	72C	-245	-83	162
	4	73A	-228	-60	168
	4	73C	-126	-7	119
	4	74A	-218	-71	146
	4	74C	-217	-53	164
	9	71A	-294	-126	168
	9	71C	-284	-164	120
	9	72A	-288	-110	178
	9	72C	-299	-155	144
	9	73A	-212	-91	121
	9	73C	-251	-85	165
	9	74A	-498	-257	241
	9	74C	-241	-137	104

Fig. 3. Table showing system performance using 4 combined circuits (71, 72, 73 and 74) after system refurbishment.

3.1.2 Current density

The design current density for existing reinforced concrete structures is 20 mA/m^2 of steel surface. Based on data retrieved from a large number of operating CP systems, in some cases the actual operating current density was significantly lower than the actual design current density.

The required current density to achieve protection and meet the protection criteria as defined in the applicable standards is related to chloride content, concrete resistivity, the level of corrosion activity, exposure conditions and steel density. The only possible method to determine the required current density is to perform a current injection test at one or several areas of the structure using a temporary power supply unit. The results from the current injection test can be obtained immediately by measuring the potential shift of the embedded steel after a short period of time from impressing CP current.

4 Separate circuits 1999					
Zone	Section	Ref	IO	24h OFF	24h Decay
			1999	1999	1999
71	4	71A	-257	-80	177
71	4	71C	-266	-40	226
72	4	72A	-265	-93	172
72	4	72C	-255	-83	172
73	4	73A	-295	-82	213
73	4	73C	-261	-62	199
74	4	74A	-253	-90	163
74	4	74C	-280	-102	178
71	9	71A	-328	-144	184
71	9	71C	-275	-110	165
72	9	72A	-230	-91	139
72	9	72C	-272	-128	144
73	9	73A	-308	-110	198
73	9	73C	-328	-76	252
74	9	74A	-289	-117	172
74	9	74C	-280	-100	180

Fig. 4. Table showing system performance using 4 separate circuits based on the original design.

Zone	Design Current @20mA/m² of steel surface area mA	Initial Jun 2016 mA	Feb 2018 mA	Jun 2018 mA
1	504	250	200	150
2	504	300	240	220
Current reduction				
Z1 mA	20	9.92	7.93	5.95
Z2 mA	20	11.9	9.52	8.73

Fig. 5. Table showing current reduction after 2 years of system operation.

The outcome from this test may in some cases reduce the design current density from the default 20 mA/m² of steel surface to a lower figure resulting in a major reduction of anode requirement, reduction of power supply unit capacity and consequently major cost reductions for installing the ICCP system. Below is a typical example of a current injection test showing data of the potential shift after a few minutes following current injection. This test was performed to select the optimum spacing for the anodes.

Pier 30, Pile 6 MRB					
L	Natural Potential	IO	Potential Shift	A	T
	mV	mV	mV		
1	-364	-525	-161	●	●
2	-362	-484	-122	●	●
3	-332	-601	-269	●	●
4	-345	-634	-289	●	●
5	-344	-620	-276	●	●
6	-365	-650	-285	●	●
7	-425	-522	-97		●
1	-329	-460	-131	●	●
2	-268	-458	-190	●	●
3	-309	-541	-232	●	●
4	-390	-638	-248	●	●
5	-390	-648	-258	●	●
6	-422	-598	-176	●	●
7	-443	-591	-148		●
1	-287	-394	-107	●	●
2	-281	-460	-179	●	●
3	-312	-511	-199	●	●
4	-356	-551	-195	●	●
5	-350	-527	-177	●	●
6	-380	-538	-158	●	●
7	-429	-514	-85		●

Fig. 6. Table showing current injection test results used to determine anode spacing.

3.2 Anode installation

In Australia, large sections of the CP systems protecting wharves and bridges are installed in tidal and splash zones and have anode design current densities not exceeding 110 mA/m². However, at most of these CP installations, there is evidence of acidification problems occurring in the tidal and splash zones. While in some cases, current dumping (due to poor system zoning design) may have contributed to higher current densities and acidification problems, it appears that in the majority of cases, water ingress to the anode due to poor anode encapsulation detail has been the major contributing factor causing acidification.

Grout acidification has been mainly associated with ribbon anode installed in tidal and splash zones. There is not enough information regarding grout acidification associated with discrete anodes in concrete. The problem is not as evident for discrete anodes as acidification may occur within the discrete anode hole.

There is a direct correlation between grout acidification and poor cathodic protection system zoning. If a CP circuit is located in a combined zone of atmospheric zone and tidal exposures, in many cases acidification has been evident in the tidal zone. A possible cause is current dumping, which is a result of water ingress to the anode level causing localised high current discharge in locations exposed to water.

Recent research at the University of New South Wales (7) confirms that the anode embedment methods currently in use by the industry in Australia, often combined with poor workmanship, allow water ingress to the ribbon anode, and these are the primary causes of acidification. The research also confirms that the application of a cementitious waterproofing coating (which minimises water ingress to the ribbon anode), can limit or stop the occurrence of grout acidification problems in existing ICCP installations.

The pictures below show a cross section of a reinforced concrete element protected by 4 ribbon anode strips and subjected to an accelerated CP application equivalent to 15 years of system operation at 20 mA/m^2 of steel surface. Figure 7 picture 1 shows the impact of water ingress and grout acidification for two anodes and figure 7 picture 2 indicates that proper encapsulation of the anode preventing water ingress eliminates the grout acidification problem caused by external water ingress.

Picture 1

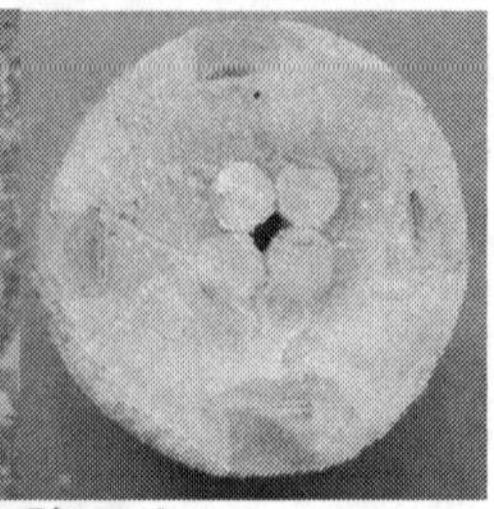

Picture 2

Picture 3

Fig. 7. Picture 1 shows the impact of water ingress and grout acidification for 2 anodes. Picture 2 shows that proper encapsulation of anodes eliminates grout acidification. Picture 3 shows ribbon anode installed for bridge piles with additional protection to eliminate water ingress to anodes.

3.3 Junction boxes

Junction boxes which are installed in areas of water exposure are normally specified with Ingress Protection (IP) suitable for such exposure. It is the author's experience that regardless of the IP protection of any enclosure, such protection is likely to be compromised during the design life of the CP system when the enclosure is installed in an area susceptible to water exposure.

For junction boxes installed in areas of potential water exposure, the following causes, or a combination of these causes can result in failure:

- Movement in the structure causing physical damage to the junction boxes and conduits and thus allowing water ingress.
- Failure of the rubber seal cover of the junction boxes; and
- Failure of the sealant applied around the conduit entries.
- Damage to cables/terminals connections in junction box.

In terms of reducing maintenance requirements associated with junction boxes, eliminating junction boxes altogether in areas where the junction boxes will be susceptible to water ingress over the design life of the CP system, or alternatively, permanently sealing the junction boxes with epoxy material or other suitable products should be considered during the design stage.

3.4 Power supply units

Various types of control systems have been installed in Australia in recent years. These control systems range from manually operated systems, to highly advanced systems with full remote monitoring and control capabilities including remote facilities for depolarisation testing and various levels of alarm functionality.

Generally, it has been the more basic, heavy duty manually operated systems which have been more reliable in comparison to remote control systems with a high level of remote control functionality.The capacity of a CP system to deliver continuous cathodic protection current to a structure is the key and most important requirement of a control system. It is the author's opinion that after one year following system commissioning, minimal and less frequent CP system adjustments are normally required to maintain optimum system operation and compliance with current standards.

The author's experience is that regular functional checks of current delivery to a CP system, in conjunction with half-yearly/yearly testing and adjustment of the system (including an inspection of the structure), is sufficient for the adequate long-term monitoring and maintenance of the CP system.

Failure to select the optimum control system for a particular structure may cause the following:

- Problems with delivering cathodic protection current to the structure, and thus reducing the capability of the system to provide corrosion protection to the structure;

- Frequent parts replacement, and;
- Excessive costs for monitoring and maintenance.

With the advancement in technology in the past two decades, it has often seemed logical for asset owners to install increasingly sophisticated control systems for the monitoring of their CP systems. However, most of the commercial systems that are currently available (while offering various communication and remote testing functions), lack the required durability, can be complex to operate, and often require a high level of technical support for maintenance and repairs. This has led to an increase in the cost of monitoring and maintenance of CP systems to a level beyond what would be normally acceptable for asset owners over the long-term. The major maintenance costs are normally associated with replacement of the power supply unit, in cases of major water damage to junction boxes or for the repair of major grout acidification problem.

Assessing the need and the benefit of such systems against the complexity of maintaining and operating them must be carried out for each individual structure. Such systems may be suitable for large and complex CP installations located in areas where they can be easily serviced. However, for relatively basic and simple CP installations or CP installations in remote locations, it is likely that such systems will add no value to the efficiency of a long term and cost-effective maintenance program.

The primary and most essential function of a control system is to provide continual delivery of current to the structure to ensure protection at all times. It is the author's opinion that a CP system should have simple functions which can allow the asset owner's maintenance staff to easily carry out all functional checks without the need for any specific software knowledge.

It is the author's experience that an optimum control unit for a CP system should consist of the following components:

- A basic and heavy-duty modular power supply unit allowing replacement of components when required without any special programming.
- An interruption facility providing the means for performing instant OFF measurements of the reference electrodes.
- A data logging facility (typically required for large systems only).
- Reliable web-based remote monitoring or standard SCADA connectivity for functional checks only (current for each circuit). This is required for installations in remote locations where such functions cannot be carried out regularly by the asset owner's maintenance staff.

All system components must be configured to enable the DC power outputs to operate irrespective of any hardware or software failure of the data logging or remote monitoring equipment.

Recent improvements in solar technology have allowed the delivery of solar cathodic protection current to small ~~to~~ and medium size structures and this eliminates the need for 240 VAC mains power and the use of high cost and high maintenance transformer rectifier power supply units. The use of solar power to deliver low voltage CP current combined with the use of basic, heavy duty and robust DC power supply units may offer a new approach for major cost reductions and simplification of operating impressed current CP systems.

4 Conclusions

Impressed current cathodic protection technology for steel in concrete has now reached maturity and can be utilised as a standard and reliable technique for the long-term corrosion protection of structures suffering from chloride induced corrosion. However, engineers and consultants operating in the field of cathodic protection must ensure that the optimum design, materials selection, installation and monitoring activities are carried out properly and professionally.

Various aspects of cathodic protection of steel in concrete structures are detailed in international and Australian standards and are available to assist owners, consulting engineers and contractors to correctly design, install, test, commission, monitor and maintain impressed current cathodic protection systems.

The proper design of CP systems and the selection of heavy duty, reliable components and control systems combined with the establishment of an efficient maintenance and monitoring program for the structure will ensure that the maintenance of the CP system becomes a routine aspect of the overall maintenance of the asset and that permanent and cost effective corrosion protection is achieved for the structure.

The failure to design simple and heavy duty ICCP systems for the corrosion protection of reinforced concrete structures will open the door for the preference of less effective galvanic anode systems which cannot achieve the applicable protection criteria and are more costly and less effective for the long-term corrosion protection of infrastructure assets.

References

1. NACE International, Impressed Current Cathodic Protection of Reinforcing Steel in Atmospherically Exposed Concrete Structures. SP0290-2007 (formerly RP0290)] (2007)
2. The European International Standard, Cathodic Protection of Steel in Concrete. ISO 12696:2016] (2016)
3. Standards Australia, AS 2832.5 – 2008, Cathodic Protection of Metals, Part 5: Steel in Concrete Structures (2008)
4. A. Cheaitani, Review of cathodic protection systems for concrete structures in Australia, lessons learnt and future directions. *Australian Corrosion Association Conference, Adelaide, Australia* (2015)

5. I. Godson, M. Polos, H. Mansour. Cold Tea Bridge (B01366) – A case study on the remediation of a bridge structure using hybrid CP system. *8th Australian small bridges conference, Gold Coast, Australia,* (2017)
6. N. Honeyfield, A. Cheaitani. Review of the cathodic protection system for wharves 4 & 5 at Port of Brisbane after 15 years of operation. *Australasian Corrosion Association Auckland, New Zealand (2016)*
7. M. Cheytani, Grout Acidification of ribbon anode in impressed current cathodic protection systems in concrete, MPhil *Thesis UNSW, Australia* (2017)

MATEC Web of Conferences **289**, 03004 (2019) https://doi.org/10.1051/matecconf/201928903004
Concrete Solutions 2019

Zinc layer anode applied on reinforced concrete structures

Xavier Hallopeau[1], *Eugène Moucadeau*[1], *Roberto Giorgini*[2], *Olivier Lesieutre*[1], *Arnaud Meillier*[1], *Camille Annede-Villeau*[1],

[1]Freyssinet International & Cie, Technical Department, 92506 Rueil Malmaison Cedex, France
[2]Corrpre, P.O.Box 387, 2740 AJ Waddinxveen, Netherlands

Abstract. The application of zinc layer anode to reinforced concrete structure has been well known for many years. This galvanic cathodic protection system was specified being a corrosion control system for general steel in concrete as well as civil structures. A good way to determine the dimensioning elements of the galvanic protection system, is to carry out a pilot test being monitored in situ, and exposed under on site conditions including local concrete remediation or repair works. This paper presents an analysis of the current distribution and steel polarization data that have been collected from a car park project in the French Alps.

1 Introduction

1.1 Purpose

The structure investigated was a car park in a ski resort in the Alps (France). This car park, built in 1979, covers 9 half-levels with a capacity of around 460 cars.

The structure is made entirely of reinforced concrete with a classical framework: columns, beams and slabs. The slabs are made with prefabricated slabs. The beams were pre-fabricated and keyed at the head of the columns.

1.2 Project

The corrosion survey, visual inspection with corrosion potential mapping and chloride content analysis following sample coring, revealed that the soffit of the slabs and the beams showed localized zones of corrosion depending on the parking levels.

Rebar corrosion can be triggered by a poor surface applied waterproofing coating, followed by chloride ingress due to de-icing salts, water build-up and cracks as shown in Figure 1, due to significant mechanical stress of concrete elements.

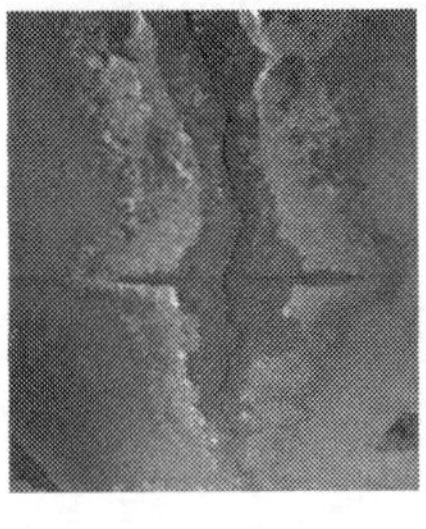

Fig. 1. Crack on soffit of slab

The car park rehabilitation project involves:
- Remediation of the surface applied sealing system preventing moisture and chlorides penetrating the concrete slab,
- Additional reinforcement avoiding concrete beam deflection,
- Installation of a galvanic cathodic protection system for corrosion control of steel in concrete.

2 Cathodic protection system

2.1 Test area

Cathodic protection zones considered for the pilot tests were as follows:
- Beam area: protection of all the rebars of the 6.5m beam and the first 20cm of reinforcement within the slabs adjacent to the beam, as shown in Figure 2.

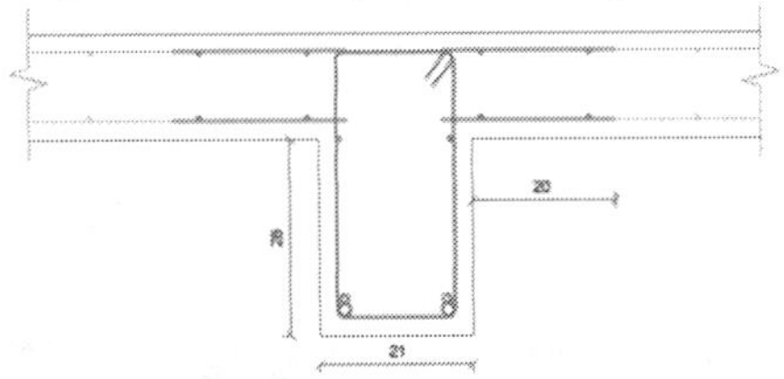

Fig. 2. Beam description

- Protection of the crack-oriented steel within the slab or deck as shown in Figure 3.

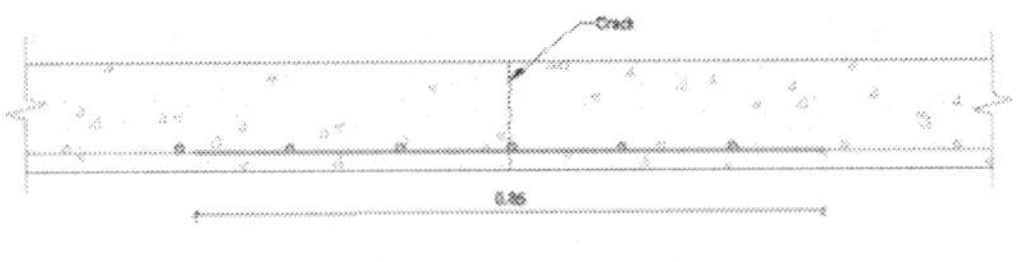

Fig. 3. Slab description

* Corresponding author: xavier.hallopeau@freyssinet.com

 https://doi.org/10.1051/matecconf/201928903004

2.2. Selection of the anodes

The cathodic protection system defined for these pilot tests was a galvanic system (SACP) using 450 µm zinc sheets (3150g zinc/m²) called a Zinc Layer Anode (ZLA) [1, 2].

These anode strips are stuck on the concrete surface thanks to an ion-conductive adhesive.

2.3 Application of the SACP pilot tests

Prior to application, the concrete surface was prepared by grit blasting.

The electric steel continuity shown in Figure 4 (longitudinal reinforcement, stirrup, top bars, bottom bars), was checked for a resistance of less than 1 ohm according to EN ISO 12696 [3]. In some cases, steel bars were welded to the existing reinforcement to restore electrical continuity.

Fig. 4. Rebar electrical continuity test

In each CP zone, 5 m² of concrete area, one ERE-20 electrode (Mn/MnO_2) and five Titanium/MMO (Mixed-Metal-Oxide) potential decay probes were embedded in the concrete to monitor the throwing power and performance of the CP system at different locations.

Each electrode was placed in a borehole at a distance of 1 to 4cm from a steel bar. Each electrode was connected to copper cable type XLPE/XLPE 2.5mm² blue color for reference electrodes and yellow for potential decay probes.

On the beams, one strip of ZLA 450µm was applied on each vertical side as shown in Figure 5 and 7.

On the crack area, one strip of ZLA 450µm was applied on both sides of the crack shown in Figure 6 and 8.

Fig. 5. ZLA on beam

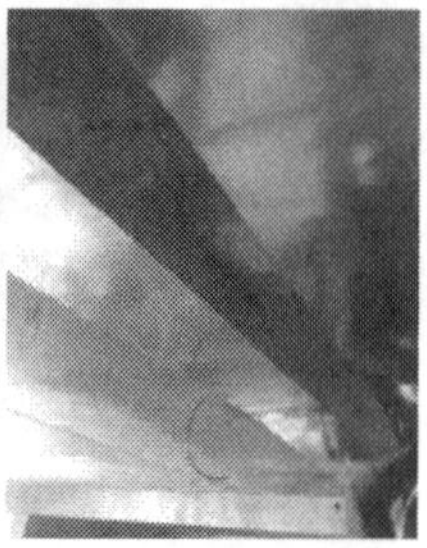

Fig. 6. ZLA on slab soffit

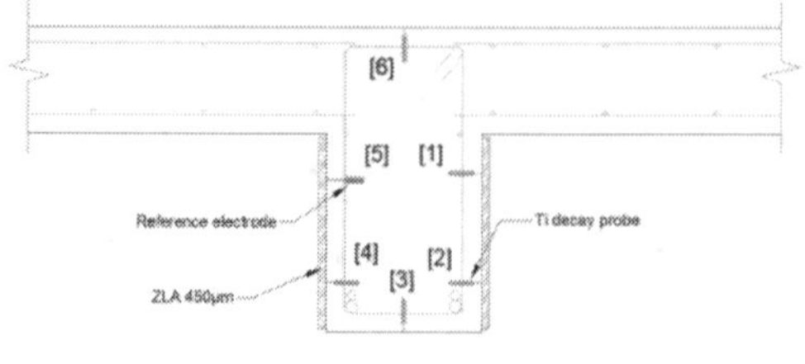

Fig. 7. Anode and monitoring description on beam

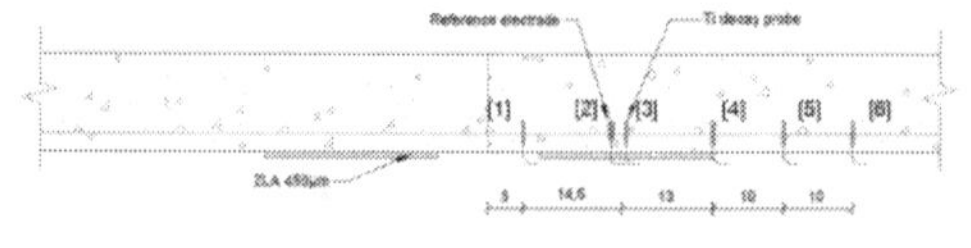

Fig. 8. Anode and monitoring description on slab soffit

The electrical continuity of the zinc layer at the overlaps was provided by plastic nails. The continuity was checked. For the connection of the anodes to the data-logger box, positive connections were made on site and were installed on the zinc sheet surface by using plastic nails. Each positive connection was connected to red copper cable type XLPE/XLPE 2.5mm².

All pilot test components were connected to copper XLPE/XLPE 2.5mm² cables, which are also connected to a voltage logger in the monitoring cabinet, shown in Figure 9.

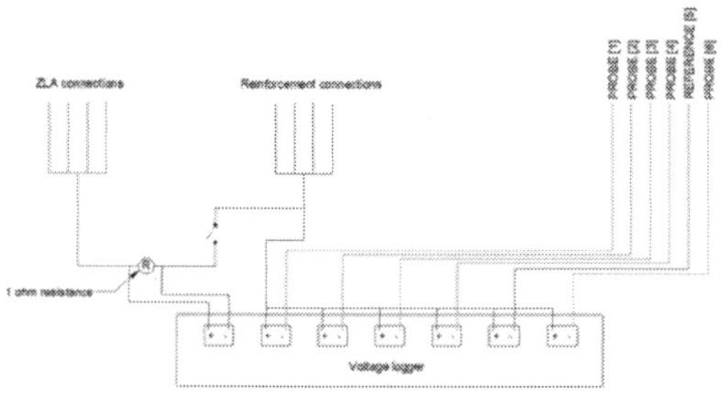

Fig. 9. Electrical scheme

Each test area was equipped with a data logger operating during the whole test period. The cables: anodic, cathodic and reference electrodes and potential decay probes, were all connected to this box. Each box was equipped with:

- Individual zinc strip to steel connections.
- Programmable data logger capturing and logging potentials and currents
- Temperature recorder.

2.3 Data recording

The data being measured included:

- Natural potential
- Polarization time
- Current per zone
- Potential ON
- Potential OFF at 0.5 to 1 second

 https://doi.org/10.1051/matecconf/201928903004

- Depolarization
- Temperature

In addition to the above-mentioned data being collected, potential mapping of the concrete surface adjacent to the zinc strips was performed using a Ag/AgCl, KCl 0.5M reference electrode.

3 Results obtained on the beam area

3.1 Polarization effect on beam

The following Table 1 shows the potential decay results after 21 days of the electrodes placed in the beam and the slab adjacent to the beam.

Measures	E_{native} 12/06/2018	E_{ON} 12/06/2018	E_{ON} 02/07/2018	$E_{Instant-OFF}$ 02/07/2018	$E_{OFF+24h}$ 03/07/2018	24h Decay
[1] - Dprobe (mV)	-370	-427	-554	-529	-388	141
[2] - Dprobe (mV)	-387	-467	-603	-566	-414	153
[3] - Dprobe (mV)	-175	-186	-418	-400	-241	159
[4] - Dprobe (mV)	-221	-235	-456	-406	-211	195
[5] - Ref (mV)	-626	-688	-889	-848	-670	178
[6] - Dprobe (mV)	-265	-280	-397	-390	-247	143
Current zone (mA)	0	269	136	0	0	

Table 1. Potential shift, current and depolarization decay at 24 hours for one zone (beam)

The results show a significant effect of cathodic protection of the beam reinforcement:
- the instant-off to 24h potential decay values,
- and the initial IR-drop when switching off the CP system.

The potential decay measured with the electrodes and probes makes it possible to validate the performance of the EN ISO 12696 standard, criteria b, i.e. a depolarization decay greater than 100mV over up to 24 hours after the system shutdown. It indicated sufficient corrosion control by the galvanic CP system.

These results validate the layout of the surface applied anode on the test beam and confirm the ability of the CP system to give sufficient corrosion control of the steel bars in the top of the beam (probe no. 6).

3.2 Effect of beam polarization on closest slab soffit

Potential mapping was performed on the surface of the soffit adjacent to the beam to evaluate the performance of the system shown in Figure 10.
The area mapping details were:
- Support: soffit slab adjacent to the beam
- Zone of mapping: 1.00m x 0.40m
- Grid of the mapping according to the direction of the beam: Axis-Y: 5 cm, Axis-X: 10 cm., 86 measurements
- Electrode used: $Cu/CuSO_4$

Fig. 10. Potential mapping area

Table 2 shows the mapped potentials of the CP-system being on after 21 days polarization and Table 3 shows the off potential values of the soffit surface area after 24 hours depolarization.

Table 2. Polarization effect (ON)

X \ Y (Column 9-K)	5cm	10cm	15cm	20cm	25cm	30cm	35cm	40cm
100cm	-422	-418	-426	-533	-553	-546	-552	-546
90cm	-383	-410	-474	-511	-527	-525	-539	-535
80cm	-344	-337	-346	-388	-374	-399	-508	-520
70cm	-346	-324	-321	-324	-339	-346	-374	-400
60cm	-365	-336	-330	-326	-330	-335	-337	-327
50cm	-379	-334	-335	-321	-302	-291		
40cm	-389	-358	-338	-315	-307	-294	-286	-284
30cm	-460	-403	-344	-306	-298	-294	-292	-282
20cm	-504	-436	-365	-326	-303	-302	-301	-287
10cm	-508	-478	-399	-344	-320	-301	-293	-272
0cm	-506	-584	-402	-327	-292	-283	-270	-255

Table 3. Depolarization

X \ Y (Column 9-K)	5cm	10cm	15cm	20cm	25cm	30cm	35cm	40cm
100cm	-430	-420	-437	-544	-552	-547	-550	-552
90cm	-466	-421	-470	-515	-530	-527	-544	-542
80cm	-370	-349	-360	-400	-473	-500	-512	-527
70cm	-364	-333	-323	-329	-350	-366	-376	-401
60cm	-387	-337	-335	-327	-324	-328	-337	-329
50cm	-392	-353	-324	-330	-322	-322		
40cm	-413	-365	-333	-318	-312	-300	-306	-284
30cm	-495	-387	-355	-342	-330	-332	-316	-310
20cm	-542	-450	-359	-336	-321	-326	-320	-296
10cm	-547	-510	-412	-351	-325	-317	-300	-274
0cm	-338	-318	-418	-336	-304	-292	-276	-263

The potential mapping carried out confirms that the arrangement of the ZLA strips on the beam did not affect or protect the steel of the soffit surface.
Reinforcement in this zone would therefore be considered to be out of the influence of the cathodic protection system and therefore were not protected against corrosion.

These results do not allow validation of the ZLA disposition on the test beam, for the protection of the lateral parts adjacent to the slab intrados beams.

3.3 Polarization results after adding ZLA on slab

In order to obtain polarization effects for the slab soffit close to the beam, just a half layer (12,5 cm) of ZLA 450 µm was applied on the slab soffit as shown in Figure 11.

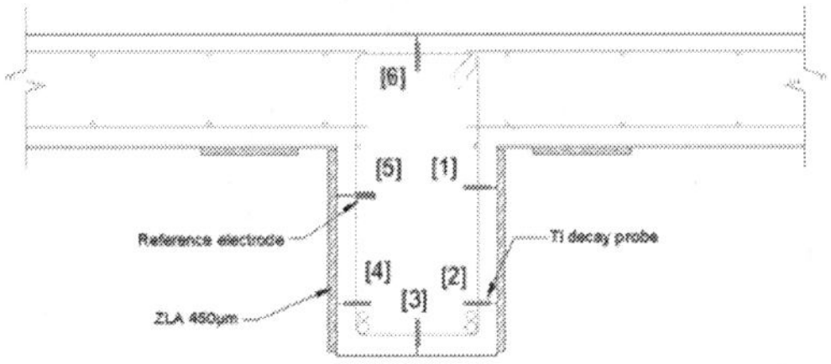

Fig. 11. ZLA layer on slab soffit

Table 4 shows the relative polarization effects based on the new anode configuration measured by the electrodes while the current has increased.

Table 4. Polarization results

Measures	E_{ON} 30/08/2018 2 ZLA strips	E_{ON} 30/08/2018 2+2x1/2 ZLA strips
[1] - Dprobe (mV)	-497	-504
[2] - Dprobe (mV)	-552	-551
[3] - Dprobe (mV)	-579	-575
[4] - Dprobe (mV)	-526	-531
[5] - Ref (mV)	-819	-828
[6] - Dprobe (mV)	-192	-210
Current zone (mA)	84	141

According to previous measurement on permanent reference electrode installed into the beam, surface potential measurements were carried out. Table 5 and 6 show the results of the On potentials mapped before and after the new configuration being installed

Table 5. Potential before installation

x \ y	5cm	10cm (half strip ZLA 450µm)	15cm	20cm	25cm	30cm
90cm	-457		-495	-522	-531	-542
80cm	-455		-397	-454	-481	-501
70cm	-472		-354	-333	-348	-350
60cm	-484		-343	-300	-293	-296
50cm	-532		-342	-271	-277	-271
40cm	-529		-328	-285	-274	-271
30cm	-536		-293	-268	-264	-264
20cm	-505		-290	-276	-260	-266
10cm	-493		-327	-281	-268	-262
0cm	-507		-305	-267	-248	-243

Column 9-K

Table 6. Potential after installation

x \ y	5cm	10cm (half strip ZLA 450µm)	15cm	20cm	25cm	30cm
90cm	-637		-556	-547	-551	-563
80cm	-610		-460	-553	-515	-518
70cm	-620		-405	-485	-375	-376
60cm	-650		-400	-361	-317	-315
50cm	-850		-380	-322	-305	-288
40cm	-804		-356	-312	-292	-282
30cm	-760		-337	-308	-281	-280
20cm	-764		-328	-292	-278	-280
10cm	-657		-320	-290	-281	-265
0cm	-720		-342	-300	-257	-251

Column 9-K

The potentials mapped confirm the positive effect of the additional ZLA strip being installed on the soffit, potentials adjacent to the new anode shifted more than 40 mV.

https://doi.org/10.1051/matecconf/201928903004

4 Results obtained for the cracked area

4.1 Polarization effect on slab

The following Table 7 shows the results after 15 days polarization of the cracked soffit area.

Table 7. Potential after connection, soffit

Measures	E_{native}	E_{ON}	E_{ON}	$E_{instant\ OFF}$	$E_{OFF\ (24h)}$	24h Decay
[1] - Dprobe (mV)	-222	-487	-444	-432	-208	223
[2] - Ref (mV)	-467	-771	-740	-686	-444	242
[3] - Dprobe (mV)	-232	-586	-557	-492	-220	272
[4] - Dprobe (mV)	-226	-426	-411	-399	-202	197
[5] - Dprobe (mV)	-210	-388	-372	-365	-198	167
[6] - Dprobe (mV)	-214	-311	-295	-293	-186	107
Current zone (mA)	0	47	36	0	0	

Depolarization readings show a significant effect for the soffit area, even at a 20cm distance from the zinc strip (probe 5 and 6).

4.2 Effect of the ZLA throwing power

Tables 8 and 9 show the mapped potentials of the neighboring zones to the zinc strip.

The potential mapping details of the cracked area shown in Figure 12 are:

- Support: slab soffit adjacent to the ZLA
- Scope of mapping: 1.00m x 0.40m
- Grid of the mapping according to the direction of the ZLA: Axis-Y: 5 cm, Axis-X: 10 cm, 86 measures
- Electrode used: $Cu/CuSO_4$

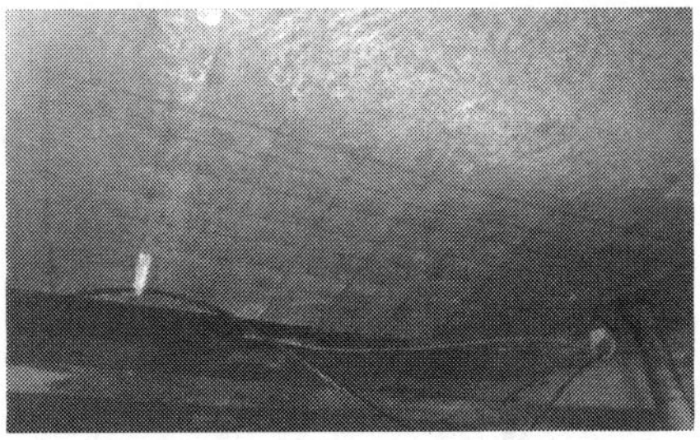

Fig. 12. Potential mapping area

The potentials mapped confirm the throwing power of the ZLA up to a distance of 20cm.

5 Discussion

Polarization readings showed the effective corrosion control of ZLA for the surface areas located on the beam and the adjacent surface area on the soffit.

Considering the current outputs and temperatures over four months' time during winter time, we clearly see a temperature dependence of the current output as shown in Figure 13, we can expect that during summer time this temperature effect on anodic current should be in the same range or greater.

Table 8. Potential mapping before connection

X \ Y (ZLA 450um)	5cm	10cm	15cm	20cm	25cm	30cm	35cm	40cm
100cm	-238	-253	-243	-236	-231	-226	-227	-215
90cm	-248	-231				-253	-239	-235
80cm	-262	-257	-253	-251	-252	-252	-249	-233
70cm	-285	-277	-279	-267	-260	-252	-250	-227
60cm	-292	-282	-283	-278	-277	-266	-259	-244
50cm	-284	-287	-284	-277	-277	-270	-252	-229
40cm	-274	-264	-272	-268	-266	-259	-248	-235
30cm	-262	-273	-266	-271	-265	-258	-243	-223
20cm	-260	-262	-273	-269	-252	-250	-233	-217
10cm	-282	-265	-276	-243	-266	-265	-239	-230
0cm	-265	-267	-255	-259	-253	-241	-231	-225

Table 9. Potential mapping after connection

X \ Y (ZLA 450um)	5cm	10cm	15cm	20cm	25cm	30cm	35cm	40cm
100cm	-477	-439	-400	-374	-356	-340	-335	-307
90cm	-472	-439				-345	-338	-322
80cm	-474	-426	-398	-375	-367	-365	-337	-313
70cm	-460	-416	-400	-386	-363	-342	-330	-312
60cm	-470	-429	-397	-376	-370	-353	-335	-308
50cm	-460	-436	-400	-387	-370	-352	-337	-315
40cm	-466	-420	-409	-384	-368	-352	-337	-305
30cm	-466	-429	-397	-392	-365	-350	-330	-300
20cm	-450	-437	-403	-395	-365	-345	-320	-292
10cm	-521	-429	-414	-373	-353	-339	-311	-300
0cm	-460	-430	-392	-373	-349	-328	-315	-300

The current output for the beam and adjacent soffit area is 40 mA coming from 4.875m^2 ZLA surface area, which gives a current density of 8.2 mA/m^2 ZLA.

Considering the cracked soffit area, the readings gave a current output of 8 mA based on 1.95 m^2 ZLA, which is 4.1 mA/m^2 ZLA.

This difference is due to the coupling ratio, in fact the steel rebar surface area is larger for the beam, 5.54 m^2 than for the soffit surface area which is 2.3 m^2.

By calculating the zinc anode consumption rates with Faraday's law, one can estimate the service life of the system.

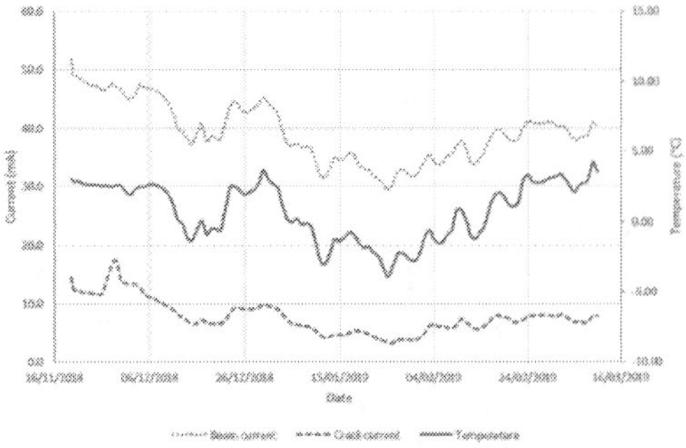

Fig. 13. Anodic current (mA) versus temperature (°C)

The following table 10 shows the service life for both configurations according to an electrochemical capacity for the Zinc of 11.2 kg/A.year, efficiency of 90%, utilisation factor of 80% and a Zinc weight of 3.15 kg/m², assuming constant current output.

Table 10. Life time expectation

Configuration	Anodic current density mA/m²	Life time (years)
Beam	8,2	24,7
Soffit slab crack	4,1	49,4

The steel current density obtained after this trial phase was found to be in good agreement with polarization levels, respectively 7.22 mA/m² for the steel of the beam area and 3.47 mA/m² for the steel of the fractured slab area.

This current density matches well with the values given in the EN ISO 12696 for cathodic protection.

Over time, we obtained the confirmation that the system is working well, as the polarization readings indicate values exceed the 100 mV criteria in the vicinity of the zinc layer anode.

5 Conclusion

Before implementing a cathodic protection on a job site, it is relevant to perform some tests on a 1:1 scale to take into account all the existing parameters.

Under these conditions, the ZLA solution has been confirmed to show test data suggesting an estimated service life of over 20 years.

References

1. R. J. Kessler, R. G. Powers, I. R. Lasa, Paper N° 234, *NACE Corrosion Conference* (1997)
2. A.J. Van Den Hondel, J. Gulikers, R. Giorgini, A. Van Den Hondel, *ICCRRR 2018*, **199**, 05004 (2018)
3. ISO EN 12696:2016, Cathodic Protection of Steel in Concrete

Residual protection on a wharf structure following application of impressed current cathodic protection

Shamir Bhuiyan, *David* Law* and *Peter* Nicholls

School of Engineering, RMIT University, Melbourne, VIC, 3001, Australia

Abstract. Recent studies have suggested that residual protection is afforded to structures following the application of impressed current cathodic protection (ICCP), in some cases for significant periods of time, while others for periods of only a few days. This study reported the findings of the de-activation of a 20 year old ICCP system installed on a 50 year old structure. The ICCP was de-activated for 84 days and the steel potentials at locations on the front pile cap and front wall were monitored via the installed reference electrodes. An adjacent water anode system was also de-activated for 48 hours during the initial de-activation period to observe the impact on the steel potentials. The results showed that out of 42 reference electrodes, 17 achieved a 100 mv decay within 24 hours and 10 had more positive instant off potentials than -150 mV. Furthermore, all displayed a positive shift in potential following deactivation of the ICCP system for a period of time, indicative of residual protection, with 22 displaying this positive shift for the whole 94 days of the trial.

1 Introduction

Reinforced concrete is one of the most widely employed construction materials, being utilized for structures including bridges, dams, high rise blocks, power plants, docks etc. The reinforcing steel is normally passive in concrete due to the high pH environment provided by the Portlandite, sodium and potassium hydroxides and C-S-H gel formed by the hydration of the cement [1]. This leads to the formation of iron oxides and hydroxides on the steel surface which prevent the oxidation of the reinforcing steel, and in doing so create a passive oxide layer on the steel which prevents corrosion [2]. However, when chlorides enter the concrete and reach a sufficient concentration at the rebar, known as the critical chloride threshold or C_{crit}, they can disrupt this passive layer and initiate corrosion of the steel [3, 4]. Once corrosion has been initiated the steel will start to corrode which will eventually lead to cracking and spalling of the concrete and significant loss of section in the reinforcing bars, until the structural integrity of the structure is compromised. Chloride induced corrosion of reinforced concrete structures is one of the major causes of deterioration worldwide, with repair and maintenance costs of billions of dollars required to maintain the operation of these critical parts of the infrastructure.

One of the most effective methods of providing protection for reinforced concrete structures is Impressed Current Cathodic Protection (ICCP). Other methodologies can be employed such as; patch repair, realkalisation, sacrificial cathodic protection, hybrid cathodic protection and chloride extraction [5]. However, ICCP remains the primary methodology for corrosion protection, particularly in aggressive marine environments. ICCP protects the steel by applying a direct current to polarise the steel to more negative values, re-establishing passivity in the steel. This is achieved by a kinetic reduction in the over potential of actively corroding steel resulting in a decrease in the corrosion current density and consequently of the dissolution of the steel [6-8]. In addition there are a number of secondary mechanisms which also contribute to the long term protection of the steel [9]:

- The production of hydroxyl ions
- The breakdown of oxygen gas
- The repulsion of aggressive anions

Recent studies have reported that residual protection has been observed on structures where ICCP systems have been de-activated for periods of time, or decommissioned [10, 11]. A study of ten structures in the United Kingdom showed that the steel remained passive for three years after the ICCP was decommissioned [12, 13]. Similar findings have also been reported on two marine structures in Australia [14]. However, residual protection has not been observed in all reinforced concrete structures, with corrosion having been re-initiated within as little as three days [14]. The reasons for these variations in performance have been attributed to a number of factors relating to the operation of the ICCP system or the secondary effects. These include charge passed, charge density, duration of ICCP operation, system design, local environments including tidal cycles, moisture content, chloride concentrations and the chemistry of the electrolyte. Hence, significant

* Corresponding author: David.Law@rmit.edu.au

 https://doi.org/10.1051/matecconf/201928903005

questions remain concerning the effects, causes and duration of residual protection from ICCP systems. This is of particular importance with many systems reaching the end of their design lifetimes. To further investigate these effects, this paper reports a study on a reinforced concrete wharf constructed in 1969 in southern Australia with an ICCP system that was installed in 1999.

2 Methodology

The study was conducted on Berth 6 at the Port of Portland in southern Victoria, Australia. Berth 6 was constructed in 1969 as part of the expansion of the Port, with the adjacent Berth 5 constructed in 1986. An ICCP system was installed on both Berth 6 and Berth 5 in 1999, due to deterioration caused by chloride induced corrosion of the reinforcing steel. As such the ICCP system has been in operation for almost 20 years.

The ICCP system consists of slotted anodes in the transverse pile headstock beams, internal anodes within the front curtain wall beam and soil anodes to protect the rear retaining wall. The slotted anodes are mixed metal oxide coated titanium ribbon located within narrow horizontal slots. The ribbon was embedded using a cementitious grout. The internal anodes were mixed metal oxide ceramic tubular anodes installed into horizontally drilled holes within the curtain wall beam. The anodes had a venting system to enable gases generated by the anodic reaction to diffuse into the centre of the porous anodes and to then vent to the atmosphere via a network of nylon tubes. The design was to allow the application of relatively high current densities to the anode in order to prevent deterioration of the backfill grout. The soil anodes were cast high silicon chromium iron anodes installed within vertically drilled holes behind the rear wall and backfilled with metallurgical coke breeze.

The slotted and internal anodes are powered by a transformer rectifier (TR) unit located in the electrical room towards the east end of the berth. The system is divided into eight operating zones with eight separate AC/DC controllers within the primary TR unit. The soil anodes are powered by zone 1 within the primary TR and a second TR unit to provide a maximum output of 20 Amps. A total of 52 Ag/AgCl permeant reference electrodes were embedded at representative locations throughout the reinforced concrete elements. Three in zone 1 to monitor the rear wall, 7 in Zone 2 to monitor the rear transverse beam and 7 in each of zones 3 to 8. These were distributed across the six zones with 7 each in the tidal zone, the splash zone and the atmospheric zone of the front pile cap and the front wall. All cables are wired back to the primary TR unit. As built drawings of the system and the operating zones are shown in Figures 1 & 2. In addition, a water anode system protects the steel piles within Berth 5. This is expected to provide some current to the lower elements in Berth 6 (i.e. those located at or below mid-tide) of the ICCP system at high tide.

With the permission of the Port Authority, the ICCP system was decommissioned for a period of 84 days. This time period was not prolonged any further as it was deemed sufficient for this study whilst minimising the impact on the structure. The water anode system was de-activated 2 hours following the de-activation of the ICCP system for a 48-hour period before re-activation, due to operational requirements. Once the ICCP system was disconnected, the steel potentials were monitored using a data-logger at 5 second intervals over the initial 5 minutes to record the instant-off data. Subsequently data points were recorded at 5 minute intervals in the initial 48-hour period when the water anodes were de-activated and then at 1-hour time intervals for the remainder of the de-activation period. Monitoring of the system commenced on the 30th of November 2018 and the ICCP was re-activated on the 22nd of February 2019.

The reference electrodes in zone 1, monitoring the rear wall and in zone 2 monitoring the transverse rear beam were not operational. As such, data was logged for the 42 reference electrodes in zones 3-6 for the front wall and front pile.

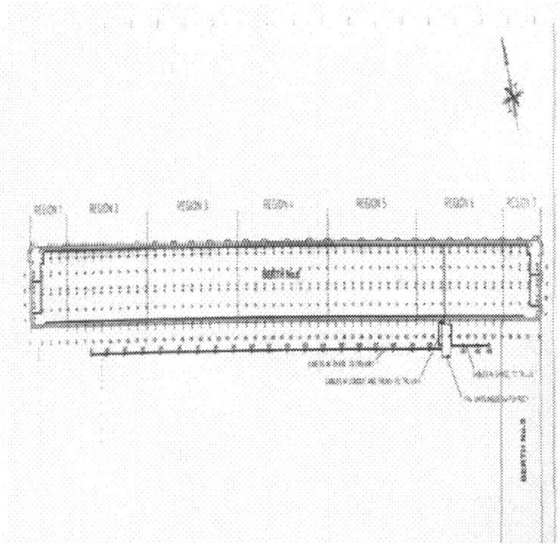

Fig. 1. As built drawing: Zonal Diagram Berth 6

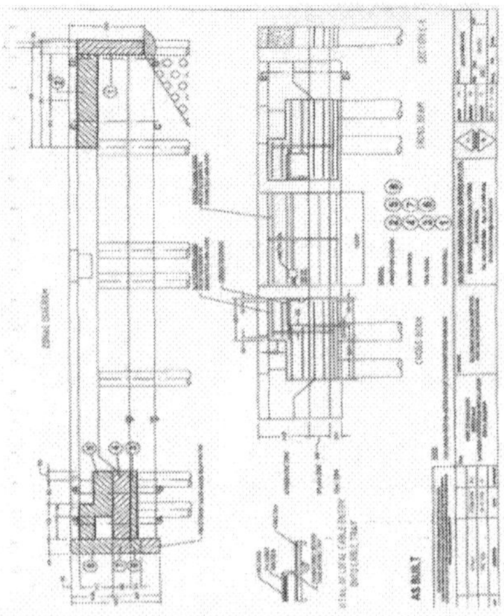

Fig. 2. As built drawing: Operating Zones Berth 6

3 RESULTS

The system operates in accordance with AS 2832.5 [15], which considers the steel being protected from corrosion when one of the following criteria are met:

1. A potential decay over a maximum of 24 hours of at least 100 mV from the instantaneous off potential. This is known as the potential decay criterion.
2. A potential decay over a maximum of 72 hours of at least 100 mV from the instantaneous off potential subject to a continuing decay and the use of reference electrodes (not potential decay sensors or pseudo reference electrodes) for the measurement extended beyond 24 hours. This is known as the extended potential decay criterion.
3. An instantaneous off potential more negative than -720 mV with respect to a silver/silver chloride reference electrode. This is known as the absolute potential criterion.
4. A fully depolarised potential, or a potential which is continuing to depolarise over 72 hours after the ICCP system has been switched off which is consistently less negative than -150 mV with respect to silver / silver chloride reference electrode. This is known as the absolute passive criterion.

It should be noted that a number of different standards used worldwide, in particular ISO 12696:2016 [16], upon which other standards (including AS 2832.5) are based, acknowledge that the above criteria are empirical in nature and allows for the use of corrosion risk assessment in order to assess the effectiveness of an ICCP system, should none of the criteria be met.

The on potential, instant off and 24-hour decay data are given in Table 1, together with the final potential before re-activation and the shift in this potential from the instant off potential. Decay curves from the tidal (zone 3), splash (zone 4) and atmospheric (zone 5) reference electrodes for front pile caps are given in Figure 3 and for the tidal (zone 6), splash (zone 7) and atmospheric (zone 8) reference electrodes for the front wall are given in Figure 4. These decay curves are typical for all reference electrodes in the front pile caps and the front wall across all zones.

The data for reference 25 showed a positive potential of 325 mV which varied from +2000 to -3000 mV over the course of the 84 days. It was concluded that the data indicated that the reference electrode was not operating correctly, and this reference electrode was discounted from the analysis.

The decay curves clearly differentiate between the tidal, splash and atmospheric zones. The tidal zone potential trends are generally the most negative and the atmospheric zone always the most positive, with those in the splash zone being between the two other zones in most cases, though occasionally being more negative than the tidal zone. The tidal cycle is clearly evident on the reference electrodes located in the tidal and splash zones and is also evident on some of the reference electrodes in the atmospheric zone, Figure 5.

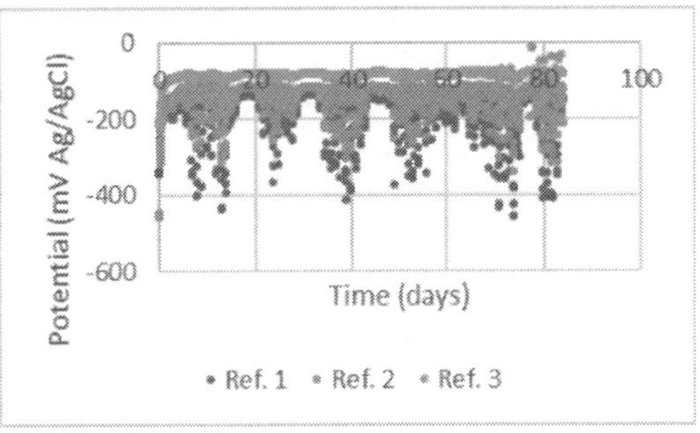

Fig. 3. Decay curves for reference electrode 1 (tidal, zone 3), reference electrode 2, (splash, zone 4) and reference electrode 3 (atmospheric, zone 5), Front Pile Caps

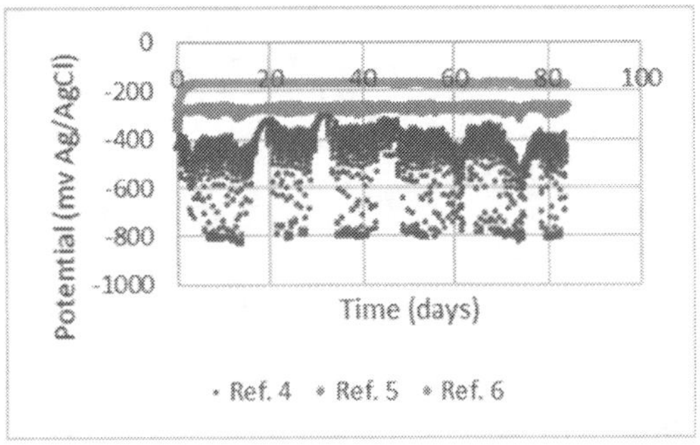

Fig. 4. Decay curves for reference electrode 4 (tidal, zone 6), reference electrode 5, (splash, zone 7) and reference electrode 6 (atmospheric, zone 8), Front Wall

The decrease in potential associated with the tidal cycles is generally of the order of 50 mV, though on occasion this does increase significantly, with potentials more negative than -800 mV being observed in the electrodes in the tidal zone. These falls in potential are consistent with those observed in previous studies and are attributed to the saturation of the concrete resulting in oxygen depletion resulting in the very negative potentials observed rather than the re-establishment of corrosion [17].

In addition to the normal tidal cycles there are a number of periods where the potentials remain relatively constant. E.g. days 17-20 and 30-33, Figure 3 and 4. These periods are associated with the mooring of ships in the berth. During these periods it is hypothesised that the berthed ships provide a protection against the tidal action of the waves and hence reduce the impact of the tides and the saturation of the concrete.

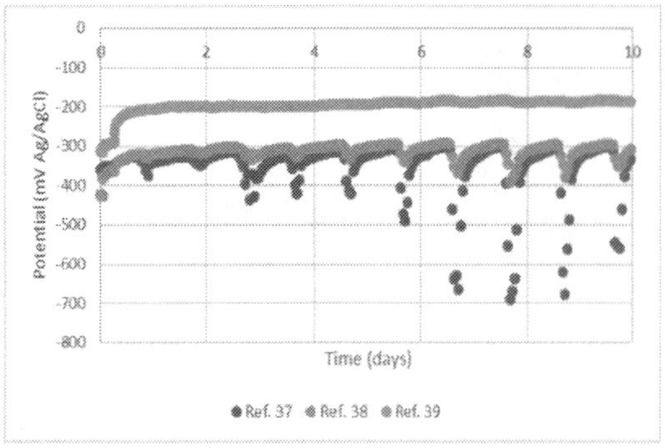

Fig. 5. Decay curves for reference electrode 37 (tidal, zone 3), reference electrode 38, (splash, zone 4) and reference electrode 39 (atmospheric, zone 5), Front Pile Caps, Day 0 to Day 10

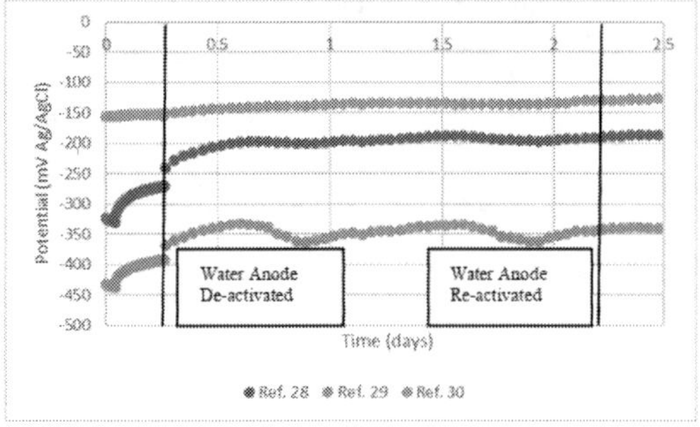

Fig. 6. Decay curves for reference electrode 28 (tidal, zone 6), reference electrode 29, (splash, zone 7) and reference electrode 30 (atmospheric, zone 8), Front Wall, Water Anode Re-Activation Period

Table 1. Location and potential data for reference electrodes (Ag/AgCl)

Ref	Zone	Location	On potential	Instant Off Potential	24 Hour Off Potential
1	3	T/FPC	-343	-295	-185
2	4	S/FPC	-456	-319	-176
3	5	A/FPC	-209	-203	-111
4	6	T/FW	-434	-434	-462
5	7	S/FW	-354	-345	-280
6	8	A/FW	-266	-264	-199
7	3	T/FPC	-464	-425	-331
8	4	S/FPC	-704	-431	-271
9	5	A/FPC	-302	-295	-201
10	6	T/FW	-689	-677	-641
11	7	S/FW	-505	-460	-378
12	8	A/FW	-465	-447	-238
13	3	T/FPC	-711	-669	-593
14	4	S/FPC	-681	-563	-383
15	5	A/FPC	-835	-670	-403
16	6	T/FW	-595	-595	-615
17	7	S/FW	-567	-467	-280
18	8	A/FW	-822	-794	-681
19	3	T/FPC	-210	-210	-205
20	4	S/FPC	-480	-476	-422
21	5	A/FPC	-414	-397	-268
22	6	T/FW	-730	-707	-490
23	7	S/FW	-375	-358	-168
24	8	A/FW	-146	-148	-145
25	3	T/FPC	N/A	N/A	N/A
26	4	S/FPC	-774	-691	-521
27	5	A/FPC	-430	-379	-239
28	6	T/FW	-331	-316	-196
29	7	S/FW	-438	-425	-350
30	8	A/FW	-156	-156	-136
31	3	T/FPC	-611	-389	-163
32	4	S/FPC	-514	-399	-293
33	5	A/FPC	-431	-372	-214
34	6	T/FW	-499	-478	-386
35	7	S/FW	-360	-356	-302
36	8	A/FW	-317	-322	-272
37	3	T/FPC	-361	-354	-341
38	4	S/FPC	-427	-381	-320
39	5	A/FPC	-316	-302	-206
40	6	T/FW	-615	-613	-595
41	7	S/FW	-302	-292	-236
42	8	A/FW	-144	-144	-115

Ref	Zone	Location	24 Hour Decay	Final Off Potential	Final Potential Shift
1	3	T/FPC	110	-145	150
2	4	S/FPC	143	-121	198
3	5	A/FPC	92	-79	124
4	6	T/FW	-28	-430	4
5	7	S/FW	65	-262	83
6	8	A/FW	65	-173	91
7	3	T/FPC	94	-221	204
8	4	S/FPC	160	-73	358
9	5	A/FPC	94	-166	129
10	6	T/FW	36	-597	80
11	7	S/FW	82	-340	120
12	8	A/FW	209	-97	350
13	3	T/FPC	76	-415	254
14	4	S/FPC	180	-295	268
15	5	A/FPC	267	-256	414
16	6	T/FW	-20	-539	56
17	7	S/FW	187	-258	209
18	8	A/FW	113	-348	446
19	3	T/FPC	5	-192	18
20	4	S/FPC	54	-279	197
21	5	A/FPC	129	-211	186
22	6	T/FW	217	-221	486
23	7	S/FW	190	-62	296
24	8	A/FW	3	-127	21
25	3	T/FPC	N/A	N/A	N/A
26	4	S/FPC	170	-238	453
27	5	A/FPC	140	-207	172
28	6	T/FW	120	-192	124
29	7	S/FW	75	-323	102
30	8	A/FW	20	-112	44
31	3	T/FPC	226	-342	47
32	4	S/FPC	106	-245	154
33	5	A/FPC	158	-214	158
34	6	T/FW	92	-363	115
35	7	S/FW	54	-143	213
36	8	A/FW	50	-263	59
37	3	T/FPC	13	-302	52
38	4	S/FPC	61	-282	99
39	5	A/FPC	96	-189	113
40	6	T/FW	18	-478	135
41	7	S/FW	56	-205	87
42	8	A/FW	29	-102	42

T- Tidal, S- Splash, A – Atmospheric, FPC - Front Pile Cap, FW – Front Wall

3.1. Discussion

An analysis of the decay potentials after current switch-off reveals that 17 out of the 41 reference electrodes satisfy the decay criterion having a potential decay over a maximum of 24 hours of at least 100 mV from the instantaneous off potential, with a further 5 almost achieving this with a decay of greater than 90 mV. There is no clear pattern in location of the reference electrodes achieving the 100 mV decay criterion. There are two each in the tidal, splash and atmospheric zone for the front wall, and two in the tidal zone, five in the splash zone and four in the atmospheric zone for the front pile caps. These correspond to two in zones 3, 6, 7 and 8, three in zone 5 and four in zone 4. However, considering those with a decay of greater than 90 mV this increases those in the atmospheric zone of the front pile cap to six, which may indicate that protection is being achieved more readily in the front pile caps than in the front wall. Furthermore, three of the five reference electrodes are located in zone 5, bringing the total located in zones 4 and 5 to 10 out of 22. A possible explanation is that this is the central section of Berth 6, which is protected whenever a ship is moored at the berth, irrespective of the ship length. This could translate into a more sheltered environment within these zones which is less aggressive with respect to being susceptible to tidal action and chloride deposition, which translates to less aggressive environment with regards to corrosion of the reinforcing steel.

Based on the 24-hour criterion, two locations actually demonstrate a negative decay, Ref. 4 and Ref. 16, both in the tidal zone of the front wall in zone 6. This indicates that protection is not being achieved in this location. When considering the extended potential decay, 72 hours, there is no change in the number of reference electrodes achieving the protection criteria, with Ref. 4 and Ref. 16 both still displaying negative decays.

Only one achieves a more negative instant off potential than -720 mV, Ref. 18, in the atmospheric zone of the front wall in zone 8. However, 10 reference electrodes achieve the absolute criterion of -150 mV. These include six reference electrodes that did not achieve any of the other three criteria. Five of these are in the atmospheric zone and one in the splash zone. This can be attributed to the local environment. Those electrodes in the atmospheric zone are influenced less by the tidal pattern, as observed in Figs 3-5 and have more positive potentials. Thus, they are more likely to achieve the absolute passive criterion but given their more positive instant off potentials, they are less likely to achieve the 100 mV decay criterion.

The decay curves also clearly show the impact of the de-activation of the water anode, Figure 6. At the point of de-activation there is a distinct positive shift in the reference electrode potentials, generally of the order of 10 – 25 mV. This shift is evident in the tidal and splash zones for most reference electrodes, as would be expected as these are in direct contact with the water. In most instances a smaller increase is observed in the atmospheric zone. Shifts of up to around 50 mV are observed in some cases, e.g. Ref 37, Figure 5. The larger shifts observed are associated with those reference electrodes displaying a more negative instant off potential. It is hypothesised that this is due to the exact location of these electrodes. While the tidal, splash and atmospheric zones are designated in the reports, the actual location is not provided and it may be that these electrodes are in direct contact with the sea water, even though they are nominally in the atmospheric zone. The positive shifts observed are attributed to the cessation of the cathodic protection afforded by the water anodes, in effect being an "instant off" for the water anodes.

Given the point in time at 0.25 days from the de-activation of the ICCP system, this shift is a contributing factor in the 24 hour and 72-hour decay values. If this was subtracted from the 24-hour decay values, a further 4 reference electrodes would not achieve the 100 mv criterion. It is interesting to note that the re-activation of the water anodes after a 48 hour off period has no discernible effect on the decay pattern for any of the electrodes, whatever the location. As such there is no evidence that the water anodes are providing any contribution to the "residual" protection to the steel when they are operating and the ICCP system is de-activated. If the initial positive shift from the de-activation of the water anodes is subtracted from the final potential measured, only an additional 3 electrodes would have values more negative than the instant off potentials.

The shape of the potential decay versus time plot can also be used to provide an indication of the corrosion activity occurring on the reinforcing steel following de-activation of the ICCP system [17]. Re-activation of corrosion in the steel is characterised by a potential shift to more negative values. A strong indicator of activation in this manner is a return to the steel native values (pre ICCP value). This re-activation behaviour can be distinguished from negative excursions caused by the tidal variations as being a sustained negative trend in the steel potentials independent of the tidal action from which the steel potential recovers to its original or more positive values at the end of the tidal cycle. It is interesting to note that 39 out of the 41 operational reference electrodes give a more positive potential value at the conclusion of the 84 day trial than their instant off value, including Ref 4 & 6 which gave a negative shift at 24 and 72 hours. One electrode, Ref. 33, in zone 5 in the atmospheric zone of a front pile cap has an identical value to that recorded after 24 hours after the instant off, and Ref. 31, in zone 3 in the tidal zone is the only reference electrode to display a final potential more negative than the 24 hour decay value, though this is still more positive than the instant off potential by 47 mV. All electrodes moved to a more positive potential following the de-activation of the ICCP system compared to their instant off value at some point in the 84 days. If this decay trend is used as an indication of passivity of the reinforcing steel, then this would indicate that residual protection is being afforded to the steel at all locations at least for a period following de-activation of the ICCP system.

An analysis of the decay over the trial period identifies only two electrodes that display a sharp fall in potential, suggesting re-initiation of corrosion Figure 7 [14]. These are Ref. 19 (71 days) and Ref. 31 (15 days), both of which are located in the tidal zone of the front wall cap in zone 3. However, a further 15 show a slight downward trend in the last 15-20 days of the trial, though this may be an effect of the tidal variations. Thus 22 show no evidence of re-activation at any point during the trial. With all the reference electrodes in the atmospheric zone indicating that protection has been maintained for the full 84 day period.

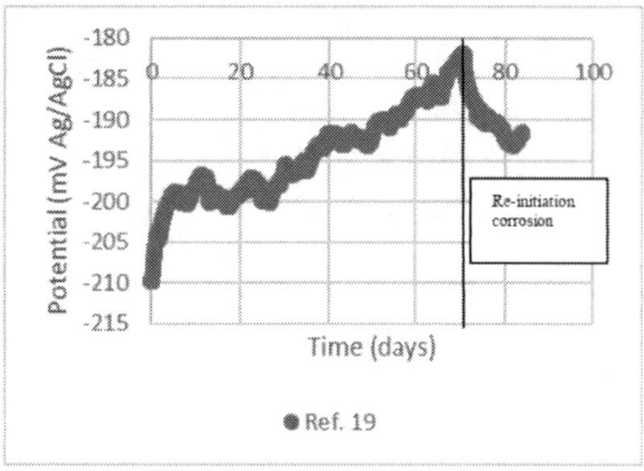

Fig. 7. Decay curves for reference electrode 19 (tidal, zone 3),

In order to further understand the behaviour of the residual protection observed, additional data is being sought with regard to the protection current applied over the lifetime of the ICCP system. At present, data has only been obtained for the period 2007-2009. The anode currents were reported to vary between a maximum of 26.5 Amps and a minimum of 16.0 Amps across all zones both rear and front with a mean current of 20.4 Amps recorded, with a total current of 367 Amps. For the front zones the maximum was 26.5 Amps and the minimum was 18.5 Amps, with a mean current of 23.6 Amps. Additional data covering the full operational life and the steel quantities, to determine the current densities is required to enable a greater knowledge of factors affecting the level of residual protection that can be achieved on structures protected by ICCP systems. The system was designed with a maximum current density of 20 mA/m^2 of steel renforcement surface area. While the steel area is not known for Berth 6 based on other berths at Portland it is estimated that the current density is likely to be between 0.5 and 2.5 mA/m^2 [14] However, the results do show that residual protection can be achieved when ICCP systems are de-activated.

4 CONCLUSIONS

Based upon the potential data from the de-activation of the ICCP system at Berth 6 at Portland the following conclusions were noted;

1. A total of 17 of the 41 reference electrodes achieved a 100 mv decay in 24 hours, satisfying the decay criterion for protection. A further 5 had a potential decay of greater than 90 mV.
2. Only one achieved a more negative instant off potential than -720 mV.
3. A total 10 reference electrodes achieved the absolute criterion of -150 mV. These include six reference electrodes that did not achieve any of the other criteria. This is attributed to their location where lack of moisture resulted in an instant off potential from which it was not possible to achieve a 100 mV decay.
4. The de-activation of the water anode resulted in a positive shift of up to 50 mV depending upon the location of the reference electrode.
5. The re-activation of the water anode had no discernible effect on the electrode potential.
6. A total of 39 out of the 41 operational reference electrodes gave a more positive potential value at the conclusion of the 84-day trial than their instant off value.
7. Based upon the shape of the potential decay curves over the 84-day trial, residual protection was afforded to all the locations for at least some period during the trial.
8. Only two electrodes display a sharp fall in potential, characteristic of the re-initiation of corrosion, while 22 show no evidence of the re-initiation of corrosion over the trial based on the shape of the decay curve.

ACKNOWLEDGEMENTS

The authors wish to acknowledge the assistance of the Port of Portland, Eldridge Electrical and Bruce Ackland in the undertaking of the ICCP trial at the Port of Portland.

REFERENCES

1. Bertolini, L., et al., *Corrosion of Steel in Concrete: Prevention, Diagnosis, Repair, 2nd Edition*. 2013: Wiley & Sons.
2. Chaussadent, T. and R. Dron, *Influence of hydroxyls, carbonates and chloride ions on the pitting corrosion of steel*. Materiales de construcción, 1992. **42**(226): p. 49-62.
3. Glass, G. and N.R. Buenfeld, *Chloride-induced corrosion of steel in concrete,*. Progress in Structural Engineering & Materials, 2000: p. 448-458.
4. Angst, U., et al., *Critical chloride content in reinforced concrete - A review*. Cement and Concrete Research, 2009. **39**(12): p. 1122-1138.
5. Robery, P., M. Gower, and S. El-Belbol, *Comparison between cathodic protection and other electrochemical repair techniques*, in *7th international conference on structural faults and repair concrete and composites,*. 1997. p. 231-237.
6. Glass, G.K. and N.R. Buenfeld, *On the Current-Density Required to Protect Steel in*

Atmospherically Exposed Concrete Structures. Corrosion Science, 1995. **37**(10): p. 1643-1646.

7. Jones, D.A., *Application of Electrode Kinetics to Theory and Practice of Cathodic Protection.* Corrosion Science, 1971. **11**(6): p. 439-&.
8. Page, C.L. and G. Sergi, *Developments in cathodic protection applied to reinforced concrete.* Journal of Materials in Civil Engineering, 2000. **12**(1): p. 8-15.
9. Glass, G.K. and J.R. Chadwick, *An Investigation into the Mechanisms of Protection Afforded by a Cathodic Current and the Implications for Advances in the Field of Cathodic Protection.* Corrosion Science, 1994. **36**(12): p. 2193-2209.
10. Glass, G.K., W.K. Green, and J.R. Chadwick, *Long term performance of cathodic protection systems on reinforced concrete structures*, in *UK Corrosion 91*. 1991: Manchester;UK.
11. Presuel-Moreno, F., A.A. Sagues, and S.C. Kranc, *Steel activation in concrete following interruption of long-term cathodic polarization.* Corrosion, 2005. **61**(5): p. 428-436.
12. Broomfield, J.P. and J.S. Tinnea, *Cathodic protection of reinforced concrete bridge components*, in *Report ID No: SHRPCUWP92618*,. 1992.
13. Christodoulou, C., et al., *Assessing the long term benefits of Impressed Current Cathodic Protection.* Corrosion Science, 2010. **52**(8): p. 2671-2679.
14. Law, D.W., P. Nicholls, and C. Christodoulou, *Residual protection of steel following suspension of Impressed Current Cathodic Protection system on a wharf structure.* Construction and Building Materials, 2019. **210**: p. 48-55.
15. AS2832.5, *AS 2328.5 Cathodic Protection of Metals; Part 5 Steel in Concrete*, in 2008, Australian Standards.
16. ISO:12696:2016, *Cathodic protection of steel in concrete*. 2016.
17. Bhuiyan, M.A., et al., *Investigation of Residual Protection of Steel following Application of Protective Current.* Construction and Building Materials, 2018. **162**: p. 503-511.

Cathodic protection of reinforcement in concrete – experience and development over 30 years

Rob Polder[1,2*] , and *Willy* Peelen[3]

[1]RPCP, Fluwelensingel 106, 2806 CH Gouda, The Netherlands
[2]Delft University of Technology, Civil Engineering and Geosciences, Delft, The Netherlands
[3]TNO Technical Sciences/Structural Reliability, Delft, The Netherlands

Abstract. This paper presents developments over 30 years in the field of cathodic protection of steel reinforcement in concrete in The Netherlands and elsewhere. From the late 1980s major developments have been: application to large numbers of precast elements corroding due to mixed-in chloride with drilled in titanium anodes and conductive coatings; analysis of working life of systems and components and end-of-life considerations; application to prestressed structures; new anode types including galvanic systems with associated life and design considerations; numerical modelling and preventative applications. Presently, CP has become a fully accepted method of securing safety and serviceability of buildings and infrastructure. Major successes and lessons learned will be presented. Technical and non-technical developments are highlighted and some recent innovative CP systems are discussed.

1 Introduction

More and more reinforced concrete structures develop corrosion of the embedded steel. This is due to long-term exposure to aggressive influences such as chlorides from sea water and de-icing salts or to the effects of mixed-in chlorides; in some cases aggravated by carbonation [1, 2]. Older structures are more sensitive to corrosion due to lower cover depths and lower concrete quality than that which is used nowadays. Steel reinforcement corrosion causes cracking and spalling of concrete and steel cross section loss, compromising serviceability and eventually structural safety. Consequently, repair and protection of concrete structures has become a major industry in the past 30 years. However, in many cases conventional methods of concrete repair have been shown to be ineffective or not durable [3, 4]. Cathodic protection (CP) on the other hand, has widely been found to be an effective and durable method for protection of steel in concrete. This paper reports on the history and experience with CP of concrete structures, based on the authors' involvement in The Netherlands and documentation from elsewhere.

CP of concrete structures was developed in the USA in the 1970s [5] and introduced in Europe in the 1980s [6]. In the USA, a large number of concrete bridge decks was damaged by corrosion due to chloride ingress from de-icing salts. Various types of anodes, including cast iron, coke breeze in asphalt and slotted systems with various mortars, were applied with varying results; many of these early anode types have been discontinued [7]. Next, new anode materials became available, such as activated titanium and conductive coatings, variants of which are still widely used today; sacrificial (galvanic) anodes were introduced in the late 1990s. From about 1985 in the UK [6], Italy [8], Norway [9, 10], Denmark, Switzerland [11, 12] and The Netherlands [13] application of CP to concrete increased. Increasing numbers of systems are being installed more recently in Germany, France, Belgium and Switzerland. Long term performance and an overview of interventions (repairs) of large numbers of systems have been analysed [14, 15].

2 Principles

Reinforcing and prestressing steel in concrete are normally passivated due to the high alkalinity in the pore solution, caused by dissolved potassium and sodium hydroxides and buffered by solid calcium hydroxide at pH values above 13. With passivation, corrosion is negligible and very low rates of oxidation and reduction occur. In aerated concrete potentials of about +0 to -100 mV versus a saturated calomel electrode (SCE) are present. Lowering of the pH due to carbonation, a reaction with carbon dioxide from the atmosphere, or the presence of chloride ions above a certain threshold concentration cause depassivation and initiation of corrosion. Subsequently anodic (oxidation) reactions and cathodic (reduction) reactions are strongly accelerated versus the passive state, respectively described by the following equations:

$$Fe \rightarrow Fe^{2+} + 2\,e \qquad (1)$$

$$2\,H_2O + O_2 + 4\,e \rightarrow 4\,OH^- \qquad (2)$$

* Corresponding author: robpolder@robpolder.demon.nl

With chloride induced corrosion localised or pitting attack develops and anodic sites may be small; locally the potential drops several hundreds of millivolts and the pH drops to 2-3 [16, 17]. Cathodic potentials remain in the more positive ranges; strong potential gradients develop between anodic and cathodic regions.

The basic principle of cathodic protection is shifting the steel/concrete interface potential difference to more negative values, slowing down oxidation reactions (eq.1) and accelerating reduction reactions (eq.2). In concrete, shifts of a few hundred millivolts are sufficient to move corrosion rates to negligible levels [8]. This is due to favourable effects of current flow: increased cathodic reactions produce hydroxyl ions and increase the pH at the steel/concrete interface and negatively charged chloride ions migrate away from the steel/concrete interface. The current is injected into the concrete from an electrode installed on the concrete surface or in the cross section, called the anode, causing the interfacial potential to shift. At this anode, oxidation reactions take place that consume hydroxyl ions or that oxidise materials at the anode/concrete interface described by the following equations:

$$4\,OH^- \rightarrow 2\,H_2O + 4\,e \tag{3}$$

$$M \rightarrow M^+ + e \tag{4}$$

Reaction (3) may cause acidification and possibly loss of anode/concrete bond by dissolving the hardened cement paste; reaction (4) oxidises galvanic anode materials or carbon particles in conductive coatings. Possible side effects at the cathode are accumulation of alkali ions due to ionic migration; and possibly hydrogen evolution according to an additional cathodic reaction:

$$2\,H_2O + 2\,e \rightarrow H_2 + 2\,OH^- \tag{5}$$

Hydrogen evolution at the surface of prestressing steel may cause its embrittlement. Reaction (5) can be avoided by limiting the current density at the prestressing (or duct) surface. As this reaction only occurs at very negative potentials, typically -1100 mV versus SCE, monitoring the steel potential can be used as a safety measure. At more positive potentials than about -1000 mV, only reaction (2) is possible and hydrogen evolution does not occur. High alkali levels may provoke alkali-silica reaction in concrete with reactive aggregates. Both of these adverse side effects occur only at relatively high current densities. For more information on principles and effects see [1]. All reinforcing steel and other metals embedded in the concrete have to be electrically continuous and connected to the cathode circuit; otherwise they will not be protected or may even develop accelerated corrosion due to stray current effects.

3 Practical aspects

Two basic types of CP systems have been used: impressed current CP (ICCP) and galvanic (sacrificial) CP (GCP). The main components are illustrated in Figure 1. In ICCP a low voltage DC power is the source of current flow. The anode consists of a material that is not (or only slowly) consumed, such as activated titanium (covered with noble metal oxides, Mixed Metal Oxide, MMO) or a conductive coating.

Activated titanium has the shape of a mesh embedded in a cementitious overlay installed on the concrete surface; or of fine mesh strips in boreholes in the concrete cross section or in slots cut in the concrete surface; holes and slots are filled with a cementitious grout. Conductive coatings are polymers filled with carbon particles, applied to the concrete surface and usually covered with a (normal) top coat. Primary anodes are metal wires that feed the current into the actual anode material; they should be spaced closely enough to avoid large potential drops in the anode material. Cementitious overlays and conductive coatings should be durably bonded to the concrete surface for good electrical contact. Achieving good bond requires good surface preparation and proper application and curing are essential.

In ICCP systems, the power source and the anodes are connected through isolated copper cables and anode-copper connections. Cable cross sections should be large enough for limited potential drop. Cables need to be redundant in order to avoid system failure due to (single) cable interruptions, e.g. by vandalism; at least two cables should feed anodes as well as cathodes, with connections as far apart as possible within a zone. Power sources are fed from the normal grid or by solar or wind energy, the latter two backed up by batteries.

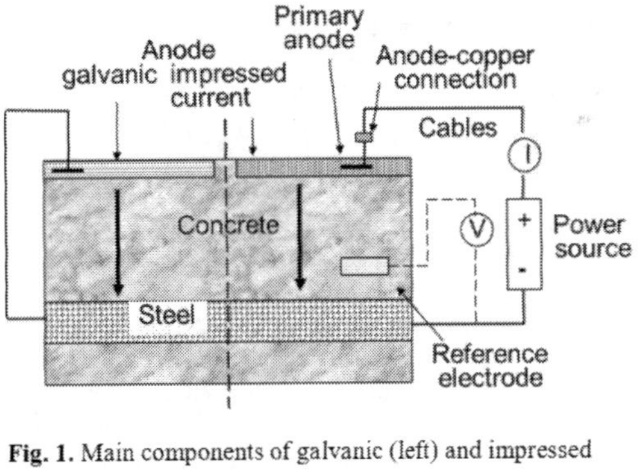

Fig. 1. Main components of galvanic (left) and impressed current (right) CP systems.

In GCP systems the power originates from the electrochemical potential difference between the anode material and the steel. Corrosion of a sacrificial material, normally zinc, provides the current according to

$$Zn \rightarrow Zn^{2+} + 2\,e \tag{6}.$$

A GCP system can be made of zinc sheet with an appropriate activator and a conductive adhesive on the concrete surface; or by embedding zinc probes with activator inside cavities or boreholes in the cross section. Activators are needed to keep the zinc corroding; they are proprietary formulations based on alkalinity, complexing agents or halides. Zinc anodes are connected to the reinforcement using copper wires.

A CP system can be divided into separately controlled parts called zones, if current densities between different parts of a structure are anticipated to be different or for practical reasons. Current densities are affected by differences in concrete electrical resistivity due to different exposure to precipitation or different cement types having been used [13, 18]. Practical reasons can apply to avoid large zones with poor controllability or a low capacity of available power units. Basic design of CP is based on a current density of 20 mA/m^2 of steel surface, as a rule of thumb. It can be debated if this should concern all steel in the entire concrete cross section or only the steel closest to the surface. Experience shows that most CP systems operate at much lower current densities, typically of 1 – 5 mA/m^2 (steel surface). However, the current density per unit of concrete surface area can be quite different. Buildings usually contain less than 1 m^2 of steel surface, typically 0.5 m^2 per m^2 of concrete surface. The steel/concrete density of civil engineering structures can be much higher, in particular in areas with higher stresses, typically 1 or more. In practice, impressed CP current is kept as low as possible by applying a low voltage to limit system degradation and negative side effects, while maintaining sufficient protection. In galvanic systems, the current cannot be externally controlled and the zinc consumption rate is determined by concrete resistivity and steel density, which will determine anode working life. In hybrid systems, in which an impressed current is first applied for some time and then are switched to galvanic mode, anode life is governed by both principles.

Preventive application requires lower (design) current densities, in the range of 0.2 – 2 mA/m^2, because preventing corrosion is easier than stopping it [8, 19]. Compared to curative systems, less anode material is needed and current distribution is more favourable.

Protection quality is tested by interrupting the current for several hours to one day or more and monitoring the steel potential using a reference electrode (RE). For ICCP, the current is switched off at the power source. For GCP, usually a special section is made that allows switching off; in the other parts of such a system the anode is directly connected to the steel at multiple spots. Upon interrupting the current, the steel potential shifts to more positive values. This test is called a depolarisation test. A steel potential depolarisation (shift) of at least 100 mV between instant off (which eliminates ohmic drop) and up to 24 hours is the most widely used criterion for sufficient protection [19]. Multiple RE's should be embedded at representative locations in each zone. Regular monitoring consists of depolarisation testing at least twice a year and a visual inspection once a year. Monitoring usually starts as part of a maintenance contract between the owner and the company that installed the system. However, in a substantial number of cases monitoring has not been carried out or has been neglected since the end of the original contract (see below). Increasing numbers of systems are remotely monitored or controlled using dataloggers and modern wireless connection technology. Nowadays more and more systems are being monitored by independent parties, after termination of the original monitoring contract. This implies that proper documentation of the as built system must be available.

4 Experience

As mentioned, application of CP in concrete started in Europe by 1985, in the UK and Italy. The first CP system in The Netherlands was installed in 1986/7 on a concrete cycle lane of the Stadium Viaduct in Rotterdam [13]. It protected 150 m^2 of concrete surface in two zones; the anodes were carbon filled polymer cables with a copper wire core with a cast cementitious overlay. A few projects involved this type of anode that was later taken off the market. In 1989, activated titanium anodes in the shape of wires were installed in precast concrete elements in an office facade with corrosion due to chloride ingress from sea water mist. 2500 precast cantilever beams with mixed-in chloride, supporting galleries and balconies in two apartment buildings in Tilburg had titanium strips in holes drilled longitudinally in the beams in 1990. Current distribution between bars in the cross section was studied in the laboratory and between beams in the field [20]. Internal cover measurement was used to avoid anodes and steel being too close or even short-circuited; a distance of 15 mm was used as minimum distance. Where steel was located too close to the borehole, a polymer isolation pipe was placed around the titanium strip. Two similar projects with 1000 cantilever beams with drilled in titanium strip anodes were made in Groningen in 1997 and 1998. On the beams and roof slabs of the top floors a conductive coating was used instead of drilled in strips. Carbon based conductive coatings were introduced as anode material and were applied to several buildings with mixed-in chloride. A series of seven buildings with columns, beams and slabs with mixed-in chloride and moderate concrete quality were protected with a conductive coating between 1993 and 2000 [13]. The long term performance of some of these early systems is discussed below.

The durability and repair of precast concrete elements with mixed-in chloride developed into a major problem in The Netherlands. In many cases conventional repair was found not to be effective: corrosion and cracking of concrete reappeared within a few years after such repairs. In contrast, CP was found to be an effective and durable solution.

Based on the notion that uniform current distribution was important and would be mainly governed by the electrical resistivity of the concrete, the resistivity of materials used for repairs associated with CP should be similar to that of the parent concrete. This was adopted by the European Standard on CP of concrete [19]. Exposure to rain and solar radiation, e.g. due to orientation of facades, was also considered to have an effect on resistivity and it was suggested as a basis for dividing systems into multiple zones.

CP companies and knowledge institutes cooperated to set up a Recommendation published in 1996 by CUR (an organisation roughly comparable to the Concrete Society in the UK) [21]. It addressed curative (not preventative) application to reinforced (not prestressed) concrete and was a typical guideline including technical considerations, test procedures and limiting values. Between 1987 and 1995 CP was applied to 16 structures, mainly buildings with mixed-in chloride, either with activated titanium anode strips in boreholes or with conductive coatings on the concrete surface. This was the pioneering period of CP in The Netherlands.

In the following years, roughly 1996-2004, application of CP on bridges and other infrastructure developed. In one project, corrosion had induced widespread delamination in a thick wall of a basement of a bridge. Using titanium mesh and shotcrete was rejected due to the possibility of the delaminations hindering current flow to the reinforcement. Subsequently a system based on drilled-in titanium strip anodes was designed. However, during installation, reinforcing bars appeared to be positioned at irregular and unexpected intervals, and many additional holes had to be drilled to avoid short circuits between anode strips and steel bars. In addition, steel continuity was largely absent, such that all rebars had to be connected. Both issues increased the cost of the system significantly. With hindsight, the preliminary investigation should have looked more carefully for the location of rebars and for continuity.

In another case corrosion had developed in the edge beam of a bridge deck above the abutments, due to joints leaking de-icing salt solution. A previous round of superficial repairs did not stop corrosion. Making deeper repairs would have required lifting the deck and taking the bridge out of traffic. CP was selected as a more user-friendly solution. The complication was that post-tensioning anchors and tendons were located in the edge beam and concern was raised for overprotection, causing hydrogen evolution and possibly hydrogen embrittlement of the prestressing steel. The problem was addressed by consulting Pietro Pedeferri, who explained that monitoring the potential of the prestressing steel ducts and maintaining it above (more positive than) a safe absolute potential value of -900 mV versus SCE would preclude hydrogen evolution [8]. This value includes a significant safety margin compared to potentials where hydrogen evolution would really occur (typically -1100 mV), as mentioned above. Subsequently CP was installed using a conductive coating; monitoring the potentials around prestressing ducts and anchors showed them to be well in the safe region. This system worked for almost 20 years until it was replaced by a new conductive coating system in 2015.

The approach outlined above for CP of prestressed concrete and the same safe absolute potential value were later adopted in the European Standard EN 12696, that was published in 2000 [19]. In addition to supporting CP of prestressed concrete, this Standard also includes preventive application, Cathodic Prevention (CPre), which had been introduced by Pedeferri [8]. The Standard has a performance based approach, rather than being a technical guideline.

Around 2000 galvanic anodes were introduced for CP in concrete in The Netherlands, based on the sacrificial action of zinc. Zinc sheet with an ion-conductive and activating adhesive was applied to various buildings and bridges.

Between 1997 and 2004 a total of 79 structures had CP installed, including buildings with mixed-in or penetrated chloride, bridges and parking structures exposed to de-icing salts and swimming pools. In some cases a single structure contained more than one CP system, e.g. on the facades and on a basement floor.

By 2004, CP had become a well-accepted method for effective long-term corrosion protection of concrete structures. Involved companies, although relatively small and with economic ups and downs, collectively sponsored a research project by TNO. It comprised setting up a database of CP cases in the country and further development of numerical models for CP [22]. The database was updated a few times. At the end of 2010, a total of 150 concrete structures in The Netherlands had CP installed, including 105 buildings (apartment blocks, offices, parking and industrial buildings) and 45 bridges and other infrastructure, totalling 85,000 square meters of protected surface [15]. In addition. 65,000 m^2 of ground floor elements with mixed-in chloride had been provided with CP.

Out of 150 CP systems, anodes were conductive coating in 65 cases, activated titanium in 50 cases and 35 had galvanic anodes. Good documentation was obtained for 105 systems, usually by the company that installed it and carried out maintenance. Out of 105, 50 cases had been operating for ten years or more and 55 cases for less than ten years. The remaining 45 cases were not documented and/or not monitored, which included most of the galvanic systems.

The survey and analysis of the findings showed a mixed picture of a low rate of overall system degradation and some weak details [15]. Weak details included accelerated corrosion of poorly insulated anode-copper connections; and some cases of degradation of primary anodes in conductive coating systems, both particularly in older ones. It appears that primary anodes should be made of truly inert materials, e.g. noble metals like platinum; less resistant materials are prone to degradation. Some cases of power unit or reference electrode failure were noted. Over time, CP companies improved these weak points, as was borne out by later systems. Just a few cases of overall anode degradation were found. However, local degradation of conductive coatings due to water leakage from overlying parts was an important cause of the need for (some cases of) maintenance. Conductive coating anodes on buildings with relatively sheltered facades were found to last as long as 15 to 20 years. This is shown by the seven buildings mentioned above. By 2013, so 13 to 20 years after their installation, these systems did not comply to the 100 mV depolarisation criterion anymore and were in need of maintenance. However, corrosion or concrete damage were completely absent. Another project with drilled in titanium strip anodes and conductive coating on the top floor showed good performance of the titanium system after 17 years; however, the conductive coating part did not work properly anymore. Activated titanium systems performed well over at least 20-25 years in many well-documented cases; working lives of 25 and more years appear very well possible. This experience is in agreement with a recent study on long term performance of CP systems on motorway structures [14], which showed that conductive coating anodes may work for a very long time, also in infrastructural works and despite visible deterioration of the coating.

5 Market and recent developments

In 2015 the database on CP systems was updated and an overview is presented. By ultimo 2014, a total of 230 structures had CP with some multiple systems installed, counting a total of 250 CP systems. Figure 2 presents annual and cumulative numbers. Structures comprised 105 buildings, 97 bridges and 44 other structures. Out of these, 103 had conductive coating anodes, 57 titanium and 58 galvanic anodes (29 had other anodes or two types). Owners were The Ministry of Infrastructure (65 systems), lower governments (41) housing companies (48), private collectives (24) and other owners (69). Increasing numbers of cases involved protecting corroding bridge abutments, preventive applications and systems with ground bed anodes; some are illustrated in more detail below.

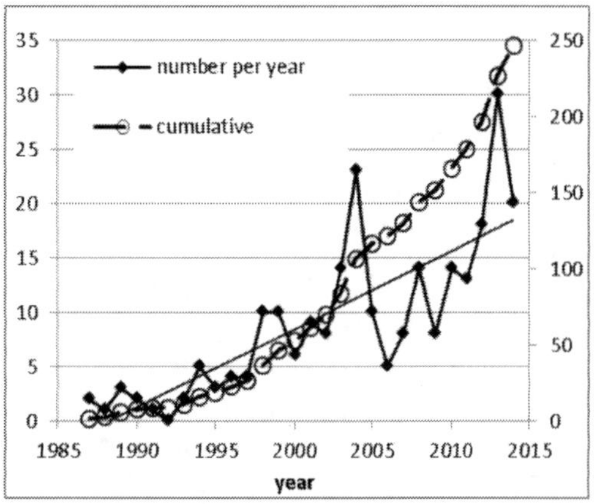

Fig.2. Number of CP systems installed in The Netherlands per year (left Y-axis), cumulative number (right Y-axis) and trend line for number per year.

CP was installed in 30 bridges of similar build up with prefabricated post-tensioned deck girders; for more background and details see [23]. Many beam heads had developed corrosion and spalling due to leakage of overlying expansion joints. Based on existing damage, 1300 beam heads were protected in 2013-2014. CP was applied with a conductive coating (without top coat) on all beam heads with damage plus undamaged adjacent ones, comprising about one-third of the total number. Two RE's were placed in every beam head with CP, one at the outer steel (for depolarisation testing) and one near the post-tensioning anchors (for monitoring the absolute potential). Subsequent operation showed that absolute prestressing anchor potentials were well on the safe side of the -900 mV criterion. A maintenance contract for 20 years was awarded, including 4-year interval visual inspection of the non-protected beam heads. In 2017 many unprotected elements were found to have developed corrosion and damage; apparently they contained chloride levels that had previously not caused corrosion damage, but they had four years later. Subsequently CP will be installed on those elements.

For a series of new bridges in a motorway around the city of Groningen, corrosion in the abutments would be highly undesired with regard to the effects of repair activities on traffic flow. Consequently, cathodic prevention was installed. Titanium strip anodes cast in mortar blocks were placed between rebar cages and concreted when casting the abutments. Similar anode blocks were placed inside the heads of precast deck beams. Because of the preventive principle, monitoring could be kept at a low level (automated checks of current flow, long interval depolarisation testing). A total of six bridges of this type were provided with CPre between 2012 and 2018.

In recent years, more and more systems are based on anodes in the ground surrounding wall-type structures such as locks and tunnel walls. One example concerns a two-tube tunnel with walls built of soil mixed with cement. Concerns for cracking and durability were countered by installing CPre, with protective current flowing from anodes placed in the ground on either side of the tunnel tubes. Numerical modelling based on ground and "concrete" resistivity was applied to calculate anode-to-anode distances while considering anode-to-wall distances allowed by the right-of-way around the tunnel. After installation and commissioning, several anodes were found to have stopped working due to later works in the area, damaging anode cables. This is a concern for systems with ground anodes.

Based on increased collaboration between CP companies, research institutes and consultants, a CP-knowledge centre was founded to promote application of CP. The knowledge centre addresses technical and non-technical issues, a.o. via working groups, and organises seminars for a wider audience. One working group set up a course for teaching technicians at level 3 of the CP personnel competence Standard [24], which has been held successfully since 2016. Another group proposed an approach to service life prediction of galvanic anodes, based on experience with current densities in practice and Faraday's law, including an efficiency factor. The benefit is that all parties involved now have a common basis for designing and offering galvanic systems for a particular working life. The basis is a model for a decreasing current demand of the reinforcement derived from practical systems: 10 mA/m^2 in the first year, 8 mA/m^2 in the second year, 5 mA/m^2 in the third year and from then on 2 mA/m^2. The efficiency factor is 0.72 (28% of the zinc does not contribute to current flow to the steel). It is intended to improve this approach based on data from GCP systems. This looks promising, as more and more long term data are being reported [26].

After 30 years of CP application, the "second life" of CP systems should be considered. In particular, older conductive coating systems may approach or have already reached the end of their working lives. The question then is how sustained corrosion protection should be achieved. Obviously, replacing the old CP system by a new one is a way. However, supposing that corrosion has not already re-activated, a new system could be designed less conservatively, for example with a lower current demand. Some information exists on residual protection after stopping the current flow [25], where positive effects were documented after short term CP operation. This deserves attention from both research and CP engineering communities.

Another long-term issue is the accumulated cost of monitoring CP systems, which may be substantial [27]. If it can be shown that after some time of CP operation re-activation of corrosion takes longer or is improbable, increasing the monitoring intervals of systems that have operated well for several years should be considered. Visual inspection remains necessary, but the interval could be increased if remote monitoring is used to guarantee current flow.

6 Overview and conclusions

In The Netherlands CP of concrete structures has seen many successful applications and considerable growth, from a few to more than 20 systems being installed per year over the past 30 years. Many different types of systems have been developed, installed and maintained and lessons have been learned. Preliminary investigation should be thorough and must, among others, establish reinforcement positioning and steel continuity. Corrosion due to mixed-in chloride can be effectively stopped with CP, which is hardly possible with conventional repair. Post-tensioned structures can be protected without the risk of damaging prestressing steel, provided a proper design is made and absolute prestressing/duct potentials are monitored. Degradation of components has occurred, but sensitive details are well-known and have been improved. Overall good working lives have been documented for many systems based on long-term field experience. Conductive coatings can remain effective for up to and above 15 years. Activated titanium systems have worked for 25 years and more. Galvanic systems provide a possibility to operate independent of power being available; an approach to predict their working life has been proposed. Installing CP on bridge joints with corrosion due to de-icing salt leakage is effective and efficient. The preventive application of CP is increasing.

The input of members of the CP-knowledge Centre (www.kb-kenniscentrum.nl) is appreciated.

References

[1] L. Bertolini, B. Elsener, P. Pedeferri, E. Redaelli, R.B. Polder, *Corrosion of Steel in Concrete: Prevention, Diagnosis, Repair,* 2nd Edition, Wiley (2013)

[2] R.B. Polder, W.H.A. Peelen, W. Courage, Mat. Corr. **63**, 1147 (2012)

[3] G.P. Tilly, J. Jacobs, *Concrete repairs – performance in service and current practice,* BRE Press (2007)

[4] J.H.M. Visser, Q. van Zon, Int. Conf. Concrete Repair, Rehabilitation and Retrofitting III, Taylor & Francis (2012)

[5] R.F. Stratfull Mat. Perf. **13** 24 (1974)

[6] J.P. Broomfield, *Corrosion of Steel in Concrete Understanding, Investigation and Repair,* 2nd Edition, Taylor & Francis (2006)

[7] J.S. Tinnea, C.B. Cryer, 1st Int. Conf. Heritage and Construction in Coastal and Marine Environment MEDACHS08 (2008)

[8] P. Pedeferri CBM **10** 391 (1996)

[9] K.A. Grefstad, Eurocorr05, (2005)

[10] O.C.N. Nerland J. Eri, K.A. Grefstad, Ø. Vennesland, Eurocorr07, (2007)

[11] Ch. Haldemann, A. Schreyer, EFC Publication **25** (1998)

[12] F. Wenk, D. Oberhänsli, Eurocorr07 (2007)

[13] R.B. Polder, HERON 43 3 (1998)

[14] C. Christodoulou, G.K. Glass, J. Webb, S. Austin, C. Goodier, Corr. Sci. **52** 2671 (2010)

[15] R.B. Polder, G. Leegwater, D. Worm, W. Courage Cem. Con. Comp. **47** 69 (2014)

[16] J. Pacheco, R.B. Polder, A.L.A. Fraaij., J.M.C. Mol, Concrete Solutions, Taylor & Francis, 147-156 (2011)

[17] M.R. Geiker, R.B. Polder, Mat. Corr., **67** 600 (2016)

[18] R.B. Polder, CBM **15** 125 (2001)

[19] EN 12696, *Cathodic protection of steel in concrete,* (2000)

[20] R.B. Polder, P.C. Nuiten, Mat. Perf. **33** 11 (1994)

[21] *Cathodic protection of reinforcement in concrete structures,* in Dutch, CUR **45** (1996)

[22] R.B. Polder, W.H.A. Peelen, F. Lollini, E. Redaelli, L. Bertolini, Mat. Corr. **60** 130 (2009)

[23] A.J. Hondel, E.L. Klamer, J. Gulikers, R.B. Polder, 4th Int. Conf. Concrete Repair, Rehabilitation and Retrofitting, Taylor & Francis (2015)

[24] EN-ISO 15257, *Cathodic protection – Competence levels and certification of cathodic protection persons : basis for certification scheme* (2017)

[25] S. Bhuiyan, D.W. Law, P. Nicholls, C. Christodoulou, CBM **162** 503 (2018)

[26] A.J. van den Hondel, A.W. van den Hondel, Concrete Solutions (2019)

[27] R.B. Polder, Y. Pan, W. Courage, W.H.A. Peelen, HERON **61** 1 (2016)

Application of CFRP as anode in cathodic protection for steel reinforced concrete – a review

Emma Qingnan Zhang*, and *Luping* Tang

Building Technology, Department of Architecture and Civil Engineering, Chalmers University of Technology, Sweden

Abstract. Impressed current cathodic protection (ICCP) is an electrochemical method to prevent or stop corrosion of steel reinforcement in concrete structures. The use of this technique is limited although it has been proven to be effective. One of the reasons is the high cost of anode material, commonly titanium anode. Hence there is a need for alternative anode materials to lower the cost and at the same time to improve the performance of the ICCP system through design. This paper reviews carbon fiber reinforced polymer (CFRP) as an anode for ICCP systems in concrete and discusses the degradation mechanism and operating conditions of the anode. It also gives an overview of the performance of CFRP as an anode material, from laboratory experiments to field operation.

1 Introduction

Chloride-induced corrosion is one of the major causes of damage or deterioration of steel reinforced concrete structures. To deal with this problem, impressed current cathodic protection (ICCP) has been proven to be an effective method to stop, slow down or prevent corrosion, even in a highly corrosive environment [1–3]. The protection of the steel is achieved by shifting the electrical potential of the steel negatively which forces the reinforcement into a passivity or immunity state. An ICCP system consists an anode system which is usually laid on the concrete surface and connected with the positive terminal of a direct voltage source.

Depending on the level of current density, in some cases, cathodic protection is referred to as "cathodic prevention". Cathodic prevention is a preventative approach when it is applied on new structures that are expected to become contaminated by chloride during their service life, using only small amount of current to suppress corrosion. According to the ISO 12696 standard, the current density for cathodic protection is 2-20 mA/m^2 (of steel surface) and for prevention it is 0.2-2 mA/m^2 [4].

For a cathodic prevention system, a small cathodic polarization of the steel should be applied early on in the beginning of the service life. The use of small current density has many advantages. The most important one is that it can reduce acid production at the anode area. Acidification at the anode area is an inevitable consequence of electrochemical reactions. The produced hydrogen will react with the anode and the cement paste, causing both anode and concrete to decompose, which will lead to a lowered protection efficiency and eventually may stop the protection [5].

As the most crucial component in a cathodic protection/prevention system, the anode material must be stable and functioning to secure the performance of such a protection system. Conventionally, the most commonly used anode material is titanium-based metal anode, because of its high current capacity, light weight and low consumption rate [6]. However, due to its high cost, it also hinders the wide application of ICCP systems.

On the other hand, in a cathodic prevention system, because of the use of low current density, the requirements of the anode material can be lowered. Therefore, as an alternative, carbon fiber reinforced polymer (CFRP) anodes have gained much interest because of their high conductivity and affordable price.

This paper is based on a literature study and reviews the usage of CFRP as anode for ICCP systems, in both laboratory experiments and field applications.

2 CFRP anodes

Carbon fiber reinforced polymer (CFRP) is a highly appreciated engineering material for its high strength, high conductivity and low weight. CFRP consists of carbon fiber threads and reinforcing polymer. Carbon fiber threads are made of thousands of carbon fiber filaments. This reinforcing polymer is usually epoxy for common engineering purpose. However, for special applications such as anodes, the reinforcing polymer can be other polymers with better electrical conductivity. For example, the resistivity of CFRP can be found between 0.015-0.05 Ω per unit length [7], [8].

Fig. 1 shows the three most common types of CFRP anodes that can be found in research for ICCP applications: woven mesh, fabric mat and rod. CFRP mesh is similar to titanium mesh which can be cast

* Corresponding author: emma.zhang@chalmers.se

directly into concrete with or without a layer of conductive overlay [9] . CFRP fabric and rod are usually used as near-surface mounted anode material together with adhesive such as epoxy [10].

Fig. 1. Image of CFRP mesh (left), CFRP fabric (middle) and CFRP rod (right)

3 Degradation mechanisms of CFRP anode systems

3.1 Durability of CFRP anode

The degradation mechanism of a CFRP anode has been investigated in simulated solutions and a concrete environment [8], [11]–[14]. Based on their results, the authors suggested that the polymer matrix was sensitive to electrochemical degradation but the carbon fiber remained with very little loss of mass even at very high current densities (up to 3 A/m^2 of anode surface). This means that CFRP materials are suitable and capable to be used as an anode material, because carbon fibers are the key component for electrical conductivity but not the reinforcing polymer. The purpose of reinforcing polymer is like a glue to make carbon fibers easy to handle and to be installed. Fig. 2 shows the images of CFRP anodes after accelerated tests of ICCP.

Fig. 2. CFRP anodes after accelerated test. (left) Image of the cross-section of CFRP. (right) Image of two single fibers with smooth surface[8]

3.2 Acidification at anode-concrete interface

To evaluate the anode system for ICCP applications, one must consider both the anode material and the anode-concrete interface, because the failure of either of these two parts can lead to the failure of the anode system. The most detrimental effect is the acidification at the anode area, which can lead to early failure or unplanned maintenance [15].

Owing to the nature of electrochemical reactions, ICCP transforms the microstructure of the concrete around the steel over time because of the migration of the ionic species and the chemical reactions [16]. The acidification at the anode-concrete interface reduces the pH value and is also associated with calcium leaching [17], [18]. Under the effect of acidification and calcium leaching, the degraded cement pastes turns into soft grey or white silica-gel-like substance, which are amorphous and have high electrical resistivity [19].

Anode degradation can cause an increase in the resistance and a loss of adhesion. This will in consequence increase the driving voltage and further accelerate the degradation process. Eventually it became impossible to maintain sufficient current for full protection to the steel reinforcement [5]. The service life of the anode system may be determined mainly by acid production due to current exchange [20]. A qualitative numerical model has been proposed to describe the acidification phenomenon under cathodic protection [21]. Accelerated tests have been conducted to investigate the relationship between the current exchange and the degradation area. Zhang and Tang suggested a non-linear relationship to describe that the total damage of the anode-concrete interface is related to both the total current exchange and the current densities [22].

4 Applications of CFRP as anode material

Only very few field operations have been reported using CFRP as ICCP anodes. Scandinavian countries have practiced their use in real construction projects as early as almost two decades ago. However, official full reports were rarely published and are difficult to find.

Vennesland et al. [23] described experimental work using CFRP mesh as anode on concrete specimens from a demolished bridge. The feeding voltage was 1 volt with a current density of 3-4 mA/m^2 and the protection criterion was achieved according to EN ISO standard 12696.

In a technical report, Mork et al. monitored and investigated a harbour structure in Honningsvaag in Norway [24]. The anode system consisted of carbon fiber mesh and cement-based mortar and the CP system was achieved with a current density of 2 to 5 mA/m^2 and a voltage of a maximum of 1.8 V. The system was functioning well as of the day of publishing, according to EN ISO 12696.

Based on these two field applications, CFRP mesh anodes have been demonstrated to be a proper alternative for cathodic prevention applications.

Studies have also been made to combine the reinforcing property together with the anode function [25], [26]. Zhu et al. [27] reported stable anode performance in accelerated ICCP tests of CFRP strip in simulated pore solutions. The stabilized potential was around 2 volts, the current densities were about 6 A/m^2 and the experiment duration was 50 days. Lee-Orantes et al. [28] performed experiments on CFRP anodes with several conductive adhered materials on reinforced concrete prisms. Nguyen reported that CFRP rod anode can operate at 128 mA/m^2 with a relatively low consumption rate, and the ultimate strength of the concrete increases by approximately 13.5% using a surface-mounted CFRP fabric [29], [30].

Based on existing research, to access the long-term performance of CFRP anodes in ICCP systems, accelerated tests have usually been adopted. To convert the results from accelerated tests to a normal current density level, the simplest rule is to apply the total charge quantity for calculation. However, according to electrochemical principles, the important effect of diffusion has been neglected because it is difficult to be quantified. The fact is that when the current density is lowered, the portion of diffusion can be increased, which helps reduce the damage of acidification and calcium leaching.

Polder and Peelen [31] described the electrochemical phenomenon under ICCP. They pointed out with low current densities (1-10 mA/m^2), the diffusion has the same order of magnitude as migration and thus the acid production is close to zero.

Other experimental works have also provided some evidence on that. Zhang et al. [22] investigated the electrochemical properties of CFRP mesh as anode in accelerated ICCP tests, with current density between 200-4000 mA/m^2 of anode area. Constant current was applied. The potential between the anode (CFRP mesh) and the cathode (steel bar) was monitored. Results showed that the potential increased dramatically after a certain period, which indicated degradation happens due to acidification and calcium leaching. The duration before potential changes was associated with current density, which was a power relation but not linear. The conclusion was that if only the total charge rule is used, it will underestimate the service life of anode systems.

Regarding the previous studies, CFRP mesh anodes are suitable and safe for cathodic prevention applications.

5 Summary and outlook

Research results have provided solid evidence that CFRP materials, especially meshes, are suitable as anode materials for cathodic prevention applications. Future investigation could focus on issues such as electrical connection. This issue is usually not highlighted or presented in research papers. However, in the authors' view, the connection points could become a weak link in the anode systems because they easily have the highest resistivity when two different materials are connected.

The connection of CFRP anode with the power supply or copper cable needs to be aided by, for example, highly conductive cement pastes. CFRP anode for ICCP systems has existed for almost 20 years. Unfortunately, still only a few field installations have been reported. This makes it hard to convince contractors and consultancy companies to use CFRP anodes instead of titanium anodes because of the lack of long-term credibility. More experiments and field installations are needed to prove the long-term durability of CFRP anode system for cathodic prevention applications.

References

[1] L. Bertolini, F. Bolzoni, A. Cigada, T. Pastore, and P. Pedeferri, "Cathodic protection of new and old reinforced concrete structures," *Corros. Sci.*, vol. 35, no. 5–8, pp. 1633–1639, 1993.

[2] N. C. Webb, "Cathodic protection of reinforced concrete," *Constr. Build. Mater.*, vol. 6, no. 3, pp. 179–183, 1992.

[3] S. 441 NCHRP, *SYNTHESIS 441 High Performance Concrete Specifications and Practices for Bridges*. 2013.

[4] ISO-12696, "Cathodic Protection of Steel in Concrete ." 2012.

[5] J. Mietz, J. Fischer, and B. Isecke, "Cathodic protection of steel-reinforced concrete: structures-results from 15 years' experience," *Mater. Perform.*, vol. 40, no. 12, pp. 22–26, 2001.

[6] R. H. Heidersbach, J. Brandt, D. Johnson, I. I. I. J.S. Smart, and J. S. Smart, *Marine Cathodic Protection, Corrosion: Environments and Industries, Vol 13C, ASM Handbook, ASM International, 2006, p 73–78* . 2006.

[7] M. Chini, R. Antonsen, Ø. Vennesland, B. Arntsen, and J. H. Mork, "A review of utilization of carbon fiber as an anodic material in cathodic protection," in *International RILEM Workshop on Integral Service Life Modelling of Concrete Structures*, 2007, pp. 395–402.

[8] E. Q. Zhang, "Durability of Reinforced Concrete under Impressed Current Cathodic Protection Failure Mechanism and Service Life Prediction," Chalmers University of Technology, 2018.

[9] E. Q. Zhang, L. Tang, and T. Zack, "Carbon Fiber as Anode Material for Cathodic Prevention in Cementitious Materials," in *Proceedings of the 5th International Conference on the Durability of Concrete Structures*, 2016, pp. 300–308.

[10] C. Van Nguyen, P. Lambert, P. S. Mangat, F. J. O'Flaherty, and G. Jones, "Near-surface mounted carbon fibre rod used for combined strengthening and cathodic protection for reinforced concrete structures," *Struct. Infrastruct. Eng.*, vol. 2479, no. May, 2015.

[11] H. Sun, S. A. Memon, Y. Gu, M. Zhu, J.-H. Zhu, and F. Xing, "Degradation of carbon fiber reinforced polymer from cathodic protection process on exposure to NaOH and simulated pore water solutions," *Mater. Struct.*, vol. 49, no. 12, pp. 5273–5283, 2016.

[12] H. Sun, L. Wei, M. Zhu, N. Han, J.-H. Zhu, and F. Xing, "Corrosion behavior of carbon fiber reinforced polymer anode in simulated impressed current cathodic protection system with 3% NaCl solution," *Constr. Build. Mater.*, vol. 112, pp. 538–546, 2016.

[13] J.-H. Zhu, L. Wei, M. Zhu, H. Sun, L. Tang, and F. Xing, "Polarization Induced Deterioration of Reinforced Concrete with CFRP Anode," *Materials (Basel).*, vol. 8, no. 7, pp. 4316–4331,

2015.

[14] H. Sun *et al.*, "Recycling of carbon fibers from carbon fiber reinforced polymer using electrochemical method," *Compos. Part A Appl. Sci. Manuf.*, vol. 78, pp. 10–17, 2015.

[15] R. B. Polder, G. Leegwater, D. Worm, and W. Courage, "Service life and life cycle cost modelling of cathodic protection systems for concrete structures," *Cem. Concr. Compos.*, vol. 47, no. 0, pp. 69–74, Mar. 2014.

[16] P. Pedeferri, "Cathodic protection and cathodic prevention," *Constr. Build. Mater.*, vol. 10, no. 5 SPEC. ISS., pp. 391–402, 1996.

[17] R. B. Polder, W. H. A. Peelen, B. T. J. Stoop, and E. A. C. Neeft, "Early stage beneficial effects of cathodic protection in concrete structures," *Mater. Corros.*, vol. 62, no. 2, pp. 105–110, 2011.

[18] R. B. Polder, T. G. Nijland, W. H. A. Peelen, and L. Bertolini, "Acid formation in the anode/concrete interface of activated titanium cathodic protection systems for reinforced concrete and the implications for service life," in *15th International Corrosion Congress (ICC), Granada, Spain, 22-27 September, 8* , 2002.

[19] E. Q. Zhang, L. Tang, D. Bernin, and H. Jansson, "Effect of the paste-anode interface under impressed current cathodic protection in concrete structures," *Mater. Corros.*, vol. 69, no. 8, pp. 1104–1116, Aug. 2018.

[20] M. G. Ali, Rasheeduzzafar, and S. S. Al-Saadoun, "Migration of ions in concrete due to cathodic protection current," *Cem. Concr. Res.*, vol. 22, no. 1, pp. 79–94, 1992.

[21] W. H. A. Peelen, R. B. Polder, E. Redaelli, and L. Bertolini, "Qualitative model of concrete acidification due to cathodic protection," *Mater. Corros.*, vol. 59, no. 2, pp. 81–89, 2008.

[22] E. Q. Zhang, Z. Abbas, and L. Tang, "Predicting degradation of the anode–concrete interface for impressed current cathodic protection in concrete," *Constr. Build. Mater.*, vol. 185, pp. 57–68, Oct. 2018.

[23] Ø. Vennesland, R. Haug, and J. H. Mork, "Cathodic protection of reinforced concrete - A system with woven carbon mesh," in *Concrete Repair, Rehabilitation and Retrofitting - Proceedings of the International Conference on Concrete Repair, Rehabilitation and Retrofitting, ICCRRR 2005* , 2006, pp. 305–307.

[24] J. H. Mork, S. Mayer, and R. Åsheim, "The performance of cathodic protection on harbor and jetty of Honningsvåg Norway ." 2007.

[25] L. Tang, E. Q. Zhang, Y. Fu, B. Schouenborg, and J. E. Lindqvist, "Covercrete with hybrid functions - A novel approach to durable reinforced concrete structures," *Mater. Corros.*, vol. 63, no. 12, pp. 1119–1126, 2012.

[26] P. Lambert, C. Van Nguyen, P. S. Mangat, F. J. O'Flaherty, and G. Jones, "Dual function carbon fibre fabric strengthening and impressed current cathodic protection (ICCP) anode for reinforced concrete structures," *Mater. Struct.*, vol. 48, no. 7, pp. 2157–2167, 2015.

[27] J.-H. Zhu, G. Guo, L. Wei, M. Zhu, and X. Chen, "Dual Function Behavior of Carbon Fiber-Reinforced Polymer in Simulated Pore Solution," *Materials (Basel).*, vol. 9, no. 2, p. 103, 2016.

[28] F. . Lee-Orantes and C. Torres-Acosta, A.; Martínez-Madrid, M.; López-Cajún, "Cathodic Protection in Reinforced Concrete Elements, using Carbon Fibers Base Composites," *ECS Trans*, vol. 3, pp. 93–98, 2007.

[29] C. V Nguyen, P. S. Mangat, P. Lambert, F. J. O. Flaherty, and G. Jones, "Dual function carbon fibre strengthening and cathodic protection anode for reinforced concrete structures," in *Concrete Repair, Rehabilitation and Retrofitting III*, 2012, pp. 1179–1185.

[30] C. Van Nguyen *et al.*, "The Performance of Carbon Fibre Composites as ICCP Anodes for Reinforced Concrete Structures," *Int. Sch. Res. Netw. ISRN Corros.*, vol. 2012, 2012.

[31] B. P. Rob and H. A. P. Willy, "Service life aspects of cathodic protection of concrete structures," in *Concrete Repair*, CRC Press , 2011, pp. 117–136.

Insights from corrosion current measurements on corrosion mechanisms in reinforced concrete and on the evaluation of other corrosion data

Ulrich Schneck[1,*]

[1]CITec Concrete Improvement Technologies GmbH, 01156 Dresden, Germany

Abstract. During the past years Tafel polarization measurements have been implemented into the scope of measurements of CITec corrosion diagnosis projects. This has created a vast database of different and corresponding corrosion parameters, such as chloride and water content in the rebar vicinity, open circuit potential, electrolyte resistance, polarization resistance (from galvanostatic pulse and linear polarization) and corrosion current from Tafel polarization measurements. Although general limitations in using these methods on macro cell systems such as reinforced concrete are known, the comparative assessment of these data has led to a better understanding of the corrosion behaviour and of specific circumstances of the structures which deviated partly from usual expectations. For instance, a low polarization resistance at high chloride content will not result necessarily in a high corrosion current, if the reinforcement in the wider vicinity of the test location is similar active, and cathodic rebar areas are either very distant or retarded by very wet concrete. So the extended range of corrosion testing gives a more precise evaluation of the corrosion situation and permits a tailored repair and maintenance concept to be found. It has also been found that the Stern-Geary equation which is often used to calculate corrosion current densities and material loss of the reinforcement from linear polarization (LPR) measurements, doesn't seem very feasible if used on reinforced concrete structures, as there appears to be a dominant influence of macro cell corrosion over the corrosion model of a homogenous mixed electrode (for which the Stern-Geary equation applies), and the true corrosion current densities may be either larger or (very often) much smaller than those calculated from Stern-Geary. This is not a new observation, and the findings will be discussed for several project cases.

1 Introduction

The corrosion behaviour of the reinforcement in concrete can also be assessed by enhanced electrochemical methods: AC impedance, galvanostatic pulse, linear polarization or Tafel polarization. With the help of a potentiostat and a measurement cell containing a reference- and a counter electrode (setup according to fig 1), these methods force the rebar – acting as a working electrode – to change its potential from the open circuit value according to the method specifications and to get additional corrosion relevant data: electrolyte resistance, polarization resistance and corrosion current. These data provide enhanced qualitative and also quantitative information which are of better accuracy than those of a potential survey only. This can help to substantially improve the interpretation of the data of a full scale potential survey, even if certain circumstances mainly regarding the LPR and the Tafel polarization are not ideal – e.g. if spots of different, closely neighboured corrosion behaviour may be touched by a single polarization measurement.

Apart from standard laboratory potentiostats, some commercial measuring devices are available for use on reinforced concrete structures.

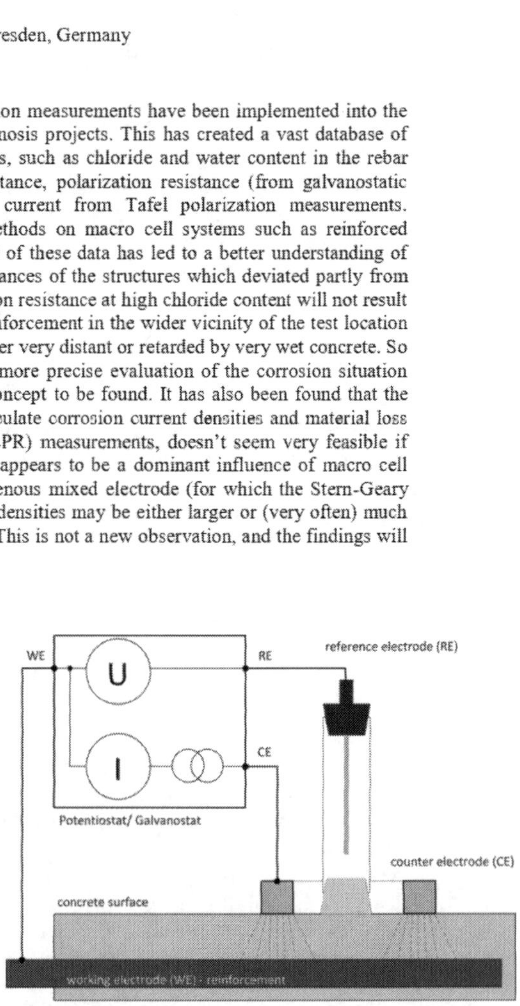

Fig. 1. Measurement setup for corrosion testing with a potentiostat

The equipment used for the measurements (fig 2) of this publication has been described in [1], and meanwhile the control software is able to do numerical and graphical data assessment on the measurement curves. The counter electrode is being operated without guard ring.

Both methods are potentiodynamic measurements, where a shift by either 10 to 20 mV (LPR) or about 200 mV (Tafel) from the open circuit potential (OCP) is

* Corresponding author: ulrich.schneck@citec-online.com

forced into a cathodic direction first, and with a slow speed of ca. 0.5 mV/s a further shift of the potential until the same respective distance from the OCP into the anodic range is made – see fig 3.

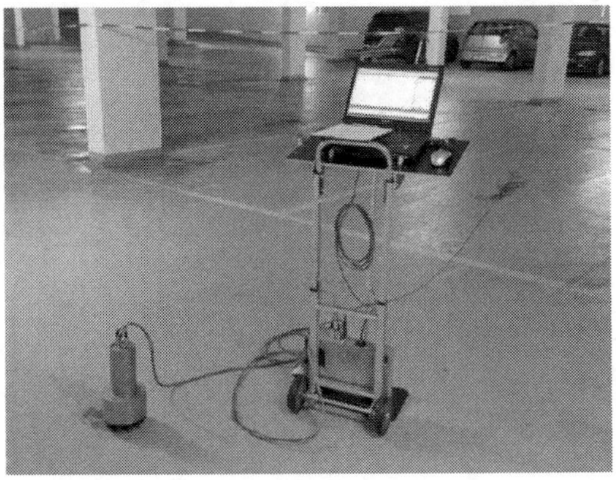

Fig. 2. Corrosion measurement system devices used for the project cases of this publication

During the whole polarization time the required current is recorded and displayed in an E/I diagram for the LPR or in an E/ log I diagram for the Tafel polarization.

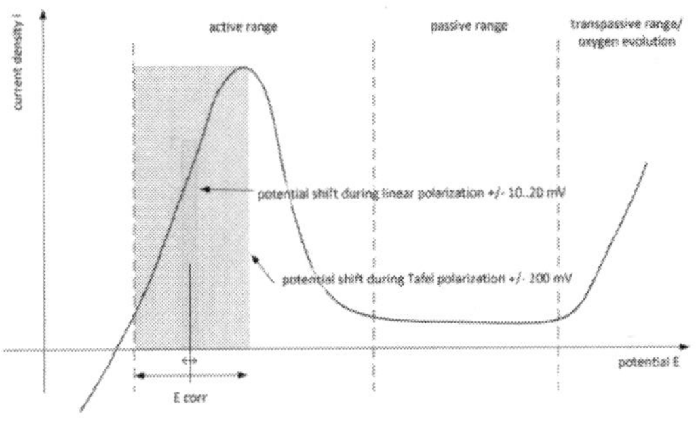

Fig. 3. Polarization scheme for LPR and Tafel measurements on the current density-potential plot

The background consideration for the LPR is that at such a low deviation from the OCP as the mentioned 10 to 20 mV, it can be expected to stay within a linear part on the current-potential curve, so the total polarization resistance can be calculated as

$$R_P' = \Delta E / \Delta I \qquad (1)$$

where R_P' contains the electrolyte resistance R_Ω and the polarization resistance R_P, which is the transfer resistance between metal and electrolyte respectively between rebar and concrete. The measurement curve will give notice if linear conditions applied or not – see figs 4 and 5.

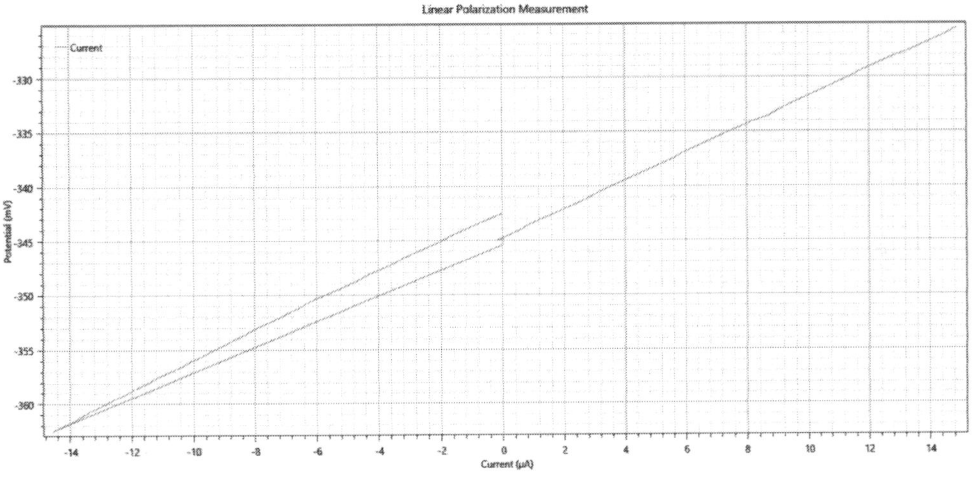

Fig. 4. LPR within the apparent linear current range

The Tafel polarization measurement, which is named after Julius Tafel, an electrochemist, who introduced it about 1905, is meant to measure the corrosion current at the OCP. This is normally not possible as there's no access into a corrosion element, and the balance of all currents (anodic, cathodic) is Zero on the outside. Typically, the measurement results into a curve according to fig. 5, where the Tafel slopes b_c (cathodic slope) and b_a (anodic slope) can be fitted. In a graphical assessment, both slopes meet over the OCP, where the externally applied current is zero. The intersection point represents the corrosion current at the OCP (or free corrosion potential E_{corr}), which cannot be measured directly. If related to the polarized rebar area, it results in the corrosion current density, which corresponds to the material loss at the time of investigation. More information about this measurement technique can be found in [2].

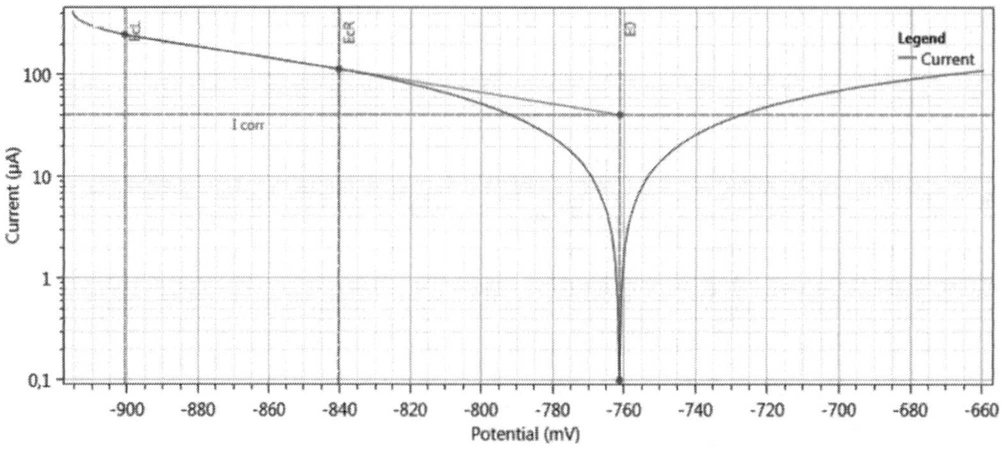

Fig. 5. Typical measurement curve of a Tafel polarization with approach for the corrosion current assessment at the free corrosion potential

For obtaining the correct corrosion current, it is further necessary to do an IR correction, because the recorded data contain a deviation $\Delta E_{IR} = I * R_B$. The control software of most potentiostats will allow such an IR correction – either by an interruptive method during the measurement or in a separate, later correction of the whole data set. Our own experience with a laboratory potentiostat and its software have shown that the deviation between the uncorrected and the IR corrected results are less than 15%, in most cases less than 10% (where the IR corrected current is smaller, because the related slopes are steeper). Considering the much larger uncertainties in selecting an appropriate electrode surface, no IR correction has been made in the following assessment, also because the uncorrected result is on the "safe side".

2 Numerical evaluation of the LPR according to Stern-Geary

The theory of the assessment and numeric calculation of corrosion current densities is explained comprehensively in several publications, such as [3-5]. Referring to [4], the approach is summarized as following: Close to the corrosion potential E_{corr}, the current curve i_S (E) can be expressed in a slope equation

$$i_S(E) = i_{corr} * \frac{E - E_{corr}}{B} \quad [\mu A/cm^2] \qquad (2)$$

where B is a constant containing the Tafel slope gradients b_a and b_c and can be calculated as

$$B = \frac{b_a * b_c}{2{,}303\,(b_a + b_c)} \qquad (3)$$

In the direct vicinity of E_{corr}, where the current density-potential curve can be considered to be a straight line, the gradient of the current equation equals the reciprocal polarization resistance

$$\frac{di_S(E)}{dE} = \frac{i_{corr}}{B} = \frac{1}{R_P} \quad [cm^2/\Omega] \qquad (4)$$

and according to Stern-Geary the corrosion current density i_{corr} can be calculated at E_{corr} from the polarization resistance R_P' as

$$i_{corr} = \frac{B}{R_P} \quad [\mu A/cm^2] \qquad (5)$$

For corroding reinforcement in concrete, B can be experimentally obtained as 0.026 V. It is important that both current density and polarization resistance need to be related to the polarized rebar area and are specific values. Another important assumption for the feasibility of the Stern-Geary equation is that the corrosion element is a mixed electrode, having the anodic and cathodic part almost on the same location (which corresponds to micro corrosion as it applies for the general current density-potential plot). Under these conditions, corrosion current densities calculated or measured from Tafel plots equal the values calculated from LPR measurements.

In reinforced concrete structures we often have macro element corrosion, especially where chloride attack occurs; anodic and cathodic rebar sections may be separated and located in distant areas, which is a different case in principle.

3 Comparative assessment of corrosion current densities from various projects

For this assessment 70 LPR and Tafel polarization data couples from 6 recent projects have been used. They represent the entire set of data per project, and most of the measurement locations have been chosen due to apparent high corrosion activity from the potential survey, but also in locations which exhibited rather passive behaviour. The data of following project cases were used:

Underground car park in Munich (UG-1): Large facility (ca. 2,000 m² per floor) that was closed down during a conversion from an office building into a hotel at least 2 years before doing the survey. At parts of the floor slab very large concrete covers < 70 mm and distinguished areas of either negative or (rather) positive (> 0 mV vs CSE) potentials with high or low chloride content accordingly; some significant reductions of rebar cross section visible.

Underground car park in Berlin (UG-2): At the time of survey the structure was ca. 45 years old, partly covered by a court area with lawn/ plants and a dwelling/ shopping building. There were severe signs of rust staining and moist concrete due to leaks in joints, numerous cracks; level -1 with frequent change of cars, level -2 with tenant parking and only few car movements.

Multi-storey car park in Braunschweig (CP-1): At the time of survey this was a less frequented structure, but previously heavily affected by de-icing salts; the car park was part of an apartment building, generally penetrated by chlorides, but only moderate corrosion damage because of meanwhile dry concrete (humidity values of ≤ 2.8 % in rebar vicinity).

Underground car park in Kassel (UG-3): Small, one-level structure with very slim columns in single foundations. Chloride ingress through brick pavement into the soil and by capillary suction into the concrete of the foundations and columns; chloride accumulation on the surface of the columns close above ground and some areas of high corrosion activity

Storm surge barrier at the North Sea (SSB): At the time of survey this ca. 40 year old marine structure with reinforced concrete slabs ca. 40 x 8 m was used as a bridge for pedestrians and cyclists on top of the construction, which are exposed to salty environment. Moderate chloride ingress (< 1% by cement mass in rebar vicinity) found and only few negative potentials < -200 mV vs. CSE in areas of lower concrete cover.

Confined spare area under pedestrian way in Abu Dhabi (CSA): Hollow space (w x h = 2 x 0-2 m) underneath an elevated pedestrian way, underpassing a bridge, exposed to salty ground water from the backfill and condensation humidity inside the confined space area. Severe chloride concentration was observed mainly in the walls toward the backfill (up to 3% by cement with no gradient in 8 cm depth), but no considerable signs of corroded rebar or loss of cross-section.

The measurement cell has been placed in all cases on a rebar crossing, and the approach for determining the related reinforcement surface was to consider a possible minimum and a maximum surface. The minimum was assumed to be the length of the 2 crossing rebars directly under the counter electrode with a diameter of 60 mm, multiplied by the rebar circumference and a surface factor of 1.2 for the ribs and other deviations from the ideal cylinder. The respective maximum surface was calculated from the double rebar length, assuming a spread of the polarization at 45° and a concrete cover of 30 mm as well as a surface factor of 2 for considering corrosion pits. Within that bandwidth the true, but unknown reacting rebar area should be situated, and for a very safe approach the maximum corrosion current density will result when dividing the corrosion current by the minimum surface.

Fig 6 shows all data couples R_P' (LPR)/ I_{corr} (Tafel) as the received absolute values from the different projects, and in fig 7 a regression curve is fitted in with a remarkably good correlation factor R^2 (0.967). For the comparative data in fig 8 the assumed rebar areas have been implemented, and it has been referred to the mentioned minimum area (cylinder under the counter electrode with d = 60 mm, rebar factor 1.2). This way, both specific parameters R_P' (LPR) and icorr (Tafel) result in relatively high values, which can be seen as the "safe" side for i_{corr} (Tafel). On the other hand, the i_{corr} values calculated by the Stern-Geary equation with B = 26 are relatively low. Naturally the regression curve of the i_{corr}/ R_P values by Stern-Geary needs to show the Stern-Geary equation.

In table 1 the respective minimum, maximum and average i_{corr} values can be seen per project and in total, resulting either from Tafel polarization or the Stern-Geary equation, including the percentage of the i_{corr} by Tafel of the respective i_{corr} by Stern-Geary.

Figs 9 and 10 show the dependency of the i_{corr} values on the observed chloride contents in the rebar vicinity. Especially large differences between the 2 i_{corr} approaches (and extremely high corrosion current densities calculated by Stern-Geary) can be seen for the underground car park Kassel (UG-3), the storm surge barrier (SSB) and the confined space area (CSA). Furthermore, these differences get especially visible at chloride contents of less than 2 %.

3 Discussion

As mentioned before, the question "how feasible LPR measurements and the calculation of corrosion current densities by the Stern-Geary equation are for reinforced concrete" has been raised before, such as in [6]. There, the background has been described in the light of the difference between the theory of mixed electrodes (with even corrosion on small areas – to which the equation by Stern and Geary refers to) and galvanic elements with the resulting macro element cells, as they are typical for the reinforcement in concrete. In laboratory experiments the effect and influence of separated anodes (carbon steel) and cathodes (stainless steel) in a solution as well as the location of the reference- and counter electrodes during the measurements have been investigated. One finding has been a general underestimation of the true corrosion current density by the calculation of i_{corr} from R_P', as parts of the applied current from the counter electrode will be discharged on other surface areas than the intended measurement location.

The measurements discussed here have been done under practical conditions with many varying, partly unknown parameters that cause difficult conditions and uncertainties for the assessment: concrete permeability, concrete humidity, chloride content, concrete cover, rebar surface and constitution of the corrosion products on the rebar surface. Regardless, both LPR and Tafel polarization measurements have been made, and a first observation is that the data couples of the absolute values seem to fit into an apparent scheme (see fig. 8). This indicates some credibility, but a true comparison can be made only when converting these data into specific values.

Furthermore, the feasibility of the Tafel polarizations on reinforced concrete structures needs to be discussed, as the large potential shift of ca. 200 mV in both directions can change the corrosion conditions and hence the received result. By experience, corrosion active surfaces with corrosion products other than the passive layer will exhibit an almost instant, reproducible return on the OCP, which indicates that the change of the conditions by the external polarization is rather negligible. In such situations, the zero current during the Tafel polarization will be seen almost on the OCP where the measurement has been started.

The other mentioned uncertainties come into effect when introducing an area on the rebar, and the choice to take a minimum surface (as would apply for a measurement with a guard ring) results in a maximum current density for the Tafel polarization, so these values are – including the skipped IR correction – on the safe side. This approach has the reciprocal effect on the i_{corr} values in the Stern-Geary equation, as the also maximized specific R_P' values are the lower part of the fraction. All this cannot consider the observations made in (Angst, Büchler 2015), but takes the concrete as a "grey box". It is next discussed how feasible the i_{corr} data appears with respect to the project cases:

Underground car park in Munich (UG-1): Here the Tafel-i_{corr} values were on average twice as high as the LPR-i_{corr} values which appears to be plausible as the potential survey made smaller areas of significant negative potentials visible. These were not caused by wet concrete (the whole slab was rather dry after ca 2 years without parking service), so the negative areas (for which the high corrosion currents and current densities were measured), certainly showed chloride induced macro cell corrosion with galvanic elements between small active/ anodic and large passive/cathodic rebar sections.

Underground car park in Berlin (UG-2): The Tafel-i_{corr} values were on average around 75 % of the LPR-i_{corr} value level. Structural problems and water ingress were clearly dominating problems with chloride induced corrosion, so no considerable macro cell corrosion with apparent large cathodes and small anodes could be stated. The high water content determined in parts of the floor slabs will have caused a larger spread of polarization forced by the counter electrode and might have been limiting cathodic reaction by a reduced presence of oxygen. The assumption of referring to a rebar area only in the dimension of the counter electrode was possibly too conservative and a true corrosion activity less than calculated by Stern-Geary seems plausible.

Multi-storey car park in Braunschweig (CP-1): The Tafel-i_{corr} values were very low and on average around 56 % of the LPR-i_{corr} value level. This is also plausible as a generally high chloride content was found across the whole floor surface which would have made most of the upper layer of the reinforcement active. In the same way, the availability of passive reinforcement for the cathodic reaction was limited. In the observed moderate to dry concrete the electrolyte resistances could increase which would result in a further retarding of corrosion activity.

Underground car park in Kassel (UG-3): Here the Tafel-icorr values were in a plausible range between passive (0.07 µA/cm²) and increased active (1.34 µA/cm²), which corresponded to the measured potentials, to the analysed chloride contents and also to the visual appearance of the rebar surface (from blank to heavily corroded). The respective i_{corr} values calculated from the LPR were on average 6 times higher, where no reason can be found, especially as very high corrosion rates were calculated at locations with low chloride content (see fig 10). The corrosion situation was not a typical macro cell case as the column feet were placed in single foundations with not much rebar surface available as cathode.

Storm surge barrier at the North Sea (SSB): The Tafel-i_{corr} values were generally in the passive range ($\leq$ 0.20 µA/cm²) and on average around 27 % of the LPR-i_{corr} value level. Basically, almost no potential differences were measured; the R_P' values were high (2,200 to 7,200 Ohm) and the I_{corr} values low (5 to 23 µA), so there was no reason to expect de passivation and corrosion activity. Since no macro cell corrosion was likely and the reinforcement was visually in a perfect passive state, no explanation for the increased corrosion rates calculated by the Stern-Geary equation could be found.

Confined spare area under pedestrian way in Abu Dhabi (CSA): Regardless of the known high chloride contents the observed corrosion rates from Tafel polarization stayed at a moderate to low level, while the respective i_{corr} values received from LPR and Stern-Geary seemed to be extremely high – relatively (more than 800 % of i_{corr}, Tafel) and absolutely (> 17.5 µA/cm²). Apparently the wet atmosphere caused a lack of oxygen and hence a cathodic control of the corrosion activity similar to submerged concrete in a marine environment. Consequently, concrete deterioration and rebar corrosion were found to be at a very low level.

Among the discussed project cases one (UG-1) was in line with the results of [5], where a macro cell with small anode and large cathode resulted in higher corrosion rates than the LPR evaluation by the Stern-Geary equation which is meant to characterize a homogeneous mixed electrode. In the other 5 projects it was the opposite case that the corrosion rates determined by the Tafel polarization were lower than those of LPR and Stern-Geary. In 3 of these 5 projects (CP-1, UG-3, CSA) a cathodic control of the otherwise anodic behaving areas was obvious, so its retarding influence could be demonstrated.

4 Conclusions

Regardless of all uncertainties and imperfections, the Tafel polarization seems feasible for the practical assessment of corrosion current densities on reinforced concrete structures, which includes a "safe" approach to do no IR correction to the measurement curve and to relate the I_{corr} to a rebar area not greater than the counter electrode. This seems to provide more plausible results than the calculation of i_{corr} from polarization resistances by the Stern-Geary equation as it could be proven by other project measurement data including the visual inspection of the reinforcement.

An interpretation of the observed differences between the Tafel and LPR assessment is that galvanic elements of macro cells influence the corrosion behaviour on a specific measurement location in either an accelerating way (on small anodes in contact with large cathodes) or a retarding way (limited cathodic reaction by missing or very high water content in the concrete). From another point of view, a measured low R_P or R_P' value is first an indicator if the respective location is active or passive. If active, it depends also on the "outside" conditions (the capability of adjacent cathodic areas) how high the corrosion rate will be. The feasibility of this conclusion should be further investigated as the apparent functional relationship between the data couples of the absolute R_P' and I_{corr} values as per fig 7 would contradict this approach. It should be mentioned that both measurements contain the IR drop and hence contain the total polarization influenced by the ohmic resistance and the polarization resistance.

If verified, this observation could be used in corrosion surveys: the corrosion rate could be calculated according to Stern-Geary and be compared to the result of the Tafel polarization. If the Tafel corrosion rate at a location of apparent corrosion activity (either because of increased chloride content or because of visible corrosion products on the rebar surface) is significantly lower than the current density resulting from the Stern-Geary equation, a cathodic control of the corrosion is likely and should be considered when specifying maintenance and repair actions. In the other case (higher corrosion current densities from the Tafel polarization), the typical macro cell corrosion is indicated.

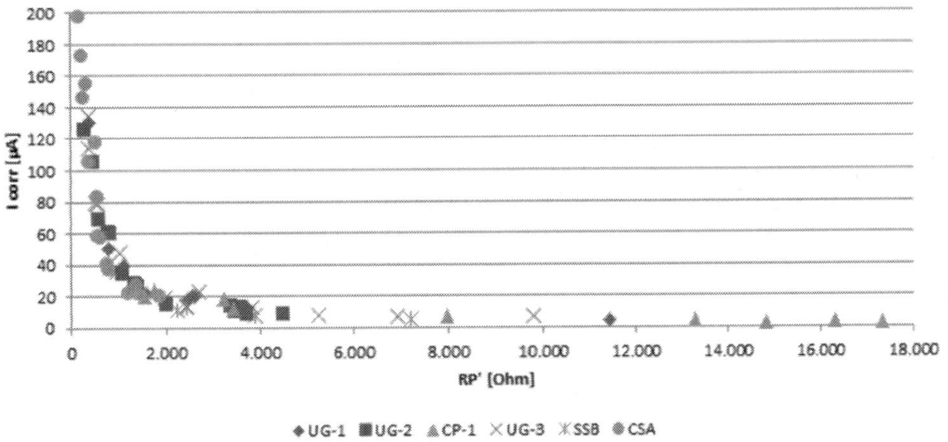

Fig. 6. Data couples of R_P' (LPR) and Icorr (Tafel) as absolute values of the different projects

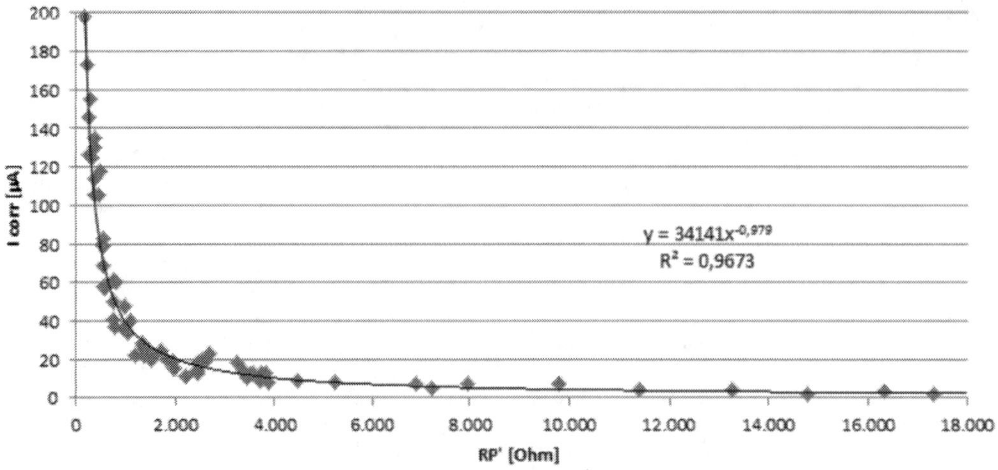

Fig. 7. Unified data couples of absolute R_P' (LPR) and Icorr (Tafel) values and regression function

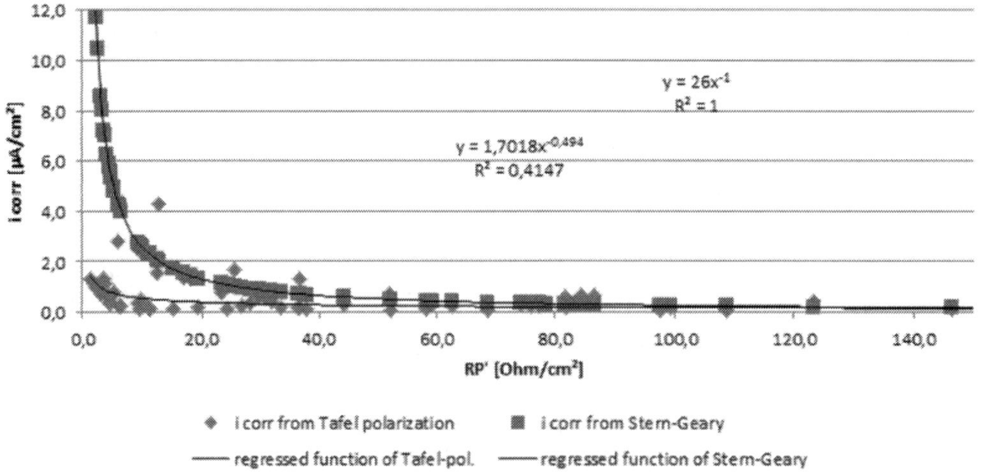

Fig. 8. Specific i_{corr} values from Tafel polarization and by Stern-Geary, related to specific R_P' values

Table 1: comparison of the min, max and avg i_{corr} values per measurement method and project

project	UG-1	UG-2	CP-1	UG-3	SSB	CSA	total
data couples R_P/ I_{corr}	9	16	10	14	5	16	70
results from Tafel polarization and assumed minimum rebar area [µA/cm²]							
min i_{corr} Tafel	0.13	0.20	0.04	0.07	0.08	0.13	0.04
max i_{corr} Tafel	4.31	2.79	0.44	1.34	0.20	1.31	4.31
avg i_{corr} Tafel	1.17	0.85	0.17	0.50	0.15	0.55	0.61
results from Stern-Geary equation, referring to the same rebar area [µA/cm²]							
min i_{corr} SG	0.07	0.26	0.08	0.27	0.24	1.72	0.07
max i_{corr} SG	2.03	4.31	0.92	8.04	0.78	17.62	17.62
avg i_{corr} SG	0.61	1.13	0.32	3.05	0.58	6.81	2.59
percentages of the corrosion current density from Tafel polarization compared to Stern-Geary							
min Tafel/SG	146%	56%	39%	13%	22%	5%	5%
max Tafel/SG	221%	93%	76%	26%	32%	12%	221%
avg Tafel/SG	192%	75%	56%	17%	27%	8%	57%

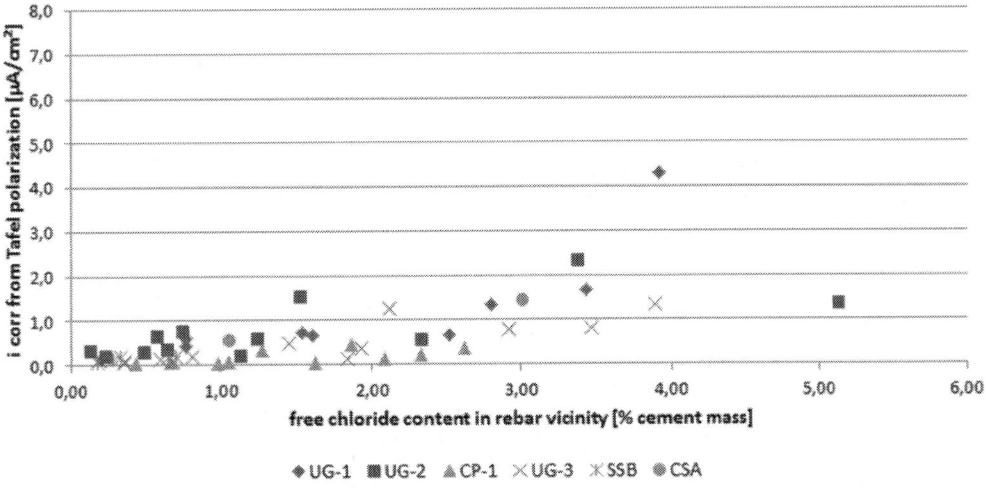

Fig. 9. i_{corr} values from Tafel measurements as function of the related chloride content in rebar vicinity

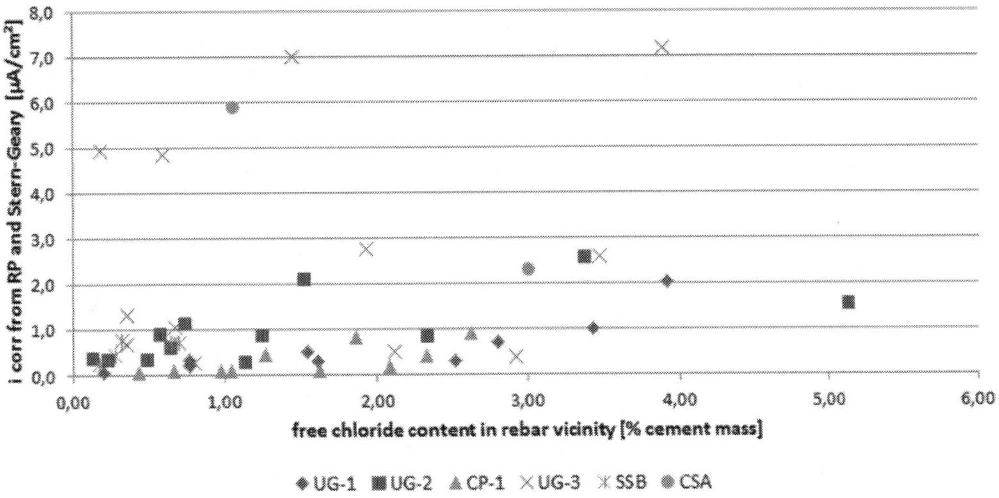

Fig. 10. i_{corr} values c as function of the related chloride content in rebar vicinity

5 References

1. K. Ahlborn, F. Berthold, W. Vonau, H. Grünzig, U. Schneck, H. Jahn, J. Köhler, *Modular Corrosion Measurement System (CMS) for electrochemical NDT.* 4th International Conference on Concrete Solutions, Proc. (2011)
2. A. J. Bard, L.R. Faulkner, *Electrochemical Methods. Fundamentals and Applications 2nd Edition.* Wiley, New York (2001)
3. E. Wendler-Kalsch, H. Gräfen, *Korrosionsschaden-kunde*, Klassiker der Technik (1998)
4. B. Elsener, D. Flückiger, H. Wojtas, H. Böhni,, *Methoden zur Erfassung der Korrosion von Stahl in Beton – FA 85/88.* ETH Zürich (1996)
5. J. Broomfield, *Corrosion of steel in concrete – understanding, investigation and repair 2nd Edition.* Taylor & Francis (2007)
6. U. Angst, M. Büchler, *On the applicability of the Stern-Geary relationship to determine instantaneous corrosion rates in macro-cell corrosion.* Materials and Corrosion 2015, 66, No. 10 (2015)

Instant performance verification after electrochemical chloride extraction by enhanced corrosion testing

Ulrich Schneck[1,*]

[1]CITec Concrete Improvement Technologies GmbH, 01156 Dresden, Germany

Abstract. Electrochemical chloride extraction (ECE) is meant to re-establish the corrosion protection of concrete for the embedded reinforcement by removing chloride non-destructively and by enhancing the alkalinity of the rebar surrounding concrete. Both effects depend on various parameters, such as concrete cover, rebar spacing, chloride profile (especially if chloride ingress is deeper than the outside rebar layer) and concrete permeability. Often these parameters require long or multi-stage treatments, which basically can achieve any desired target level of chloride profile and impressed charge, but become a costly solution after a while. The acceptance criteria mentioned in CEN TS 14038-2 clause 8.6 refer to the achieved chloride content and to the amount of impressed charge, which are the conventional, easy measurable, but not direct parameters for evaluating the corrosion activity. A third parameter – the re-measurement of potentials for assessing (intended) low potential gradients and more positive average potentials – requires some weeks to months of depolarization and evaporation of water, before such a measurement can be applied successfully. A promising approach for an instant performance testing after an ECE treatment has been made on several occasions with follow-up measurements of electrolyte resistance, polarization resistance and corrosion current. Convincing changes towards significantly lower corrosion activity could be obtained (and compared to known classified values) – regardless of sometimes high residual chloride and very wet concrete. These data could be verified when re-assessed after some weeks, so enhanced corrosion measurements seem to be a useful tool for either establishing that the designed treatment time has been sufficient or to check on possible earlier termination of the treatment during a running ECE.

1 Introduction

During the past 8 years enhanced corrosion measurements (AC impedance, galvanostatic pulse measurement, linear polarization and Tafel polarization) have been increasingly used in corrosion survey projects of CITec, and the device applied for these measurements has been described in [1]. After collecting encouraging experiences, especially in clarifying scenarios that were difficult to evaluate on the basis of the "traditional" test methods potential survey plus chloride and water content [2], it was a logical step to use them also for the assessment of ECE applications.

General explanations about ECE are given for instance in [3]. It has been observed from many projects, that much higher amounts of residual chloride than 0.4 %, related to the cement mass can be tolerated even in the rebar vicinity with proven elimination of corrosion activity. This information was received mainly by potential surveys that demonstrated significantly more positive values in the treated areas than before ECE and a low deviation from potentials in adjacent, chloride-free areas. Unfortunately such conditions cannot be found immediately after terminating ECE, where the strong cathodic polarization enforces instant-off potentials < -800 mV vs. CSE, and in the usually very wet concrete it lasts some 3 to 6 months until equalized, stable moisture conditions allow a useful assessment of potential measurements.

For having a solid evaluation of the success of the treatment, respective measurement results must be available directly at its termination and not just many weeks after. The 3 case studies of this paper give good confidence in a promising approach.

2 Short descriptions of the enhanced corrosion measurements

For the enhanced electrochemical measurements a potentiostat/ galvanostat is needed, which is a standard testing device in electrochemistry [1]. In addition to the voltmeter and the reference electrode which are used for potential measurements on the reinforcement, it has a current source and a counter electrode for triggering changes to the working electrode/ reinforcement (fig. 1). The following electrochemical test methods have been applied at different stages of the ECE:

AC impedance: With this method it is possible to obtain the electrolyte resistance of the concrete in a 2-electrode setup at a frequency of ca. 1 kHz.

* Corresponding author: ulrich.schneck@citec-online.com

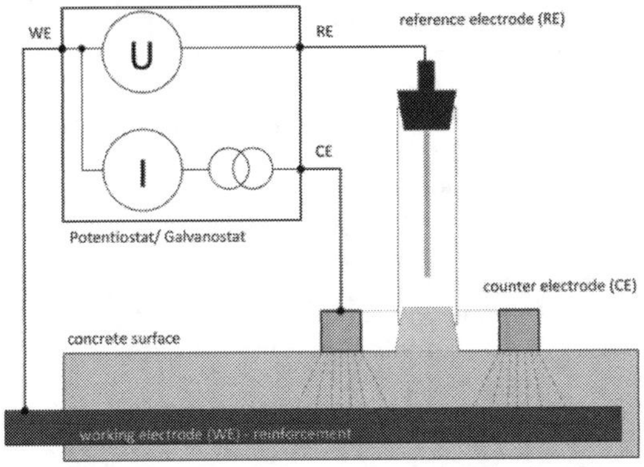

Fig. 1. Corrosion measurement setup with a potentiostat

Galvanostatic pulse measurement (GPM): Here, a small anodic constant current is applied via the counter electrode, and the forced shift of the potential at the working electrode is measured. The change of potential is to be divided into the IR drop, which equals the electrolyte resistance R_B, and the "true" polarization resistance R_P. Details of the application of this method on reinforced concrete structures can be read in [2], [4]. There exist various approaches for the measurement setup; here, the anodic pulse has been at -100 µA over 60 s.

Linear polarization measurement (LPR): The potential is shifted slowly in a small range around the open circuit potential (OCP) (max. ±20 mV at ca. 0.5 mV/s scan rate). This is considered to exhibit a linear dependency between potential and current. The total polarization resistance RP' (including electrolyte and polarization resistance) is the forced shift of potential divided by the resulting shift of current.

Tafel polarization measurement (TPM): In this method, a much wider potential shift is done, usually ca. ±200 mV around the open circuit potential. The measurement curve is displayed in a log I/ E-graph, and at the two visible straight sections the so called Tafel slopes will be fitted. At their intersection with the OCP, the corrosion current I_{corr} at the corrosion potential can be determined. If no IR correction is done, the I_{corr} is by experience ca. 10 to 15% higher than the true value. More information about this method is available in [5] and [6]. For obtaining corrosion rates from the I_{corr} value, a related rebar area has to be assumed, which is very tricky in practical cases. The approach taken here considers the range between a minimum and a maximum area, referring either to a rebar length directly under the counter electrode (l = 6 cm) or under an angle of 45° (which results at a cover of 3 cm in a rebar length of 12 cm), multiplied by the number of rebars, their perimeter and a surface factor of between 1.2 for the ribs and 2 for a surface additionally covered with corrosion pits.

3 Project cases and results

In this case study, 3 different ECE applications are presented, where data of enhanced corrosion measurements were available from dust sampling locations before, directly after and mostly also some weeks or months after terminating the ECE. All numeric results are summarized in table 1 at the end of this paper, along with the graphical presentation of the electrochemical data in dependency on the chloride content in rebar vicinity.

Hollow box girder floor slab: This is a typical, very beneficial application field for ECE, as chloride has penetrated into the mainly dry concrete of the floor slab from leaking drainage pipes that are meant to remove surface water from the bridge deck. Along with the repair of the drainage system, ECE can remove the chloride smoothly and without restricting the use of the bridge. Multi-stage applications, which might be necessary in case of deep chloride ingress and low concrete cover, usually do not cause technological or administrational conflicts. With the eliminated origin of chloride ingress, the corrosion protection in a floor slab can be restored by ECE durably, and no other, long-term or additional steps are required.

Fig. 2. ECE electrode layout on the box girder floor slab

As long as tendon ducts do not have large defects, ECE can be applied safely also in pre-stressed, post-tensioned slabs (Faraday Cage). This has been the case in the project introduced here; furthermore some small areas of delaminated concrete were present; in one of them a test location has been specified and surveyed.

Fig. 3. Floor slab area after removing the ECE electrodes and while running the first corrosion measurements

All test locations have been surveyed with dust sampling, OCP and impedance measurements, GPM, LPR and TPM before ECE, directly after removing the electrodes and 3 months later. As can be seen in table 1, the chloride values in the rebar vicinity did not get reduced dramatically during the one-stage, 7-week application, but a very high amount of chloride has been analysed from the re-usable electrodes – ca. 100 g/m². The data obtained from the one test location in a delaminated area (yellow marked) did not deviate much from the other 3 locations, so the ECE seems to have worked there in a similar performance than on the other, undamaged areas.

Track bed of a subway: In a subway station considerable amounts of chloride and rebars showing pitting corrosion have been found in the track bed area. The source of the chlorides remained unclear; one assumption was that in came from acidic cleaning agents which had been used over many years on the tiles of the adjacent wall.

Fig. 4. ECE in the track bed of a subway under traffic

After a corrosion diagnosis and extensive testing of the ECE equipment for its feasibility under these special circumstances (no influence of the running equipment on the automated train control, and that the electrodes and cables remain safely in their position also at full speed of passing trains), the application was run as designed in up to 3 stages of 6 weeks each per sub-area and intermissions of also 6 weeks between the applications.

Fig. 5. View of the ECE electrodes and their fixing to the rails

The resulting data are also to be found in table 1. The amount of removed chloride was in the same range as on the box girder floor slab (100 g/m²), but the initial chloride contents as well as the amount of impressed charge were much higher. The end chloride level was at 3 of the 4 shown test locations very low, but in one (ref 3) 1.67 % residual chloride was analysed. The R_B, R_P and I_{corr} values are on a similar "good" level as in the other test locations.

Wall section of a highway tunnel: It was a special issue of this ECE application that it ran as part of a major repair project in the tunnel. This way, there was a strict schedule limiting the time of application and the accessibility of the wall section during the application. Furthermore, the power supply of the ECE was connected to the temporary power supply of the construction site, which caused frequent shut-downs during the first of two parts/ areas and required many site visits to replace fuses and to reboot the process. During the second part, high-capacity UPS devices were added that could filter voltage peaks and compensate short power interruptions even for the DC power supplies. In the first part, only 3.5 of possible 6.5 weeks operation time could be achieved, but in the second part the ECE worked during the available 6.5 weeks without interruptions.

Fig. 6. ECE electrodes on the tunnel wall

Fig. 7. ECE control centre and power supply

The ECE was applied as a one-stage treatment to wall areas at the tunnel entrance with high local chloride contents and relatively low concrete cover (ca. 30 – 40 mm, regularly > 60 mm). The general repair concept included a surface protection that prevents further chloride ingress.

Again, all data of the tunnel wall application can be seen in table 1. Its first part worked just 3.5 weeks net, and low amounts of removed chloride and impressed charge resulted from this. In test location ref 2 some reduction from the initial 5.35 % chloride content can be seen, but 3.7 % remained in an analysis 2 months after the scheduled termination of the treatment. Unlike ref 1 and 3, the corrosion measurement data of ref 2 showed no significant changes towards passivity. The open circuit potential was found to be rather positive, similar to the test locations ref 1 and 3.

3 Discussion

Generally, the data obtained from the enhanced corrosion measurements appear to be plausible and fit into expected schemes: R_P', LPR ≥ R_P', GPM, as fig 8 shows (at least in the interesting range of resistances up to 3,000 Ohm). The I_{corr} values fit quite well in an exponential relationship with the R_B values from AC impedance and the R_P values from GPM (figs 9 and 10).

It should be mentioned that these data have been taken on very different structures at different stages of the ECE, which were tested under widely varying conditions. In another relation, the electrochemical data are displayed versus the respective chloride content in rebar vicinity (figs 11 to 13).

In fig 11 it can be seen that due to the partly intensive wetting and strong cathodic polarization, extremely negative potentials were measured. Apparently these influences did not affect the other measurements (AC impedance, GPM, LPR, TPM), which is an indicator for the feasibility of these measurements under conditions that won't allow a useful interpretation of potential measurement data or where varying moisture conditions make the potential assessment very difficult.

The fit of the R_B and R_P data over the chloride values is not that excellent as over the I_{corr} values, very likely because the dust sampling has a larger "spread by coincidence" and does not match exactly to the test location of the measurement cell.

Figs 14 and 15 show the dependency of the assumed corrosion rates on the chloride contents, and this refers to the conversion of the obtained, absolute I_{corr} values into specific corrosion rates according to the remarks made in the introduction of the TPM. One has to be aware of the uncertainties of that step, but with the suggested corridor between estimated minimum and maximum rebar areas the true corrosion rate should be covered reasonably.

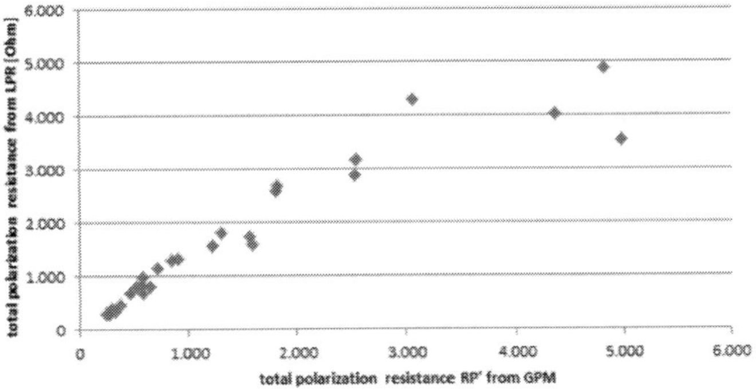

Fig. 8. Total polarization resistances R_P' from LPR versus R_P' from GPM

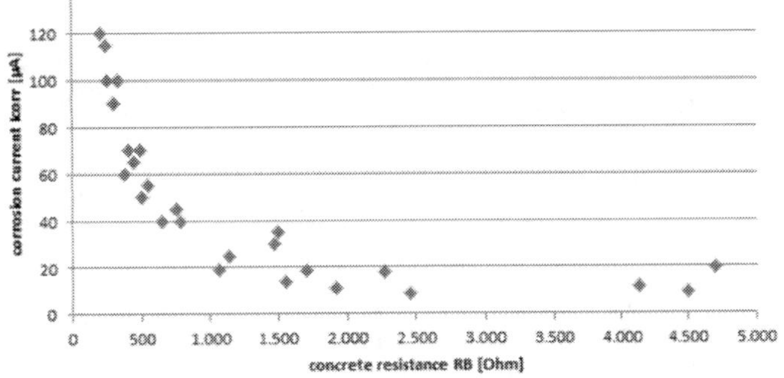

Fig. 9. Corrosion currents I_{corr} versus electrolyte resistances R_B

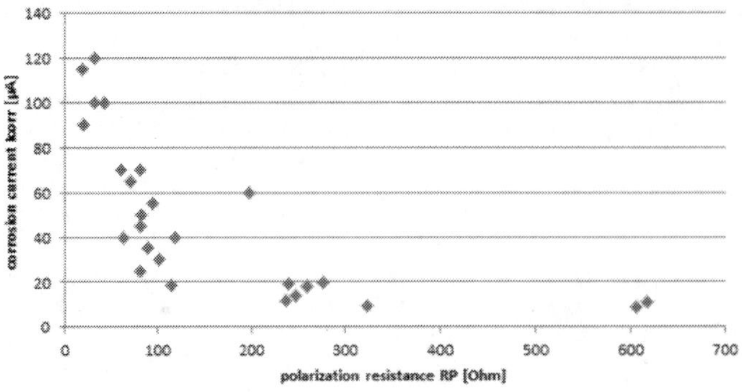

Fig. 10. Corrosion currents I_{corr} versus polarization resistances R_P

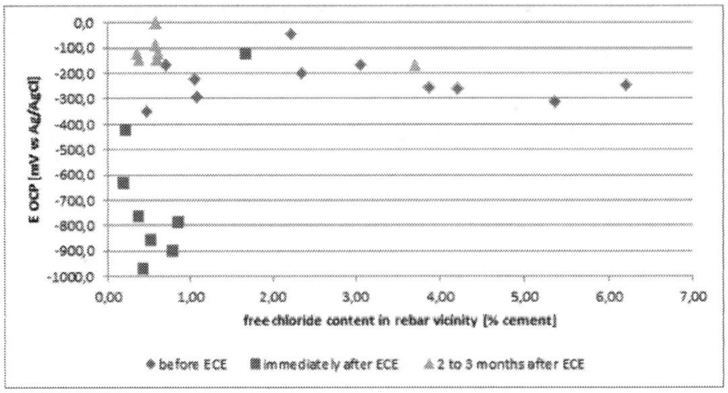

Fig. 11. Open circuit potentials versus chloride content in rebar vicinity at different stages of ECE

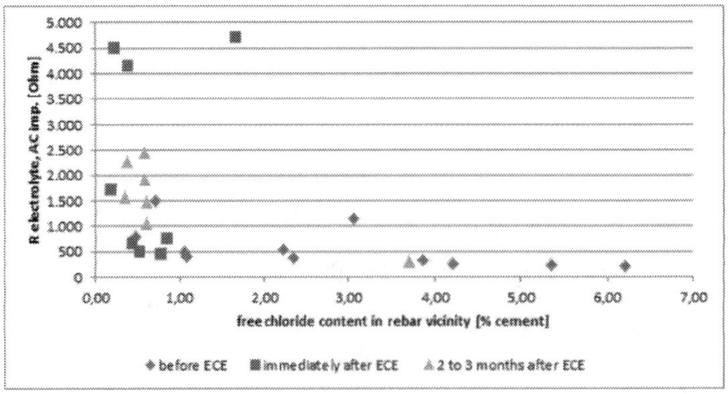

Fig. 12. Electrolyte resistances versus chloride content in rebar vicinity at different stages of ECE

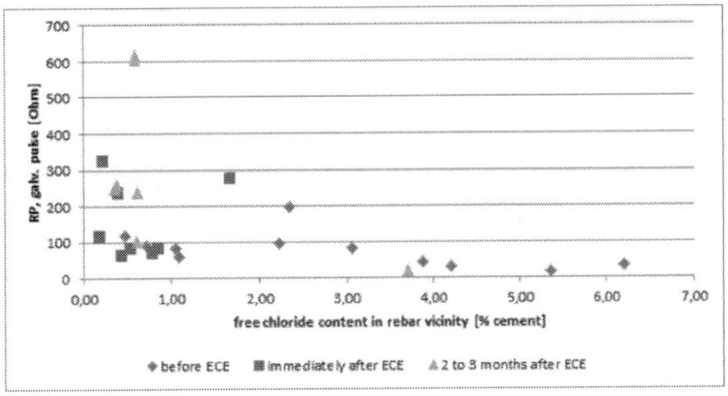

Fig. 13. Polarization resistances versus chloride content in rebar vicinity at different stages of ECE

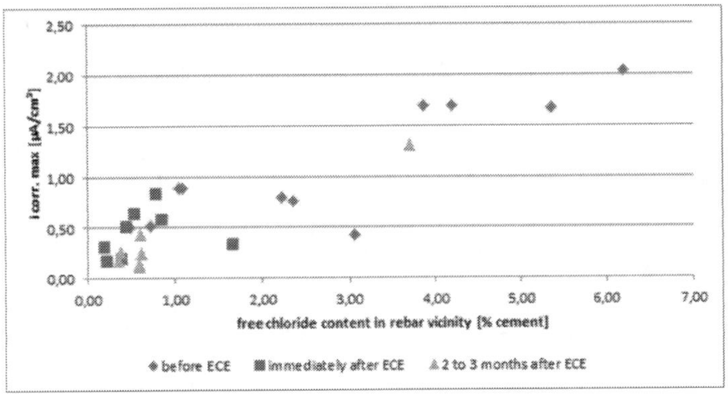

Fig. 14. Assumed maximum corrosion rates versus chloride content in rebar vicinity at different stages of ECE

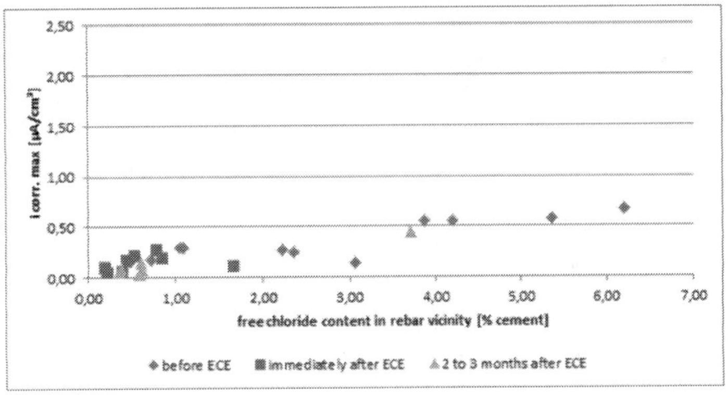

Fig. 15. Assumed minimum corrosion rates versus chloride content in rebar vicinity at different stages of ECE

In figs 16 and 17, the minimum and maximum i_{corr} values are displayed against the respective R_B and R_P values The 4 regression lines have a good fit to the data points ($R^2 > 0.8$), so it is possible to extract some limit values for R_B and R_P that may characterize passive behaviour. As passive corrosion rates, 0.1 or 0.2 μA/cm² are commonly considered. In [7] these two values are assigned to the type of measurement cell; without a guard ring 0.2 μA/cm² is suggested, which has been the case for the measurements discussed here. Table 2 summarizes the results:

Table 2. Limit R_B and R_P values for passive condition

$i_{passive}$ μA/cm²	R_B Ohm			R_P Ohm		
	min	max	avg	min	max	avg
0.20	700	2,800	**1,750**	90	340	**215**
0.10	1,700	6,000	3,850	200	700	450

With these average R_B and R_P limit values and the average i_{corr} values, the evaluation of the measurement results of table 1 is made in table 3. There, an extended assessment scheme is used with limit values, partly specified by experience, which classify the data into "good", "intermediate" and "poor". The ECE results of the 3 project cases can now be discussed as follows:

Hollow box girder floor slab: Despite that in 3 of 4 test locations the target chloride content of 0.4 % cement mass in rebar vicinity could not be met according to CEN TS 14038-2 and the impressed charges were rather low (between 300 and 650 Ah/m²), the electrochemical data indicated that passive conditions were achieved for the polarization resistance and the corrosion rate.

In this project, all data were measured before, directly after and 3 months after the ECE, and a similar development of the electrochemical data can be seen in all test locations: directly after the ECE the electrolyte resistances R_B are at least in the same range as before, the polarization resistances RP have partly dropped, while the I_{corr} values have remained or dropped slightly under apparent critical measurement conditions (between 6 and 15 % water content). The data obtained 3 months later indicate that this can already be interpreted as signs of a successful treatment as then all corrosion rates dropped by 65 to 80 % and were found in the passive range. The R_B and R_P values were also measured at levels indicating passivity according to table 2.

Track bed of a subway: In 3 of the 4 test locations (ref 1, 2, 4) the target chloride was reached, and due to the multi-stage treatments, the impressed charge was well above 1,000 Ah/m². The R_B, R_P and i_{corr} values of ref 2 were not in the specified range of passivity yet, but the respective OCP certainly indicated cathodic polarization, so the corrosion behaviour could be expected to be in the safe passive range in the near future. Ref 3 exhibits still a high residual chloride (after > 6 % of initial chloride content), but surprising positive OCP along with sufficiently high R_B and R_P values. Considering the experiences from the box girder floor slab case, sound passivity can be expected also here within the next months.

Wall section of a highway tunnel: The data of this ECE application in table 1 show the effect of the insufficient treatment time on the results: in ref 1 and 2 only little charge could be impressed (the vertical arrangement of the electrodes cannot support good conditions for wetting the concrete even with an implemented water supply). Consequently, the water content remained rather low, and especially in ref 2 more than 3.7 % chloride remained in the rebar vicinity. However, some effect of corrosion protection seemed to be achieved, as the potentials were at a quite positive level and did not show considerable differences between the test locations. Because of the time restrictions on this site it was decided to provide additional corrosion protection by discrete zinc anodes in the areas of initially very negative potentials.

4 Conclusions

The use of enhanced corrosion measurements for the instant evaluation of the success of an ECE treatment seems feasible and can support the usual evaluation by chloride content determination and impressed charge quite well, especially if the target chloride values could not be met.

The observed similar increase of R_B and R_P along with the substantial drop of corrosion currents and of the assumed corrosion rates during 3 months after should be verified in future projects. If there is systematic, reliable development, this knowledge can be used for the extrapolation of the corrosion data and also for determining a suitable, early end point of treatment.

However, it is not advised to run an ECE under a tight schedule, especially if the conditions are difficult (chloride behind outer rebar layer, low concrete cover, chance of technical problems). If the targets of the ECE treatment are not met for any reason and there is no opportunity for repeated treatments, the ECE cannot be successful, which is not a disadvantage of the technology but the consequence of an improper design.

5 References

1. K. Ahlborn, F. Berthold, W. Vonau, H. Grünzig, U. Schneck, H. Jahn, J. Köhler, *Modular Corrosion Measurement System (CMS) for electrochemical NDT*. 4th International Conference on Concrete Solutions, Proc., Dresden: Balkema, Rotterdam, 2011
2. U. Schneck: *Instant and solid corrosion diagnosis on reinforced concrete structures with galvanostatic pulse measurement.* Concrete Solutions 5th International Conference on Concrete Repair, Proc., Belfast: Balkema, Rotterdam, 2014
3. U. Schneck: *Electrochemical chloride extraction*. In M. Grantham (ed): Concrete Repair – A practical guide, p. 147-168: Taylor & Francis, Milton Park, Abington, 2011
4. R. Bäßler, A. Burkert, T. Frølund, O. Klinghoffer: *Usage of portable equipment to determine the corrosion state of concrete structures*. In M. Raupach, B. Elsener, R. Polder, J. Mietz (eds.): EFC-Publications Number 38 - Corrosion of reinforcement in concrete. p159 - 169: Woodhead Publishing Ltd., Abington, 2007
5. U. Schneck: *Insights from corrosion current measurements on corrosion mechanisms in reinforced concrete and on the evaluation of other corrosion data.* Concrete Solutions 7th International Conference on Concrete Repair, Proc., Cluj Napoca: MATEC Web of Conferences 2019
6. A. Bard, L. Faulkner: *Electrochemical Methods. Fundamentals and Applications*, John Wiley and Sons Inc., 2001
7. J. Broomfield, *Corrosion of steel in concrete – understanding, investigation and repair 2nd Edition.* Taylor & Francis, Milton Park, Abington, 2007

Table 1: Results of the enhanced corrosion measurements in comparison to the other process data of all 3 project cases

project	ref	stage	cover mm	chl. max % cement	chl. rebar % cement	water rebar % concrete	chl. removed g/m²	charge Ah/m² rebar	weeks	cycles	E_{OCP} mV vs SE	R_B Ohm	$R_{P,GPM}$ Ohm	$R_P'{}_{,GPM}$ Ohm	$R_P'{}_{,LPR}$ Ohm	I_{corr} µA	i_{corr}, max µA/cm²	i_{corr}, min µA/cm²
box girder floor slab	1	before ECE	19	2,35	2,35	2,96	98	665	7,0	1	-197,7	382	198	580	696	60	0,77	0,25
		after ECE		0,86	0,86	12,44					-790,7	760	82	842	1.310	45	0,57	0,19
		ECE + 3m		0,59	0,59	1,68					-87,5	2.460	606	3.066	4.307	9	0,11	0,04
	2	before ECE	20	1,05	1,05	3,46	98	665	7,0	1	-225,3	490	82	572	848	70	0,89	0,29
		after ECE		0,79	0,79	15,34					-898,6	450	71	521	819	65	0,83	0,27
		ECE + 3m		0,62	0,62	4,15					-122,3	1.060	239	1.299	1.803	19	0,24	0,08
	3	before ECE	35	1,05	0,48	3,32	98	505	7,0	1	-346,9	790	118	908	1.321	40	0,51	0,17
		after ECE		0,52	0,44	6,45					-970,9	650	63	713	1.144	40	0,51	0,17
		ECE + 3m		0,36	0,35	3,72					-120,0	1.560	248	1.808	2.586	14	0,18	0,06
	4	before ECE	36	1,86	1,09	3,75	98	302	7,0	1	-293,0	407	61	468	689	70	0,89	0,29
		after ECE		1,22	0,54	6,17					-857,9	500	83	583	978	50	0,64	0,21
		ECE + 3m		1,25	0,59	4,34					0,0	1.920	617	2.537	3.190	11	0,14	0,05
trackbed	1	before ECE	38	3,87	3,87	3,97	67	1.215	13,0	2	-257,0	327	42	369	475	100	1,70	0,55
		after ECE		0,59	0,39	4,31					-765,0	4.134	237	4.371	4.021	11	0,19	0,06
	2	before ECE	40	4,20	4,20	4,37	103	1.950	16,0	3	-261,0	258	31	289	410	100	1,70	0,55
		after ECE		0,19	0,19	4,20					-635,0	1.710	114	1.824	2.690	18	0,31	0,10
	3	before ECE	42	6,21	6,21	5,17	103	1.525	16,0	3	-248,0	205	32	237	300	120	2,04	0,66
		after ECE		1,67	1,67	4,88					-127,0	4.702	277	4.979	3.553	20	0,33	0,11
	4	before ECE	44	5,78	3,06	4,09	103	1.423	16,0	3	-168,0	1.140	81	1.221	1.577	25	0,43	0,14
		after ECE		0,42	0,22	5,14					-423,0	4.500	324	4.824	4.882	10	0,16	0,05
tunnel wall	1	before ECE	38	1,56	0,72	1,62	37	219	3,5	1	-168,0	1.500	90	1.590	1.600	35	0,51	0,18
		ECE + 2m		0,61	0,61	2,92					-145,0	1.470	101	1.571	1.730	30	0,44	0,15
	2	before ECE	33	5,35	5,35	2,85	37	131	3,5	1	-313,0	250	18	268	300	115	1,67	0,58
		ECE + 2m		3,71	3,71	3,01					-165,0	303	20	323	348	90	1,31	0,45
	3	before ECE	39	2,82	2,23	n.a.	101	441	6,5	1	-46,0	550	95	645	820	55	0,79	0,28
		ECE + 2m		0,57	0,38	n.a.					-145,0	2.270	259	2.529	2.875	18	0,26	0,09

classification of corrosion current densities
measurement cell without guardring

high (> 1.0 .. 10.0 µA/cm²)
moderate to low (> 0.2 .. 1.0 µA/cm²)
passive (< 0.2 µA/cm²)

 https://doi.org/10.1051/matecconf/201928903009

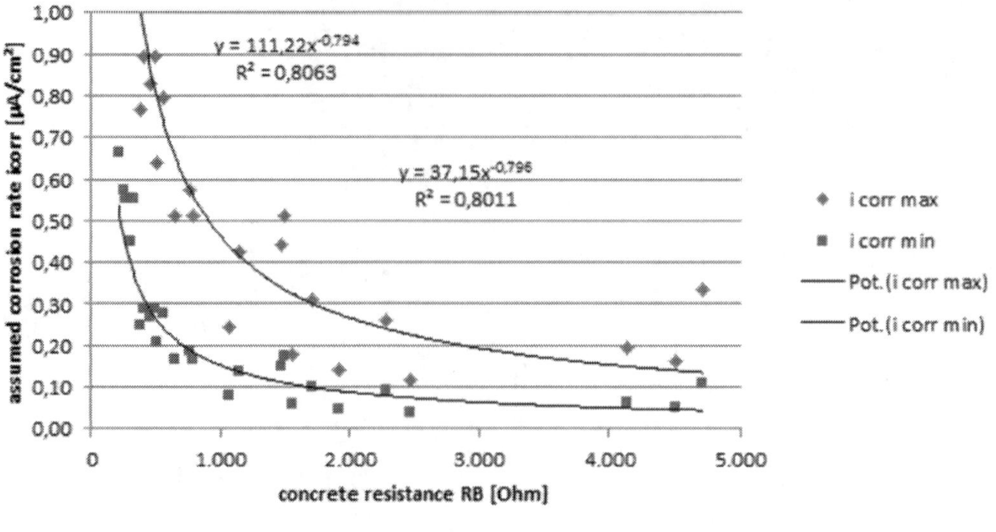

Fig. 16. Regression curves of the assumed maximum and minimum corrosion rates versus the concrete resistances

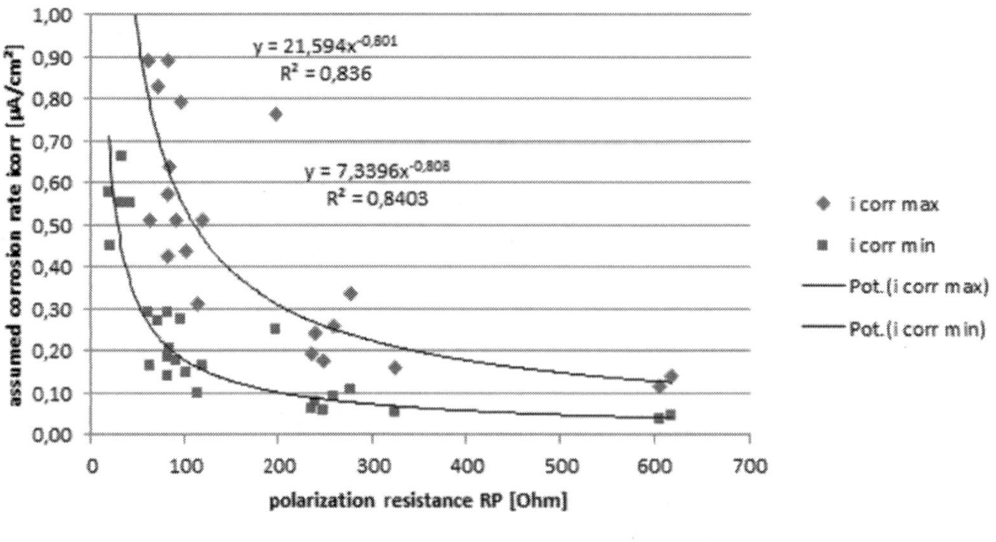

Fig. 17. Regression curves of the assumed maximum and minimum corrosion rates versus the polarization resistances

 https://doi.org/10.1051/matecconf/201928903009

Table 3. Evaluated data of the measurements after ECE

project	ref	stage	chl. rebar	water rebar	charge	E_{OCP}	R_B		R_P		i corr, avg	I corr	I corr
			% cement	% concrete	Ah/m² rebar	mV vs SE	Ohm	increase	Ohm	increase	µA/cm²	reduction	end to 3 m
box girder floor slab	1	after ECE	0,86	12,44	665	-790,7	760	199%	82	41%	0,38	-25%	
		ECE + 3m	0,59	1,68		-87,5	2.460	644%	606	306%	0,08	-85%	-80%
	2	after ECE	0,79	15,34	665	-898,6	450	92%	71	87%	0,55	-7%	
		ECE + 3m	0,62	4,15		-122,3	1.060	216%	239	291%	0,16	-73%	-71%
	3	after ECE	0,44	6,45	505	-970,9	650	82%	63	53%	0,34	0%	
		ECE + 3m	0,35	3,72		-120,0	1.560	197%	248	210%	0,12	-65%	-65%
	4	after ECE	0,54	6,17	302	-857,9	500	123%	83	136%	0,42	-29%	
		ECE + 3m	0,59	4,34		0,0	1.920	472%	617	1011%	0,09	-84%	-78%
track bed	1	after ECE	0,39	4,31	1.215	-765,0	4.134	1264%	237	564%	0,13	-89%	
	2	after ECE	0,19	4,20	1.950	-635,0	1.710	663%	114	368%	0,21	-82%	
	3	after ECE	1,67	4,88	1.525	-127,0	4.702	2294%	277	866%	0,22	-84%	
	4	after ECE	0,22	5,14	1.423	-423,0	4.500	395%	324	400%	0,11	-62%	
tunnel wall	1	ECE + 2m	0,61	2,92	219	-145,0	1.470	98%	101	112%	0,30	-14%	
	2	ECE + 2m	3,71	3,01	131	-165,0	303	121%	20	111%	0,88	-22%	
	3	ECE + 2m	0,38	n.a.	441	-145,0	2.270	413%	259	273%	0,18	-67%	

	chl. rebar	water rebar	charge	E_{OCP}	R_B	R_P	i corr, avg
good/ suitable	< 0,4	< 3,5	> 600	> -200	> 1750	> 215	< 0,2
intermediate	0,4 .. 1,0	3,5 ..4,5	400 .. 600	-200 ..-300	1750 .. 1000	215 .. 100	0,2 ..1,0
poor/ insufficient	> 1,0	> 4,5	< 400	< -300	< 1000	< 100	> 1,0

Galvanic chloride extraction by an embedded zinc anode: Ion distribution mapped by laser induced breakdown spectroscopy (LIBS)

W. Schwarz[1)], *Gerd Wilsch*[2)], *A. Pichlhöfer*[1)], *G. Ebell*[2)], *T. Völker*[2)]

[1]CAS Composite Anode Systems GmbH, Vienna, Austria
[2]BAM – Bundesanstalt für Materialforschung und -prüfung, Berlin, Germany

Abstract. An important aspect with regard to the service life of zinc based galvanic anodes and the durability of the corrosion protection of steel in concrete is the "galvanic chloride extraction". Chloride ions move in the electric field generated by the current, flowing between the galvanic anode and the cathodic steel. Migration leads to an accumulation of anions, e.g. chloride ions, at the anode and depletion of chlorides near the steel rebar surface. The ion migration was studied on steel reinforced concrete specimens admixed with 3 wt.% chloride/wt. cement and galvanically protected by a surface applied embedded zinc anode (EZA). The zinc anode was embedded and glued to the concrete surface by a geo-polymer based chloride free binder. The EZA was operated over a period of 1 year and the ion distribution between anode (EZA) and cathode (steel reinforcement) was studied by laser induced breakdown spectroscopy (LIBS) after 5 months, 7 months and 12 months. The results show that chloride ions efficiently migrate in the direction of the zinc-anode and accumulate there. Chloride distribution in the EZA correlates with the distribution of zinc ions generated by the anodic dissolution of the zinc anode in the binder matrix. The microstructure of the binder matrix and its interface to the zinc-anode are studied by REM/EDX – preliminary results will be reported.

1 Introduction

Galvanic corrosion protection of steel in concrete is based on the formation of a galvanic element when a metal which is less noble than cast iron steel, in direct contact with the concrete overlay, is electrically connected to the steel rebars. The reinforcing steel is protected from corrosion as long as sufficient galvanic current flows between the galvanic anode and the steel reinforcement. Most commonly, zinc is used as the sacrificial anode material. The galvanic element formed corresponds to a conventional zinc/air battery that is becoming popular again as an alternative source of energy.

Galvanic corrosion protection was first employed to protect a bridge deck in Illinois in 1977 within the cooperative highway research program, with mixed results [1]. A problem with the initially applied sacrificial anodes was that their protection current decreases with time, and they eventually become passive, so most systems have a relatively short useful life [2].

In the 1990's, sacrificial anode systems based on sprayed zinc anodes, zinc foil glued to the concrete surface (zinc hydrogel system) and zinc mesh pile jackets around bridge columns filled with sea water were starting to be evaluated and used for the protection of bridge structures [2 – 5].

The efficiency of galvanic corrosion protection depends on the lasting activity of the zinc anode. Deposition and agglomeration of the anodic products like zinc hydroxide and zinc hydroxychlorides or contact with calcium hydroxide in the pore solution may passivate the zinc anode surface. The service time of the zinc anode may be limited by self-corrosion which increases with the activation of the zinc anode and may reach up to 70% of the zinc consumed during operation.

A galvanic zinc anode system composed of a zinc mesh embedded into a proprietary mortar that solidifies into a solid electrolyte, installed on various civil structures (bridges, marine harbour piles, parking decks, tunnel entrances) protects the steel reinforcement reliably and efficiently from corrosion for up to 11 years [6]. The solid electrolyte of the embedded zinc anode system (EZA) is based on a tecto-alumosilicate-binder (TAS) containing additives that prevent passivation of the zinc anode, assure high and durable galvanic activity of the zinc anode and high and durable adhesion towards the concrete overlay.

Previous studies of concrete overlay samples drawn from field installations indicated chloride migration and chloride fixation near the zinc anode [7], however verification and quantification on samples drawn from field installations proved to be difficult as the original chloride distribution in the concrete and the embedding mortar are difficult to evaluate.

* Corresponding author: schwarz@cas-composite.com

Migration of chlorides in concrete in an applied electric field is well established and employed e.g. for rapid determination of diffusion constants in concrete [8-10]. The electric field generated by the galvanic current, is estimated to be in the range of 300 – 600 mV, estimated from the open-circuit potential between zinc-anode and steel reinforcement, and is much lower than the electric fields usually applied (about 1 Ampere /m^2) for electrochemical chloride extraction [11]. Migration is only slightly influenced by the applied electric field at high electric fields (e.g. 5 – 10 Volts, about 1 Ampere/m^2) and mainly controlled by the current [12]. The influence of the electric field on migration at very low field strength is not well studied.

Therefore, the performance of an EZA embedded into a chloride free embedding TAS binder (shrinkage reduced embedding zinc activating cement – SEZAC) was studied and evaluated under controlled laboratory conditions. The influence of the galvanic field on the ion distribution, especially on the chloride distribution in the concrete overlay and the embedding binder could be evaluated and visualized by LIBS – laser induced breakdown spectroscopy. Results are reported in this paper; evaluations by REM and EDX are under way and will be reported soon in CCR (Cement and Concrete Research).

2. Experimental

2.1. Galvanic Anode/Mortar Probe

Table 1: steel reinforced mortar - test plates

dimensions	26 x 17 x 6 cm	
steel reinforce-ment	Ø 10 mm spacing: e 17 connecting steel	4 bars à 240 mm 4 mm & 6 mm
mortar	SikaMonotop 412 N water sodium chloride	13,85 kg 2,0 kg 0,242 kg 0,84 wt.% Cl/kg fresh mortar
reference cell	ERE 20	Type: Mn/MnO$_2$

Note: SikaMonotop 412 N mortar was chosen as being certified for CP use. The cement content is rather high (640 kg/m^3 fresh mortar).

Embedding binder for zinc mesh: geopolymer based 2-component SEZAC binder admixed with ground marble filler. Mixing ratio component A (binder), component B (aqueous high alkali potassium silicate), Filler (0,2 – 0,5 mm ground marble):

A : B : Filler = 100:50:104

Specific weight of wet binder: 1,84 kg/liter

Zinc-mesh: 99% zinc alloy, mesh size 5 mm – 7 mm, wire diameter 1,1 mm; 2,7 kg zinc/m^2 mortar surface, 1,1 kg zinc-mesh per plate embedded into 0,750 kg SEZAC binder.

The EZA probes were stored at 99% rh at RT for 1 week for initial hardening and then stored at 80% rh for an additional week.

2.2. Operation of the EZA Galvanic Anode System

The EZA probes were stored in a climatized compartment at 75% rh. Current between the steel reinforcement and the zinc anode was measured over a 0,1 Ohm shunt and transformed into voltage with an operational amplifier. Steel potentials were measured with a 20 MΩ input impedance. Data were recorded by a MODAC data logger. Additionally, steel potentials were measured manually with an Ag/AgCl reference cell placed onto the surface of the mortar probe.

2.3. Laser Induced Breakdown Spectroscopy (LIBS)

The LIBS technique was used for the investigation of the ion transport processes . LIBS - is a combination of laser ablation and plasma generation by a short (ns) high energy (several mJ) laser pulse and the analysis of the emitted radiation by optical spectroscopy. LIBS can capture all elements of the periodic table simultaneously through the underlying measurement principle [13]. An overview of the potential of LIBS for the investigation of building materials can be found in the literature [13-17]. For the investigation in this paper the FiberLIBSlab system from SECOPTA analytics was used (see fig. 1).

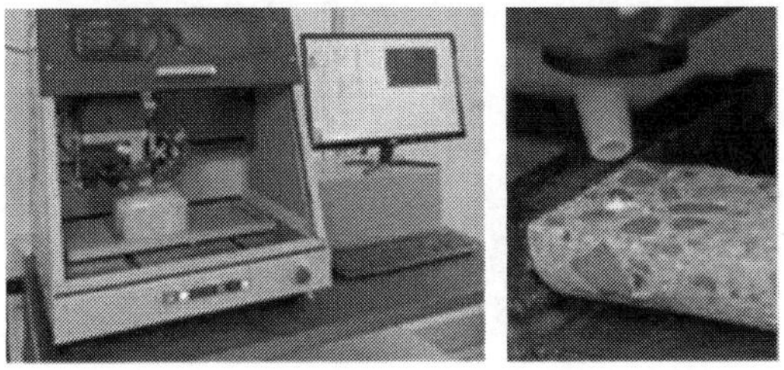

Fig. 1. Overall view of the LIBS laboratory system (left). Detailed view on the sample surface with plasma and nozzle of helium feeder (right).

For the quantitative analysis, a calibration of the system with 15 reference samples based on cement (CEM I) with a concentration range of chloride of 0,05 to 2,5 wt% was performed. The limit of detection for chlorine in concrete is 0,04 % regarding to the mass of cement. A scanner is used to obtain a two-dimensional element mapping.

All the elements listed in table 2 were recorded simultaneously for each measuring point. The multielement information can be used for exclusion of aggregates, rebars or points belonging to the zinc anode by defining limits for the LIBS-signals in the software.

Table 2: main parameters of the used LIBS system

Plasma generation	Detection of radiation
NdCr:YAG-laser: 1064 nm, 3 mJ, 1.5 ns, 100 Hz passive Q-switch	spectrometer: Czerny-Turner 1200 lines/mm, 2048 pixel temperature stabilised
Ø laser spot: 80 µm	NIR 750 nm - 950 nm UV 230 nm - 500 nm
power density: 40 GW/cm²	elements: NIR: Ca, O, K, Na, Cl, C, S UV: Ca, Si, Mg, Al, Fe, Ag, Ti
process gas: integrated scanner:	helium A_{max} = 14 cm x 17 cm

The galvanic anode/mortar probes were cut along the broad side for performing the LIBS measurements on the cross-section as shown in fig. 2:

Fig. 2. Galvanic anode/mortar probe cut broadside for LIBS measurements

On the sample an area of about 85 mm by 65 mm measurements were taken with a spatial resolution of 0.5 mm x 0,5 mm. As result the element distribution was plotted as multicolor image over the measured area (e.g. figure 4). Three samples were prepared after 5 months, 7 months and 12 months of galvanic operation.

3. Results

3.1. Galvanic Performance

The galvanic current between the EZA mounted on the mortar probe stored in a climatized chamber (75% rh, 20 – 25°C) and the steel reinforcement was monitored. The current flow over one years' time is shown in figure 3.

The galvanic current stabilized after about 8 months of operation at 7 – 10 mA/m². Galvanic operation resulted in a significant shift of the steel potentials towards positive values (operational data are summarized in table 3). 24 h depolarisation according to EN 12696 after 12 months of operation yielded a value of 138 mV.

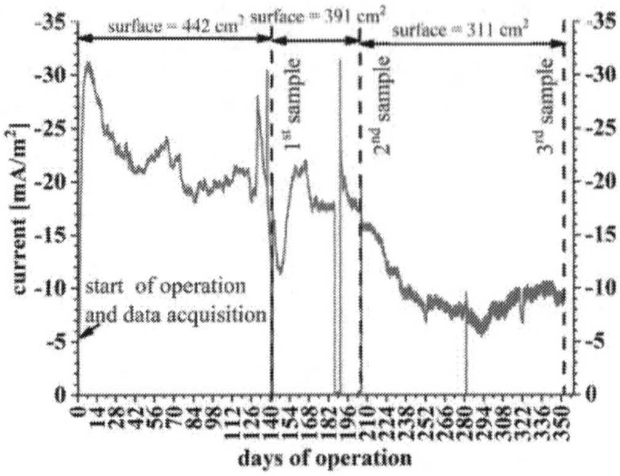

Fig. 3. Galvanic current flow over time at 75% rh and RT.

Table 3. steel potentials before and during galvanic operation of EZA mounted on the mortar probe during 24 h depolarisation measurements (off-potentials)

Time of galvanic operation [month]	0	5	12
24 h off-potential [mV]	-374	-246	-172
Charge passed [kC]	0	261	476

After 12 months of operation, no visible corrosion of the zinc anode could be detected (figure 2).

3.2. LIBS measurements

LIBS measurements were performed on samples cut after 5, 7 and 12 months of galvanic operation of EZA mounted on the mortar probe.

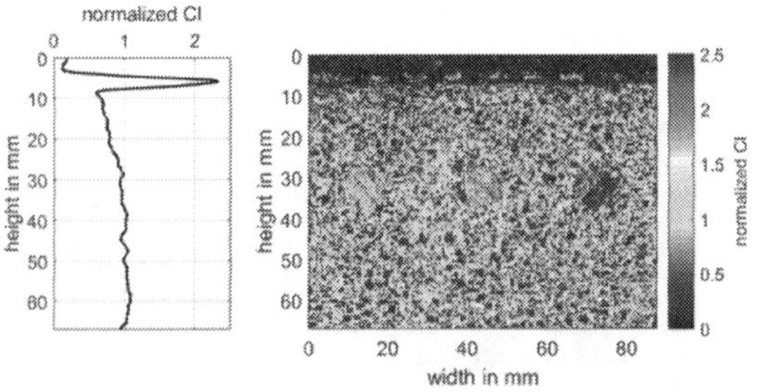

Fig. 4. Color coded chlorine distribution determined by LIBS in the cross section of EZA/mortar probe after 7 months of galvanic operation (right) and depth profile of chlorine (left). The chloride content is normalized to the chloride content of the mortar in areas not exposed to the electric field generated by the galvanic current.

After 7 months of operation, chloride migration towards the anode was significant, creating a depletion zone between the steel reinforcement (that was not yet visible after 5 month of galvanic operation) and the zinc-anode. The chloride accumulated near the zinc-anode – there was no chloride in the binder material itself detectable (figure 4) and the chloride gradient between steel reinforcement (cathode) and anode increased significantly after 5 months additional galvanic operation (figure 5).

Note: the circular area with an apparent red circle in it (o), between the two rebars on the centre right originates from a cut Ag/AgCl reference cell, the red circle is due to the KCl electrolyte.

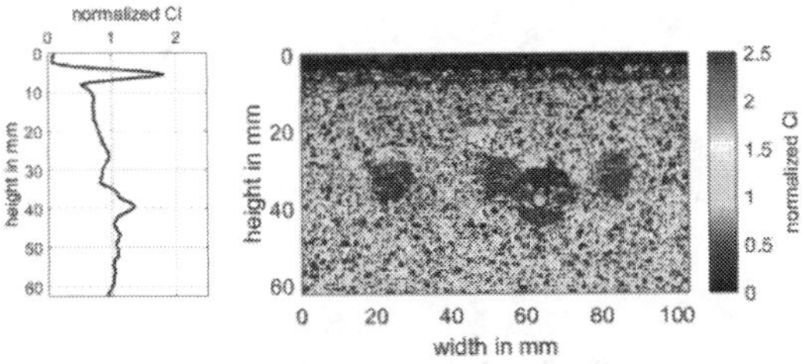

Fig. 5. Color coded chlorine distribution determined by LIBS in the cross section of EZA/mortar probe after 12 months of galvanic operation (right) and depth profile of chlorine (left).

The software of the LIBS allows, as described in chapter 2.3., the exclusion of the areas that contain zinc, iron and aggregate as shown in figure 6 & 7: In the area of Zn-exclusion, chloride concentration was zero – indicating that the chloride was linked to zinc products accumulated in the vicinity of the zinc-anode, such as zinc-hydroxide.

The given quantitative chloride values (Fig. 6) should only be regarded as guide values. As already mentioned, the LIBS system was calibrated with cement-based reference samples (CEM I). A deviation from the material can lead to errors in quantification due to matrix effects (e.g. here the binder).

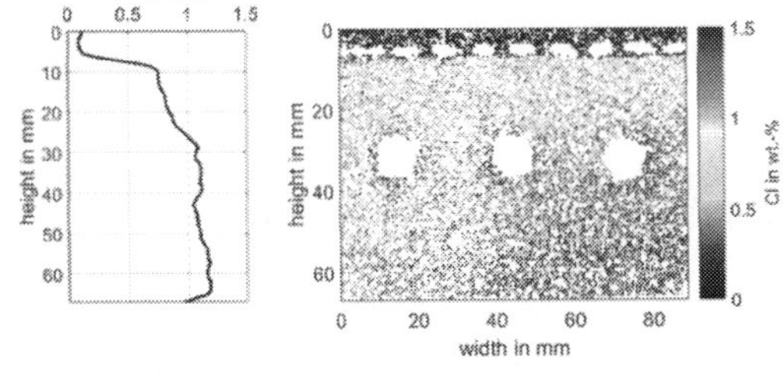

Fig. 6. Chloride distribution determined by LIBS in the cross section EZA/mortar probe after 7 months of galvanic operation: Zn, Fe & aggregate excluded/deducted (white).

This data supports the hypothesis that chloride forms with the zinc-hydroxide a zinc-hydroxy-chloride, e.g. zinc chloride hydroxide monohydrate (Simonkolleite). The nature of the mineral phases formed near the zinc anode is currently being investigated by REM and EDX and will be published soon.

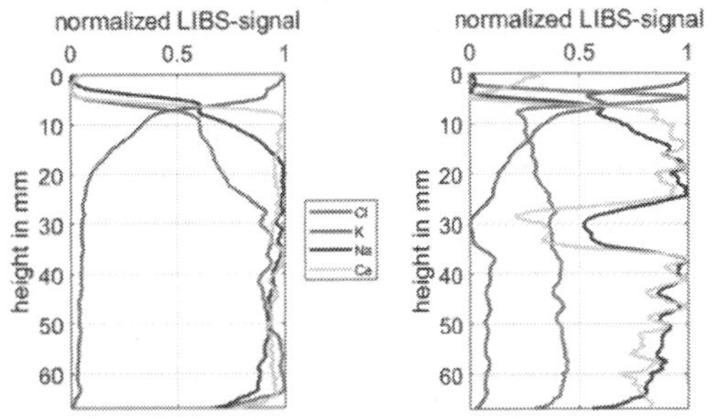

Fig. 7. Ion distribution determined by LIBS in the cross section EZA/mortar probe after 7 months of galvanic operation: Zn, Fe & aggregate excluded/deducted (left graph) in comparison with the ion distribution without exclusion (right graph).

The distribution of chloride, potassium, sodium and calcium under the exclusion of aggregate, iron and zinc in the cross-section of the EZA-mortar probe is shown in figure 7: The chloride depletion zone is clearly visible (left graph in fig. 7). Sodium and potassium migrate towards the cathode – the binder is rich in potassium, free in sodium and nearly calcium-free – calcium is detected only in the mortar phase and in the aggregates in the binder phase (figure 7, ion distribution without exclusion). The absence of calcium in the binder is essential for the functioning of the SEZAC binder as zinc-activating cement.

4. Summary and outlook

The evaluation of elemental distribution in the concrete/binder matrix by LIBS allows the investigation and graphical representation of the changes in the ion distribution caused by the applied electric field.

The results show that there is a significant accumulation of chloride in the vicinity of the zinc-anode directly linked to the local presence of zinc compounds resulting from the anodic reaction of zinc. There is no chloride in the binder surrounding the zinc-anode indicating that the chloride may be bound by the anodic products of the zinc anode.

A significant chloride gradient showing decreasing chloride concentration from the steel reinforcement towards the zinc anode indicates chloride extraction into binder/zinc-anode placed on the mortar surface. The results indicate that significant chloride extraction may occur on a time scale of a magnitude larger than the time period of observation (12 month).

However, one has to take into account the high cement content of the mortar (700 kg/m^3) that will chemically absorb significant amounts of chloride and therefore significantly impede chloride migration. Therefore, chloride migration during galvanic corrosion protection in probes with a cement content of the mortar more common to field situations – max. 300 kg/m^3 - should be studied.

The results indicate that the chloride is chemically bound in the vicinity of the zinc anode therefore impeding back diffusion of chloride into the mortar/concrete matrix. The identification of the mineral phases and their chemical composition of the zinc-compounds in which chloride is bound are important and are underway. The results will be reported soon.

5. References

1. J.L. Kepler, D. Darwin and J.R. Locke, *Evaluation of Corrosion Protection Methods for Reinforced Concrete Highway Structures*, Structural Engineering and Engineering Materials SM Report No. 58, University of Kansas Center for Research Inc., Lawrence, Kansas, May 2000
2. Y. P. Virmany, G.G. Clemena, *Corrosion Protection-Concrete Bridges*, Report No. FHWA-RD-98-088, Federal Highway Administration, Washington, D.C., 1998
3. R.J. Kessler, R.G. Powers, and I.R. Lasa, *Un update on the long-term use of cathodic protection of marine structures*, Corrosion 2002, paper 02254, NACE International
4. S. Szabo, I. Bakos, *Cathodic Protection with Sacrificial Anodes*, Corrosion Reviews 24: 2006, pp. 231 – 280
5. S.J. Bullard, S. Cramer and B.Covino, *Final Report – Effectiveness of Cathodic Protection*, SPR 345. Report No. FHWA-OR-RD-09-18, National Energy Technology Laboratory, Oregon, 2009
6. W. Schwarz, A. Pichlhöfer, A. van den Hondel and H. Esteves, *Maintenance and repair of steel reinforced concrete structures by galvanic corrosion protection – field experiences over 10 years, ICCRRR 2018 Cape Town, South Africa, November 19-21, 2018, in* MATEC Web of Conferences 199, 05005 (2018)
7. W. Schwarz, F. Müllner, A.J. van den Hondel, *Maintenance and repair of steel reinforced concrete structures by simultaneous galvanic corrosion protection and chloride extraction, fib Symposium 2016, Cape Town, South Africa 21-23 November 2016*
8. A. Delagrave, J. Marchaud, and E. Samson, *Prediction of Diffusion Coefficients in Cement-Based Materials on The Basis of Migration Experiments*, Cement and Concrete Research, Vol. 26, No. 12, pp. 1831-1842, 1996
9. C. Andrade, Calculation of chloride diffusion coefficients in concrete from ionic migration measurements, Cement and Concrete Research, Volume 23, Issue 3, May 1993, Pages 724-742
10. L. Tang, H. E. Sørensen, *Precision of the Nordic test methods for measuring the chloride diffusion/migration coefficients of concrete*, Materials and Structures, Vol. 34, October 2001, pp 479-485
11. B. Elsener, U. Angst, *Mechanism of electrochemical chloride removal*, Corrosion Science, Volume 49, Issue 12, December 2007, 4504-4522
12. E. Samson, J. Marchand, K.A. Snyder, Calculation of ionic diffusion coefficients on the basis of migration test results, Materials and Structures, Vol 36, April 2003, pp 156 – 165
13. Cremers, D.; Radzemski, L., 2006: *Handbook of Laser-induced Breakdown Spectroscopy*, John Wiley & Sons Ltd (2006).
14. G. Wilsch, F. Weritz, D. Schaurich, H. Wiggenhauser, *Determination of chloride content in concrete structures with laser-induced breakdown spectroscopy*, Construction and Building Materials 19 (10) (2005) 724 – 730. doi:10.1016/j.conbuildmat.2005.06.001.
15. G. Wilsch, D. Schaurich, H. Wiggenhauser, *Imaging laser analysis of building materials - practical examples*, AIP Conference Proceedings 1335 (1) (2011) 1315 – 1322. doi:10.1063/1.3592085.
16. G. Wilsch, T. Eichler, S. Millar and D. Schaurich, *Laser Induced Breakdown Spectroscopy (LIBS) - alternative to wet chemistry and micro-XRF*, CRC Press 2014 (2014) 611 – 615. doi:10.1201/b17394-94
17. S. Millar, C. Gottlieb, T. Guenther, N. Sankat, G. Wilsch, S. Kruschwitz, *Chlorine deter-mination in cement-bound materials with Laser-induced Breakdown Spectroscopy (LIBS) - A review and validation*, Spectrochimica Acta Part B: Atomic Spectroscopy, Volume 147, September 2018, p. 1-8

A fresh look at depolarisation criteria for cathodic protection of steel reinforcement in concrete

Nikita Rathod[1], *Peter* Slater[1], *George* Sergi[2], *Gamini* Seveviratne[2] and *David* Simpson[2]

[1]School of Chemistry, Birmingham University, Edgbaston, Birmingham B15 2TT, UK
[2]Vector Corrosion Technologies Ltd, 27a Upper High Street, Cradley Heath, West Midlands B64 5HX, UK

Abstract. Criteria for the successful application of cathodic protection (CP) for steel reinforced concrete have been fixed for decades and form part of ISO EN12696. The most used criterion is the achievement of 100 mV depolarization over a period not exceeding 24 hours after discontinuation of the applied current. Although more empirical than theoretically based, the criterion has served the CP industry well. It does, however, exclude any systems that may not always achieve that level of depolarization but have been shown to offer adequate protection, and so there is a need to explore ways of assessing depolarisation data more effectively. On a fundamental level, non-linear polarisation, as described by the Butler Volmer equation, relates corrosion rate to polarisation for a given applied current density and shows that at low current densities, estimated corrosion rates can be shown to be still insignificant at less than 100 mV polarisations. This paper explores the use of non-linear polarisation as an additional supportive criterion based on the measured 24-hour depolarisation level for a known applied current density and tests its applicability in the laboratory and in the field. It speculates that a reducing apparent corrosion current density trend in combination with a depolarised potential moving in a more noble direction is likely to be a suitable alternative criterion, where 100 mV depolarisation is not achieved.

1 Introduction

Both ISO EN 12696 : 2016 [1] and NACE SP0290 : 2007 [2] have a set of criteria that need to be satisfied to specify that a cathodic protection system is deemed to be adequately protecting the steel reinforcement. Meeting the criteria is believed to be an indication that the system is delivering satisfactory cathodic current to either passivate the reinforcing steel or reduce its corrosion current to low and non-destructive levels.

Polarising any metal to a near immunity state is recognised as a possible way of achieving protection hence ISO EN 12696 requires, as one of its criteria, achieving an instantaneous-off potential of -720 mV vs Silver-Silver chloride reference electrode, 0.5M KCl. In saturated concrete this level of polarisation is achieved easily so the standard also tries to ensure that the steel potential does not fall much below this value, especially for prestressing steel so it sets a limit of -900 mV vs Silver-Silver chloride reference electrode, 0.5M KCl, in order to avoid hydrogen gas evolution on the steel surface.

Unlike steel in aggressive aqueous or soil environments, in concrete, steel forms a protective oxide film owing to the alkalinity of the cement matrix so corrosion, if caused by chlorides, only occurs once the pitting potential is exceeded. To repassivate the steel, it is not essential to polarise the steel to immunity but to the repassivation potential which requires considerably lower cathodic current [3]. Both Standards, allow for this lower polarisation but only by an empirically determined criterion of 100 mV depolarisation (excluding the IR drop) over a period of 24 hours. Such a level of polarisation can be assumed to reduce the corrosion rate roughly by an order of magnitude. In concrete, 24 hours is rarely enough time to achieve full depolarisation of the steel owing primarily to a slow migration of oxygen to the steel surface, especially in partly saturated concrete cover. ISO EN 12696 deals with this possibility by allowing a depolarisation period longer than 24 hours and up to several days provided the level of depolarisation exceeds 150 mV.

ISO EN 12696 hints that the criteria are empirical by stating in Note 4 that *"Criteria, a), b) and c) (relating to the polarisation/ depolarisation criteria) are not necessarily supported by theoretical considerations, but are a non-exhaustive, practical series of criteria to indicate adequate polarization which will lead to the maintenance or re-establishment of protective conditions for the steel within the concrete"*.

A NACE International Publication [4] attempted to validate the 100 mV criterion both theoretically and empirically with a level of success in the case of buried pipelines. The document concluded that a 100 mV depolarisation of steel in concrete over a maximum of 24 hours is likely to equate to a total depolarisation of 150 mV over a longer period, which justifies the ISO EN 12696 specification of this higher depolarisation level for periods longer than 24 hours. According to Bennet and Mitchell [5], this level of depolarisation may be required in most aggressive conditions based on a suggestion that

* Corresponding author: nikitar@vector-corrosion.com

https://doi.org/10.1051/matecconf/201928903011

the cathodic Tafel slope of steel in concrete may be as high as 150 mV/decade which represents an order of magnitude reduction in the corrosion rate.

Apart from these depolarisation criteria, very little else is suggested in the standards that can aid in the validation of a cathodic protection system. There are suggestions of the range of current density (2-20 mA/m^2) that would normally be applied to steel in a CP system, but the actual applied current is normally adjusted post commissioning in an impressed current CP system (ICCP) to achieve the required depolarisation levels.

NACE SP0290 assumes that CP is no longer required if the steel potential is less negative than -200 mV vs copper/copper sulphate electrode (CSE) and ISO EN 12696 states in Note 5 *"As an investigative criterion it may be considered appropriate to seek a steel/concrete potential in a fully depolarized structure after the cathodic protection system has been switched off for a long period (typically 7 days or longer) less negative than −150 mV with respect to Ag/AgCl/0.5 M KCl"*.

It is reasonable to suggest that at the instance when the steel potential is shown to be passive there would be no obvious need to apply further polarisation other than to prevent re-initiation of corrosion with time so a cathodic prevention level current density, suggested as 0.2-2 mA/m^2 in ISO EN 12696, would suffice. Often, the levels of polarisation will be less than 100 mV in such instances.

There are also numerous cases when the depolarisation criteria cannot be met either because current distribution is variable owing to localised irregularities in resistance, or, more commonly, because the current output of the protective system is limited such as when galvanic cathodic protection systems are used or aging ICCP anode systems can no longer adequately deliver the required current. It may be justifiable to maintain a lower current density if it can be shown that adequate protection is maintained. In such cases, it is essential that alternative ways of determining whether the steel reinforcement is adequately protected at any given time are identified.

This paper reports on ways that analysis of site data, such as the relationship between the applied current density to the level of depolarisation and/or the trend of the depolarised potential with time can help assess the adequacy of the level of protection. Site data are supported by laboratory studies where corrosion rates estimated by non-linear polarisation, as defined by the Butler Volmer equation, were compared to equivalent corrosion rates determined by linear polarisation.

2 Site Data

Three bridge abutment sites in the UK which had galvanic anodes embedded in concrete for controlling the corrosion of the steel reinforcement were monitored over a period of between 15 and 19 years. The results were analysed carefully, observing trends and apparent corrosion rates to analyse their performance.

2.1 Site-1, M53 Motorway Bridge, South Clatteridge

Chloride-laden water sprayed from adjacent traffic of the motorway was causing corrosion concerns of the base of two adjacent abutments of a bridge (Fig. 1). Corrosion potentials were indicative of a low and variable degree of corrosion but there were no signs of corrosion-induced cracking. Two lines of galvanic anodes were embedded in holes broken away at a spacing of approximately 300 mm, 11 No. Size-1 cylindrical anodes in Abutment-1 (Fig.1) and 8 No. Size-2 cylindrical anodes (double current output compared to Size-1 anodes) in Abutment-2.

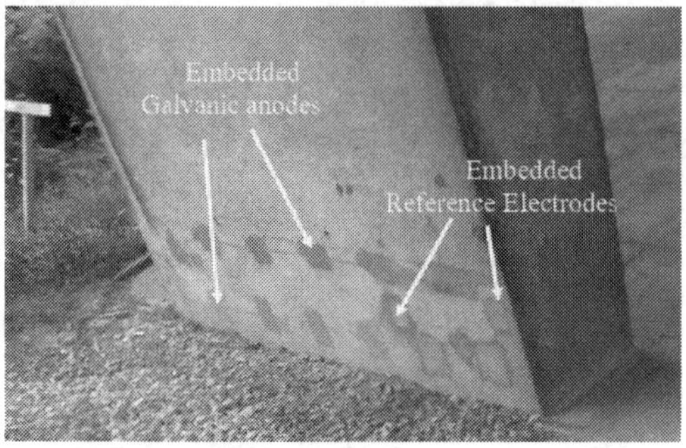

Figure 1 Abutment-1 of a bridge on M53 showing two lines of 11 galvanic anodes and two reference electrodes.

The anodes were connected along a chain of double insulated cable and the two end points of the chain taken to a junction box/monitoring unit at the back of the abutment. Two embeddable silver/silver chloride 0.5M KCl reference electrodes were also inserted in holes at anode midpoint locations. The cables of these and from two connections to the steel reinforcement were also taken to the monitoring box which enabled connection and disconnection between the anodes and the steel reinforcement. This allowed monitoring of the current flowing between the anodes and the steel, the on and instant-off potentials of the steel by the embedded reference electrodes and the depolarisation potentials over 24 hours after switching off the current.

Results were collected over a period of 15 years. The anodes are still in operation.

2.2 Site-2, Ivy Street Bridge Abutment, Leeds

Failure of the waterproofing joints above the abutment was allowing chloride-laden water to run down the abutment walls and settle at their base (Fig. 2). These had penetrated the cover concrete of the walls causing significant corrosion of the steel reinforcement.

Following some localised repairs, an array of Size-2 galvanic anodes were embedded in cored holes across the whole wall in a grid configuration at a spacing of 300 mm. Two small areas of 9 anodes (3 x 3), one at the top end of the abutment (Zone-1) and one at the base of the abutment (Zone-2) were not connected directly to the steel reinforcement but, via a cable, were taken to a junction box/monitoring unit where an external controlled connection was made to the steel reinforcement. Two embeddable reference electrodes were positioned, one in the centre of 4 anodes and one in-between 2 anodes in each monitored area in order to record potentials at various stages.

As in Site-1, results were collected over a period of 15 years with the anodes still in operation.

Figure 2 Abutment wall of bridge protected by an array of galvanic anodes showing Monitored Zones 1 and 2.

2.3 Site-3, Groby Bridge, Leicestershire

The third site was a bridge in Leicestershire which, owing to the failure of the joints, had suffered from chloride-induced corrosion, cracking and spalling of the cover concrete [6]. All affected concrete was removed up to about 25 mm behind the steel, exposing all the steel and enabling its cleaning. Galvanic anodes were then attached on the steel around the periphery of the patch at a spacing of approximately 600 mm before reinstating with proprietary repair mortar. Two adjacent areas were set up in a way as to allow monitoring of the current delivered to the steel by each individual anode via a junction box (Fig. 3).

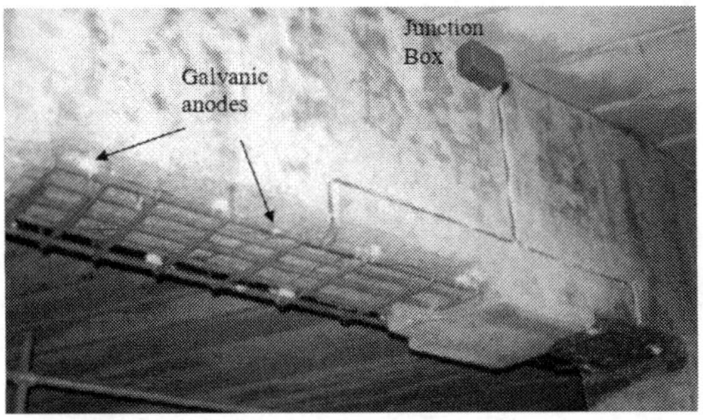

Figure 3 Galvanic anodes attached to steel reinforcement as part of a patch repair with cables from individual anodes taken to a junction box to aid monitoring of their performance.

As with the other two sites, the current was monitored with time. As no reference electrodes were embedded within the patch repairs, the steel potentials were measured from the surface in a grid configuration both while the current was on and after 24-hour depolarisation following interruption of the current. The IR-drop was estimated from instant-off readings on two marked points on the surface away from the anodes, but this was often less than 10 mV.

Results were collected over a period of 19 years. As in the case of the other two sites, monitoring is continuing.

3. Results from the monitored sites

The variation of current density by steel area delivered by all anodes over the equivalent area in Site-1, including 300 mm beyond the edge of the line of anodes, is shown in Figure 4. Overall, the current density has been decreasing with time with the larger Size-2 anodes constantly delivering a higher current to the reinforcement.

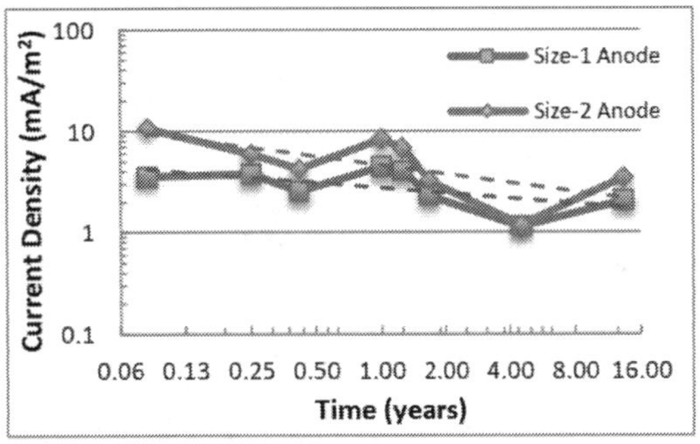

Figure 4 Current density by steel area delivered by the anodes with time to the steel reinforcement in Site-1

A similarly decreasing current density was observed in Site-2 (Fig. 5); this time, even though the anodes were the same size in both zones, a higher current density was observed throughout in the lower Zone-2. It is believed to have been influenced by the wetter and higher chloride-containing environment of the zone at the base of the abutment as water had tended to remain there for longer periods. It is also evident that the current was reducing at a faster rate in Zone-1, suggesting a partial drying of the abutment at the higher level. The apparent sudden drop in current density at around 2 years was caused by a lower temperature (see also Fig. 6).

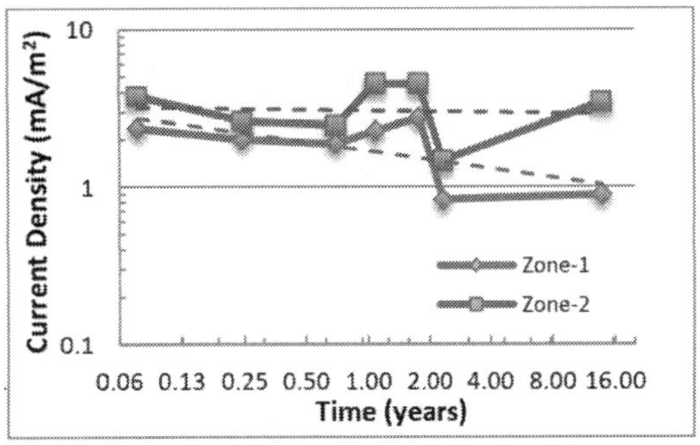

Figure 5 Current density by steel area delivered by the anodes with time to the steel reinforcement in Site-2.

The mean current density of the 12 anodes within the two adjacent patch repairs in Site-3 was also seen to be decreasing with time (Fig. 6). As more frequent readings were taken at the early stages of the monitoring, the effect of temperature on the current density is seen very clearly, summer readings being significantly higher than winter readings.

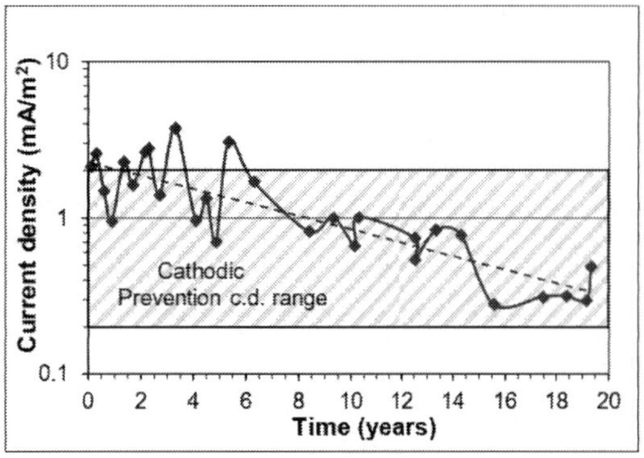

Figure 6 Mean current density by steel area delivered with time by the anodes to the steel reinforcement in Site-3

4 Analysis of Site Results

Analysis of the results has revealed some very interesting relationships and trends. Plotting the logarithm of current density against the depolarisation at the equivalent time indicates a strong relationship as shown in Figures 7 and 8. In Site-1 (Fig. 7), where two different size anodes were utilised on similar adjacent abutments, the slope of the plots appear to be almost identical, the only obvious difference being the higher depolarisations, mainly beyond 100mV up to a high of 145 mV, achieved by the larger Size-2 anodes which were able to deliver a higher current throughout. Depolarisation achieved with the smaller, Size-1, anodes was overall less and lay within a range of 70 mV and 120 mV.

The relationship between current density and depolarisation in Site-2 was found to be significantly different for the two zones (Fig.8). Somewhat lower overall current densities attained in Zone-1 had achieved significantly higher depolarisation values (70-115 mV for current densities of 0.85-2.75 mA/m^2) compared to Zone-2 (60-95 mV for current densities of 1.5-4.5 mA/m^2).

Such logarithmic, non-linear, polarisation behaviour is described by the Butler Volmer equation (eq. 1). The formula should allow utilisation of the depolarisation and current density values to determine an apparent corrosion current density.

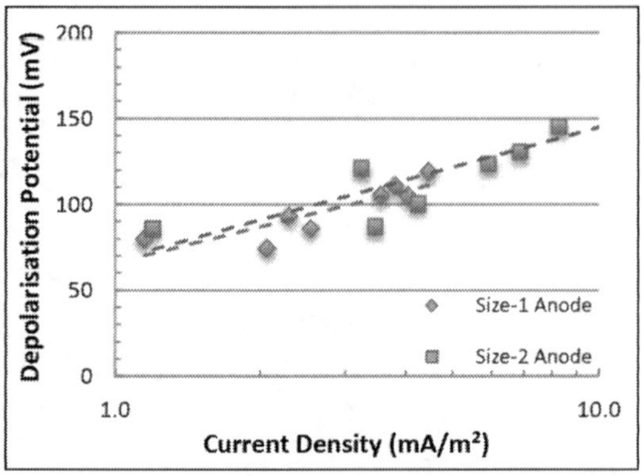

Figure 7 Relationship between current density and depolarisation potential for Site-1.

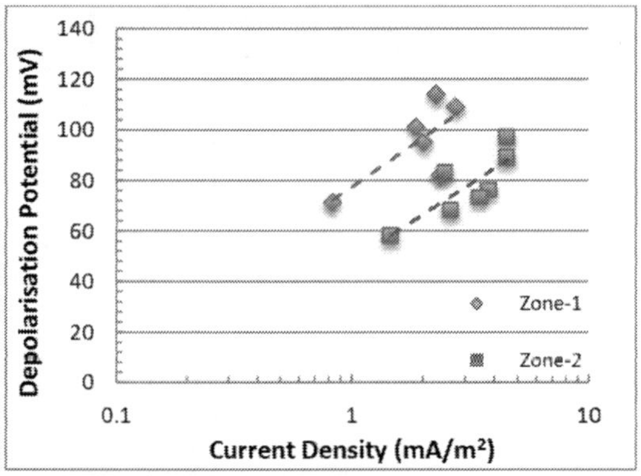

Figure 8 Relationship between current density and depolarisation potential for Site-2

$$i_{corr(app)} = \frac{i_{appl}}{exp\left(\frac{2.3\eta}{\beta a}\right) - exp\left(\frac{-2.3\eta}{\beta c}\right)} \quad (1)$$

Where,

$i_{corr(app)}$ = apparent corrosion current density,
i_{appl} = applied current, i.e. current density,
η = potential shift, i.e. depolarisation potential,
βa = Anodic Tafel Constant (assumed as 120 mV),
βc = Cathodic Tafel Constant (assumed as 120 mV).

For Site-1, the calculated apparent current density values are plotted against time in Figure 9. Within experimental error, $i_{corr(app)}$ appears to be unchanged for both abutments over the whole timescale, ranging within 0.25-0.65 mA/m^2.

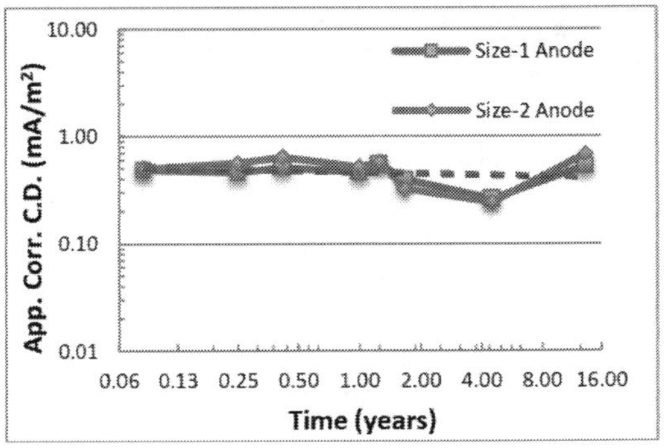

Figure 9 Variation of calculated apparent corrosion current densities with time in Site-1

In Site-2, the results reveal something different (Fig. 10). While the results of Zone-2 appear very similar to Site-1 (0.55-0.95 mA/m^2), Zone-1 shows a decreasing trend with time, $i_{corr(app)}$ falling from around 0.5mA/m^2 to less than 0.1 mA/m^2.

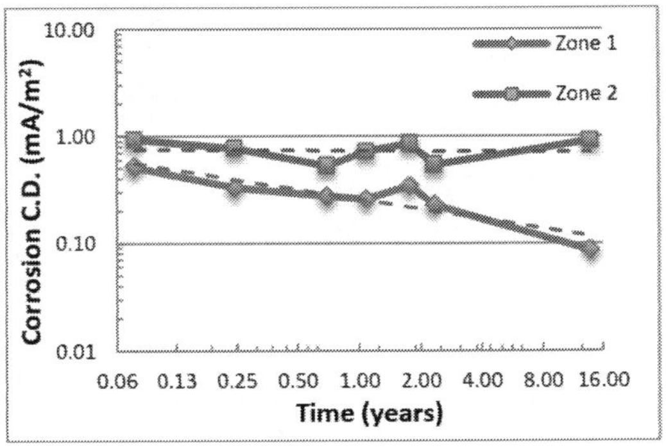

Figure 10 Variation of calculated apparent corrosion current densities with time in Site-2

As was observed in Figure 8, a low current density was inducing high depolarisation values. This is indicative of low corrosion levels of the steel which, with time, as seen in Figure 10, became more passive with the environment possibly becoming drier. In all the other conditions (Figs 9 & 10), the level of corrosion remained relatively stable.

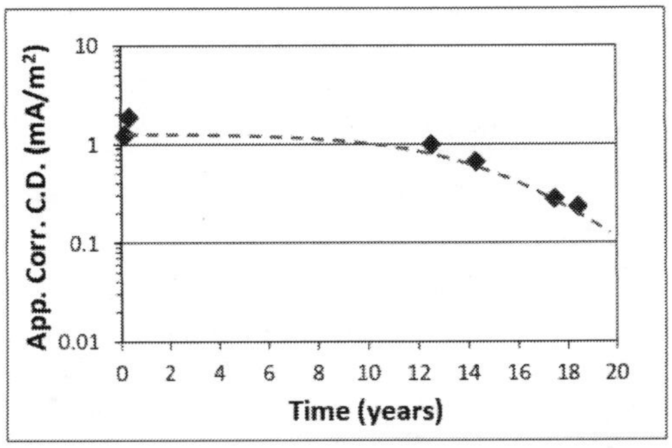

Figure 11 Variation of calculated apparent corrosion current densities with time in Site-3.

In Site-3, the average apparent corrosion current density also reduced with time, finally falling to around 0.2 mA/m^2 (Fig. 11). This correlates well with the 24-hour depolarised corrosion potentials which became more noble with time (Fig. 12).

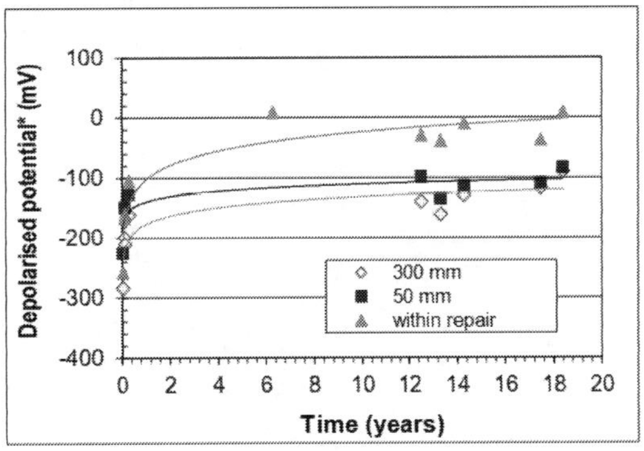

Figure 12 Depolarised potential with time within and up to 300 mm from the edge of the repair

Combining all the values and plotting the apparent corrosion current density against the depolarised potential*, i.e. the steel potential 24 hours after disconnection of the anodes, a reasonable correlation is observed (r = 0.74) with a cathodic slope of 150 mV/decade, close to the assumed cathodic Tafel slope of 120 mV/decade (Fig. 13). Two points on the graph appear to fall outside the 95% confidence intervals. In fact, both these points were obtained when depolarisation potentials exceeding 150 mV at current densities above 20 mA/m^2 during an externally induced applied current. The two points exhibiting lower corrosion potentials were the same two conditions prior to the induced current in a corrosive state. These were not part of the three site conditions described but part of a separate site trial involving a two-stage CP process [7].

If a depolarised potential of -150 mV is assumed to signify passivity of the steel reinforcement, the equivalent $i_{corr(app)}$ value appears, from Figure 13, to be around 0.6 mA/m^2 but owing to the variability of the results could arguably lie within a range of 0.2-2 mA/m^2 (95% confidence interval).

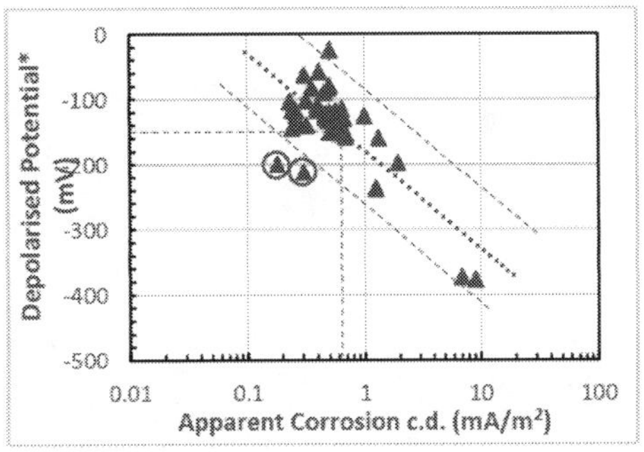

Figure 13 Relationship between apparent corrosion current density and 24-hour depolarised potential.

*All potentials are vs silver/silver chloride 0.5M KCl reference electrode

5. Results from laboratory investigations

Experiments were run in the laboratory in an attempt to verify the relationship between applied current and resulting depolarisation, as observed in the site results. Specimens of the sort depicted in Figure 14 and reported elsewhere [7, 8] were employed for the studies.

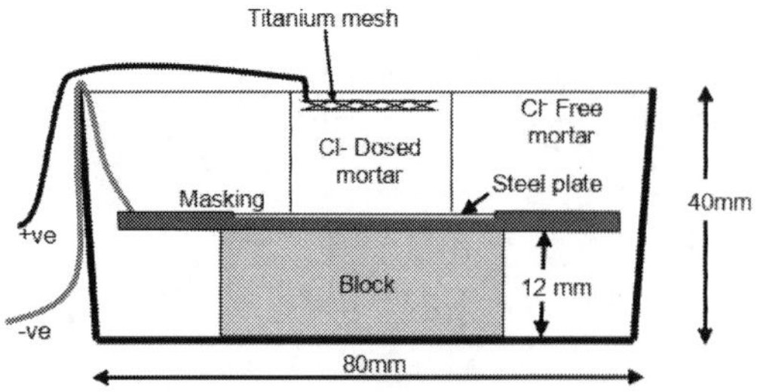

Figure 14 Schematic diagram of laboratory steel-mortar specimens

Steel plates in mortar, partially dosed with 2-4% chloride, were pre-corroded to various levels by applying anodic polarisation to various degrees, as detailed in an earlier publication [8]. They were then cathodically protected over several weeks at constant current densities ranging between 20 mA/m^2 and 50 mA/m^2 (steel surface area) until repassivation of the steel was achieved. The current was interrupted on a weekly basis. The data recorded included the current density, the 24-hour depolarisation potential and the 24-hour depolarised potential. Linear polarisation was also performed while the steel plates were in a 24-hour plus depolarised state.

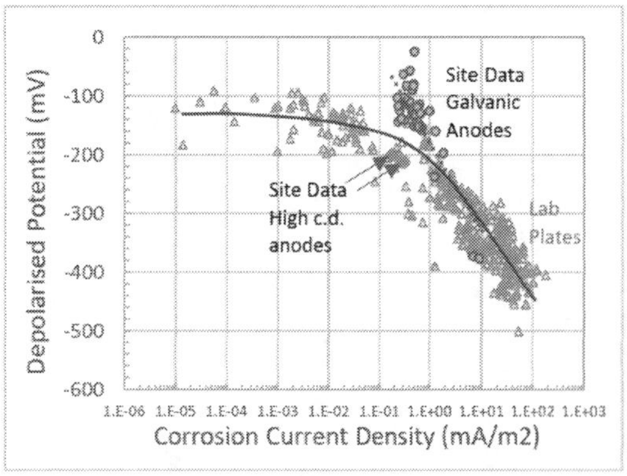

Figure 15 Apparent corrosion current density versus depolarised potential for all laboratory samples compared to site data from Fig. 13.

A whole mass of results obtained from all the tested samples, irrespective of level of pre-corrosion, applied current density and chloride dosing status, when plotted in a single graph appear to fall, within experimental error, on a single curve (Fig. 15). At the higher corrosion condition and lower current deliveries, the points appear to form an extension of the straight line obtained with the site results (Fig.13). As $i_{corr(app)}$ approaches what may be considered as passivation, as indicated by the depolarised potentials, the line curves away and eventually flattens horizontally with the depolarised potential remaining relatively constant at around -150 ± 50 mV over a range of at least four orders of magnitude. It is revealing that the two points from the high current high polarisation data (Fig. 13) also appear to fall on the curve in Figure 15. A reason for this behaviour is suggested below.

It is often assumed that activation control is in operation during cathodic polarisation of steel in concrete. For this to be true, oxygen migrating to the cathodic site must be able to replace the consumed oxygen as indicated by the most likely cathodic reaction (eq. 2).

$$\frac{1}{2}O_2 + H_2O + 2e^- \rightarrow 2OH^- \qquad (2)$$

This may be true at low cathodic current densities, depending on factors such as porosity and level of saturation of the concrete, but as the cathodic current increases, the reaction will gradually change to diffusion control and depend on the speed of oxygen replenishment at the cathode.

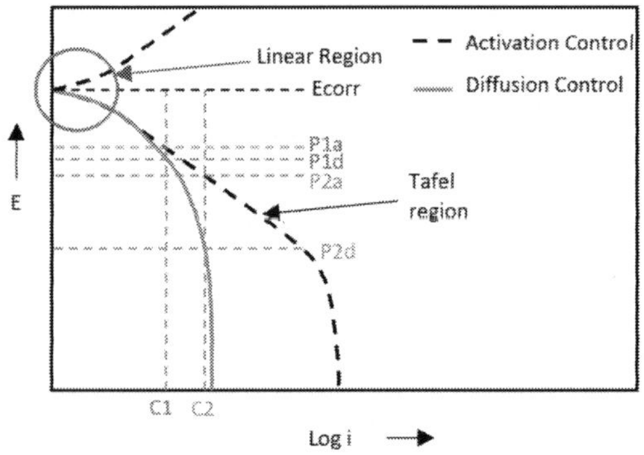

Figure 16 Schematic of cathodic polarisation showing the difference between activation control and diffusion control.

The different behaviour between activation control and diffusion control is illustrated in Figure 16. While the level of polarisation away from Ecorr is not very different when the applied current, C1, is low (P1a for activation control and P1d for diffusion control), the difference is greatly exaggerated at the higher current, C2, where the level of polarisation under activation control is much lower (Ecorr to P2a) than that achieved under diffusion control (Ecorr to P2d). This exaggerated level of polarisation under diffusion control is translated to a lower $i_{corr(app)}$ as calculated by the Butler Volmer equation (eq. 1) since the equation assumes Tafel behaviour and, therefore, activation control. Alternatively, or in addition, the cathodic behaviour of highly passive steel is altered considerably as current flowing through a very dense passive film is greatly restricted and is essentially unchanging over a wide range of polarised potentials leading to very high polarisation levels with slight increases in applied current.

In the case of the site results, with depolarisation levels between 60 mV and 140 mV achieved with current densities in the region of 0.8-8 mA/m^2 (Figs 7 & 8), polarisation may be mainly under activation control leading to the logarithmic relationship obtained in Figure 13. The laboratory specimens were subjected to higher current densities (20-50 mA/m^2) where diffusion control is more likely to be prevalent leading to the reduced apparent corrosion current densities seen in Figure 15. As the cathodic current in the experiments passivated the steel surface to ever greater levels, polarisation became much easier to achieve [3] leading to the apparently minute corrosion current densities observed.

That polarisation becomes easier with enhanced passivation can be seen in Figure 17. The level of measured depolarisation increased significantly as more cathodic charge was delivered to the steel at a constant current density of 50 mA/m^2. Both sets of specimens were pre-corroded with an anodic current density of 20mA/m^2 over two weeks. The more corroded samples containing 4% chloride by weight of cement showed a slower response, but both became more passive over the total delivered charge of 120 kC/m^2 as indicated by the depolarised potentials (Fig. 18). An increasing alkali concentration and reducing chloride level at the steel surface caused by polarisation are thought to reduce the corrosion risk [3,8].

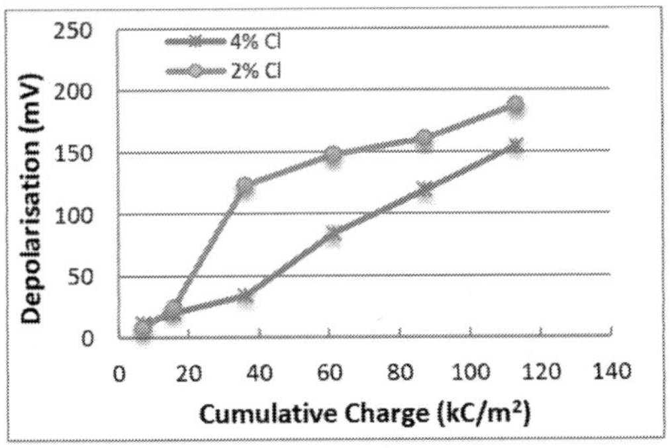

Figure 17 Increased level of depolarisation achieved with cumulative applied charge at a constant cathodic current density of 50 mA/m^2.

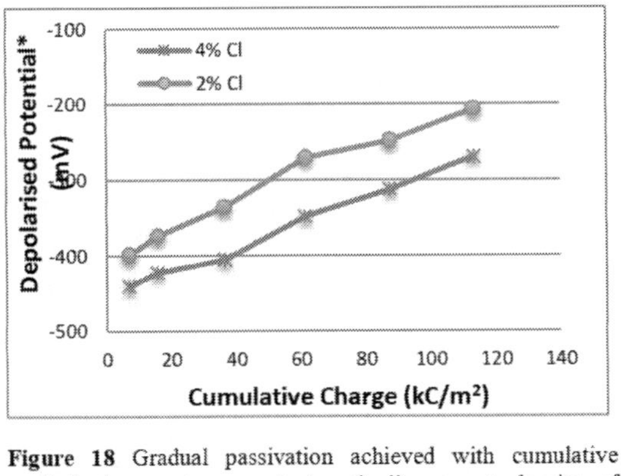

Figure 18 Gradual passivation achieved with cumulative applied charge at a constant cathodic current density of 50 mA/m^2.

Corrosion current densities determined by linear polarisation measurements on the steel plates just after depolarisation were compared with $i_{corr(app)}$ over the straight-line region of Figure 15.

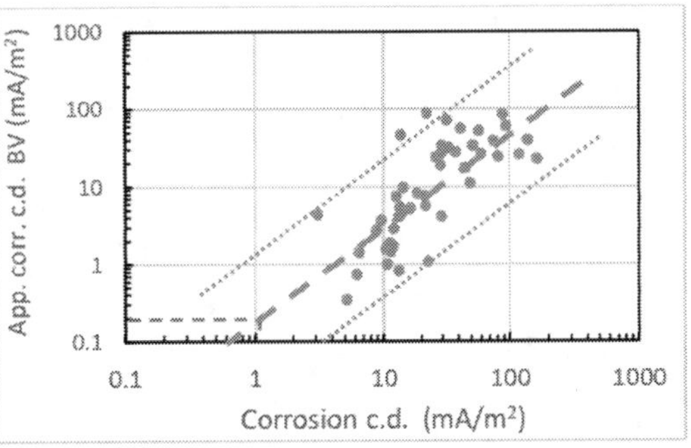

Figure 19 Correlation between corrosion current densities determined either by linear polarisation (Stern & Geary equation) or by non-linear polarisation (Butler Volmer equation).

The results (Fig. 19) suggest that a correlation exists (r = 0.75) between the two sets of calculated values but, at the lower corrosion current densities the two sets diverge. At icorr = 1 mA/m^2, as determined by linear polarisation, considered to be the limit where passivity of the steel is demonstrated [9], the equivalent $i_{corr(app)}$ value is nearer 0.2 mA/m^2. Although this value is lower than 0.6 mA/m^2 that appears to be the critical apparent corrosion current density in Figure 12, the two set of results certainly overlap if the 95% confidence intervals are considered.

Any values calculated in this way are only likely to be accurate within an order of magnitude, especially if error factors such as total affected area of steel during polarisation and variation in localised resistivity are considered. It is encouraging, however, that for a single monitored area (Figs 9-11), where the assumed steel area receiving current is essentially constant and the potential is always measured from a fixed position, even if the calculated $i_{corr(app)}$ value is assumed to be erroneous, the trend appears to be realistic. Such a trend could be used as an indicator of the performance of the cathodic protection system.

As the depolarised potential also follows a trend which can indicate lowering corrosion activity (Figs. 12 & 18), it is suggested that a combination of $i_{corr(app)}$ and the depolarised potential viewed together as trends with time can be very helpful in assessing the performance of the protective system. Both these parameters can be easily determined from measurements that are already taken during monitoring of the system, viz., 24-hour depolarisation and depolarised potential in combination with the applied current density.

6 Conclusions

It is suggested from analysis of site and laboratory results in this study that a combination of an apparent corrosion current density, $i_{corr(app)}$, easily calculated from a combination of applied current density and 24-hour depolarisation for the equivalent current density can be a useful indicator of the performance of the cathodic protection system.

A constant or reducing $i_{corr(app)}$ trend with time in combination with an upward trend (less negative) of 24-hour depolarised potential can be viewed as an indicator that the system is in a good operating condition achieving adequate protection of the steel reinforcement.

These parameters can be used in order to decide if protection is deemed to be adequate when the depolarisation criteria of ISO EN 12696 are not met.

The results further indicated that adequate polarisation leads to a gradual decrease in the corrosion rate of the reinforcement and can, if sufficient charge is delivered to the steel at an appropriate current density, lead to passivation of the steel.

References

1. ISO BS 12696:2016 "Cathodic protection of steel in concrete.
2. NACE SP0290:2007 "Impressed Current Cathodic Protection of Reinforcing Steel in Atmospherically Exposed Concrete Structures".
3. P Pedeferri "Cathodic Protection and Cathodic Prevention" Construction & Building Materials, **10**, 5, (1996) pp. 391-402.
4. NACE International Publication 35108 "One Hundred Millivolt (mV) Cathodic Polarisation Criterion", 2008
5. J.E. Bennett, T.A. Mitchell, "Depolarization Testing of Cathodically Protected Reinforced Steel in Concrete" MP 30, 12 (1990): pp. 20-25.
6. G Sergi "Ten-year results of galvanic sacrificial anodes in steel reinforced concrete" Materials and Corrosion, **62**, 2 (2011) pp 98-104.
7. G Sergi & A M G Seneviratne "Extending residual service life of steel reinforced structures by the use of galvanic anodes" *Concrete Solutions*, M Grantham, P A M Basheer, B Magee & M Soutsos Eds, CRC Press, Taylor & Francis Group, 2014, pp. 229-235.
8. N Rathod, G Sergi, G Seneviratne & P R Slater "A suggestion for a two-stage corrosion mitigation system for steel reinforced concrete structures" *EUROCORR 2008*, ICE Krakow, 9-13 September 2018.
9. C. Andrade, C. Alonso, J Gulikers, R. Polder, R. Cigna, O. Vennesland, M. Salta & A. Beharinaivo "Test methods for on-site corrosion rate measurement of steel reinforcement in concrete by means of the polarisation resistance method" Materials and Structures (2004) **37**, 273, pp. 623-643.

Monitoring and testing corrosion and responsive behaviour in galvanic systems.

Christian Stone[1], Gareth Glass[1], Adrian Roberts[1], and Nigel Davison[1]

[1]*Concrete Preservation Technologies, Long Eaton, Nottinghamshire, UK*

Abstract. Testing methods for corrosion prevention in structures have, historically, been based on the steel polarising from the protective current. Modern galvanic systems aim to protect the steel by restoring the protective alkaline environment, rather than through polarisation. Since 2012 standard guidelines allow for alternative testing to be used [1]. Scientific testing of galvanic systems will be discussed looking at the nature of constant potential galvanic systems and comparing them to the traditional constant current impressed systems, looking in particular at unique anode behaviours such as low median current during times of low corrosion risk and responsive behaviour during corrosion hazards. This work lays out a protocol for the evaluation and testing of existing galvanic corrosion protection (CP) systems in reinforced concrete through the lens of various real-world studies. Data taken from over a decade of monitored systems are used to support the protocol. Modern testing strategies such as 2-dimensional relative potential mapping and corrosion rate monitoring will measure the steel's corrosion. Galvanic currents rise and fall with corrosion risk providing protection while ensuring a long life to the design.

1. Introduction:

Cathodic protection (CP) in reinforced concrete structures has been used to control corrosion on bridge structures since the 1950s [2] [3]. The question of how to test and monitor these structures has been an area of concern from early on in part because many systems have no inbuilt monitoring devices. Existing galvanic systems are reaching or surpassing their estimated design life and knowledge of the ongoing corrosion is essential to the health of the protected structures. New systems also must be actively monitored to assess the effectiveness of the corrosion management.

Unlike typical impressed current systems which run at a constant current and aim to polarise the steel in order to halt corrosion [4], galvanic systems are electrochemical and hold a constant voltage. They supply current relative to changes in moisture, ion concentration and heat [5], with the aim of maintaining the naturally protective concrete environment to manage corrosion.

The monitoring and testing protocol of constant current impressed current systems is well known [6]. However, given it is designed to measure polarisation rather than corrosion it is ill-suited to be a comprehensive protocol for constant voltage systems. This work will cover a scientific approach to monitoring the risk of corrosion both for constant current and constant voltage systems, both impressed current and galvanic, whether they be low maintenance simple patch repairs or intensively monitored systems for larger projects. Data-driven measurements are essential to ensuring that the corrosion in the area is prevented or managed in a satisfactory manner.

2. Low maintenance systems:

Patch repairs and low-risk projects may not need the intensive monitoring of a bridge pillar or car park. For these system designs, a simpler testing and monitoring protocol is used. This section constitutes the complete practice for low maintenance anode systems. One or many of the following surveying methods may be necessary based on the needs of the system and the client.

2.1 Visual and delamination Surveys:

All projects benefit from visual and delamination surveys in and around areas of corrosion risk and patch repairs at intervals agreed upon with the client that is consistent with the usage of the concrete structure. During the visual inspection, areas of moisture, staining, vandalism, cracking, exposed rebar, and discolouration should be noted. Any source of rust staining should be investigated. Cracks forming parallel to the rebar and spalling will be of particular concern. Visual surveys can be taken from a short distance or using binoculars if access is difficult.

Touching distance visual surveys are necessary if delamination is suspected or every six years (as noted in BD 63/17 [6]) for highway structures. These tests may include a "hammer tap" test for delamination in areas of concern found in the visual survey. These timelines can be adjusted according to access and risk, based on discussions between manufacturers, the appropriate agencies, and official inspectors.

https://doi.org/10.1051/matecconf/201928903012

2.2 Corrosion potential measurements:

Measuring steel potentials against the potentials of a standard reference electrode (half-cell) is a well-established technique. In general measurements more positive than approximately -200mV against $Cu/CuSO_4$ (copper/copper sulphate) are considered to be in the area of small corrosion risk and measurements more negative than -350mV against $Cu/CuSO_4$ (CSE) are considered to be in the area of high risk of ongoing corrosion [7] [8]. Steel potentials can give a good approximation of the corrosion risk of steel, but they are affected by various other factors. Below are the shifts we expect to see in the potentials due to different environments in the concrete:

Tab. 1. Typical potential ranges measured in concrete using a Copper Sulphate reference electrode (V) [8]

dry concrete	0 → +0.2
dry, carbonated concrete	0 → +0.2
humid, carbonated concrete	-0.4 → +0.1
humid, chloride free concrete	-0.2 → +0.1
wet, chloride contaminated concrete	-0.6 → -0.4
water saturated concrete without Oxygen	-1.0 → -0.9

Different reference electrode materials will give different absolute potential measurements. It is therefore important to make a note of which is used during data collection.

Tab. 2. Relative potential against the standard hydrogen electrode for different half-cell materials (V) [8]

Relative potentials against the standard hydrogen electrode for different half-cell materials (V)	
Copper/Copper Sulphate saturated (CSE)	+0.318
Calomel (Hg/Hg_2Cl_2) KCl saturated (SCE)	+0.241
Silver Chloride (Ag/AgCl) KCl saturated	+0.199

Tab. 3. ASTM Standard potentials and their corrosion risk [9]

Potential vs. SCE (mV)	Potential vs. CSE (mV)	Probable Corrosion
>-123	>-200	Low (10%) risk
-123 to -273	-200 to -350	Intermediate risk
<-273	<-350	High (90%) risk
<-421	<-498	Severe corrosion

Manganese dioxide is also often used as the in-situ reference electrode given its stability, high pH and low chloride content.

2.3 Potential Maps:

A good way to reduce the effects of many of these potential shifts is to move away from taking individual potential measurements. Detailed potential maps consisting of corrosion potential measurements at close intervals (typically 50mm) may be undertaken at defined monitoring areas (limited in size) that are representative of a concrete structure or have a high risk of corrosion (ASTM 2015 [9]).

Corrosion potential or "half-cell" measurements can be taken with a wheel-based electrode or with a hand held electrode at set points between the CP anodes. The potentials are typically measured with reference to the potential of the reinforcement or, when not practical, with respect to another set position half-cell electrode [10]. This would reduce the need to damage concrete to expose the steel. These tests will be undertaken along the steel furthest from the sacrificial anodes, this steel is a relatively high risk given the lower protective current or using a line or two-dimensional grid (drawn onto the concrete) and will show localised changes in potential where anodes are present, as shown in Figure 1. This allows any possible anodic locations to be observed in the reinforcing steel as well as corroding galvanic anodes. Anodes indicate areas of active corrosion. Steel anodes represent actively corroding steel. Further investigation is then needed to see if this corrosion is in a controlled state or whether this is cause for concern.

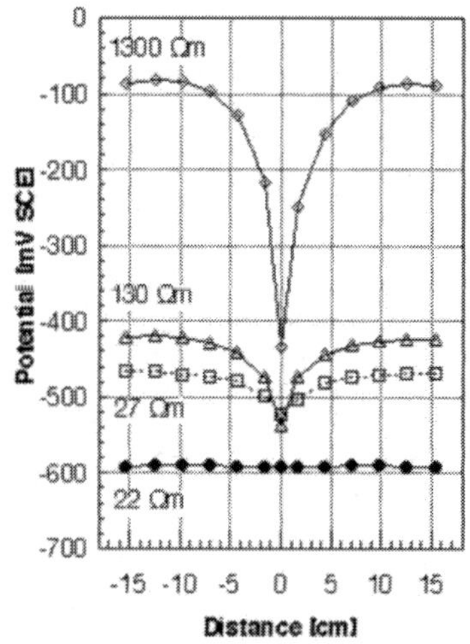

Fig. 1. half-cell potential model of anodic steel at different electrolyte resistivities in very wet, chloride containing mortar. Cover depth 20mm, anode 0.5cm. in very wet, chloride containing mortar. Rilem [8]

A detailed map of anode sites and historic potential map data are useful tools to see changes over time and to spot new or growing anodic sites for further investigation. Special attention should be given to the areas furthest from the anodes, as they will have the lowest protective current.

If no prior data is available, a visual assessment and an initial half-cell risk assessment over the structure or over representative areas should be used to find small areas of specific risk. These areas will be half-cell mapped alongside more typical areas and tracked throughout the remaining life of the structure.

To ensure good data is collected:

- Target steel locations within and immediately adjacent to any patch repairs.

- ICCP systems should be turned off if possible, testing after 24-hours of depolarisation is preferred.
- There must be no coating or surface other than bare concrete.
- The voltmeter shall have at least 10MΩ input impedance to limit the effect of meter impedance on the result.
- The surface must be wet to reduce the effect of moisture variation (see Table 1) and the half-cell electrode must make contact via a wet sponge.
- The sponge should be washed or replaced to avoid contaminants from the electrode affecting the measurement (sulphate and to a lesser extent chloride react with hydrated cement and change the potential).
- Use non-destructive techniques, like cover-meter mapping, to find the rebar if measurements are to be taken targeting specific steel elements. When steel locations are known a fast survey can be taken along the steel furthest from the sacrificial anodes.
- The calibration of all portable electrodes should be checked in a lab, and on-site against each other before use.
- Readings in an area specifically selected for monitoring must be recorded over the smallest practical distance (typically 50mm) to spot corrosion and increase the sensitivity of results. The size of the grid should be chosen so that there will be adequate points between known sacrificial anodes to identify an anode on the steel. A typical grid would have no more than 50-100mm between measurement sites. These areas should be tested and tracked through accurate records for comparison during each scheduled evaluation.
- These regular test zones may be relatively small areas (typically 0.2-1m^2) that are at risk of corrosion and/or reprehensive of the structure.

The large amount of data collected per square meter when looking for anodes makes this technique only suitable for assessing limited, high-risk areas within a structure or failing that a few select representative sample locations of the structure. It is very useful however to ensure that what is being seen is corrosion, especially local pit corrosion and not a difference in moisture. Relative potentials show anode locations as sharp local changes in potential. A wheel anode increases the ease of this data collection.

When doing a fast survey, the steel areas further from the anodes will be the most vulnerable to corrosion. Figure 2 (below) shows (1) typical steel rebar reinforcement with (2) regularly spaced anodes. Location (3) shows a desired minimal testing of the steel between the anodes which get the least protection and therefore at the highest risk. If a local anode is suspected, location (4) shows how two further measurements can ensure that the drop is a localised anode on the steel.

The criterion for success is simply that there are no anodic sites measured between or immediately around existing anodes. If anodic sites are found, they should be monitored over time to look for growth and any change in potential.

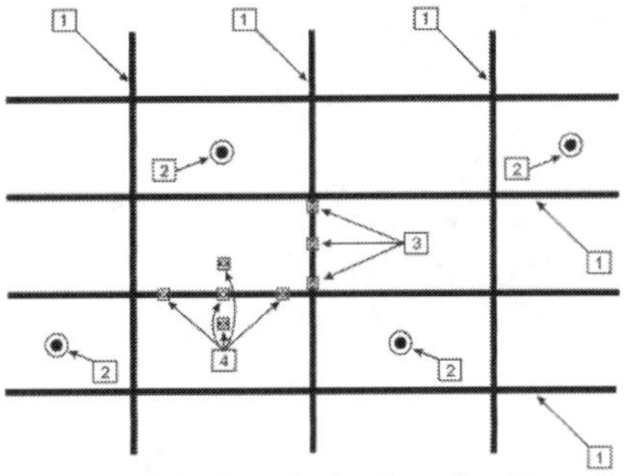

Fig. 2. Typical schematic showing rebar (1), anodes (2), primary half-cell testing (3) and secondary testing (4).

An example of a potential map is shown below:

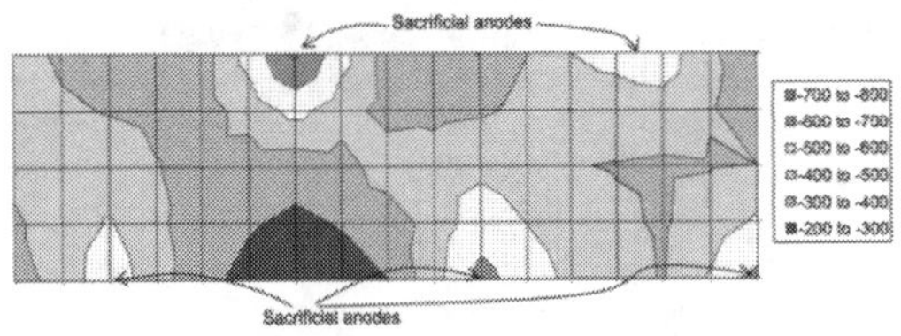

Fig. 3. Potential map of a structure containing embedded sacrificial anodes measured on a 50 mm grid. [8]

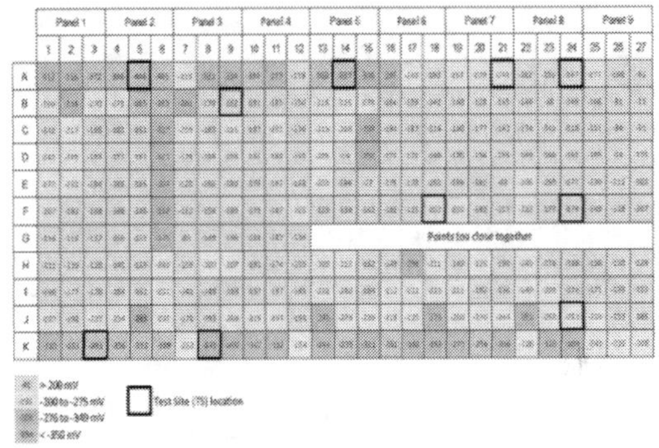

Fig. 4. Data taken from a bridge underpass before treatment. It can be used as a comparison for future half-cell tests. Areas that are red are at a higher risk of being an anode and would need further investigation.

In cases where a coating is to be applied, small coin sized areas can be ground down to give access or at least three embedded half-cell electrodes can be placed in the relatively high corrosion risk areas between anodes. Measurements can then be taken in the future between these electrodes, preferably using an electrode embedded at a lower risk site as a reference electrode.

Moisture differentials between reference and measurement electrodes can cause weak artificial anodes to occur on the map. This can be minimised by using reference electrodes in the same environment as the measurement and noting where mortar boundaries and other moisture boundaries are on the maps. Wetting the concrete surface and allowing it to permeate before measuring should also be standard practice to minimise this issue.

2.4 Current measurements

This may be used on constant voltage ICCP and galvanic systems. Protective currents delivered by one or more galvanic anodes may be measured by installing a resistor (typically 1-10 Ohms) in the circuit between the anode(s) and the steel and then measuring the voltage across that resistor. A sensitive voltmeter, with a resolution of at least 0.1mV, should be used to measure directly across the resistor to give the best data. Anodes with a higher current density are said to be more "active".

Galvanic anodes and constant voltage ICCP anodes often initially deliver a high current. The initial galvanic current off an anode is typically 3mA. The current decreases year-on-year after repairs due to many factors including; the concrete drying, the restoration of the concrete surrounding the steel, the creation of a benign and more resistive concrete environment, the build-up of resistive products around the anode, the reduction in galvanic anode size, and the galvanic anodes finding a local electrochemical equilibrium. The question of how we know the anodes are still active is one that troubles many in the field observing older systems.

During times of increased moisture or higher temperature the current output of the anodes increases. This is also when the steel is most vulnerable to corrosion. Significant transient increases in current have been measured. This is termed responsive behaviour because the anode responds to the corrosive nature of the environment [5]. We can use this behaviour to see if the system is still active.

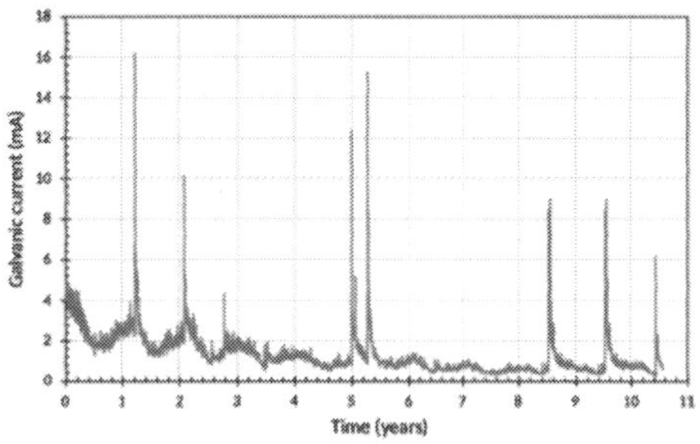

Fig. 5. Galvanic current from an anodic array over more than 10 years, showing the galvanic current response to flooding (tall peaks) and rainfall (smaller peaks) events.

Dry periods of low protective current are usually not a sign of the anodes depleting or disappearing. Galvanic systems work on a constant potential model so when the concrete is dry and stable, the steel is protected by the concrete. The anodes corrode less and maintain a lower protective current.

The higher initial current powers the migration of chloride ions away from the steel surface, the re-passivation of the steel, and a lowering of the pH in the concrete around the reinforcement by the creation of hydroxide ions at the steel interface. To test whether the CP system is still "active, test the current flow after rain or during the day vs at night to see the responsive behaviour to increased ionic diffusion.

There is no minimum current criterion. In a completely benign concrete environment, the steel is protected without the application of any current.

ICCP systems have been turned off and yet years later corrosion has not yet become an issue because the environment has been restored and the steel is passive [11].

Figure 6 shows the anode current in response to a flooding event several years after installation. We can see that the anodes respond sharply to the changed environment and increased corrosion risk. The current then reduces as the concrete dries out over many weeks. Currents are higher in the zone with the highest moisture content.

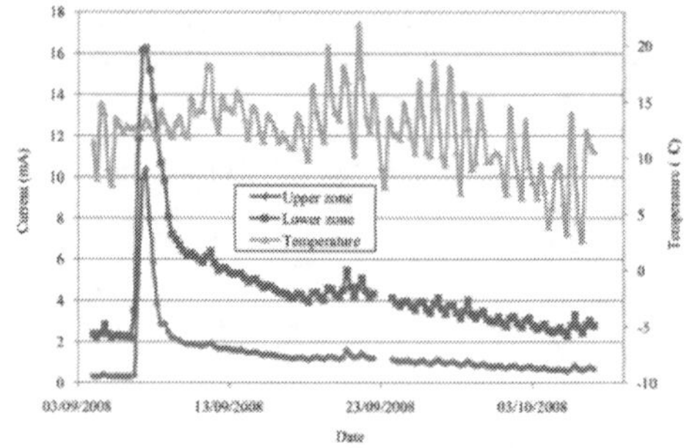

Fig. 6. Galvanic anode current response to flooding of Whiteadder Bridge, Northumberland, UK. [5]

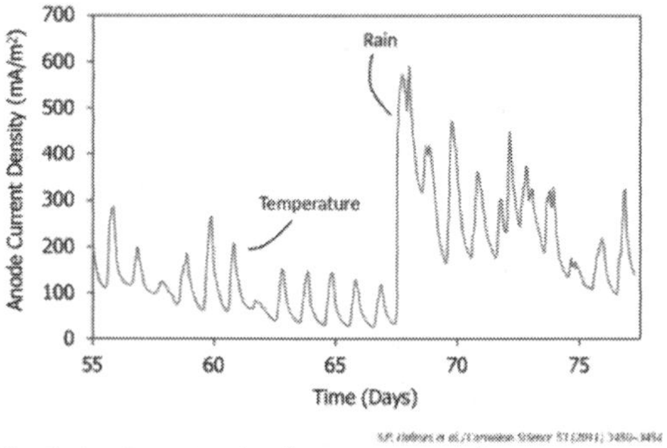

Fig. 7. Anode current density response to daily temperature oscillations and rainfall.

Figures 7 and 8 show how the current responds to temperature. Again, the current responds to the temperature dependent corrosion risk. The delay in response is attributed to the temperature recordings being atmospheric and the concrete taking a finite time to heat up.

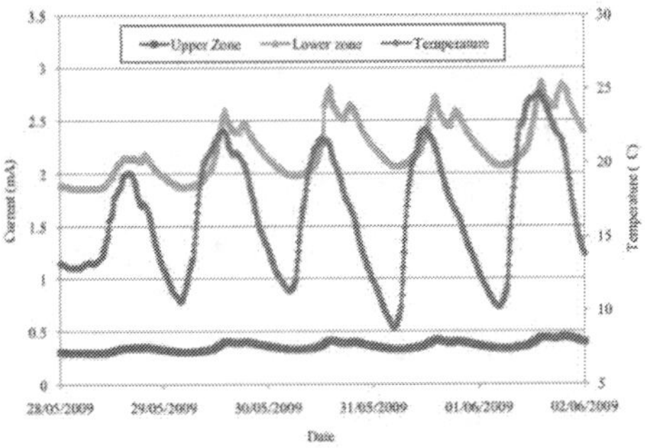

Fig. 8. Anode current response to temperature. Current and temperature data from a 5-day period in May/June 2009. [5]

The above constitutes the complete practice for low maintenance anode systems. What follows is an additional set of methods for intensively monitored, long-lifetime systems.

3. Intensively monitored systems with long design lives:

The ISO BS EN 12696:2016 Standard has the following non-exhaustive criteria for achieving protection. No more than one must be met to meet the standard

a) an "Instantaneous OFF" potential more negative than -720mV with respect to Ag/AgCl/0.5M KCl
b) from "Instantaneous OFF" a potential decay of at least 100mV over, at most, 24hrs
c) potential decay over an extended period (typically over 24hrs) of at least 150mV after "Instantaneous OFF". Use a reference electrode for any measurement over 24hrs.

This data is taken via a polarisation test. Although this test is not supported by theoretical considerations it is seen as a non-exhaustive, practical standard. These criteria point towards the polarisation of the steel being adequate to achieve protection. In the case of galvanic and constant-voltage impressed current systems, protection is as a result of the re-establishment and maintenance of a protective concrete environment and this test methodology would be inappropriate.

An alternative set of performance verification criteria can be found in the notes of the ISO BS EN 12696:2016 standard and is permitted if none of the non-exhaustive list can be met. This entails a further assessment of corrosion. The following indicate a low risk

a) steel/concrete potential less negative than -150mV with respect to Ag/AgCl/0.5M KCl in a fully depolarised structure after the cathodic protection has been turned off (typically 7 days or more)
b) corrosion rates of less than $2mA/m^2$ or preferably less than $1mA/m^2$ to indicate passive steel.
c) a combination of a rising corrosion potential and falling corrosion rate.

The rest of this work will cover how to test to this combined standard. The focus will be on galvanic systems as these are the least understood, though all tests can be equally applied to constant-voltage impressed current systems that would also aim to restore the environment and benefit from this alternative standard.

3.1 Polarisation decay and corrosion rate monitoring systems:

Corrosion rate data should be collected from areas of the structure that are most at risk from further corrosion-induced deterioration. A visual inspection may identify areas that are most susceptible to further chloride contamination. Potential mapping may be used to identify areas of active corrosion. Areas historically known to be at high risk occur in the parent concrete adjacent to areas of concrete patch repair. A range of locations for monitoring should be selected to cover all risks.

3.1.1 Polarisation decay measurements:

When a current is applied to a metal, the potential changes. This potential shift is known as polarisation. By interrupting the current, the potential decays. The potential decay allows the calculation of a conservative estimate of the polarisation which in turn provides an indication of the passivation of the steel rebar.

A typical arrangement used to measure the potential shift in reinforced concrete that is treated with a hybrid electrochemical treatment is shown in Figure 9. With zinc anodes buried in cavities in the concrete. A reference electrode is located within the segment to determine the steel potential shift induced by the galvanic current. The rest of the anode system (the blue circuit in Figure 9) is used to minimize edge effects and confine the current from the anode segment to the steel below the segment (often termed a guard ring effect).

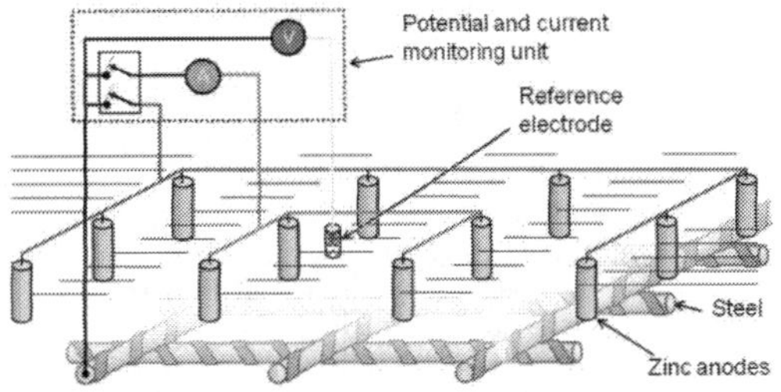

Fig. 9. Arrangement used to obtain data for corrosion rate calculation.

Polarisation potentials should be measured in the following way

- Remove galvanic link to turn off the system and instantly measure the steel/concrete potential, if the potential is more negative than -720mV than it has passed, and no further test is needed
- Turn off for up to 24 hours and measure the steel/concrete potential. A shift in potential from the first measurement of 100mV is the minimum ISO standard, if passed no further measurement is needed

- Turn off and measure the potential drop from the instant off potential over a longer time period, 150mV is the ISO standard.

This method is not based on theoretical models but is a known standard that has been used for some time. It measures the energy stored in the steel, or polarisation, which makes corrosion less energetically favourable and is dependent on the current supplied to the rebar by the anodes. For responsive systems such as hybrid galvanic anodes, once the concrete has dried and the steel environment re-alkalised the current needed to maintain the steel will often be too low to meet this standard, although the steel is being adequately protected.

3.1.2 Corrosion rate measurements:

If positive results are not possible with the previous test an alternative, a more scientifically rigorous measure, the corrosion rate, can be calculated. The basic method for this follows.

In Figure 9, the anode array is segmented so that the current density from a small segment of anodes (the green circuit in Figure 9) can be measured. This is used to determine the current density delivered to the steel by these anodes. The reference electrode should be located to see the average effects of the current (not the extremes due to variations in current distribution).

Rather than using an ammeter, insert a resistor in the galvanic circuit for each anode surrounding the reference electrode. After a few minutes, to let the current settle, use a voltmeter set in parallel with the resistor to measure the current off the segment of anodes. Using knowledge of the local steel density to calculate the applied current density per m^2. From this the corrosion rate can be calculated. This relationship is given by the following equation.

$$i_{appl} = i_{corr}\left(exp(2.3\Delta E/\beta_c) - exp(-2.3\Delta E/\beta_a)\right) \quad (1)$$

i_{appl} is the applied current density, i_{corr} is the corrosion rate, ΔE is the steel polarisation potential shift and β_a and β_c are constants. This equation builds the basis for measuring the corrosion rate from the polarisation potential shifts. Values of β_a and β_c of 120mV were used in producing the current guidelines on the interpretation of corrosion rate data.

The following table can be used as a reference to analyse the results of this analysis.

Tab. 4. Corrosion rates and their relative risks taken from BRE Group [12]

Corrosion rate [mA/m^2]	Risk
≤2	Very Low - Passive Steel
2-5	Low/Moderate
5-10	Moderate/High
>10	High

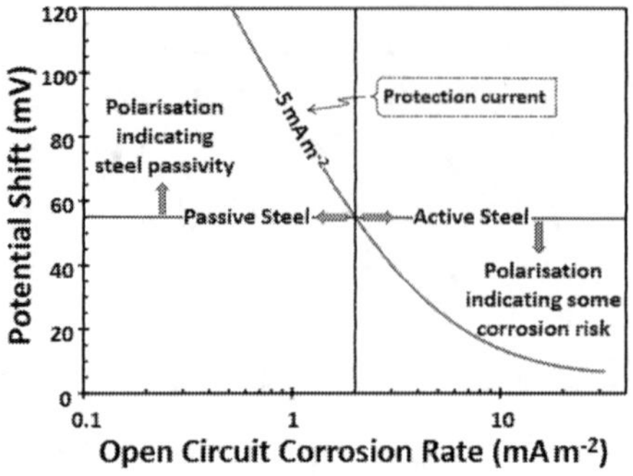

Fig. 10. Corrosion rate calculated as a function of potential shift at a current density of 5 mA/m^2 together with an example of its interpretation [13]

A large polarisation potential shift is induced by a small current density when the steel is passive. In the example in Figure 10, a potential shift of 55mV or more induced by a current density of 5 mA/m^2, would indicate that the steel is passive [13]. A falling corrosion rate and a rising corrosion potential also indicate that the system is providing protection and meets the alternative criterion allowed in ISO BS EN 12696:2016.

3.2 Carbonation and chloride profiling:

If an area appears to be corroding and the historic data shows the issue is getting worse, the last investigation before doing a destructive analysis of the area is testing the chemical make-up of the concrete around the rebar. Chloride sample standards are covered in ISO EN 14629:2012 and BRE Digest 444 Part 2:2000.

In areas of particular concern, following the assessment above, a semi-destructive chloride ion profile, can be used to test for the ingress of pH lowering ions near the target steel rebar. This will be done by drilling a 25mm hole into the concrete and collecting the dust from different depths. The first 5mm of concrete cover is discarded. Samples are then collected in 20 or 25mm depth increments up to the rebar depth. These samples are then sealed and sent off to a lab for chemical analysis. Chloride amounts at the depth of and by the rebar of 0.4% or lower by weight of cement show a negligible to low probability of corrosion occurring as long as the concrete has not been carbonated. Ideally, the weight percentage will be 0.3% by weight of concrete or less (Bridge Advice Note 35 [14]).

Carbonation may be tested by applying phenolphthalein to a freshly taken core of concrete. If the result quickly shows a pink hue, then the concrete is still strongly alkaline, and carbonation has not reduced the pH to a dangerous level. If several minutes pass before the change in colour it may be a sign of some carbonation.

If a concrete structure has failed to meet this standard, then those high-risk areas will be in need of possible repair and/or further protection.

Rebar depth can and will vary over a structure. Technical drawings or a measurement in another location should not be taken as an accurate indication of the true rebar depth. Places with less concrete cover will be more likely to undergo higher levels of corrosion.

4 Conclusion:

Laid out above are the different tests that can be used to monitor old and new CP systems and how they meet with current ISO standards. Low galvanic current during dry periods should not be mistaken as a sign of the system underperforming if the corrosion risk remains low and the system responds to hazards with increased current output.

Using the alternative performance standards established in ISO EN BS 12696:2016 galvanic and constant-voltage impressed current systems can be held to a clear standard for the testing of corrosion risk.

Actions to take if a corrosion risk is identified include installing more galvanic anodes or applying a temporary electrochemical treatment (using the galvanic anodes, if possible) and attaching a temporary power supply (for hybrid systems). The options available will depend on the galvanic system used.

5 References

1. ISO: EN BS 12696 (2016).
2. R. F. Stratful, *Progress report on inhibiting the corrosion of steel in a reinforces concrete bridge,* Corr. (Jun. 1959).S. Beamish, *Corrosion protection for reinforced concrete highway structures*, [online] http://www.corrosionprevention.org.uk/Web%20Docs/Presentations%20for%20website/Manchester/Highway%20Structures%20-%20Sam%20Beamish.pdf (Sept. 2018)
3. H. Esteves, R. Brueckner, C. Atkins, T. Gerrard, and U. Hammer. *Case studies of cathodic protection installations.* [book auth.] J. Broomfield P. Chess. *Cathodic Protection of Steel in Concrete and Masonry, 2nd Ed.* CRC Press, (2013).
4. S. Holmes, G. Wilcox, P. Robins, G. Glass, and A. Roberts. *Responsive behaviour of galvanic anodes in concrete and the basis for its utilisation.* 10, Corr. Sci., **53**, 3450-3454. ISSN 0010-938X (2010).
5. BD63/17, *Highway Structures: Inspection and Maintenance of Highway Structures.* Highways England, (2017).
6. P. Bamforth. *Enhancing reinforced concrete durability Guidance on selecting measures for minimising the risk of corrosion of reinforcement in concrete.* Concrete Society, (2004).
7. B. Elsener, C. Andrade, J. Gulikers, et al. *Half-cell potential measurements—Potential mapping on reinforced concrete structures.* 7, RILEM, Materials and Structures, **36**, 461–471 (Aug. 2003).
8. ASTM C876-15, *Standard Test Method for Corrosion Potentials of Uncoated Reinforcing Steel in Concrete.* (2015).
9. C.S. TR60, *Electrochemical tests for reinforcement corrosion.* Concrete Society, (2004)
10. C. Christodoulou, G. Glass, J. Webb, S. Austin, C. Goodier, *Assessing the long term benefits of Impressed Current Cathodic Protection,* Corrosion Science, **52**, Issue 8, 2671-2679. (2010).
11. BRE Digest 434, K. Hollinshead, *Corrosion of reinforcement in concrete: Electrochemical monitoring.* (Nov. 1998).
12. G.K. Glass, A.C. Roberts, N. Davison, *Assessment of cathodic protection applied to above ground reinforced concrete*, Materials and Corrosion, **70**, 3, 503-510 (2019).
13. DMRB:3-3, BA 35/90, *Inspection and repair of concrete highway structures* (1990)

Case studies of hybrid and galvanic systems on bridge structures

Christian Stone[1], *Michel Donadio*[2], *Christian Christodoulou*[3], *and Nigel Davison*[1]

[1]*Concrete Preservation Technologies, Long Eaton, Nottinghamshire, UK*
[2]*Sika, Refurbishment, Le Bourget, Hauts-de-France, France*
[3]*AECOM Ltd, Birmingham, U.K*

Abstract. Corrosion of steel in bridge structures is a growing issue worldwide. Case studies of galvanic and hybrid systems show the diverse range of challenges and corrosion hazards that are faced during the design of these systems. Data from these case studies show the effectiveness of this approach. Hybrid systems have been utilised in controlling corrosion during refurbishment work on bridges for over a decade. Galvanic systems have been around much longer. By utilising these systems individually or together there is great flexibility in design. This allows intensive concrete restoration from a hybrid system alongside the maintenance of repairs in lower risk areas from galvanic anodes. This paper presents case studies covering hybrid and galvanic systems applied to bridge structures. Included is a prestressed concrete bridge in a tidal environment, we believe to be the first of its kind fitted with a hybrid system. Tidal considerations and flooding are key challenges in corrosion mitigation. Data shows the systems are particularly effective in these transitory environments reacting rapidly to increased risk.

1 Introduction

Throughout the world, bridge structures are exposed to moisture, salt, and other corrosion hazards. NACE estimated in 2016 that corrosion costs the world $2.5 trillion (USD) every year and corrosion prevention and management could save up to $875 billion (USD) every year [1]. The implementation of sacrificial zinc-rich anodes in rebar-reinforced concrete structures was one of the cited examples of cost-effective corrosion management.

With many of the world's busiest commuter cities being located on coastlines where corrosion risks are particularly high, effective corrosion management and monitoring are essential to avoid more costly repairs. Galvanic systems have been used for two decades to increase the lifetime of repaired concrete on corroding structures [2]. New hybrid galvanic systems restore the concrete environment around the rebar minimising the need for concrete replacement which saves significant labour and reduces the time the structure is impacted by the work.

It has been over a decade since some of the first hybrid systems were first installed on bridge structures [4, 5] in Scotland. The challenges, designs, and results of such systems will be covered here, including the monitoring and testing of this and more recent structures fitted with hybrid and galvanic patch anodes. With the characteristic responsive behaviour alongside falling galvanic currents and rising steel potentials the effectiveness of these systems can be shown.

2 Structures

This work will be discussing the following structures:

2.1 SH16, Whau river bridge

Fig. 1. Aerial view of the existing Whau River Bridge, looking south-west (Corbett & Watterson, 2016 [6])

A 182m long, eight-span bridge bringing traffic across the Whau River. Constructed in stages from the 1950s through to the 1990s, connecting Auckland to the city's western outskirts. The bridge is part of the SH16 which carries an average of 90,000 vehicles per day through an environmentally sensitive coastal reserve [7].

The superstructure consists of reinforced concrete supports with cast reinforced deck beams and deck slabs.

The substructure varies across the bridge, the result of many expansions and alterations.

2.2 SH16, Oakley causeway bridge

Fig. 2. Aerial view of the existing Causeway Bridge, looking south-west (Corbett & Watterson, 2016 [6])

A 75m long, five-span bridge that brings SH16 traffic over the Oakley inlet waterway close to the Whau River bridge. Originally constructed in the 1950s this structure, like the Whau River bridge, has undergone widening and other alterations into the 1990s. This structure includes pre-tensioned concrete piles from both the 1960s structures and 1990s widening [7].

2.3 B6461, Whiteadder bridge

Fig. 3. Whiteadder bridge, Scotland, taken 10 years after installation

A 90m single carriageway bridge structure constructed from steel girders supported on bearings directly above 4 trapezoidal reinforced concrete piers, the deck is constructed from reinforced concrete beams spanning horizontally between the four primary steel girders. The unique shape of the piers leads to steel density increasing towards to bottom. Built in 1973, it carries motor traffic over a tributary flowing into the river tweed which is prone to flooding [8].

2.4 H6 canal bridge, Milton Keynes

Fig. 4. Abutment shown from opposing bank 4 years after patch repairs

A 1970s 14m bridge carrying traffic via a dual carriageway over the canal in central Milton Keynes. Reinforced abutments support the concrete deck that spans the waterway. Leaking bridge joints resulted in contamination and corrosion in the abutments.

3 Assessment

This covers the assessment of both the SH16 causeway and river bridges. The structures underwent the following tests:

- Both SH16 structures above underwent full durability assessments utilizing mathematic models to predict the condition in 50 years' time.
- Visual inspection;
- Chloride depth profile sampling;
- Colorimetric carbonation depth testing;
- Compressive strength testing;
- Cement content testing;
- Reinforcement continuity;
- Resistivity of the concrete;

The Whau river bridge piles exhibited cracking, delamination, spalling and section loss. Chloride ion concentrations varied from 0.02 to 0.32% by weight of concrete, the standard allows for 0.06% or 0.03% by weight of concrete for plain and pre-stressed steel respectively [2]. Chloride concentrations were concluded to be the major cause of deterioration in this structure. These levels alone are a high enough risk to warrant an intervention.

Colorimetric testing showed no risk of carbonation at the depth of the steel, typically finding carbonation depths from 1-3mm with only one outlier measurement of 23mm. The reinforcement depth varied from 40-80mm in the piles depending on the age of the concrete.

The Oakley causeway bridge piles were in a similar state of deterioration as the Whau River bridge with chloride concentrations ranging from 0.02 to 0.25% by weight of concrete. Calculations using predictive modelling based on Fick's second Law [ASTM C1556 –

04 and NZTA Report 337] which gives an idea of the time to corrosion initiation. These confirm the chloride levels are most likely the cause of the corrosion on both structures.

Whiteadder bridge and the H6 canal bridge had suffered severe delamination with exposed steel visible in areas. Both bridges underwent the following tests to assess the structures:

- Visual inspection;
- Chloride depth profile sampling;
- Colorimetric carbonation depth testing;
- Cement content testing;
- Reinforcement continuity;
- 2D potential mapping;

Whiteadder bridge showed substantial spalling and delamination on the piers from what appeared to be chloride ingress caused by leaking deck joints. Salt deposits were visible on the surface in some areas. The high concentration of chlorides was confirmed through testing a variety of locations by drilling in sections to the depth of the steel and collecting dust for analysis. The highest chloride ion content was found in two locations on a pier on the west side of the river. It measured 1.68% chloride by weight of cement, a high risk of chloride-induced corrosion. The half-cell potential map confirmed the high risk of corrosion and gave information that would allow the targeting of known, local anodic areas within the structure.

Visual inspection of the H6 canal bridge showed visible signs of chloride induced pitting on the exposed steel and the source of chloride appeared, from staining patterns, to be a leaking bridge joint allowing salt from the road above to be carried down through seepage. Through chemical analysis, it became clear that chloride was the instigator of the corrosion. Being by a canal the environment was very humid leading to widespread corrosion. The 2D potential map showed large anodic areas in need of protection.

Fig. 5. Chloride pitting on the steel broken out during the repair of H6 canal bridge

4 Corrosion prevention systems

4.1 Galvanic systems

Galvanic systems utilise the electrochemical potential difference between two dissimilar metals to drive the protection of the steel. It is well known that anodic metals have more available ions and are therefore more energetically favourable to corrode. By adding a material that is less noble than steel to the concrete environment (commonly zinc) and connecting it to the reinforcement via a wire, charge can flow and a dissimilar metal couple is created which pushes the steel into a cathodic, protected state. The less noble material becomes an anode and corrodes sacrificially to supply electrons to the cathode protecting the steel during times of increased corrosion risk.

At the anode the material will tend to undergo oxidation reactions:

$$Zn \rightarrow Zn^{2+} + 2e^- \quad (1)$$

$$Zn^{2+} + 2H_2O \rightleftharpoons Zn(OH)_2 + 2H^+ \quad (2)$$

The electrons produced by this dissolution polarise the steel, protecting it from corrosion. These electrons fuel reduction reactions at the steel surface:

$$O_2 + 4e^- + 2H_2O \rightarrow 4OH^- \quad (3)$$

Oxygen and water in the pores of the concrete react on the rebar surface to produce alkaline hydroxide ions. [2]

Steel is not just protected by its polarisation; new concrete has a naturally high pH which protects the steel from corrosion. Carbonation and chloride ions decrease the pH environment of the steel making the steel vulnerable to anodic corrosion. An effective way of managing corrosion is to restore and maintain the concrete around the steel. The cathodic steel uses electrons to create an alkaline hydroxide environment. These ions restore the pH of the concrete. Once the pH gets high enough the steel will passivate, and corrosion will be managed. [14]

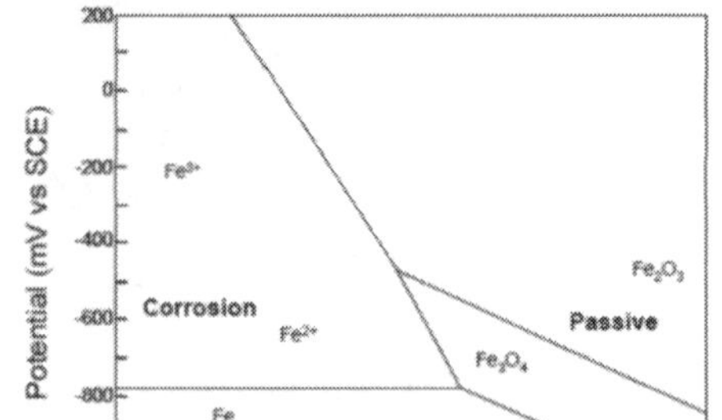

Fig. 6. Simplified Pourbaix diagram of iron

The anode and steel system act as a cell with a driving voltage defined by the electrochemical galvanic potential difference between the two metals. This cell produces a current based on the resistivity of the concrete electrolyte and the corrosion of the anode. The resistivity is dependent on the moisture levels in the concrete and the anodic corrosion is dependent on the presence of corrosion hazards such as water, oxygen, heat and chloride. Therefore, the cell is producing the most current during times of high corrosion risk. The anodes can be said to work reactively in response to hazards [8]. When galvanic anodes are installed it is common to soak the concrete. This gives the system an initial surge of current to halt ongoing corrosion. As the anode corrodes to protect the steel it is, like any cell slowly depleted, making lifetime calculations using predicted currents a key part of system design.

Galvanic systems are very effective for maintaining systems such as patch repairs and low chloride content concrete. They are however limited in the power they can provide by the surface area of the anode and the resistivity of the concrete. Galvanic anodes are typically used on the edges of patch repairs or in arrays throughout low-risk or repaired concrete. Though galvanic anodes can and do restore the concrete environment, their limited current makes them a poor choice in heavily contaminated areas. These areas are best served by impressed current or hybrid systems.

4.2 Impressed current systems

Impressed current cathodic protection (ICCP) systems will not be covered here in detail. ICCP systems have commonly been used to polarise steel in contaminated concrete by driving a current larger than that typically seen in galvanic systems through titanium ribbons, mesh or inert discrete anodes [15] reversing the direction of the current produced by corrosion. This reduces the need for replacing much of the concrete.

ICCP systems are typically held at constant current utilising a transformed and regulated mains electricity connection to provide protection for the lifetime of the system. The systems need frequent adjusting and monitoring; too high a current and the acid formed at the inert anode will start to locally dissolve the concrete, too low and the steel will not be adequately polarised. The running costs of these systems sometimes lead to them being turned off. It has been shown that when turned off the steel depolarises, but corrosion does not initiate for some time. This is due to the restoration of the steel environment [16, 17, 18].

The current driven to the steel generates a hydroxide bath through reduction at the steel interface, restoring the pH and the positively charge anodes pull harmful chloride ions away from the reinforcement. This restored environment keeps the steel passive.

4.3 Hybrid systems

Hybrid anodes were developed as the best of both worlds, combining a short, high intensity impressed current phase to restore the concrete environment with the long term, low maintenance responsivity of a galvanic system. Rather than using inert ICCP anode, hybrid systems supply the initial current through the same galvanic anodes as the second protective phase. The in-situ anodes then maintain the restored environment for the lifetime of the system. This protection comes without the need for the permanent housing of a large and complicated powering system and the corresponding large running cost and vandalism risk.

Like galvanic systems, discrete zinc anodes are embedded into pre-drilled cavities within the concrete. Hybrid anodes are arranged in regular arrays spaced according to the density of the steel reinforcement used. The anodes comprise zinc cylinders 25-220mm long with a coating designed to keep the anode active during its design life. Like ICCP systems these anodes have a titanium wire connector, better suited to delivering an impressed current.

During the initial impressed phase, power is typically driven from a 9V power supply providing the 50 kC/m^2 of steel surface area charge needed to restore passivity [3, 4] The current is often much larger than ICCP systems which are limited to 110mA but given the coatings typically used on the anodes and the short time scale of this phase, acid is not able to accumulate.

After adequate charge has been provided to the steel, typically a few weeks, the power source is removed, and the anodes are connected directly to the steel to start galvanic protection. During the galvanic phase, currents will vary responding to moisture and other corrosion risk factors.

The combination of hybrid and galvanic systems into highly contaminated and low contamination zones respectively allows for the creation of low running cost corrosion management systems that react to the corrosion hazards in the structure's environment and can be effectively monitored with embedded reference electrodes over the lifetime of the structure.

5 System Design

5.1 SH16 Whau river and Oakley causeway bridges

These structures had a variety of steel depths on the piers as well as other challenges to factor in during the design phase. A major factor in the choice of the cathodic protection system was using a system that could be installed in a short amount of time and with access constraints dependent on the nature of tides in the marine environment. Hybrid anodes are well suited in part due to their fast and straight-forward installation with a simple and easy wiring design.

Pre-stressed steel structures offer many advantages for a large concrete structure but can suffer from hydrogen embrittlement when protected with an impressed current. If the current is too high and the steel potential is driven too low the steel will start to embrittle which could put the entire structure at risk. Care must be taken providing a high current to pre-stressed steel but as long as polarisation rates do not stray more negative than -900mV vs SCE for an extended period of time there is little risk of embrittlement. Galvanic anodes are naturally limited in voltage by the dissimilar metal couple and do not suffer from this problem in concrete.

As these structures are exposed to seawater in a tidal area there are three major zones to consider on the piles:

- Atmospheric zone; from the soffit to 1m above the highest tide mark;
- Splash zone; from 0.5m above the average high tide to 1m above the highest tide;
- Tidal zone; from mean low water springs to 0.5m above the average high tide;

Protection below this point is in this case was not deemed necessary, in part due to the sparsity of oxygen in a marine environment limiting corrosion.

Hybrid anodes, with capacities exceeding 300kC, manufactured using 18 mm zinc cylinders and length 165-220mm were then installed into predrilled holes using an activated proprietary ready-to-use mortar around the anodes and a high-quality low-shrinkage repair mortar to cap the hole. The titanium connector wires were embedded in chases between the anodes to allow a current to be impressed via a junction box housed on the abutments of each bridge. This junction box also contained a connection to the steel reinforcement and a data logging system.

A manganese Oxide reference electrode was used to record continual monitoring data, measuring the steel potential, anode current, and corrosion current density from polarisation resistance. These are used to calculate both the steel polarisation and corrosion rate which can be used to assess the corrosion risk.

5.2 Whiteadder bridge

Regular flooding, splash zones, steel density changes, and tidal effects all play a part in the unique challenges of designing this system. Leaking from the joints also allowed, during the winter, chloride rich water to seep into the piers from above. It was determined that the design and the monitoring should be in two zones. An upper 'dry' zone and a lower 'wet' zone.

Cylindrical zinc hybrid anodes of 18 mm diameter, ranging from 25 to 220 mm in length, with an integrated titanium connector wire, were installed with a separation of 400mm for the lower zones and 500mm for the upper zone over the surface of 4 reinforced piers. The anodes were placed closer in the lower zones to combat the higher density of steel in that area. Reference electrodes were embedded in upper and lower zones on the piers in 12 locations. Ten rows of five hybrid anodes were installed in the monitored area. The upper zones were known to be dry, outside of flood or tidal splash zones. The lower zones would be in areas with more moisture. A wireless modem powered by a battery which could be recharged from a solar panel allowed remote access via a desktop PC. Data samples were taken every 6 hours.

5.3 H6 canal bridge

Galvanic anodes were embedded about 400mm apart around the patch repairs in the parent concrete on the abutment. The anodes were directly connected to the reinforcement. The anode capacity and density over the abutment depended on the risk, with a higher density located adjacent to the repair areas and in wet areas arising from the leaking joints. The deck joints were not repaired and continued to leak after the anode installation.

6 Results

6.1 Initial testing of Whau River and Oakley causeway bridge

Following installation, the hybrid systems were connected to the power source and energised to deliver the 50kC/m² of charge to the steel reinforcement. Data showing the polarisation of the reinforcement up to 50kC/m² of steel surface area and the current applied to a single zone are shown below:

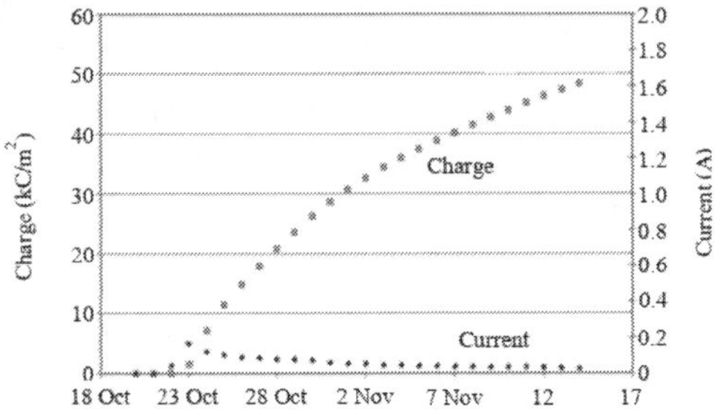

Fig. 7. Typical charge density and current applied to a zone [7]

After completing the impressed current phase, the system is tested using a depolarisation test as stated in the ISO 12696:2016 standard. Below is shown the depolarisation test data for atmospheric and splash zones:

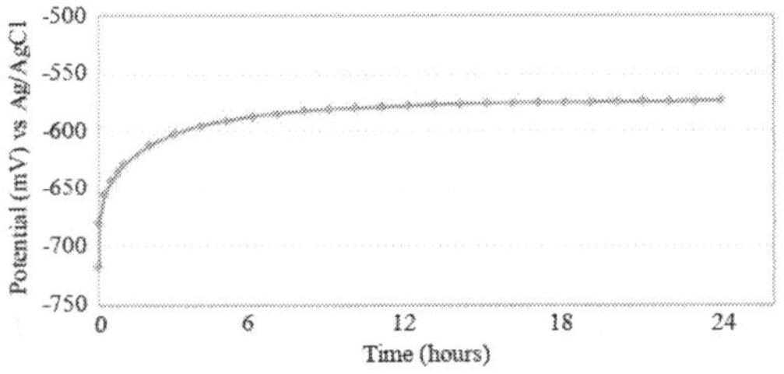

Fig. 8. Depolarisation taken in an atmospheric zone [7]

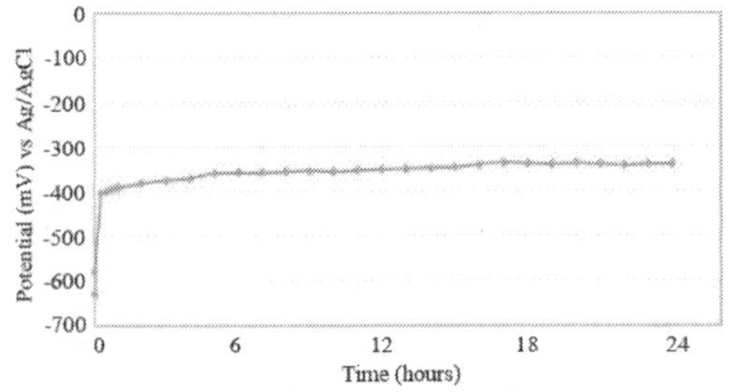

Fig. 9. Depolarisation taken in a splash zone [7]

It can be seen that a depolarisation greater that 100mV over 24hrs has been achieved for the both zones. The steel is therefore seen to have sufficient cathodic protection. In the splash zone, a greater polarisation potential of 240mV is measured over 24hrs. In these zones, the steel has been passivated during the impressed current phase and the steel is completely protected by the anodes.

Tidal zones may not be able to be assessed in the same manner given the changing conditions from the sea water level rising and falling during the depolarisation test and complications caused by oxygen depletion when there is a variety of steel potentials. However, the data indicates quite clearly the responsive nature of the anodes to corrosion hazards.

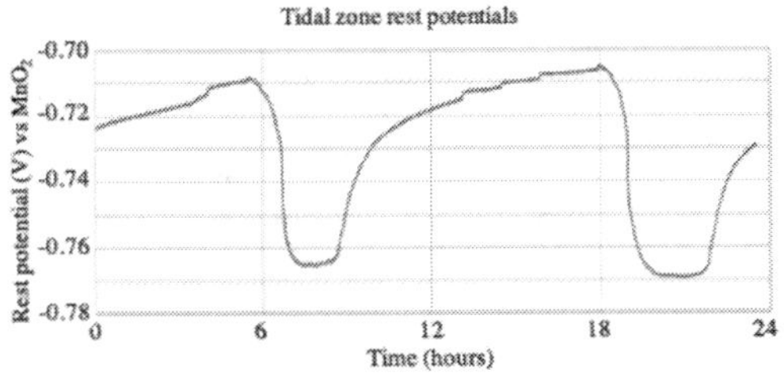

Fig. 7. Rest potentials in a tidal zone [7]

As shown above, the drop seen in potential during low and high tide is 40-65mV over approximately 9 hours across the piers. This changing environment will limit the ability to do standard depolarisation tests. Instead, calculated corrosion rates for the areas will be used.

The Bulter-Volmer equation gives a way of calculating a corrosion risk that can be assessed as a corrosion risk factor. Corrosion rates greater than $2mA/m^2$ are typically associated with some corrosion risk. Though, higher but falling corrosion rates in an area also show protection in the standard ISO12696:2016, because of this it is recommended to keep a log of data for historical comparison.

Given the measured corrosion rates fall below the threshold for corrosion risk and clearly rising steel potentials in the tidal zones, protection can be said to have been achieved.

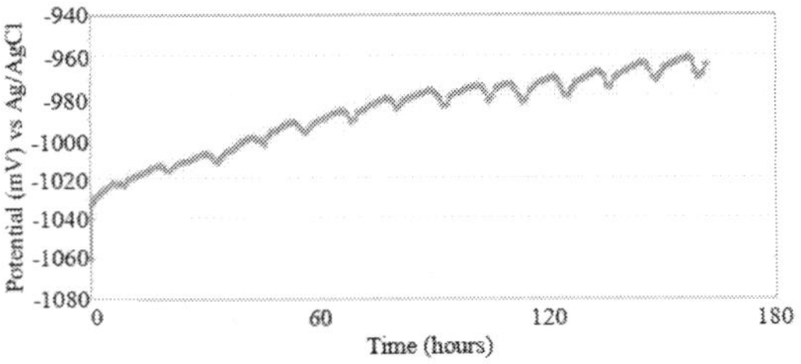

Fig. 8. Depolarisation of a tidal zone over 160 hours [7]

6.2 Monitoring of H6 canal bridge and Whiteadder bridge

Patch protection and hybrid systems tend to have different intensities of monitoring. The H6 canal bridge project was installed as a low maintenance system and therefore had no in-built reference electrodes. The Whiteadder project, by contrast, had continual monitoring taking data several times a day to allow off-site employees to monitor the system remotely.

After 4 years, the H6 canal bridge underwent a visual inspection and a surface half-cell potential measurement using a calibrated copper sulphate electrode with a steel connection to the local reinforcement. Examination of the structure shows that there was no concrete delamination or cracking despite the continued contamination with salt water from the failed joint on the bridge above:
Potential testing was undertaken at small intervals along the concrete to determine if those anodes could be identified after 4 years of operation. A half-cell electrode was placed on the concrete surface at regular intervals in the vicinity of the installed galvanic anodes embedded in the host concrete to generate a local 'potential map'.

When potential measurements are taken with respect to the reinforcement, a large negative potential indicates an anodic area. Therefore, when potential maps are taken, the most negative areas should be the sacrificial galvanic anodes that were embedded. Areas of negative potential away from the anodes will point towards possible corroding anodic steel, these areas will be likely exposed to increased chloride or moisture levels.

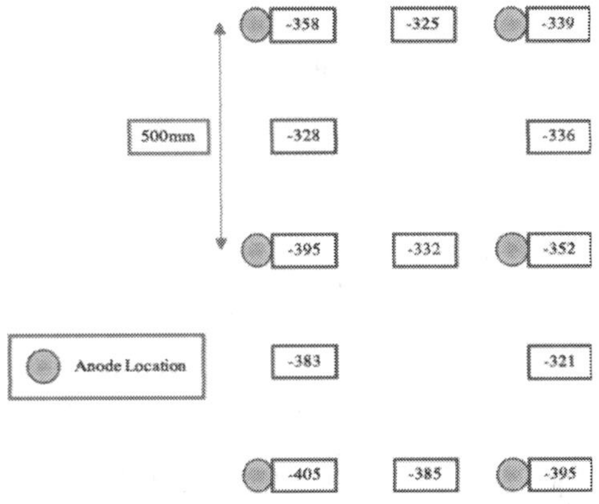

Fig. 9. Half-cell potential data from Test Area 1 (in mV v CSE)

It can be seen that the galvanic anodes can be clearly identified by measuring surface half-cell potentials i.e. the most anodic areas are associated with the installed galvanic anodes – this confirms that the anodes continue to generate current and are polarising the steel such that this can be measured on site. This also confirms that the steel has the least negative potential and as it is relatively cathodic and being protected by the sacrificial anodes.

A close interval half-cell potential test was undertaken at 50mm centres along the edge of the large repair in the centre of the bridge and data was obtained as follows;

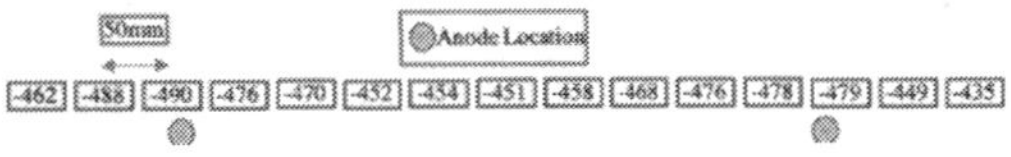

Fig. 10. Very Close Interval Half-cell potential data from Test Area 2 (in mV v CSE)

The half-cell potential rises adjacent to the anode and drops off moving away from the anode location indicating the sacrificial anodes are active and offering protection to the more cathodic steel.

A further area was tested around the large central patch, as follows;

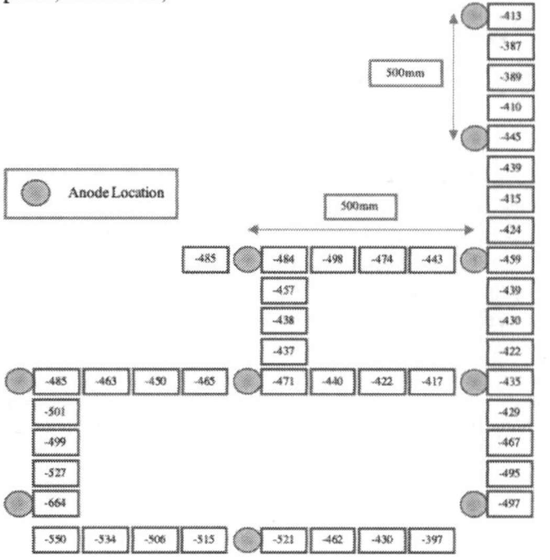

Fig. 11. Half-cell potential map from Test Area 3 (in mV v CSE)

Data from test area 3 was obtained from the bottom of the bridge structure adjacent to the large patch repair and shows that the anodes again can be identified from surface half-cell potential measurements. In addition, one of the anodes at the base of the structure is at a very negative potential (-664mV) indicating greater activity in an area where moisture levels are higher. It is clear that potentials generally get more negative on moving towards the bottom of the structure where there is more moisture and sacrificial anode activity is higher. This demonstrates the responsive behaviour of sacrificial anodes which generate current when the moisture content, and therefore the tendency for higher corrosion risk increases. The anode systems are functioning well to control corrosion and respond to moisture levels from both the canal and salt-laden water from the bridge above.

Whiteadder bridge was one of the first structures in the world to be fitted with a hybrid anode system. After 8 years AECOM did an independent study of the site over 2 years [4]. A waterproofing membrane added after installation limits the ability to do effective potential mapping. However, reference electrodes were placed in both upper 'dry' zones and lower 'wet' zones with data collected several times a day. Polarisation tests were done throughout the life of the project.

A visual inspection was carried out in March 2014, 8 years after installation. Although there was discolouration on the piers arising from the leaking deck joints above, no visible defects were found.

The initial current generated by the anodes after the end of the impressed current phase is quite high because of the moisture in the embedding mortar. The current then reduces as the mortar dries until it reaches a steady, low, maintenance current, this steady state current represents the work done to maintain the system once the local concrete environment is restored. Like the bridges in New Zealand, there is a clear difference between the 'wet' lower and 'dry' upper zones. A larger current is drawn in the 'wet' zone due to the lower resistivity of the concrete which protects against the increased risk from the moisture.

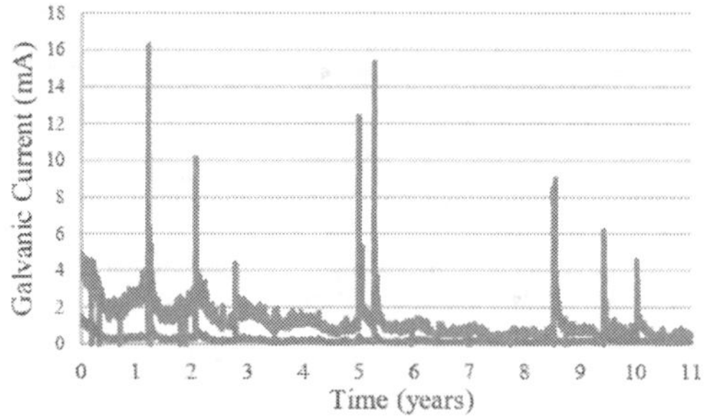

Fig. 12. Galvanic currents in upper and lower zones of Whiteadder bridge over 11 years

This is the most recent data for Whiteadder. Large peaks, an order of magnitude greater than the steady-state, can be seen at irregular intervals over the 11 years of data. These peaks are reactions to the river flooding with the rise mainly seen in the lower zone. The system is working well to protect the steel increasing the current reactively during a period of high risk until the moisture levels and resistivity drop back to normal.

A noticeable swell in the galvanic current can be seen every year. The swell is highest in the summertime and lowest during the winter months. This is the result of temperature variation during the year increasing the ionic mobility and with it, the work needed to maintain the concrete. This responsive behaviour shows that the system is actively working to protect the steel.

Galvanic systems aim to maintain the environment rather than polarising the steel, therefore corrosion current density calculations are used to get an accurate impression of the state of corrosion. The following table shows the first eight years of average current output for both upper 'dry' zones and lower 'wet' zones.

Tab. 1. Average current output (mA) for the first 8 years of operation [4]

Year	Average Anode Current Output (mA)	
	Upper Zone	Lower Zone
1	0.5	2.52
2	0.42	2.22
3	0.27	1.8
4	0.17	1.29
5	0.13	0.95
6	0.16	1.39
7	0.12	0.72
8	0.11	0.74
8-year average	0.24	1.45

Both the upper and lower zones have falling average anodic currents with 8-year averages. When polarisation tests were used to calculate corrosion rates, the lower zone falls well within the design current density of 2mA/m^2 and met the threshold for satisfying the low corrosion risk criteria. The upper zone has a natural variance depending on the depolarisation over the 6 test sites.

Tab. 2. Corrosion rate measurements using the area protected by the monitoring zone as 0.7m^2, an approximate steel density of 1.95m^2 and 2.05m^2 per m^2 of concrete for upper and lower zones respectively over an 8-year period [4]

	Corrosion Rate (mA/m^2)	
Date	Upper Zone	Lower Zone
19/04/07	1.89	0.67
27/08/07	2.48	0.66
20/10/07	1.74	0.43
06/03/08	2.38	0.49
12/05/09	2.88	0.47
04/06/13	2.07	0.45
20/06/14	2.47	0.46
12/03/15	1.60	0.24

Further proof of adequate protection is given by the migration of the steel potentials towards more positive values, a sign of the steel passivity. The dips correspond to the same flooding events seen before which caused responsive behaviour from the anodes. Given that after 8 years the corrosion rate was calculated at 1.60mA/m^2 coupled with the rising potential of the steel, it was determined that adequate protection had been given.

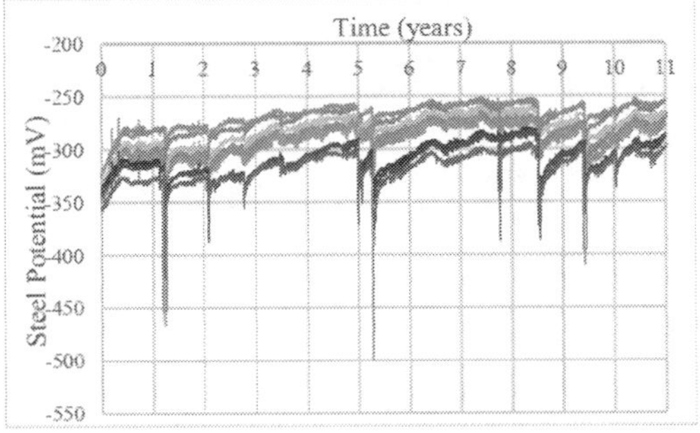

Fig. 13. Steel potentials data over 11 years [4]

There is a strong correlation between the anode current and steel potential values. As the potentials tend more positive the anodic current tends towards lower values as the demand is reduced by the benign environment of the steel. During the flooding events, near instantaneous reactions are observed to both the anodic current and the steel potential, showing that even after many years the galvanic system remains active and responds to higher corrosion risk by preferentially corroding and polarising the steel.

7 Conclusions

The responsive behaviour of galvanic anodes makes them uniquely suited to the transitory environments found on bridges or other river and marine structures worldwide. Whether the water source be from humidity, a canal, flooding, rainfall, or salinated from salt laden water seeping through leaking joints, galvanic anodes respond to the corrosion risk to protect the steel reinforcement.

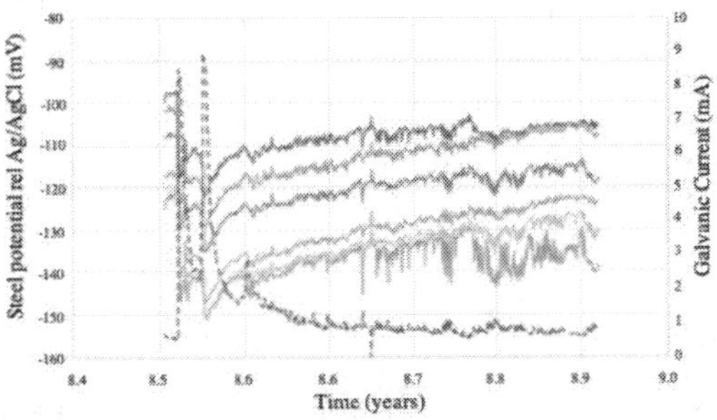

Fig. 14. Response of steel potential and galvanic current to a flooding event [4]

This reactive, self-regulating system therefore only provides high currents in response to corrosion hazards increasing the lifetime of the system. Whiteadder bridge was designed for a 30-year lifetime and was guaranteed for 20 years but is currently on track to have a usable lifetime in excess of 50 years [4, 8]

On the earliest installed hybrid system, we see this responsive behaviour active over its first 11 years of life, with calculations showing that the anodes have many decades of lifetime remaining. This is due to the anodes having very low corrosion currents when there is little moisture present and little risk of corrosion. The lowest corrosion rates, in fact, were calculated from data recorded in the highest risk areas.

Galvanic systems are well suited to dealing with sudden changes in environment such as flooding, rainfall, and tides, instantly increasing protection during the event and maintaining adequate protection during the high risk drying out period.

On the H6 canal bridge we see that anodes react locally to risk, only active in areas with high moisture to prevent the steel corroding. Using this knowledge, structures may be designed to use larger capacity anodes in higher risk areas, increasing the design life of the project efficiently.

The SH16 bridges are the first utilisation of a hybrid system on a pre-stressed structure. Care must be taken during the short impressed current phase not to exceed the hydrogen embrittlement threshold for an extended period. However, due to the fixed driving voltage of the galvanic anodes during the galvanic phase there is no risk of hydrogen embrittlement.

Concrete environment restoration, temporal and local responsivity alongside a fixed safe voltage for prestressed structures make hybrid and galvanic anodes a great solution for corrosion on bridges and waterways. The simple installation and straightforward implementation are well suited to projects with limited and tidal access and the reactive nature of these anodes also allows for extended design lifetimes. Whiteadder bridge, the oldest structure is on-track to provide protection past its design life by 20 or more years.

8 References

1. G. Koch, J. Varney, N. Thompson, et al, *International measures of prevention, application, and economics of corrosion technologies study*, NACE International, Impact.nace.org/documents/Nace-International-Report.pdf (2016).
2. C. Christodoulou, G. K. Glass, J. Webb, V. Ngala, S. Beamish, P. Gilbert, *Evaluation of galvanic technologies available for bridge structures, In: Structural Faults and Repair*, 12th International Conference and Exhibition. Engineering Technics Press, Edinburgh, UK (2008).
3. G. K. Glass, A. C. Roberts, N. Davison, Hybrid corrosion protection of chloride-contaminated concrete, Proceedings of the Institution of Civil Engineers-Construction Materials, **161**, 163-172 (2008).
4. W. Dodds, C. Christodoulou, C. Goodier, *Hybrid anode concrete corrosion protection – independent study*, Proceedings of the Institution of Civil Engineers-Construction Materials, Ahead of Print, https://doi.org/10.1680/jcoma.16.00024 (2016).
5. G. K. Glass, A. C. Roberts, N. Davison, *Assessment of cathodic protection applied to above ground reinforced concrete,* Materials and Corrosion, **70** (3), 503-510, (2019).
6. T. J. Watterson, P. W. Corbett, *Widening of existing bridges on State Highway 16 in Auckland.* Proceedings of the ICE - Bridge Engineering. **169**. 1-12. (2015).
7. C. Christodoulou, R. Cobbs, P. Corbett, and M. Elliot, *Service life extension of state highway 16 bridges – New Zealand's first hybrid corrosion protection application*, MATEC Web of Conferences **199**, 05002 (2018).
8. C. Christodoulou, W. Dodds, B. Jackson, et al, *Hybrid galvanic corrosion protection of Whiteadder Bridge – 10 years on*, unpublished (2019).
9. Concrete Society, Technical Report 60, Electrochemical tests for reinforcement corrosion, Surrey, UK (2004).
10. U. Angst, B. Elsener, C.K Larsen, O. Vennesland, Critical chloride content in reinforced concrete - A review. Cement and Concrete Research, 39, 1122-1138 (2009).
11. ASTM C1556 – 04
12. NZTA Report 337 – Deterioration of Prestressed Concrete Bridge Beams.
13. S. P. Holmes, *The response of galvanic protection current to environmental changes*, EUROCORR Nice, France, Paper 8254 (2009)
14. G. K. Glass, A. C. Roberts, N. Davison, *Pit realkalisation and its role in the electrochemical repair of reinforced concrete*, Journal of Corrosion Science and Engineering, **9**, Paper 10, (2007).
15. ISO. 12696. *Cathodic Protection of Steel in Concrete*, (2016).
16. G. K. Glass, J. R. Chadwick, Corrosion Science, **36**, 12, 2193-2209 (Dec. 1994).
17. C. Christodoulou, G. K. Glass, J. Webb, S. Austin, C. Goodier, *Assessing the long term benefits of impressed current cathodic protection*, Corrosion Science, **52**, 2671-2679 (2010).
18. C. Christodoulou, A. Sharifi, S. Das, C. Goodier, *Cathodic protection on the UK's Midland Links motorway viaducts*, Proceedings of the Institution of Civil Engineers: Bridge Engineering, **167**, 43 – 53 (2013).

Results on the long-term performance and effectiveness of a galvanic CP system applied on a lightweight concrete bridge with prestressed steel as determined by depolarisation values and current densities

A.J. (Hans) van den Hondel MSc

Vogel Cathodic Protection, Zwijndrecht, The Netherlands

Abstract. This paper gives insight in the research and investigations which were undertaken on a post tensioned light-weight concrete box girder bridge in the Netherlands. This project started in the first half of 2013 and measurements are still executed regularly. Due to long-term leakage of a longitudinal joint between 2 parts of the bridge, chlorides had penetrated into the concrete up to the level of the reinforcement and beyond, causing severe corrosion of the steel, spalling of the concrete cover and eventually causing danger for the shipping underneath the bridge due to falling lumps of concrete. To solve this problem, Galvanic Cathodic Protection (GCP) was applied in 2 stages. Firstly, in 2013, zinc based anode strips were applied on the soffit of 2 post tensioned lightweight bridge segments for the purpose of testing and monitoring. After obtaining good results from the test, in 2014/2015 the remaining 29 bridge segments were repaired and protected with the same GCP system as well. This paper reviews subsequent potential decay and current densities measurements, which have provided valuable information on GCP performance and throwing power. The results of anode performance at different locations and depths are reviewed. Concrete resistivity has shown a major influence on the effectiveness of the installed GCP system.

1 Introduction

1.1 The bridge

The bridge in question is the Neerbossche Bridge which spans a canal at Nijmegen, The Netherlands (Figure 1).

Fig. 1. The Neerbossche Bridge over the Maas-Waal canal.

The bridge is a post tensioned double box girder bridge from 1982, of which the structure above the canal is constructed with light-weight concrete (1650 kg/m^3). The in-situ cast concrete segments have a length of about 3.4 meters. For each box girder, the part of the bridge located between the piers comprises 33 segments of which the 2 segments above the piers are constructed with conventional normal weight concrete. So each box girder is comprised of 31 light-weight concrete segments above the canal. The soffit of the light-weight concrete sections of the southern box girder showed serious damage and consequently maintenance was required in the short term in view of structural safety as well as possible danger for the shipping underneath the bridge due to falling lumps of concrete.

1.2 The project

The damage in the bridge was extensively investigated and reported [1] in a preliminary phase and it was concluded that the damage was limited to the concrete surface at the soffit of the northern cantilever of the span of the southern bridge (see Figure 2). Due to long-term leakage in the longitudinal joint between the northern and southern bridge, chlorides originating from the frequent application of de-icing salt during winter periods had penetrated into the concrete up to the level of the reinforcement and beyond. As a consequence, the mild steel reinforcement had started to corrode which eventually resulted in the development of significant tensile stresses in the concrete, leading to cracks, delaminations and spalling at multiple locations.

The corrosion induced damage amounted to approximately 44 m^2, which was about 22% of the total surface of the soffit. At other locations, the likelihood of corrosion was demonstrated by corrosion potential measurements which showed values down to -450 mV ($Cu/CuSO_4$) and -350 mV (Ag/AgCl; see Figure 3).

Fig. 2. Location of the damaged cantilever (southern bridge).

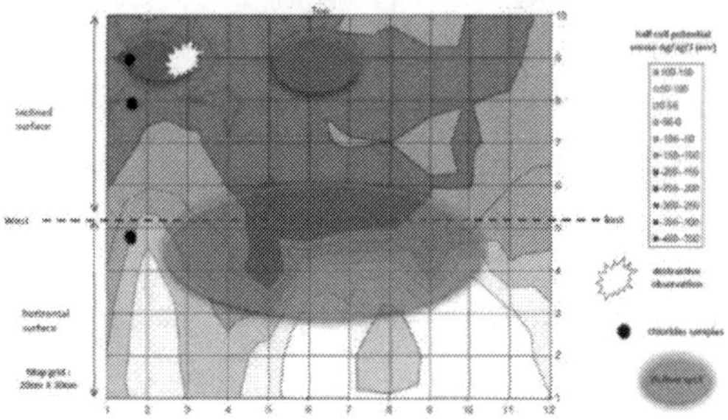

Fig. 3. Corrosion potential measurements at the cantilever in the southern bridge. The grey areas denote hollow spots and the white flash a destructive investigation point.

As a long-term solution to this problem, cathodic protection (CP) was considered to present a viable and economical option. As the structure was post tensioned, Galvanic Cathodic Protection (GCP) was in favor to be applied in order to avoid the risk of hydrogen embrittlement in the high quality steel tendons. In the summer of 2013, the steel reinforcement in 2 bridge segments was protected with zinc based strip anodes, for testing purposes applied in small size foils which could be connected to the reinforcing steel on and off separately. Also, a large number of reference electrodes (titanium decay probes, referred to as RE's) were embedded in the concrete of these 2 segments in order to monitor the effectiveness of the CP system, not only within the protected area but also outside the protected area.

With these 'remote' RE's the so-called throwing power of this CP system could be established. For the same reason RE's were installed at different depths in the concrete cover as well as in repaired areas in order to study the corrosion behavior of the steel in these areas due to the action of the CP system.

In view of the good results that were obtained in this test, the suitability of this GCP system was proven, the full scale project on the remaining 29 bridge segments was executed in the winter of 2014/2015.

1.3 The application of the GCP system

The highlights of the full scale project in 2014/2015 were reported extensively in 2016 [2], with respect to the accessibility of the soffits 10 meters above the water level, the application of shotcrete using a light-weight mortar with a high cement content and no additives for the concrete repairs and the application of the GCP system (Figure 4-6).

The anode material used for the protection of the steel reinforcement was a surface mounted zinc anode. This so-called Zinc Layer Anode (ZLA) is applied as a foil on the surface and consists of a 250 μm thick zinc sheet provided with a 750 μm thick ion-conductive adhesive.

Fig. 4. Access to the soffit at the damaged cantilever.

Fig. 5. Repair of the damaged locations with shotcrete.

Fig. 6. Application of the zinc anode in 25 cm strips.

The complete zinc surface, including the edges, was covered with a 2 mm waterproof cement-based coating in order to prevent the zinc from direct contact with moisture due to condensation and resulting atmospheric corrosion.

1.4 Scope of the paper

This paper will focus on the new data and insights from the still ongoing research and investigation on the effectiveness of the GCP system over time, by measuring potential decay values and current densities.

2 Developments in time of the effectiveness of the GCP system

2.1. The test segments

The original test provisions from 2013 on the 2 segments (see Figure 7; [2]), with a large amount of RE's and the narrow zinc strips (with a width of 83 mm), were kept unchanged, as much as possible, during the full scale project in 2014/2015 in order to be able to continue the monitoring of the GCP system and to start research of the long-term effects of this type of GCP systems with a surface mounted anode.

Since 2013 depolarisation measurements of the CP system in different seasons have been carried out, using all the RE's in and remote from the protected area, as well as in depth. The amount of protection current being generated by each zinc strip was measured and monitored in order to predict the service life of the (sacrificial) anodes. Also, other system parameters, such as the resistance in all the components, were monitored. These measurements were carried out during almost 5 years and show interesting results, expanding the knowledge on GCP and the long-term effects on potential decay values, (protective) current densities and the throwing power of galvanic anodes, as well as the influence of repair mortars and the moisture content of the concrete on the behavior of installed GCP systems.

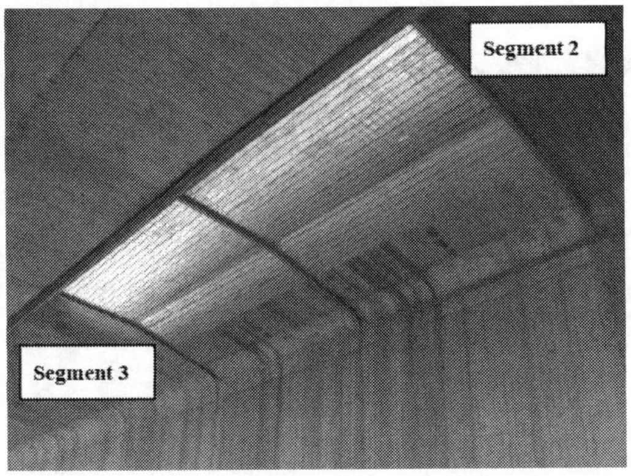

Fig. 7. Test locations with the zinc anode material.

2.2 Data and properties of the test segments

Two bridge segments (of 33) were selected in 2013:

1. Segment 2 with 4% of concrete damage, a concrete cover between 29 mm and 45 mm (μ = 37 mm) and corrosion potential measurements with values from 0 mV down to -270 mV ($Cu/CuSO_4$)
2. Segment 3 with 33% of concrete damage, a concrete cover between 25 mm and 46 mm (μ = 38 mm) and corrosion potential measurements with values from -30 mV to -350 mV (at the location of the damage)

The reinforcement of the soffit was placed in 2 layers, crosswise, with rebar's of Ø12 mm and Ø16 mm in diameter. The carbonation depth as measured in the soffit varied between 2 mm to 11 mm, however high chloride contents were measured, up to 3% on cement weight and up to depths of over 50 mm.

The CP system at the test segments was applied on the horizontal part of the cantilever and the adjacent 50 cm of the inclined segment (thus over a total width of 2 meters). Each test segment was in this way covered with 6.2 m^2 of ZLA (see Figure 7).

2.3 Effectiveness of the installed GCP system

The measurements were carried out using the titanium decay probes (RE's), embedded in the concrete of the first 2 segments, in and remote from the protected area.

From the start in 2013, measurements showed that the applied GCP system was effective. The depolarisation (potential decay) values of the 1st layer of reinforcement mostly met the code-based CP criteria (ISO-EN 12696; [3]), i.e. a minimum of 100 mV depolarisation after 24 hours or 150 mV after complete depolarisation (see Figure 8 & 9). The potential decay values of the 1st layer decreased in time: with 52% in 6 years at segment 2 and with 17% in the first year and 60% in the next 4.5 years after replacement of the anode material at segment 3.

Furthermore it was found, from the RE's placed at different depths in the concrete cover, that in segment 2 the 2nd layer of reinforcement is protected as well. However, in segment 3 the results are far less positive and the potential decay values decreased rapidly in the first year after the replacement of the anode material in 2014/2015 Since then the results are stabilized (with 74% decrease in 4.5 years, to an average (24 hours) value of 70 mV).

Since 2016 the code based CP criteria are no longer met at all 4 RE's, although it should be noted that 2 of the RE's are placed in repaired locations (RE7 & RE8). The same is applicable to the remaining 2 RE's placed at the 1st layer of reinforcement in repaired areas of segment 3 (RE9 & RE10).

Segment 2												
Date:	28-8-2013	22-11-2013		28-1-2014	4-12-2014	20-3-2015	16-10-2015	10-8-2016	22-9-2017	24-5-2018	1-5-2019	
Session:	1st	3rd		4th	7th	8th	9th	10th	11th	12th	13th	
	Depolarisation											
	(mV)	(mV)	(mV)	(mV)	(mV)	(mV)	(mV)	(mV)	(mV)	(mV)	(mV)	
RE 1	240	255	461	376	157	263	300	170	170	171	161	1st layer of reinforcement
RE 2	337	344	387	452	183	240	304	177	174	166	163	
RE 3	217	211	241	214	83	109	101	106	94	84	81	
RE 4	231	217	225	226	101	155	123	120	96	84	91	
RE 5	146	151	160	131	61	87	95	109	105	109	105	2nd layer of reinforcement
RE 6	157	158	165	173	77	130	121	116	119	120	115	
RE 7	187	188	237	234	108	139	141	137	137	134	133	
RE 8	224	229	320	271	116	190	173	152	139	125	118	
RE 9	270	334	481	324	226	289	184	199	161	144	148	Most neg.potential
RE 10	101	104	163	148	72	104	2	48	50	45	41	Remote from CP area: 20 / 40 / 60 cm
RE 11	64	69	90	90	49	60	31	33	33	30	28	
RE 12	47	48	74	69	37	45	13	8	26	23	21	
	24	24	168	144	23	74	240	24	24	24	24	Hours of depol.
	100 mV	100 mV	150 mV	150 mV	100 mV	150 mV	150 mV	100 mV	100 mV	100 mV	100 mV	Threshold value

Fig. 8. Depolarisation values of segment 2.

Segment 3														
Date:	28-8-2013		22-11-2013		28-1-2014	10-9-2014	4-12-2014	20-3-2015	16-10-2015	10-8-2016	22-9-2017	24-5-2018	1-5-2019	
Session:	1st		3rd		4th	6th	7th	8th	9th	10th	11th	12th	13th	
	Depolarisation													
	(mV)	(mV)	(mV)	(mV)	(mV)	(mV)	(mV)	(mV)	(mV)	(mV)	(mV)	(mV)	(mV)	
RE 1	134	155	152	164	152	62	77	210	185	161	117	134	126	new anode material — 1st layer of reinforcement
RE 2	154	177	184	202	219	104	91	268	196	145	86	95	97	
RE 3	112	133	120	232	231	135	112	293	240	171	119	126	122	
RE 4	58	64	57	99	149	77	101	320	192	137	86	95	96	
RE 5	130	147	122	137	141	60	75	313	221	94	66	72	69	2nd layer of rein-forcement, partly in repair locations
RE 6	74	85	89	114	144	66	33	310	188	98	59	70	66	
RE 7	58	65	67	77	89	53	56	247	211	87	59	65	64	
RE 8	51	53	53	66	70	50	50	190	216	91	79	89	80	
RE 9	51	57	52	73	74	25	36	202	221	122	74	81	82	in repair locations
RE 10	31	39	36	46	70	37	38	209	213	122	80	93	90	
RE 11	285	333	294	309	533	307	191	363	180	323	254	300	294	Most neg.potential
RE 12	87	102	101	150	138	86	82	81	33	56	38	54	52	Remote from CP area: 20 / 40 / 60 cm
RE 13	56	65	57	85	84	57	54	38	21	39	27	35	36	
RE 14	25	49	29	72	60	41	40	26	-1	27	21	27	26	
	24	79	24	168	144	26	23	74	240	24	24	24	24	Hours of depol.
	100 mV	150 mV	100 mV	150 mV	150 mV	100 mV	100 mV	150 mV	150 mV	100 mV	100 mV	100 mV	100 mV	Threshold value

Fig. 9. Depolarisation values of segment 3.

From the start they showed less depolarisation than similar RE's in non-repaired locations and since 2017 the code based criteria are no longer met.

In general, the potential decay values of the 2nd layer were less compared to the ones located in the 1st layer of reinforcement. As an average, during 6 years, 64% to 97% of the potential decay values of the 1st layer were measured at the 2nd layer of reinforcement in segment 2 and 57% to 97% in segment 3. It should be noted that the 2nd layer showed less concrete damage compared to the 1st layer and levels of chloride contamination were lower.

The best results in segment 3 were clearly obtained for the reinforcement showing the most negative natural potentials prior to installation of the CP system (RE11 with a 4.5 years average of 293 mV (after 24 hours of depolarisation)). In segment 2, the potential decay values of the similar reference electrode (RE9) were also more than satisfactory with a 6 years average of 212 mV.

Finally, the measurements also showed that the CP system was not only effective in the protected area but also at a limited distance from the protected area. At the start in 2013 and during the first year after installation, the measurements showed that the code-based criteria were met at least up to 20 cm adjacent to the protected area. In time, the potential decay values in this area decreased considerably and since December 2014 the code based criteria are no longer met at 20 cm from the anode material.

2.4 Development in time of potential decay values

The measured values of the depolarisation in the protected area decreased over time (Figures 10 and 11). At the start in 2013 the measurements showed higher values indicating that the applied GCP system complied with the code-based ISO-EN12696 criteria ([3]) at almost all locations where RE's are installed, resulting in sufficient protection of the reinforcement.

After 17 months (in December 2014) 4 out of 5 RE's (80%) at the 1st layer of reinforcement in segment 2 still met the code-based protection criteria (compared to 100% compliance in 2013). In October 2015, more than 2 years after the installation of the GCP system, this is reduced to 3 out of 5 RE's (60%) at the 1st layer. Since then the results at different measuring moments (up to 6 years after the installation) are stabilized and 60% to 100% of the RE's still meet the protection criteria.

The RE's placed at the 2nd layer of reinforcement in segment 2 have behaved more or less in the same way. After 17 months, 2 out of 4 RE's (50%) still met the code based protection criteria (compared to 100% compliance in 2013). In October 2015 this was further reduced to 25%, but since then the measurement results are back to 100% compliance.

In segment 3, the reaction of the RE's is less favorable. Starting with a compliance of 80% in 2013, after 17 months this reduced to 60% (3 out of 5 RE's) at the 1st layer of reinforcement. At the 2nd layer of reinforcement in segment 3, only 1 RE (25%) met the code based protection criteria at the start in 2013. After 17 months these criteria are no longer met at any of these RE's (0%).

At segment 3, a part of the zinc anode, as well as the majority of the RE's, were replaced during the full scale project in 2014/2015 due partly to repair mortar disbondment in the test phase (not corrosion related). As a consequence, the last measurements with the original anode material dated from December 2014. After repairing the disbonded areas with shotcrete and replacing parts of the zinc anode the system reacted immediately and showed good results again. In March 2015, the GCP system (with partly new zinc anode material) met the code based criteria again, indicating sufficient protection of the reinforcement at all locations where RE's are installed, i.e. at both the 1st and 2nd reinforcement layer and in the repaired areas. Moisturizing of the concrete and the new installed anode material can be the explanation for this observation.

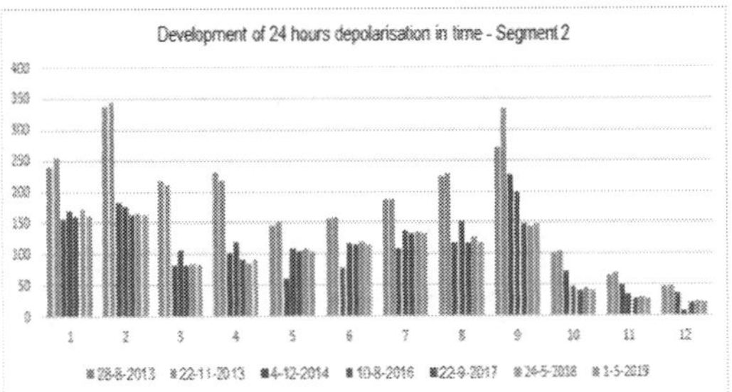

Fig. 10. 24 hours depolarisation values of segment 2.

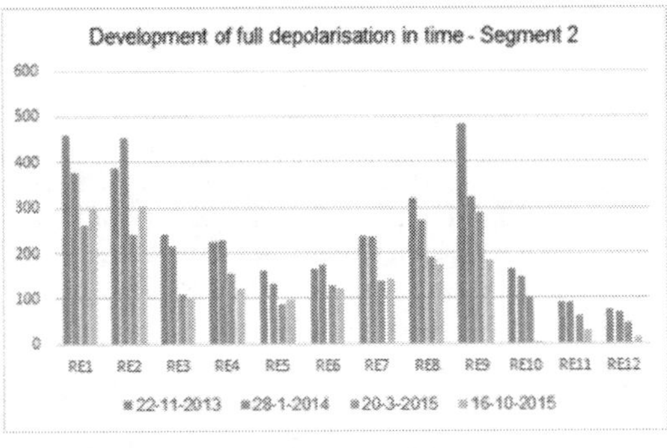

Fig. 11. Full depolarisation values of segment 2.

In August 2016, all 7 RE's (100%) at the 1st layer of reinforcement in segment 3 still met the code based protection criteria, including the RE's in the repaired areas. In September 2017 this was reduced to 43% (3 out of 7 RE's) and the same results were obtained in 2018 and 2019. At the 2nd layer of reinforcement none of the RE's (0%) were, in 2016, still in compliance with the criteria and since then the potential decay values have further reduced. Drying out of the concrete and polarisation of the steel are considered to be the explanation for this observation.

The best result in segment 3 is, in 2019, still coming from the reinforcement with the most negative natural potentials prior to CP installation (RE11).

At the areas located more remote from the anode, i.e. adjacent to the protected surface, similar observations were found. At the start in 2013 and during the first year after installation the protected area stretched out at least up to 20 cm adjacent to the anode material, meeting the code based criteria. In time the potential decay values of the RE's at 20 cm, 40 cm and 60 cm decreased with approximately 50% and in 2019 the code based criteria are not met at any of these RE's. The positive side effect due to the so called 'throwing power' was no longer measured after 2 years, even at a distance of 20 cm from the anode material. Drying out of the concrete is considered to be the explanation for this observation.

Finally, the RE's placed in repaired areas (segment 3; RE 7-10) showed, after having disbonded areas from the test phase repaired and zinc anodes replaced (December 2014), sufficient depolarisation at all locations. The most probable explanation for the negative outcome of the potential decay values at these RE's during the first 17 months lies in disbondment of the repair mortar. Since 2017, less than 3 years after the replacement, none of the RE's (0%) placed in repaired areas is still in compliance with the criteria. Drying out of the concrete and polarisation of the steel are considered to be the explanation for this observation.

2.5 Development in time of the current density

The current output readings were carried out for each separate zinc foil (22 strips per segment with a width of 83 mm and a total zinc surface of 0.28 m^2 for each strip) at 5 moments in time. The total current output was measured as well for the complete anode system per segment (with a total zinc surface of 6.2 m^2).

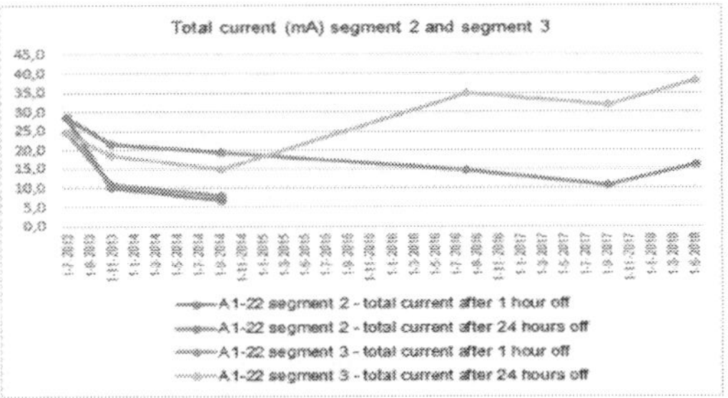

Fig. 12. Development of the total current output in time for segments 2 and 3.

The measured current outputs decreased over time (see Figures 12-14), with a strong accent in the first months. However, in 2018 the current outputs increased again.

At the start in 2013 the total current output amounted to 28.6 mA (equal to 4.6 mA/m^2_{anode} and roughly 7.7 mA/m^2_{steel}) and 24.8 mA (equal to 4.0 mA/m^2_{anode} and roughly 6.7 mA/m^2_{steel}) for segment 2 and 3, respectively. After 14 months the total current output was reduced (measured 1 hour after switching off the GCP system) to 6.8 mA (equal to 1.1 mA/m^2_{anode}, roughly 1.8 mA/m^2_{steel} and 76% reduction) and 8.1 mA (1.3 mA/m^2_{anode}, roughly 2.2 mA/m^2_{steel} and 67% reduction). After 3, 4 and 5 years the current output was measured after switching off the system for 24 hours. In segment 2 a total current output of 14.6 mA was measured in 2016, 10.8 mA in 2017 and 16.3 mA in 2018 (equal to 2.4, 1.7 and 2.6 mA/m^2_{anode} respectively and roughly 4, 2.8 and 4.3 mA/m^2_{steel}).

After replacement of parts of the zinc anode in segment 3 in December 2014, the current output increased. After switching off the system for 24 hours 35 mA was measured in 2016, 32 mA in 2017 and 38 in 2018 (equal to 5.7, 5.2 and 6.2 mA/m^2_{anode} respectively and roughly 9.4, 8.6 and 10.2 mA/m^2_{steel}).

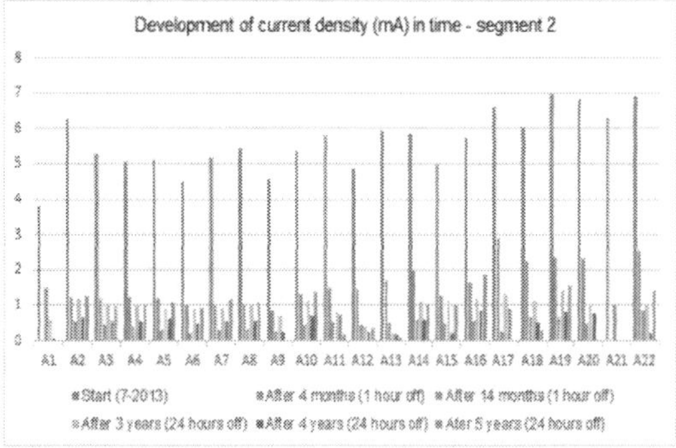

Fig. 13. Development of current output in time for segment 2.

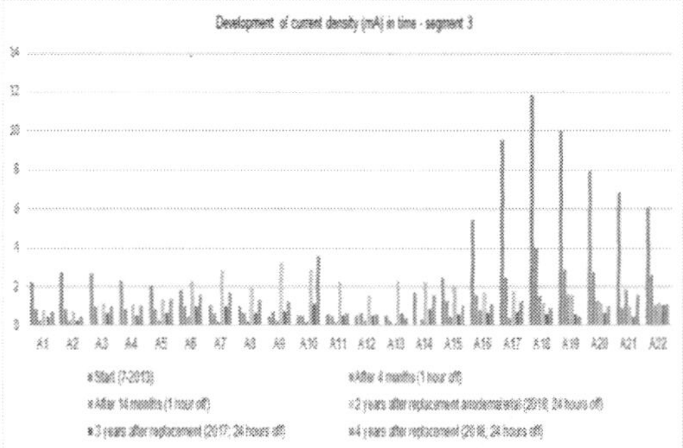

Fig. 14. Development of current output in time for segment 3.

During the first year the current output for the separate zinc foils (22 strips per segment) varied strongly in segment 3 (from 0.5 - 11.8 mA/strip; see Figure 14). The parts of segment 3 with patched areas (from zinc strip A5 to A16 and with the larger patched areas from A9 to A13; see Figures 15-16; [5]) showed much lower current outputs compared to the zinc strips in non-patched areas. The disbonded repair mortar caused a higher resistance and subsequent lower current densities at these locations.

Fig. 15. Repair locations and RE's in segment 3.

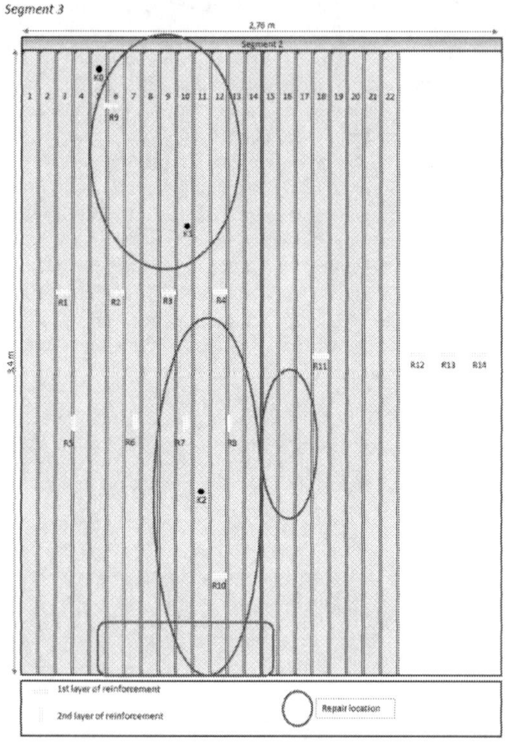

Fig. 16. Anode strips, RE's and repair locations in segment 3.

As mentioned above, the current output in segment 3 increased significantly after having the disbonded areas from the test phase repaired and the majority of the zinc anodes replaced in December 2014. At the same time the current distribution over the zinc strips became more equally spread (varying from 0.3 - 1.2 mA/strip in September 2017 ($\approx$ 0.28 m^2 for each strip) and 0.3 - 3.6 mA/strip in May 2018; see Figure 14).

The zinc strips applied in segment 2, which has less patched areas, showed more equally distributed current outputs from the start: varying from 3.8 - 7.0 mA/strip in July 2013 to 0.1 - 0.9 mA/strip in September 2017 and 0.0 – 1.9 mA/strip in May 2018; see Figure 13).

To compare the current output of each zinc strip vs the potential reading of respective RE's is complicated due to the fact that the RE is affected by the current output of all neighboring zinc strips.

2.6 Development of the concrete resistance in time

The measurements of the concrete resistance were carried out by measuring the resistance between the reinforcement and each individual RE and anode strip for each segment.

The measured values of the concrete resistance on these specific locations increased strongly over time (see Figure 17). At the start in 2013 the measured resistances at the RE's in segment 2 varied between 1 to 9 kΩ (with an average (μ) of 3.4 kΩ), in segment 3 in the patched areas between 0.4 to 1 kΩ and between 1 to 6 kΩ in non-patched areas (μ_{total} = 2.2 kΩ). After 14 months (2014), it increased substantially and the resistances varied at that moment from 7 to 212 kΩ in segment 2 (μ= 45 kΩ) and in segment 3 from 2 to 5 kΩ in patched areas and from 4 to 130 kΩ in non-patched areas (μ_{total}= 26 kΩ). The highest resistances in both segments were measured at RE's placed adjacent to the protected surface. Drying out of the (uncovered) concrete is considered to be the explanation for this observation.

During the next 3 years, the measurements showed a stabilized situation in segment 2. After 4 years (2017) the resistance at the RE's within the protected area (RE1 – RE9) showed an average of 19 kΩ (compared to 18 kΩ in 2014 and 16 kΩ in 2016 in the same season of the year). In 2018 this average was 16 kΩ again. In segment 3 the majority of the RE's were replaced in Dec. 2014.

	Segment 2						
	12-7-2013 start	10-9-2014 14 months	20-3-2015 20 months	10-8-2016 3 years	23-9-2017 4 years	24-5-2018 5 years	increase of resistance
	Resistance						
RE's	(kΩ)	(kΩ)	(kΩ)	(kΩ)	(kΩ)	(kΩ)	(kΩ)
RE 1	1,96	14	23	14	17	14	12
RE 2	2,59	11	16	12	15	12	9
RE 3	1,86	10	17	10	12	10	9
RE 4	2,60	19	16	14	18	16	13
RE 5	2,06	15	26	13	14	12	9
RE 6	1,33	7	12	6	7	5	4
RE 7	2,33	19	16	11	13	11	9
RE 8	1,84	13	21	12	13	11	10
RE 9	5,43	56	84	52	64	56	50
RE 10	8,57	212	55	92	110	90	82
RE 11	3,08	48	33	28	34	29	26
RE 12	6,83	116	90	74	94	78	71
	RE's vs Reinforcement						
average:	3,4	45,0	34,1	28,2	34,2	28,7	25
Anode strips	(kΩ)	(kΩ)		(kΩ)	(kΩ)	(kΩ)	(kΩ)
A1	0,217	3,80		0,383	3,5	12,2	loose contact
A2	0,112	0,135		0,151	0,200	0,181	0,069
A3	0,129	0,154		0,167	0,226	0,205	0,076
A4	0,136	0,170		0,176	0,232	0,216	0,080
A5	0,137	0,182		0,260	0,238	0,220	0,083
A6	0,152	0,206		0,190	0,242	0,236	0,084
A7	0,128	0,160		0,163	0,196	0,179	0,051
A8	0,125	0,158		0,157	0,205	0,187	0,062
A9	0,148	0,225		0,420	1,5	7,2	7,1
A10	0,126	0,158		0,143	0,184	0,166	0,040
A11	0,116	0,159		0,146	0,188	0,410	0,294
A12	0,138	0,577		0,518	0,700	0,640	0,502
A13	0,112	0,171		0,583	1,4	5,7	5,6
A14	0,121	0,199		0,200	0,533	0,413	0,292
A15	0,139	0,286		0,169	2,4	0,430	0,291
A16	0,122	0,181		0,205	0,225	0,172	0,050
A17	0,108	0,678		0,263	0,246	3,7	3,6
A18	0,113	0,194		0,196	0,475	1,6	1,487
A19	0,106	0,180		0,164	0,208	0,192	0,087
A20	0,057	0,348		0,443	0,266	22	21,9
A21	0,113	0,601		45	29	49	break in wiring
A22	0,105	0,191		0,238	1,5	0,244	0,1
	Anode strips vs Reinforcement						
average: kΩ	0,13	0,25		0,25	0,568	0,465	2,09
A_{tot}-K_s			12,2 Ω	47 Ω	16,7 Ω	18 Ω	
calculated: Ω	5,4	9,7		10,0	14,7	15,2	$1/R_t = \Sigma 1/R_n$

Fig. 17. Measured concrete resistances in segment 2.

In 2013, the measured resistances at the anode strips in segment 2 varied between 57 to 217 Ω (μ= 130 Ω), in segment 3 in the (large) patched areas between 0.8 to 1.4 kΩ and between 41 to 606 Ω in non-patched areas (μ_{total}= 432 Ω). After 14 months, these resistances were increased as well, varying at that moment from 135 to 678 Ω in segment 2 (μ= 253 Ω) and in segment 3 from 1.2 to 3.1 kΩ in patched areas and from 0.2 to 0.8 kΩ in non-patched areas (μ_{total} = 1kΩ). The highest resistances were measured at (partly) disbonded patch repairs in segment 3.

From 2016 to 2017 the resistances increased furthermore to an average of 568 Ω in segment 2. Drying out of the concrete is considered to be the explanation for this observation. However, in 2018 the resistance of most of the anode strips in segment 2 decreased again to an average of 465 Ω. The measured high resistances of 6 (of the 22) anode strips in this segment needs to be further investigated. Oxidation of the (anode) contacts or defects in the wiring are expected to have caused this deviation.

In segment 3 the measured resistances changed substantially after having the disbonded areas from the test phase repaired and the majority of the zinc anodes replaced in December 2014.

2.7 Polarisation of the anode material

The polarisation of the anode was measured in order to determine the condition of the anode material. Measurements were carried out 3 and 4 years after installation of the GCP system in segment 2 and after 3 years (with the new anode material) in segment 3, by measuring the potential between the anode and each individual RE, as well as between each anode strip and the reinforcement.

The measured values show that the anode material was still in good condition after 3 and 4 years respectively (see Figures 18 and 19).

The measured potentials of the older anode strips in segment 3 (roughly anode strips A1-A3 and A18-A22) differ slightly from the anode material that was replaced in December 2014 (roughly A4-A17). At this moment the reason for this observation is uncertain but could have been caused by the simultaneous replacement of the majority of the RE's in this segment in 2014. This will be investigated in 2019.

		Segment 2					
		10-8-2016					
		measured anode potential after switching off CP system: RE (Ti*)					
		on	off after 3 minutes	off after 30 minutes	off after 1 hour	off after 3 hours	polarisation anode after 3 hours
Segment 2		(-mV)	(-mV)	(-mV)	(-mV)	(-mV)	(mV)
RE 1	Anode - M2	448	923	961	974	991	68
RE 2	Anode - M2	431	906	937	947	958	52
RE 3	Anode - M2	247	767	809	824	846	79
RE 4	Anode - M2	249	774	817	831	850	76
RE 5	Anode - M2	246	778	817	830	853	75
RE 6	Anode - M2	263	784	822	836	858	74
RE 7	Anode - M2	263	733	771	784	803	70
RE 8	Anode - M2	311	787	822	832	849	62
RE 9	Anode - M2	468	912	935	945	963	51
RE 10	Anode - M2	212	763	811	830	864	101
RE 11	Anode - M2	127	685	737	760	799	114
RE 12	Anode - M2	79	637	691	716	763	126
RE (Ti*)	Zinc anode						

Fig. 18. Anode polarisation in segment 2 (vs Ti*), 3 years after installation.

		Segment 2			Segment 3		
		July 2013	September 2017		July 2013	September 2017	
		natural potential	measured anode strip potential after switching off CP system: steel		natural potential	measured anode strip potential after switching off CP system: steel	
Anode strip	Reinfor-cement	before starting up	off after 30 minutes	off after 24 hours	before starting up	off after 30 minutes	off after 24 hours
		(-mV)	(-mV)	(-mV)	(-mV)	(-mV)	(-mV)
A1	Steel	1005	596	668	956	573	697
A2	Steel	985	642	776	920	412	517
A3	Steel	960	616	757	921	464	588
A4	Steel	943	579	723	904	398	509
A5	Steel	955	596	743	890	377	493
A6	Steel	923	597	740	881	407	530
A7	Steel	916	621	766	831	303	412
A8	Steel	937	615	757	752	180	262
A9	Steel	910	580	707	763	181	254
A10	Steel	909	595	739	769	107	144
A11	Steel	918	584	725	744	176	232
A12	Steel	896	526	673	832	174	227
A13	Steel	900	537	651	696	281	360
A14	Steel	916	615	758	781	284	373
A15	Steel	895	608	746	843	219	298
A16	Steel	904	636	784	774	351	456
A17	Steel	922	629	777	805	346	458
A18	Steel	904	616	740	838	449	570
A19	Steel	930	636	783	808	348	444
A20	Steel	930	647	793	821	614	736
A21	Steel	927	break in wiring		880	421	419
A22	Steel	953	640	771	860	691	740

Fig. 19. Anode strip polarisation (vs steel) in segments 2 and 3, 4 years and 3 years after installation respectively.

3 Conclusions and recommendations

3.1 General

Since 2013 potential decay values, current outputs and resistances were measured during a period of almost 6 years on 2 selected bridge segments in which a large amount of RE's were installed and on the surface narrow zinc strip anodes. These measurements showed interesting results on the effectiveness of the installed GCP system with surface-mounted zinc strip anodes (ZLA), on the development of the long-term potential decay values and (protection) current outputs and on the throwing power of galvanic anodes. The concrete resistance, being influenced by the moisture content of the concrete and the mortars used for repairs, was identified to have a major influence on the effectiveness of the installed GCP system and its development in time.

Monitoring of the installed CP system will continue and will be further reported in the future.

3.2 Effectiveness of the installed GCP system

From startup, the measurement results have demonstrated that the applied GCP system resulted in an effective solution. The depolarisation values complied with the code-based protection criteria of (ISO-EN12696; [3]) at almost all of the locations where RE's were installed.

The 2nd reinforcement layer was protected as well (in non-patched areas), however with substantially less potential decay values compared to the 1st layer.

The CP system was not only effective in the anode covered area but also more remote from this area, at least up to 20 cm adjacent to the zinc anodes during the first year after installation.

From the start, the RE's located in patched areas showed less depolarisation compared to similar RE's in non-patched areas and since 2017 the code based criteria have not been met at these locations. It is concluded and recommended that the effectiveness of a CP system should not be determined by RE's placed in (large) repairs.

3.3 Potential decay values over time

Depolarisation values gradually decreased over time (at the 1st layer of reinforcement 52% in almost 6 years) and part of the installed RE's then showed values which did not comply with the code based protection criteria any more: predominantly the readings from RE's installed at the depth of the 2nd layer of reinforcement. Drying out of the concrete and polarisation of the steel are considered to be the explanation for this observation.

In December 2014, i.e. during the full-scale project, some zinc strip anodes needed to be replaced because of the disbondment of repair mortar during the test phase. After replacing the zinc strips and repairing the disbonded areas, the system reacted immediately and showed good results again at all locations with installed RE's in this area. Moisturizing of the concrete and new installed anode material are considered to be the explanation for this observation. During the following 4.5 years the potential decay values decreased again (with 60% at the 1st layer of reinforcement) due to drying out of the concrete.

The effectiveness of the GCP system to protect steel located at a greater distance from the protected area decreased over time as well (with approximately 50% in 4 years). After 2 years this positive side-effect due to the so called 'throwing power' was no longer measured with the RE's placed at a distance of 20 cm from the anode material. Drying out of the concrete is considered to be the explanation for this observation. It is concluded that the effectiveness of a GCP system adjacent to an area protected with surface-mounted zinc anodes, is limited in dense and/or dry concrete and declining in time. Therefor it is recommended not to overrate this side-effect if specific test results (mainly with regard to the concrete quality and the surface environment) on the effectiveness of the protection of steel reinforcement remote from anode material are not available.

3.4 Current outputs over time

The current outputs decreased strongly over time, with a strong accent in the first months. At the start, in 2013, the total current output was 4 - 4.6 mA/m^2_{anode} (equal to roughly 6 - 8 mA/m^2_{steel}), which was reduced to 2.4, 1.7 and 2.6 mA/m^2_{anode} in 2016, 2017 and 2018 respectively (equal to roughly 4, 3 and 4 mA/m^2_{steel}). Drying out of the concrete and polarisation of the steel are considered to be the explanation for this observation.

After replacement of parts of the zinc anode in segment 3 the current output increased. At the same time the current distribution over the zinc strips became more equally spread.

In the first year, current outputs at large patched areas were substantially lower compared to non-patched areas. This was mainly caused by the disbondment of (parts of) the repair mortar in the test phase. After repair of these areas the current output was more equally distributed over the anode material and variations in the current leveled with the concrete resistance at the subsequent zinc strips.

References

1. J. Gulikers, R. Giorgini, A.J. van den Hondel, Application of a surface-applied cathodic protection system on a light weight concrete bridge - Part I: Condition assessment, intervention options and trial investigations, *Proceedings of Concrete Solutions – Thessaloniki,* CRC Press (2016)
2. A.J. van den Hondel, R. Giorgini, J. Gulikers, Application of a surface-applied cathodic protection system on a light weight concrete bridge - Part II: Developments in time of the effectiveness by potential decay values and current densities, *Proceedings of Concrete Solutions – Thessaloniki,* CRC Press (2016)
3. ISO-EN12696:2016, Cathodic protection of steel in concrete (Geneva, ISO (2016))
4. CUR Recommendation 45, Cathodic Protection of reinforcement in concrete constructions (*in Dutch*, Gouda, CUR (1996)).
5. A.J. van den Hondel, Execution of concrete repairs with the use of galvanic CP - Neerbosschebrug Nijmegen (*in Dutch*, Utrecht, RWS (2015))

Behaviour of column constructed with FRP tubes filled with concrete

Ted Donchev[1,*], *Vineta Srebrenkoska*[2], *Diana Petkova*[1], *Mohammad Albkeirat*[1], and *Hikmatullah Akhundzada*[1]

[1] Faculty of Science, Engineering and Computing, Kingston University, London, UK
[2] Faculty of Technology, University of Goce Delcev in Strip, Republic of North Macedonia

Abstract. This paper investigates the behaviour of Concrete-Filled FRP Tubes (CFFT) as columns. The experimental programme consists of preparing and testing one steel column acting as control sample and three columns made with GFRP. The GFRP tubes were produced by filament winding method where the amount and orientation of the fibres was changed. The tubes had dimensions of 1000x100 mm (length x diameter) and were filled with C25/30 concrete. The columns were tested under compression and the load was applied at a pace rate of 0.5 mm/min. It was found that the GFRP tubes can efficiently confine the concrete and could be used as alternative material to steel tubes. The steel and GFRP samples developed a high level of strain throughout the testing. The GFRP sample with fibre orientation of 90° failed by FRP rupture, whereas the remaining samples failed by buckling. The orientation of the fibres at 90° was more efficient than orientation of fibres at 45° in terms of increasing the ultimate capacity. The GFRP samples displayed lower ultimate capacity compared to steel samples with same wall thickness, but increasing the wall thickness of the GFRP columns increased the ultimate load accordingly.

1 Introduction

Reinforced Concrete (RC) structures require periodic maintenance and monitoring in order to sustain their design life and durability. The structures built in close proximity to a marine environment, bridges and car-parks are under higher risk of corrosion. The risk of corrosion for Concrete Filled Steel Tube (CFST) is significantly higher as the steel tubes are directly exposed and the outside layer of steel is not protected by alkalinity of concrete [1]. Over the past few decades, Fibre Reinforced Polymer (FRP) composites have been used in the construction industry as an alternative to steel reinforcement. FRP materials provide several advantages over steel such as, high tensile strength, light weight, higher corrosion and chemical resistance [2]. Although the initial construction cost of FRP RC members is relatively higher; the total life cycle cost is expected to be less [3].

The behaviour of Concrete Filled FRP Tubes (CFFT) have been investigated in the literature for new columns as alternative to CFST and steel RC columns [4], [8-14]. In this method, FRP tubes are used as formwork to contain the fresh concrete which results in significant cost reduction [5]. However, the primary reason for the use of FRP tubes is confinement of concrete which eliminates the use of longitudinal and transverse reinforcement [6]. The orientation of the fibres in CFFT tubes dictates whether it acts as longitudinal or transverse reinforcement [4]. It is reported that CFFT tubes improve the ductility of columns and increase the shear resistance of the CFFT columns [7,8].

The axial compressive behaviour of CFFT columns have been experimentally investigated over the past two decades [4], [8-14]. FRP tubes produced by the filament winding method, are reported to deliver strength, ductility and large energy absorption [9,10]. Experimental and analytical studies were conducted to develop equations for predicting compressive strength and failure strain of carbon and glass FRP composite tubes [11]. The behaviour of filament-wound CFRP columns built at different angles of winding were experimentally and analytically presented [12]. FRP tubes could be effectively used to confine high and ultra high-strength concrete (HSC and UHSC) and to achieve a highly ductile compressive behaviour [13]. Carbon CFFT columns tested under eccentric loading are less effective in terms of confinement compared to steel helices. However, addition of FRP bars inside the concrete increases the axial load and deformation at maximum loads compared to RC columns [4]. The parametric study on carbon FRP tubes shows that fibre orientation, FRP bar reinforcement ratio and confinement ratio has a direct impact on $P-M$ (axial load – moment) interactions of CFFT columns.

This paper presents the experimental results of the glass CFFT columns tested under axial loading. The experimental variables are the fibre orientation and wall thickness of the FRP tubes. The results are analysed and compared with a steel CFST column. This research is aimed to enhance and reinforce the current understanding of confining concrete compression elements.

[*] Corresponding author: t.donchev@kingston.ac.uk

2 Experimental programme

2.1. Details of specimens

The experimental programme consisted of a total of four specimens, three made of glass FRP and one specimen made of steel. The specimens were cylindrical in shape and had dimensions of 1000 x 100 (height x diameter). Table 1 shows the specimen designation and details. The wall thickness of the first three samples were identical whilst the wall thickness of the fourth sample was three times thicker. The GFRP tubes were filament wound with various winding angles with respect to the longitudinal axis of the tube. Fig 1 shows the filament winding process for different orientation of the fibres.

Fig. 1. Filament-winding process for production of GFRP tubes at different angles (a) 45° (b) 90°

Table 1. Details of test specimens.

Sample Designation	Winding angle (°)	Number of layers	Wall thickness (mm)
Steel	N/A	N/A	3.6
FRP-45	45	5	3.6
FRP-90	90	6	3.6
FRP-Mix	10/60/90	14	10.8

2.2. Material properties

2.2.1 Concrete

The concrete used for this research was batched and mixed at Kingston University's concrete laboratory. Natural aggregates with maximum diameter of 10 mm were used in the mix. The compressive strength of the concrete was determined after 28 days, and on the day of testing by testing a series of concrete cubes and cylinders. The characteristic compressive strength of the concrete for all the samples was C25/30 on the day of testing. The tubes were filled in layers and the concrete was well-compacted with a needle vibrator. The concrete mix design is shown in Table 2.

Table 2. Concrete mix design.

Materials	Amount
Cement (kg)	339
Water (kg)	210
Sand (kg)	667
Coarse aggregates (kg)	1184

2.2.2 Steel and GFRP

The circular hollow section used in this experiment was manufactured by Parker Steel. These tubes were cold formed, made with mild steel and with a yield strength of 235 MPa. The type of glass fibre used for production of GFRP tubes was 185 OCV with tensile strength of 11 N per fibre. The fibres were wrapped at different angles on a mandrel (core) with a size of FI 100 mm x 1110 mm. Three types of resin system were mixed at specific ratio for creating the tubes namely, epoxy resin LY 1135, Araldite 917 and Ascc.960. The fibres were wrapped at a speed of 21 m/min. The tubes were cured at a temperature of 80 – 100°, for a duration of 4 – 5 hours.

2.1. Instrumentation and testing procedure

The specimens were instrumented with nine strain gauges and four linear variable differential transformers (LVDTs) to monitor the behaviour of the columns throughout the testing. The strain gauges were mounted on the outer surface of the tubes. Three strain gauges were installed at the mid-point of the tubes and the remaining six strain gauges were installed at the top and bottom end of the tubes. These strain gauges were evenly spaced around the perimeter of the tube and created an angle of 120 degrees. The LVDTs were mounted on the top and bottom of the load-plate in order to monitor the axial deformation of the columns. The instrumentation of the column is shown in Fig 2.

The samples were tested under monotonic axial compression using a Denison 1000 kN compression machine. The top and bottom ends of the columns were capped with high-strength steel plates, in order to uniformly distribute the load and prevent localised crushing. The application of the load was displacement-controlled with a speed of 0.5 mm/min. The data was recorded at a rate of 10 data/sec using data loggers. Fig 3 shows the test setup for testing the columns.

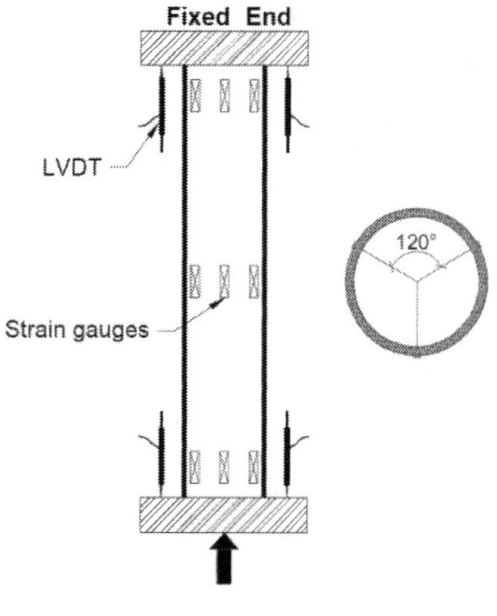

Fig. 2. Column instrumentation.

Fig. 3. Test setup for testing columns under compression.

3 Test results and discussion

3.1. Load-displacement

The load-displacement graph for the steel and GFRP samples is presented in Fig 4. At the initial stages of the loading, the columns adjusted in the testing frame which is observable from the initial negative curvature in the graph.

The steel sample displayed an elastic-plastic behaviour throughout the testing. This sample showed a linear behaviour from the start of the loading until reaching its obvious plastic behaviour at a load over 750 kN. The yielding occurred at a load of 774 kN with approximately 7mm axial displacement. A sudden drop in the load could be observed after the yield point. The capacity of this column started to gradually decrease when it reached its plastic limit.

The FRP-45 column displayed a very similar behaviour to the steel column. This sample reached its maximum load at a displacement of around 11 mm during its linear stage. Although, normally, FRP materials only display linear behaviour, here, due to the composite action with the concrete and the buckling effect, non-linear behaviour is evident above 330 kN as shown in Fig 4. This sample maintained a significant amount of its peak-load at the beginning of the plastic stage.

The performance of the FRP-90 column was relatively different from the steel and FR-45 sample. It displayed a higher degree of deformability throughout the testing. The FRP-90 column displayed a linear behaviour and the failure initiated at a load of 618 kN with a displacement of 26 mm. The failure was relatively brittle with rupture of GFRP walls of the tube and no buckling was observed.

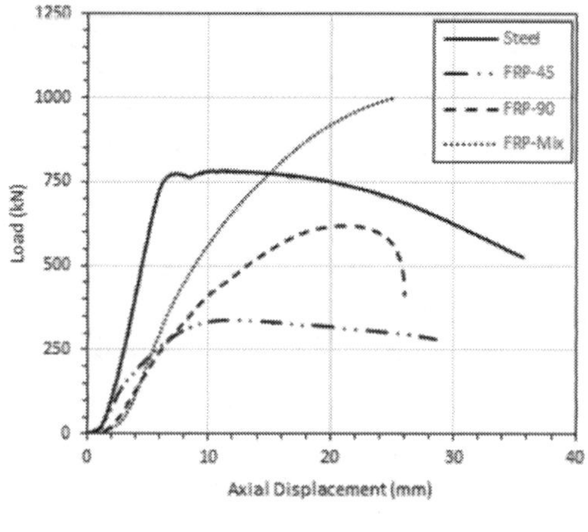

Fig. 4. Load vs axial displacement for all samples.

The FRP-Mix sample which had a wall thickness of 10.8 mm displayed the highest maximum load and exceeded the capacity of the testing frame. A linear behaviour up to a load of 250 kN is observable from Fig 4. However, the slope of the curve is getting gentler which points towards failure of the column. The column is expected to probably reach to 10% of its current maximum load at the failure point. The column was intact after the testing and no sign of damage was visible by visual inspection.

3.2. Observed failure modes

The failure modes for the steel and GFRP columns are shown in Fig 5. Two types of failure modes were observed for the tested columns; buckling and rupture of

FRP fibres. The steel and FRP-45 samples failed by mid-span buckling. Both of the columns maintained a significant amount of load with large amounts of deformability. The FRP-90 sample failed by rupture of FRP fibres at the mid-span. The column failed abruptly and split into two pieces.

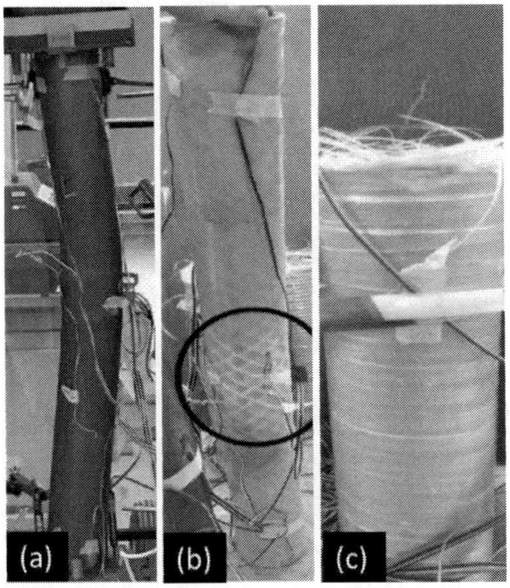

Fig. 5. Failure mode of all samples (a) steel (b) FRP-45 (c) FRP-90.

3.3. Ultimate load

The ultimate failure load for all the samples is shown in Fig 6. Although, the FRP-45 and the FRP-90 sample had the same wall thickness, there was a significant variation in their respective ultimate load. The FRP-90 sample was more effective in confining the concrete and displayed almost double the load compared to the FRP-45 sample. It is evident that the fibre orientation (winding angle) plays a significant role in the failure mode and ultimate load of CFFT. The GFRP tubes with similar wall thickness to the steel tube, displayed a lower ultimate load. The ultimate capacity of the FRP-90 sample was about 21% lower than the steel sample. The GFRP-Mix sample, which had a wall thickness of three times higher than steel sample, displayed the highest ultimate load which reached loads of in excess of 1000 kN.

The efficiency of GFRP tubes with identical cross-sectional area to steel tubes is relatively low. The CFST and CFFT tubes displayed significantly higher capacity compared to normal RC columns; therefore, the CFFT tubes could be used as an alternative to RC columns. Future design of such elements have to include consideration for appropriate fire protection, such as intumescent paint or encasing them in fire resistant materials. By increasing the wall thickness of the GFRP tubes, similar behaviour to steel tubes in terms of ultimate load and ductility could be achieved. The GFRP tubes have the potential to be used as alternative to steel tubes in corrosive environments.

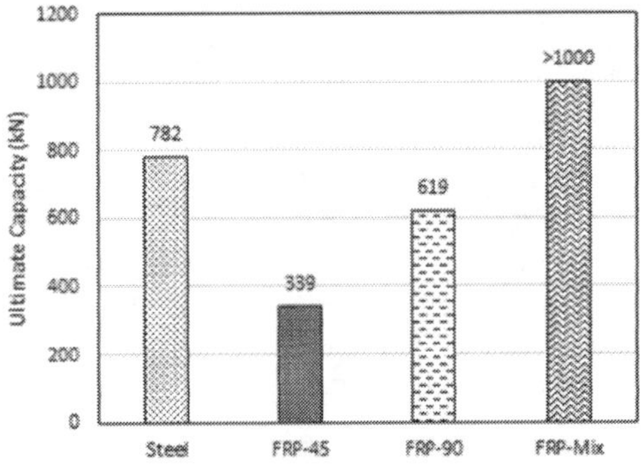

Fig. 6. Ultimate load for all samples.

4 Conclusions

This paper has presented the experimental results of testing CFST and CFFT tubes in compression. The influence of the fibre orientation and the wall thickness of the GFRP was investigated in this study. The following conclusions can be drawn:

1. The GFRP tubes produced by the filament winding method, could be used as an alternative material to steel tubes. The GFRP tubes with the same wall thickness to steel tubes displayed a lower ultimate load but achieved a substantial level of deformability. The difference in ultimate load of the GFRP and steel sample was around 21%.
2. The orientation of the fibre (winding angle) in GFRP tubes dictated the ultimate load and failure type. The capacity of FRP-90 sample was approximately two times higher than the FRP-45 sample. The orientation of fibres at 90° depicts stirrups in RC columns which successfully confine the concrete.
3. By increasing the wall thickness of the GFRP tubes, the ultimate load was increased accordingly. The FRP-Mix sample exhibited higher load and greater level of deformability throughout testing.

References

1. C.L. Page, *Mechanism of corrosion protection in reinforced concrete marine structures Nature.* (1975)
2. C. Pantelides, M. Gibbons, L. Reaveley, J. Compos. Constr. **17**(3), (2013)
3. H. Tobbi, A.S. Farghaly, B. Benmokrane, ACI Struct. J. **111**(2), (2014)
4. M.N.S. Hadi, Q.S. Khan, M.N. Sheikh,. Constr Build Mater. **122**, (2016)
5. T. Ozbakkaloglu, D.J. Oehlers. J Compos Constr. **12**(4), (2008)

6. A.Z. Fam, S.H. Rizkalla, *J. Compos. Constr.* **6**(2), (2002)
7. D. Lillistone, C.K. Jolly, Struct Eng. **78**(23/24), (2000)
8. A. Mirmiran, M. Shahawy, H. Samaan, H. El Echary, J.C. Mastrapa, O. Pico, J Compos Constr. **2**(4), (1998)
9. A. Mirmiran, M. Shahawy, American Society of Civil Engineers, New York, (1995)
10. A. Mirmiran, M. Shahawy, ACSE Journal of Structural Engineering. **123**(5), (1997)
11. M. Saafi, H.A. Toutani, A. Li, ACI Material Journal, **96**(4), (1999)
12. W.K. Hong, H.C. Kim, Can. J. Civ. Eng., **31**(2), (2004)
13. T. Vincent, T. Ozbakkaloglu, Compos Part B Eng **50** (2013)
14. Q.S. Khan, M.N. Sheikh, M.N.S. Hadi, Comps Part B. **133** (2018)

Implications upon using exterior post-tensioning tendons to retrofit a low strength concrete waffle slab

Andrei Faur[1,*], *Virág* Jacint[1]

[1]Department of Structures, Faculty of Civil Engineering, TUCN, 25, G. Barițiu Street, 400027, Cluj-Napoca, Romania

Abstract. A common problem when casting reinforced concrete structures is to have a lower concrete compressive strength as opposed to what has been assumed in design. The deflections that may occur under the serviceability conditions can exceed the maximum allowed. Although less frequent nowadays, there is still an ongoing demand to offer simple and practical retrofitting solutions for such situations. To cast a new concrete layer above an existing waffle slab facing these problems is sometimes, only by appearance, the simplest retrofitting method. The major disadvantages reside in the growth of the overall steel consumption and overall weight of the structure. Additional reinforcement is needed for the new slab and at the interface of the two concretes. Interventions that are not always simple and accessible to the builder may still be needed to satisfy the strength demands of the already cast elements. Therefore, a post-tensioning system placed bellow the existing concrete seems to be a better solution. The present study refers to a real-case investigation on a 5-level residual building where a low concrete class was identified for several waffle slabs. Casting an overall top concrete slab or applying an exterior post-tensioning system were the two main retrofitting solutions offered by the designer. Both solutions are compared in terms of structural performance and cost efficiency.

1 Introduction

The present paper refers to a real-case investigation upon finding the proper solution for retrofitting several waffle-slabs. The problem appeared at the construction site, where lower concrete strengths were obtained after casting the first slabs. Concerns coming from the builder about the whole structure being compromised started to emerge. This was mainly because at the time this problem was identified, the first three floors were already built. Therefore, a third-party designer offered the expertise to inspect the current situation and to propose a retrofitting solution if necessary.

The preliminary inspection revealed that the concrete did not achieve the prescribed strength at the time the in-situ core samples were tested. All the collected concrete cores had a diameter of 60 to 65 mm and a length to diameter ratio of approximately 1 (i.e. 3 to 4 samples having 60 to 65 mm in diameter and length were collected from each structural element). The cores were tested under laboratory conditions after 8 days, and the procedures from NP 137:2014 [1] and EN 13791:2007 [2] were used to determine the in-situ characteristic concrete strength.

The concrete strength loss was detected not only for slabs, where the concrete reached as far as the equivalent of two concrete classes below the design prescription, but in all the concrete elements. By the time the first retrofitting solutions were proposed, a new batch of in-situ cores were tested, revealing a slight raise of the compression strength. This was assumed to be caused by the ongoing curing process, mainly because the time difference between the two measurements was approximately one month, and the weather was favourable during this time (i.e. relatively high temperatures combined with normal humidity levels). Anyway, it was concluded that only the already cast waffle-slabs needed to be retrofitted, and this to fulfil the serviceability requirements. The retrofitting solutions are discussed in the following.

1.1 A brief description of the building

The building in discussion is conceived mainly as residential and combined with a few office areas and parking lots at the basement floor. The structural system consists of reinforced concrete waffle-slabs supported by perimeter walls, beams and columns. The building has six openings along the longitudinal direction and a single opening along the transversal direction. It has a total of six storeys with unsymmetrical setbacks along the longitudinal direction, the first three floors being cast at the time the investigation started. The distance between the transversal axes is of 12 meters, while the distances between the longitudinal axes vary between 2.8 and 8 meters. Therefore, there are two regions for each floor where the waffle-slab has an area of 8x12 meters (i.e. 96 m^2) between the axes, and consequently between the support elements. A 3-D representation of the structural system is presented in Fig. 1, and the first-floor layout in Fig. 2.

* Corresponding author: andrei.faur@dst.utcluj.ro

 https://doi.org/10.1051/matecconf/201928904002

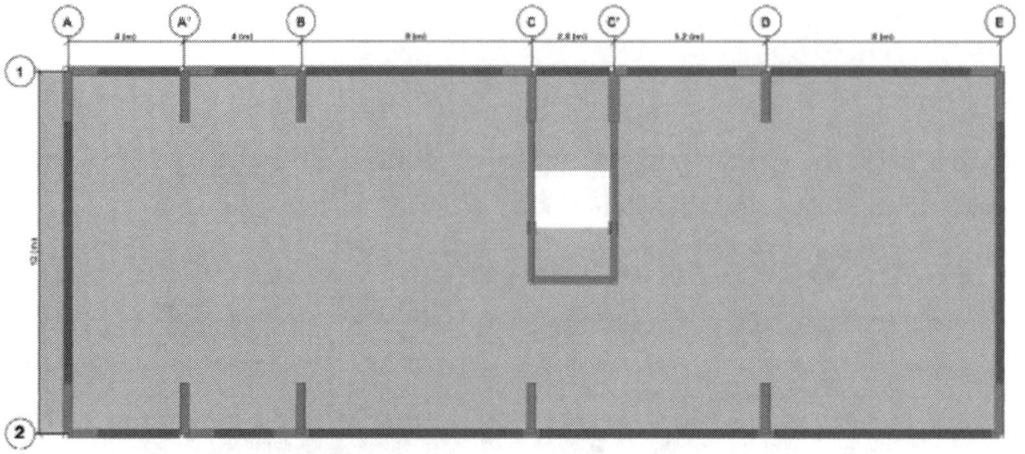

Fig. 1. First floor layout.
Fig. 2. 3-D representation of the structural system.

The first two floors have no setbacks. At the third floor the slab stops at gridline A', and from here it starts to withdraw at gridline B and at the middle between gridlines B and C for the fourth and the fifth floor, respectively. The edges of the slabs were strengthened with reinforced concrete beams in order to prevent excessive deflections.

The storey height is equal to 3.06 m, except for the basement floor, where the height is 3.73 m.

The building is located in Cluj County, Romania.

1.2 The structural system

Based on the building's architecture, the floors were conceived as waffle-slabs, except for the areas near the vertical supports (i.e. walls and columns) and along the strips following the axes A', B, C and D. No supporting ribs were used in these areas, but 30 cm thick solid slabs were cast instead. The overall depth of the waffle-slab is also 30 cm.

The reinforced concrete walls are arranged as vertical piers of 30 cm thickness. These are depicted in Fig. 2 with red colour. Rectangular section beams or confinement beams are placed on the perimeter to sustain the slab edges and are depicted in the same Fig. 2 with blue colour. The structure for the staircase is made of frame elements (i.e. columns and beams) and filled with masonry panels.

The ramps and the pedestals are designed as 18 cm reinforced concrete flat slabs.

The concrete class assumed in design is C16/20 and the steel for the reinforcement has a yield strength of 345 kPa and a tensile strength of 510 kPa.

2 Preliminary tests and structural expertise

At the time the structure was built up to the third floor (i.e. the foundations, the ground floor, the semi-basement floor and the ground floor's top slab) it was found that the concrete in some elements had a lower compression strength than that established by design. The cause of this problem is still under investigation. It was revealed based on the in-situ and laboratory tests that almost all the structural elements were affected. The compression strength of the in-situ cubic concrete samples and their equivalent concrete class are summarised in Table 1:

Table 1. In-situ samples test results.

Structural element	$f^{is}_{ck,cube}$ [MPa]	Equivalent concrete class
Basement Walls	16.19	C16/20
Basement Top Slab	10.43	C8/10 to C12/15
Semi-basement Walls	9.91	C8/10
Semi-basement Top Slab	11.73	C8/10 to C12/15
Ground Floor Walls	13.82	C12/15
Ground Floor Top Slab	≈10	C8/10

where $f^{is}_{ck,cube}$ is the concrete characteristic cube strength of the in-situ samples. All the mentioned values were determined accordingly with the Romanian standard for the assessment of in-situ compressive strength in concrete structures NP 137:2014 [1], and the European standards Eurocode 2 [3] and EN 13791:2007 [2].The structural technical expertise was performed based on these tests. The capacity of the structural system having the new concrete properties was established by running several numerical analyses. The computations were done through the use of ETABS [4] analysis program. Both the vertical and the horizontal (i.e. the territory of Romania is fully regarded as an active seismic region) types of analyses were performed for both the ultimate and the serviceability limit states. The results revealed the necessity to retrofit the waffle-slabs and two solutions were proposed to the builder as being the most efficient:

- by pouring a new a new concrete layer over the existing slabs, or
- by placing a post-tensioning system below the existing slabs.

Both retrofitting solutions were proposed because the analyses showed excessive vertical deflections of the slabs in the large spanning areas (i.e. between axes B-C and D-E). The maximum allowed deflections for the serviceability state were supposed to be exceeded by far in the future if no structural interventions are made immediately to the slabs. Moreover, the shear and the flexural forces developed in some slab ribs exceeded the ultimate limit state conditions under the new concrete strength values. Therefore, not only the stiffness but also the strength of the waffle-slabs has to be increased as quickly as possible. Both retrofitting solutions are discussed in the following. The builder excluded the demolition starting from the basement top slab, even if a C8/10 class concrete is not suitable for constructional purposes, claiming he was behind works.

3 Numerical analyses

All the numerical models were developed in ETABS [4] and SAP2000 [5]. Several 3-D models representing the complete structure were preferred against simpler 2-D representations. These were intended to capture the interaction between different elements to both vertical and lateral loads. Simpler numerical models were still used but only for finding the mechanical characteristics of the waffle slabs.

3.1 Analysis of the initial state of the slabs

One of the most delicate parts of the expertise was the evaluation of the waffle-slab mechanical properties. One of the first studies referring to the orthotropic behaviour of the slabs were those of Timoshenko and Woinowsky-Krieger [6]. While being mathematically rigorous, the authors of the mentioned publication restrict their theory to simple geometries and boundary conditions. Even so, the shape functions for finding the deflected shape of the shells are too complex to be used in a finite element context. Other authors proposed simpler shape functions [7], and finite element formulation for specific waffle-slab shell elements were developed since the 80's [8]. Therefore, the analytical approach is considered to be too time consuming, while other methods requiring high finite element knowledge are far to impractical.

In order to simplify the analysis, it was decided to model the waffle-slab as a flat slab with similar stiffness and strength properties. The SAP2000 [5] analysis program was used for this purpose. The equivalent thickness of the slab was determined following several trials of comparing the vertical deflections of two numerical models:

- The first model – consisted in a flat slab having the dimensions of 8 m × 12 m and 205 mm in thickness;
- The second model – consisted in a real representation of the waffle-slab with 0.75 m spaced frame elements as ribs (0.15 m × 0.30 m are the cross-sectional dimensions) and a 8 mm top slab, and having the same 8 m × 12 m dimensions as the first model.

Moreover, the 8 mm slab used in the second model was placed at the top of the frame elements. Plate-Thin shell elements (i.e. supports only bending moments and transverse forces [5]) were used in both models. The in-plane behaviour was neglected for simplifying the computation and for not overestimating the slab's rigidity. Anyway, the infinite diaphragm hypothesis is assumed for all the 3-D models and, therefore, such simplification is justified.

The openings for both models were chosen with respect to the large spanning areas of the slabs (i.e. the regions where the maximum vertical deflections are likely to occur). Pinned supports were placed along the perimeter to simulate the boundary conditions of the real slab. The finite element analyses were performed after the discretisation of the area elements and the same vertical displacement was obtained for a 205 mm thickness of the flat slab. The difference between the results was under 1% (i.e. the vertical displacement for the flat slab model was 14.6 mm, and 14.5 mm for the beam-slab model). The deformed shapes of both models are presented in Fig. 3, where u_z represents the vertical displacement. A vertical load of 10 kN/m^2 was used for both analyses.

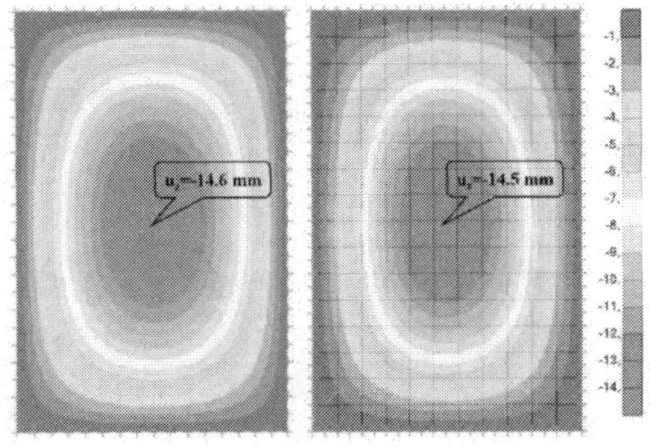

Fig. 3. Vertical displacement of the equivalent flat slab model (left) and the frame-slab model (right).

https://doi.org/10.1051/matecconf/201928904002

Based on the obtained results, it was concluded that similar stiffness properties can be obtained if the slab is modelled with simple area finite elements of 205 mm in thickness.

It is mandatory that the slab deflections must be verified for the serviceability limit states (SLS) under these new conditions of concrete strength loss. The total deformation including creep may be calculated by using an effective modulus of elasticity for concrete, as indicated in Eurocode 2 [3]:

$$E_{c,eff} = \frac{E_{cm}}{1+\varphi(\infty, t_0)} \tag{1}$$

where: $E_{c,eff}$ – effective modulus of elasticity of concrete; E_{cm} – secant modulus of elasticity of concrete; and $\varphi(\infty, t_0)$ – final value of creep coefficient. The age of concrete at the time of loading t_0 was taken as 28 days and 21000 MPa was considered as the secant modulus of elasticity for the C8/10 concrete class. The secant modulus for each structural element was taken accordingly with the data presented in Table 1. The final values of the creep coefficient for the waffle-slabs were estimated as 2.65 for the C16/20 concrete class, 2.85 for C12/15, and 3.35 for C8/10. Therefore, an approximatively four-time reduction in the total flexural stiffness of the slabs (i.e. the product of the effective modulus of elasticity and the moment of inertia) results for the SLS analysis.

The loads were applied on slabs and on frames. The permanent loads, except the self-weight of the structural elements, were taken as 1.45 kPa and were applied on slabs. The live loads were taken as 2.70 kPa for the residential areas and 3.70 for the office areas but including the self-weight of the movable partitions. Based on these loadings and stiffness properties, the maximum SLS deflections of the slabs at the quasi-permanent combinations are presented in the following figures:

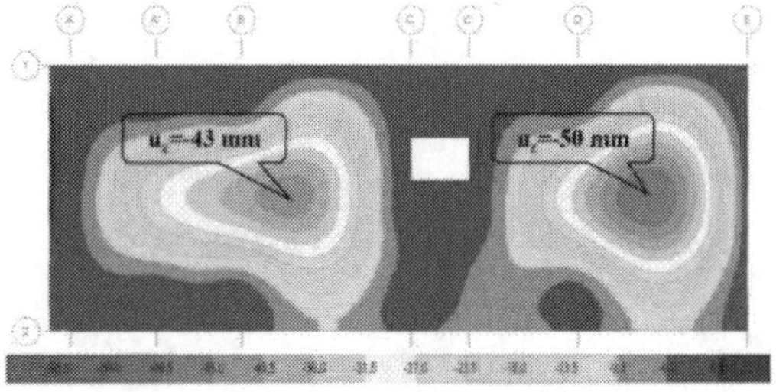

Fig. 4. Slab deformed shape – basement floor.

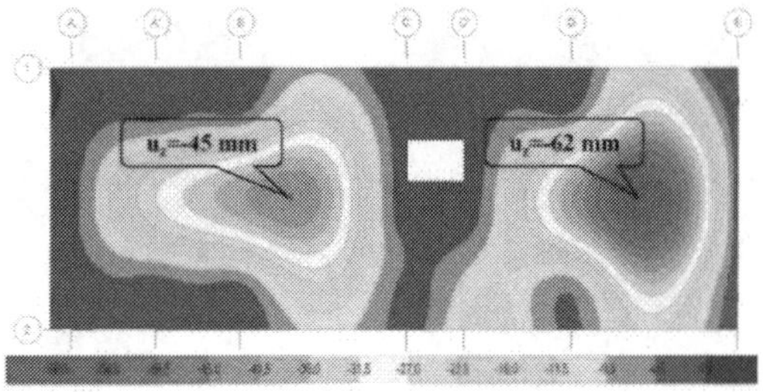

Fig. 5. Slab deformed shape – semi-basement floor.

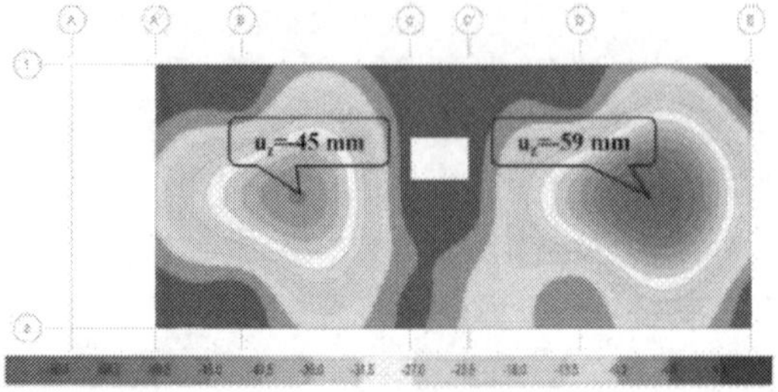

Fig. 6. Slab deformed shape – ground floor.

If the maximum deflection is restricted to span/250 (i.e. for the 8 m spans between axes B-C and D-E the maximum allowed deflection is 32 mm), then the presented deflections exceed this value by far. The slab deflection between axes D and E at the basement floor is almost twice higher than the maximum allowed (i.e. 93.78% higher). Retrofitting of the slabs is mandatory in this situation.

Based on the reinforcement plans and the ultimate limit state (ULS) analysis it was found that the element forces in some regions are greater than the yield forces, but do not exceed the resisting moments, shear forces, or the resisting axial forces. Therefore, no other structural interventions beside the two proposed retrofitting solutions were predicted at this stage.

3.2 The top concrete layer model

The simplest solution to overcome the serviceability deflections of the slabs seems to be realized by pouring a concrete layer above the existing waffle-slabs. The rigidity should increase and, therefore, the deflections should drop below the admissible values. Following a series of numerical computations, it was found that the minimum requirements are achieved if the reinforced concrete top layer has at least 60 mm in thickness and the concrete class is C30/37.

The retrofitted slabs were modelled in ETABS [4] with layered shell elements (i.e. a type of area object composed of multiple overlaid shells, see [4]). As in the previous case, the existing waffle-slab was considered as an equivalent flat slab of 205 mm in thickness. The supplementary 60 mm slab of higher concrete class was placed above, and perfect bond was considered between the two concretes. This last assumption will be respected by providing enough reinforcement crossing the interface. The increase in stiffness resulted by over-concreting was estimated as being 2.5 to 3 times higher, depending on the existing concrete strength and the creep.

It should be mentioned that the stiffness has been evaluated based on the same principles that led to the finding of the equivalent thickness of the flat slab: the vertical deflection of the initial and the retrofitted slabs were compared under the same load, boundary and geometry conditions (i.e. 8 m × 12 m simply supported slabs loaded with a uniformly distributed vertical load of 10 kN/m).

The maximum SLS deflections of the slabs at the quasi-permanent combinations for the new layered slabs are presented in Fig. 7, Fig. 8 and Fig. 9. It may be observed that the maximum deflections are lower than in the absence of the top concrete layer. The maximum allowed deflection of 32 mm is nowhere reached (i.e. the highest resulted value of 23 mm being 28% lower than the admissible limit). Therefore, the SLS deflection criteria are fulfilled if the existing waffle-slabs are retrofitted with a top reinforced concrete layer of 60 mm in thickness and C30/37 concrete class.

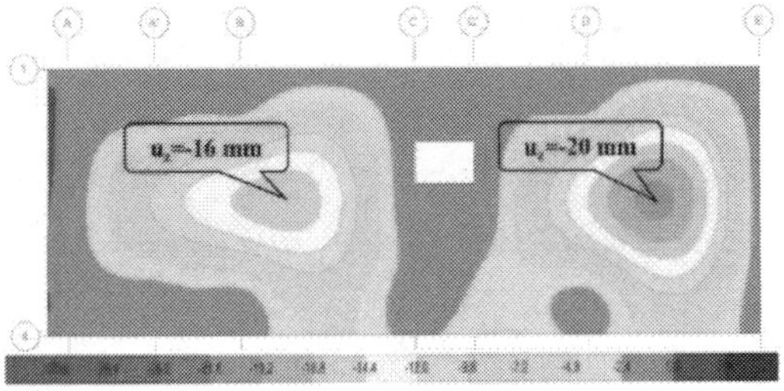

Fig. 7. Over-cast slab deformed shape – basement floor.

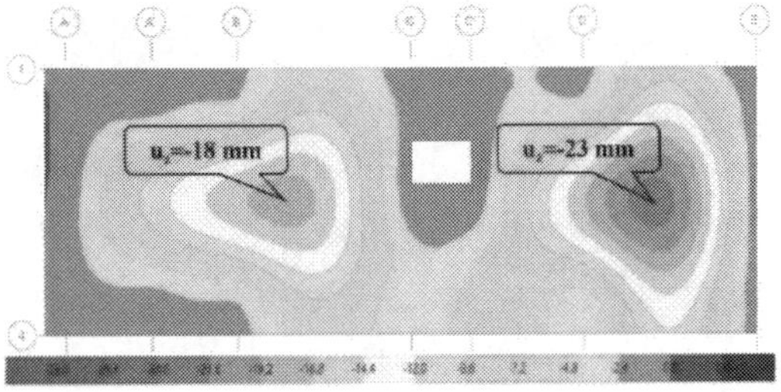

Fig. 8. Over-cast slab deformed shape – semi-basement floor.

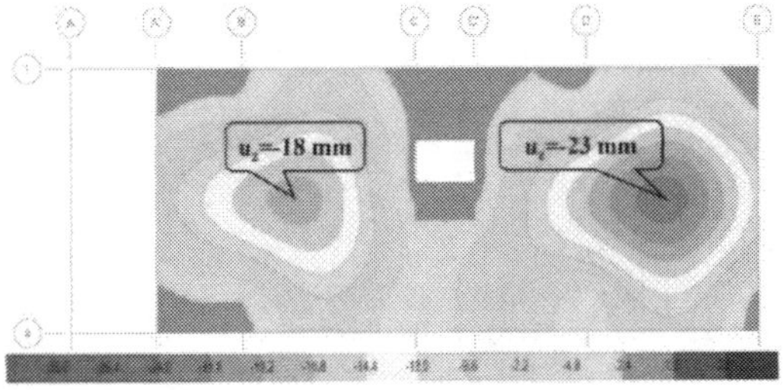

Fig. 9. Over-cast slab deformed shape – ground floor.

One problem with this retrofitting solution was observed during the analysis. It referred to the ultimate limit state (ULS) verifications. Some of the existing slab ribs, especially those located near the edges parallel to axes 1 and 4, were subjected to forces that were higher than these rib elements were capable to overcome. Therefore, the design flexural moments and shear forces cannot be undertaken if no supplementary structural interventions are made.

These forces become almost twice higher than in the absence of the top concrete layer. Such a force redistribution is supposed to be caused by the slab's overall stiffness growth, thus directing the forces towards the supports. While this phenomenon seems to be favourable for low concrete strengths (i.e. causing a force redistribution towards the middle of the slabs where sufficient reinforcement is provided), an initial design error cannot be excluded. Whatever the case, it is certain that in order to fulfil the ULS criteria, several slab ribs need to be retrofitted for flexural and shear strengthening.

3.3 The post-tensioning system model

The inconveniences caused by the necessity for supplementary interventions beside over-concreting led to other retrofitting solutions to be considered. One of the most appealing implies the use of exterior post-tensioning. If the slabs can be subjected to counter-deflections through post-tensioning while verifying the SLS conditions, then no other major interventions to the existing structure may be necessary.

Following the previous considerations, it was necessary that the exterior post-tensioning tendon profile and the minimum tensioning force to be found. It was concluded that the post-tensioning is necessary to be placed only below the slab regions having the larger spans (i.e. between and, at the same time, oriented parallel with axes B-C and D-E). The vertical component of the tensioning force causing the lifting of the slab (i.e. generating the vertical counter-deflection) depends upon both the tendon's eccentric position to the existing slab and the magnitude of the tensioning force.

The maximum distance of the tendon to the bottom of the slab was set to 250 mm at the middle, and to 200 mm at the quarter distance on both sides of the 12 m bay. These values were established due to some architectural limitations regarding the clear floor height (i.e. the builder agreed a maximum 300 mm ceiling drop to permit a proper protection against fire and humidity, while separating walls were additionally provided bellow the areas where tendons require maximum eccentricities).

The stabilized post-tensioning force is also limited to 70% of the yield force of the tendons (i.e. calculated after all the post-tensioning losses have been consumed). If using tendons with a cross-sectional area of 150 mm^2 and a characteristic tensile strength of 1770 MPa, the final stress caused by post-tensioning is:

$$\sigma_p = 0.7 \times f_{p0.1k} \cong 1075 MPa \qquad (2)$$

where σ_p – the final post-tensioning stress; $f_{p0.1k}$ – the characteristic 0.1% proof-stress of post-tensioning steel.

The exterior post-tensioning effect is modelled by using frame elements as tendons. The deflecting devices are also modelled with frame elements. Even if there are studies presenting different modelling techniques for the post-tensioning [9], the following simplifications were considered:

- each tendon is divided between the deflection devices into four individual frames;
- the frames between the deflection devices have the bending stiffness released at both ends and the torsional stiffness released at only one end (see Fig.

10), thus simulating a cable-type behaviour while avoiding unpredictable mechanisms or instabilities;

- the post-tensioning force is induced as a thermal load by applying a minus temperature change, and thus producing an axial contraction of the tendons.

A numerical model example having the frames and the deflecting devices positioned bellow the slabs is presented in Fig. 10.

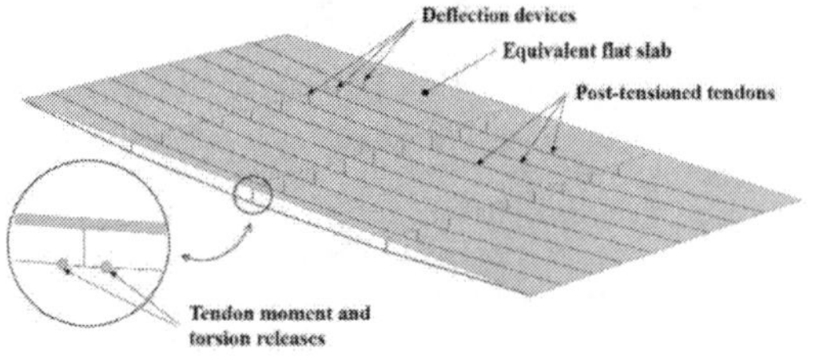

Fig. 10. Exterior post-tensioned slab numerical model.

It should be mentioned that the cable-type behaviour is obtained even if the shear stiffness of the frames is maintained. In the presence of the shear stiffness and the mentioned end releases, a truss-type behaviour is generated instead. However, by neglecting the friction between tendons and deflecting devices, and due to the symmetry of the applied loading (i.e. the temperature load produces equal axial retractions of both ends of the frame), only tension forces (i.e. axial with no compression) are developed. The necessary temperature change is found by the following equation:

$$\Delta t = \frac{\sigma_p}{\alpha_t E} \quad (3)$$

where Δt – is the imposed temperature change; α – is the coefficient of thermal expansion, and E – is the modulus of elasticity of the tendon material. Having the final post-tensioning stress set to 1075 MPa, and taking E = 195000 MPa and α_t = 0,0000117 1/°C (i.e. °C – Celsius degrees), the resulted temperature change is 470°C.

The deflection devices were determined with respect to the estimated forces induced by post-tensioning: the vertical component causing the lifting of the slab, and the horizontal component coming from the friction with the tendons. Therefore, the deflection devices are designed as steel frames subjected to axial, bending and shear forces. Coated and grouted strands are proposed in order to minimize these frictions, but also to reduce the risk of corrosion. Based on these considerations, the design suggested the use of square hollow cross-section steel bars as deflection devices, having: the cross-sectional dimensions of 80x80x5 mm (i.e. 80 mm total depth and width, and 5 mm thickness), and the steel yield strength of 275 MPa.

The analysis of the post-tensioning applied in the lack of any other vertical load (i.e. also excluding the self-weight of the existing slabs) generated the deflected shapes presented in Fig. 11, Fig. 12 and Fig. 13. It should be observed that slab counter-deflections are present to all floors. This confirms the efficiency of the proposed modelling technique to include the exterior post-tensioning effect. The self-weight of the slabs is excluded from this particular analysis merely to emphasise the uplifting effect, and it doesn't correspond to a realistic loading situation by its own.

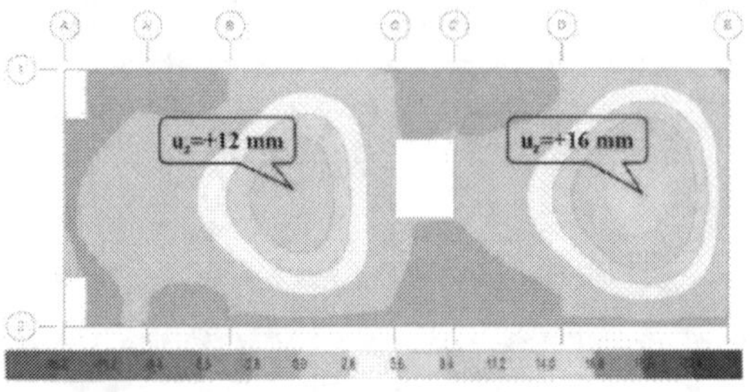

Fig. 11. Pure post-tensioned slab deformed shape – basement floor.

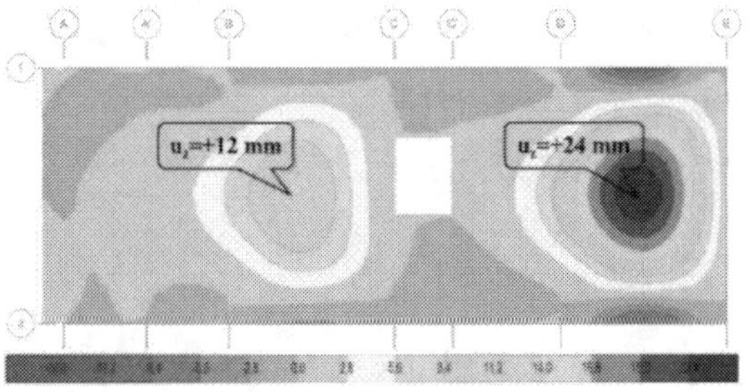

Fig. 12. Pure post-tensioned slab deformed shape – semi-basement floor.

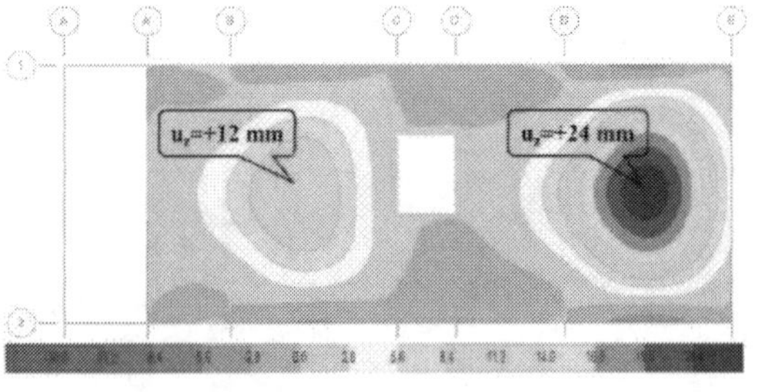

Fig. 13. Pure post-tensioned slab deformed shape – ground floor.

If the vertical loads are added (i.e. including the self-weight of the slabs), then the deflected shapes presented in Fig. 14, Fig. 15 and Fig. 16 are obtained. The permanent and live load values were taken as in the previous analyses. The shown relative vertical displacements of the slabs are lower or at least equal with the maximum allowed, thus indicating the exterior post-tensioning as a viable retrofitting solution.

Moreover, the verifications performed at the ULS suggest that no supplementary structural interventions are needed beside those involving the installation of the post-tensioning system (i.e. components of the deflection and anchorage devices).

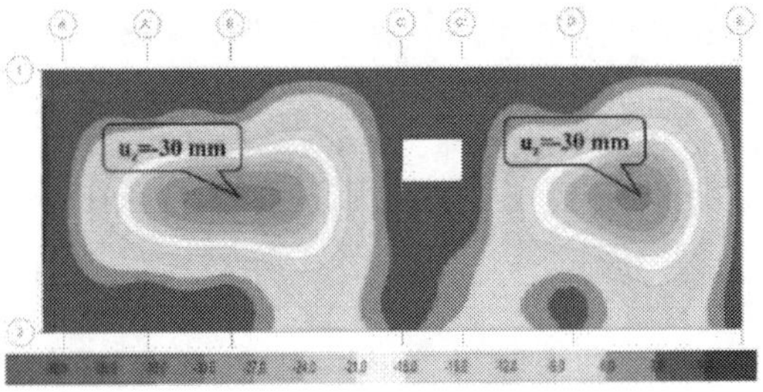

Fig. 14. Post-tensioned slab deformed shape – basement floor.

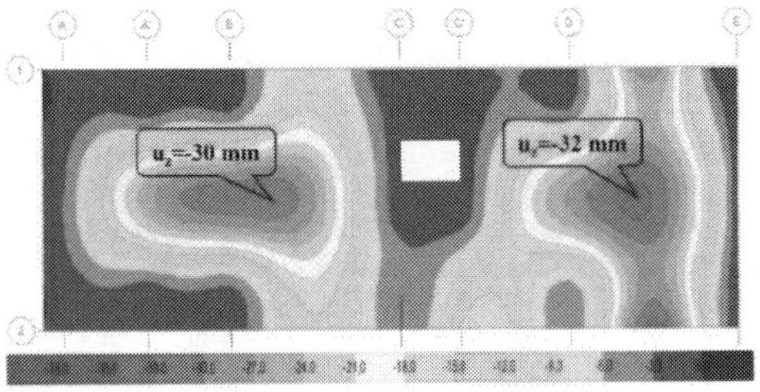

Fig. 15. Post-tensioned slab deformed shape – semi-basement floor.

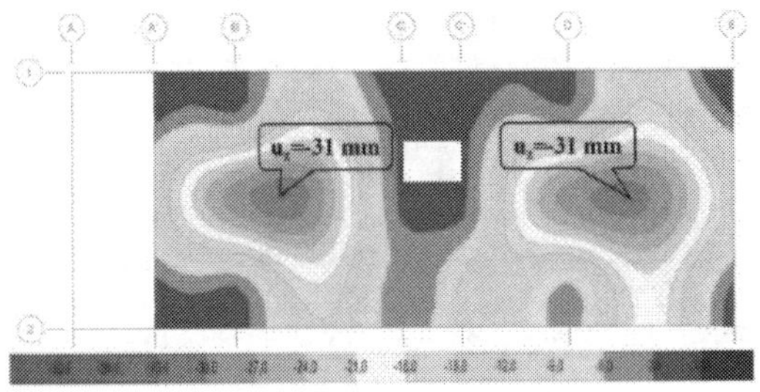

Fig. 16. Post-tensioned slab deformed shape – ground floor.

3.4 Comparison of the analytical results

If both retrofitting solutions are compared in terms of slab deflections, it is clear that the highest stiffness is obtained for the top concrete layer solution. The deflections obtained from the in-situ initial state model ($u_{z,IN}$), the top concrete layer model ($u_{z,CL}$), and the post-tensioning model ($u_{z,PT}$) are presented in Table 2. The values are reported to the admissible SLS deflections ($u_{z,ADM}$), also. Only the slab deflections that are measured between axes B-C and D-E are reported.

The highest deflections are obtained for the semi-basement top slab, between axes D-E. Even if the concrete strength is lower at the ground floor top slab, the live loads at the semi-basement floor are higher (i.e. office area destination) and this causes higher deflections. More tendons will be needed in these areas to compensate the slab deflections, as it will be seen in the following.

Table 2. Numerical analyses results comparison.

Slab	Initial state deflection $u_{z,IN}$ [mm] ($u_{z,IN}/u_{z,ADM}$)		Top concrete layer deflection $u_{z,CL}$ [mm] ($u_{z,CL}/u_{z,ADM}$)		Exterior post-tensioning deflection $u_{z,PT}$ [mm] ($u_{z,PT}/u_{z,ADM}$)	
	B-C	D-E	B-C	D-E	B-C	D-E
Basement Top Slab	-43 (1.34)	-50 (1.56)	-16 (0.5)	-20 (0.63)	-30 (0.94)	-30 (0.94)
Semi-basement Top Slab	-45 (1.41)	-62 (1.94)	-18 (0.56)	-23 (0.72)	-30 (0.94)	-32 (1.00)
Ground Floor Top Slab	-45 (1.41)	-59 (1.84)	-18 (0.56)	-23 (0.72)	-31 (0.97)	-31 (0.97)

It should be mentioned that neither analysis showed the necessity to retrofit the perimeter beams. This is because there is enough reinforcement provided in order to overtake the exterior forces (i.e. counting separately on the supplementary weight of the top concrete layer or the vertical component of the post-tensioning force). The flexural and the shear capacity of these beams were estimated with the help of the Section Designer module (i.e. sectional analysis software integrated in SAP2000 [5]) and the recommendations of Eurocode 2 [3], respectively. In the regions were post-tensioning should be provided, the perimeter beams have 300x600 mm the cross-sectional dimensions. Four rebars of 25 mm in diameter are placed at both the upper and the bottom sides of the beams and through the whole length. Vertical stirrups spaced at 150 mm of 8 mm in diameter are also provided through the whole length of the beams. Under the mentioned material strengths (i.e. C8/10 concrete class and reinforcement having the yield strength of 345 kPa and the tensile strength of 510 kPa), the following bending and shear capacities are obtained:

- Bending yield moment = 335 kNm;
- Bending ultimate moment = 407 kNm;
- Design shear resistance without shear reinforcement = 57 kN;
- Design shear resistance provided by the transversal reinforcement for a concrete strut angle of 45 degrees = 75 kN;
- Design shear resistance limited by crushing of the compression struts = 233 kN.

The maximum bending moment obtained from the analyses is lower than 200 kNm, while the maximum external shear force does not exceed 70 kN.

4 Retrofitting solutions

The analyses performed for the ULS offer the stress distribution to be used in design for both the considered retrofitting solution. As mentioned before, the top concrete layer solution implies additional reinforcement for the new concrete and at the interface of the two concretes, while if exterior post-tensioning is used, then supplementary devices will have to be designed and mounted as deflection devices and end anchorages.

4.1 The top concrete layer retrofitting solution

The new concrete layer should be cast above all the existing slabs, and not just over the areas where the deflections do not meet the SLS requirements. The decision came following the intention to ensure a rigid support for all the coming floor layers.

Eurocode 2 [3] standard specifies that the shear at the interface between concretes cast at different times should be taken by the reinforcement crossing the interface. The reinforcement is calculated based on the shear force distribution. The equation upon which the reinforcement at the interface is calculated, is given bellow:

$$\underbrace{\frac{\beta \cdot V_{Ed}}{z \cdot b_i}}_{v_{Edi}} = \underbrace{c \cdot f_{ctd} + \mu \cdot \sigma_n + \rho \cdot f_{yd}\left(\mu \cdot \sin\alpha + \cos\alpha\right)}_{v_{Rdi}} \le 0.5 \cdot \nu \cdot f_{cd} \qquad (4)$$

where: v_{Edi} – is the design value of the shear stress in the interface; v_{Rdi} – is the design shear resistance at the interface; β – is the ratio of the longitudinal force in the new concrete area and the total longitudinal force either in the compression or tension zone, both calculated for the section considered; V_{Ed} – is the transverse shear force; z – is the lever arm of composite section; b_i - is the width of the interface; c and μ – are factors which depend on the roughness of the interface; f_{ctd} – is the design tensile strength of the concrete; f_{yd} and f_{cd} – are the design yield strength of reinforcement and the design value of concrete compressive strength, respectively; ρ – is the ratio of the reinforcement crossing the interface and the area of the interface or joint; ν – is a strength reduction factor (see Eurocode 2 [3]); and α – is the angle of the interface reinforcement.

The distribution of the reinforcement crossing the interface is obtained based on the shear force diagrams. A proper anchorage length is provided in both concretes (i.e. the existing concrete and the new top concrete layer), thus increasing steel reinforcement quantities. A detail of the interface reinforcement is shown in Fig. 17. The total steel quantity crossing the interface for all the slabs is estimated to approximately 1500 kg.

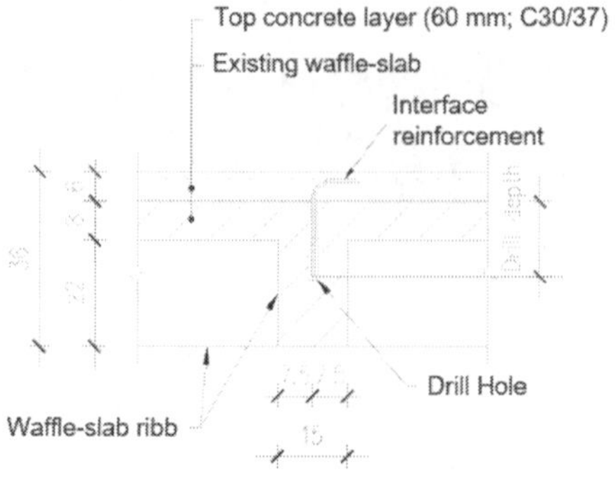

Fig. 17. Interface reinforcement detail.

The longitudinal reinforcement in the new concrete layer is found thanks to the bending moment diagrams resulting from the ULS analysis. The obtained longitudinal reinforcement ratio for the top concrete layer is approximately 0.005, thus meaning approximately 9200 kg of steel reinforcement for all the retrofitted slabs.

Based on the same ULS analysis it was found that supplementary punching shear reinforcement is necessary near the piers supporting the slabs. This is a consequence of having a low strength concrete and a supplementary dead load coming from the additional concrete layer. Steel anchor rods were used as punching shear reinforcement. These anchor rods were designed to be mounted into drilled holes crossing the whole existing slab and fixed with injection mortar or a resin-based grout. Bolts and special designed steel plates are use as anchorage devices, and a detail of this retrofitting solution is shown in Fig. 18. The slab in these areas near the piers is flat (i.e. no ribs are provided) and has a thickness of 300 mm without over-concreting. The total steel quantity for the punching shear retrofitting solution for all the slabs is estimated to approximately 1400 kg.

Another consequence of having a diminished concrete strength is the reduction of the shear resistance, especially of the capacity limited by crushing of the compression struts (i.e. named $V_{Rd,max}$ in Eurocode 2 [3]). Here, the capacity is directly related to the design value of concrete compressive strength and the width of the concrete element. Therefore, the ULS analysis showed that the induced shear forces can no longer be overcome by some of the slab ribs.

The proposed retrofitting solution for these elements consists in fixing a U channel steel profile above the slab ribs and combined with exterior stirrups.

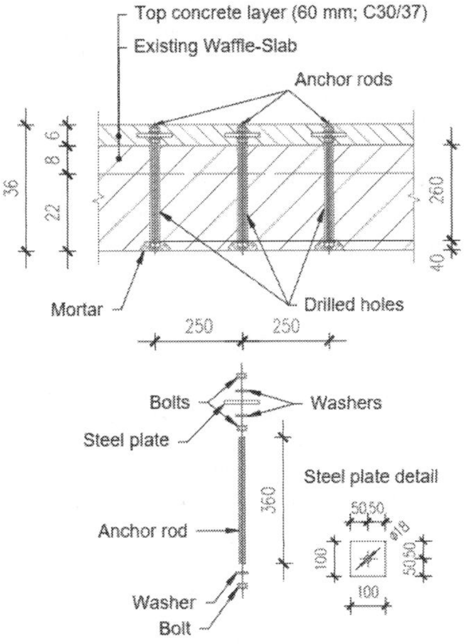

Fig. 18. Punching shear retrofitting solution.

A steel plate is placed under the steel profile to protect the existing concrete. Steel angles with L profiles are provided at the corners of the slab ribs for the same protective reason. All steel elements are welded. For simplicity, all the shear force is considered to be taken by the steel profile and, therefore, it is designed accordingly. A detail of this retrofitting solution is presented in Fig. 19. The total steel quantity for all the slabs, including the steel profiles, is estimated to approximately 2200 kg. A total of 39 slab ribs require interventions of this kind.

Finally, a total steel quantity of 14300 kg is estimated to be needed for this retrofitting solution and a total concrete volume of 70 cubic meters (i.e. the volume of the new concrete layer).

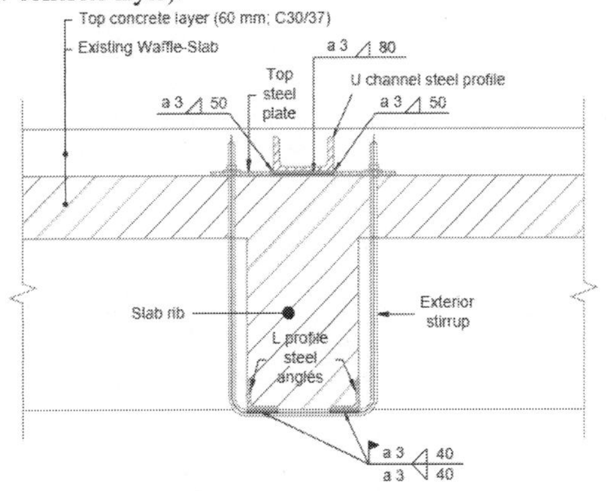

Fig. 19. Punching shear retrofitting solution.

4.2 The post-tensioning retrofitting solution

Despite all the appearances, the proposed retrofitting solution using exterior post-tensioning is actually much simpler. All the tendons were placed under the existing slabs and anchored into the exterior concrete beams. The tendon layout is ensured with steel deflection devices that are designed for a maximum of 1 m radius of the strands. Moreover, the ULS analysis showed no risk in having exterior forces that are higher than the structural elements are capable to overcome. A representation of the tendon layout is shown in Fig. 20.

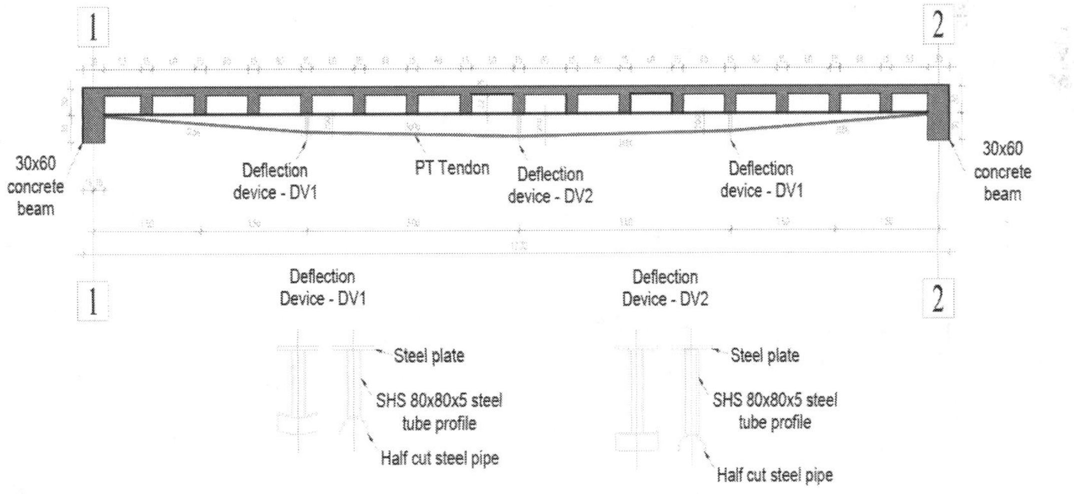

Fig. 20. Exterior tendon profile.

The tendons are designed to be placed under the slab ribs and spaced at 0.75 m (i.e. also the distance between the ribs). The number of tendons gets higher in the areas where the maximum deflections were obtained during the analysis. These areas are located at the semi-basement and at the ground floor, and between axes D and E. A plan representation of the tendons for each floor is offered in Fig. 21, Fig. 22 and Fig. 23. As seen, a total of 62 tendons are needed, each having the length of approximately 12.350 m.

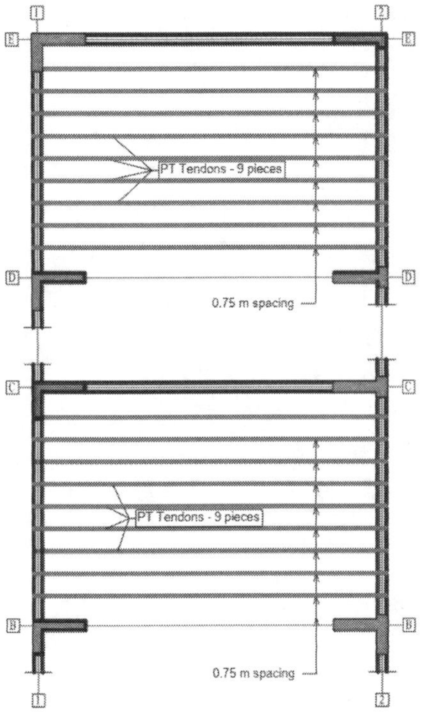

Fig. 21. Positioning of tendons – basement floor.

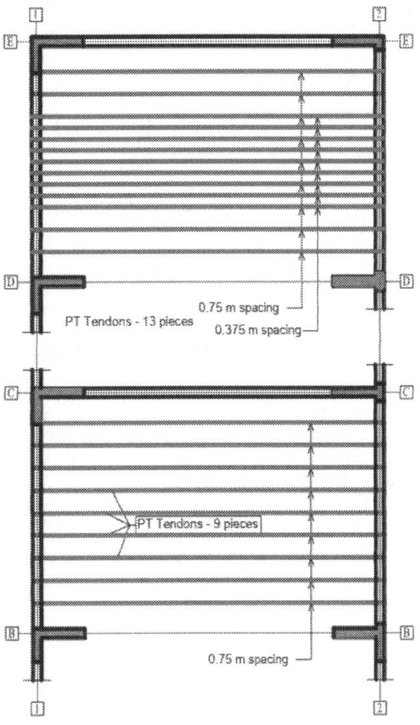

Fig. 22. Positioning of tendons – ground floor.

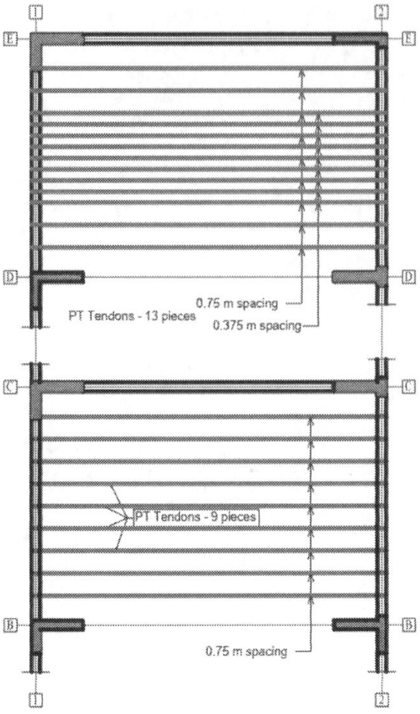

Fig. 23. Positioning of tendons – semi-basement floor.

A total of 124 pieces of deflecting devices DV1 and 62 pieces of deflecting devices DV2 are needed. The total steel consumption that is estimated for this retrofitting solution is estimated at about 1100 kg. Of course, no additional concrete layer is needed and only the areas where the maximum deflections were encountered must be retrofitted.

Due to the high-tension forces in the tendons and the low concrete class of the perimeter beams, supplementary anchor devices are needed at both ends. These devices are meant to reduce the tensions induced in the support elements caused by the post-tensioning, and therefore should be made from thick steel plates. The dimensions are established based on the tension forces from the strands. The resulted quantities were contained in the estimated consumptions presented previously.

4.3 Comparison of the proposed retrofitting solutions

Except for the simplicity of the solution, it is of great importance to compare the overall steel consumptions of the two retrofitting solutions. It should be mentioned that these quantities are estimations, but no significant differences are expected to be encountered in practice.

Therefore, the steel consumption for the top concrete layer solution is 13 times higher than post-tensioning solution. A cost estimation is considered not so relevant, merely because it is difficult to estimate the final impact upon sales while dealing with a smaller living space due to post-tensioning. Based on the major difference regarding the steel consumption and by the fact that no additional concrete layer and no supplementary structural interventions are needed for the post-tensioning solution, is assumed that this last-mentioned retrofitting solution is also the costliest efficient.

5. Conclusions

Two retrofitting solutions were studied for a multi-storey building having waffle-slab floors. The necessity to retrofit the slabs came as a consequence of detecting poor concrete strengths in the already cast elements. The SLS analysis showed that the maximum allowed deflections are highly exceeded, while the SLU analysis did not suggested any necessary structural interventions when analysing the in-site situation. Therefore, the retrofitting solutions were adopted and focused merely on the stiffness enhancement of the slabs to long period loadings.

The top concrete layer solution was considered simpler at first, but complex structural interventions have been suggested after performing the SLU analysis. The necessary thickness of the top concrete layer is 60 mm, and the analysis showed that the infrastructure has the capacity to overcome the resulting additional weight.

The exterior post-tensioning solution is considered to be the most efficient. The overall steel consumption is 13 times lower than obtained by over-concreting. If the intension is increasing the stiffness, then it is considered that the presented post-tensioning solution is even more cost-efficient than using carbon fibre. Therefore, exterior post-tensioning should be used as an efficient retrofitting solution to overcome the SLS excessive deflections.

References

1. Normativ privind evaluarea in-situ a rezistenței betonului din construcțiile existente, indicativ NP 137, (2014).
2. BS EN 13791: 2007: Assessment of in-situ compressive strength in structures and pre-cast concrete components.
3. EN 1992- 1- 1: 2004: Eurocode 2: Design of concrete structures - Part 1- 1: General rules and rules for buildings. Incl.
4. Computers and Structures Inc., *ETABS - Extended Three Dimensional Analysis of Building Systems - Nonlinear v16.0.0.*, California, USA, (2016).
5. Computers & Structures Inc., *CSI Analysis Reference Manual For SAP2000®, ETABS®, and SAFE*, (2016)
6. S. Timoshenko and S. Woinowsky-Krieger, *Theory of Plates and Shells*, 2nd ed, McGraw-Hill, N.Y. USA, (1987).
7. I. Chowdhury and J. P. Singh, *Analysis and design of waffle slabs with different boundary conditions*, Indian Concr. J., vol. **86**, no. 6, pp. 43–52, (2012).
8. W. Kanok-Nukulchai and A. K. Gilani, *A macroelement for waffle slab analysis*, Comput. Struct., vol. **15**, no. 2, pp. 117–122, (1982).
9. A. Faur, C. Mircea, and M. Păstrav, *A Modeling Technique for Precast Concrete Frames with Hybrid Connections*, Acta Technica Napocensis: Civil Engineering & Architecture, vol. **55**, no. 3, 2012.

Hybrid fibre-reinforced geopolymer (HFRG) composites as an emerging material in retrofitting aging and seismically-deficient concrete and masonry structures

Ernesto Guades [1,*], *Henrik* Stang [1], *Gregor* Fisher [1] and *Jacob* Schmidt [1]

[1]Department of Civil Engineering, Technical University of Denmark, Lyngby, Denmark

Abstract. Fibre-reinforced polymer (FRP) systems have recently become popular in repairing concrete or masonry structures because of their inherent advantages. In spite of these benefits, FRPs have drawbacks having low fire resistance, poor environmental sustainability and incompatibilty with the substrate concrete. The effort to address these issues has led to the development of an emerging strain hardening cementitious (SHC) material using an inorganic polymer known as hybrid fibre-reinforced geopolymer (HFRG) composites. Compared with cement-based SHC composites, HFRG has better bond performance to concrete substrates, higher fire resistance, greater corrosion durability and helps to reduce CO_2 emissions. This paper reviews the recent development of HFRG composites as an emerging repair material. Literature reveals that flowability of a fresh HFRG mixture decreases with increasing fibre content though still workable up to 2% fibre volume. Fibre synergy could result in 10–181% higher flexural toughness of geopolymer composites than when just using mono fibres. The application of HFRG composites to RC beams increased displacement ductility by to 263%. To date, there has been no reported field application of HFRG as a repair material though mono-fibre FRG has been field-applied as a strengthening material in large-diameter sewer RC pipes, RC culverts, RC sewerage manholes and dam surface improvement.

1 Background

Many existing buildings and bridges in Europe are in need for rehabilitation to enhance and extend their service performance and life. Retrofitting of these concrete and masonry structures and elements has traditionally used cement mortar patching or steel jacketing, however the use of fibre-reinforced polymer (FRP) systems have recently become popular because of their inherent benefits. FRPs, derived from combining of fibre and organic polymer (i.e. epoxy), are known to perform better than traditional repair systems due to their high specific strength, relative ease and speed of application and high corrosion resistance. The performance of such systems has been thoroughly investigated in the past decades and recently some in–situ applications have been reported [1–2]. In spite of these advantages and reasonably successful application, FRP method suffered drawbacks owing to low fire resistance, poor environmental sustainability, incompatibility with the substrate concrete and being relatively expensive [3–4]. The quest to overcome the limitations of FRP systems has led to the development of a relatively new strain hardening cementitious (SHC) material using an inorganic polymer known as fibre-reinforced geopolymer (FRG) composite. Instead of using cement as the matrix, FRGs utilize aluminosilicate powder such as fly ash (FA), metakaolin (MK) or ground granulated blast-furnace slag (GGBS) and are synthesized through the process of geopolymerization [5]. Geopolymers are considered to be an environmentally-friendly alternative for cement-based composites since their production does not necessarily emit CO_2. In addition, they have good bond strength to the concrete substrates [6], lower creep and shrinkage and greater durability in aggressive environments [7].

The reinforcement in FRG composites can be in the form of short mono-fibres or hybrid fibres. Compared to the first, hybridization offers superior engineering properties because the presence of one fibre enables the more efficient utilization of the potential properties of the other fibre [8]. Consequently, hybrid FRG (HFRG) composites are undoubtedly an ideal strengthening material for structures, especially when situated in aggressive environments. HFRG composite is an emerging strengthening material and is relatively new in this application, therefore an effort is made to gather relevant information from the previous studies. This paper presents the recent information on HFRG from literature sources. The discussion is concentrated on the concept of fibre hybridization and its effect on the fresh and hardened state properties of the geopolymer composites. In addition, laboratory-based findings and field application performance of FRG-strengthened concrete and masonry structures are discussed. It is expected that the knowledge provided in this paper could

* Corresponding author: ergua@byg.dtu.dk

contribute in the development of effective FRG repair system guidelines and leading to their global acceptance.

2 HFRG material constituents and fibre hybridization

Fly ash, MK and GGBS are the most common precursor material in producing geoplymer. FA is an industrial by-product of burning coal or lignite deriving from electricity generating plants. The particle size, curing time, curing temperature and the alkaline solutions are considered major parameters that affect the mechanical strength of the fly ash-based geopolymer. MK is formed by a controlled calcination process of the clay mineral kaolinite that contains high SiO_2 and Al_2O_3 [9]. The chemical solutions, curing system and the characteristics of the MK used greatly affect the initial reaction process and performance of the MK-based geopolymer. On the other hand, GGBS is a non-metallic material from heating iron ore, coke and limestone to a temperature between 1500°C to 1600°C [10]. The ores, fluxing stone and impurities in the coke charged into the blast furnace are the factors affecting the composition and properties of GGBS. Among these binding materials, MK-based geopolymer has been shown to have less impurities, exhibits good mechanical properties and easier control on the Si/Al ratio [11]. However, MK is relatively expensive since this material requires manufacturing through the process of calcination, as compared to FA and GGBS that are abundantly available in landfill sites. In the geopolymerization process, alkaline solutions are required to activate either FA, MK or GGBS. The most common alkaline activator that is suitable for FA, MK or GGBS is a combination of sodium silicate (Na_2SiO_3) and sodium hydroxide (NaOH).

Short fibre reinforcement used in HFRG composites for structural applications are usually formed and fabricated from steel (ST), polyvinyl alcohol (PVA), polyethylene (PE), polypropylene (PP), carbon, glass and basalt materials. They can be classified as high modulus (i.e., steel and carbon) and as low modulus (i.e., PVA, PE, PP, glass and basalt). The high modulus fibre usually achieve its optimal response at small to medium crack openings while the low modulus fibre would develop its capacity only at large cracking [12]. If hybridized, high and low modulus fibres are anticipated to produce a composite material with excellent crack-mitigating performance. In fibre–reinforced concrete, theoretically macro-fibres increase the composite toughness by bridging macro-cracking while micro-fibres enhance the response to micro-cracking prior to the formation of macro-cracks [13]. If combined, it is expected that micro-fibres will enhance the pull–out capacity of the macro-fibres resulting in composites with high strength and toughness [12].

3 Workability of fresh HFRG composites

Figure 1 shows the effect of different hybrid fibre combinations on the workability of fresh geopolymer composites. Attempt to quantify the effect of fibre hybridization on the workability of fresh FRG composites have been done in [14]. This study combined two steel fibres (i.e., spiral + hooked end) and found that, like in a mono fibre reinforcement, increasing its combined fibre content decreased the workability of the mixture. Combining different materials (i.e., steel + PE), however, does not necessarily result in the same trend since an optimum flow value can be obtained at a combined fibre volume content of 1.5%. Increasing the amount of PE fibre in (steel + PE) combination at a fixed fibre ratio at 2% reduced the fluidity of mixes containing them because of its relatively lower density [14–15]. Regarding the effect of the steel/PE fibre ratio at a fixed volume fraction, it can be observed from Figure 1 that decreasing the amount of steel (i.e., increasing the

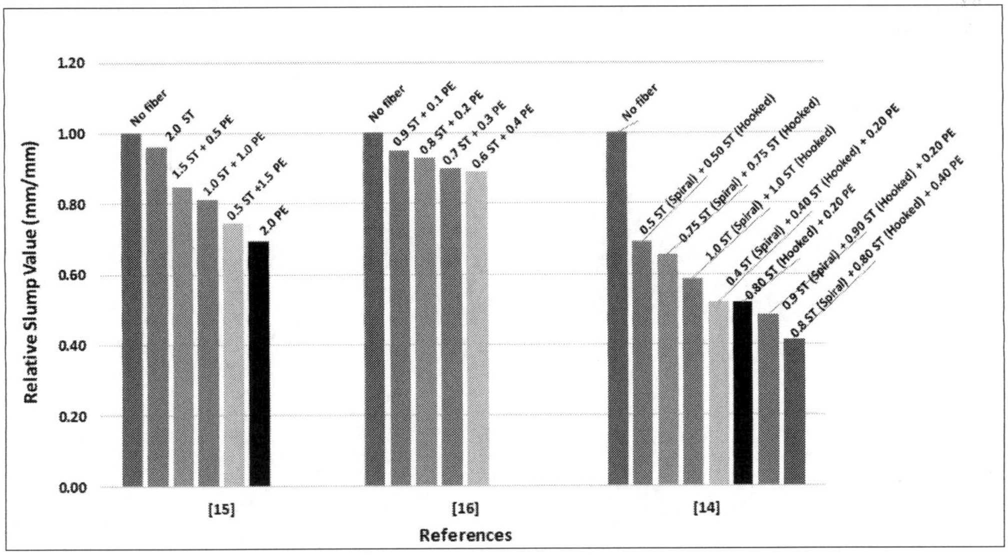

Fig. 1. Workability characteristics of HFRG composites.

PE fibres) would generally drop the flowability of the HFRG composites. This finding was also observed in [15] and [16] whereby the higher the steel fibre volume used in a fixed (steel + PE) combination, the higher the workability found. Meanwhile, Gao et al. [17] characterized the slump flows of samples with 1% fibre volume content at different long/short fibre ratios (i.e., 80/20, 60/40, 40/60 and 20/80) and reported that higher long/short fibre ratio exhibited a relatively low slump. The information on the effect of fibre hybridization on the flowability of fresh FRG composites is rather limited, and apparently additional studies in this aspect need to be conducted due to its complexity. FA-based FRG composite is observed to be more workable than slag-based composite because of its inherently smooth spherical shape bringing a ball-bearing effect that reduces internal friction in the mixture. Other factors that influenced the workability condition of fresh FRG composites include fibre types and smoothness, fibre mechanical properties, alkaline activator properties, and other admixtures [7,17].

4 Mechanical properties of HFRG Composites

Figures 2–5 summarize the different studies utilizing hybrid fibres and the effect of hybridization on the mechanical properties of FRG composites. It should be noted that the hybrid combinations included in the figures (details shown in Tables 1 and 2) are the ones producing the highest strengths or toughness, nevertheless other combinations can be found in the individual concerned studies. Moreover, the values indicated for the studies of [15] in Figure 5 are from the fracture toughness calculation instead of the flexural toughness. It can be observed from Figures 2-3 that compressive and tensile strengths of hybrid composites are typically in between the envelopes of the composites utilizing mono fibres. Hybridization does not play a significant contribution to enhance the strengths of composites under compressive and tensile loadings especially if the synergy is between fibres of different materials (i.e., steel combined with PP, PE or glass fibres). The closest compressive strength variation between the hybrid and mono fibre composites with highest strength is around 2% with 10% PP + 90% steel combination [16]. On the other hand, the closest difference in tensile strength is 7% when using synergy of (0.5 PP + 1.0 Steel, % by volume) fibres compared to 1.5 Steel mono fibre [18]. However, there is a tendency that hybridization becomes significant if the combination is on the length of fibres with same material type. For instance, Gao et al. [17] found out that combining 60% of long steel fibre and 40% of short steel fibre (% by weight in total steel fibre combination) yields an increase in compressive strength of 5–9% compared when only using long or short steel fibre alone. Accordingly, hybridization presents a synergetic effect by constraining the cracks from both micro and macro intensities, resulting in the occurrence of an optimum compressive strength.

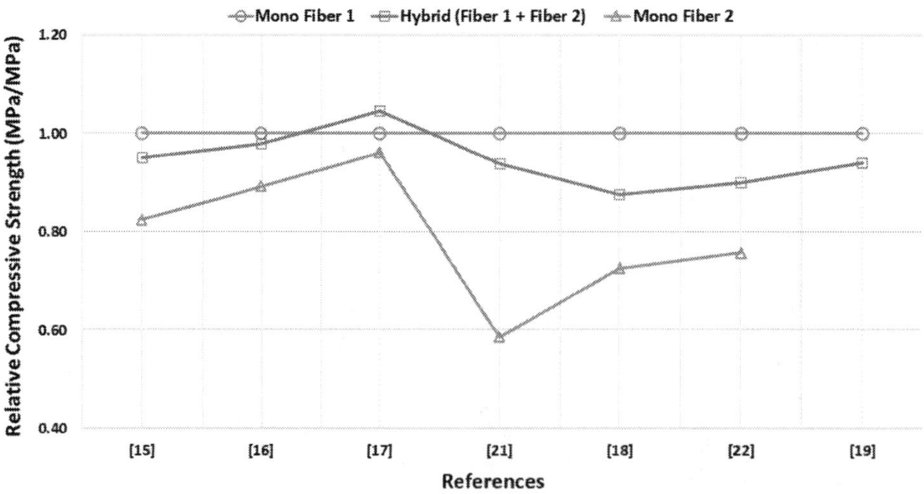

Fig. 2. Relative compressive strength of HFRG composites.

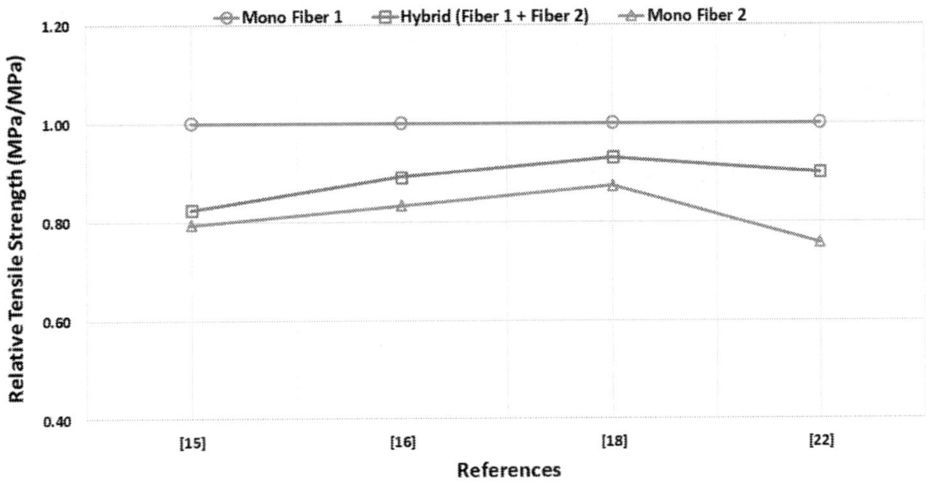

½

Fig. 3. Relative tensile strength of HFRG composites.

Table 1. Fibre combinations of FRG composites for compressive and tensile properties.

Mechanical properties	References	[a]Mono fibre 1	[a]Hybrid	[a]Mono fibre 2
Compressive strength	[15]	2.0 ST	0.5 PE + 1.5 ST	2.0 PE
	[16]	1.0 ST	0.1 PP + 0.9 ST	1.0 PP
	[17]	1.0 ST Long	0.6 ST Long + 0.4 ST Short	1.0 ST Short
	[18]	1.5 ST	0.5 PP + 1.0 ST	1.5 PP
	[19]	1.6 ST	1.3 ST + 0.3 Glass	1.6 Glass
	[21]	1.0 ST	0.2 PP + 0.8 ST	1.0 PP
	[22]	3.0 ST Long	3.0 (ST Long + 3ST Short)	3.0 ST Short
Tensile strength	[15]	2.0 ST	1.0 PE + 1.0 ST	2.0 PE
	[16]	1.0 ST	0.1 PP + 0.9 ST	1.0 PP
	[18]	1.5 ST	0.5 PP + 1.0 ST	1.5 PP
	[22]	3.0 ST Long	3.0 (ST Long + ST Short)	3.0 ST Short

[a]% by volume

The effect of hybridization in the flexural strength of FRG composites (Figure 4) is positive as reported in other studies. Combining 30% PP and 70% steel fibres (% by volume in total PP and steel fibre combination) in composites outperformed 100% steel mono fibre composites by an 18% increase in terms of flexural strength performance [16]. Similarly, the blend of (0.3 glass + 1.3 steel, % by volume) fibre in the composites brought a flexural strength increase of about 7% compared to using composites reinforced by steel mono fibre [19]. This increase is mainly attributed to the glass and steel fibre synergy that interconnected the micro and macro cracks in the tensile zone of the composites resulting in effectively controlling crack proliferations. These results indicated that the highest flexural strength can still be achieved by combining two or more materials of different type (i.e., steel + PP of glass) with proper proportioning. One can observe in Figure 4 that hybridization does not really deliver a positive effect to increase the flexural strength when using the same fibre material (i.e., long + short steel fibre combination). Observing Figures 4 and 5, it seems that the effect of fibre hybridization is most substantial in improving the flexural toughness of the FRG composites rather than its strength. Toughness of the composites as repair material is extremely important especially if the structure is under dynamic loadings (e.g., seismic, impact or blast) and therefore hybrid fibres are suitable in these applications. Figure 5 demonstrates that all studies reported an excellent toughening performance provided by HFRG, regardless of the method of hybridization. Although the flexural strength of most of the HFRG composites reflected in Figure 4 is intermediately located in the mono fibre system composites, their strain hardening behaviour has been greatly improved. Due to hybridization, high-stiffness or high strength fibres (i.e., steel fibres) maintain their capacity to increase the flexural strength of the composite while low stiffness fibres (i.e., PVA, PP or PE) maintain their ability to increase the deflection capacity of the composite [20], thus increasing their toughness. The maximum increase of the flexural toughness from the mono fibre composites is around 181% when a combination of (2.0 PE + 0.5 ST, % by volume) is used. On the other hand, the synergy of (0.5 PP + 1.0 ST, % by volume) provided a minimum toughness increase of about 14%. Hybridization using (steel + PVA) yields a better toughening performance than steel + PP combination because mono PVA fibres are tougher than mono PP fibres. It is worth noting that by properly selecting the ratio, combining long and steel fibres in composites can increase the flexural toughness by almost 100% compared to using long fibre alone [22].

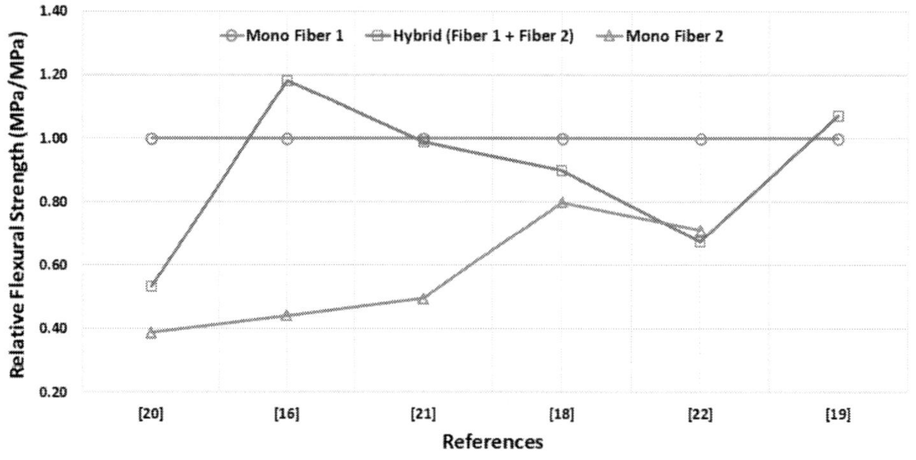

Fig. 4. Relative flexural strength of HFRG composites.

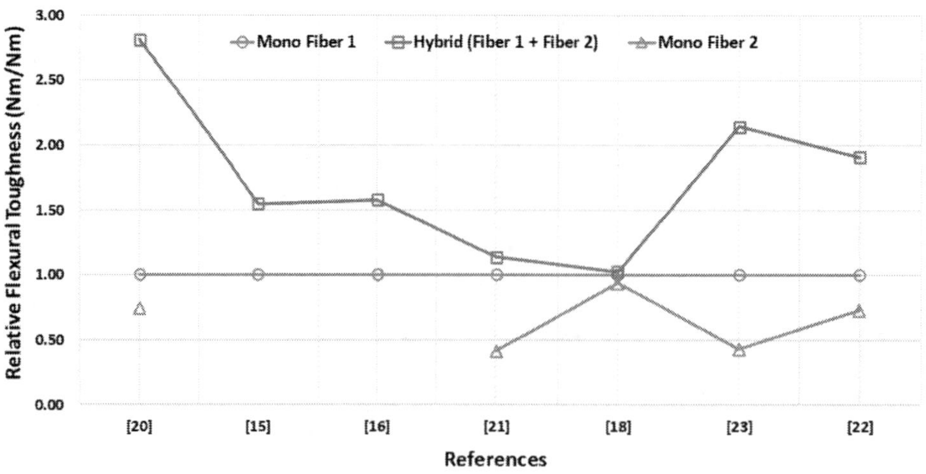

Fig. 5. Relative flexural toughness of HFRG composites.

Table 2. Fibre combinations of FRG composites for flexural properties.

Mechanical properties	References	[a]Mono fibre 1	[a]Hybrid	[a]Mono fibre 2
Flexural strength	[16]	1.0 ST	0.3 PP + 0.7 ST	1.0 ST
	[18]	1.5 ST	0.5 PP + 1.0 ST	1.5 PP
	[19]	1.6 ST	1.3 ST + 0.3 Glass	1.6 ST
	[20]	2.5 ST	1.0 PE + 1.5 ST	2.5 PE
	[21]	1.0 ST	0.2 PP + 0.8 ST	1.0 PP
	[22]	3.0 ST Long	3.0 (ST Long + ST Short)	3.0 ST Short
Flexural toughness	[15]	2.0 ST	1.5 PE + 0.5 ST	2.0 ST
	[16]	1.0 ST	0.3 PP + 0.7 ST	1.0 ST
	[18]	1.5 PP	0.5 PP + 1.0 ST	1.5 ST
	[20]	2.5 PE	2.0 PE + 0.5 ST	2.5 ST
	[21]	1.0 ST	0.2 PP + 0.8 ST	1.0 PP
	[22]	3.0 ST Long	3.0 (ST Long + ST Short)	3.0 ST Short
	[23]	2.0 PVA	1.0 PVA + 1.0 ST	2.0 ST

[a]% by volume

5 Laboratory-based studies on HFRG-strengthened elements

To date there are limited studies characterizing the behaviour of concrete or masonry elements strengthened with HFRG composites. Krishnaraja and Kandasamy [24] investigated the flexural performance of RC beams strengthened using PVA, PP and steel-based FRG composites. Their study revealed that both mono PVA fibre and (PP + steel) fibre combinations improved the initial cracking load response of the repaired concrete by 4.9% and 7.6%, respectively. However, using (PVA + steel) fibre in combination and mono PP reduced the initial cracking load capacity of the repaired beam by 5.2% and 8.6%, respectively. When it comes to the ultimate load of strengthened RC beams, they observed a 7–12% value increase when using either mono or hybrid fibres. Their study also revealed that this strengthening process could provide excellent energy absorption capacity, 71–100% higher than its un-strengthened counterpart. Among fibres, (PVA + steel) blend offered the highest energy absorption response of around 1,490 N-m, a result due to a better toughening effect of PVA fibre. The strengthened RC beam exhibited a better ductility performance showing an increase of displacement ductility between 67–263%, using (PP + steel) fibre combination as the most ductile strengthening fibre material.

There are also reported studies on FRG as repair material, however they focused more on the mono fibre as the reinforcement. For instance, Al-Majidi et al. [25] investigated the performance of a 2.4 m long RC beam repaired by 25 mm and 50 mm thickness PVA-based FRG composites. They reported that both the initial cracking and ultimate loads of repaired RC beams exhibited relatively higher values than their unrepaired counterparts. This study also showed that the initial cracking load capacity of the strengthened beam without exposing the whole beam to corrosion was substantially enhanced by 41 and 63% for 25 mm and 50 mm thickness repaired layers, respectively. They also stated that the difference between a 25 mm and 50 mm repair layer in their initial cracking load values (i.e., 41 and 63%, respectively) is more obvious than in its ultimate load values (i.e., 15 and 14%, respectively). Majidi et al. emphasized that the typical failure mode of the FRG-strengthened RC beam is flexural fracture and cracks propagating at the mid-span, indicating that a strong and effectively bonded interface was developed between the geopolymer repair material and concrete substrate. Likewise, Sial and Khan [26] studied the flexural response of RC beams with 60 and 90 mm thickness overlays of PVA-based FRG with or without additional steel reinforcement. Two common findings observed in the studies of [25–26] can be inferred; firstly, different overlay thicknesses apparently improved the initial cracking and ultimate load performances of the repaired beam. Secondly, the effect of increasing layer thickness is only significant in the initial cracking load improvement but not in the ultimate load value. The latter reported an increase in initial cracking load of the fortified RC beam of 63 and 86% when using 60 and 90 mm thickness repair overlays, respectively. These values are relatively higher compared to that observed in the study of [24] most probably because the former used thicker repair overlays than the latter. Nevertheless the recorded values in the ultimate load increase of the repaired RC beams of the two studies are more or less comparable. Donnini and Corinaldesi [27] investigated the behaviour of a concrete column strengthened using a glass or steel fibre reinforced geopolymeric matrix system, and compared it with that of an FRP jacketing and fibre-reinforced cement-based matrix (FRCM) system. The result revealed that the FRG repair system enhanced significantly the compressive strength of the concrete column. Using a 30 mm thickness of glass-based or steel-based FRG could increase the strength to 227 and 289%, respectively. It was found that the FRG system exhibited better compressive strength than an FRP-epoxy wrapping system, the strength of the former was 63–86% higher than the latter. FRG strengthening method also outperformed FRP-cement mortar wrapping system in terms of compressive strength by 61–69%. However, comparing with the ductility behaviour of the three different systems, this study showed that concrete columns strengthened by FRG composite systems are less ductile than when using FRP and FRCM systems.

6 Field applications of HFRG composites

There has been no documented field application of HFRG as a repair material although mono fibre-reinforced geopolymer composites have already been used in the rehabilitation of water culverts, pipes and dams. For instance, FRG mortar was spray-applied to repair a fully-deteriorated reinforced concrete pipe (RCP) main sewer in Texas, USA ([28]. The selected test pipe for the demonstration was a 49 m. long, 1.5 m diameter RCP and was located at a depth of 7.6 m. The FRG mortar that served as lining was a water-activated micro fibre reinforced fly ash-based geopolymer having a compressive, tensile and flexural strength of 55 MPa, 5.2 MPa, and 8.3 MPa; respectively. The lining was applied at an approximate thickness of 84 mm total, which was accordingly well above the design value of 48 mm. The lining process was concluded by smoothing the sprayed surface of the repaired pipe manually using a trowel. Although the repair method using FRG has been successfully demonstrated in deteriorated RCP, Matthews and his associates recommended that a shaker table or a vibrator be used in the future practice to minimize voids and that a need for continuous studies on the finished coating be undertaken.

Similarly, several demonstration projects on the use of FRG in repairing deteriorated culverts and pipe structures have been reported. The rehabilitation projects included large diameter concrete pipes, concrete sewerage manholes, rectangular concrete box culverts and storm water corrugated culvert [29]. The diameter of these concrete pipes ranged between 1.35–1.5 m and a length of around 53–854 m long. All of these structures were repaired using a spraying-technique of FRG mortar having a thickness in the range of 25–50 mm.

FRG composites were used to rehabilitate a dam that served as a water power plant in Thüringen, Germany [30]. The dam which is primarily made of concrete and dump stone was observed to have cracks on the concrete block wall with spalling and leaky joints. Repair was undertaken in the year 2011 to the dam, having a height of about 17 m and width of 3.5 m using a sprayed FRG composite system. The repair mortar contained 2 % by volume of PVA fibre (12 mm long and a diameter of 40 μm) with fly ash as the geopolymer precursor. A total mortar layer thickness of around 30–50 mm and 120 mm were sprayed on the surface wall and deteriorated joint of the dam, respectively. The final step in the repair process was to manually level the last layer of the sprayed mortar to achieve a smooth surface. No monitoring device was installed in this project, however visual inspection of the retrofitted zone was performed 4 years after its completion. They observed branching and a fine pattern of several cracks above the block joint, nevertheless the FRG-strengthened structure still performed according to expectations. Although the initial performance of the project was satisfactory, Müller and Mechtcherine highlighted that a monitoring system should be installed to collect detailed information of the crack behaviour for further improvement of the system.

7 Conclusions

The present study has performed a review on the development of HFRG composites as an emerging material in strengthening concrete structures. Literature reveals as expected that the flowability of fresh HFRG mixtures decreases with increasing fibre content though workable mixes are reported up to 2% fibre volume. It is known that fibre hybridization in geopolymer composite is more notable in increasing its flexural properties than its compressive or tensile strength. The compressive or tensile strength of geopolymer reinforced by hybrid fibre is between the strength envelopes of the composites with mono fibres. Combining 20% glass and 80% steel fibres or 30% PP and 70% steel fibres result in a flexural strength increase of 7–18%. Fibre synergy could result in a 10–181% higher flexural toughness of geopolymer composites than when using mono fibres. RC beams strengthened by HFRG exhibited a better ductility performance showing an increase of displacement ductility between 67–263%, with (PP+ steel) fibre synergy as the most ductile strengthening fibre material. To date, there has been no reported field application of HFRG as a repair material though, mono-fibre FRG has been field-applied as a strengthening material in large-diameter sewer RC pipes, RC culverts, RC sewerage manholes and dam surface improvement.

Acknowledgements

This project has received funding from the European Union's Horizon 2020 Research and Innovation Programme under the Marie Sklodowska-Curie grant agreement No. 713683.

References

1. J. Strydom, E. du Plessis, L. Pieters, *ICCRRR MATEC Web. Conf.* **199**, 1–9 (2018)
2. E.J. Guades, Recoletos Multidiscip. Res. J. **4**:1, 45–60 (2017)

3. F. Bencardino, A. Condello, Constr. Build. Mater. **117**, 171–181 (2016)
4. J. Donnini, V. Corinaldesi, *ICCRRR MATEC Web. Conf.* **199**, 1–5 (2018)
5. J. Davidovits, *Geopolymer '99 I Conf.* (1999)
6. S. Hu, H. Wang, G. Zhang, et al., Cem. Concr. Compos. **30**:3, 239–244 (2008)
7. N. Ganesan, R Abraham, S.D. Raj, Constr. Build. Mater. **93**, 471–476 (2015)
8. A. Bentur, S. Mindess, *Modern Concrete Technology Series:* Taylor & Francis (2009)
9. C. Kuenzel, T.P. Neville, S. Donatello, et al., Appl. Clay. Sci. **83–84**, 308–314 (2013)
10. N.I. Ishak, M.A.M. Johari, S.F.S. Hashim, *AIP Conf. Proc.* **1892**, 1–8 (2017)
11. H.Y. Zhang, V. Kodur, S.L. Qi, Constr. Build. Mater. **55**, 38–45 (2014)
12. N. Banthia, N. Nandakumar, Cem. Concr. Compos. **25**, 3–9 (2003)
13. C.P. Ostertag, J. Blunt, *Proc. CBM-CI Intl. Workshop* 73–90 (2008)
14. M.Z.N. Khan, Y. Hao, H. Hao, et al., Cem. Concr. Compos. **85**, 133–152 (2018)
15. Y. Alrefaei, J.G. Dai, Constr. Build. Mater. **184**, 419–431 (2018)
16. C.P. Devika, R.N. Deepthi, Int. J Sci. Res. **4**:7, 130–135 (2015)
17. X. Gao, Q.L. Yu, R. Yu, et al., Mater. Struct. **50**:165, 1–14 (2017)
18. M. Mastali, A. Dalvand, A.R. Sattarifard, et al., Compos. Part B-Eng. **151**, 186–200 (2018)
19. N.P. Asrani, G. Murali, K. Parthiban et al, Constr. Build. Mater. **203**, 56–68 (2019)
20. F. U. A. Shaikh, M. Maalej, P. Paramasivam, Constr. Build. Mater. **21**, 1088–1097 (2007)
21. P. Sukontasukkul, P. Pongsopha, P. Chindaprasirt, et al., Constr. Build. Mater. **161**, 37–44 (2018)
22. M.H. Al-Majidi, A. Lampropoulos, A.B. Cundy, Compos Struct. **168**, 402–427 (2017)
23. F.U.A. Shaikh, Constr. Build. Mater. **43**, 37–49 (2013)
24. A.R. Krishnaraja, S. Kandasamy, Arch. of Civ. Eng. **63**:4, 173–189 (2018)
25. M.H. Al-Majidi, A. Lampropoulos, A.B. Cundy, et al., Constr. Build. Mater. **164**, 603–619 (2018)
26. S.U. Sial, M. I. Khan, Performance of strain hardening cementitious composite as strengthening and protective overlay in flexural members, ICCRRR MATEC Web. Conf. **199**, 1–8 (2018)
27. J. Donnini, V. Corinaldesi, *ICCRRR MATEC Web. Conf.* **199** (2018) 1–5
28. J. Matthews, A. Selvakumar, S. Vaidya, ASCE Pract. Period. Struct. Des. Constr. **20**:4, 1–6 (2015)
29. https://infrastructure.milliken.com (Date accessed: 2 May 2019)
30. S. Müller, V. Mechtcherine, *ICCRRR MATEC Web. Conf.* **199**, 1–6 (2018)

Experimental testing of medium scale GFRP reinforced concrete frames

George Hopartean[1,*], *Ted* Donchev[1,*], *Diana* Petkova[1], *Costas* Georgopoulos[1], *Mukesh* Limbachiya[1] and *Noel* Parnada[1]

[1]School of Engineering and the Environment, Faculty of Science, Engineering and Computing, Kingston University, London, UK

Abstract. Fibre reinforced polymers (FRP) have been used as strengthening for existing RC structures for many decades. Lately, there has been a lot of interest in using FRP as internal reinforcement in beams, slabs and columns. One potential area of application could be reinforced concrete frames internally reinforced with GFRP bars. With limited research in this direction, the objective of this publication is to assess the behaviour of glass FRP (GFRP) reinforced concrete frames under reversed cyclic lateral in plane loading and to analyse the seismic performances of such elements. For the purpose of this paper, experimental testing of two 1/3 scaled down frames is conducted in displacement-controlled mode with the loading history according to ACI 374.1-05. The control sample is reinforced with conventional steel reinforcement and the results obtained are compared with the sample reinforced with GFRP bars. In summary, observations on the sample behaviour at specified drift ratio such as load-displacement behaviour, envelope curves and energy dissipation are presented.

1 Introduction

The use of fiber reinforced polymer (FRP) reinforcement as a viable alternative to steel in reinforced concrete structures has been the main interest for many researchers in the last decades. One of the main characteristics of the FRP bars is attributed to their non-corrodible nature. Consequently, there have been many successful applications reported in the literature in the use of FRP bars as main reinforcement in new reinforced concrete (RC) elements, such as beams, slabs and columns. However, very limited research has been conducted to study the behaviour of frames subjected to seismic loading.

In 2004, Said and Nehdi [1] investigated the performance of steel-free GFRP-reinforced beam-column joints. Two full-scale beam column joints, one reinforced with steel and the other with GFRP, were tested under reversed cyclic loading. The results were acceptable in terms of drift demand. Similarly, Mady et. al. [2] in 2011 investigated the seismic behaviour of beam-column joints reinforced with GFRP bars and stirrups. Five full scale column beam joints were tested under seismic loading and the results indicated the feasibility of using GFRP bars and stirrups as internal reinforcement in beam column joints. Moreover, similar studies [3,4] indicate that concrete frames reinforced with GFRP are able to withstand significant lateral displacement. Aliasghar-Mamaghani et al. [5] suggest that frames reinforced with GFRP bars show higher strength than frames reinforced with steel bars under seismic loads. Current publications [6] for GFRP reinforced concrete frames argue that the existing design codes have little or no seismic provisions for these types of structures due to lack of data and research in this area.

This paper presents the experimental results of 1/3 concrete frames reinforced with bent GFRP bars subjected to reversal cyclic loading. It represents a continuation of research conducted at Kingston University as published at FRPRCS 14 [7].

2 Experimental program

2.1. Test specimens

Two 1/3 scale down reinforced concrete frames were constructed and tested in the department of Civil Engineering at Kingston University, London.

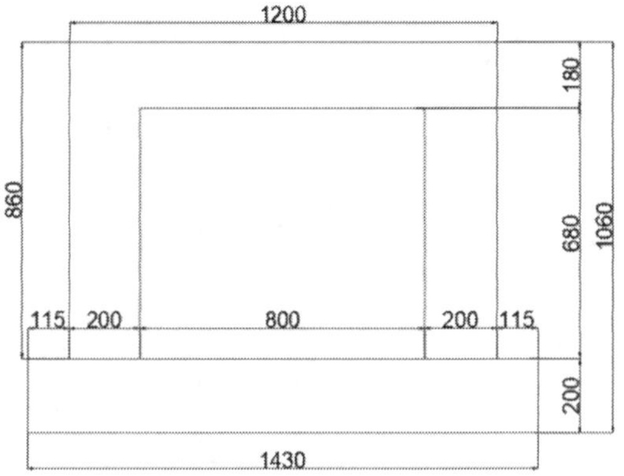

Fig. 1. Overall dimensions of both specimens

The control sample was a frame reinforced with conventional longitudinal steel and steel stirrups (SS)

while the second frame was reinforced with longitudinal GFRP bars and steel stirrups (GS).

The test specimens were of identical geometry and reinforcement details. Figure 1 shows the overall dimensions of the specimens, which were 1200 mm by 860 mm. The column heights were 860 mm with a cross section of 120x200 mm while the beam length was 1200 mm with a cross section of 120x180 mm. The foundation beam with a cross section of 200x300 mm was reinforced with 8 bars of 10 mm diameter so as to ensure sufficient stiffness is provided to the frame.

Figure 2 shows the reinforcement details of both specimens. The columns were reinforced with 8 rebars of 6 mm diameter while the beam was reinforced with 4 rebars of 6 mm in diameter. The stirrups were 4 mm diameter.

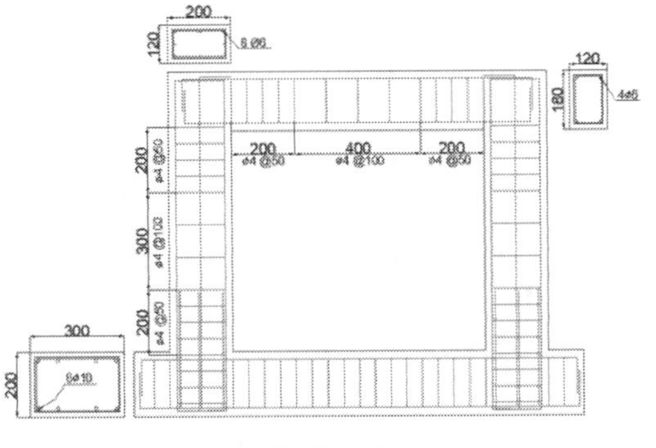

Fig. 2. Reinforcement details of specimens

2.2. Material properties

The specimens were constructed in the laboratory by using normal-weight mixed concrete with maximum aggregate size of 10 mm. The frames were cast and wet-cured in a horizontal position for 28 days. The targeted 28-day concrete compressive strength was C30/37.

Two types of bent reinforcing bars were used for building the frames: conventional steel and GFRP bars as shown in the Figure 3.

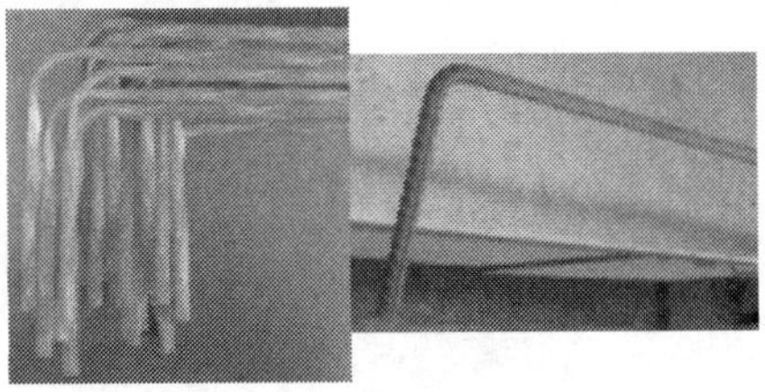

Fig. 3. Bent reinforcement bars

All the bars used in the experiment were anchored by using 90-degree hooks. The mechanical properties of the GFRP bars provided by the manufacturer are listed in Table 1 along with the properties of the steel bars.

Table 1. Mechanical properties of GFRP bars provided by manufacturer

Bar type	Bar diameter (mm)	Modulus of elasticity (GPa)	Tensile strength (MPa)
GFRP #6	>6	>40	>900
Steel #6	6	200	500

2.3. Test set-up and instrumentation

The experimental program consisted of testing two frames up to failure under reversed cyclic loading. The specimens were bolted to the strong reaction frame with the anchorage bolts from the RC foundation. The lateral load was applied at the beam level in the in-plane horizontal direction using a screw jack, electric motor and invertor as shown in Fig. 4.

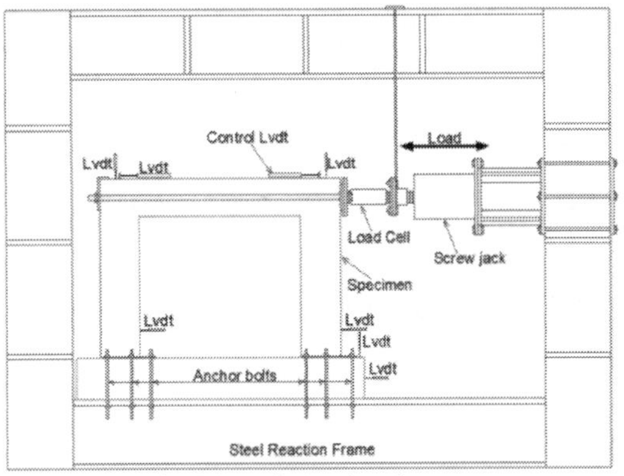

(a) Schematic drawing of set up

(b) Photo of set up

Fig. 4. Test setup: (a) Schematic drawing of set up; (b) Photo of set up

The pushing and pulling loading mechanism were possible by using high strength studs which were attached along the middle longitudinal axis of the beam and were secured at both ends with thick steel plates.

The load applied to the specimens in both directions was monitored in real time through a 500 kN load cell that was attached to the jack. Six linear variable displacement transducers (LVDTs) were attached to measure the displacement of the frame and two LVDTs to monitor the movement of the foundation (if any).

Strain gauges were installed on the reinforcing bars and linear wire gauges (PL-60-11 type) on the concrete. The acquisition rate was 10 readings per second.

The loading procedure consisted of one displacement-controlled phase, which was based on ACI-374.1:05 [8] as shown in Figure 5. During this phase, three fully reversed cycles were applied at each drift ratio.

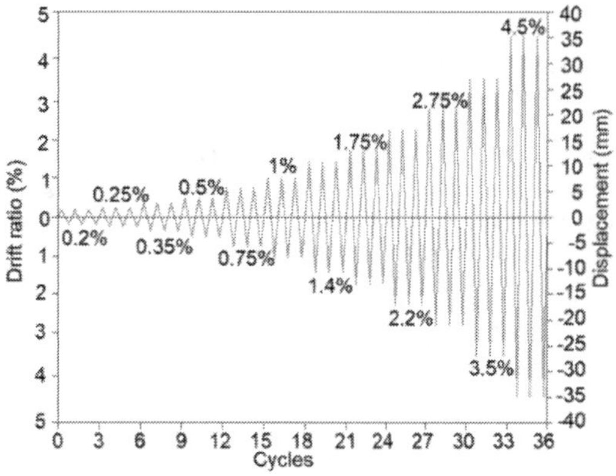

Fig. 5. Loading history

3 Test results and observations

Figure 6 illustrates the load-displacement hysteretic curves and photos for both specimens at the last drift.

The envelope load-displacement relationship is indicated with a red dashed line on the hysteretic graph. Both specimens, SS and GS, showed almost symmetric responses in pushing and pulling directions as shown in the figure 6. They were both tested up to failure.

At the drift ratio of 1.75% which is at 13.5 mm displacement, the SS specimen recorded a peak load of 71 kN in the pushing direction. In the pulling direction it reached its maximum load of 68 kN at a drift ratio of 2.2% which is at 16.9 mm displacement. In general, the SS sample exhibited lower resistance in pulling than in pushing and with a difference of 3 kN at maximum load. The maximum displacement the specimen reached was 32 mm in both directions.

The GS specimen reached the maximum load of 69 kN and 63 kN respectively in the pushing and pulling directions at the same drift ratio of 3.50% which is at 26 mm displacement. The GS sample showed a slightly higher resistance in pulling than in pushing up to a drift ratio of 1.75%. After that point, the resistance changed, becoming higher in pushing than in pulling mode. The difference at maximum load between the directions was only 6 kN. The maximum displacement the specimen reached was 40 mm in both directions.

The crack formation and development was almost symmetrical in pushing and pulling directions for both specimens.

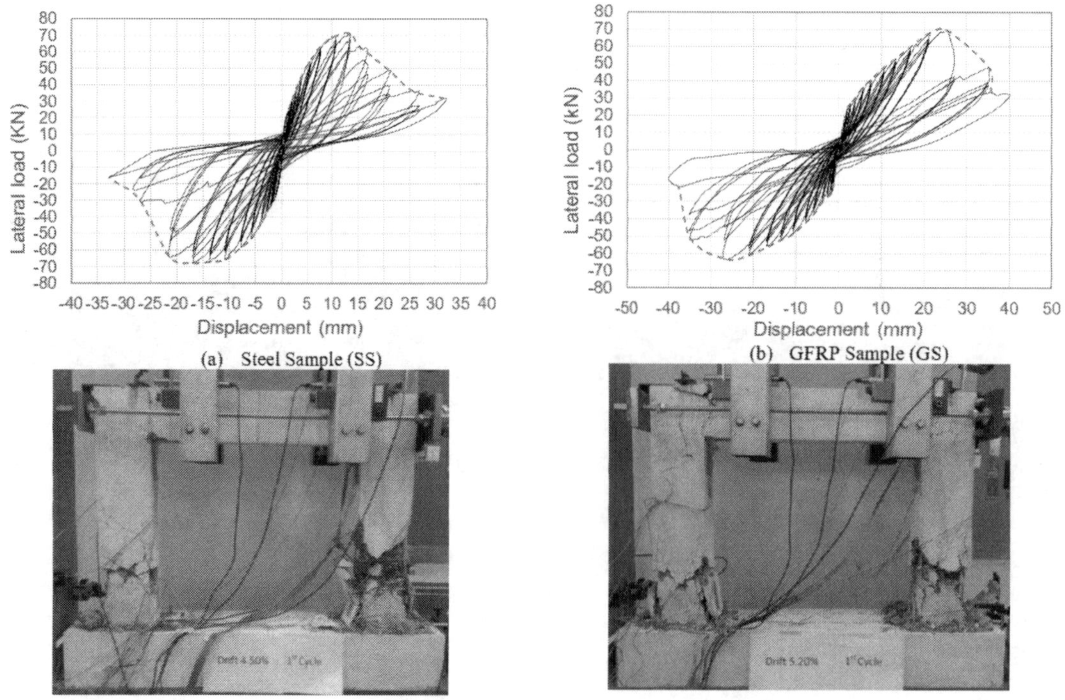

Fig 6 Hysteretic curves and photos of specimens at the last drift: (a) SS sample; (b) GS sample

4 Analysis and discussion

Figure 7 shows the maximum average load from pushing and pulling directions at different amplitudes and it was calculated as the average of the maximum load from both directions from the 1st cycle at each amplitude.

It can be noted that the SS sample is stiffer than the GS sample. Although the SS specimen had a slightly higher load, the GS specimen withstood loading at higher drift ratios.

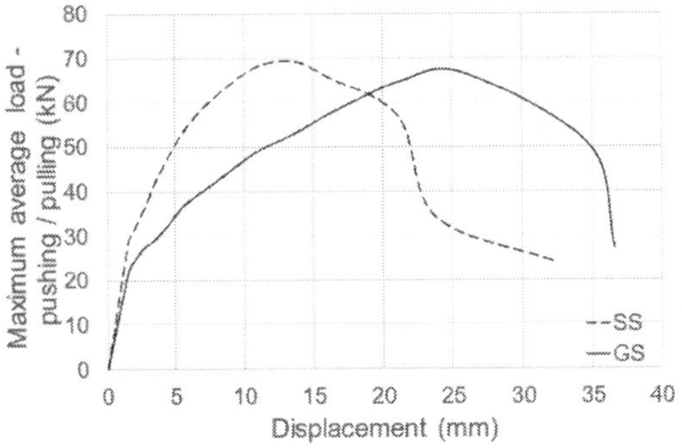

Fig. 7. Maximum average load from pushing and pulling

It can be observed that while the SS specimen reached its maximum load at 1.75% drift, the GS specimen had the peak at a drift of 3.50%. So, at around 25 mm in displacement when the GS sample reached its maximum load, the loading capacity of the SS sample had dropped to half of its ultimate load.

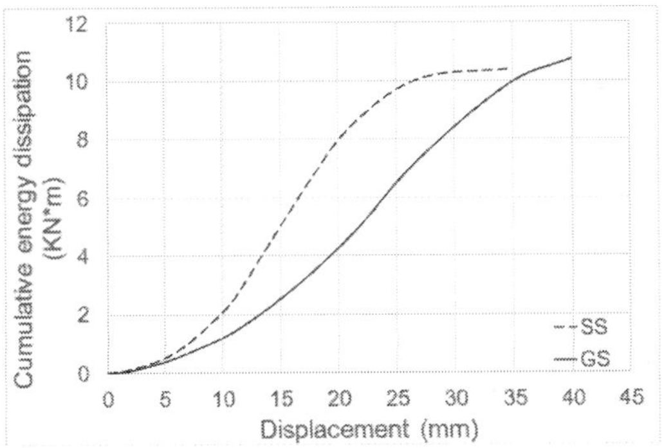

Fig. 8. Cumulative energy dissipation for both samples

Energy dissipation is an important factor in designing structures for earthquake requirements. In seismic areas, moment resisting systems, like RC frames, need to have the ability to dissipate the energy input from ground motion. Lateral load resisting systems with good energy dissipation characteristics are allowing to produce fast decrease of vibrations generated from initial impact.

Figure 8 shows a comparison of the cumulative energy dissipation for the samples. The cumulative energy dissipation was calculated by summation of the enclosed area in the hysteretic loops in successive load-displacement cycles. For each cycle, the energy dissipation is calculated as the area that the loop encloses in that particular load displacement graph.

It can be noted that the energy dissipation of the SS specimen is higher than for the GS specimen and up to a drift of 3.5% it is almost double. After that point, the energy dissipation of the SS sample starts to decrease considerably and at almost 35 mm displacement, it has almost the same amount of cumulative energy dissipation as the GS sample with a tendency for the GS sample to reach higher values.

5 Conclusions

Based on the experimental results, the following conclusions can be drawn:

1. Both tested specimens were able to sustain drift ratios higher than 2.5%.
2. The SS specimen is stiffer than the GS specimen
3. The SS specimen recorded the maximum load of 71 kN in the pushing direction at a drift ratio of 1.75%.
4. The GS specimen reached the maximum load of 69 kN in the pushing direction at a drift ratio of 3.50%.
5. The energy dissipation of the SS specimen was higher than that of the GS specimen up to 35 mm displacement and then decreased above this drift.

The conducted experiments indicate that GFRP reinforced frames are developing approximately same loading capacity as the control frame but the ultimate values of load turned up to be at higher displacement. Therefore, at the displacement of 25 mm, when the GS sample reached its ultimate load, the load bearing capacity of the SS sample is already reduced to approximately 50% of its ultimate capacity. During an earthquake event, the behaviour of a GFRP reinforced frame could allow bigger displacements than the conventional RC frames before reaching ultimate load values, this way allowing for better distribution of stresses.

The authors would like to express their gratitude for the GFRP bent bars provided by ATP construction composites, Italy.

References

1. A.M. Said, M.L. Nehdi, *Use of FRP for RC frames in seismic zones: Part II. Performance of steel-free GFRP-reinforced beam-column joints*. In: Applied Composite Materials (2004), pp. 227–245
2. M. Mady, A. El-Ragaby, E. El-Salakawy, *Seismic behavior of beam-column joints reinforced with GFRP bars and stirrups*. In: Journal of Composites for Construction **15(6)** (2011), pp. 875-886
3. M. Kazem Sharbatdar, M. Saatcioglu, B. Benmokrane, *Seismic flexural behavior of concrete connections reinforced with CFRP bars and grids*. In: Composite Structures (2011)

4. S.K. Ghomi, E. El-Salakawy, *Seismic Behavior of Exterior GFRP-RC Beam–Column Connections: Analytical Study*. In: Journal of Composites for Construction **22(4)** (2018), pp. 04018022
5. M. Aliasghar-Mamaghani, A. Khaloo, *Seismic behavior of concrete moment frame reinforced with GFRP bars*. In: Composites Part B: Engineering, (2019), pp. 324–338
6. M. Hasaballa, E. El-Salakawy, *Anchorage Performance of GFRP Headed and Bent Bars in Beam-Column Joints Subjected to Seismic Loading*. In: Journal of Composites for Construction **22(6)** (2018), pp. 1-14 04018060
7. G. Hopartean, T. Donchev, D. Petkova, C. Georgopoulos, M. Limbachiya, Z. El Kashik, *Experimental investigation of FRP reinforced RC frames under Cyclic in-plane loading,* Proceedings of the International Conference FRPRCS14, Belfast (2019)
8. ACI-374.1:05, *Acceptance Criteria for Moment Frames Based on Structural Testing and Commentary*. In: American Concrete Institute 05 (Reapproved) (2014)

Experimental study regarding the influence of fibre to matrix compatibility on general performance of Fibre Engineered Cementitious Materials (FECM)

Anamaria Mircea[1,2*], *Călin* Mircea[2], *Henriette* Szilágyi[2], *Cornelia* Baeră[3] and *Andreea* Hegyi[2]

[1]N.I.R.D. URBAN-INCERC, Cluj-Napoca Branch, 400524, Calea Florești 117, Cluj-Napoca, Romania
[2]Technical University of Cluj-Napoca, Civil Engineering Faculty, 400020, C-tin Daicoviciu 15, Cluj-Napoca, Romania
[3]N.I.R.D. URBAN-INCERC, Timișoara Branch, 300223, Traian Lalescu 2, Timișoara, Romania

Abstract. Fibre Engineered Cementitious Materials (FECM) represent composites with similar overall performance as Engineered Cementitious Composites (ECC), namely developing strain hardening behaviour under loading, which generates the material capacity of high deformability. The pattern of multiple microcracks successively developed under increasing loading is proved to be the key of material self-consolidating potential and ability to support loads after the first crack occurrence. The matrix to fibre compatibility is considered to be one essential parameter controlling the multiple micro-cracking pattern (MC) and consequently, the strain hardening effect in the material. Factors like fibre type and reinforcement percent in the mixture represent sensitive variables, with major influence for matrix to fibre compatibility and overall performance of the composite. Cement based materials, whose compositional heterogeneity traditionally represents a lack in their regular usage, can be valorised and designed to produce the width controlled cracking typology, beneficial for material behaviour. This paper presents an experimental study on the fibre to matrix compatibility effect in the FECM design and producing process. Several types of dispersed reinforcing typologies for FECM development are experimentally tested and analysed. The results confirm the importance of matrix to fibre compatibility in enhancing superior material performance: physical, mechanical and even durability (Self-Healing potential evaluation).

1 Introduction

Fibre Engineered Cementitious Materials (FECM) are a strain hardening type of fibre reinforced composite designed, cast, evaluated and eventually tailored for a better overall physical and mechanical performance at Incerc Cluj-Napoca, in the last years. The compositional design of FECM considered the ECCs concept [1] as a starting point of their development, namely the ability to develop high deformability under loading, due to the controlled microcracking mechanisms under loading. Consequently, this unique mechanical performance is achieved, distinct from the classical cement-based composites and best emphasised under tensile loading. Brittle failure, as a specific feature of classic cementitious composites, is thus avoided, opening the path to a new typology of building materials, characterised by the strain hardening performance under loading.

The matrix to fibre compatibility is proved to be one essential parameter controlling the multiple micro-cracking pattern (MC) and consequently the strain hardening effect in the material, which is one of the most important compositional design considerations [1, 2]. Fibre type and also the dispersed reinforcement percentage of the cementitious matrix represent essential factors, influencing the material performance in developing the strain hardening, metal like behaviour, under loading.

2 Fibre Engineered Cementitious Materials (FECM) – design approach

2.1 Strain hardening behaviour

The ECC design approach is essentially related to the development of a succession of restricted width cracks (generally bellow 100 μm) under load increase This cracking pattern, assimilated to the High-Performance Fibre-Reinforced Cementitious Composites (HPFRCC) and particularly to the Strain Hardening Cementitious Materials (SHCM), contradicts the general, concrete typical failure mode, produced via one sudden crack with large opening and associated with a significant strength drop, (Fig. 1). The intermediate material typology, namely Fibre Reinforced Concrete (FRC) develops a quasi-brittle, tension softening response under loading. Small quantity fibres, specific to FRC, may have a positive effect in counteracting drying shrinkage cracking and also sudden, brittle failure, but still, one crack formation would enlarge under loading and will not provide the deformability of the material.

* Corresponding author: anamaria.mircea@incerc-cluj.ro

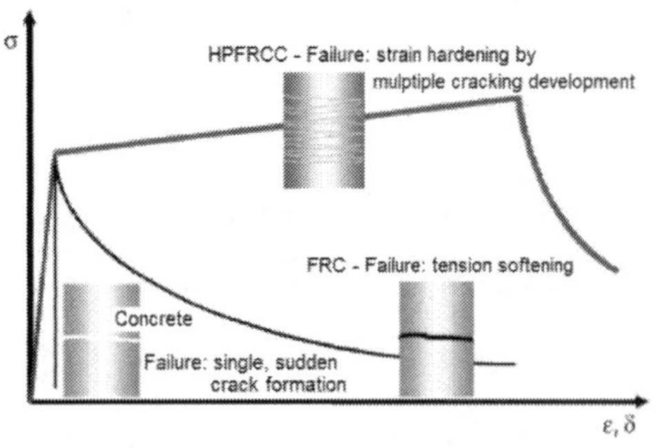

Fig. 1. Failure typology under uniaxial loading for cementitious materials: concrete, FRC and HPFRCC [1, 3]

In order to achieve the strain hardening typology [4, 5, 6, 7, 8, 9, 10], two criteria have to be simultaneously satisfied: a) the strength criterion; b) the energy criterion. The first criterion determines the micro cracks' debut, from the inherent flaws (derived from the heterogeneity of the cementitious matrix with dimensions ranging from nano- to millimetre scale), before the tensile load (σ_{cs}), exceeds the maximum bridging capacity of the fibre (σ_0): $\sigma_0 \geq \sigma_{cs}$. The second criterion considers an energetic balance within several simultaneous mechanisms occurring within the matrix, like absorption, sliding and bridging, debonding and friction [2]. It entirely depends on the fibre to matrix interface, and thus their compatibility, which generates the steady state cracking pattern, namely the constant, small width along the crack length, as opposed to the unstable crack formation (the Griffith crack, with crack opening varying over their length).

It is generally proven that the fibre to matrix compatibility emphasised through their interface, depending on matrix and fibre particular, intrinsic, characteristics is essential for the material strain hardening behaviour under loading [1, 2].

The current experimental procedure validates the importance of these parameters.

2.2 FECM development

The development of FECM at Incerc Cluj-Napoca, as specified above, had as a starting point Li's ECCs and later on, considering Snoeck et al's [1, 2, 11, 12] observations for mix proportions and mixing technology, successive compositional adjustments were introduced, also based on local typology of raw materials [13, 14].

The main goal was developing FECM mixes with strain hardening behaviour by using easily available local raw materials in order to facilitate future potential technological transfer to the Romanian construction industry. In the same time, the high fly ash content of the FECM mix design, increases substantially the potential of the concept, considering the large fly ash production on Romanian territory and the very little amount of its use in the construction industry, opposing its conversion into waste [15, 16].

At the same time, the evaluation of the possible integration of Polypropylene fibres of local source (PP hyb), as a reliable alternative to the lubricated Polyvinyl Alcohol (PVA) fibres, as dispersed reinforcement of the cementitious matrix, represents an important step of the research [13].

3 Experimental procedure – materials and methods

The experimental procedure is elaborated in order to support a pertinent comparative analysis among several FECM mixes, where the variable is the fibre type in order to evaluate this parameter influence on the composite strength (flexural and compressive) and also the strain hardening potential by the means of a four point bending test (4PB). The influence of fibre content is also considered.

3.1 Backgrounds of research

The first trials to introduce the PP fibres (of local source), as potential replacement for the lubricated PVA fibres, already proved to be efficient [1, 2, 17], but less accessible from the economic point of view and showed initial, early age, good results (Fig. 2, a), [14].

a)

b)
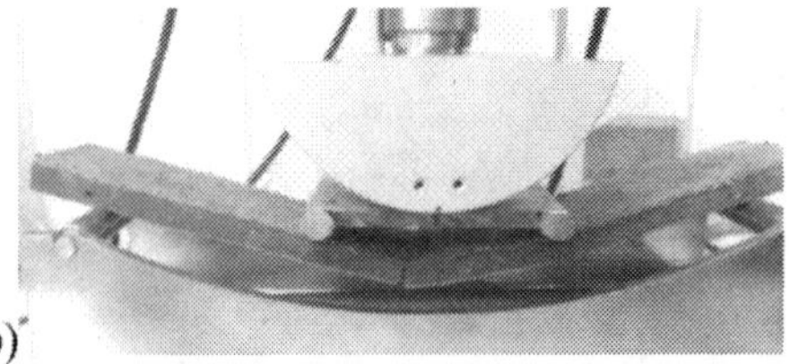

Fig. 2. Mid-span maximum deflection under 4PB test performed on: a) early age (7 days) specimens [14]; b) 28 days specimens

The 4PB performed on PP-FECM coupons specimens at the age of 28 days did not confirm the expected results (Fig. 2, b), as the material became brittle and rigid, developing just a limited number of cracks (max. 2) that widened considerably under loading, leading to failure. Still, the specimens did not break completely, the bridging effect of fibres preventing integrity loss.

The current study includes some adjustments regarding the mix design, in terms of raw materials and secondary addition changes. All changes are described below, in the raw materials paragraph.

3.2 Raw materials

The raw materials used for the FECM mixes are mainly provided by local producers.

The ECC binding system represents a combination of cement (C) and a high fly ash (FA) content. The specific ratio FA/C, which usually can vary for ECC mixes from 1.0 to 1.2 or even higher [1, 2, 13, 17], was 1.2 for the FECM mixes of the current study.

The cement used was the regular Ordinary Portland Cement (OPC), namely CEM I 42.5 N (Fig. 3, a). The class F Fly-Ash (FA), provided by a local power station (Fig. 3, b) is characterised by the XRF analysis presented in Table 1.

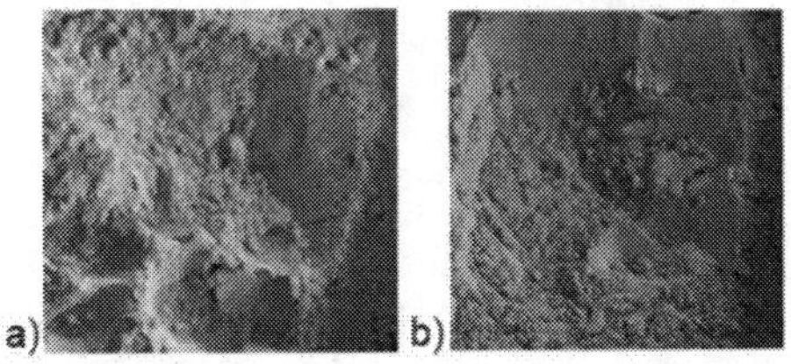

Fig. 3. The binding system: a) Cement 42.5 N, (C) b) Class F Fly Ash, (FA)

Table 1. XRF analysis data for the fly ash samples.

FA Oxides, %					
SiO_2	**Al_2O_3**	**Fe_2O_3**	**CaO**	**MgO**	**SO_3**
53.61	26.16	7.58	2.42	1.49	0.26
Na_2O	**P_2O_5**	**TiO_2**	**Mn_2O_3**	**L.O.I.**	-
0.59	0.12	1.04	0.08	3.57	-

The polymeric fibres used were of two types: the highly performing Polyvinyl Alcohol (PVA) fibres lubricated (by 1.2% mineral oil) and provided by Kuraray, Japan. The PVA-FECM is presently developed as a control mix, in respect with the PP-FECM would be evaluated further on regarding the relevant material features previously mentioned. The second type of fibres were Polypropylene (PP_hyb), namely a combination of two types of PP fibres, provided by a local producer.

PVA fibre (Kuralon K-II fibre RECS15x8) characteristics, in accordance with the data sheet provided by the producer were as follows: diameter 5.0±3.0 dtex, cut length - 8.0±1.0 mm), tensile strength 12.0±2.5 cN/dtex, converted to 1560±325 MPa, Elongation - 6.5±1.5%, Modulus - 320 cN/dtex.

PP1 fibres – RoWhite represents 100% Virgin Polypropylene, with modest mechanical properties and usually used for shrinkage prevention; the basic characteristics are as follows: shape - monofilament, Density - 910 kg/m^3, tensile strength - 300 MPa, ut length - 6 mm, colour - white, Absorption - none. The general conformity is specified in accordance to EN 14889-2 [18].

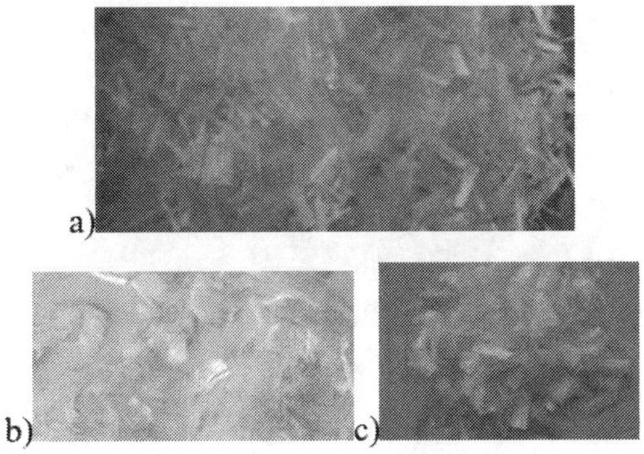

Fig. 4. Polymeric fibres: a) PVA Kuralon K-II fibre (RECS15x8) (Kuraray, Japan); b) PP RoWhite (PP1); c) ECONO NET (PP2)

PP2 fibres – Econo Net represents 100% Virgin Polypropylene, with increased mechanical properties (for impact and wear resistance) and also used for shrinkage prevention; the basic characteristics are as follows: shape fibre network, density - 910 kg/m^3, tensile strength - 90-110 ksi (620-758 MPa), cut Length - 19 mm, colour - white, high acid resistance, absorption - none. The general conformity is specified in accordance with ASTM C-1116 [19].

The PP1/PP2 ratio, 1/3, is considered optimum for the PP_hyb_FECM mix design.

The fibre content in the mix was generally 2% (by volume). A supplementary PP_hyb_FECM mix was developed considering a higher fibre content, namely 2.9% (by volume), in order to evaluate the influence of tis parameter on the overall material performance.

The Aggregate is represented by fine silica sand (FSS) provided by the traditional producer Bega Minerale SA (Cluj, Romania); the maximum grain size was 0.3 mm (Fig. 4, a). The FSS aggregate replaced the initially used silica sand (SS), provided by same company and with the maximum grain size of 0.5 mm. This change was performed for a better packing of the matrix and a superior MC development under loading, in accordance to Snoeck [2].

The mix packing optimisation included another replacement, related to the lime powder, used as *secondary addition* to the composite [Cracovia]. Initially, inert limestone filler (L) was added to the mix in order to control the potential bleeding and to generally improve the fresh state characteristics (fibre balling and bleeding prevention, homogeneity and the creamy aspect of the mix), and the hardened characteristics as well (packing of the matrix, fine particle content increase, and extra source for calcium carbonate ($CaCO_3$) precipitation and consequently increase of crack closing efficiency (Self-Healing potential of the material)).

https://doi.org/10.1051/matecconf/201928904005

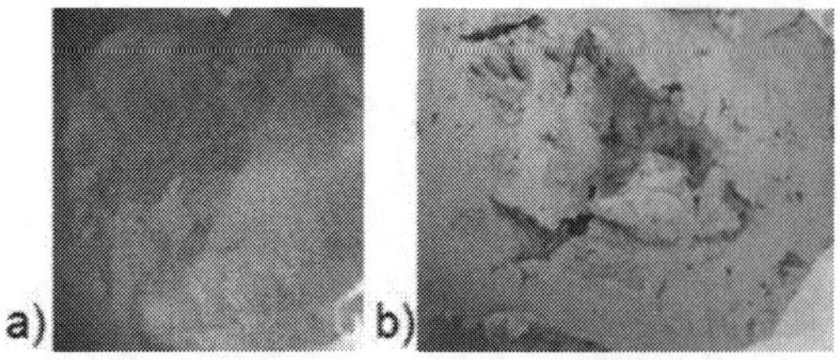

Fig. 4. Raw materials: a) Aggregate, Fine Silica Sand (FSS); b) Limestone Slurry (SL)

The *slurry lime* (Fig. 4, b) a lime paste with determined water content, provided as a waste of a local industry, was tested and considered proper to replace the lime powder addition. The total water adjustment in the mix was performed considering the actual water content of the lime paste.

The low water to binder ratio (W/B), essential for material performance, was maintained as low as possible by means of the *polycarboxylate superplasticizer* Glenium 51 (HRWR), (BASF).

3.3. Mix design

Table 2 presents the mix proportions of the FECM mixes (PVA- and PP_hyb-) namely the specific ratios with respect to the cement (C) content (C = 1.0), for optimum emphasising of the design process. Three FECM mixes are considered for the current study: the reference, namely the PVA-FECM, containing 2% (by volume) PVA lubricated fibres; 2.0 PP_hyb-FECM, containing 2% (by volume) PP_hyb fibres; and the mix with increased content of PP, namely 2.9 PP_hyb-FECM, containing 2.9% (by volume) PP_hyb fibres.

Table 2. Mix proportions of FECM.

Components	FECM Mixes		
	FSS-SL-PVA-FA/C (1.2)	FSS-SL-PP(2.0)-FA/C (1.2)	FSS-PP(2.9)-FA/C (1.2)
C*	1.0	1.0	1.0
FA	1.2	1.2	1.2
B	2.2	2.2	2.2
FSS	0.8	0.8	0.8
SL	0.18	0.18	0.18
W	0.65	0.65	0.65
HRWR	0.021	0.021	0.021
FIBRE	2.0%	2.0%	2.9%
W/C	0.67	0.67	0.67
W/B	0.30	0.30	0.30
Liq./B	0.31	0.31	0.31

**C = 1.0;*
Supplementary Symbols: B = C + FA (Binding system); FSS = Fine Silica Sand; SL = Slurry Lime; W – Water; HRWR- Superplasticizer; Liq. = W+ HRWR;

For a better relevance of the study, the matrix mix proportions were kept similar for all the composites, the fresh state aspect allowing for this to happen.

4 Experimental tests: results and discussions

4.1 Mechanical properties

The classic mechanical characteristics, the flexural and compressive strengths were determined according to EN 196-1 [20] and EN 1015-11 [21] methodologies, at 28 days and also at 56 days; later age testing proved to be relevant for the high content fly ash composites, due to the fly ash's late pozzolanic reaction. 40 x 40 x 160 mm³ prismatic specimens, cured according to cementitious materials: removal from the mould 24 hours after casting and kept immersed in water, at T (21 ± 2)°C till the testing age, were used for the testing.

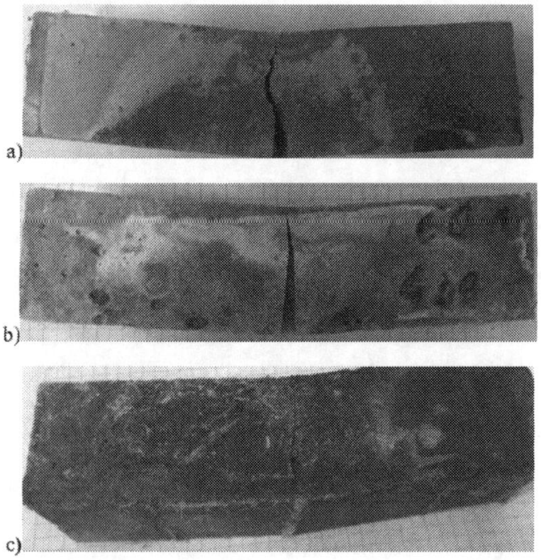

Fig. 5. PVA/PP – FECM, 3PB failure cracking pattern: a) FSS-SL-PVA-FA/C (1.2); b) FSS-SL-PP (2.0)-FA/C (1.2); FSS-PP(2.9)-FA/C (1.2)

Tables 3 and 4 presents the results, in terms of strength and also the strength gain (%), from 28 days to later age of 56 days. The graphical representation of the results is presented in Fig. 6, (Flexural test) and Fig. 7 (Compression test).

Table 3. Mechanical strengths: 3PB Flexural Resistance.

FECM Mix (PVA&PP)	3PB Flexural resistance (MPa)		
	28 d	56 d	Strength gain (%)
FSS-SL-PVA-FA/C (1.2)	23.8	24.1	1.1
FSS-SL-PP(2.0)-FA/C (1.2)	13.6	14.8	8.8
FSS-SL-PP(2.9)-FA/C (1.2)	14.0	15.9	13.2

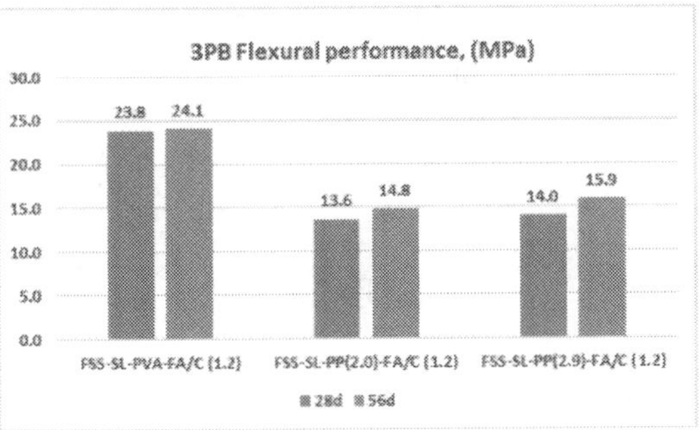

Fig. 6. PVA/PP – FECM, Graphical 3PB Flexural Resistance

The best results were provided, as expected, by the PVA-FECM mix, which developed an impressive flexural strength at 28 days. The value of 23.8 MPa was slightly improved at 56 days - 24.1 MPa. The PP-FECM showed modest results in comparison to the PVA-FECM (designated as control mix for the current evaluation): a flexural strength reduced by 42.8% for the PP_hyb (2.0%) mix, and by 41.1% for the PP_hyb (2.9%) mix. The PP-FECMs also showed a modest flexural strength gain from 28 to 56 days, of 8.8% and respectively 13.2%, but slightly better than the 1.1%, of the PVA mix.

The cracking evaluation (Fig. 5) for the PVA composite showed the visible multiple cracking network in the critical zone. For the PP composites, the failure crack tended to be more singular; still, the bridging action of the fibres prevented the integrity loss and ensured the raised flexural strengths of 14 – 16 MPa, which exceeded the normal expectations of a FRC.

The flexural strength evaluation leads to the main conclusion that the extra addition, of almost 1% (by volume) of PP fibres, is not efficient, not producing a substantial improvement in the overall flexural performance of the material.

Table 4. Mechanical strengths: Compression

FECM Mix (PVA&PP)	Compression resistance (MPa)		
	28 d	56 d	Strength gain (%)
FSS-SL-PVA-FA/C (1.2)	52.5	70.6	34.4
FSS-SL-PP(2.0)-FA/C (1.2)	44.4	75.7	70.8
FSS-SL-PP(2.9)-FA/C (1.2)	47.3	77.3	63.5

The compressive results collected at 28 days showed almost similar performance of all the PVA / PP-FECMs: 15.6% and 10% reduced strengths of the PP mixes with respect to the PVA control mix. The 56 days compressive strength evaluation showed a consistent strength gain for both of the PP mixes, of 70.8% and 63.5% with respect to the 28 days evaluation. Opposing this, the PVA mix recorded a rather modest increase, of 34.4%, proving that generally it reaches the maximum bearing capacity around 28 days. On the other hand, the PP mixes showed a common, later compressive strength gain, with respect to the PVA composite; this behaviour can be attributed only to fibre to matrix physical, mechanical and chemical interaction and bond, clearly distinct from PP to the PVA fibres. Once again, the 2.9% of PP fibres in the mix failed to prove its efficiency, considering the modest strength gain: 6.2% (28 days) and 2.1% (56 days).

The pattern failure prevented the integrity loss due to fibre bridging effect.

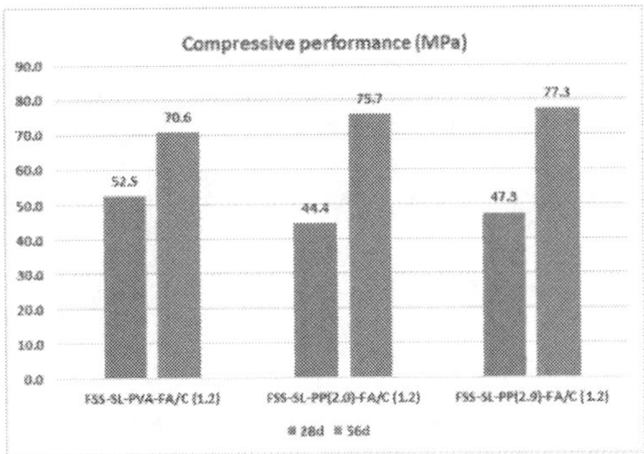

Fig. 7. PVA/PP – FECM, Compressive Strength

The flexural strength evaluation leads to the main conclusion that the extra addition of almost 1% (by volume) of PP fibres is not efficient, not producing a substantial improvement.

4.2 Microcracking pattern (MC)

The composite capacity of developing successive steady state cracks, with narrow opening, is directly connected to the strain hardening, metal like behaviour and therefore to the capacity of deformability, without sudden bearing capacity drop or integrity loss. The multiple cracking (MC) capacity at a certain age is considered to be best evaluated by the flexural 4 point bending test (4PB), performed on coupon specimens (Fig. 8).

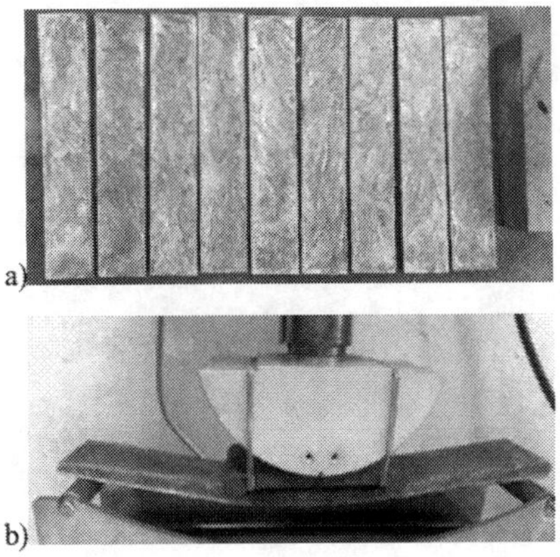

Fig. 8 4PB for PVA – FECM: a) coupon specimens; b) Coupon developing high deformability under loading

The 4PB test is currently performed on 200x40x8 mm³ (Fig. 8) coupon specimens at the age of 28 days, considered to be relevant for the evaluation.

The specimens were identically cast and cured as the 3PB specimens. The testing device used an adapted 4PB displacement control UNIFRAME MINI (Fig. 2, Fig. 8, b). The crack network was analysed considering the general aspect, the crack numbers and their openings (minimum, mean and maximum values); the crack typology (with or without branches) was also considered. The evaluation was performed on samples containing three coupon specimens / mixture, by using the microscopic visual analyses, namely a stereomicroscope Leica S8 APO, equipped with digital camera Leica DMC2900. The LAS Analysis Bundle software was used to process the data acquired via optical microscopic measurements. The analysis results are summarised in Table 5.

Table 5. Microcracking potential (MC): 4PB testing at characteristic ages of 28

FECM Mix (PVA&PP)	MC interval			
	Crack number (n)	Crack opening		
		W_{max} (µm)	W_{med} (µm)	W_{min} (µm)
FSS-SL-PVA-FA/C (1.2)	11-13	48	29	10
FSS-SL-PP(2.0)-FA/C (1.2)	2-3	58	20	34
FSS-SL-PP(2.9)-FA/C (1.2	2-3	58	20	34

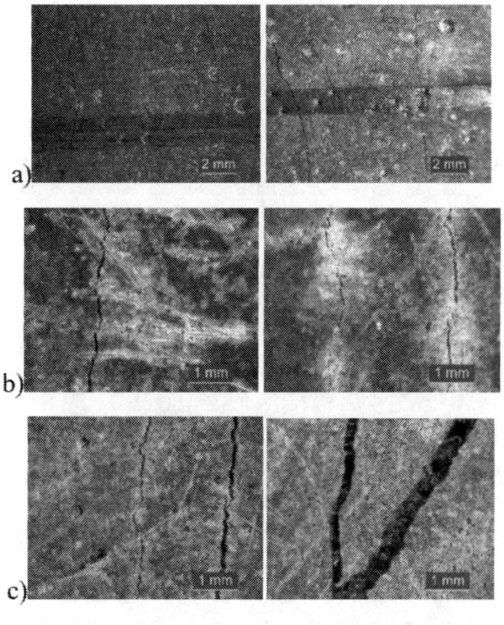

Fig. 8 Microscopic analysis of 4PB cracking pattern: a) FSS-SL-PVA-FA/C (1.2); b) FSS-SL-PP (2.0)-FA/C (1.2); FSS-PP(2.9)-FA/C (1.2)

It was noticed that MC capacity generally degrades over time [2], as the material hardens. Previous studies on PP_hyb-FECM mixes (with local PP fibres) [13, 14], despite of the encouraging early age results (3 and 7 days), showed a significant drop in the 28 days MC capacity (Fig. 2).

The current evaluation was performed directly at the age of 28 days and did not show significant improvement of the PP_hyb-FECMs. The supplementary fibre addition (2.9% instead of the typical 2.0%) did not bring an effective, quantifiable beneficial effect to the composite behaviour. The MC potential was quite scarce, 2-3 cracks in comparison to 11-13 cracks of the lubricated PVA, proving a superior compatibility between PVA and the cementitious matrix, ensuring the strain hardening effect. As expected, brittle failure was avoided for all mixes.

4 Summary and conclusions

The experimental procedures for mechanical evaluation of different types of FECMs developed and optimised with mainly local raw materials offer encouraging results. The overall performance of FECM, with PVA r PP fibres as dispersed reinforcement, 2% (by volume) as optimal fibre content in the mix, proved the viability of the material and its capacity of being tailored in accordance to the necessary specific characteristics, imposed by the potential domain of direct applicability. The study experimentally proved the increased impact of the matrix to fibre compatibility on the complete set of characteristics and material performance. This aspect is essential for design optimising of the composites.

At the same time, the study confirmed that 2% fibres (by volume) represented the optimum fibre content in the ECC typology of materials. An extra addition of fibres involves increased costs, more mixing difficulties etc., and it does not bring an effective superior performance, which might justify it.

As a supplementary point, the study showed increased potential for the fly ash (FA) valorisation, as main addition to cementitious composites, as cement substitute, with positive overall effect in the matrix synergetic processes and, consequently, for the FA waste prevention as main research goals of Incerc Cluj. [16].

Acknowledgements

The current experimental program was carried out in the framework of the PhD. Research Program of the first author at the Technical University of Cluj-Napoca. All the experimental work has been carried out in the laboratory of NIRD. URBAN-INCERC, Cluj-Napoca Branch, Romania and was supported by the Research Programmes for Smart Specialization, Sustainable Territorial Development, Environment Preservation and Resilience of Building Heritage – "CONCRET", PN 18 35 04 03: "Research for valorising the inert, hydraulic late or pozzolanic mineral additions in innovative cementitious materials for resilient structures, in the context of implementing the "Circular Economy" concepts in Romania and PN 19 33 04 02: "Sustainable solutions to ensure health and safety of the population in the open innovation context and environmental protection", financed by the Romanian Government.

References

1. V.C Li, *Engineered Cementitious Composites (ECC) Material, Structural, and Durability Performanc,.* in Nawy E. (Editor), Concrete Construction Engineering Handbook, Chapter 24, CRC Press (2008), pp. 1-78
2. D. Snoeck, *Self-healing and microstructure of cementitious materials with microfibres and superabsorbent polymer*, PhD. Thesis, Ghent University, Ghent, Belgium (2015)
3. L. Vandewalle et al., *RILEM TC 162-TDF: test and design methods for steel fibre reinforced concrete: sigma-epsilon-design method—final recommendation*, Mater. Struct. **36** (2003), pp. 560–567
4. V.C. Li, C.K.Y. Leung, *Steady state and multiple cracking of short random fiber composites*, ASCE Engineering Mechanics **188** (1992) 11, pp. 2246-2264
5. V.C. Li, C. Wu., S. Wang, A. Ogawa, T. Saito, *Interface Tailoring for Strain-Hardening Polyvinyl Alcohol - Engineered Cementitious Composites (PVA-ECC)*, ACI Materials Journal **99** (2002) 5, pp. 463-472
6. V.C. Li, H. Stang, H. Krenchel, *Micromechanics of Crack Bridging in Fibre-Reinforced Concrete*, Materials and Structures **26** (1993) 8, pp. 486-494
7. V.C. Li, S. Wang, C. Wu, *Tensile strain-hardening behavior of polyvinyl alcohol engineered cementitious composites (PVA-ECC,)* ACI Materials Journal **98** (1997) 6, pp. 483-492
8. V. Mechtcherine, F. de Andrade Silva, M. Butler, D. Zhu, B. Mobasher, S.-L. Gao, E. Mäder, *Behaviour of strain-hardening cement-based composites under high strain rates*, Journal of Advanced Concrete Technology **9** (2011) 1, pp. 51-62
9. E.-H. Yang, *Designing added functions in Engineered Cementitious Composites*, University of Michigan, Ann Arbor (2008), pg. 293.
10. E.-H. Yang, Y. Yang, V.C. Li, *Use of High Volumes of Fly Ash to Improve ECC Mechanical Properties and Material Greenness*, ACI Materials Journal **104** (2007) 6, pp. 303-311
11. V.C. Li, E. Herbert, *Robust Self-Healing Concrete for Sustainable Infrastructure*, Journal of Advanced Concrete Technology **10** (2012) 6, pp. 207-218
12. D. Snoeck, N. De Belie, *Mechanical and self-healing properties of cementitious composites reinforced with flax and cottonised flax, and compared with polyvinyl alcohol fibres*, Biosystems Engineering **111** (2012) 4, pp. 325-335
13. C. Baeră, H. Szilagyi, A. Lăzărescu, *Developing Fibre Engineered Cementitious Materials with Self-Healing abilities (SH-FECM) by using polyvinyl alcohol fibres and supplementary powder addition*, Wit Derkowski et al. (Eds.), Proceedings of the fib Symposium 2019 held in Kraków, Poland (2019), pp. 387-294
14. C. Baeră, H. Szilágyi, C. Mircea, A. Hegyi, D. Snoeck, *Fibre engineered cementitious materials (FECM) optimisation by the use of hybrid polypropelene fibres and supplementary powder addition*, Proceedings of the 17th International Multidisciplinary Scientific GeoConference & EXPO – SGEM 2017, Albena, Bulgaria (2017), pp. 169-176
15. A. Lăzărescu, C. Baeră, H. Szilagyi, *A Review on the Development of Fly Ash-Based Geopolymer Materials Using Romanian Local Raw Materials*, Proceedings of the EUROCOALASH 2019, Dundee, Scotland, UK (2019), pp. 145-156
16. H. Szilagyi, C. Baeră, A. Hegyi, A. Lăzărescu, *Romanian resources of waste and industrial by-products as additions for cementitious mixtures*, Proceedings of the 18th International Multidisciplinary Scientific GeoConference SGEM 2018, 30 June – 9 July, Albena, Bulgaria (2018), pp. 325-332
17. C. Baeră, *Self-Healing capacity of Advanced Engineered Cementitious Composites*, PhD. Thesis, Technical University of Cluj-Napoca, Cluj-Napoca, Romania (2016)
18. ASRO, *SR EN 14889-2:2007 – Fibres for concrete - Part 2: Polymer fibres - Definitions, specifications and conformity*, Romanian Standards Association, Bucharest, Romania (2007)
19. ASTM, *C-1116 – Standard Specification for Fiber-Reinforced Concrete and Shotcrete*, American Society for Testing and Materials, USA (2003)
20. ASRO, *SR EN 196-1:2016 Methods of testing cement. Determination of strength*, Romanian Standards Association, Bucharest, Romania (2016)
21. ASRO, *SR EN 1015-11:2002/A1:2007 - Methods of test for mortar for masonry. Determination of flexural and compressive strength of hardened mortar*, Romanian Standards Association, Bucharest, Romania (2007)

Influence of the surface modification by sanding of carbon textile reinforcements on the bond and load-bearing behavior of textile reinforced concrete

Cynthia Morales Cruz[1,*] and *Michael* Raupach[1]

[1]Institute of Building Materials Research, RWTH Aachen University, Schinkelstr. 3, Aachen 52062, Germany

Abstract. In the context of the application of carbon Textile Reinforced Concrete (TRC) layers for the durable repair of building surfaces, uniaxial tensile tests on rectangular TRC samples were carried out to compare the bond and load-bearing behavior of an epoxy-impregnated carbon textile and its surface modified version. The aim of the surface modification, consisting of a subsequent coating with epoxy resin and sanding with quartz sand, is the improvement of the composite material regarding crack width reduction and an increase of the load-bearing capacity. A total of 15 series were examined and the parameters: reinforcement type, orientation and ratio were varied. In addition, long-term load tests were conducted. An optical 3D-video measuring system in combination with a DIC-software was used, which allowed the analysis of the process of crack formation during the entire testing time. With the surface modified reinforcement the formation of approx. 1.5 times the number of cracks with averagely 33 % smaller crack widths and up to 50 % smaller crack spacings were observed, regardless of the ratio of reinforcement. The residual behaviour of the series subjected to a permanent load of 1500 MPa over 1000 h showed no reduction of the tensile stress compared to short-term tests.

1 Introduction

Due to its specific properties Textile Reinforced Concrete (TRC) can be used for purposes ranging from the strengthening of existing structures to new load bearing constructions. Because of its high tensile strength and the non-corrosiveness of textile reinforcements (for example carbon fibers), TRC can be much thinner than steel-reinforced concrete. Additionally, TRC has a higher crack distribution ability than steel reinforced concrete. When it cracks, the crack distribution is finer; therefore the cracks are thinner, making TRC especially suitable for the repair of cracked building structures.

At the Institute of Building Materials Research (ibac) of the RWTH Aachen University, "DURTEX" was developed as an alternative to the usual repair methods for concrete and steel reinforced concrete structures. DURTEX consists of a textile reinforced mortar/concrete layer and a delamination material (see figure 1) [1, 2]. The concept enables the permanent protection of cracked building surfaces with rear water pressure and cyclically moving cracks with a crack width opening between Δw_{op} 0.2 and 0.6 mm [2]. The de-bonding material (see Fig. 1, ①) impedes the bond between substrate and mortar/concrete (see Fig. 1, ②) in the area of the crack and thus leads to an increase in the free expansion length of the TRC. The textile reinforcement (see Fig. 1, ③) embedded in the mortar/concrete allows the realization of a thin reinforced layer and ensures that the crack width opening Δw_{op} of the substrate crack can be bridged and is distributed over several cracks with single crack widths $w_i < 0.1$ mm in the TRC (see Fig. 1.)

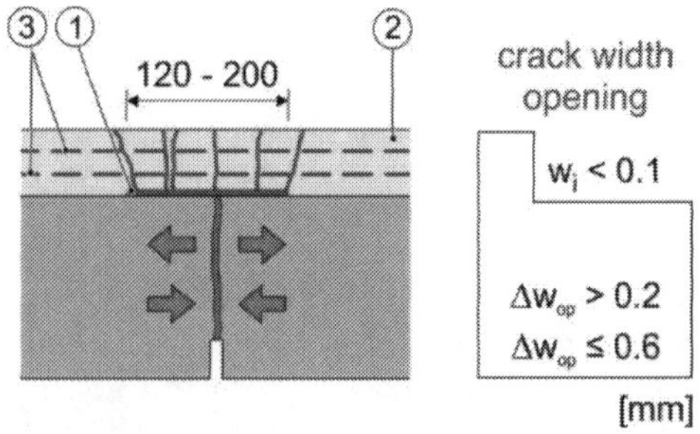

Fig. 1. Construction and function of carbon TRC repair system based on [1, 2]

While in [3] the overall performance of the system is investigated, this present work deals with the influence of surface-modified textile reinforcements on the load and bonding behaviour of TRC. In order to keep the test setup simple, but still allow the transfer to the repair system, which so far has been tested in complex crack bridging tests (see e.g. [4]), uniaxial, deformation-controlled tensile tests were carried out on rectangular TRC sample bodies. In previous studies [5], an improved composite and load-bearing behaviour of textile concrete with textile reinforcements modified under laboratory conditions was demonstrated. In this current work,

* Corresponding author: morales@ibac.rwth-aachen.de

however, only commercially available materials were investigated. In the context of these studies, the influence of the reinforcement type as well as the reinforcement ratio and test direction of the textile reinforcement in combination with a commercially available repair mortar were investigated. Furthermore, endurance tests and overlap attempts were carried out.

2 Experimental

2.1 Materials

In this study, the carbon textile reinforcement solidian GRID Q85/85-CCE-21 was investigated in two variations (reference and sanded). The reference reinforcement (hereafter referred to as: Q85a) was impregnated with an epoxy resin (see Fig. 2, top). The sanded version (see Fig. 2, bottom) was subsequently coated with an epoxy resin and sprinkled with a quartz sand with a grain size of 0.3 - 0.8 mm by the manufacturer solidian GmbH (hereafter referred to as: Q85b).

Fig. 2. Carbon textile reinforcement Q85a (top) and Q85b (bottom), close-up of the intersection point

The properties of the reinforcement according to the manufacturer are summarised in Table 1 [6]. The abbreviation "CCE" stands for "Carbon-Carbon-Epoxy" and describes the production materials used, in this case carbon rovings were used in both warp and weft direction. The rovings were impregnated with a hot-curing epoxy resin to improve the bond properties between the rovings and the mortar/concrete matrix. The rovings of the textile reinforcement had an axial distance of 21 mm and a fibre cross-section area of 85 mm²/m or 1.81 mm² per roving in both warp and weft direction. The fineness of the rovings was 3200 tex and the characteristic tensile stress was 2500 MPa. In the warp direction the roving had an average tensile stress of 3300 MPa and in weft direction 3550 MPa. The modulus of elasticity was greater than 220 GPa for the warp direction and at least 205 GPa for the weft direction.

Table 1. Properties of the textile reinforcements Q85a according to the Manufacturer´s data sheet [6]

property	unit	value	
		warp	weft
distance between rovings	mm	21	
roving cross-section	mm²	1,81	
cross-section (reinforcement)	mm²/m	85	
average tensile strength (based on roving tests)	MPa	3300	3550
modulus of elasticity of the reinforcement	GPa	> 220	> 205

For the production of the rectangular TRC samples the polymer modified, cement-bonded repair mortar (RM) StoCrete TG 203 from StoCretec GmbH with a maximum grain size of 2 mm, which is usually used as a concrete substitute in the repair of concrete and reinforced concrete components, was used. The manufacturer's information on properties is given in Table 2. The compressive strength after 28 days is 60 ± 5 MPa, the modulus of elasticity is 25 MPa. [7]

Table 2. Properties of the mortar StoCrete TG 203 according to the Manufacturer´s data sheet [7]

property	unit	value
fresh mortar density	kg/dm³	2.2
tensile bond strength (28 days)	MPa	> 2.0
compressive strength (28 days)		> 55 - 65
modulus of elasticity, static (28 days)	GPa	25

2.2 Characterisation methods – textile reinforcement

To quantify the surface modification, the parameters roving cross-section area, roving circumference and tensile strength were determined.

2.2.1 Roving cross-section area and circumference

Using light microscopy, the cross-section of the warp and weft rovings (3 samples each) were examined on polished sections (see Fig. 3). The mean cross-section

https://doi.org/10.1051/matecconf/201928904006

area of the warp Q85a rovings was approx. 4.0 mm^2 with an average circumference of approx. 10.1 mm. While the average cross-section area of the weft Q85a rovings was approx. 4.8 mm^2 with a circumference of approx. 11.9 mm. Due to the wrapping with the knitting thread, the warp roving is normally more compact than the weft thread and thus has a smaller circumference and a 15 % smaller bonding surface to the mortar matrix. When comparing the calculated cross-section areas with the manufacturer's specifications, it becomes clear that the epoxy resin matrix accounts for up to 62 % of the composite roving cross-section area. The images show cavities in the interior of the Q85a rovings that were formed during impregnation. These can lead to an uneven activation of the filaments during tensile loading.

As a result of the surface modification, the warp roving circumference was increased from 9.5 mm by about 20 %, so that a larger bonding area was available. A similar increase of the circumference, namely from 10.4 to 12.9 mm, was measured for the weft roving.

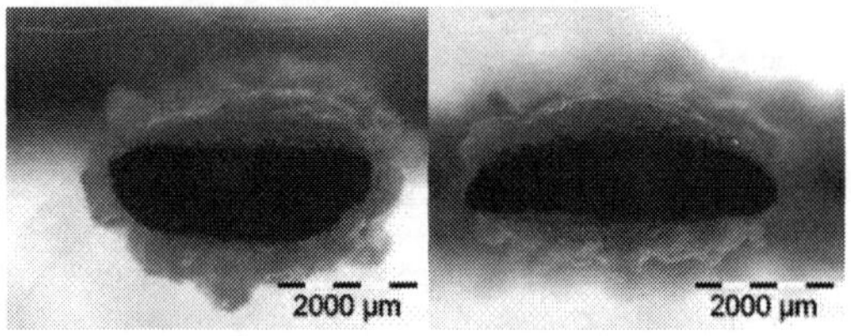

Fig. 3. Cross-section image of warp (left) and weft (right) rovings of the textile reinforcement Q85b

2.2.2 Roving tensile-strength test

Roving tensile tests on the Q85b reinforcement were carried out to verify that the tensile strength of the Q85b reinforcement was not reduced or damaged during the surface modification. Furthermore, these values are necessary to make a statement on the ratio of efficiency of the reinforcement in TRC. For this purpose 10 individual tests were carried out. Therefore, 350 mm long rovings in warp direction were taken from the reinforcement Q85b and glued into steel hulls. Steel hulls with a length of 75 mm and an outer diameter of 25 mm were used to transfer the load from the testing machine/bearing to the samples without transverse pressure (see Fig. 4, left).

The preload was set to 150 N, which corresponds to approx. 2 % of the failure load. The tests were performed displacement controlled with a rate of 2 mm/min until failure. One DSA-sensor, with a measuring length of 50 mm, was fixed directly onto the roving's surface to measure the deformation but removed by approx. 50 % of the failure load to prevent damaging the sensor.

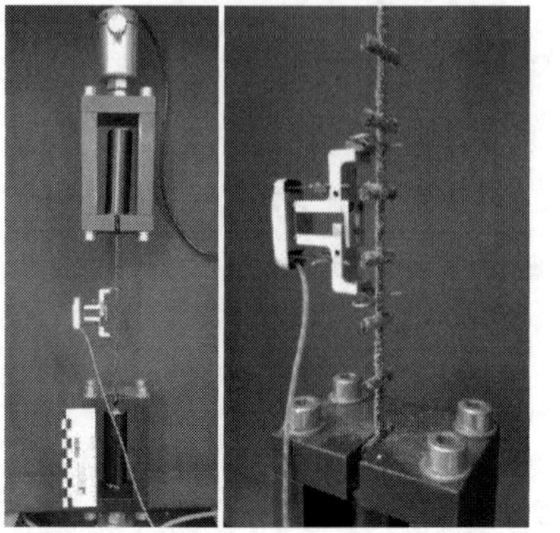

Fig. 4. Roving tensile strength setup

The modulus of elasticity was determined within the linear range (between 10 % and 30 % of the tensile strength) of the calculated stress-strain curves. The mean roving tensile strength of the Q85b reinforcement in the warp direction was determined to be 3530 ± 200 MPa and the mean modulus of elasticity to be 221400 ± 22700 GPa. These values show good agreement with those given by the manufacturer.

2.3 Test variations

For the main part of this work, rectangular TRC samples 1000 mm long, 100 mm wide and 20 or 30 mm thick reinforced with either 1 or 2 layers were produced at ibac and tested (see Table 3). The parameters reinforcement type, ratio of longitudinal reinforcement, number of layers and test direction were varied. Furthermore, samples were prepared to determine the residual load bearing capacity after a permanent load of more than 1000 hours.

The series designation are based on the following key:

T – Q85x – A – L – D – N

with:

- T type of test
  - DK: short-term load
  - DS: permanent load
- x reinforcement variation
  - a: without sand (reference)
  - b: sanded
- A number of longitudinal rovings: 4 or 5
- L number of layers: 1 or 2
- D test direction: 0° (corresponds to the warp rovings) or 90° (corresponds to the weft rovings)
- N sample number

 https://doi.org/10.1051/matecconf/201928904006

Table 3. Testing matrix

Test type	Reinforcement type	Testing direction	No. of layers	No. of rovings	No. of samples
DK	Q85a	0 °/ 90 °	1	4 / 5	6
			2	5	3
	Q85b		1	4 / 5	
			2	5	
		0 °	2	4	
DS			2	4 / 5	

2.4 Manufacturing process

Before concreting, the textile reinforcement was cut to a length of approx. 1200 mm and a width corresponding to the selected number of rovings using an angle grinder.

The mortar was mixed according to the manufacturer's specifications (0.13 parts by weight of water were added to the dry mortar mixture). Due to the consistency of the mortar, the samples were produced exclusively by lamination technique explained in [8]. This involves alternately applying a layer of mortar, a layer of textile and then another layer of mortar – this procedure is repeated until the respective number of layers is reached. The formwork construction used allows the textile reinforcement to be clamped so that floating can be avoided and the reinforcement is centered in the test sample. All samples were demoulded after approx. 24 h and then stored in the climate-controlled room at 22 °C and 53 % RH for at least 28 days until testing.

2.5 Test procedure

2.5.1 Tensile strength test – rectangular TRC samples

The samples were clamped with steel plates in the upper and lower load introduction area over a length of 250 mm each. A contact pressure of 100 or 150 Nm was applied on the 20 or 30 mm thick samples via 4 threaded rods M16 each with hexagonal nuts (see Fig. 5). A tightening torque of e.g. 100 Nm and a clamping surface of 250 x 100 mm² resulted in a surface pressure of approx. 3.8 MPa on the sample in the clamping area, which was significantly below the compressive strength of the mortar. The free test length was 500 mm long and the measuring range was 400 mm.

To increase friction and thus prevent the samples from slipping out of the load introduction area and to avoid stress peaks due to small unevenness of the sample surface caused by the manufacturing process, 3 mm thick rubber mats were clamped between the sample and the steel plates (see Fig. 5, right).

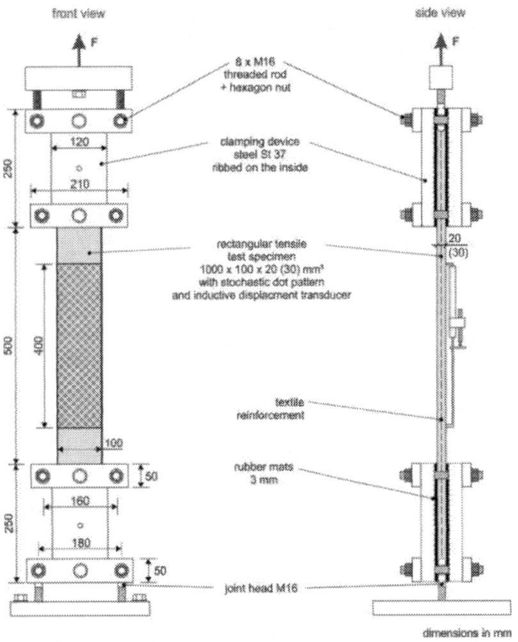

Fig. 5. Schematic view of the test setup for rectangular TRC samples [3]

After the sample was clamped in the testing machine, a linear variable displacement transducer (LVDT) was attached on the back of the sample between two longitudinal rovings.

The tensile strength tests were performed in a universal testing machine Z150 TL from Zwick/Roell, precision class 1 according to DIN EN ISO 7500-1, with a maximum load of 150 kN. The tests were carried out displacement-controlled with a rate of 2 mm/min until sample failure.

The test load, the piston displacement and the deformation of the displacement transducer were measured with a measuring rate of 10 Hz via the testing machine, as were the deformations of the sample in the measuring range using the GOM Correlate software with a frequency of 0.2 Hz.

2.5.2 Tensile strength test – deformation measurements

Additionally to the LVDT, an optical deformation measurement system was used in order to be able to analyse the individual crack widths and the crack development over the entire test time.For this purpose, the ARAMIS System was used. ARAMIS is a contact-free, optical 3D deformation measuring system that evaluates 3D coordinates under static or dynamic load [9]. In addition the digital image correlation software, GOM Correlate, was used for the deformation analysis.

A stochastic pattern of 400 x 100 (L x B) mm² was applied to the sample's cast side using a graphite spray prior to the test. This stochastic pattern is used for photogrammetric recording and evaluation of the length changes and crack movements of the sample. Before starting the test, the system was calibrated with a corresponding measuring volume of 400 x 330 x depth of field in mm³ (see Table 4). The calibration parameters are listed in Table 4.

Table 4. Parameter for the ARAMIS measurement

Parameter	unit	value
calibrated 3D space (measuring volume)	mm³	400 x 330 x 400
facet geometry	pixels	19 x 19
camera resolution		2448 x 2050
aperture of the lens	-	5,6
focal length of the lens	mm	23
max. calibration deviation	pixels	0.050
max. scale deviation		0.031

The first image captures the undeformed state of the sample (reference image). During the tensile test further images in different load stages of the sample, so-called deformation stages, were recorded. ARAMIS compares these images with each other and calculates the deformation over defined object characteristics in the stochastic pattern by means of rectangular image sections (facets) [9]. The selected facet had a size of 19 x 19 pixels. This corresponds approximately to 3 x 3 mm² with the measuring volume used. An overlap of 3 pixels resulting in a distance of 16 pixels between two facet centers was selected. The measuring inaccuracy parallel to the testing direction is 8 µm for the chosen measuring volume, facet size and camera resolution.

At the end of the test, the evaluation was carried out manually using the images in the individual deformation stages based on [10]. Firstly, the main deformation (parallel to the test direction) was displayed in the computational mask (see Fig. 6, green surface) and scaled to 1.5 % to make the cracks visible. Afterwards, facet points were set at the crack banks and their distance was calculated by the software in the next work step. Subsequently, the crack spacings and the individual crack widths for each crack and all stages were evaluated at three different points (see Fig. 6, top left). Figure 6 shows the evaluation of the measured crack widths at three points of one crack at the final deformation stage before sample failure.

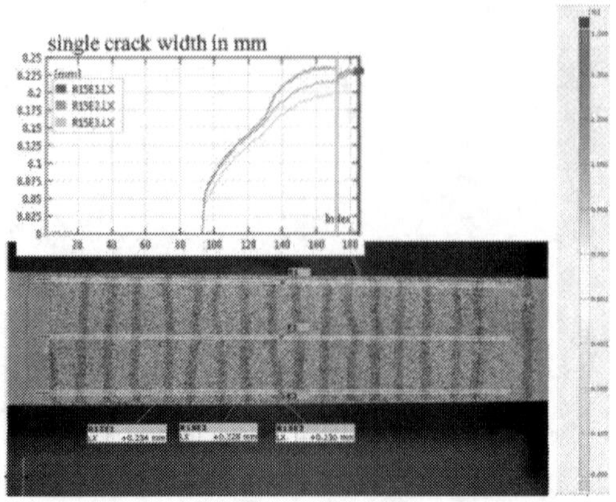

Fig. 6. Crack evaluation method with the GOM Correlate software; sample DK-Q85b-5-2-0-1

2.3.3 Permanent load

At the age of 28 days, two series were stored for 42 days (1000 h) at ca. 23 °C and 55 % RH and were subjected to a stress of 1500 MPa (dead load) which e.g. corresponds to 43 % of the short-term strength of the reference series DK-Q85b-5-2-0 and 46 % of the short-term strength of the reference series DK-Q85b-4-2-0. Crack widths and crack spacings were determined shortly before the end of permanent load using a crack magnifier and a crack width scale. The values determined, in particular the crack widths, are numerically lower than the evaluation with the ARAMIS software. However, the reading accuracy of the crack magnifier was 0.05 mm. After long-term loading, the residual load of the sample was tested. The crack measurement with the ARAMIS system was omitted since the characteristic crack distribution was developed with the dead loads and no further crack distribution was expected after the crack formation in the permanent load.

3 Results

The aim of the experiments was to quantify the influence of surface modification of the carbon textile reinforcements Q85a and Q85b on the bond and load behavior of TRC, in particular the parameters crack spacing, crack width, number of cracks and tensile strength at fracture.

3.1 Evaluation method

At least 3 test samples per series were tested (s. Table 4). All stresses in the stress strain curves or stress deformation curves are related to the textile cross-section area. The tensile strength σ_t was determined as follows:

$$\sigma_t = \frac{F}{A_t} = \frac{F}{n \cdot A_r} \quad (1)$$

F machine force in kN
n number of rovings in testing direction
A_t textile cross-section area in mm²
A_r roving cross-section area in mm²

The strain of the TRC was calculated from the recorded change in length of the LVDT in relation to the measuring length of 400 mm. The deformations in the stress-deformation diagrams correspond to the piston displacement, since the displacement transducers for some samples were removed at about 80 % of the failure load and therefore not enough measurement data were available to show the strains to fracture (see Fig. 8, top). In some series, e.g. Fig. 8, top, the displacement transducers were not removed, so that complete stress-strain curves are displayed.

In the framework of this study, the stress strain curves (see e.g. Fig. 7) were divided into different sections based on the OH-Model [11] and approximated using a linear function (see Fig. 7, black line). Point B was determined which corresponds to the initial cracking stress. The stress and strain range of crack formation in state IIa (between point B and C) was also determined. Additionally, the elongation at sample failure (point D) was evaluated with ARAMIS. At this point the number of cracks, crack width and crack spacing was evaluated (see Table 4).

The theoretical initial cracking stress in Table 4 was determined on the basis of the stress strain curves, related to the simplified mortar cross-section area. This was 2000 mm² for single layer reinforced samples and 3000 mm² for double layer reinforced samples. It is apparent from table 4 that the Q85a reinforcement, with the exception of the series DK-Q85-5-1-0, shows a much higher first crack stress than the Q85b in the TRC. However, if the textile cross-section measured in section 2.2.1 is taken into account the results are more or less consistent with the increase of roving circumference.

In state IIa, as in the OH-model, a slight increase in tensile stress during crack formation can be observed. The gradient factor for the Q85a reinforcement was between 1.0 and 1.17, whereas for the sanded reinforcement it was between 1.15 and 2.13.

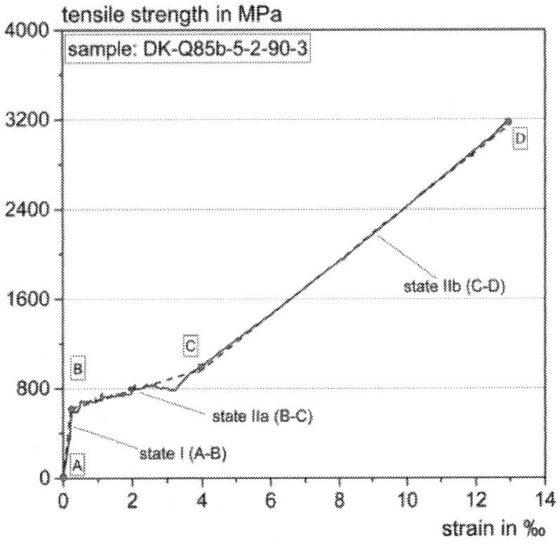

Fig. 7. Stress strain curve of the DK-Q85b-5-2-90-3 sample (blue line) and the linearized stress-strain diagram (black dotted line) according to the OH-Model [11]

3.2 Summary of the results

All samples failed by exceeding the tensile strength of the reinforcement (rupture), with the exception of the overlap tests and the inclined reinforcement tests, where failure could be observed by delamination and/or complete textile pull-out. Longitudinal cracking was observed (i.e. for 2 layers reinforcement) after completed crack formation (in state IIb). Exemplary stress strain curves of the TRC samples with the Q85a and Q85b reinforcement are shown in Fig. 8.

Table 4. Evaluation results of the tensile strength tests of TRC samples with the textile reinforcement Q85a and Q85b

series	initial crack stress [MPa]	tensile strength [MPa]	no. of cracks	crack width [mm]	crack spacing [mm]
DK-Q85a-4-1-0	4.98 ± 0.25	3004 ± 252	10 ± 1	0.473 ± 0.013	35.5 ± 2.5
DK-Q85a-4-1-90	4.66 ± 0.31	3451 ± 388	10 ± 1	0.484 ± 0.070	41.2 ± 3.4
DK-Q85a-5-1-0	4.69 ± 0.40	3275 ± 154	13 ± 2	0.384 ± 0.085	33.6 ± 1.3
DK-Q85a-5-1-90	4.83 ± 0.28	2509 ± 171	13 ± 2	0.331 ± 0.080	32.0 ± 6.1
DK-Q85a-5-2-0	4.33 ± 0.13	2952 ± 115	12 ± 2	0.519 ± 0.061	34.8 ± 3.7
DK-Q85a-5-2-90	4.10 ± 0.56	3421 ± 113	13 ± 1	0.459 ± 0.011	31.7 ± 0.5
DK-Q85b-4-1-0	3.33 ± 0.37	3272 ± 176	19 ± 1	0.282 ± 0.042	20.6 ± 1.3
DK-Q85b-4-1-90	3.38 ± 0.29	3683 ± 102	19 ± 3	0.296 ± 0.030	20.5 ± 3.0
DK-Q85b-4-2-0	2.73 ± 0.17	3249 ± 232	17 ± 2	0.209 ± 0.059	24.1 ± 3.8
DK-Q85b-5-1-0	4.34 ± 0.27	3192 ± 122	18 ± 0	0.304 ± 0.012	22.8 ± 0.7
DK-Q85b-5-1-90	3.31 ±1.30	3596 ± 277	20 ± 1	0.222[1)] ± 0.083	20.5[1)] ± 1.4
DK-Q85b-5-2-0	3.43 ± 0.17	3506 ± 190	17 ± 3	0.376 ± 0.029	24.4 ± 3.5
DK-Q85b-5-2-90	3.68 ± 0.93	3257 ± 59	19 ± 1	0.283 ± 0.010	20.9 ± 1.2
DS-Q85b-4-2-0	-	3505 ± 179	20 ± 1	0.106[2)] ± 0.014	20.8 ± 1.0
DS-Q85b-5-2-0	-	3684 ± 52	19 ± 1	0.136[2)] ± 0.015	21.6 ± 0.3

[1)] crack width and crack spacing not evaluated immediately before fracture; [2)] measured at a 1500 MPa

The results of the tensile strength tests of TRC samples including the crack analysis are found in Table 4.

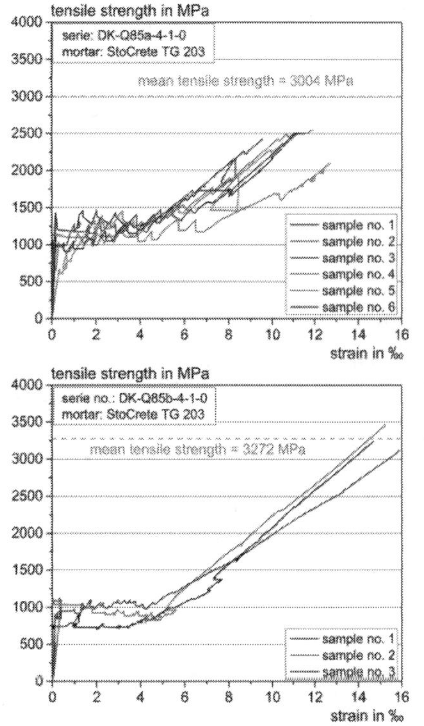

Fig. 8. Stress strain curves of the DK-Q85a-4-1-0 (top) and DK-Q85b-4-1-0 (bottom) - samples

4 Discussion

4.1 Influence of the testing direction

For the Q85a reinforcement no clear tendency regarding the testing direction could be determined even though a higher tensile strength as well as a better bond behavior of the weft rovings can be expected due to the larger surface area as well as the higher roving tensile strength. The results are partly contradictory. As expected, the samples with four and ten rovings in the weft direction show higher fracture stresses. With a higher reinforcement ratio (five rovings) the exact opposite was observed. Possible reasons here are scattering of the Q85a reinforcement.

In the case of Q85b reinforcement, the expected dependence of the tensile strength due to the testing direction was confirmed for samples with one reinforcement layer. The samples with reinforcement in weft direction presented higher tensile stresses than the samples with reinforcement in warp direction (see Table 4). The difference arises during the manufacturing process where the warp rovings are often slightly damaged and therefore have a lower load carrying capacity.

4.2 Influence of the reinforcement type

In the case of the Q85a reinforcement, it could be shown that the number of cracks increases with the ratio of reinforcement, while the crack spacings and the associated crack widths become smaller. This coincides with the observations of [12, 13]. The mean crack spacing for all Q85a samples is between 32 and 40 mm. This corresponds approximately to 1.5 to 2 times the mesh size of the textile reinforcement.

It is apparent from table 4, that the Q85b reinforcement leads to a higher tensile strength. Additionally, there is less scatter in the values of tensile strength, number of cracks, crack width and crack spacing. Since the measured textile tensile strength of the Q85b reinforcement is the same as that of the rovings, the authors assumed that the load-bearing capacity of the reinforcement was thus fully utilised.

The stress strain curves with the Q85b reinforcement show significantly smaller stress drops ($\Delta\sigma_{t,m} < 100$ MPA), so called reinforcement deficits, than with the Q85a reinforcement ($\Delta\sigma_t \sim 200$ MPA) during the crack formation stage, which can be attributed to the improved bond behavior (see Fig. 8). The improved bond compared to the Q85a reinforcement also results in a higher number of cracks with smaller crack spacing and correspondingly smaller crack widths. The mean crack spacing of approx. 21 mm corresponds to the mesh size of the textile reinforcement (see Table 1). This means that the maximum possible number of cracks was reached.

A comparison of the results of the crack width evaluation of samples with one reinforcement layer is presented in Fig. 9. The samples with the Q85b reinforcement show on average 50 % more cracks and up to 50 % smaller crack spacing than with the Q85a reinforcement. Additionally, the crack widths of the TRC samples with the Q85b reinforcement are on average 33 % smaller than with the Q85a reinforcement. For both reinforcement types, longitudinal cracks were observed with increasing ratio of reinforcement, i.e. two reinforcement layers and five rovings. These occurred in the sample free length after completed crack formation. Since no spalling occurred and the longitudinal cracks did not lead to failure, the authors assume that the chosen mortar covering was sufficient.

There is a clear correlation between the number of cracks and the strain in state IIa especially with the Q85b reinforcement (see Fig. 10). With fewer cracks, the respective crack widths or strains get larger. Also with increasing number of cracks the local strains become smaller. Thus the state IIa with the Q85b reinforcement falls generally smaller.

Fig. 11 shows the correlation between the elongation at fracture and the mean crack spacing of the TRC samples determined with ARAMIS. The higher the mean crack spacing, the less textile elongation in the cracks was recorded.

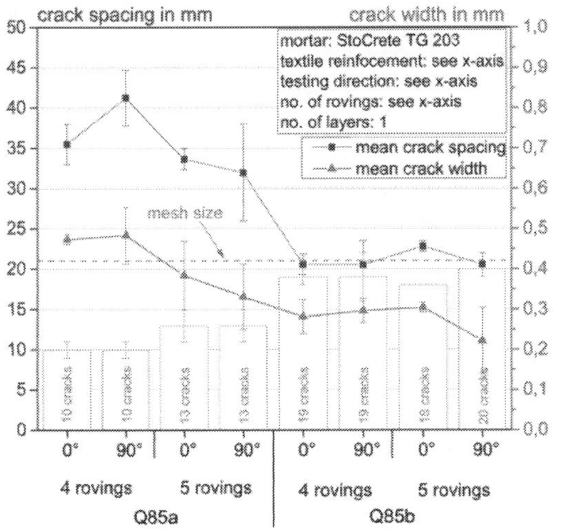

Fig. 9. Analysis of the no. of cracks, crack spacing and crack width evaluation; one reinforcement layer

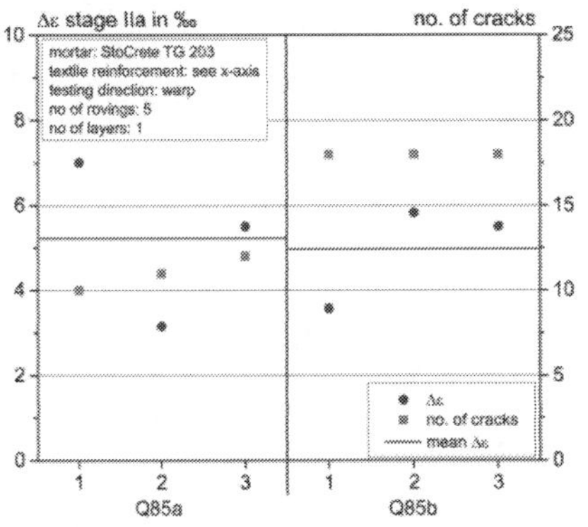

Fig. 10. Correlation between number of cracks and strain range in state IIa (crack formation)

Due to the poorer bond of the Q85a reinforcement and thus the uneven strain distribution, the textile elongations in the cracks were locally exceeded and premature failure of individual rovings occurred. This resulted in lower elongation at fracture. The authors are of the opinion that the improved bond behaviour of the Q85b reinforcement to the mortar resulted in a better load distribution within the reinforcement cross-section. This led to a higher tensile strength of the Q85b reinforcement.

4.2.1 Visual inspections

In addition to the mechanical experiments on TRC, visual inspections of the reinforcement surface after the tensile strength tests were performed and evaluated. In Fig. 12, top, the bond between the mortar matrix and the Q85a reinforcement after sample failure is recorded. It is visible that the mortar does not adhere well to the smooth reinforcement surface. Only residues of cement are visible. The mortar adhesion was practically non existent, so that the reinforcement was easily detached.

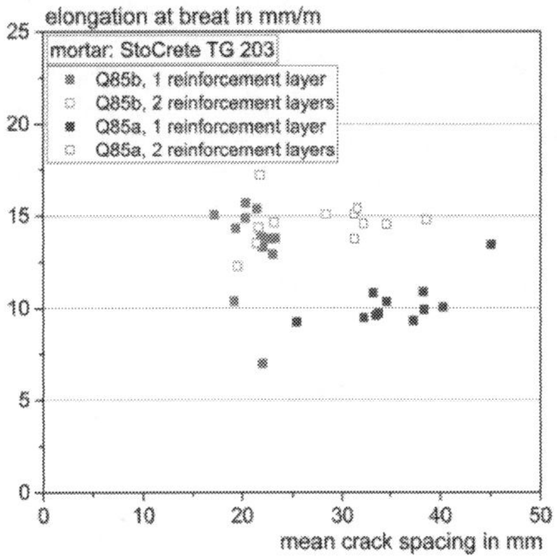

Fig. 11. Dependence of elongation at fracture on mean crack spacing

Fig. 12, bottom, shows that the surface modification with sand has a good bond to the mortar matrix since mortar residues remain on the roving after sample failure. It was partially found that the additional coating does not bond well with the Q85a reinforcement since it was detached during the test. In comparison to the Q85a reinforcement, the Q85b reinforcement has a significantly better bond to the mortar and is difficult to detach from the fractured TRC samples.

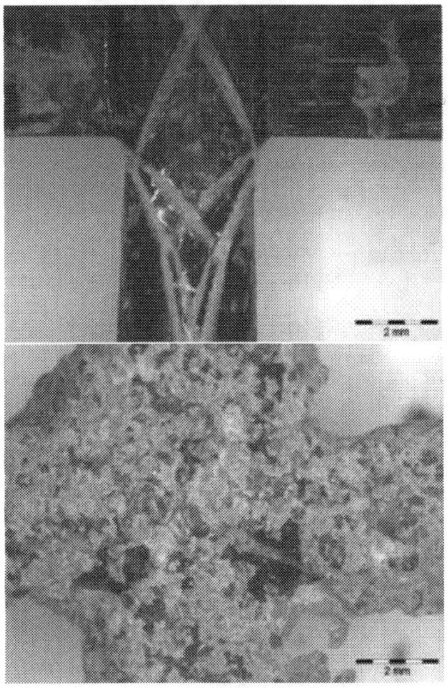

Fig. 12. Carbon textile reinforcement Q85a (top) and Q85b (bottom) after tensile test

4.3 Influence of a permanent load

Fig. 13 shows the stress-strain curves of the DS-Q85b-5-2-0 series of a TRC sample with two layers reinforced after 42 days (1000 h) at 43 % of the short-term strength. The reference series DK-Q85b-5-2-0 had an average tensile strength of 3506 MPa. The reference series DK-Q85b-4-2-0 had an average tensile strength of 3249 MPa. Compared to the reference series, the samples reinforced with four or five rovings showed an 8 % or 5 % increase in tensile strength. A possible explanation could be the even load distribution on the filaments under continuous stress. This coincides with the observations of [14]. Since the Q85a reinforcement was not subjected to permanent loading, no statements on the influence of the modification can be made. Since the increase in strength is very small, it can be concluded that the carbon filaments were generally well bonded to each other and the load was evenly transferred.

Since the characteristic crack distribution was developed during application of dead loads, the stress displacement curves in Fig. 13 are approximately linear. The crack spacings and crack widths in Table 4 were measured shortly before the end of the permanent load with a measuring magnifier.

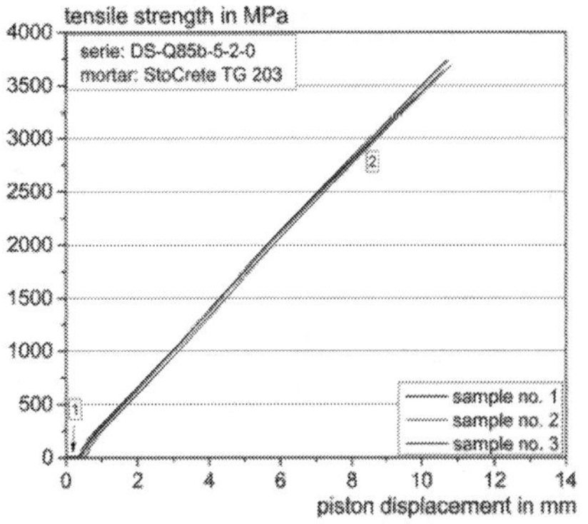

Fig. 13. DS-Q85b-5-2-0 stress piston displacement curves of the residual tensile strength

5 Conclusions and outlook

The aim of this experimental study was to determine the influence of the surface modification by sanding of textile carbon reinforcement with (Q85b) and without surface modification (Q85a). The following conclusions can be drawn from the experimental study:

- The results of the uniaxial tensile strength test on rectangular samples show that the surface modified Q85b textiles results in up to 1.5 times the number of cracks with up to 50 % smaller crack spacing and on average 33 % smaller crack widths then the reference material Q85a.
- In addition, lower initial cracking stresses (based on simplified concrete area), smaller reinforcement deficits and higher fracture stresses were observed with the surface modified textile Q85b.
- With regard to the test direction, higher tensile strength, smaller crack widths and crack spacing tend to be achieved with the weft Q85b reinforcement. This effect is not pronounced for the Q85a reinforcement.
- Light microscope pictures do not indicate any damage of the surface modified Q85b textiles at fracture.
- Higher tensile strengths were measured with the Q85b reinforcement. This is attributed to the improved composite bond behavior and thus better load distribution.
- As a result of a permanent load, no reduction of the tensile stress at fracture was determined. Instead, a small increase in the tensile strength of the reinforcement of 5 or 8 % compared to the short-term strength due to an even load distribution on the filaments was observed.

Undergoing investigations include temperature tests with rectangular test samples with both as short-term and permanent loads. In addition, cyclic tests on cracked textile concrete samples are planned to further investigate the durability of the bond between textile and mortar.

References

1. Büttner, T.; Raupach, M.: Büttner, T.: Raupach, M.: Des Bauwerks neue Kleider. Funktionsprinzipien und Einsatzmöglichkeiten von Textilbetonschichten zum Schutz von Bauwerken. In: Bauen in Bestand 35, No. 6, p. 70-75, 2012.
2. BAW-MITEX (2019): Merkblatt Flächige Instandsetzung von Wasserbauwerken mit textilbewehrten Mörtel- und Betonschichten (MITEX) – yellow print -- Bundesanstalt für Wasserbau.
3. Morales Cruz, Cynthia: Crack-Bridging Carbon Textile Reinforced Concrete Protection Layers. PhD thesis, Civil Engineering Faculty, Rheinisch-Westfälischen Technische Hochschule RWTH Aachen University, (to be published)
4. Morales Cruz, C.; Raupach, M.: Textile Reinforced Concrete Layers For a Durable Protection of Building Surfaces: Structural Faults & Repair, 15th International Conference and Exhibition, London. 8-10 Juli 2014
5. Morales Cruz, C.: Improving the Bond Behavior of Textile Reinforcement and Mortar Through Surface Modification. In: Brameshuber, W. (Hg.): 11th International Symposium on Ferrocement and Textile Reinforced Concrete 3rd ICTRC (Ferro-11). RILEM publications. 2015, p. 215–223
6. https://www.solidian.com/en/products/flat-reinforcement/, last access on 14.07.2019

7. https://www.stocretec.de/de/produkte/produktprogramm/produktdetail_131011432.html, last access on 17.07.2019
8. Brameshuber, W.: Recommendation of RILEM TC 232-TDT: Test Methods and Design of Textile Reinforced Concrete: Uniaxial Tensile Test: Test Method to Determine the Load Bearing Behavior of Tensile Samples Made of Textile Reinforced Concrete. In: Materials and Structures Vol. 49 (2016), No. 12, p. 4923–4927
9. https://www.gom.com/metrology-systems/aramis.html, last access on 10.07.2019
10. Morales Cruz, C. ; Raupach, M.: Investigations on controlled crack distribution of textile-reinforced protective layers with optical 3D deformation analysis. In: Restoration of Buildings and Monuments Vol. 20 (2014), No. 5, p. 29-40
11. Ohno, S. ; Hannant, D. J.: Modeling the Stress-Strain Response of Continous Fiber Reinforced Cement Composites. In: ACI Materials Journal Vol. 91 (1994), No. 3, p. 306–312
12. Hinzen, M.: Einfluss von Kurzfasern auf die Frisch- und Festbetoneigenschaften sowie das Tragverhalten von Textilbeton, RWTH Aachen, PhD thesis, 2014
13. Voss, S.: Ingenieurmodelle zum Tragverhalten von textilbewehrtem Beton, RWTH Aachen, PhD thesis, 2008
14. Lorenz, E. ; Schütze, E. ; Weiland, S. : Textile concrete - properties of composite material. In: Concrete and reinforced concrete construction Vol. 110, Special (2015), p. 29-41

Corrosion protection of steel reinforcement by using surface applied corrosion inhibitors

Engin Cueneyt Seyhan[1,*] *Fred* Goodwin[2], and *I-Wen (Evan)* Huang[2]

[1]BASF Construction Chemicals, Regional Business Segment Management Europe, Piccardstr. 11, 86159 Augsburg, Germany
[2]BASF Construction Chemicals, Global Corrosion Competency Center, 23700 Chagrin Blvd, 44122 Beachwood, USA

Abstract. Surface applied corrosion inhibitors (SACI) are widely used to mitigate the corrosion process of steel reinforcement in concrete. But they remain controversial as to their effectiveness and the ability to compare materials from different manufacturers and technologies. They are applied onto the surface of hardened concrete and penetrate towards the steel reinforcement. This paper discusses the corrosion inhibition performances of a dual-phase corrosion inhibitor and field surveys through corrosion monitoring the selected structures. The efficiency of the corrosion inhibitor is assessed by selected testing procedures in laboratories.

1 Concrete deterioration and cost of corrosion

Concrete is the most widely used construction material in the last two centuries. Local availability of ingredients such as sand and cement, ease of shaping and high compressive strength made concrete the best alternative to masonry for the building of new structures. As this man-made stone could not provide adequate tensile strength, engineers had to incorporate steel to overcome this deficiency. The advent of reinforced concrete led to innovative uses of concrete for all types of structures and made reinforced concrete the material of choice for designers and builders. However, it is a disturbing truth that corrosion of the steel represents a major problem for the longevity of the structures concerned. Although the cement matrix provides physical protection and the highly alkaline environment keeps steel reinforcement in a passive state, it has been observed that corrosion of the reinforcing steel has initiated in almost every structure sooner or later.

Over the past 50 years, corrosion has emerged as the most significant cause of damage to reinforced concrete structures. Apart from the considerable expense of refurbishment work, the loss of access or use also results in significant cost. Considering the five major sectors of infrastructure: utilities, transportation, manufacturing industry and government, the total annual cost of corrosion-related repairs throughout the world is estimated at US$2.5 trillion, corresponding to about 3.4% of the world's Gross Domestic Product (GDP) (2013) [1]. Corrosion repairs to reinforced concrete structures such as roads, bridges, marine structures, industrial buildings, commercial and residential buildings account for a significant share of the total direct cost of corrosion. The indirect cost to the community of corrosion damage is hard to estimate but, based on a life cycle analysis of corrosion maintenance, repair, and rehabilitation for bridges in the USA, the cost in terms of lost productivity, traffic delays etc. could be as high as the direct costs [2].

2 Corrosion of steel reinforcement

Corrosion may be defined as the conversion of a refined metal (in this case steel) into its more stable oxides, with the consumption of oxygen and water. These electrochemical reactions take place at two distinct locations on a metal component, the anode (1) and the cathode (2) [3].

$$Fe \rightarrow Fe^{+2} + 2e^{-} \quad (1)$$

$$2e^{-} + H_2O + \frac{1}{2}O_2 \rightarrow 2OH^{-} \quad (2)$$

Concrete provides a natural shield, protecting the steel reinforcement against corrosion. The highly alkaline environment created by the cement matrix surrounding the reinforcement bars results in the formation of a passive layer around the steel. This passive film is only a few nanometers thick and is composed of hydrated iron oxides with a varying degree of Fe^{2+} and Fe^{3+} [4]. This passive layer is an impermeable oxide film which prevents further corrosion reactions by keeping oxygen and water away from the metal surface. This protective film is maintained if the highly alkaline environment provided by the concrete matrix persists. However, this passive condition cannot be maintained over the course of time because of deterioration occurring within the concrete. In chloride attack, chloride ions are transported through the concrete to the steel. This disrupts the passive layer and leads to corrosion. In addition, carbonation due to exposure to carbon dioxide dissolved in water reduces the pH of the cement matrix, which also resulting with breakdown of the passive layer and furthermore increases

* Corresponding author: engin.seyhan@basf.com

the free chlorides by releasing the bound ones. Once the passive layer breaks down, corrosion reactions start, rapidly attacking the reinforcement bars. Water and oxygen are the only ingredients which are needed for the reactions to continue. Depending on the corrosion rate, the reinforcement bars push off the concrete cover due to volume expansion of the corrosion product and eventually lose their bond to the concrete, which compromises the structural integrity.

3 Corrosion protection of steel reinforcement

Corrosion of steel in concrete is an electrochemical process. A corrosion cell requires both an anodic site, where the metal is dissolved, and a cathodic site, where oxygen is reduced, and hydroxyl ions are produced; it also requires electronic conduction paths (through the reinforcing steel) and ionic or electrolytic conduction paths (through the pore-electrolyte). To reduce the rate of corrosion, a reduction in the rate or elimination of the associated processes is required [5].

2.1 Use of silanes in concrete protection

Chloride ions diffuse into concrete along with the water and deteriorate the protective passive layer around the steel reinforcement and initiate or accelerate the corrosion process. Water is the main chemical compound along with oxygen, which are both consumed in corrosion reactions. Therefore, restricting water ingress and keeping chloride ions out of the concrete matrix reduces the corrosion rates significantly.

Silanes are a diverse class of chemicals of organo-silicon molecules of various functions and sizes. The small size of silane molecules means that they can penetrate deep into the concrete, where they can form chemical bonds with the hydrated components of the cement matrix due to their silica structure. After reacting with concrete, silane molecules repel the water and stop further chloride ions from penetrating the concrete [6,7,8]. Any water already present in the concrete can evaporate through open pores since this water-repellent layer created by silanes is considered 'breathable'.

Alkyltrialkoxysilanes are composed of an alkyl group that serves as a water repellent and three alkoxy groups for chemical reactivity and bonding, all joined to a silicon atom. The effectiveness of silanes depends on the type of alkyl group in size and branching, the molecular weight, and the reaction rate and surface tension of the silane liquid [9]. It is critical to select the suitable alkyl chains for formulating the silane blends for achieving the best performance on concrete substrates. Besides the water repellency, the resistance to carbonation by water repellency is the key for long term performance.

2.2 Corrosion inhibitors

NACE International defines corrosion inhibitor as "a chemical substance that, when added in small concentration to an environment, effectively decreases the corrosion rate". Based on the electrochemical mechanism of action alone, all inhibitors fit into one of three broad classes: (a) anodic, (b) cathodic and (c) mixed, depending on whether they affect the anodic reaction, the cathodic reaction, or both [10]. These inhibitors are chemical compounds, which can be carried to the steel reinforcement in different ways. They can be added to concrete admixtures and mixed directly with the fresh concrete. Unfortunately, this method is limited to fresh concrete and only relevant for new structures. In the European construction industry, the refurbishment of existing structures accounts for a large part of activities. In view of this fact, scientists have developed various types of surface-applied corrosion inhibitors (SACI) to stop existing corrosion processes in hardened concrete.

Silanes are used as an effective corrosion inhibitor due to their excellent hydrophobicity on concrete surfaces. Besides silanes, amines (e.g. aminoalchohol, aminocarboxylates) and sodium monofluorophosphate (MFP) are known SACIs for protecting steel reinforcement. However, the efficiency of these inhibitors on reinforced concrete are limited under certain conditions. For instance, MFP shows very little effectiveness on the corrosion rates of steel in both non-carbonated and carbonated concretes [11]. Several experimental studies and field surveys revealed that the amine-based inhibitors (aminoalcohol and aminocarboxylates) also have very limited impact on inhibiting corrosion [12,13,14,15].

3 Dual phase corrosion inhibitor

In the light of research carried out over the past two decades, BASF scientists succeeded in combining silanes and selected corrosion inhibitors to develop a new type of surface-applied corrosion inhibitor. This product is a clear liquid with combination of high-quality blended silanes and selective corrosion inhibitors (including latent phase inhibitors) and is designed to break the corrosion mechanism in reinforced concrete structures in various conditions. Silane technology allows deep penetration into the concrete and chemical bonding, which avoids wash out. In addition, the high alkali resistant nature of silanes leads to long service life when used on concrete surfaces. While the silane backbones of the product guarantee high water repellency and the exclusion of water and chloride ions, selective corrosion inhibitors stop the corrosion reactions and help the reinforcement bar to restore its passive layer. The presence of latent phase inhibitor lies dormant within the concrete until activated by moisture, which then penetrates the surface through cracks in concrete and sustains effective corrosion protection even after the silane barrier breaks down. This spectacular performance leads to a significant extension of service life and the maintenance-free period and reduces refurbishment costs.

3.1 Working mechanism

Dual phase corrosion inhibitor employs dual functionality. The silane base (alkylalkoxysilane) provides similar benefits to the water-repellent sealers, and the integral corrosion inhibitors are carried into the concrete along with the silane. The latent phase inhibitors remain inactive in the concrete until the water repellency diminishes over time, or until the concrete cracks. They then become mobile and are carried deeper into the concrete by moisture. Dual phase inhibitor has a surface tension roughly 1/3 that of water, and low viscosity to improve penetration into concrete. Its special blend of silanes provides a balance between drying time and penetration over a wide temperature range, as well as lower VOC content and a higher flash point than many traditional silane-based inhibitors. The result is a very fluid material with viscosity close to water, which can be applied via spray, brush or roller.

The most observable property of dual phase inhibitor is water repellency on the concrete surface. The silane component of dual phase inhibitor provides the same benefit with the added feature of its ability to act as a corrosion inhibitor. The alkylalkoxysilane structure of the dual phase inhibitor protects concrete from penetration of liquid water, but it keeps concrete permeable to water vapor. Thus, while keeping additional water from penetrating into the concrete, it will allow any moisture contained in the concrete to evaporate. Dual phase inhibitor utilizes multiple silane functionalities to create different effects on the concrete substrate. These include the formation of a hydrophobic shield as well as a corrosion inhibition effect.

Creation of the hydrophobic properties begin when the silane hydrolyzes to form silanol and alcohol. Next the silanol group condenses, forming a bond to hydroxides on the concrete surface. Fig. 1 [16,17] depicts the chemical reactions associated with this process.

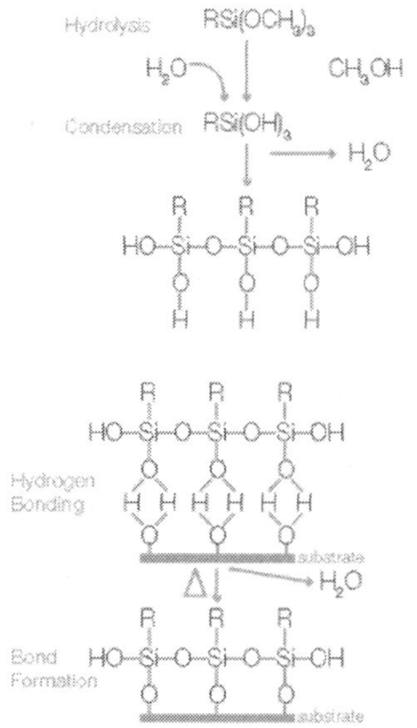

Fig. 1. Reaction of alkyl alkoxy silane with water (hydrolysis) to produce silanol and alcohol followed by condensation and bonding to the pores of the concrete substrate.

The R in the figure represents the hydrophobic portion of the chemical compound. CH_3OH is the methanol produced when dual phase inhibitor reacts with water. The substrate in this case is the pores in the concrete. This series of reactions leads to the formation of the hydrophobic layer (in molecular level) observed on and in the surface of the concrete treated with dual phase inhibitor.

Alkylalkoxysilanes can react within the concrete to reduce the conductivity of the electrolyte. The reaction increases the resistivity of the concrete immediately adjacent to the reinforcing steel which in turn slows down the electrochemical reactions for the corrosion process.

3.2 Experimental studies

Independent testing of dual phase corrosion inhibitor has shown it to be very effective at reducing active corrosion in steel reinforced concrete. The combination of the silane blend and the latent corrosion inhibitors serves synergistically to reduce active corrosion and prevent the onset of additional corrosion activity.

3.2.1 ICCET testing laboratories

An experimental study to determine the corrosion protection of reinforced concrete structures from chloride ion attack and carbonation by surface applied dual phase corrosion inhibitor was done in Eduardo Torroja Institute (ICCET) laboratories in Spain. The study was conducted on standard mortar (acc. EN 196-1) specimens with embedded steel reinforcement. The specimens were exposed to induce corrosion either by admixed chloride ions or a 100 % CO_2 atmosphere, after which their surface was treated with the inhibitor. Steel corrosion was monitored via electrochemical tests to determine corrosion potential, corrosion rate, and electrical resistance of the hardened mortar specimens.

The first phase of the experimental study was determining the efficiency of the inhibitor in case of chloride induced corrosion. This test was designed to determine the efficiency of the corrosion inhibitor with admixed chloride ions. This entailed adding 2 wt% (cement) Cl- in the form of $CaCl_2$ to the mixing water. Corrosion rate, corrosion potential, and resistivity measurements were taken 24-hours after mixing until the measurements verified the onset of corrosion. The inhibitor was then applied onto the specimens.

The corrosion rates observed for the specimens prepared with admixed chlorides are shown in Fig. 2. The blue lines represent the steel in the untreated reference specimen and the green and orange lines represent the specimens treated with the inhibitor.

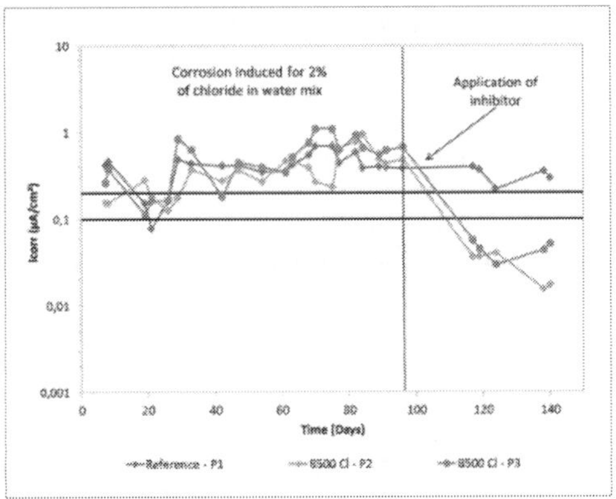

Fig. 2. Mean steel corrosion rate for specimens bearing chloride ions in the mixing water.

The data for the first period depicted in Fig. 2 (corrosion induced by 2 % admixed chloride by weight of cement), i.e., the corrosion rates prior to applying the inhibitor, confirmed these three specimens are pre-corroded. At day 95, random specimens with active corrosion were selected and treated with the inhibitor. It was then noticed that the corrosion rate declined drastically below 0.1 μA/cm^2 (steel de-passivation threshold) and remained below that value until the end of the tests. On the other hand, the control specimen (blue line) continued to exhibit mean corrosion values of 0.4 μA/cm^2.

In addition to corrosion currents, electrical resistance of the specimens was also investigated. The concrete electrical resistance curves before and after surface treatment in Fig. 3 shows that electrical resistance rose steeply 24 hours after application.

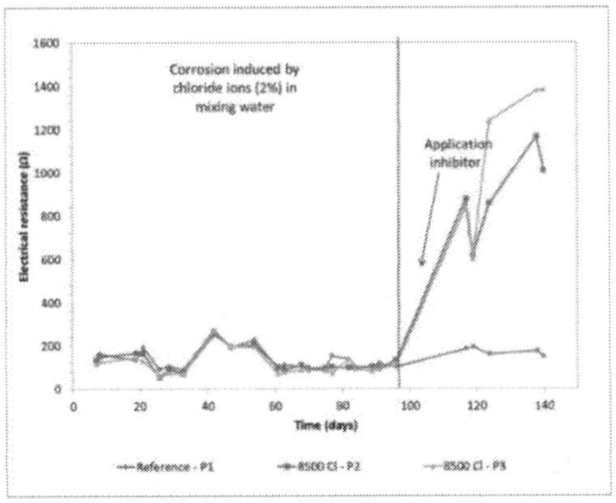

Fig. 3. Mean electrical resistivity rate for specimens bearing chloride ions in the mixing water.

The second phase of the experimental study was determining the efficiency of dual phase inhibitor in case of carbonation induced corrosion. The corrosion rate for the specimens exposed to accelerated carbonation in a 100 % CO_2 atmosphere is plotted in Fig. 4.

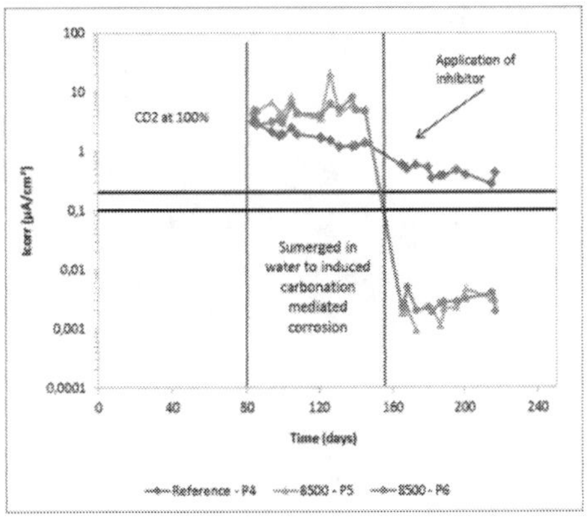

Fig. 4. Mean steel corrosion rate for specimens exposed to accelerated carbonation.

After 80 days of exposure and immersion in water, the steel embedded in the specimens exhibited high corrosion rates (5 μA/cm^2). Once the product was applied, however, the values declined to under 0.1 μA/cm^2 steel de-passivation threshold. These values continued another 60 days until the end of the test.

Fig. 5 shows electrical resistance of concrete before and after corrosion treatment. Here initial electrical resistance was very high due to prior specimen exposure to accelerated carbonation, which raised this parameter substantially. After treatment, resistance rose even further, to over 10.000Λ.

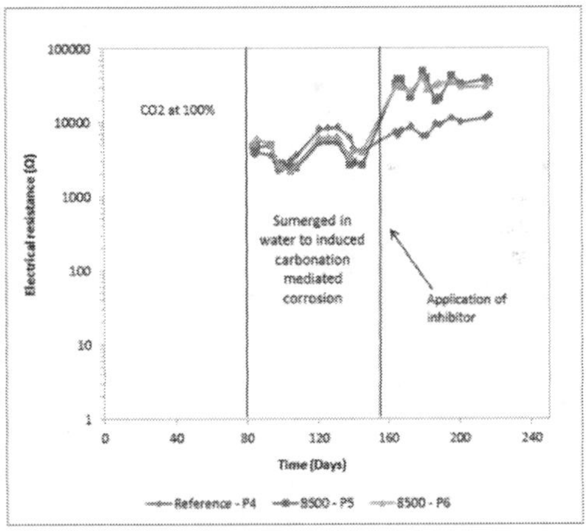

Fig. 5. Mean electrical resistivity for specimens exposed to accelerated carbonation.

3.2.2 ASTM G 109

ASTM G109 is a test method for "determining the effects of chemical admixtures on the corrosion of metals in concrete" [18]. The specimens, which were built according to this test standard were modified for corrosion inhibitor testing, shown in Fig. 6.

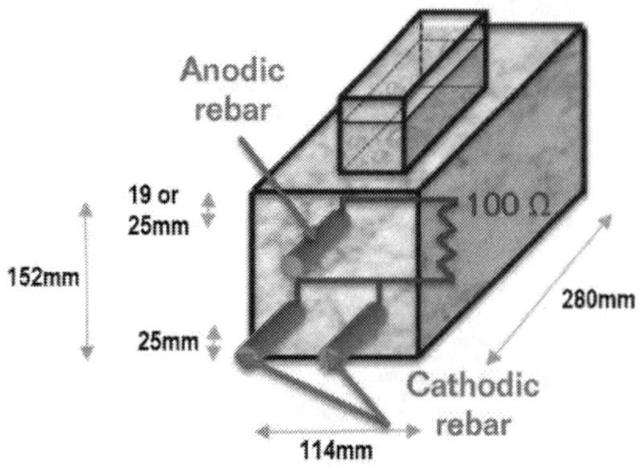

Fig. 6. ASTM G109 specimens modified for SACI testing.

The configuration for ASTM G109 is used, but instead of the test material being admixed into the concrete, the dual phase corrosion inhibitor is applied to the top surface of the hardened concrete specimen and allowed to dry before installing the ponding dam and coating the sides of the specimen with epoxy.

Before applying the inhibitors, the sample surface was abrasive blasted to ICRI CSP 5-6 after curing. Then they were treated with different silane-based surface applied corrosion inhibitors including dual phase corrosion inhibitor and an amino-functional silane and ponded with 3% sodium chloride (NaCl) solution one week after the SACI treatment. In this testing, both corrosion inhibitors worked well, as shown by the total coulombs of corrosion current between the upper and lower bars, Fig. 7.

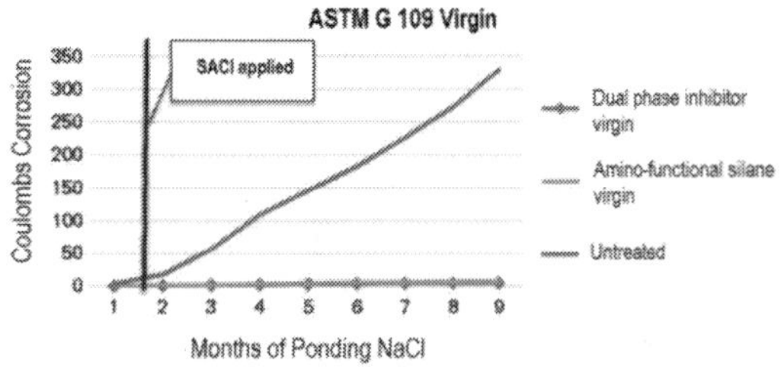

Fig. 7. ASTM G109 specimens treated with dual phase inhibitor, amino-functional silane, and an untreated control specimen.

As is easily seen in Fig.7, there is no appreciable difference between the two treated specimens after nine months of ponding. Both inhibitors reduced the corrosion rate drastically. In order to test the dual functionality and the efficiency of the latent phase inhibitors, another series of tests have been done by using modified ASTM G 109 specimens. Three specimens were cracked before the treatment with the corrosion inhibitor, and an additional three specimens were cracked after treated with the dual phase inhibitor. The crack widths were approximately 0.3mm for both pre (before treatment) and post-cracked (after treatment) samples. Specimens were ponded with 3% NaCl for two weeks followed by drying at 50% relative humidity to simulate the accelerated corrosion conditions. During ponding with chlorides, corrosion current was measured between the upper and the lower bars. The accumulated corrosion current demonstrated that latent phase inhibitors in the new material reduces the corrosion rates significantly, represented in Fig. 8. This proves the dual functionality of the new material enhanced with latent phase inhibitors.

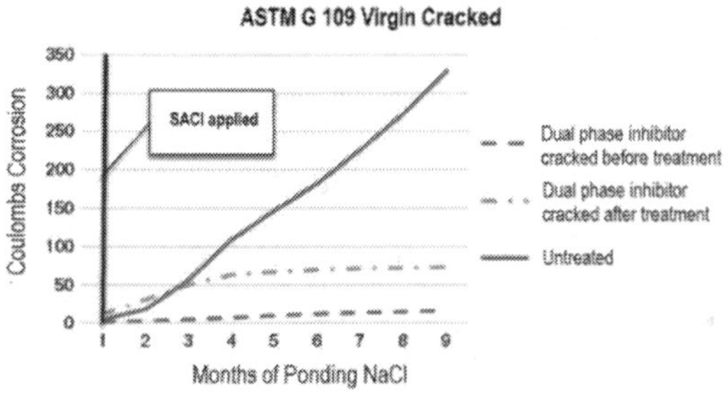

Fig. 8. ASTM G109 specimens treated with dual phase corrosion inhibitor before and after cracking, and an untreated control specimen.

4 Summary and outlook

Corrosion of reinforcing steel is the leading cause of deterioration of concrete structures around the world. Guidance exists for design and construction practices to produce durable structures; however, corrosion-induced deterioration continues due to failure to follow good trade practices. Concrete requires maintenance, especially to correct issues from cracking. Use of SACIs is a maintenance technique that can be applied proactively to existing structures as well as reactively to mitigate corrosion. Silane-based materials reduce chloride ingress and allow concrete drying to increase resistivity, both reducing corrosion. Incorporating latent corrosion inhibitors into silane-based surface treatments combines the benefits resulting from silane treatments and the corrosion inhibiting effects in concrete with cracks before, or even after treatment.

Experimental studies have revealed the significant high performance of dual phase inhibitor when compared to traditional corrosion inhibitors. ASTM G109 tests show that it performs well in both uncracked and cracked samples. In addition, the efficiency of the latent phase inhibitor is proved on the post-cracked specimens. In addition to superior performance obtained in the ASTM G109 test program, the inhibitor efficiency has been also proved in the experimental studies performed in Torroja Institute. Dual phase inhibitor not only performed outstandingly in the specimens subjected to a high amount of chloride ions, but also in the carbonated specimens. This guarantees that reinforced concrete elements can be protected effectively against both chloride and carbonation induced corrosion.

References

1. G. Koch, J. Varney, N. Thompson, O. Moghissi, M. Gould, J. Payer: *International measures of prevention, Application, and Economics of Corrosion Technologies Study,* NACE International Report, Houston, (2016)
2. FHWA-RD-01-159 Report, Corrosion Costs and Preventive Strategies in the United States, (2001)
3. BRE Digest 444 Part 1, Corrosion of Steel in Concrete, Durability of Reinforced Concrete Structures, BRE Center for Concrete Construction, (2000)
4. L. Bertolini, B. Elsener, P. Pedeferri, R.P. Polder: Corrosion of steel in concrete, WILEY –VCH, Weinheim (2004)
5. BRE Digest 444 Part 3, Corrosion of Steel in Concrete, Protection and Remediation, BRE Center for Concrete Construction, (2000)
6. C. Christodoulou, … et al.: *Long-term performance of surface impregnation of reinforced concrete structures with silane.* Construction and Building Materials **48** (2013), pp.708-716
7. L. Zhichao, W. Hansen: *Effectiveness of Concrete Surface Treatment Materials in Reducing Chloride-Induced Reinforcement Corrosion,* Cement Concrete Comp. **69 (2016)**: pp. 49-60
8. C. D. Murray, R. Deschenes, W. M. Hale: *Durability of silane sealer in highly alkaline environment*, ACI Materials Journal **113 (2016) 4**, pp. 387-393
9. ACI 515.2R-13, Guide to Selecting Protective Treatments for Concrete, American Concrete Institute (2013)
10. A.M. Vaysburd, P.H. Emmons: *Corrosion inhibitors and other protective systems in concrete repair: concepts or mis concepts*, Cement&Concrete Composites, **26** (2004), pp. 255-263
11. V.T. Ngala, C.L. Page, M.M. Page: *Corrosion inhibitor systems for remedial treatment of reinforced concrete Part2: sodium monofluorophosphate,* Corrosion Science **45** (2003), pp. 1523-1537
12. FHWA-RD-01-097, Long-Term Performance of Corrosion Inhibitors Used in Repair of Reinforced Concrete Bridge Components, Federal Highway Administration Research and Technology (2002)
13. B. Elsener: *Corrosion inhibitors for steel in concrete*, International Congress on Advanced Materials, (2000)
14. M. Ormellese, M. Berra, F. Bolzoni, T. Pastore: *Corrosion inhibitors for chloride induced corrosion in reinforced concrete structures*, Cement and concrete Research **36** (2006), pp. 536-547
15. Y. Schiegg, F. Hunkeler, H. Ungricht: *The effectiveness of corrosion inhibitors – a field study*, 16th Congress of IABSE, (2000)
16. A.I. Karayan, Y. Cubides, H. Castaneda: *Evaluation of different surface applied corrosion inhibitors systems on reinforced concrete structures based on electrochemical techniques and standard testing,* Department of Defense Allied Nations Conference proceedings, Paper No. 2017-0000
17. X. Pan, Z. Shi, C. Shi, T.C. Ling, N. Li: *A review on concrete surface treatment Part I: Types and mechanisms*, Construction and Building Materials. **132**. 578-590. 10.1016/j.conbuildmat.2016.12.025.
18. ASTM G109-07, Standard Test Method for Determining Effects of Chemical Admixtures on Corrosion of Embedded Steel Reinforcement in Concrete Exposed to Chloride Environments, ASTM International, (2013)

Reinforced concrete corrosion prevention/reduction by hydrophobic impregnation

Michel Donadio[1], *Mirdash* Bakalli[2], *Zeno* Dan[3]

[1]*Sika Services AG, Switzerland*
[2]*Sika Technology AG, Switzerland*
[3]*Sika Europe Management AG, Switzerland*

Abstract: The prevention of corrosion in reinforced concrete structures can be achieved by different means. The aim of this paper is to present the efficiency of corrosion prevention using silane hydrophobic impregnation, through laboratory investigations as well from long term field studies. The laboratory investigation was carried out in Zürich, is based on modified ASTM C109, where the product was applied on cracked concrete beams (before cracking and before corrosion initiation; after cracking and before corrosion initiation and finally after cracking and after corrosion initiation). The long term field studies were carried out over 12-years exposure to de-icing salts in a Swiss tunnel, and after 10 years exposure to a marine zone simulation in Japan. This paper will show that the use of a silane hydrophobic impregnation can be an effective way to reduce the risk of corrosion for concrete structures, including structures exposed to chlorides in a marine environment, or from the use of de-icing salts in winter time.

1 Introduction

The recent disastrous collapse of the Genova Bridge in Italy and different investigations in various countries have shown that the state of bridges in many countries is critical i.e. in need of urgent renovation works.

The cost of corrosion [1] is now well documented by various different organizations including NACE in the USA and the World Corrosion Organization. It is estimated that a quarter of the world's annual steel production is destroyed by corrosion every year. This is equivalent to 150 million tons per year (equivalent to 5 tons/second). The W.C.O. estimates the cost of corrosion as being between 3.1 to 3.5% of a nation's GDP annually for industrialized countries (e.g. between €1.3 and 1.4 trillion in USA). As these figures reveal this corrosion is very costly for society and not only financially.

Table 1 shows very clearly that this cost of corrosion versus GDP is also unfortunately not reducing over time.

Table 1: Cost of corrosion (1) in % of GDP

Country	Year	% of GDP
US	1950	2.1
UK	1970	3.5
Japan	1974	1.2
US	1975	4.5
US	1988	3.1
US	2013	3.1

A study from COMACAC (Morocco) estimated the cost of corrosion at 5% of GDP, and in France this is estimated to cost from 1 to 2 € per day per inhabitant.

Fig. 1: Example of corrosion of reinforced concrete.

The steel reinforcement in concrete starts to corrode in the presence of moisture when the passive layer formed by the surrounding highly alkaline concrete matrix has broken down due to carbonation, or the presence of chlorides.

To initiate an electrochemical process, several elements are required to be present:

- A loss of passivity (due to carbonation or the presence of chlorides)
- moisture and oxygen
- stray current

To avoid such expense and the risk of eventual collapse of the structures due to steel corrosion, various methods to improve the durability of reinforced concrete structures are available. One of them is to prevent or to reduce the penetration of chlorides, and to dry out the concrete so that the speed of any corrosion that is initiated will be reduced.

Deeply penetrating, surface applied, hydrophobic, silane based, impregnation materials, such as the liquid type triethoxy(2,4,4-trimethylpentyl)silane (~99% active content), or the cream type (same chemical base but with ~80% active content), will reduce the moisture level [2] in the concrete. This reduction of humidity around the reinforcement bars will reduce any corrosion activity (basis of the Principle 8 Resistivity Increase of EN 1504-9 [3]). Additionally, by preventing moisture ingress, these products will also prevent the further ingress of water-borne chloride ions into the structure.

2 Methodology

In order to assess the suitability of such deeply penetrating, silane based, hydrophobic impregnations to prevent/arrest corrosion in chloride environments, this paper will summarise the results from three studies, one from laboratory investigations in Switzerland, and two from field testing, in Switzerland and Japan respectively.

2.1. Swiss Laboratory Investigations [4]

The cracked concrete beam corrosion test (CCBCT) is a laboratory time-to-corrosion test. It was developed by Paul Tourney and Neal Berke [4] and the test is adapted from ASTM G109 [5].

2.1.1 Preparation of steel reinforcing bars

To prevent crevice corrosion during the test, the reinforcing steel bars are prepared as followed:

Table 2: Reinforcing steel bar preparation

item	Description
Type:	Grey mild steel reinforcing rods, grade S235JR+AR (EN 10025-2:2004-10)
Dimension:	Diameter 12 mm, length 914 mm
Preparation:	- Removal of rust particles mechanically with a power wire brush - Degreasing the bars with acetone - Both ends (203 mm – 8 in) of the rods are fully covered with an epoxy coating - Both ends are further protected against crevice corrosion with a heat shrink tube

2.1.2 Concrete Specimens Mix Design

Production of the concrete mix design was performed according to EN 1766

Cement CEM I 42.5 N:	355 kg/m^3
Aggregates:	0-16 mm
W/C:	0.40
Flow table spread:	33-36 cm
Air content:	1.6 to 2.4 %
Curing after demoulding:	20°C 95% RH

3.1.2 Concrete Specimens

The test specimens produced are concrete beams with dimensions 150 x 150 x 750 mm (6 x 6 x 30 in), each with three of the 12 mm diameter mild steel rebars embedded. Two rebars are placed 38 mm (1.5 in) from the bottom of the beams, the third 43 mm (1.7 in) below the top surface of the beams.

No mould release agent was used during production of these beams, in order to prevent any interference with the subsequent application and penetration of the hydrophobic impregnation.

After curing the specimens in a humidity chamber for 28 days, followed by storage for a month at approximately 23°C, a 5 mm (0.2 in) deep and 3 mm (0.1 in) wide notch was sawn into the top surface of the beams. The resulting concrete cover, measured below this cut notch, was 38 mm (1.5 in).

The sides and the bottom of the beams were painted with two coats of a 100% epoxy resin coating in order to prevent / limit water evaporation, as well as to prevent contamination from spills of the chloride solution during cyclic ponding.

The bottom rebars were permanently connected to each other with 1.0 mm diameter copper wire, and the top rebars were electrically short-circuited using a 10-ohm resistor soldered onto the copper wire.

A Plexiglas dyke, 305 mm long, 114 mm wide and 51 mm deep, was attached to the top surface of the cracked beams and sealed with an elastic polyurethane sealant. The dyke was used to expose the beams to a 3.0% (5.0%) by mass solution of sodium chloride using a 2-week wet, 2-week dry ponding cycle procedure. During the wet cycle, the dyke was filled to within 20 mm off the top with sodium chloride solution. The dyke was also covered with a plastic plate in order to avoid/limit water evaporation.

The average time period between the end of the 28-day curing period and the subsequent cracking procedure was: 2 months (average time period between casting and cracking the beams: 3 months).

The average time period between the end of the 28-day curing period and the start of the first ponding cycle was: 3 months (average time period between casting the beams and the beginning of the first ponding cycle: 4 months). After concrete curing, all of the specimens were stored in the laboratory at approximately 23°C.

Fig. 1: Prepared concrete beams for corrosion testing.

2.1.3 Testing Concept

The products to be tested were applied at different intervals and to mimic different field scenarios:

- **Before** cracking (forced) and **Before** corrosion initiation (**B/B**)

 https://doi.org/10.1051/matecconf/201928905002

- **After** cracking (forced) and **Before** corrosion initiation (**A/B**)
- **After** cracking (forced) and **After** corrosion initiation (**A/A**)

2.1.4 Forced Cracking Procedure

Each beam was carefully cracked along the cut notch using flexural load bearing techniques, so that the crack extended down to the top rebar. Whilst under the cracking load, the crack was shimmed to a consistent crack size to be used throughout the whole corrosion test. For this purpose three to four 0.25 mm (0.01 in) thick stainless steel shims were inserted into the crack at intervals to hold it open once the load was released.

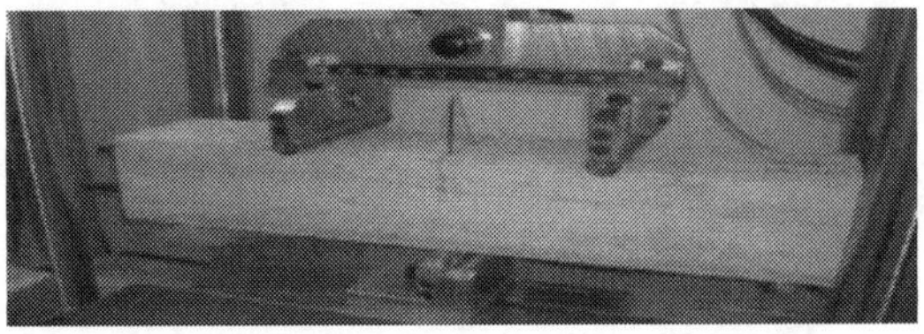

Fig. 2: Forced cracking.

2.1.5 Selection of the Hydrophobic Products

Two liquid hydrophobic products were selected for the test; the first was triethoxy(2,4,4-trimethylpentyl)silane based with ~99% active content (Product A) and the second product was an organo-functional silane with 98% active content and claiming to also contain a corrosion inhibitor (Product X).

2.1.6 Application Procedure for Hydrophobic Products

Two consumption rates were used for Product A corresponding to 300 g/m^2 which was applied in 2 coats and to 500 g/m^2 which was applied in 3 coats.

One consumption rate of 500 g/m^2 (corresponding to the typical recommended rate and applied in 3 coats, was used for Product X.

2.1.7 Ponding Procedure

Initially, the ponding exposure was performed for 19 cycles with a 3.0% by mass of sodium chloride solution, on the basis of a 2-week wet, 2-week dry ponding cycle.

Beginning with the 20th cycle, the sodium chloride concentration of the ponding solution was increased to 5.0% by mass, in order to further accelerate the CCBCT, thereby achieving even harsher test conditions.

2.1.8 Electrochemical Measurements

Different methods [7][8] of testing were used to assess the corrosion activities of the different concrete specimens:

- Corrosion Potential
- Corrosion Current
- Integral Corrosion Current

2.2. Japanese Field Study [9]

Miho Tsujita of Kagoshima University of Japan conducted a study over a 10 year period on beams placed in the simulated exposure of a shower of salted water, designed to simulate 10 years exposure in a splash zone.

Fig. 3: Test site for salt spray exposure testing at the Kagoshima University in Japan.

2.2.1 Concrete mix design for Japanese specimens

Cement:	OPC 243 kg/m^3
Sand:	River sand 912 kg/m^3
Aggregates:	Sandstone 20mm 970 kg/m-
W/C:	0.68
Air:	4.0%
Size:	100x100x150mm
Rebar:	Two deformed steel bars (Ø 10 mm)
Cover:	45 mm and 17.5 mm from the two opposite exposed sides of the specimens.
Cure:	28 days in water
Types:	1 set of specimens without cracks (NC) 1 set with cracks induced BEFORE surface treatment (BC) 1 set with cracks induced AFTER surface treatment (AC)

Note: with the exception of the two exposed surfaces, all of the other four beam surfaces were coated with epoxy resin to prevent additional side penetration of chlorides.

2.2.2 Selection of Hydrophobic & Surface Improvement Products

Table 3: Types of surface treatment materials

	Type	Aspect	Component	%
A	Silane	Liquid	Silane Alcohol	~34 ~66
B	Silane/Siloxane	Cream	Alkylalkoxysilane Ethanol	NA
C	Silane/Siloxane	Gel	Trietoxysilane Ethanol	~90 ~5
D	Siloxane	Liquid	Silane/siloxane Distilled liquid	60- 100
E	Silicate	Liquid	Super silicate particle Catalyst	30-60
F	Acrylic Silicate	Liquid	Emulsion	NA

All of the specimens were allowed to dry for 2 weeks after application at a constant temperature of 20°C.

2.2.3 Exposure tests

All of the specimens with the surface impregnated materials were exposed to simulated splash zone conditions using the salt water shower. This test facility is located in the Kurihama bay area (Yokosuka, Kanagawa prefecture, Japan) and has artificially repeated dry and wet cycles, that is, salt water shower for about four hours and then natural drying in the air for eight hours in each cycle (Fig. 3). One day consisted of two cycles and the total exposure time was 10 years.

2.2.4 Measurements

After 10 years exposure to repeated cycles of wetting with salt water and subsequent drying, the following assessments and measurements were carried out:

- Penetration depth of the tested products
- Carbonation depth
- Corrosion of the steel bars

2.3. Swiss Field Study [10]

Yves Schiegg has reported the results of a 12-year long field study aimed at investigating the corrosion behaviour of ordinary and stainless steel rebars in various types of concretes.

2.3.1 Composition of the Concrete Plate Specimens

Table 4: Composition of concrete plate specimens

Ref.	Type	W/C	Rebar
H1-H4	CEM I (reference)	0.50	Carbon Steel B500B
G1-G4	CEM I	0.35	
D1-D4	CEM I + 20% Slag	0.50	
E1-E4	CEM I + 15% Fly ash	0.50	
F1-F4	CEM I + 7% silica fume	0.50	
A1-A4	Reference + hydrophobic	0.50	
B1-B4	Reference	0.50	Galvanised carbon steel
C1-C4	Reference	0.50	Stainless Steel

2.3.2 Field Exposure Panel in the Naxberg Tunnel

The Naxberg tunnel is located on the Swiss A2 motorway, close to the Gotthard tunnel and at around 1000 m altitude. As a part of scheduled tunnel restoration work at the end of the 1990's, the opportunity was provided to place a concrete test panel in the tunnel wall at a distance of 50m from the north-facing tunnel entrance (Fig. 4). The panel consisted of 32 concrete plate specimens, each with the dimensions of 600 x 500 x 100 mm.

Due to the different positions of the plates (height / distance from road (0 to 2.4 m), the exposure class according to European standard EN 206 was expected to vary from XD3 (alternating wet/dry with exposure to chloride-containing spray water) to XD1 (moderately wet with exposure to chloride-containing mist water).

2.3.3 Test Measurements

Fig. 4: Visual state of the concrete test panels in the Naxberg tunnel in the year 2000 (left) and 2012 (right).

Various test measurements were carried out on the concrete plates from the panels during this investigation. This paper will only present some extracts of these results on

- Carbonation depth,
- Chloride profiles
- Corrosion profiles.

3 Test Results

3.1. Swiss Laboratory Investigations

Only the macrocell measurements are presented in this paper.

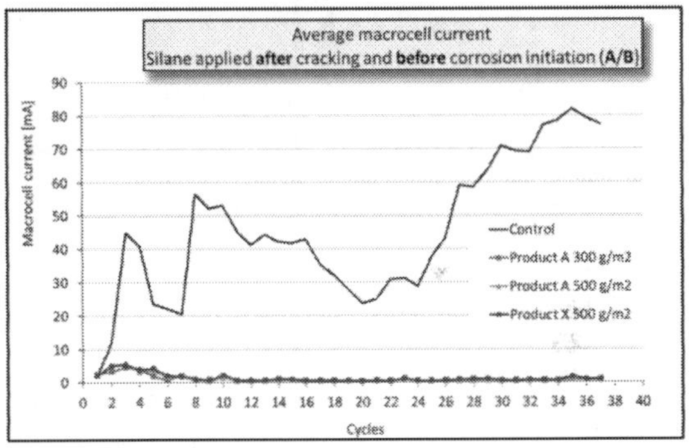

Fig. 5: Macrocell measurements – Silane applied after cracking & before corrosion initiation

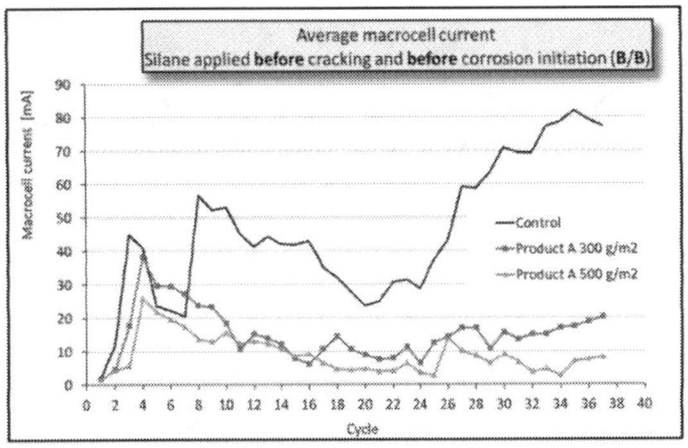

Fig. 6: Macrocell measurements – Silane applied before cracking & before corrosion initiation

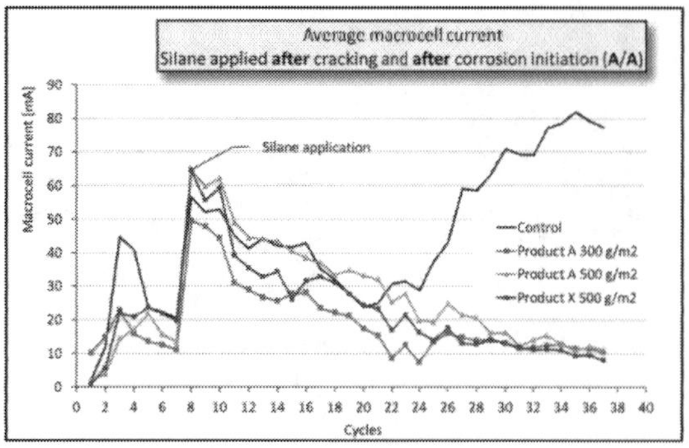

Fig. 7: Macrocell measurements – Silane applied after cracking & after corrosion initiation

3.2. Japanese Field Study

Note: The numbers in bracket in the following graphs correspond to the concrete cover.

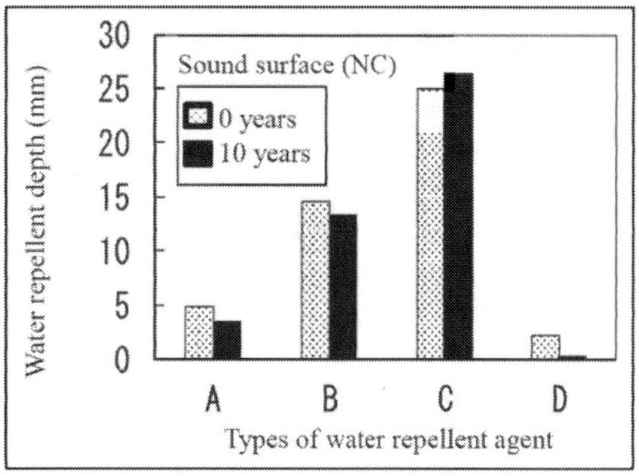

Fig. 8: Penetration depth of hydrophobic specimens at the start and at the end of the study

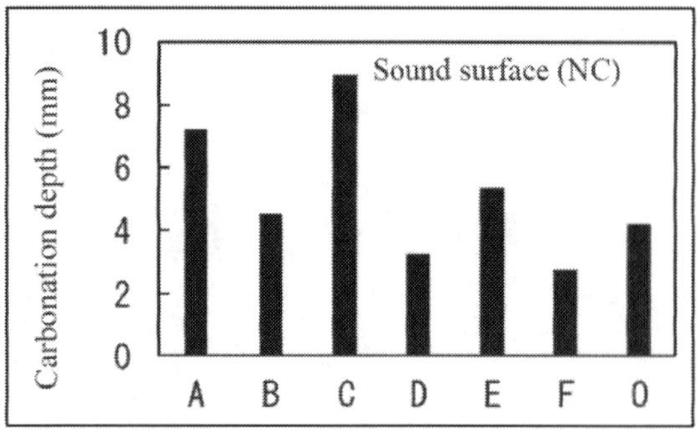

Fig. 9: Carbonation depth of treated specimens versus control (0) on non cracked concrete surfaces.

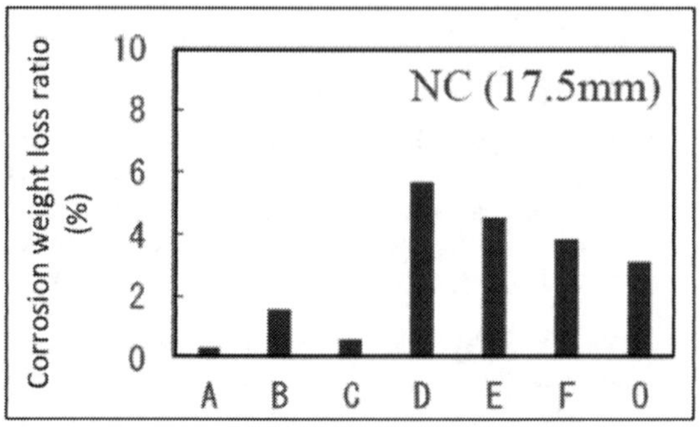

Fig. 10: Corrosion levels with low concrete cover on non-cracked surfaces.

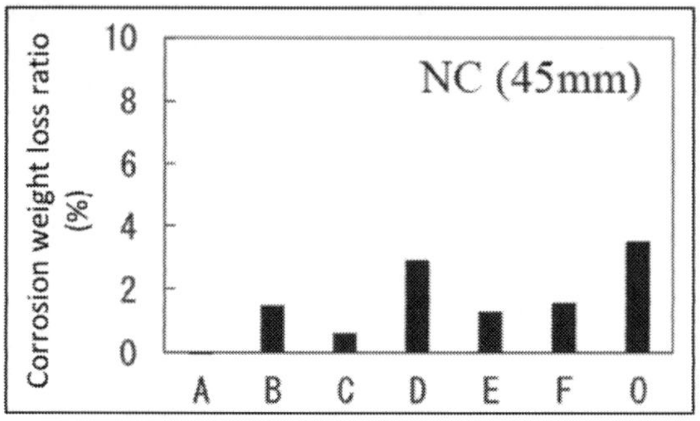

Fig. 11: Corrosion levels of normal concrete cover on non-cracked surfaces.

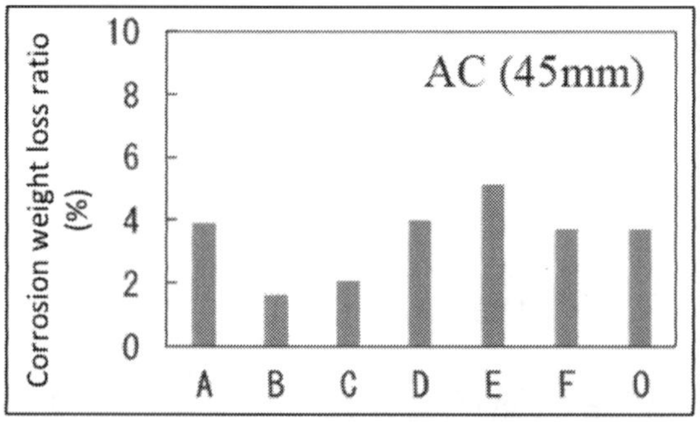

Fig. 12: Corrosion levels of normal concrete cover on cracked surfaces – surface treatment carried out AFTER cracking

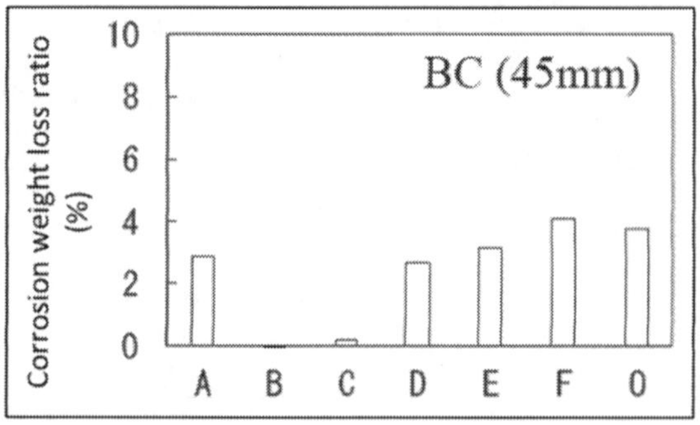

Fig. 13: Corrosion levels of normal concrete cover on cracked surfaces – surface treatment carried out BEFORE cracking

3.3. Swiss Field Study

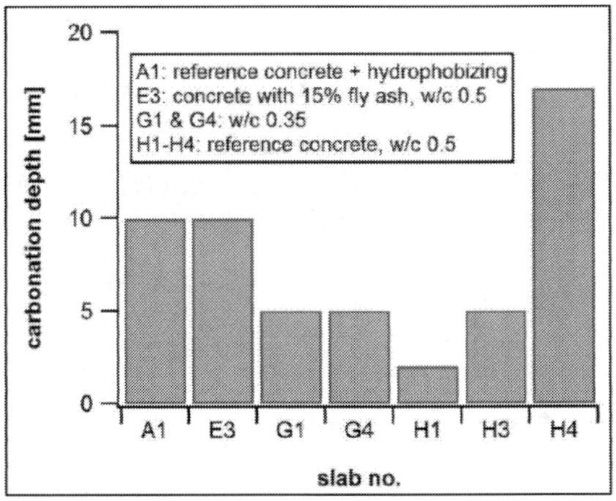

Fig. 14: Carbonation depth after 12 years exposure in the Naxberg tunnel.

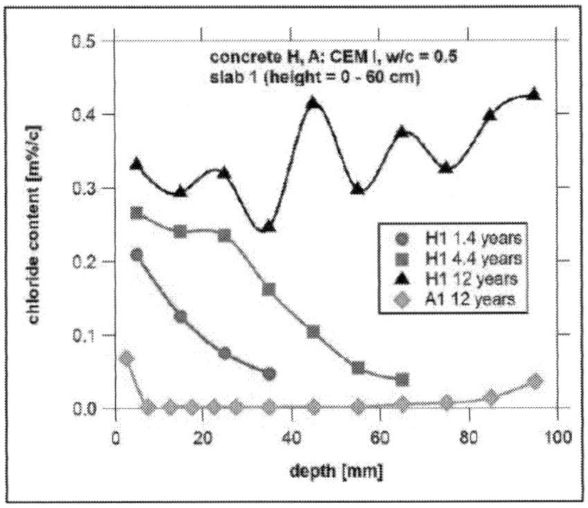

Fig. 15: Chloride profiles on the lower plate of the control specimens compared to the specimen treated with the hydrophobic impregnation.

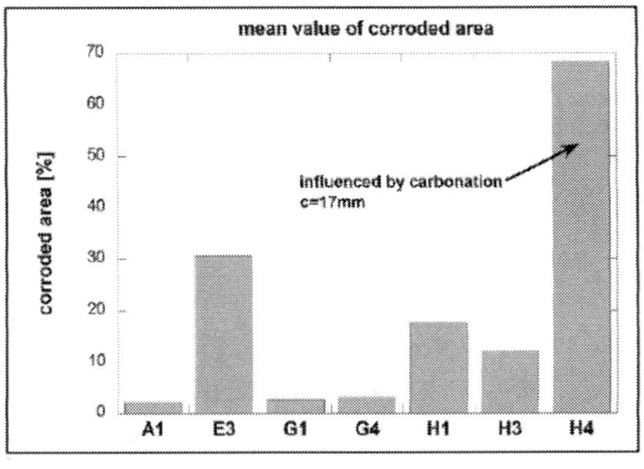

Fig. 16: Average corroded surface area of the steel rebars

4 Discussion

4.1. Swiss Laboratory Investigations

Figure 5 clearly shows that the hydrophobic impregnation provides the best increase in protection when applied after the cracks have already formed.

When the cracks happen after application of the treatment (refer to figure 6), we can also see the importance of the materials consumption (closely related to their penetration depth) in the level of protection.

Also when the treatment is carried out once corrosion is already advanced, we can also see a positive effect from the relative drying-out of the specimens on reducing corrosion activities – this is further confirmed by independent tests carried out by ZAG (Slovenian National Building & Civil Engineering Institute) [11].

The lack of difference between the two different products tested shows that the positive effect of prevention / reduction of corrosion is achieved due to the hydrophobic performance of the two products.

4.2. Japanese Field Study

Unfortunately no information was provided in the original paper about the consumption rates of the different products used.

However, from the type of product selected and the penetration levels achieved (refer to fig 8), it clearly shows that to obtain efficient protection against corrosion, even in the presence of cracks, high performance, deeply penetrating products such as the cream (B) and gel (C) will provide significant protection. This 10 years study also shows the limited efficiency of silicate based products (E & F) in arresting the corrosion.

4.2. Swiss Field Study

Unfortunately no information is provided in the original paper about the type and consumption rate of the hydrophobic impregnation used.

However, during the subsequent discussion at the EUROCORR2017 conference where the paper was presented, it was stated by the author that the product used was similar to the triethoxy(2,4,4-trimethylpentyl) silane (Product A) used in the above Swiss Laboratory Investigations.

The carbonation depth (Fig. 14) of these different specimens confirmed the results from the Japanese study, showing that hydrophobic impregnation products are not intended for arresting carbonation.

However, Figure 15 above clearly shows the benefit of using these types of products for arresting the penetration of chlorides as at the typical depth (50 mm) of steel reinforcement. The control specimens show that after 12 years of exposure chloride levels up to almost 0.4% by weight of cement, whilst the specimens treated with hydrophobic products show no chlorides penetrating the concrete at this depth.

Figure 16 shows a low level of corrosion in the treated standard concrete specimen at the same level as the concrete with a very low water / cement ratio (G).

The author of this study concludes that it is clearly of benefit to protect such exposed reinforced concrete structures with this type of hydrophobic impregnations: "After 12 years of exposure only the plates that contained stainless steel rebar or were hydrophobically treated showed no rust spots on the surface".

5 Conclusions

To reduce the costs related to steel corrosion in reinforced concrete infrastructure, these laboratory investigations and field studies have all demonstrated that treatment of concrete with deeply penetrating, hydrophobic impregnations such as the liquid product triethoxy(2,4,4-trimethylpentyl) silane, or the cream/gel equivalent version of the product, will effectively reduce chloride migration, and therefore arrest or mitigate established corrosion. Other benefits from the application of such products are the prolongation of the time to repair and increased service-life of the concrete structures. In addition the professional and periodic application of deeply penetrating, hydrophobic impregnations will also significantly contribute to increasing the durability and safety with lower risk structures.

References

1. Sources: World Corrosion Organization, NACE, COMACAC, IMIST-CNRST© (2011 - 2015)
2. E. R. Giannini, University of Texas. *Coatings and overlays for concrete affected by alkali-silica reaction*. Eds. M. Grantham, V. Mechtcherine, U. Schneck, CRC Press, pp. 823-831, (2011)
3. EN 1504-9-2008; *Products and systems for the protection and repair of concrete structures - Definitions, requirements, quality control and evaluation of conformity - Part 9: General principles for the use of products and systems*. (2008)
4. M. Donadio, Internal Sika Article, *corrosion tests on cracked concrete beams. Adapted from ASTM G109* (2018)
5. Paul Tourney and Neal Berke, Concrete International. *A Call for Standardized Tests for Corrosion-Inhibiting Admixtures*, (1993), p. 57-62
6. ASTM G109-07-2013, *Standard Test Method for Determining Effects of Chemical Admixtures on Corrosion of Embedded Steel Reinforcement in Concrete Exposed to Chloride Environments*, ASTM International, West Conshohocken, PA, (2013)
7. ASTM C876-15, *Standard Test Method for Corrosion Potentials of Uncoated Reinforcing Steel in Concrete*, ASTM International, West Conshohocken, PA, (2015)
8. *Corrosion of Reinforcement in Concrete - Monitoring, Prevention and Rehabilitation Techniques*, Eds. by M. Raupach, B. Elsener, R. Polder and J. Mietz, EFC Publication No. 38, Woodhead Publishing, (2007)
9. Mijo Tsujita et all, School of Science and Engineering, Kagoshima University, Japan. *Proceedings of Hydrophobe VIII*, Hong Kong, *Effect of surface treatment materials on corrosion protection of concrete structures subjected to chloride attack: An experimental study*. (2017)
10. Yves Schiegg et all, Technology and Research for Concrete Structures. *EUROCORR2017. Monitoring rebar corrosion propagation in concrete – Results of the Naxberg field experiment after 12 years*. (2017)
11. Report from the Slovenian National Building and Civil Engineering Institute; P859/11-440-1. *Corrosion testing and visual assessment of reinforced concrete beams treated with Sikagard®-705 L*. (2011)

Opportunities regarding the potential use of the self-cleaning concept within urban contemporary architecture in Romania

Elvira Grebenişan[1,2,*], *Henriette* Szilagyi[1], *Andreea* Hegyi[1], *Călin* Mircea[2], and *Cornelia* Baeră[3]

[1]N.I.R.D. URBAN-INCERC, Cluj-Napoca Branch, Romania
[2]Technical University of Cluj-Napoca, Civil Engineering Faculty, Cluj-Napoca, Romania
[3]N.I.R.D. URBAN-INCERC, Timişoara Branch, Romania

Abstract. Contemporary urban architecture faces two important issues: degradation of buildings, caused by exposure to various environmental factors (air and water pollution, mainly generated by the fuels combustion used for transport and heating) and also the costs for repair, cleaning and maintenance of the buildings facades. Regarding the last mentioned aspects, recent research led to development of materials with self-cleaning potential and consequently pollution reduction. Self-cleaning concrete represents a state-of-the-art material with photocatalytic properties generated by the addition in its composition of nanomaterials like TiO_2. Already known for its intrinsic photocatalytic character, TiO_2 has the ability to catalyse the decomposition of organic substances like grease and dirt, facilitating their quick removal only by rainwater action. Therefore, a building façade made of TiO_2-SiO_2-containing material develops substantial savings regarding maintenance costs, water consumption and less detergents contamination due to its intrinsic super hydrophilic effect of the surface in the presence of UV radiation, leading to easy dirt removal when water reaches it. The aim of present paper is presenting the latest stage of worldwide research regarding the obtaining of self-cleaning concrete and also the possibility of adapting the concept to the actual Romanian architecture needs, as a sustainable solution for urban pollution reduction.

1 Introduction

Nowadays the possibility of developing cementitious composites with self-cleaning properties is reported worldwide due to the photocatalytic properties of TiO_2 nanoparticles, used as an addition or as a substitute for a part of cement, in the mix.

The photocatalytic substance is a material activated by the action of sunlight or UV radiation, causing an oxidation-reduction reaction Titanium dioxide, TiO_2. This is a crystalline substance, with photocatalytic properties, that can be found in three forms of crystallization: anatase, rutile and brookite. Of all substances with photocatalytic properties, the use of TiO_2 anatase has the following advantages: it proves a high chemical stability under the action of acids and bases, it is not toxic, it has a relatively low price and has the ability to decompose organic and inorganic substances into stable compounds, oxides / salts, neutralized, thus contributing to depollution, sterilization and unpleasant odour elimination. The most important property of TiO_2, accidentally discovered in 1995, was regarding a composition of TiO_2-SiO_2, which, in the presence of UV radiation revealed the effect of surface superhydrophilicity. Another important feature of the TiO_2-SiO_2 compositions is that, unlike TiO_2 whose photocatalytic activity ceases without UV radiation, the photocatalytic effect continues hours, even days after removal of the UV source. The combination of these two photo-induced properties (photocatalytic and superhydrophilicity) is the basic mechanism of self-cleaning cementitious materials, with antibacterial, depollution, water and air purification properties [1]. In 1997, Luigi Cassar et al. present the first official publication related to the manufacture of cement materials with the self-cleaning property [2], [3].

The first experimental attempts in the topic of self-cleaning materials were performed by using enriched TiO_2 in white cement compositions. In 1996, the first relevant results were reported and in 2003 the first large-scale construction of this kind, the church Dives in Misericordia, Rome, was put into use (Fig. 1) [5].

Fig. 1. The church Dives in Misericordia, Rome, Italy

* Corresponding author: elvira.grebenisan@incerc-cluj.ro

Some reports have indicated that up to 2003, approximately 5000 constructions in Japan have used cementitious composites enriched with TiO_2 nanoparticles [4]. The latest use of this composite material was to build the famous Palazzo Italia Expo 2015, Milan, Italy. The purpose of this paper is to present the latest stage of global research on concrete with self-cleaning properties and to adapt the concept to the current needs of Romanian architecture, as a sustainable solution for reducing urban pollution.

2 Design and realization of cement composite materials with TiO_2 nanoparticles

Titanium dioxide has been used in the field of construction to induce the self-cleaning ability of various building elements, namely: building facades, exterior tiles, finishing coatings, road blocks, concrete pavements, etc., due to its chemical stability and compatibility with traditional building materials [6].
Three technologies for exploiting the properties of TiO_2 nanoparticles are currently known and used: surface coatings with TiO_2 dispersion solutions, surface coatings with a superficial cementitious composite layer including TiO_2 and the realization of entire concrete elements with TiO_2. According to the studies reported in the literature, the best photocatalytic efficiency is the introduction of nanoparticles in superficial layer, without vibration [7]. Generally, the literature recommends that the TiO_2 nanoparticles to be initially added and dried mixed to the cement, followed by the addition of hydration water. It does not chemically react with any crystallographic form of titanium dioxide, nor does a reaction occur between the photosensitive nanoparticles with the cement phases, therefore the hydrolysis hydration reactions are not influenced [1].

The particle size and distribution of the TiO_2 powder was proved to influence the cementitious composite material; consequently, at the University of Milan, Italy, research was carried out by using TiO_2 powders with micrometric dimensions (m-TiO_2) or nanometric dimensions (n-TiO_2) [1]. The obtained results showed that there are advantages and disadvantages in both cases. Thus, dispersion and distribution in the matrix is more convenient for the use of micrometric-sized granules as the nanometric ones tend to agglomerate, thereby reducing the total reactive surface available for photocatalytic reactions initiation. On the other hand, the use of nanoscale granules is advantageous, albeit more difficult, because the composite would have a better adsorption capacity of pollutant oxides, which can easily penetrate the nanoparticle agglomerations. The research carried out so far has not revealed a defining influence of particle size granulometry on the efficiency of the cement matrix enriched with TiO_2, from the point of view of the decomposition capacity of the polluting molecules and the dirt particles. Thus, it would be preferable to use nanoscale particles as they sum up both, the more efficient decomposition effect of the pollutant particles of the type of nitrogen oxides as well as their superhydrophilicity, thus decreasing the contact angle of the water and creating a uniform water film on the treated surfaces [8]. This prevents the contact between the external dirt and the surface itself, thus facilitating the washing of the pollutants and dirt particles. However, the influence of photoactivation light is neglected, because studies showed that there are differences even in the case of keeping all the initial preparation parameters, if the preparation of the composite takes place in artificial light, natural or dark light or in the same type of light but with different intensities [1], [9].

The unanimous conclusion is that, for each case, it is necessary to determine the optimum quantity of TiO_2 nanoparticles used in the composite; an additional amount of TiO_2 is not economic and very often it can negatively influence some parameters of the cement composite matrix enriched with nanoparticles TiO_2. It is also considered that, in most cases, the enrichment with TiO_2 nanoparticles of the entire mass of the concrete element is not necessary; a coating layer of composite material with photocatalytic properties is proved to be sufficient to increase element durability and to achieve the environmental purification effect.

3 Performance of cement composite materials with TiO_2 nanoparticles, in fresh concrete

A series of investigations to date have highlighted that introducing the TiO_2 nanoparticles influences the properties of fresh concrete. The first observed effect in preparing of cement mortar and concrete is the increase in water demand to achieve the standard consistency [4]. Regarding the influence of the used amount of TiO_2 nanoparticles, considering the workability point of view of, it can be said that increasing it leads to clear decrease of workability. This effect was noticed in both, simple cement paste mixes and also in composites containing slag. Thus, the sequence of used content (percentage) of nanoparticles of titanium dioxide (TiO_2) was: 0 and 2.0%; 0, 0.5, 1.0, 1.5 and 2.0%; 0, 1.0 and 3.0%; 0, 1.0, 3.0 and 5.0%; and 0, 1.0, 2.0, 3.0, 4.0 and 5.0%, in simple cement systems (Fig. 2), respectively 0, 5.0 and 10.0% in composite systems mixed with slag [6]. However, some research has indicated that cement replacement with a maximum of 1.0% TiO_2 (by weight), does not significantly affect the fluidity of the cement mortar, independently of slag adding or not to the mix [10].

When introducing different amounts of TiO_2 nanoparticles (like: 0, 5.0 and 10.0%; 0, 0.5, 1.0, 1.5 and 2.0%; 0, 5.0, 7.0 and 10.0%), it was observed that initial and final open time decreased with increasing the content of TiO_2 nanoparticles [6]. (Fig. 3).

The decrease in both, workability and open time with the increase in the amount of TiO_2 nanoparticles introduced into the cement paste can be explained by the catalytic effect of the nanoparticles on the cement hydration reaction and they function as potential storage cores of hydration products.

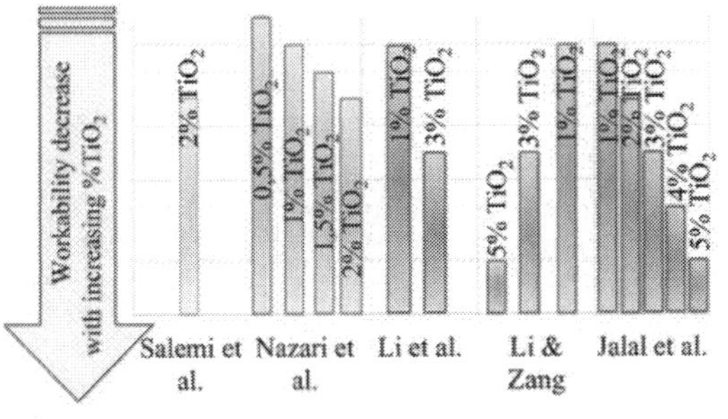

Fig. 2. Workability decreasing with increasing of TiO_2 content [processing reference 6]

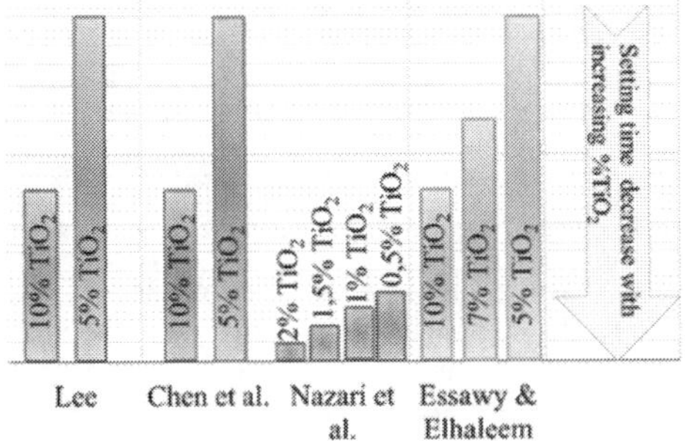

Fig. 3. Setting time decreasing with increasing of TiO_2 content [processing reference 6]

Research on matrix hardening processes has revealed significant increase of the hydration heat together with increase of the cement hydration rate [4], [10], the change of the structural orientation of the CH crystals and of their dimensions, as well [4]. Figure 4 shows the growth rate of the cement hydration heat for the pastes where different content (percent by weight) of TiO_2 nanoparticles added. Figure 5 emphasises the increase rate of the heat released by hydration for cement pastes with slag addition and various TiO_2 nanoparticles content.

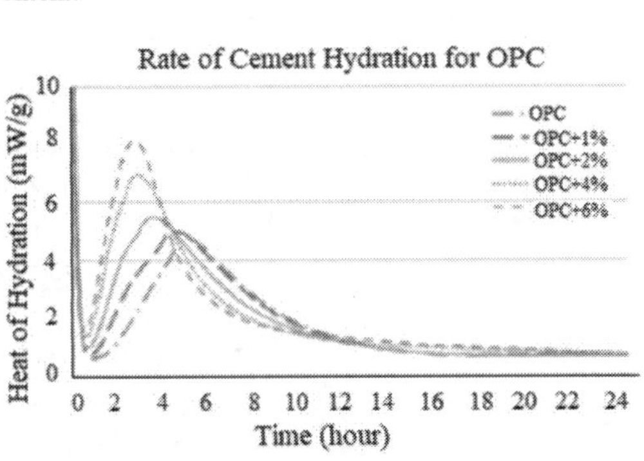

Fig. 4. Rate of heat release evolution for Portland cement pastes with different TiO_2 dosages [11]

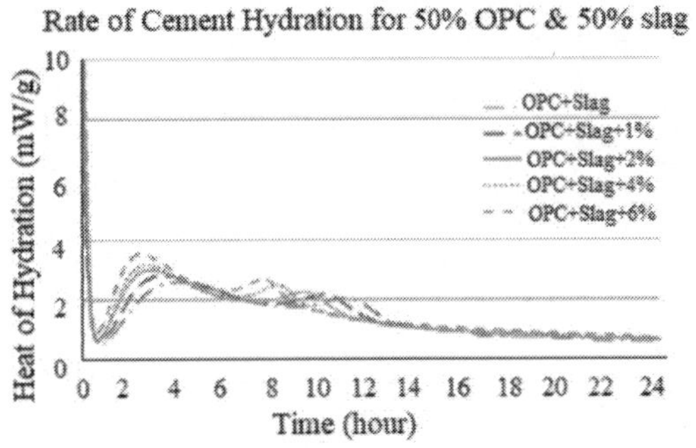

Fig. 5. Rate of heat release for cement pastes with 50% Slag and different TiO_2 dosages [11]

4 Self-cleaning performance of cementitious composite materials with TiO_2 nanoparticles in hardened concrete

According to the research done so far, in terms of the durability of the self-cleaning property as well as the overall durability of the composites, it is appreciated that the light wavelength in the visible field is sufficient to induce the photoactivation of TiO_2, but it may lead to colouring, sensitization and colour degradation through this mechanism [1]. For photoactivation is generally preferable to select waves in the UV field around the wavelength of 360 nm. The degradation mechanisms vary function to the wavelength at the time of preparation, and function to illumination amount during conditioning [1]. A large number of experimental studies have shown that compressive strength increases with increasing the nanoparticles content, the optimum content being of 1%. Other studies claimed that compressive strength decreases as the TiO_2 content increases [6]. (Fig. 6).

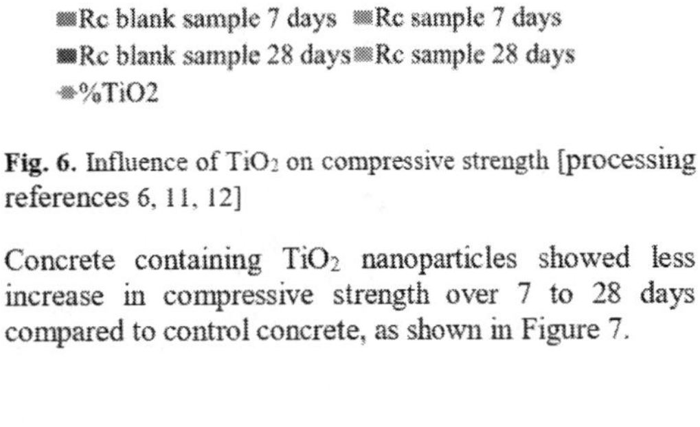

Fig. 6. Influence of TiO_2 on compressive strength [processing references 6, 11, 12]

Concrete containing TiO_2 nanoparticles showed less increase in compressive strength over 7 to 28 days compared to control concrete, as shown in Figure 7.

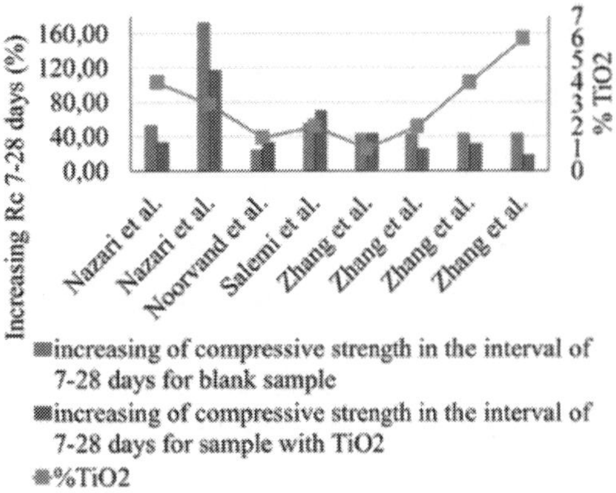

Fig. 7. Influence of TiO_2 on the increase of compressive strength in the interval of 7-28 days [processing references 6, 11, 12]

According to the literature, the early age compressive strength growth rate was higher than at older ages. Consequently, 5% (by weight) of TiO_2 nanoparticles introduced in the mixes increased the compressive strength 1, 3 and 7 days but it reduced it at the age of 28 days, at a water to binder (w/b) ratio of 0.5. A 10% TiO_2 nanoparticle addition increased the compressive strength at 1 and 7 days but it reduced it at 3 and 28 days, at a water to binder (w/b) ratio 0.5. Considering a 0.4 w/b ratio, the compressive strength increases along with the TiO_2 nanoparticles content increasing; considering the w/b ratio of 0.5 or 0.6, the compressive strength decreases with the content increase of TiO_2 nanoparticles [6].

In terms of flexural strength, the use of 1.0% and 3.0% TiO_2 nanoparticles addition in relation to the amount of cement, increased the flexural strength, while a 5% addition decreased it. The optimum TiO_2 nanoparticles addition was considered to be 1% (by weight, related to the cement content). Splitting and flexural strength increased with increasing the TiO_2 nanoparticles addition; 4% is considered the optimum percentage. High flexural and splitting resistances are also reported by other authors who established that the optimal percentage of TiO_2 nanoparticles is 1% for water curing and 2% for limewater curing [6]. In literature, the increase of fatigue strength is experimentally confirmed, 1% being the optimum TiO_2 nanoparticles addition. Cement composites developed with 10% slag addition and TiO_2 nanoparticles additions of 5% and respectively 10% (cement related content) led to compressive strength increase at 1 day and induced a decrease of the compressive strength at 28 days. Increase of the flexural strength also occurs for the mixtures containing 45% slag and 3% TiO_2 nanoparticles [6].

The increase of TiO_2 nanoparticles leads to the chlorine permeability coefficient reduction, the optimum content of nanoparticles addition being 1%, according to the reports made by some authors, and 4% being the optimal addition of nanoparticles according to experimental researches carried out by others [6]. Simultaneously with the increase of the amount of TiO_2 nanoparticles used in mix, the porosity reduction occurs. Water absorption of the composite, determined at the age of 28 days and 90 days, showed that increasing the nanoparticles content reduces the water absorption, 0.5% being the optimum addition content of TiO_2 nanoparticles. For water absorption evaluated at the age of 7 days, the 4% TiO_2 nanoparticle content is proved to develop the best results. However, in all cases, if water absorption is evaluated at the age of 2 days, the addition of TiO_2 nanoparticles tents to reduce it. In terms of aesthetic durability, experimental research has shown that keeping white or coloured (in the case of coloured matrix) is not influenced by the amount of nanoparticles addition. TiO_2 particles are characterized by a high degree of whiteness, but their relatively uniform mass distribution does not affect the possibility of staining using classical pigments. Once the composite mass was coloured, no time fades were observed due to the photocatalytic activity of TiO_2 nanoparticles [1].

5 Environmental influence of self-cleaning building

Although the influence of TiO_2 nanoparticles on the properties of fresh and reinforced concrete still has many uncertainties, the antibacterial, antimicrobial and antifungal effect of TiO_2, for some of the most common microorganisms (cladophora, chlorella vulgaris, escherichia coli, aspergillius niger), is analysed and proved [13]. Thus, TiO_2 enriched cement materials became an ally in the fight for disinfection, sterilization and purification of surfaces, water and air from sensitive public spaces (hospitals, schools, restaurants, airports, etc.). Therefore, a significant number of studies has demonstrated the ability of this semiconductor to decompose organic substances, grease and dirt, making them to be easily rainwater washed [4], [8]. Since the TiO_2 cement material surface is superhydrophilic (water no longer forms droplets but lamellas instead, which are able to easily remove decomposed compounds), the washing process is clearly more facile. Moreover, surfaces retain and regenerate their aesthetic appearance and colour under the action of UV radiation [4]. Thus, some authors reported that tests performed on 1%, 2%, 4% or 6% TiO_2 concrete surfaces indicated the self-cleaning capacity preservation even after continuous exposure to UV for 1500 hours [11]. Therefore, because of the self-cleaning capacity of these kind of surfaces, the water and detergent consumptions are reduced.

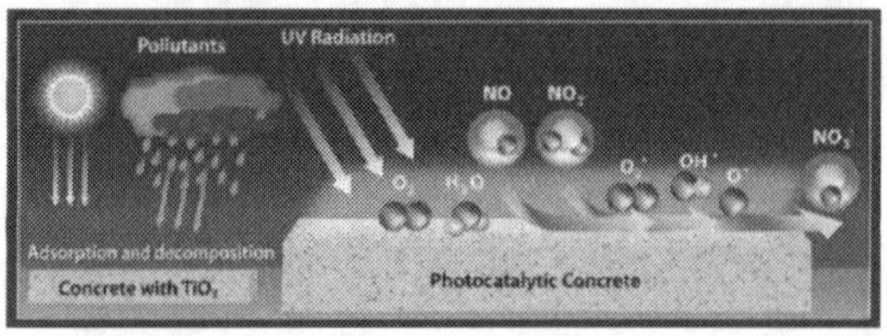

Fig. 8. Schematic of photocatalytic air purifying pavement [14]

Regarding the air pollution, Luigi Cassar called the concrete with TiO_2 "smog-eating concrete" [2]. In urban areas air pollution is mainly caused by combustion of heating and transport fuels, so the most frequent pollutants are CO, NOx, VOC and dust. The research results have shown that TiO_2 contributes to the elimination of CO, NO_x, SO_x, NH_3, aromatic hydrocarbons, VOC [8], [15]. There are authors who believe that surfaces of cementitious materials enriched with TiO_2 nanoparticles, as a result of their self-decontamination capability, represents passive forms of safety technology to increase safety of biological, radiological and nuclear accidents. Regarding to this respect, the authors recommend the use of cementitious materials enriched with TiO_2 nanoparticles for strategic military structure projects as well. The travertine durability, due to its efficient surface treatment by using a suspension of TiO_2, has been demonstrated. Thus, when exposing the travertine surface, previously TiO_2 cured, to the action of H_2S, this is converted into H_2SO_4 by the means of the photocatalysis process and then it is eliminated by evaporation [16].

Experimental research has shown that the reduction in pollution by the absorption of nitrogen oxides, sulphur, etc., and the decomposition of larger pollutants (fats, oils, dusts, microorganisms) are also influenced by the light exposure conditions, meaning the number of hours of illumination, air traffic, traffic conditions, concentration and type of pollutants. Thus, in southern Europe where the climate and lighting conditions are very favourable, the degree of absorption of nitrogen oxides can reach up to 60%. Under these conditions, namely the favourable lighting, an efficiency of 78 (± 2) % has also been recorded for the toluene removal (with a considerable speed of 100 mg/h*m^2), when compared to less favourable regions, from the lightning and climatic point of view. [1].

6 Standardization

Due to the already proven benefits of TiO_2 nanoparticulate concrete, some specific analysis methods for material evaluation have already been standardized over the past 15 years, as shown in Table 1. It is noted that most of these are Japanese and Italian national standards, some being internationally assimilated.

Table 1. Standards for methods of analyzing the photocatalytic action of TiO_2 [15].

Performance attributes	Principle of test method	Standard
Air purification effect	Nitric oxide removal	JIS R 1701-1:2004 UNI 11247:2007 ISO 22197-1:2007
	Removal Volatile Organic Compound (VOC)	UNI 11238-1, 2:2007
	Acetaldehyde removal	JIS R 1701-2:2008 ISO/CD 22197-2
	Toluene removal	JIS R 1701-3:2008
Water purification effect	Active oxygen-forming	JIS R 1704:2007 ISO 10676: 2010
Self-cleaning effect	Water contact angle change	JIS R 1703-1:2007 ISO 27448: 2009
	Methylene blue decomposition	JIS R 1703-2:2007 ISO 10678:2010
	Rhodamine	UNI 11259:2008
Biocidal effect	Antibacterial activity	JIS R 1702:2006 ISO 27447:2009
	Antifungal activity	JIS R 1705:2008
Other	Light source for test under UV irradiation	JIS R 1709:2007 ISO 10677: 2011

7 Conclusions

Today the use of TiO_2 nanoparticles is performed either by coating surfaces with TiO_2 dispersion solutions or by coating surfaces with a superficial cementitious composite layer with TiO_2 addition or by entirely producing elements of concrete with TiO_2. The latter case is the least favourable, from the point of view of the high consumption of nanoparticles, some of which remaining unexploited due to their inherent positioning in the depth of the composite mass.

The purpose of this paper was to present the latest stage of global research on concrete with self-cleaning properties and to consider some possibilities for adapt the concept to the current needs of Romanian architecture, as a sustainable solution for reducing urban pollution. On the basis of the above it can be said that:

- TiO_2 nanoparticles have water and air purification properties, antimicrobial, antifungal and anti-allergic in the presence of UV.
- Under the action of UV rays, a TiO_2-SiO_2 composition becomes superhydrophilic, so that the water that reaches the surface of this material, as a result of the oxidation-reduction reaction, forms the lamellae which facilitate the uptake of the impurities, the photocatalytic effect continuing hours even days after removal of the source UV.
- Activated by UV rays, the TiO_2 nanoparticles facilitate the decomposition of organic pollutants (fats, oils, aromatic hydrocarbons) into simpler compounds, easily washable by rainwater; they also entrain the dust particles deposited on the superhydrophilic surface.
- UV-activated TiO_2 nanoparticles convert inorganic pollutants such as nitrogen, sulphates, acids, air, water or solids into oxide compounds / salts by oxidation reduction / neutralization reactions into stable compounds, oxides / salts, thus contributing both, in the self-cleaning of urban buildings and in the depollution.
- In the fresh concrete the TiO_2 nanoparticles reduce the workability, open time and porosity.
- In terms of reinforced concrete performance: increased mechanical strength, increased frost-freeze resistance, increased resistance to abrasion and increased

resistance of concrete to chemical agents, when the optimum TiO_2 nanoparticles addition is used.

- The influence of nanoparticles on the mechanical strengths of concrete is still controversial, but it can be already established that the first requirement is, on a case-by-case basis, the determination of the optimum TiO_2 nanoparticles content, with respect to the cement.
- The literature reports agree on the beneficial effect of TiO_2 nanoparticles regarding the reduction of water and air pollution and on their self-cleaning capacity, despite the fact the mechanisms are not yet fully elucidated.
- Over the past 15 to 20 years, research regarding cementitious composite matrixes with TiO_2 addition has increased. as a direct result of the interest in achieving sustainable, aesthetic, low maintenance and improved performance, as well as identifying pollution reduction possibilities.
- The optimal compositions of nanoparticles as well as the optimal granulometric distribution for the maximum benefits from the photocatalytic effect are not yet fully identified. On the other hand, even the photocatalytic activation mechanisms are not yet fully elucidated.
- Since 2007 there were implemented some standardized methods of analysis in this topic and the trend of international assimilation is noticed.

References

1. A. Folli, *TiO_2 photocatalysis in Portland cement systems: fundamentals of self-cleaning effect and air pollution mitigation*, PhD Thesis, University of Milan, Italy (2010)
2. L. Cassar, *Nanotechnology and photocatalysis in cementitious materials*, NICOM'2, Bilbao (2005)
3. A. Mujkanović, D. Bečirhodžić, N. Merdić, *Self-Cleaning Concrete – A construction material for building cleaner world*, 20th International Research/Expert Conference "Trends in the Development of Machinery and Associated Technology" (2016), pp. 93-96
4. M. Janus, K. Zając, *Concretes with Photocatalytic Activity, High Performance Concrete Technology and Applications*, INTECH (2016)
5. Richard Meier & Parteners Arhitects LLP: Jubilee Church, https://www.richardmeier.com, accessed in July 2019
6. A.M. Rashad, *A synopsis about the effect of nano-titanium dioxide on some properties of cementitious materials - a short guide for civil engineer*, Rev. Adv. Mater. Sci. **40** (2015), pp. 72-88
7. S. Hanson, *Evaluation of concrete containing photocatalytic titanium dioxide*, PhD Thesis Department of Civil and Environmental Engineering, The University of Utah (2014)
8. E. Quagliarini, F. Bondioli, G.B. Goffredo, C. Cordoni, P. Munafò, *Self-cleaning and de-polluting stone surfaces: TiO_2 nanoparticles for limestone*, Construction and Building Materials **37** (2012), pp. 51–57
9. D.H. Chen, K. Li, *Photocatalytic Coating on Road Pavements/Structures for NOx Abatement*, Annual Project Report Submitted to Houston Advanced Research Center and Office of Air Quality Planning and Standards U.S. Environmental Protection Agency Research Triangle Park, NC 27711 (2007)
10. J.V. Sorathiya, S.G. Shah, S.M. Kacha, *Effect on Addition of Nano "Titanium Dioxide" TiO_2 on Compressive Strength of Cementitious Concrete*, International Conference on Research and Innovations in Science, Engineering & Technology At: Birla Vishvakarma Mahavidyalaya, Gujarat, India **1** (2017)
11. S.M.-H. Zhang, D. Tanadi, W. Li, *Effect of photocatalyst TiO_2 on workability, strength, and self - cleaning efficiency of mortars for applications in tropical environment*, 35th Conference on Our World in Concrete and Structures, Singapore (2010)
12. A. Hegyi, A. Lăzărescu, C. Dico, H. Szilagyi, *The effect of TiO_2 on the properties of cementitious composite materials – the current state-of-the art*, Proceedings of the 18th International Multidisciplinary Scientific GeoConference & EXPO – SGEM 2018, Albena, Bulgaria (2018), pp. 391-398
13. B. Ma, H. Li, J. Mei, X. Li, F. Chen, *Effects of Nano-TiO_2 on the Toughness and Durability of Cement-Based Material*, Advances in Materials Science and Engineering (2015)
14. Boonen E., Beeldens A.: Recent Photocatalytic Applications for Air Purification in Belgium, Coatings (2014), 4, pp. 553-573
15. Y. Ohama, D. Van Gemert, *Applications of Titanium Dioxide Photocatalysis to Construction Materials*. State-of-the-Art Report, Ed. Springer (2011)
16. A. Sassolini, A. Malizia, F. D'Amico, M. Carestia, D. Di Giovanni, O. Cenciarelli, C. Bellecci, P. Gaudio, *Evaluation of the effectiveness of titanium dioxide (TiO_2) self-cleaning coating for increased protection against cbrn incidents in critical infrastructures*, Defence Science and Technical Bulletin **7** (2014), pp. 9-17

Self-Cleaning Concrete for Landscaping Applications

Diana-Maria Mircea[1,*]

[1] University of Agricultural Sciences and Veterinary Medicine of Cluj-Napoca, 400372 Cluj-Napoca, Romania

Abstract. Negative effects of environmental pollution pose a significant risk to agriculture, water resources and human health. This can however be reduced by selecting appropriate materials in construction and landscape architecture. It is well-known that strong sunlight or ultraviolet light decomposes many organic materials in a slow natural process. Photocatalytic substances accelerate this process and when used in concrete (which is one of the most widely used construction materials), permit the treatment of pollutants close to their source by applying a self-cleaning principle: decomposing organic materials, biological materials and pollutants into molecules like oxygen, water, carbon dioxide, nitrates, and sulphates. Catalytic material activation will start due to energy that is received from sunlight (it also can be activated using artificial lights), and self-cleaning begins when this material is activated. Beside other properties, this innovative self-cleaning concrete also keeps its colour for far longer than other traditional building materials, making it a desirable solution for landscaping applications.

1 Introduction

People are more aware nowadays of the importance of preserving the environment. In a modern urban environment, green spaces often serve as isles of habitat for native species of plants, insects, birds and small animals. In densely populated areas, landscape architecture provides outdoor recreation areas within aesthetically pleasing and culturally meaningful sceneries that increase the life quality and enhance the well-being of local population.

All industrialized places and densely populated urban areas have problems with pollution. Factories are essential to provide a source of economic growth but also have negative effects on the environment. Environmental pollution poses a significant risk to agriculture, water resources and human health. This can however be reduced by selecting appropriate materials in construction and landscape architecture.

Structural and decorative concrete elements, covering a large variety of shapes and colours, are used in landscaping in order to expand the design possibilities when putting into practice new architectural solutions.

A photocatalyst is a semiconducting substance [1] which can be chemically activated by light radiation that results in an oxidation-reduction reaction. The application of photocatalysts in concrete technology is by now a well-established concept. Starting in the late 1960s [2], the research on titanium dioxide (TiO_2) began with photoelectrochemical solar energy conversion, then through the area of environmental photocatalysis, including self-cleaning surfaces, and nowadays it has emerged as an excellent photocatalyst material for environmental purification.

2 Properties of self-cleaning concrete

Landscape compositions create harmonious combinations of natural and artificial elements of the environment (Figure 1), in accordance with certain artistic expertise into a holistic spatial design able to improve the location and to develop recreational facilities, both efficient and inspired. A contribution to regeneration of polluted areas can be achieved through landscaping applications, especially if resourceful materials and methods are implemented.

Self-Cleaning Concrete is an innovative construction material that has been developed in the past decades providing the "self cleaning" ability of concrete surfaces, by retaining their light colour for longer, along with the property to remove airborne toxins, particularly nitrogen oxides. The reactions begin when the catalytic material added to concrete is activated, due to sunlight, or artificial light with similar properties (e.g. ultraviolet light).

Fig. 1. Landscape composition with concrete elements.

* Corresponding author: diana.mircea@rocketmail.com

Since strong sunlight decomposes many organic materials in a slow natural process, photocatalysts accelerate this process, and stimulate the chemical transformation without being consumed by the reaction. To increase the efficiency of photocatalysts, they must be accessible to sunlight and activated, so their presence on the surface of the material is essential [3]. When used in concrete, they decompose organic materials like oil and silt, biological materials (mould, algae and bacteria) and pollutants such as volatile organic compounds and tobacco smoke into molecules like oxygen, water, carbon dioxide, nitrates, and sulphates. The whole reaction results in a significant reduction in the concentration of pollutants. It should be noted, however, that most inorganic pollutants and stains, including rust, are not catalysed.

In the past years, these technologies have been applied to glass, ceramic, and cement-based materials. The main applications of TiO_2 (traditionally used as white pigment) include pollution remediation, self-cleaning and self-disinfecting surfaces. Along with the advantage of using solar light and rainwater as the driving force, they confer environmentally compatible properties to building materials [4] and landscaping applications. The cement industry [5], has already developed new cement products that offer the possibility to improve environmental quality by decomposing air pollutants in harmless substances, while using a photocatalytic reaction based on a patented technology.

Concrete is a material that allows creating almost every shape: simulations of natural landforms, slopes, hills, retaining walls, stairs, ramps, terraces, urban furniture (Figure 2). It allows the possibility to embed into the landscape various functional, social, cultural and entertaining elements with interesting decorative characteristics and forms.

Fig. 2. Decorative elements suitable for self-cleaning concrete.

Paving elements and concrete sidewalk plates are installed in different commercial zones, garden-parks and other pedestrian areas, being used for walk-path surfaces, in a variety of shapes and colours, which expand the design possibilities when implementing traditional or modern architectural solutions. For self-cleaning concrete pavement applications, since photocatalysts transform the pollutants into water soluble compounds, the porosity of the prefabricated concrete elements is very important.

The porosity is affected by aggregate size, water-to-cement ratio, and, in general, by the mix design of the concrete. Study results [6] emphasize that the photodegradation of nitrogen oxides (NO_x) is related to the porosity of the pavement elements; accordingly, a greater porosity results in increased ability for the removal of nitrogen oxides (NO_x). The self-cleaning effectiveness of concrete depends of the titanium dioxide (TiO_2) content of the cement. Study results [7] reveal that higher TiO_2 content in cement increases the self-cleaning activity.

Placed where air quality is greatly affected by vehicle exhaust, noise barriers and visual screens are construction elements with both functional and aesthetic properties, being part of the built environment as well as of the landscaping applications. If made of concrete containing photocatalytic cement, or prepared with the application of surface mortar coatings, or of a photocatalytic concrete cover (which is also more economically feasible [8]) it will allow the oxidation of air pollutants to occur on their surface, along with self-cleaning characteristics and impact on deteriorated air quality in urban environments.

Strategic placement and selection of plant species within an architectural project to highlight and enhance aspects of spatial design produce many daily benefits. Concrete-made interior surfaces (used for vertical gardens, plant storing arrangements), may be subject of self-cleaning characteristics (i.e. air purification) by using the photocatalytic oxidation [9], to destruct volatile organic compounds (VOCs) in indoor air.

3 Application of self-cleaning concrete elements in design - Case study

The path of a landscape project from the beginning to its end typically contains the following four major phases (each with its own agenda of tasks and issues): initiation, planning, implementation, and closure. The environmental compatibility for construction and landscaping projects is revealed through active multidisciplinary collaboration, from the design study and preliminary design phases, to execution and supervision, all the way to project completion [10]. Architectural characteristics have to be preserved, landscape design objectives have to be accomplished, and innovative materials, methods and features have to be implemented. In this regard, with respect to regulations, every design theme gives freedom of choice and allows guidance through selection of plants, decorations, shapes, structures and materials.

 https://doi.org/10.1051/matecconf/201928905004

Figure 3 presents the large application possibilities of self-cleaning concrete related to often encounter landscaping demands.

Fig. 3. Landscaping design proposals for functional and decorative applications of self-cleaning concrete elements.

4 Conclusions

The sustainable building concept focuses on the importance of an environment-friendly management of materials throughout their entire life-cycle, from resource extraction, during the fabrication phase, on-site installation, use, maintenance and post-utilization. Using sustainable building materials contributes to conservation of natural resources along with the protection of landscape and its natural balance [11]; it helps to reduce pollution, to achieve energy savings, and to provide a cleaner and healthier living environment.

The use of concrete for the decorative landscape elements is both widely applicable, as well as custom-made, due to the durability and shape adaptability qualities of this material [12]. To be effective (i.e. to react with sun light and also to reduce cost) it is more useful when cement with addition of TiO_2 is added in surface coating or in the form of concrete cover, rather than being mixed into the construction concrete.

Based on experimental evidence regarding the ability to carry out self cleaning by reaction with compounds responsible for surface finish degradation, building elements containing cement with addition of TiO_2 can maintain their aesthetic appearance over time [13]. The lasting performance of photocatalytic activity was assessed [14] during one year of outdoor exposure trial and 2000 h of accelerated ageing in a chamber with UV radiation and condensation cycles. Results showed that photocatalytic activity on concrete was unaltered.

The constructive and compositional landscaping design solutions depend by their placement in combination with vegetation, relief, walkways, patio areas, and many more locally arising particular details. Along with the addition of photocatalysts to concrete, environmentally compatible materials are created, which can reduce pollution and act like self-cleaning surfaces, also diminishing maintenance related costs.

References

1. Y. Ohama, D.V. Gemert, *Application of Titanium Dioxide Photocatalysis to Construction Materials*, State of the Art Report of the RILEM Technical Committee 194-TDP, Springer Dordrecht Heidelberg London New York (2011).
2. A. Fujishima*, T.N. Rao, D.A. Tryk, Titanium dioxide photocatalysis, *Journal of Photochemistry and Photobiology* C: Photochemistry Reviews 1 (2000) 1–21.
3. A. Beeldens, An environmental friendly solution for air purification and self-cleaning effect: the application of TiO2 as photocatalyst in concrete, *Proceedings of 10th International Symposium on Concrete Roads*, Belgian Road Research Centre Brussels, Belgium (2006).
4. J. Chen, C. Poon, Photocatalytic construction and building materials: From fundamentals to applications. *Building and Environment*, 44, (2009) 1899–1906.
5. Hanson Cement, TioCem - Reducing pollution in the urban environment, www.hanson.co.uk.
6. C.S. Poon, E. Cheung, NO removal efficiency of photocatalytic paving blocks prepared with recycled materials. *Construction and Building Materials*, Vol. 21, No. 8, (2007) pp. 17461753.
7. F. Pacheco-Torgal, S. Jalali, Nanotechnology: advantages and drawbacks in the field of construction and building materials, *Construction and Building Materials*, V. 25, pp. 582-590 (2011).
8. C.J. Churchill D.K. Panesar, Life-cycle cost analysis of highway noise barriers designed with photocatalytic cement, *Structure and Infrastructure Engineering*, Maintenance, Management, Life-Cycle Design and Performance, Vol. 9, 10 (2013) pp. 983-998.
9. J. Zhao, X. Yang, Photocatalytic oxidation for indoor air purification: a literature review. *Building and Environment*, V. 38, No.5 (2003) pp. 645-654.
10. A.T. Mircea, Selecting Environmentally Compatible Materials for Priority Integration into the Design Process, *17th International Multidisciplinary Scientific Geo Conference SGEM 2017*, Vol. 17, 62, Section Green Buildings, Technologies and Materials, Albena, Bulgaria, pp. 409-416. (2017) *doi: 10.5593/sgem2017/62/S26.052.*
11. A.T. Mircea, Study upon Applying Multi-Criteria Analysis to the Selection of Building Materials, *28th IBIMA Conference on Vision 2020: Innovation Management, Development Sustainability, and Competitive Economic Growth*, Seville, Spain (2016).
12. EN 1992-1-1:2004 Eurocod 2: Design of concrete structures - Part 1-1: General rules and rules for buildings (2004).
13. L. Cassar, Cementitious materials and photocatalysis, *Congress of International Glassfibre Reinforced Concrete Association*, Prague (2008) pp. 1-7.
14. P.M.Carmona-Quiroga, S.Martínez-Ramírez, H.A. Vilesa, Efficiency and durability of a self-cleaning coating on concrete and stones under both natural and artificial ageing trials, *Applied Surface Science*, Vol. 433, (2018), pp. 312-320.

*Renderings by the author.

Comparative study on shotcrete repair solutions of buried steel pipelines

Călin GR Mircea[1,*], *Sergiu* Blaga[1], and *Ovidiu* Dăbăcan[1]

[1]Technical University of Cluj-Napoca, Civil Engineering Faculty, Cluj-Napoca, 28 Memorandumului Street, 400114, Romania

Abstract. The research aims to optimize the rehabilitation works of steel buried pipes after their lifespan is reached. It is based on objective real situations met in the Romanian agricultural irrigation system. The study proposes an analytical approach to predict the behavior of the rehabilitated pipelines initially subjected to various corrosion rates, from low to high agressivity of the environment. Besides traditional lining with reinforced concrete, a new trend given by the relative modern structural repair mortars is considered. The time behavior of the shotcrete lining is estimated based on information given by the appropriate European norm. The results question the design philosophy, still not clarified in the field, and emphasize the superior performance of modern cementitious materials, both in terms of technical performance and economical efficiency. Future research perspectives are also underlined with regard to the new materials meeting sprayed concrete technology requirements, and design philosophy.

1 Background

1.1 General considerations

Steel pipelines are traditionally used to transport liquids or gases over long distances, during time proving reliability and cost efficiency. Such pipelines operate at high pressures and sometimes elevated temperatures. During service, pipelines endure damage due to built-in metallurgical defects, welding imperfections and interaction with internal and external surroundings.

Most pipeline systems were designed for a lifespan of 30 years but are still in service after more than 40 years. External corrosion is the most frequent cause for buried pipeline failures. It reduces strength capacity, affects structural integrity and results in shorter service life and increased vulnerability. Besides service disruption, deterioration could end in leakage and bursts that may provoke significant financial losses and serious environmental damage.

When dealing with corrosion, repair techniques must comply with the demands to prevent future corrosion, leakage sealing and restore strength performance. When affected by extended damage, pipelines repair presume full-encirclement sleeves. Spraying thin layers of concrete/mortar inside pipes is a traditional technique, which improves hydraulic properties and ensures corrosion protection. However, to control strength and cracking, steel mesh is needed: leading to thick layers and steel consumption. Thus, the technique becomes expensive, time consuming and hard to implement on long segments.

Nowadays, the new generation of discrete reinforced cementitious materials (e.g., fiber reinforced concrete, engineered cementitious materials, structural repair mortars etc.) deliver versatile alternatives. Even if they require full flow diversion, pipe cleaning and primer coating, their use ensures fulfillment of all demands and provides durability of the repaired systems as well.

1.2 The Romanian irrigation system

The water supply for Romanian agriculture is ensured by an irrigation system built mainly between 1971 and 1989 (Figure 1), which covers an area of about 29,920 km^2. The River Danube provides 85 % of the water transported through 26,700 km of buried pipelines and 10,670 km of channels.

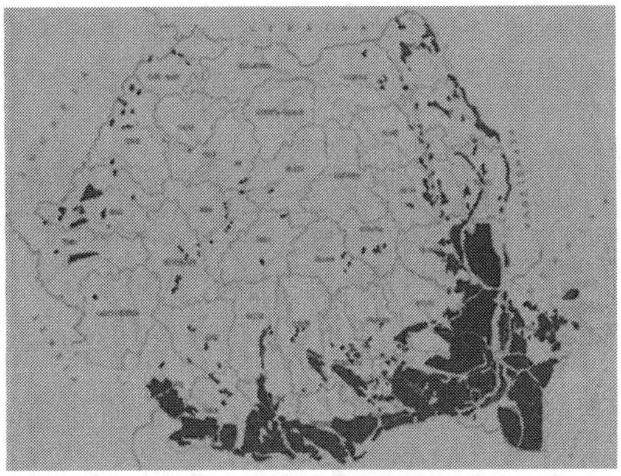

Fig. 1. Areas operated by the Romanian irrigation system.

Given the extended service and poor technical state of the system (e.g., about 75 % of the capacity was inoperable at the beginning of the decade), in 2016 a

* Corresponding author: calin.mircea@dst.utcluj.ro

governmental effort was launched to rehabilitate the public system and guarantee access to the all end users from the served zones.

2 Research significance

The study aimed to optimize the interventions made at the steel buried pipes from the Covurlui plain (Galaţi county, as shown in Figure 2). These were in service since 1985-1989 and are under a comprehensive rehabilitation program.

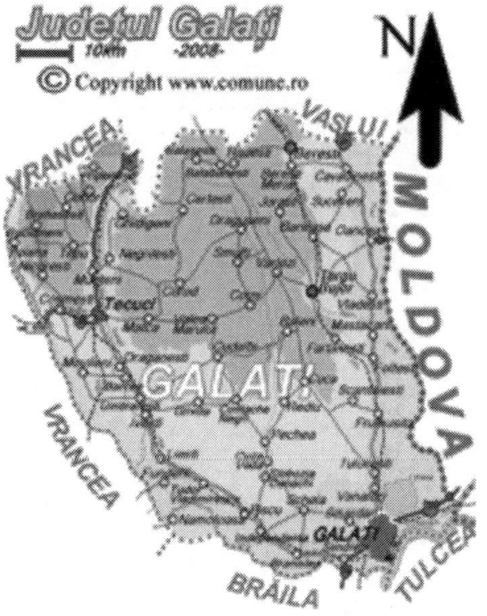

Fig. 2. Irrigation system emplacement.

In previous work, Mircea et al. [1] presented a successful rehabilitation project made by shotcrete reinforced with steel welded meshes. This research refers to the rehabilitation by sprayed concrete, through a comparative analysis. Besides the traditional reinforced concrete solution, a structural repair mortar class R4 is considered. After detailed but limited technical inspections and tests on materials, an analytical model was proposed and the solutions are objectively compared.

Various pipe diameters and functions (e.g., siphon, adduction or suction) were considered. Table 1 presents data for three characteristic pipes, analyzed forward within the paper.

Table 1. Pipe characteristics.

Characteristic / Pipe symbol	P1	P2	P3
Exterior diameter D_e [mm]	3500	3100	1800
Pipe thickness t_{ps} [mm]	10	10	10
Steel pipe yielding strength f_{spy} [MPa]	240	240	240
Modulus of elasticity E_{sp} [GPa]	200	200	200
Working pressure p_w [MPa]	0.62	0.95	0.80
Hydraulic shock pressure p_{hs} [MPa]	1.00	1.30	2.00

3 Analysis and modeling

3.1 Inspections and preliminary remarks

Besides local damage (e.g., rusted grilles and flanges, silting etc.), general electrochemical corrosion was noticed on several pipes during technical inspections. The measurements performed with ultrasonic thickness gauges revealed steel section reductions on average up to 41.7 % (i.e., 4.17 mm). However, a few zones fully corroded were also found.

According to EN ISO 12944-2:2018 [2], pipes are framed in the corrosion class IM 3.

Considering corrosion rates from low (i.e., 20 µm/year for soil without aggressivity) up to high (i.e., 250 µm/year for soil with high aggressivity), Table 2 presents an estimation of the possible pipe thicknesses after 30 years of service and processing for rehabilitation.

Table 2. Estimated pipe thickness.

Corrosion rate [µm/year]	Remaining thickness [mm]	
	gross	effective
20	9.40	9
50	8.50	8
100	7.00	6
150	5.50	5
200	4.00	3
250	2.50	2

Table 3 presents the framing of the pipe lining in exposure classes on the ground given by EN 1992-1-1 [3].

Given a design lifespan of 30 years for the rehabilitated pipes, a concrete class C 30/37 and a pre-dosed repair mortar class R4 were considered.

Table 3. Framing of shotcrete lining in exposure classes.

Risk	Exposure class
Corrosion induced by carbonation	XC 4 - cyclic wet and dry environment
Chemical attack	XA 3 - highly aggressive chemical environment

3.2 Materials properties

The results of the mechanical tests (Figure 3) made on the concrete C 30/37 and repair mortar R4 mixes at the age of 28 days are summarized in Table 4.

Table 4. Mean experimental strengths.

Mix	f_{cm} [MPa]	f_{ctm} [MPa]	τ_{bm} [MPa]
Concrete C 30/37	39.2	3.1	0.6
Structural repair mortar R4	58.4	10.3	2.3

Characteristic strengths and secant Young's modulus were determined using standard relations given by EN 1992-1-1 [3]. The results are shown in Table 5.

Fig. 3. Specimens and tests.

Table 5. Characteristic strengths and Young's modulus.

Mix	f_{ck} [MPa]	f_{ctk} [MPa]	E_{cm} [MPa]
Concrete C 30/37	31.2	2.1	33144
Structural repair mortar R4	50.4	7.2	37345

For the case of reinforced concrete, a steel mesh with characteristic yielding strength of f_{yk}=460 MPa and Young's modulus E_s=200 GPa was considered to ensure both cracking control and resistance capacity.

3.3 Basics of the analytical model

The proposed analytical model considers a perfect bonding assumption (Figure 4) between the steel pipeline and the shotcrete lining. This statement may be considered valid even for the case of concrete with much lower bonding strength (see Table 4) because the steel reinforcing meshes require numerous fixing points, that result also in improved intimate contact.

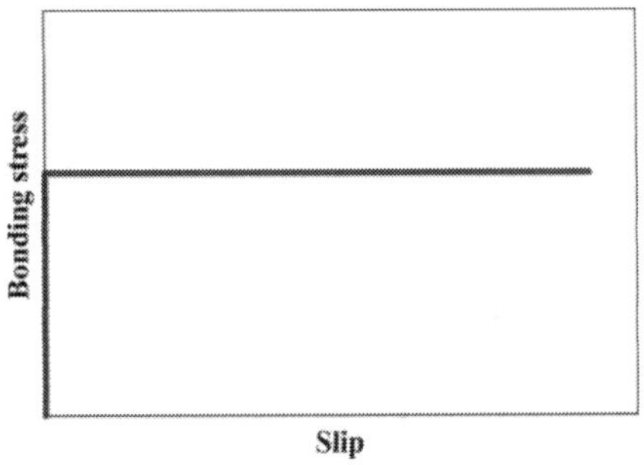

Fig. 4. Perfect bond followed by free slip assumptions.

The axial restraint degree between the initial steel pipe and the shotcrete lining is introduced [4, 5] to consider shrinkage in analysis

$$R = \frac{\text{free slip - restrained slip}}{\text{free slip}} \tag{1}$$

resulting in

$$R = \frac{1}{1 + \dfrac{A_c E_{c,eff}}{A_{sp} E_{sp}}} \tag{2}$$

where A_c and $E_{c,eff}$ are the shotcrete normal area and the effective Young's modulus of shotcrete, and A_{sp} and E_{sp} are the steel pipe normal area and modulus of elasticity. The areas depend by the direction considered in analysis, the transversal ring cross-sectional areas for the longitudinal direction, and the area of the unit strips on the radial direction.

The corresponding tensile stress in the shotcrete due to restrained shrinkage becomes

$$\sigma_{cs} = R\varepsilon_{cs} E_{c,eff} \tag{3}$$

with ε_{cs} the free shrinkage strain (i.e., autogenous and drying shrinkage are considered) and $E_{c,eff}$ calculated according to the European norm [3]. Integrated on the entire surface of shotcrete lining, the axial force results as

$$N_{cs} = R\varepsilon_{cs} E_{c,eff} A_c \tag{4}$$

The internal pressure p (i.e., p_w or p_{hs}) in the pipe of internal diameter D_i generates the axial tensile forces as follows:

- in the radial direction

$$N_r = p\frac{D_i}{2} \tag{5}$$

- in the longitudinal direction

$$N_l = p\frac{D_i}{4} \tag{6}$$

Finally, the axial forces are distributed to the steel pipe and shotcrete proportional with their axial stiffness. The total axial forces are:

- In the old steel pipe

$$N_{r(l),sp} = N_{r(l)} \frac{A_{sp} E_{sp}}{A_{sp} E_{sp} + A_c E_{c,eff}} \tag{7}$$

- In the new shotcrete lining

$$N_{r(l),c} = N_{r(l)} \frac{A_c E_{c,eff}}{A_{sp} E_{sp} + A_c E_{c,eff}} + N_{cs} \tag{8}$$

In the case of ordinary reinforced concrete lining, the influence of reinforcement increases the effective area. Therefore, the idealized cross-sectional area A_{ci} should be introduced

$$A_{ct} = \iint_{A_c} \frac{E(x,y)}{E_{c,eff}} dxdy = A_c + \left(\frac{E_s}{E_{c,eff}} - 1 \right) A_s \qquad (9)$$

where A_s is the reinforcement area.

All above relations are time dependent due to the influence of creep. Finally, standard relations [3] are applied to control the crack width and ensure water tightness of the rehabilitated ensemble. The resistance capacity must provide an allowable tensile stress higher than the hydraulic shock pressure both in the steel pipe and the shotcrete lining. If ordinary reinforced concrete is implemented, only the reinforcement may be considered, and the tension stiffening neglected.

To meet the above performance at the reinforced concrete solution, c=20 mm concrete cover was adopted, and the layer of concrete taken with a thickness t_c=50 mm. Just the quantity of reinforcement was varied. In the second solution, shotcrete with structural repair mortar class R4, the lining thickness was varied t_c.

4 SYNTHETIC RESULTS

4.1 Reinforced concrete solution

Table 6 presents the time dependent evolution of the free shrinkage strain (i.e., autogenous shrinkage cumulated with the drying shrinkage) and the creep coefficient for a relative humidity of 90 % on radial and longitudinal direction of the pipes.

Table 6. Free shrinkage strain and creep coefficient (P1/P2/P3).

Shotcrete age [years]	Free shrinkage strain ε_{cs} [mm/m]	Creep coefficient φ
Radial direction		
1	0.169 / 0.179 / 0.175	1.35 / 1.33 / 1.31
5	0.185 / 0.195 / 0.190	1.79 / 1.77 / 1.74
10	0.187 / 0.197 / 0.192	1.91 / 1.89 / 1.86
15	0.188 / 0.198 / 0.193	1.96 / 1.93 / 1.90
20	0.188 / 0.198 / 0.193	1.99 / 1.96 / 1.93
25	0.188 / 0.198 / 0.194	2.00 / 1.98 / 1.95
30	0.188 / 0.198 / 0.194	2.01 / 1.99 / 1.96
Longitudinal direction		
1	0.190 / 0.192 / 0.193	1.52 / 1.58 / 1.19
5	0.197 / 0.198 / 0.194	1.90 / 1.96 / 1.64
10	0.198 / 0.198 / 0.195	1.99 / 2.04 / 1.78
15	0.198 / 0.199 / 0.195	2.02 / 2.07 / 1.83
20	0.199 / 0.199 / 0.195	2.04 / 2.09 / 1.86
25	0.199 / 0.199 / 0.195	2.05 / 2.10 / 1.88
30	0.199 / 0.199 / 0.195	2.06 / 2.10 / 1.90

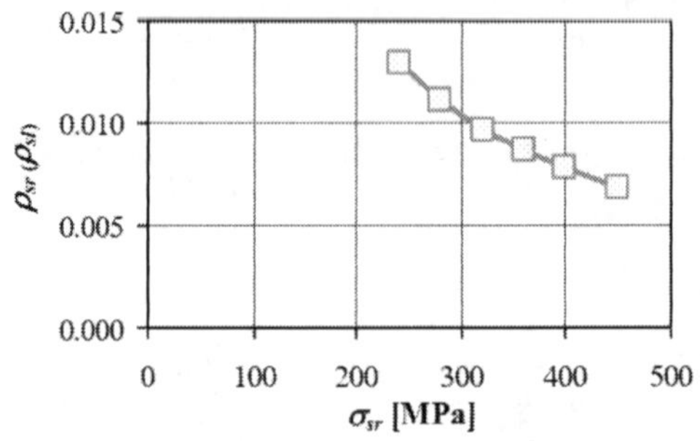

Fig. 5. Minimum reinforcing coefficient for cracking control.

Figure 5 shows the reinforcement coefficients corresponding to the radial ρ_{sr} and longitudinal ρ_{sl} directions, resulting from the control of cracking corresponding to maximum crack width of 0.1 mm and various stress values after cracking σ_s, and resistance capacity respectively. Figure 6 presents the same coefficients derived to ensure the resistance capacity.

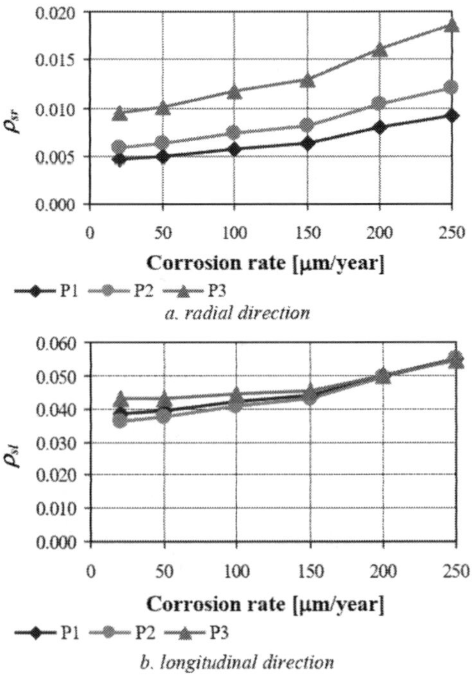

Fig. 6. Minimum reinforcing coefficient for resistance capacity.

4.2 Structural repair mortar R4 solution

Table 7 presents the time dependent evolution of the free shrinkage strain and the creep coefficient for a relative humidity of 90 %.

Table 7. Free shrinkage strain and creep coefficient (P1/P2/P3).

Shotcrete age [years]	Free shrinkage strain ε_{cs} [mm/m]	Creep coefficient φ
Radial direction		
1	0.175 / 0.183 /	1.28 / 1.27 / 1.25
5	0.190 / 0.188 /	1.50 / 1.41 / 1.38
10	0.192 / 0.191 /	1.54 / 1.45 / 1.40
15	0.193 / 0.191 /	1.56 / 1.47 / 1.41
20	0.193 / 0.191 /	1.56 / 1.48 / 1.41
25	0.194 / 0.191 /	1.57 / 1.48 / 1.42
30	0.194 / 0.191 /	1.57 / 1.48 / 1.42
Longitudinal direction		
1	0.196 / 0.192 /	1.40 / 1.29 / 1.21
5	0.199 / 0.194 /	1.64 / 1.43 / 1.26
10	0.199 / 0.197 /	1.69 / 1.47 / 1.29
15	0.199 / 0.198 /	1.70 / 1.50 / 1.31
20	0.199 / 0.198 /	1.71 / 1.50 / 1.32
25	0.199 / 0.198 /	1.72 / 1.50 / 1.32
30	0.199 / 0.198 /	1.72 /1.50 / 1.32

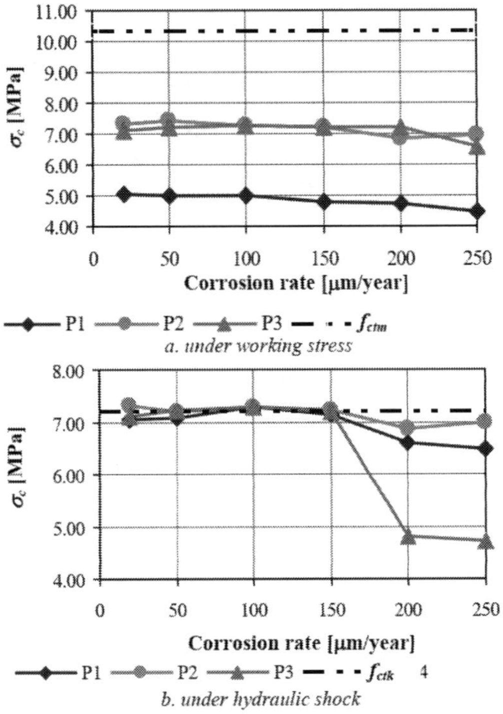

Fig. 7. Shotcrete radial tensile stress with corrosion rate.

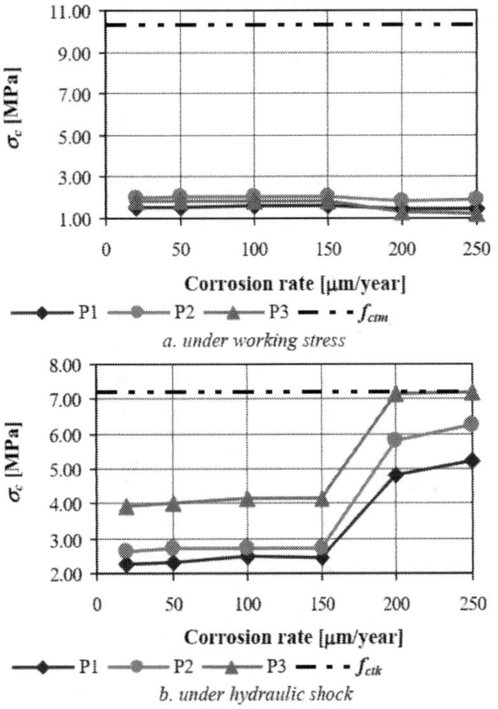

Fig. 8. Shotcrete longitudinal tensile stress with corrosion rate.

Figures 7 and 8 emphasize the tensile stresses related to the shotcrete lining depths shown in Table 8. With two exceptions, the depths resulted from the longitudinal resistance criterion, the other being related to the same criterion on the radial direction.

Table 8. Shotcrete depths for rehabilitated pipes.

Corrosion rate [μm/year]	P1	P2	P3
	t_c [mm]		
20	55	70	80
50	60	80	85
100	70	95	100
150	80	105	110
200	35	40	60
250	40	45	70

4 Final Remarks

It is evident the reinforced concrete shotcrete lining requires high work costs and low material costs, and vice versa is valid for the repair mortar R4 lining. After a global analysis, the authors are inclined to recommend the second alternative, which seems to be more reasonable from economical motivations, but not entirely.

At the present, there is not much research done on the topic. Therefore, more reliability analyses may accept lower values for the allowable stresses under a hydraulic shock (i.e., water hammer), phenomenon that is also subjected to technical equipment control. Moreover, given the relative reduced social and economical impact of such an event, a mean tensile strength is recommended to be considered as allowable stress under such an accidental situation. Thus, the acceptable shotcrete depths reach reasonable values of 20-50 mm - very competitive from both the technical and economical perspective.

More research is needed in order to quantify more accurately the long-term behavior of the rehabilitation shotcrete lining, both for ordinary/traditional reinforced concrete and structural repair mortars. Other cementitious materials should also be considered, as soon as their fresh state properties become compatible with the sprayed concrete technology.

References

1. C.G.R. Mircea, O. Corbu, E. Maier, C. Rus, *Rehabilitation of the Suhurlui Irrigation Pipeline*. Shotcrete mag., **Vol. 19**, No. 7, pp. 42-45 (2017)
2. EN ISO 12944-2: Paints and varnishes - Corrosion protection of steel structures by protective paint systems. Part 2: Classification of environments (2002)
3. EN 1992-1-1: Eurocode 2: Design of concrete structures - Part 1-1: General rules and rules for buildings (2004)
4. G.F. Kheder, R.S. Al-Rawi, J.K. Al-Dhahi, *A Study of the Behavior of Volume Change Cracking in Base Restrained Concrete Walls*. Materials and Structures, **27**, pp. 383-392 (1994)
5. S. Bernader, M. Emborg, *Risk of Concrete in Massive Concrete Structures–New Development and Experiences, in Thermal Cracking in Concrete At Early Ages*. Proc. of the RILEM International Symposium, Edited by R. Springenschmid, E &FNSpon, London, pp.385-400 (1994)

Estimation of In-situ concrete strength using drilling resistance

Serkan Karatosun[1], *Muhammet* Asan[2] and *Oguz* Gunes*

[1]Istanbul Technical University, Department of Civil Engineering, 34467 Istanbul, Turkey
[2]Gebze Technical University, Department of Civil Engineering, 41400 Gebze, Turkey

Abstract. Rapid and reliable condition assessment of reinforced concrete structures in high seismicity regions is a priority task in estimating their seismic safety. Non-destructive testing (NDT) methods may contribute to the condition assessment practice by providing fast and reliable strength estimation while causing minimal or no damage to the structure. Drilling resistance is an NDT method that has been used for mechanical characterization of natural stone and wood by measuring the force response for constant penetration rate and rotational speed. This paper focuses on the relationship between drilling resistance and compressive strength of concrete, including when it is combined with other NDT methods. Concrete cube samples produced using 6 different concrete mixtures were tested. Correlation equations were then obtained using statistical analysis. The results reveal that it may be possible to reliably estimate the compressive strength of concrete using drilling resistance method.

1 Introduction

Several major earthquakes and the resulting loss of life in the last decades drew attention to seismic safety of existing structures. Rapid and reliable condition assessment of structures in high seismicity regions is a priority task in estimating their seismic safety, especially when the quality of concrete is questionable. There are several test methods for in-situ strength assessment of reinforced concrete structures [1]. Testing samples cast concurrently during construction or cores drilled from the structure for compressive strength determination are the most frequently used conventional methods for estimating the strength of concrete in existing structures [2]. However, traditional methods have distinct drawbacks, for instance, results are not obtained rapidly, curing of concrete specimens may differ from that of the structure, and size and shape of concrete specimens may affect the strength properties [3]. Thus, many non-destructive testing (NDT) techniques have been developed to overcome these drawbacks. NDT techniques measure certain properties of concrete that can be indirectly related to strength and durability of structures. NDT methods may contribute to the condition assessment practice by providing fast and reliable strength estimation while causing minimal or no damage to the structure. The most commonly used NDT methods to estimate concrete strength are ultrasound pulse velocity (UPV) and Schmidt hammer [4].

In order to improve the reliability of assessment and the characterization of reinforced concrete structures, combining NDT methods is a strategy investigated by various researchers [5]. The main idea behind combining different NDT methods is the improvement of reliability through measurement of different unrelated characteristics that all relate to the compressive strength [6].

The primary objective of this study is to observe the relationship between drilling resistance and concrete strength, including when it is used in conjunction with other NDT methods. Statistical relationships were established between NDT results from Schmidt hammer, Silver-Schmidt hammer, ultrasonic pulse velocity and drilling resistance with those obtained from compressive strength tests. Single and multivariate regression relations were obtained for individual and combined use of NDT methods in relation to compressive strength test results.

2 Concrete constituent and specimens

A total of 17 concrete cube samples of 15 cm dimensions from 6 different concrete mixtures with the same maximum aggregate size (12 mm) were produced and tested in this study. All specimens were produced from the same constituent materials for consistency and were moist cured for 28 days after demoulding.

Table 1 presents the six different concrete mix proportions used in this study. The water to cement ratio by weight varied from 1.85 to 0.68 to vary the concrete strength. Compressive strengths of the tested concrete cube samples were in the range of 3.7 MPa to 39 MPa as shown in Table 1.

3 Experimental work

Four different NDT methods were used for this research, namely, Schmidt hammer, Silver-Schmidt hammer,

* Corresponding author: ogunes@itu.edu.tr

Table 1. Mix proportions, densities of concrete mixtures and compressive strength test results.

Mix No	Cement (kg/m^3)	W/C	N.S[a] (kg/m^3)	C.S.[b] (kg/m^3)	C.S.1[c] (kg/m^3)	Density[d] (kg/m^3)	(f_c)[d] (MPa)
1	130	1.85	552	677	647	1967	3.8
2	160	1.44	552	677	647	1979	6.9
3	190	1.16	552	677	647	2074	13.8
4	220	0.95	552	677	647	2091	15.2
5	250	0.8	552	677	647	2116	23.9
6	280	0.68	552	677	647	2192	38.8

[a] Natural Sand, [b] Crushed Sand, [c] Crushed Stone 1, [d] Mean for mixture.

ultrasonic pulse velocity test and drilling resistance method. Results obtained from individual and combined use of these NDT methods were related to those obtained from uniaxial compressive strength tests.

3.1 Compressive strength test

The compression strength of concrete cube samples was obtained destructively using a concrete compression testing machine (Fig.1). The tests were performed according to EN 12390-3 [7].

Fig. 1. Uniaxial Compressive strength test

3.2 Schmidt hammer

A Schmidt hammer, also known as rebound hammer is a device used to estimate the in-situ concrete strength in relation to surface hardness. Schmidt hammer is the most commonly used NDT technique to estimate concrete strength and its uniformity within a structure owing to its ease of use and low cost [8]. The Schmidt hammer basically measures the surface hardness from the rebound of a plunger that strikes the concrete surface (Fig. 2.a). The test procedure was performed according to EN 12504-2 [9].

3.3 Silver-Schmidt hammer

In the traditional Schmidt hammer, the rebound number is obtained in relation to the distance the hammer mass compresses a spring upon impacting the concrete surface. Internal friction and the inclination are the major factors that influence the results [10]. The Silver-Schmidt hammer is a recently developed, more advanced device that measures the rebound number through an optical method in relation to the speed of hammer mass upon impact, which is almost unaffected by friction and inclination [11]. Thus, the results from this method are relatively less dispersed [10]. This test procedure was also performed according to EN 12504-2 [9].

(a)

(b)

Fig. 2. Rebound hammer tests. (a) Standard Schmidt hammer, (b) Silver-Schmidt hammer

3.4 Ultrasonic pulse velocity

The ultrasonic pulse velocity test has been used to determine the quality of the concrete since the 1960s [4]. This test can be carried out in direct, semi-direct or indirect fashion. In the direct method, the time of flight of ultrasound pressure waves travelling across the concrete cube specimen can be measured precisely (Fig. 3). Knowing the cube dimension, the ultrasonic pulse velocity (UPV) can be easily calculated. The test procedure was performed according to EN 12504-4 [12].

Fig. 3. Ultrasonic pulse velocity test

3.5 Drilling resistance

Drilling resistance is an NDT method that has been used for mechanical characterization of natural stone and wood. Although it is not yet a well-established and standardized non-destructive testing method, the drilling resistance test is considered to have a high potential for rapid and reliable estimation of in-situ concrete strength (Fig. 4).

This method simply involves estimation of concrete strength based on the resistance of drilling a hole inside concrete [13]. The time required to drill to a certain depth was measured while keeping the force and rotational speed constant. The constant force was achieved by placing the specimen setup on an electronic scale and adjusting the force to read an approximately constant weight on the scale. A powerful drill was used to ensure that the rotational speed is not affected by the concrete strength. To this effect, the constant rotational speed is an assumption, but even if it is violated in practice, it is a bias that is valid for all specimens of similar strength, which makes the correlations still valid but specific to the drill used. The relationship between drilling resistance expressed as the time to drill a unit depth (s/cm) and compressive strength was established through regression analysis. Since the drilling is performed to a certain depth in concrete, the method is less affected by carbonation and other factors such as fire damage [14,15].

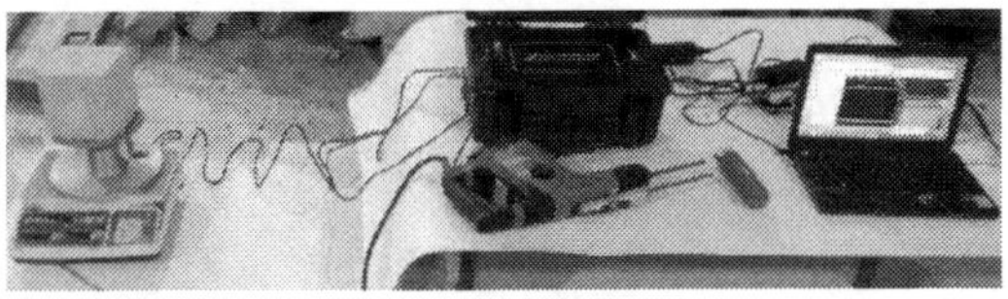

Fig. 4. Drilling resistance test

4 Test results and discussions

Four different NDT measurements were performed prior to compression tests on the specimens. Correlation relations were then derived through regression analysis on measured data from each NDT method in relation to the compression test results. These were compared with those based on the code equations in EN 13791 [16]. Additionally, multivariable regression analyses were performed by combining at least two NDT methods with the compression test. The correlation constants obtained for combined use of NDT methods gives an indication of whether the reliability is improved.

All correlation equations are compared with respect to the root mean square error (RMSE) and the coefficient of determination (R^2) as a means for reliability assessment.

4.1 Univariate regressions

In this section, correlation equations were obtained for the results of each NDT method in relation to destructive test (DT) results are presented. Table 2 provides these equations and the pertinent statistics. Plots of the correlation equations are provided in Figs. 5-9.

Table 2. Relationship between compressive strength and individual NDT results.

Method	Model	R^2	RMSE*
Rebound,R	$f_c = -33.5 + 1.80 \times R$	0.809	5.482
Rebound,Q	$f_c = -28.0 + 1.40 \times Q$	0.940	3.085
UPV	$f_c = -45.2 + 18.3 \times V$	0.820	5.326
Drilling Resistance,D	$f_c = -31.1 + 35.4 \times D$	0.768	6.050
Rebound, R	$f_c = 0.002 \times R^{2.76}$	0.822	5.296
Rebound, Q	$f_c = 0.001 \times Q^{2.80}$	0.970	2.190
UPV	$f_c = 0.040 \times V^{4.75}$	0.906	3.857
Drilling Resistance,D	$f_c = 4.960 \times D^{3.41}$	0.906	5.120

*Root Mean Square Error in unit of Mpa

Fig. 5 and Fig. 7 show the relationship between rebound numbers obtained from standard and Silver-Schmidt hammers, respectively, with the compressive strength test results. Both linear and non-linear relations in the form of power curves are shown in the figures. The coefficient of determination (R^2) for the Silver-Schmidt was found to be higher than that of the standard Schmidt for both linear and power curve fits. Hence, the Silver-Schmidt results provide relatively better correlation with the compressive strength as shown by both R^2 and RMSE values in Table 2.

The base curve provided in EN 13791 [16] is illustrated in Fig. 6. The required shift of the curve was calculated as -5.4 in according with the code relation given by Eq. 1. Coefficient k_1 is 1.48 since the number of paired test results in the study is more than 15.

$$\Delta f = \delta f_{m(n)} - k_1 \times s = \quad (1)$$

The relationship between UPV test results and compressive strength is shown in Fig. 8. EN 13791 provides polynomial base curve relation between UPV and compressive strength. Hence, a 2nd order polynomial regression curve was found for UPV (km/s) vs. compressive strength (MPa). UPV test results show similar correlation with compressive strength for both the polynomial curve and power curve fits.

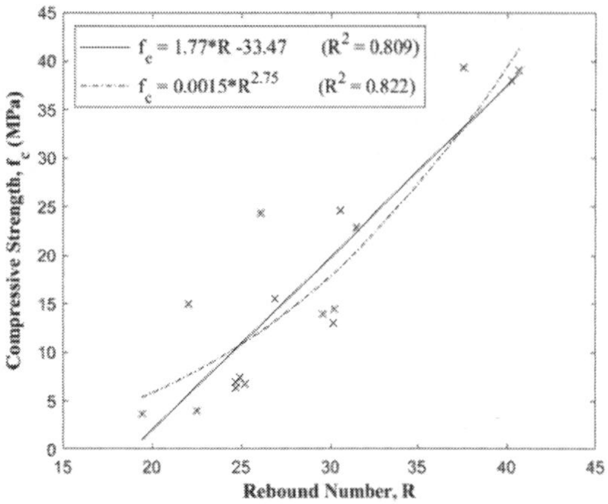

Fig. 5. Correlation between rebound number (R) and compressive strength

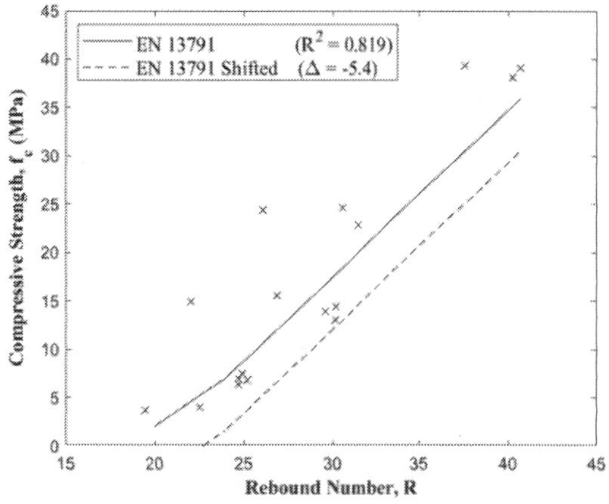

Fig. 6. Basic curve and shifted basic curve for Schmidt hammer (EN 13791)

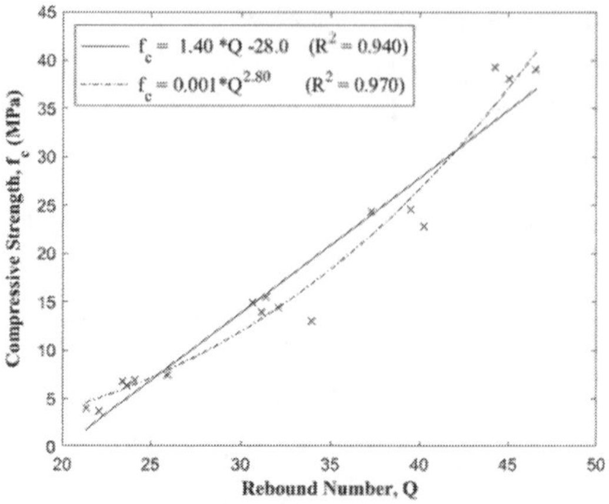

Fig. 7. Correlation between rebound number (Q) and compressive strength

The correlation between UPV and compressive strength is higher than that between standard Schmidt hammer and compressive strength in terms of R^2 and RMSE. On the other hand, the Silver-Schmidt provides more reliable strength estimation than all other NDT methods used in this study in terms of R^2 and RMSE (Table. 2) .

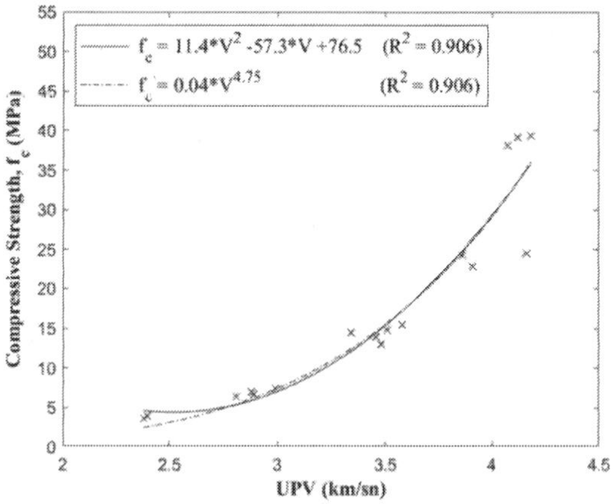

Fig. 8. Correlation between UPV (V) and compressive strength

The relationship between drilling resistance test results and compressive strength test result is shown in Fig. 9. The coefficient of determination for linear and power regression curves are 0.768 and 0.835, respectively. Furthermore, the RMSE of the power curve fit is about 15% less than that of the linear regression curve, which are 5,12 and 6,05 MPa, respectively. Consequently, the power curve fit gives a better representation of the relationship between drilling resistance and compressive strength test results.

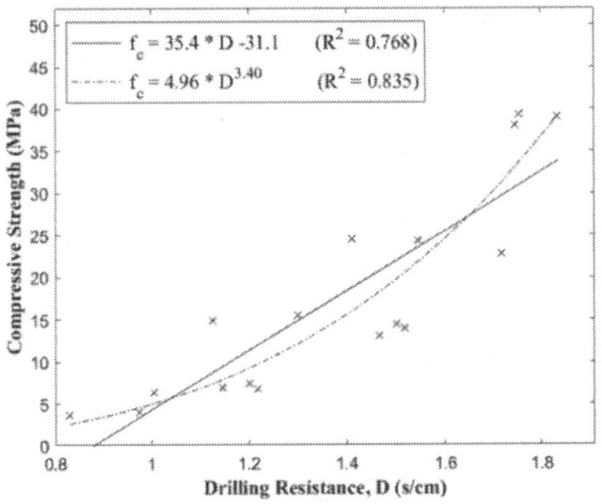

Fig. 9. Correlation between drilling resistance (D) and compressive strength

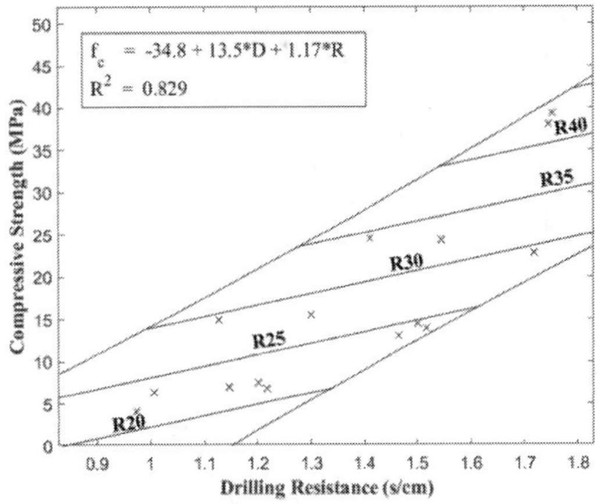

Fig. 10. Combined drilling resistance and rebound number (R) vs compressive strength with 85% confidence level boundaries

4.2 Combined regressions

Different techniques can be combined to increase accuracy and reliability of concrete strength estimation. Table 3 summarizes regression equations, R^2 and RMSE of multivariate regression analyses. Also, these correlations are plotted in Figs 10-14.

Table 3. Relationships between concrete strength and drilling resistance combined with other NDT methods

NDT	Equation	R^2	RMSE (MPa)
R	$f_c = -34.8 + 13.5 \times D + 1.17 \times R$	0.829	5.38
Q	$f_c = -29.9 + 6.3 \times D + 1.19 \times Q$	0.930	3.32
UPV	$f_c = -43.2 + 13.8 \times D + 12.2 \times V$	0.845	5.12

The relationship between the drilling resistance test - used in combination with both rebound hammer tests - and compression tests are shown in Fig. 10 and Fig. 11. Fig. 12 illustrates the relationship between drilling resistance and UPV combination with compressive strength. Multiple linear regression analyses were performed to obtain the regression curves. As can be seen in Table 3, the drilling resistance and Silver-Schmidt combination gives the best correlation with highest R^2 (0.93) and lowest RMSE (2) among all the alternative combinations.

Equations 2 and 3 were derived to estimate concrete strength by combined use of drilling resistance, UPV and standard Schmidt hammer (Eq. 2) or Silver-Schmidt hammer (Eq. 3). R^2 of these correlations are 0.912 and 0.947, respectively.

$$f_c = -47.5 - 10.1 \times D + 1.2 \times V + 12.7 \times R \quad (2)$$

$$f_c = -20.9 - 4.9 \times D + 1.9 \times V - 4.7 \times Q \quad (3)$$

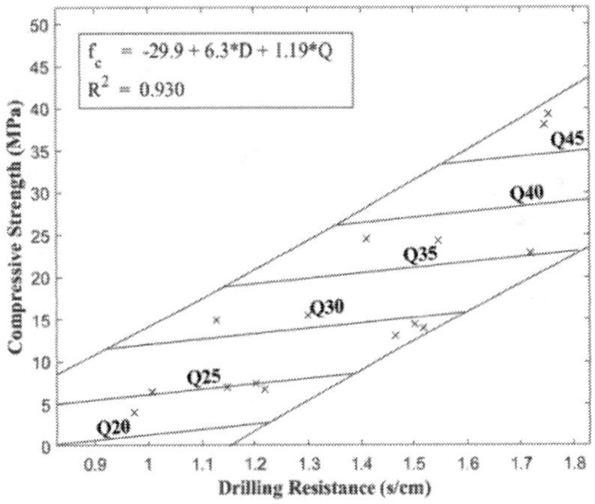

Fig. 11. Combined drilling resistance and rebound number (Q) vs compressive strength with 85% confidence level boundaries

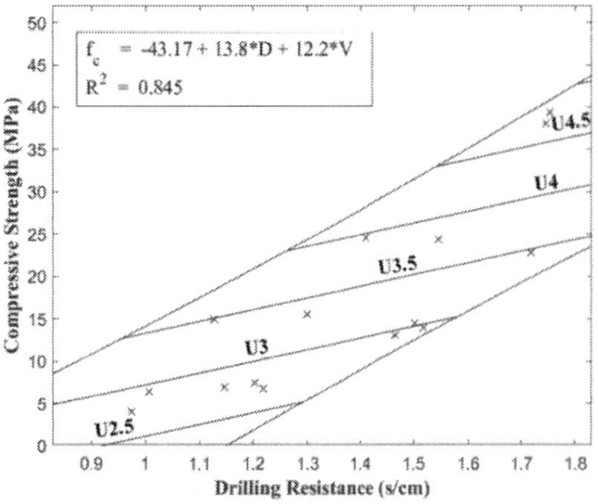

Fig. 12. Combined drilling resistance and UPV (V) vs compressive strength with 85% confidence level boundaries

4 Conclusions

In this paper, the relationship between drilling resistance method and concrete strength, including when it is combined with other NDT methods was investigated and the established correlation curves were reported. Single and multivariate regression analyses were performed for individual and combined use of NDT methods in relation to compressive strength test results. The results reveal that it may be possible to reliably estimate the compressive strength of concrete using drilling resistance method especially when it is combined with Silver-Schmidt hammer. Furthermore, combined use of the drilling resistance method with Schmidt hammer, Silver-Schmidt hammer and UPV methods increased the reliability of concrete strength estimation.

Two equations (Eq.2 and Eq.3) were derived to estimate concrete strength in relation to the combination of 3 NDT methods, namely drilling resistance, UPV, one of two rebound hammer tests. Eq. 2 which involved Silver-Schmidt as rebound hammer shows a relatively better correlation than that involved the standard Schmidt hammer (Eq. 3).

Acknowledgement: The authors gratefully acknowledge the support provided by the Scientific and Technical Research Council of Turkey (TUBITAK) under Project No. 214M054.

References

1. T. Uomoto, Non-destructive testing in civil engineering, Seiken symposium,**26**, (2000).
2. Pucinotti, Reinforced concrete structure: Non-destructive in situ strength assessment of concrete, Construction and Building Materials, **75**, 331-341, (2015).
3. S. K. Verma, S. S. Bhadauria, S. Akhtar, Review of Nondestructive testing methods, journal of construction engineering,**1**,11, (2013).
4. Malhotra VM, Carino NJ. Handbook on non-destructive testing of concrete, ASTM. 2nd ed. CRC Press; 2004.
5. S. P. Pessiki, In-place methods to estimate concrete strength. ACI 228.1R-03 report, (2003).
6. D. Breysse, Non-destructive evaluation of concrete strength: An historical review and a new perspective by combining NDT methods. Construction and Building Materials, **33**, 139-163. (2012).
7. EN 12390-3, Testing hardened concrete, Compressive strength of test specimens, CEN, Brussels, (2010).
8. B. Graeye, H. Laar, J. Eijik, W. Gijbels, L. Lauriks, On-site strength assessment of limestone based concrete slabs by combining ndt, J.B.E, **13**, 216-223, (2017).
9. EN 12504-2, Testing concrete, determination of rebound number, CEN, Brussels, (2012).
10. D. Corbett, Advancing the rebound hammer method: a new concrete test hammer, Nondestructive Testing of Materials and Structures, Rilem book series **6**, (2013).
11. M. Breccolotti, M.F. Bonfigli, A.L. Materazzi, Influence of carbonation depth on concrete strength evaluation carried out using the SonReb method, NDT&E Int.,**59**,96-104, (2013).
12. EN 12504-4, Testing concrete, determination of ultrasonic pulse velocity, CEN, Brussels, (2004).
13. O. Gunes, Turkey's grand challenge: Disaster-proof building inventory within 20 years. Case Studies in Construction Materials, **2**, 18-34, (2015).
14. R. Felicetti, the drilling resistance test for the assessment of fire damaged concrete, Cement & concrete composites, **28**, 321-329, (2006).
15. R. Nogueria, A. P. F. Pinto, A. Gomes, assessing mechanical behaviour and heterogeneity of low strength mortars by the drilling resistance method, Construction and Building Materials, **68**, 757-768, (2014).
16. EN 13791, Assessment of in situ compressive strength in structures and precast concrete. CEN, Brussels, (2007).

New warning sensors to detect corrosion risk in reinforced concrete

Mahdi Khadra[1,2,*], *Elisabeth* Marie-Victoire[2], *Myriam* Bouichou[2], *Christian* Crémona[3], and *Stéphanie* Vildaer[3]

[1]Cercle des Partenaires du Patrimoine, 29 rue de Paris, 77420 Champs-sur-Marne, France
[2]Laboratoire de Recherche des Monuments Historiques, 29 rue de Paris, 77420 Champs-sur-Marne, France
[3]Bouygues Travaux Publics, Pôle Ingénierie Matériaux, 78114 Magny-les-Hameaux, France

Abstract. Corrosion is the most frequent but also the most deleterious deterioration mechanism affecting reinforced concrete. In addition to the economic impact of the repair works, for historical concrete structures, corrosion can generate irreversible losses of original material of great cultural value. If the usual non-destructive electrochemical methods have highlighted their efficiency in evaluating on-going corrosion activity, they also have pointed out their drawbacks for accurate extrapolation and prevention. To prevent the corrosion phenomenon, by detecting the penetration of aggressive agents, a new warning sensor system has been developed. The principle of the technique is to embed thin metallic sheets (called orphan blades) in the concrete cover, at different distances from the surface to the reinforcing bars. Then the corrosion of those very reactive orphan blades is followed during the propagation of the carbonation front and/or the penetration of chloride ions using stimulated infrared thermography. The corrosion of the sensors at different depths is indicative of the ingress speed of the front and can alert about the risk of corrosion of reinforcing bars in the concrete. The purpose of this study is to present this new technique and the first results obtained in the laboratory on corroded and non-corroded sensors.

1 Introduction

Nowadays, concrete has become the most used material in construction. However, concrete structures interact with the external environment which leads to their degradation over time [1]. Atmospheric CO_2, in the presence of moisture, reacts with the main constituents of concrete [2]. This reaction, called carbonation, reduces the pH of the interstitial solution, which is initially basic (12 to 13) to a value lower than 9, which initiates corrosion of the reinforcing bars [3]. In marine environments or in areas frequently exposed to de-icing salts, chloride ions can gradually penetrate concrete structures until they reach the reinforcement, and when the concentration of chloride ions exceeds a critical threshold, particularly active corrosion can develop [3,5]. Today, a large number of degraded concrete structures have reached an alarming threshold, and the economic issue related to the rehabilitation of this stock is estimated to 2 billion US dollars over the next 10 years only for the United States territory! [11]. Corrosion also causes major problems for conservation in reinforced concrete structures listed as historical monuments [4].

To evaluate the corrosion risk or even to assess active corrosion, the most used techniques are electrochemical methods, performed from the concrete surface [5, 6, 7, 8, 9, 10]. But usually the measurements are local and depend on the environmental conditions: they can over- or under-estimate the annual corrosion risk. For many years, the concrete department of LRMH has studied an adaptation of these techniques to improve the diagnosis accuracy by implementing continuous monitoring using sensors placed on the surface or embedded in the concrete structure [12, 13]. These sensors are generally connected by wires to a data logger, remotely searchable by cell phone. This means that the installation is visible and that connection issues can appear on the long term. In addition most of these methods require a physical connection to the reinforcing bars.

To overcome these issues, but also to anticipate the corrosion before it reaches an irreversible state, a new warning system named "orphan blades" has been developed and patented by Bouygues Travaux Publics [14]. It is wireless, invisible from the surface, not based on electrochemical measurements and allows detection of the penetration of aggressive agents such as carbonation or chloride. In this paper this sensor system is explained and a laboratory study performed on reinforced concrete slabs, with sensors showing different states of corrosion, embedded at several depths is presented.

2 New sensor systems

2.1 Principle of the technique

The idea is to place ultra-thin metal sheets in the concrete cover and to monitor their corrosion during the

* Corresponding author: mahdi.khadra.ext@culture.gouv.fr

propagation of CO_2 or chloride ingress (Fig. 1) using stimulated infrared thermography. The gradual corrosion of the sheets (or orphan blades) embedded at different depths is indicative of the propagation of aggressive agents and of the risk of corrosion for the reinforcing bars.

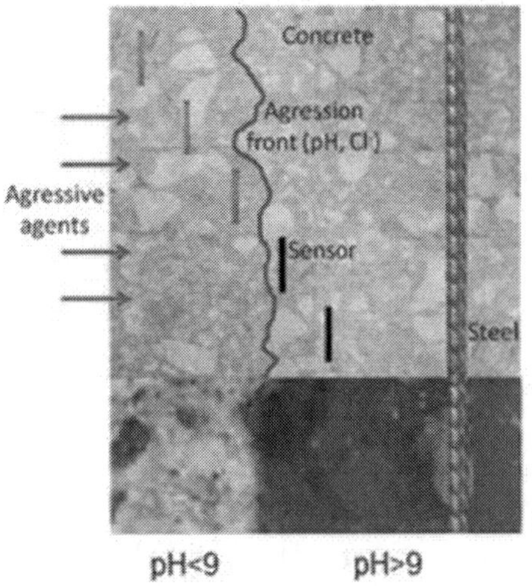

Fig. 1. Schematic principle of the technique

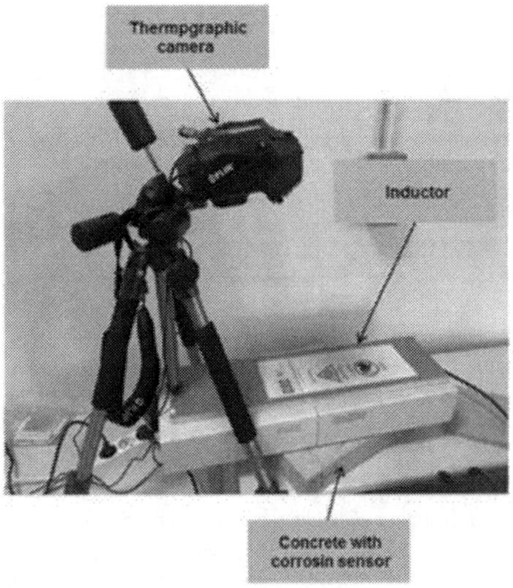

Fig. 2. Setting of the technique

In several industries, for nondestructive control of defect detection in conductive composites or metals, electromagnetic methods are widely used. Magnetic induction (Lenz's law) is one of the most used techniques. The currents induced by the coil of the inductor placed close to the metallic surface, heat the surface region. Then any surface defect disturbs locally these currents and induces local modification in the temperature of the surface.

In the idea of stimulating the orphan blades embedded in the concrete using magnetic induction, a specific inductor was developed. It uses an appropriate electrical power, which, associated with a suitable coil, generates an alternating magnetic field powerful enough to heat the metallic blades up to several centimeters into the concrete cover, by mobilizing electrons on their surface (eddy currents). The consequence is that sound metallic blades will be heated while corroded ones will behave differently. The rise of temperature of the sheets propagates to the surface and can be detected using a thermographic camera (Fig. 2).

1.2 Sensor positioning

To perform the most accurate measurement, sensors must be placed in the concrete cover parallel to the concrete surface in order to be parallel to the camera line of sight during the thermographic acquisition. The compliance of this criterion is fundamental but can be disturbed by the vibration of concrete during casting, which can modify the position or angle of orientation of the sensors. To solve this problem, a sensor support system was designed (Fig. 3). It is composed of two parts. The first part is a frame that keeps the sensor parallel to the concrete surface, and prevents any deformation. It is composed of a square slot, perpendicular to a rectangular rod that is dedicated to positioning the sensor vs the reinforcing bars. The second part, which is wedge-shaped, is intended to stabilize the system, to adjust the distance between the sensor and the reinforcing bars in the formwork, and therefore to adjust the desired depth of the sensor versus the surface. Thanks to the combination of the rod of part 1 and the wedge of part 2, the system can be attached to the reinforcing bar using a plastic clamp collar for which a central band in the rod was designed.

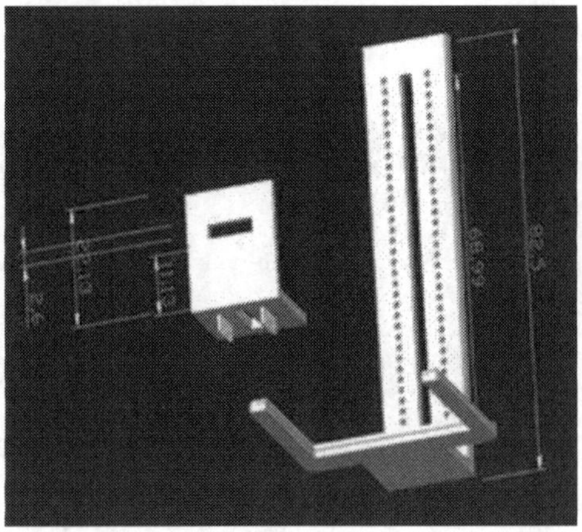

Fig. 3. Sensor support system.

The whole support system was produced using 3D printing with a specific heat inert polymer.

1.3 Data acquisition and treatment processes

For the data acquisition, a T420 FLIR thermal camera, with a 0.1°C thermal sensitivity was used. The data was acquired using the ResearchIR MAX 4.0 software by connecting the camera to a computer (Fig. 4). The measurements were carried out according to a 7 step protocol.

- Step 1: Place the specimen to be tested under the thermal camera so that the lens of the camera is parallel to the surface of the sample. The camera is equipped with a laser pointer to focus on the centre of the sensor whose position is known in the slab.
- Step 2: Connect the camera to the computer equipped with the dedicated software and adjust the parameters, such as the emissivity (ε = 0.93 for concrete), the distance between the camera and the specimen, and the ambient temperature.
- Step 3: Place the inductor on the surface of the specimen so that the centre of the coil of the inductor is placed on the centre of the sensor.
- Step 4: Turn on the inductor and heat for a pre-set time.
- Step 5: During heating, press the record button to record the video of the thermal response of the sensor.
- Step 6: At the end of the heating time, turn off the inductor and remove it from the surface of the specimen.
- Step 7: Follow the evolution of the thermal response on the computer until the total dispersion of the thermal signal on the surface of the sample; turn off recording and save the file for data processing.

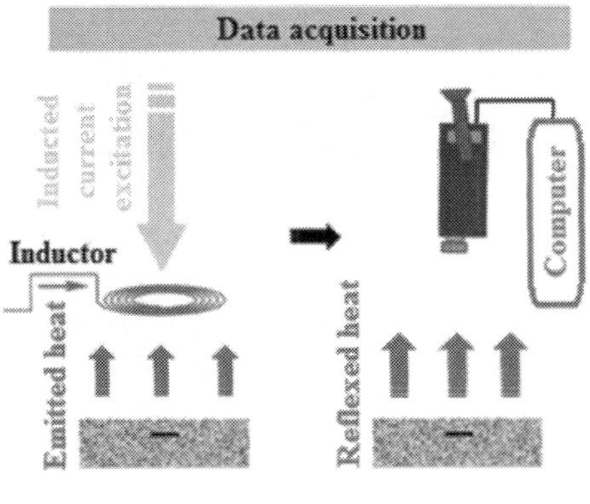

Fig. 4. Data acquisition processes

The first step in data processing is to extract the first thermal image appearing immediately at the end of the induction heating time (Fig. 5). The second step consists in framing the reflected area of the sensor on the concrete surface using the BOX tool provided by the software, and to follow the evolution of the thermal response on the diagonal line of the BOX (Fig. 6).

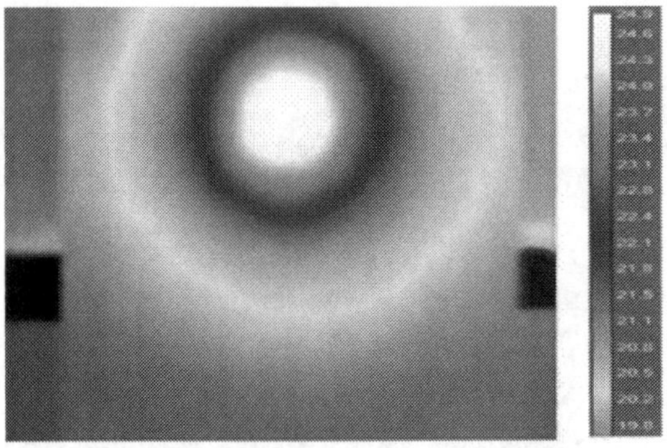

Fig. 5. Thermal apparent image in the infrared spectrum after heating.

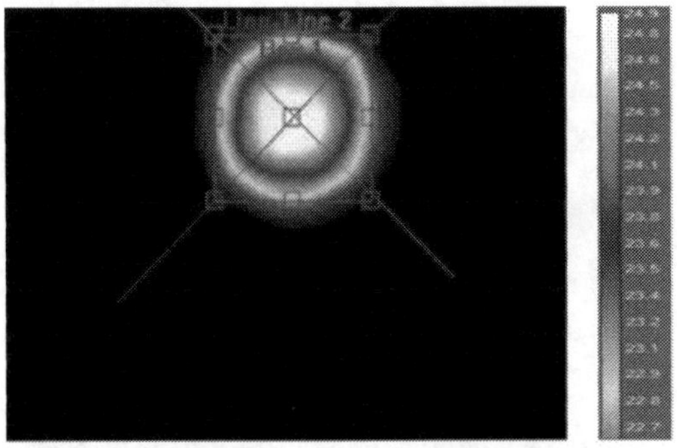

Fig. 6. Thermal apparent image in the infrared spectrum of the sensor area after heating.

The thermal responses are then represented as a regression of the maximal apparent temperature of the line versus time.

3 Experimental program and results

In order to evaluate the sensitivity of the technique regarding the corrosion state of the sensor, a series of tests was performed using corroded and non-corroded sensors placed into concrete at different depths.

3.1. Artificial corrosion

Prior to their introduction in the concrete the sensors were submitted to artificial ageing. They were first inserted in their polymer frame. Then a series of cycles of immersion (in the evening and overnight) in a hypochlorite solution with 2.6% (m/m) active chlorine, followed by drying (in the morning and the afternoon), was applied. To perform a preliminary characterization of the sensors after artificial corrosion, images in both the visible and infrared spectrum were taken for each sensor, after 6 days at room temperature (Fig. 7).

 https://doi.org/10.1051/matecconf/201928906002

Corrosion is visible to the naked eyes, but also on the IR images, the sound zones reflecting the radiation and therefore having a higher apparent temperature than the corroded parts.

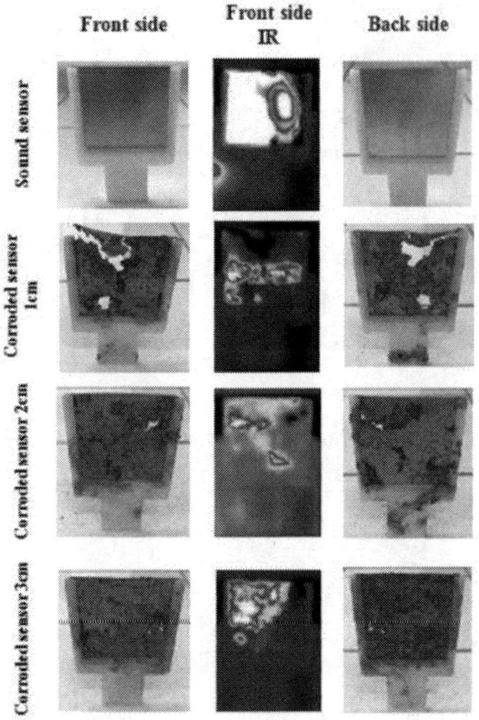

Fig. 7. Images of the sensors in visible and infrared spectrum.

3.2. Sample preparation

The concrete considered for this study was a mix of CEM I 52.5 N CP2 (275 Kg/m^3), with a water to cement ratio of 0.74 and aggregates with a max diameter of 16mm. The slabs were cast in wooden moulds of 22x17x8 cm. In each slab, only one sensor was fixed at the middle of the rebar before concrete casting, at a depth controlled thanks to the support system. Three depths were considered: 1cm, 2cm and 3cm. The formwork was filled in each case in two steps in order to optimize the vibration and the quality of the concrete around the sensors (Fig. 8). After casting, the slabs were stored for one month in an outside covered place in contact with atmospheric air.

Fig. 8. Fixation of the sensor system to the rebar before concrete casting.

3.2. Results and discussion

Prior to the measurements, the slabs were moved to a room with controlled temperature for two days, in order to stabilize their initial temperature. Measurements were then carried out according to the protocol detailed in section 1.3.

Figure 9 shows the infrared images obtained on the slabs containing either sound or corroded sensors placed at 1cm from the concrete surface, after 10 minutes of heating with a 25 KHz induction frequency. In the table, two types of images are presented: those showing the thermal response from the entire surface detected by the camera and those obtained by focusing only on the sensors.

As can be seen, the intensity of the hot spot and the elevation of the apparent temperature are higher for the sound sensor than that of the corroded sensor.

The intensity of the apparent temperature can be quantified by extracting the maximal apparent temperature on the diagonal line of the BOX. When plotting the curve of these maximal apparent temperatures versus time, sound and corroded sensors can be clearly distinguished (Fig.10), so that this parameter can be considered as a quantitative indicator of corrosion.

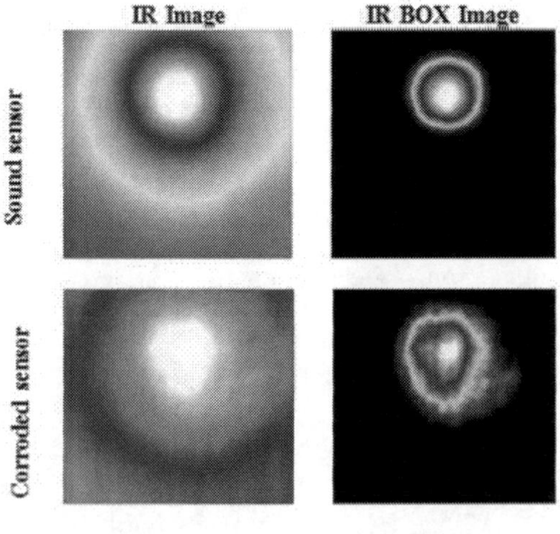

Fig. 9. Results obtained after 10 minutes of heating for the slabs containing sensors with a 1cm concrete cover.

Figure 11 gathers the infrared images obtained on the slabs containing either sound or corroded sensors placed at 2cm from the concrete surface, after 10 minutes of heating. As can be seen, the spot of the sound sensor is more defined and brighter than that of the corroded sensor. But the elevation of temperature of this sensor depth is lower than for the sensors placed at 1cm.

This is probably due to less efficient heating of the 2cm-depth sensors compared to those at 1cm-depth because the magnetic field decreases with depth, and because the thermal wave needs more time to appear on the surface of the slabs, and therefore part of the heat dissipates in the concrete by conduction.

 https://doi.org/10.1051/matecconf/201928906002

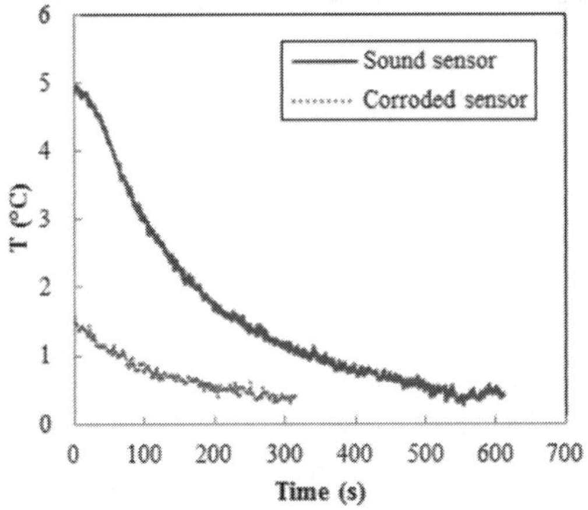

Fig. 10. Temperature versus time of the sensors with a 1cm concrete cover after10 minutes of heating.

The qualitative observations were confirmed by the quantitative data as shown on figure 12, where the maximal apparent temperature is plotted versus time. It can also be noticed that for this higher concrete cover, sound and corroded concrete can still be quantitatively distinguished.

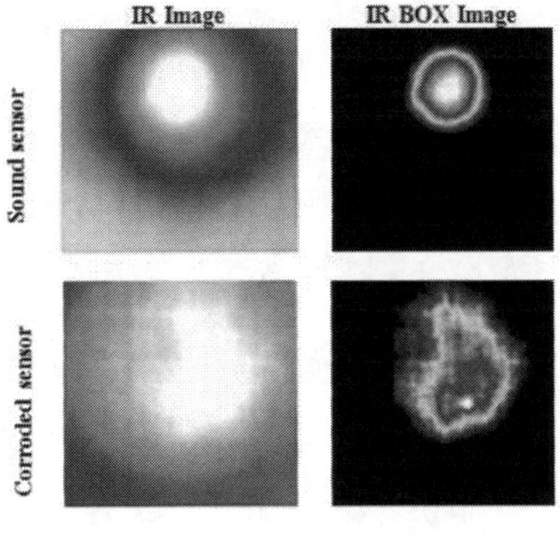

Fig. 11. Results obtained after10 minutes of heating for the slabs containing sensors with a 2cm concrete cover.

Figure 13 gathers the infrared images obtained for the sensors placed at 3 cm from the surface of the slabs. If sound and corroded sensors can be distinguished from the entire infrared images, it is more difficult from the BOX images. At the induction frequency used in this study, when the concrete cover increases, the heating of the sensor decreases and the thermal wave diffuses more in the concrete, and as a consequence, the apparent temperature at the concrete surface is reduced. Thus, the maximal temperature for the sound sensor placed at 3cm is about 1.3°C (figure 14) when it is was 2.7°C for the sound sensor placed at 2cm from the concrete surface (figure 12).

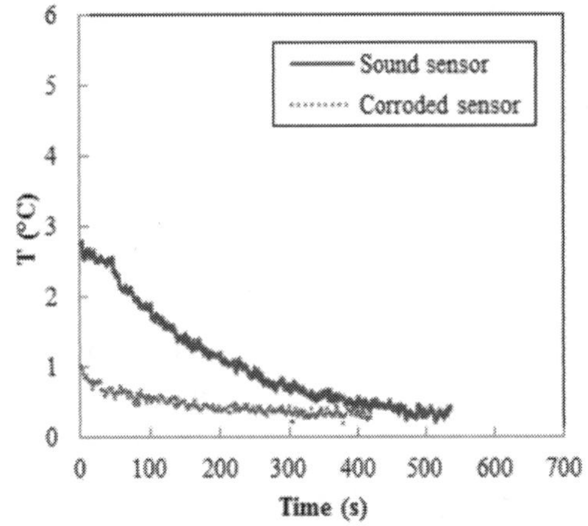

Fig. 12. Temperature versus time of the sensors with a 2cm concrete cover after10 minutes of heating.

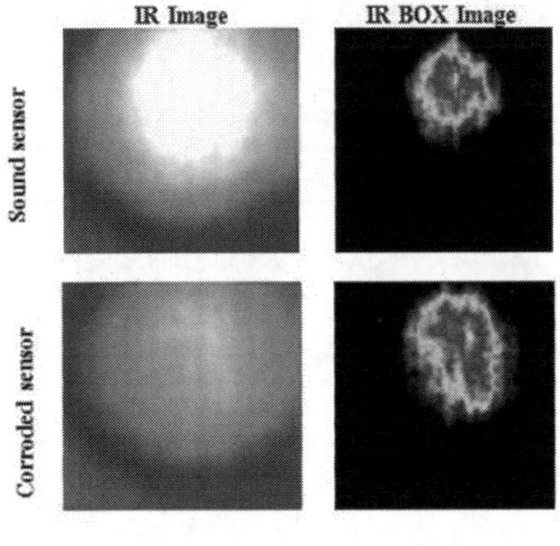

Fig. 13. Results obtained after10 minutes of heating for the slabs containing sensors with a 3cm concrete cover.

When comparing the curves of temperature versus time for the sound and corroded sensors with a 3 cm concrete cover, it can be observed that the maximal temperature difference is about 0.5°C (Fig. 14). This cannot be considered sufficient to distinguish sound and corroded sensors.

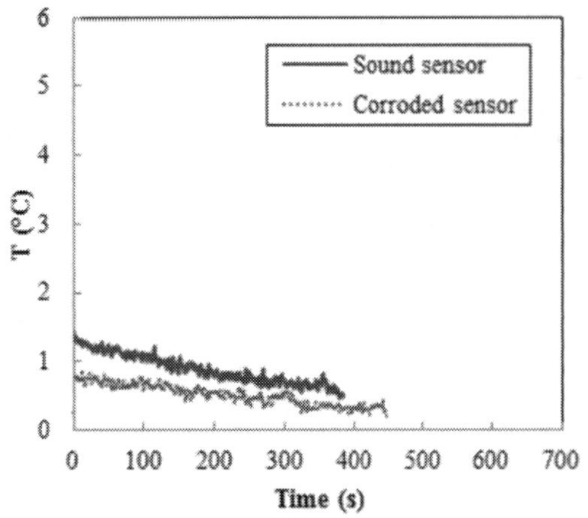

Fig. 14. Temperature versus time of the sensors with a 3cm concrete cover after10 minutes of heating.

4 Conclusions and outlook

A new warning system designed to anticipate corrosion of reinforcing bars in concrete was designed. The principle of this system is to place ultra-thin metal sheets, named orphan blades, at different depths in the concrete cover till the depth of the reinforcing bars and to monitor their corrosion versus time by stimulated infrared thermography. When aggressive agents such as carbonation or chlorides penetrate the concrete, the orphan blades will gradually corrode, their thermal signature will be modified, and therefore will warn about the progression of the risk of corrosion.

In this paper, a comparison between sound and corroded sensors embedded at three different depths in a concrete with a high water to cement ratio, susceptible to be more sensitive to corrosion was presented. The results show that whatever the concrete cover, sound and corroded sensors can be qualitatively distinguished, even when the sensors are not totally corroded. But dealing with quantitative analysis of the data, if the temperature differences between the two states (sound or corroded) were significant for the sensors embedded at 1 and 2 cm from the concrete surfaces, it was less evident for the sensors placed at 3 cm. This is probably due to a less efficient induction heating when the depth increases and to the diffusion of the thermal wave into the concrete before it reaches the concrete surface.

As a conclusion, the orphan blades system appears to be a very promising technique to anticipate corrosion of reinforcing bars in concrete and constitutes an interesting alternative to the usual electrochemical measurements.

Further studies on sensors corroding directly in the concrete or on the influence of the heating time and environmental parameters are on-going.

A new inductor with a higher heating frequency (33 KHz) is also under development. The first results obtained on sensors embedded at 3cm and 4cm from the concrete surface show that the maximal heating temperature is significantly higher than those obtained with the inductor used in this study. Therefore the technique could be applied to reinforced concrete with higher concrete covers.

References

1. J.P., Broomfield, *Corrosion of steel in concrete: understanding, investigation and repair*, ed. C. Hall, London: E & FN SPON. 238 (1997)
2. M. Thiery, Modélisation de la carbonatation atmosphérique des matériaux cimentaires, Ecole Nationale des Ponts et Chaussées, PhD. Thesis (2005)
3. A. Bentur, S. Diamond, N.S. Berke, *Steel corrosion in concrete: fundamentals and civil engineering practice.* Modern concrete technology 6, ed. C. Hall, London : E & FN SPON (1997)
4. E. Marie-Victoire, M. Bouichou, T. Congar, R. Blanchard, Concrete cultural heritage in France - Inventory and state of conservation, *In* 4th ICCRRR : International Conference on Concrete Repair Rehabilitation and Retrofitting, October 5-7, 2015, Leipzig, Germany, 8p.
5. A. Dollet, G. Taché, Méthodes de diagnostic du béton armé, Anticorrosion et durabilité dans le bâtiment et le génie civil, S. Audisio and G. Béranger, pp. 659-668. Presses polyt. Et univ. Romandes ed (2010)
6. M. Raupach, Chloride-induced macrocell corrosion of steel in concrete-theoretical background and practical consequences. Construction and Building Materials, Vol. 10, 5, pp. 329-338 (1996)
7. M. Raupach, P. Schiessl, Monitoring system for the penetration of chlorides, carbonation and the corrosion risk for the reinforcement. Construction and Building Materials, Vol. 11, 4, pp. 207-214 (1997)
8. P. Schiessl, Corrosion measuring cell. Patent, US5015355 (1991)
9. J.-F. Lataste, C. Sirieix, D. Breysse, M. Frappa, Electrical resistivity measurement applied to cracking assessment on reinforced concrete structure in civil engineering, NDT & E International, Vol. 36, No. 6, pp. 383-394 (2003)
10. E. Marie-Victoire, M. Bouichou, H. Jourdan, Water-repellents as alternative carbonation-induced corrosion treatments for reinforced concrete cultural heritage. In XIV DBMC: 14th international conference on durability of buildings materials and components, 29-31 May 2017, Ghent University, Belgium, p.399.
11. FHWA, Corrosion Costs and Preventive Strategies in the United States, FHWA-RD-01-156 (2002)
12. G. Villain, J.P. Balayssac, M. Fares, S.M. Sbartaï, J.F. Lataste, X. Dérobert, Comparaison de profils de taux de saturation obtenus par plusieurs méthodes

non destructives électrique et électromagnétiques, actes Cong. Diagnobéton, Marrakech (MA), (2016)

13. M. Fares, Y. Fargier, G. Villain, X. Dérobert, O. Coffec, S. Palma-Lopes, Détermination du profil de teneur en eau dans le béton d'enrobage par inversion de mesures de permittivité au moyen de sondes capacitives, actes cong. Diagnobéton, Toulouse (2014)
14. Patents FR2955178A1, EP2521906B1, CA2786449C, WO2011083142A1, AU2011204622B2, MA33918B1, HRP20140053T1, Method for measuring corrosion in a concrete building, Bouygues Travaux Publics (2010-2014)

MATEC Web of Conferences **289**, 06003 (2019) https://doi.org/10.1051/matecconf/201928906003
Concrete Solutions 2019

Comparison of 2D and 3D ultrasonic pulse echo imaging techniques for structural assessment.

Lubos Misak[1] and *David* Corbett[2], *Michael* Grantham[3]

[1]Proceq SA, Proceq Europe Department, 8603 Schwerzenbach,Swizerland
[2]Proceq SA, Product Development Department, 8603 Schwerzenbach,Swizerland
[3]Leeds University and Sandberg LLP, London, UK

Abstract: The use of ultrasonic waves in the medical field is quite advanced, especially in the area of three-dimensional (3D) imaging of the human body and is quite common. Also, in Non-Destructive Testing of metal (for example weld checking), flaw detection with reliable 3D imaging is already in use. In the past 1 -2 years, the construction industry has made significant advances in two-dimensional (2D) and three-dimensional (3D) imaging of concrete structures, for finding problems, or irregularities inside, or just for confirming the quality of the construction work. This paper describes the principles of the different types of scan, A Scan, B Scan etc and how the scans can be combined to make 3D images. The advantages of multi head array systems in providing more detail and revealing depth information are shown. This paper will show the advantages of this development and how it can ease workflow.

1 Introduction

The use of ultrasonic waves in the medical field is quite advanced, especially in the area of three-dimensional (3D) imaging of the human body and is quite common. Also in Non-Destructive Testing of metal (for example weld checking), flaw detection with reliable 3D imaging is already in use.

In the past 1 -2 years, the construction industry has made a big step in two-dimensional (2D) and three-dimensional (3D) imaging of concrete structures, for finding problems, or irregularities inside, or just for confirming the quality of the construction work.

This has been made possible by progress in the electronics industry (smaller, faster microchips, high resolution of screens) and the possibility to handle a big volume of data, with powerful software algorithms to produce and display two-dimensional (2D) and three-dimensional (3D) objects.

This paper will show the advantages of this development and how it can ease workflow.

2 Recent Research

Khairi et al (1) discussed the theory behind ultrasonic tomography and illustrated it with a number of applications cross-industry, but including its application in concrete and mortar and in evaluation of damage to structures.

Scott (2) discussed the use of ultrasound computed shear wave tomography for the inspection of Nuclear Structures to detect hidden defects.

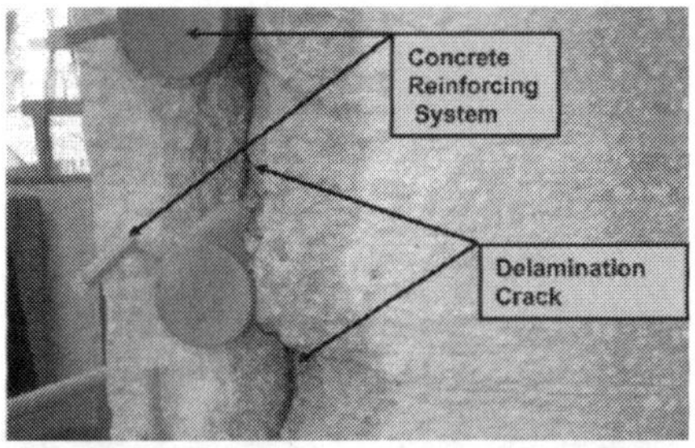

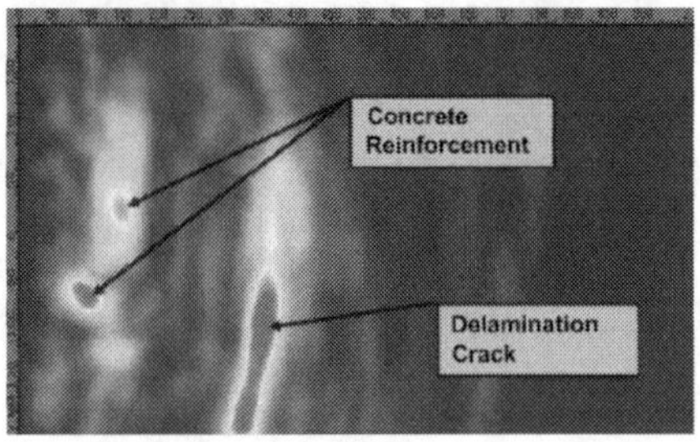

Fig 1. Delamination in a nuclear containment structure wall (top) and an ultrasound tomographic scan (bottom) - after Scott (2).

Schabowicz, (3) discussed the principle of the technique, illustrating its use on laboratory test specimens and also on a cracked, defective foundation slab, suspected of being too thin and being found by tomography to be non-compliant with its design.

Choi et el (4) described a comprehensive visualization scheme to characterize reinforced concrete structures, based on integrated ultrasonic tomography and 3-D computer vision. Their paper describes a recently developed hybrid aircoupled/contact ultrasonic transducer system which enables the generation of internal tomographic images that characterize the interior condition of structures. for example the presence of voiding defects, Based on tomographic cross-sectional slice images, 3-D volumetric internal images were built up to obtain full stereoscopic analysis of full-scale reinforced concrete elements.

Haach and Ramirez (5) described the application of ultrasonic tomography for the qualitative assessment of concrete. They conducted several experimental tests with variations in concrete composition, presence of voids, arrangement and frequency of transducers on concrete prisms. The results showed that ultrasonic tomography was more efficient than conventional ultrasonic tests. Analyses of several transducer arrangements revealed a large number of measurements did not necessarily return better images.

2 A-scan principles

The classical measurement with Ultrasonic transducers (sender – receiver) looks like this:

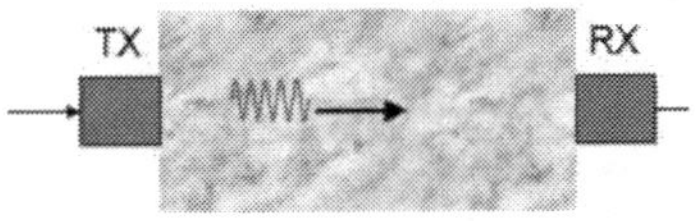

Fig 2 Normal UPV direct transmission

An ultrasonic signal is from transmitter transducer to the receiver transducer through the material and results in an A-scan.

The A-scan presentation is a display of the ultrasonic signal in which the x-axis represents the time and the y-axis the amplitude

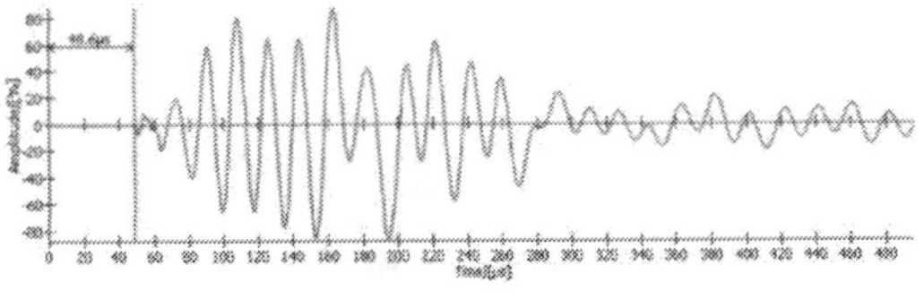

Fig 3. Typical UPV Trace showing signal amplitude vs time of flight. The 48.6 μs represents the time for the first receipt of signal.

This procedure requires the use of a couplant to ensure a good contact between the transducer and the material (concrete, brick, mortal, ..) and the measurement is usually taken directly, semi-directly, or indirect (on the surface).

3 B-scan principles

The B-scan presentation is an image of the results of an ultrasonic examination showing a cross section of the test object perpendicular to the scanning surface and parallel to the reference direction. The cross section will normally be the plane through which the individual A-scans have been collected.

The B-scan is principally a series of A-scans in a line in one direction. For example, in one axis (X-axis) this would be a **B-scan**.

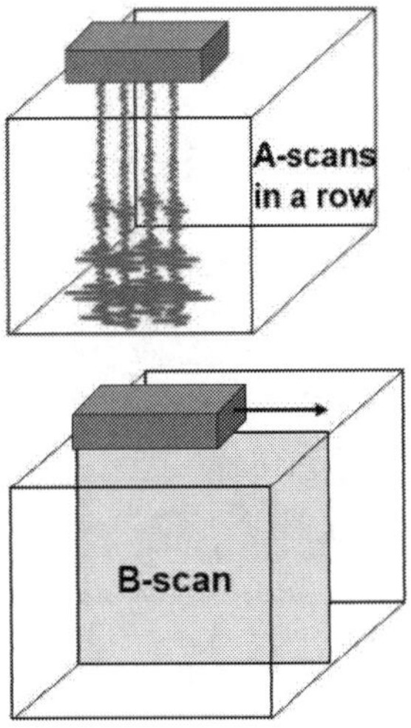

Fig. 4 The "B" Scan principle

4 C-scan, D-scan, P-scan, Volume scan

The same scan in the perpendicular axis is often referred to as a **D-scan.**

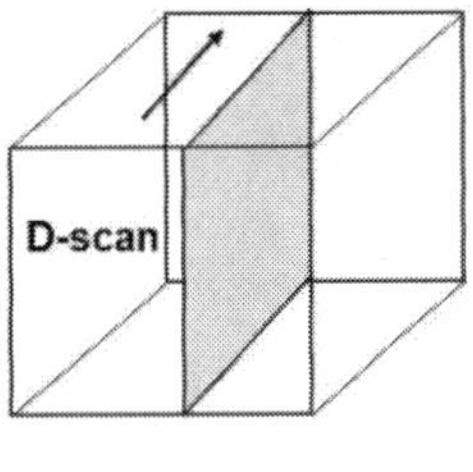

Fig 5 The "D" Scan Principle

The C-scan presentation is an image of the results of an ultrasonic examination showing a cross section of the test object parallel to the scanning surface. The evaluated parameter in the image is the amplitude of the signal.

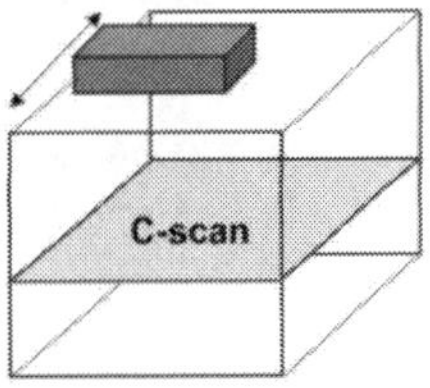

Fig 6 Typical "C" Scan

P-scan (time slice) and **Volume scan.** A projection view of several B-Scans or C-Scans. Once we have a volume scan we can look at gated slices in each of the three dimensions, at any depth we choose.

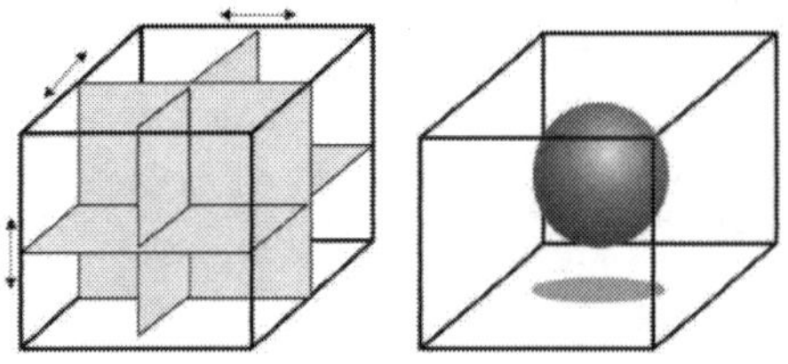

Fig 7. "P" Scan (Left) and Volume Scan (Right)

5 DPC transducers and Array Probe.

The development of DPC (Dry Point Contact) transducers have been a significant advance and have helped to have contact with the material without using couplant.

These transducers are used in pulse echo mode in which the transmitter and receiver are on the same side – which is an advantage, if there is access only from one side (tunnels, subways, bridges, ...). This technique allows reflections from internal flaws to be obtained (object detection inside), or to detect the thickness of the object from the back-wall echo.

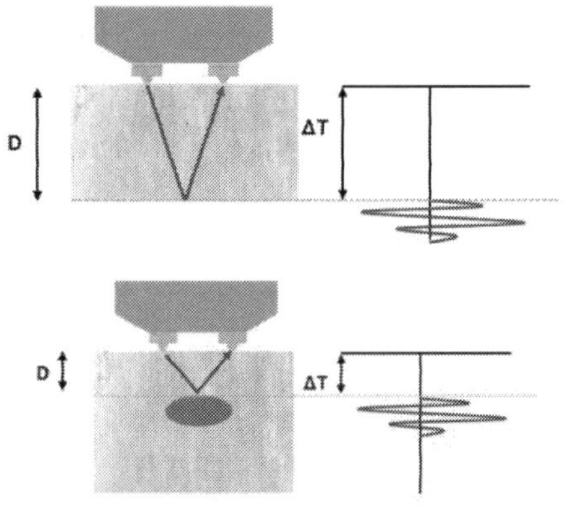

Fig 8. Typical "one side" UPV scan where reflections are used to see defects or the back wall in a structure.

The technique is further improved by using multiple channels in an array transducer, enabling faster scanning of bigger areas.

The working principles are, that one channel transmits and the others are listening, each receiving a separate A-scan, then the second channel transmits and the others are listening and so on. This continues in a cycle and the resulting A-scans are used to create a B-scan in real time.

The principle is shown in Figure 8 where the clear advantages in time and resolution using a multi-scanning system are obvious.

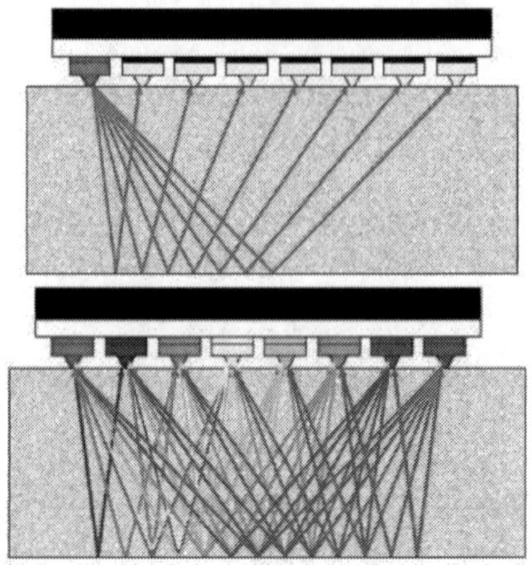

Fig 9. Multi scanning array as used in advanced ultrasonic tomography.

It is also possible to combine array transducers together to double the scanning surface. This reduces the scanning time.

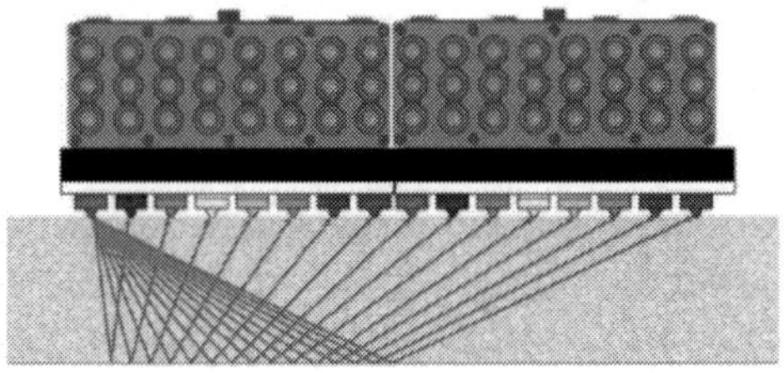

Fig 10. Combining two transducer arrays to speed up scanning.

6 Comparison of 2D and 3D scans.

Having access to both 2D and 3D scans provides the user with more information about the test object. The 2D images are in the X and Y axis (B-scans), and provide depth information about objects. Combination and interpolation of B-scans allows 3 D images to be created which then allow the user to look at the amplitude information of the entire scan area at a chosen depth. This provides information about the extent of objects or defects.

In Fig 11a we can see two objects, very clearly. To determine the extent of such objects it is possible to use the 3D scan set to the depth of the object in question and in the following images – 11b, 11c and 11d which show information at different depths, it is clear to see that it extends throughout the length of the scan and is almost certainly a pipe of sorts. (In fact, the object in this case was a tendon duct.) Accurate depth information is determined by detailed analysis of the A-scan at any particular location.

https://doi.org/10.1051/matecconf/201928906003

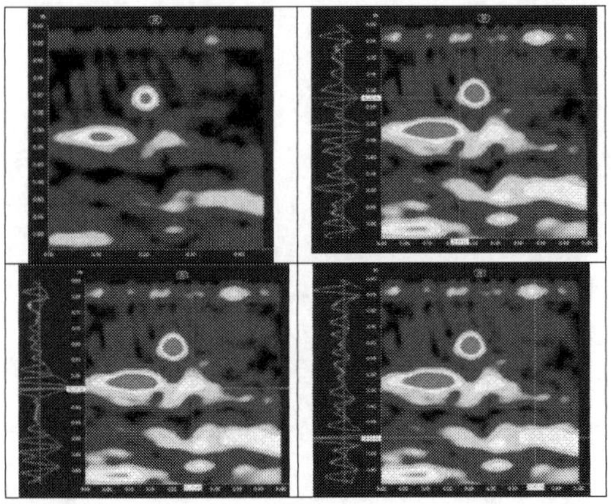

Fig 11 clockwise from top left: a,b,c,d

With the 3D scan, we have a volumetric model of scanned structure inside of the object, which we can see and rotate to get a visual appreciation of the spatial relationship between objects detected, even quite complex ones as in fig 12.

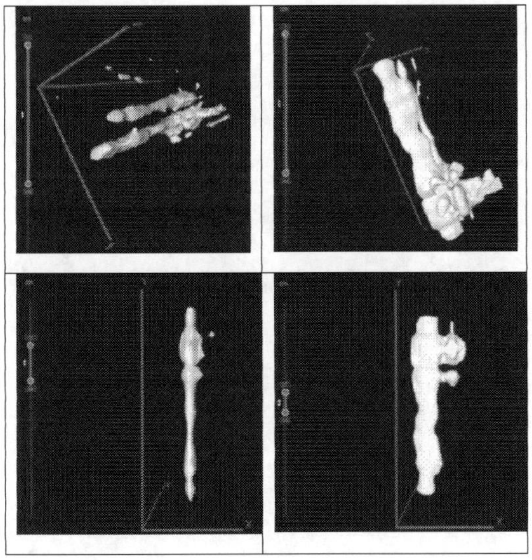

Fig 12. Complex object visualized and rotatable in 3D

Also, in this type of representation it is important to be able to blend out sections of the scan or increase the gain at various depths. For example, rebars are not so easily detected by ultrasonic pulse echo instruments, but they can be made visible by increasing the gain and lowering the pulse velocity. However, these actions have the effect of distorting the location of deeper objects.

7 Conclusions

Ultrasonic pulse echo scanning has developed significantly in recent years, but the main issue remains in interpretation of the images. Both 2D and 3D images are useful to the engineer for assessing a structure. Each provides specific information that is essential in trying to make sense of the scanned structure. The advanced hardware and software technique for ultrasonic testing is making it possible to create better images of the object for technicians and engineers, but also make simple to understand 3D renders which are much easier to explain to non-NDT specialists. The latest advances are even making use of techniques such as Augmented Reality (AR) which allow the scan to be viewed in-situ on a tablet and even to be able to draw in the position of bars etc. on the structure surface using a special marker which is visible on the tablet and can then mark the positions of bars from the virtual image.

Fig 13. Graphic depicting the future of internal visualization of structures.

The 2D pictures are more suitable for detail analysis, for specialists, where they can put more energy into the picture, check the A-scan and go for details but for non specialists the power of visualizing and exploring inside a structure in an understandable way is a huge advance. Undoubtedly, the technique will continue to advance and better, improved techniques and visualisations will become available.

References

1. M. T. M. Khairi, S. Ibrahim, M. A. M. Yunus, M. Faramarzi, G. P. Sean, J. Pusppanathan, A. Abid, Measurement, **146,** 490-523, (2019).
2. D.B. Scott, Energy Conversion and Management. **74,** 582-586, (2019).
3. K. Schabowicz, Archives of Civil and Mechanical Engineering, **14**, 295-303, (2014).
4. H. Choi a, Y. Hamb, J S. Popovics, Construction and Building Materials **123**, 384–393 (2016)

5. V.G. Haach, F. C. Ramirez, Construction and Building Materials **119** 61–70 (2016)
6. D. Corbett, One Belt One Road Concrete Technology Forum, Singapore (2018)
7. D. Corbett, Pundit 250 Array Application (2019)

Effectiveness of two field methods of saturating near surface concrete on the water permeability of *in situ* concrete

Kai Yang[1], ***Muhammed Basheer***[1*], *Sreejith Nanukuttan*[2], *Yun Bai*[3], *Adrian Long*[2]

[1]School of Civil Engineering, University of Leeds, Leeds, UK
[2]School of Natural and Built Environment, Queen's University Belfast, Belfast, UK
[3]Civil, Environmental and Geomatic Engineering, University College London, London, UK

Abstract. Determining the water permeability of concrete in structures remains a challenge because of difficulties in removing the influence of its moisture content. Saturating concrete with water could be one option, but this is not easy to achieve on site. This paper reports a testing programme carried out to assess the reliability and effectiveness of two field saturation methods, viz. vacuum saturation and ponding. The water permeability test results after applying the vacuum saturation and ponding were compared with that obtained after incremental immersion. It was found that ponding was unable to remove the influence of moisture, whilst vacuum saturation was effective for wet concretes. The results obtained from the electrical resistance measurements after incremental immersion suggested that the water permeability of concretes can be accurately determined by carrying out *in situ* permeability tests if the near surface region up to a depth of 25 mm is fully saturated.

1 Introduction

Concrete has the reputation as a strong and durable material [1, 2]. However, the occurrence of durability problems and some catastrophic failures of reinforced concrete structures since 1970's have caused deep concerns to civil engineers in their ability to design and construct durable reinforced concrete structures [3]. Since then numerous approaches have been proposed to assess the durability of concrete.

The permeability of cover concrete has been considered as a key parameter to assess the durability of concrete structures and numerous methods for measuring the *in situ* permeation properties of normal concretes have been proposed [4-6]. However, they are not effective for distinguishing the permeation characteristics of very low permeability concretes, such as those typically associated with high-performance concrete (HPC), because of their low sensitivity to variations in permeation characteristics. Against this background, a new air permeability test and a water permeability test were developed to ensure that the differences between HPCs can clearly and reliably be identified [7].

For most field permeability assessments, a technical barrier is that the results are significantly affected by the moisture content of the concrete [8, 9]. An investigation by Yang *et al.* has indicated that to remove the influence of moisture on air permeability of HPCs, the relative humidity in the near-surface region (from surface to 20 mm depth) should be less than 60% [7] as opposed to the value of 80% proposed by Basheer and Nolan for NCs [10]. As this moisture condition is not easy to reach for HPCs on site, an alternative could be to measure their permeation properties by carrying out *in situ* water permeability tests.

In situ water permeability tests are usually carried out when the concrete is in a fully saturated state. However, relatively few publications have been found to deal with the influence of moisture on water permeability tests. For the Clam Water permeability test [11], the test area is saturated for 24 hours by ponding water before carrying out the test, but this approach is not effective to achieve the saturated state for dense concretes (e.g. w/c<0.5). Whiting attempted to saturate concrete using a similar technique, as part of the on-site Rapid Chloride Permeability Test [12]. However, it can be noted that only normal concretes were tested in their research. In the case of HPCs, almost no investigations on the effect of moisture on *in situ* water permeability tests can be found. Furthermore, studies do not give any detailed information on the effectiveness of vacuum saturation and, hence, the preconditioning procedures are not fully understood. Therefore, the *in situ* procedures for vacuum saturation have to be developed for HPCs.

In this study, three saturation regimes were selected, including vacuum saturation, ponding and incremental immersion. The vacuum saturation and ponding are considered as possible approaches for field use, whilst the incremental immersion is considered as the 'reference' method, as the previous work [13] has shown that reliable water permeability results can be obtained after incremental immersion of HPCs.

2 Experimental programme

* Corresponding author: Muhammed Basheer, School of Civil Engineering, University of Leeds, Email: p.a.m.basheer@leeds.ac.uk

2.1 Materials and concrete mix proportions

In the context of this research, HPCs refer to concretes with low permeation properties. Details of mix proportions of the three HPCs are given in Table 1. CEM-I cement conforming to BS-EN 197 and two supplementary cementitious materials (SCMs), viz. microsilica (MS) and pulverised fuel ash (PFA), were used in this study. The PFA, conforming to BS-EN 450, was obtained from Kilroot Power station in Northern Ireland, UK and microsilica, conforming to BS-EN 13263-1, was used in slurry form which was supplied by Elkem. A polycarboxylic acid based superplasticiser was used to maintain the required consistence.

Table 1. Concrete mix proportions and both fresh properties and compressive strength.

Concrete	PFA	PC	MF
Water (kg/m^3)	145	145	145
Portland cement (kg/m^3)	388	485	449
Microsilica (kg/m^3)	0	0	36
PFA (kg/m^3)	97	0	97
Sand (kg/m^3)	668	689	652
Coarse aggregate (kg/m^3)	1150	1150	1150
Superplasticiser (% of binder content)	1.4	1.3	1.5
Air content (%)	0.6	1.0	1.6
Slump (mm)	220	225	240
28 day compressive strength (MPa)	81.3	81.8	84.2

The fine aggregate was medium graded natural sand and the coarse aggregate was crushed basalt with 10mm and 20mm size proportioned at equal mass. Before casting, the aggregates were dried in an oven at 105 (±5)°C for 24 hours followed by cooling to 20 (± 1)°C for one day to control the moisture content.

2.2 Preparation of specimens

The concrete was mixed in accordance with BS 1881: part 125 and the fresh concrete was assessed for slump and air content according to BS-EN 12350-2 and BS-EN 12350-7 respectively. For each concrete mix, 410×100×250 mm blocks were cast for carrying out the water permeability test and 100mm cubes for determining the compressive strength at the age of 28 days. The slab specimens contained an electrode array for resistance measurements. After casting, the specimens were covered with wet hessian and placed in a constant temperature and relative humidity room at 18 (± 2)°C and 60 (±10)% respectively so that all specimens had similar environmental conditions during their initial period of curing. The fresh properties and the compressive strength values for each concrete are reported in **Table 1**.

2.3 Test methods

2.3.1 High pressure water permeability test

A high pressure water permeability test setup was used in this study, details of which and the test procedure are available in Yang *et al.* [7]. To carry out the test, the test head was clamped on the surface of the specimen, which was then connected to an air compressor for pressurising the testing system. During measurements, the pressure was maintained at 7 bar by advancing a piston through a cylinder. The volume of water within the cylinder was recorded every minute. Each measurement took 120 minutes.

2.3.2 Electrical resistance measurement

Electrical resistance measurements were carried out in order to monitor variations in the degree of saturation within the cover region. The stainless steel rods were placed at depths of 10, 15, 20, 25, 30mm from the surface, which is the same as that used by McCarter *et al.* [14]. The changes in electrical resistance with time across each electrode pair were measured by an LCR meter and recorded by a data logging system.

2.3.3 Vacuum saturation setup

Figure 1 illustrates the vacuum saturation setup. The design concept of the vacuum saturation set-up used in this study was based on Whiting's work [12] and that used in some other surface permeability tests [15]. The vacuum pressure was controlled by a regulator and the change in vacuum level in the chamber was monitored using a vacuum pressure gauge.

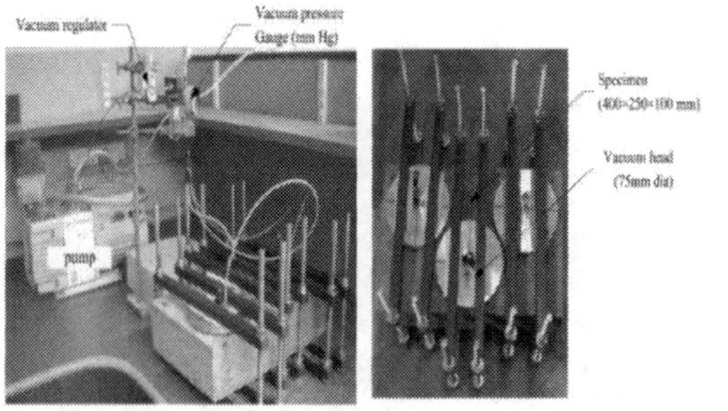

(a) Vacuum set up (b) Vaccum ring

Fig. 1. The vacuum saturation set-up.

Two levels of vacuum (40 and 240mm Hg) and two duration of application (3 and 6 hour) were chosen as the test variables. These are denoted as V40-3, V240-3, V40-6 and V240-6 whilst reporting the results. After setting up the vacuum saturation system, the vacuum pump was switched on and the vacuum pressure was adjusted to the desired level of vacuum for the two duration of vacuum application using the regulator. The pressure level was monitored by the gauge and adjusted, if necessary. At the end of the two specified periods of vacuum application, water was admitted into the chamber. After filling the saturation head, vacuum was released so that the specimen would admit water for another 40 hours. The water permeability test was then performed at this location.

3 Results and discussion

3.1 Investigation of the effectiveness of vacuum saturation

3.1.1 The possibility of achieving the 'steady-state' after vacuum saturation

Figure 2 shows the water flow into the concrete blocks after the four different vacuum saturation regimes at five different locations. The behaviour of the water flow was examined before attempting further interpretation because the new water permeability test is based on the steady-state flow theory. The duration to reach a 'steady-state' as well as the rate of flow were determined, the procedure of which is similar to that reported in the previous publication [6]. It is evident that the relationship between the volume of in-flow and time is non-linear in all cases, especially during the initial period. Further, there was a test outlier in both V40-3 and V240-3 and two test outliers in V240-6. These data indicate that either the test location was not properly saturated or the ring was not properly fixed on the test surface, hence they were not used in any further analysis.

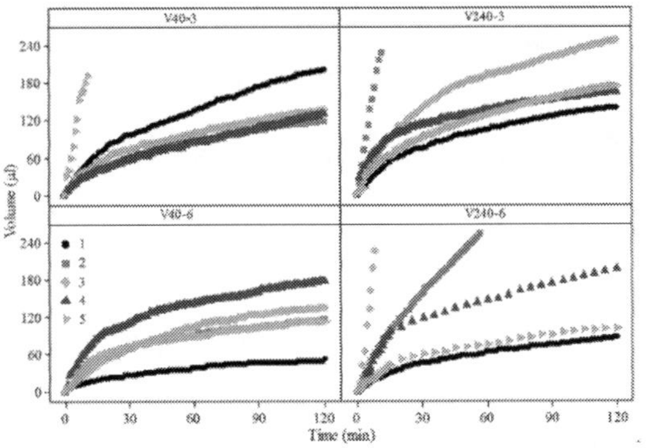

Fig. 2. Water permeability test data after vacuum saturation of the MF concrete

The flowrates at different test duration were estimated using the regression analysis of the volume of water versus time. These results indicated that the flowrates decreased as time increased, and stabilised after 60 minutes. Therefore, the flowrates for all tests were estimated based on the regression analysis of the data after 60 minutes.

3.1.2 The effect of vacuum pressure and duration of vacuum application

The effect of vacuum pressure and duration of vacuum on the steady-state flowrate was analysed through a 2^2 factorial experiment. Table 2 summarises the results of the statistical analysis and Figure 3 shows the main effects of the factors and the interaction between vacuum pressure and vacuum duration. As can be seen from the table, only the effect of vacuum duration is significant, whilst the others do not remarkably affect the flowrate. In Figure 3, it can be seen that the vacuum duration had a negative contribution on the response, meaning that a higher level leads to a lower flowrate. This can be expected because longer duration can remove the air effectively and more pores can become saturated, hence giving a lower flowrate. Furthermore, the increase of the vacuum pressure can be seen to cause an increase in the flowrates.

Although Figure 3 demonstrates the existence of interaction between the vacuum pressure and the duration of vacuum application, Table 2 shows that this is not significant. Therefore, the vacuum level and its duration can be decided based on their main effects. In other words, a lower vacuum pressure and longer vacuum duration are preferable for field applications to achieve the steady state of flow.

Table 2 The analysis of variance of the 2^2 factorial experiment

Factor	Sum of square	DoE	Mean square	p-value*
Vacuum pressure	0.0062	1	0.0062	0.637
Duration	0.147	1	0.147	0.036
Vacuum pressure×Duration	0.0258	1	0.0258	0.342
Error	0.316	12	0.0263	
Total	0.522	15		

*Note: p-value < 0.01 means highly significant, 0.05 < p-value < 0.01 means significant, p-value > 0.05 means non-significant.

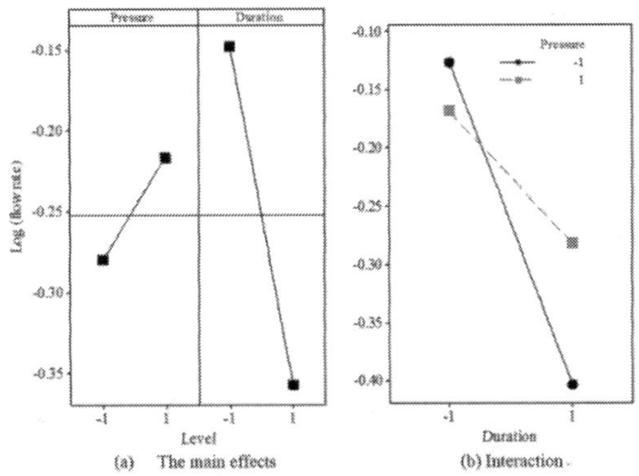

Fig. 3. The plots of main effects and interaction

3.2 Investigation of the effectiveness of ponding

Ponding the test area with water was another approach that was investigated to saturate concrete on site and, hence, its influence on water permeability results was assessed. The test head that was described in section 2.3.3 was clamped on the concrete surface and water was filled in the head (termed as ponding with water). After 48-hours of ponding, the new water permeability test was carried out at the conditioned region and the volume of water transported into the concrete at different duration determined. From these data the flowrate was obtained, as was done for the vacuum saturation method. The flowrate corresponding to the steady state is presented in Figure 4, along with those from the four combinations of vacuum saturation method and the incremental immersion of test specimens for 6 and 10 days. [The incremental immersion, principally reported by researchers from Queen's University Belfast [16], was a method to saturate concrete samples in the laboratory. The incremental immersion method is able to remove air in specimens by leaving one surface exposed to the ambient whilst water is absorbed through other surfaces; this enables almost 100% degree of saturation. According to results obtained previously [11, 17], this method could give results similar to that obtained from the vacuum saturation method.

It can be seen from Figure 4 that the flowrate after ponding is roughly three times of what was obtained after the vacuum saturation. This is because ponding is known to be effective only to saturate the near surface region, typically up to 1 to 2 mm [18], which is lower than the effective depth in the water permeability test (typically 25 mm). Therefore, it can be concluded that ponding is not an effective saturation method for the steady state water permeability tests on HPCs.

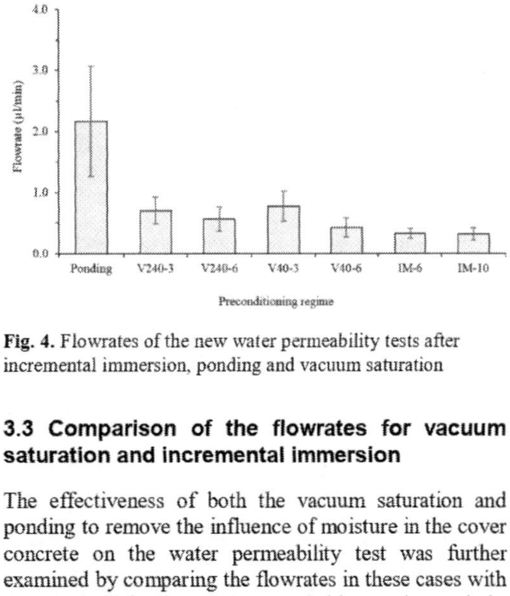

Fig. 4. Flowrates of the new water permeability tests after incremental immersion, ponding and vacuum saturation

3.3 Comparison of the flowrates for vacuum saturation and incremental immersion

The effectiveness of both the vacuum saturation and ponding to remove the influence of moisture in the cover concrete on the water permeability test was further examined by comparing the flowrates in these cases with that obtained for the two incremental immersion periods, viz. 6 days and 10 days (Figure 4). A comparison between the two duration of incremental immersion was also made in order to establish whether or not the duration of immersion had any significant effect on the flowrate. The mean values in Figure 4 are cross-compared using the least significant difference (LSD) [19] and the results are summarised in Table 3. The reason for comparing the mean flowrates of both the ponding and the vacuum saturation conditions (V-240-3, V240-6, V40-3 and V40-6) with that for the incremental immersion for 6 days (IM-6) in this table is given below.

In Figure 4, three features can be identified. Firstly, the flowrates after incremental immersion are the lowest (IM-6 and IM-10 in Figure 4) and no obvious difference existed between 6 days and 10 days of immersion. The data in Table 3 show that the difference between the means for these two test conditions is not significant. Therefore, the effectiveness of other saturation methods can be assessed by comparing with the flowrate for either of these two; hence, hereafter the comparison is made with data from incremental immersion for 6 days (IM-6).

Secondly, there is noticeable difference in flowrate between incremental immersion and ponding (Figure 4). The data in Table 3 highlights that this difference is highly significant.

Thirdly, as the vacuum level was decreased from 240 mmHg to 40 mmHg and the duration of vacuum application was increased from 3 hours to 6 hours, the flowrate decreased. Further, the 40 mmHg-6 hour vacuum saturation regime gave similar flowrate to that of the incremental immersion and the data in Table 3 confirms that the difference between the mean flowrates for these two test conditions was not significant. However, this is not the case for other treatment combinations.

Therefore, it has been concluded that vacuum saturation with the application of the lower vacuum pressure (40 mmHg) and the longer duration (6 hours) is adequate to remove the influence of variations in moisture on the steady state water permeability test and ponding for 48 hours is not sufficient to achieve this.

Table 3 Summary of cross-comparison of the flowrates determined from the new water permeability tests after different preconditioning regimes

Saturation method	Comparison	Mean difference	DoE	P-value*
Ponding	Ponding vs IM-6	0.802	1	< 0.0001
V240-3	V240-3 vs IM-6	0.327	1	0.0036
V240-6	V240-6 vs IM-6	0.227	1	0.0347
V40-3	V40-3 vs IM-6	0.368	1	0.0013
V40-6	V40-6 vs IM-6	0.092	1	0.3441
IM-6	IM-6 vs IM-10	0.025	1	0.7961

*Note: p-value < 0.01 means highly significant, 0.05 < p-value < 0.01 means significant, p-value > 0.05 means non-significant.

3.4 Confirmation of the effectiveness of the proposed vacuum saturation regime

Despite the similarities in flowrates between vacuum saturation and incremental immersion, there is a potential risk to draw an improper conclusion. To see if the effect of mixes on water flowrates determined by using the vacuum saturation method is similar to that from the incremental immersion method, the flowrates for the two methods of saturating three concrete mixes were obtained and compared in Figure 5. It can be seen that the flowrates after vacuum saturation are similar to those after incremental immersion, albeit the vacuum saturation offering slightly higher value for all the mixes. However, the t-test [19] for comparing the flowrates (Table 4) shows that the difference between the flowrates for the two precondition regimes is not statistically significant. That is, the proposed vacuum saturation regime is sufficient to achieve steady state permeability values similar to that for the incremental immersion.

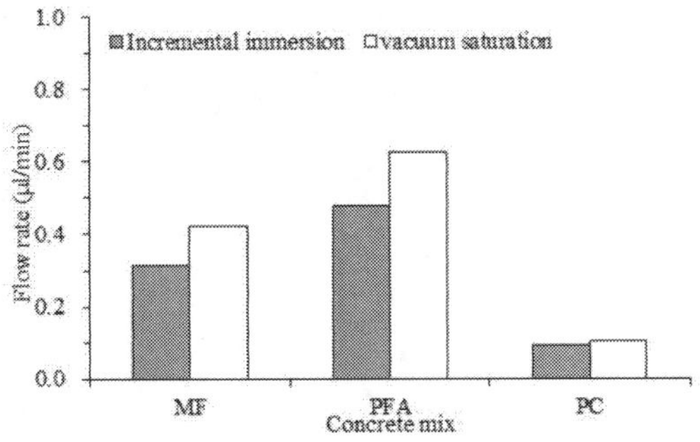

Fig. 5. Comparison of flowrates after saturating by incremental immersion and vacuum saturation for three different concrete mixes

Table 4 Summary of the results of the paired t-test between incremental immersion and vacuum saturation for PFA and PC

t-test	PFA incremental immersion Vs Vacuum saturation	PC incremental immersion Vs Vacuum saturation
	0.087	0.053
σ^*_d	0.338	0.274
$t_{0.05}/(n-1)^{0.5}$	0.953	0.953
$t_{0.01}/(n-1)^{0.5}$	1.676	1.676
$A_{0.05}$	0.322	0.261
$A_{0.01}$	0.566	0.459
Conclusion	Non-significant	Non-significant

Although the objective of the research reported in this paper was not to compare the water permeability of different HPCs, but to evaluate the preconditioning regimes for carrying out the steady state water permeability tests on HPCs, it is important to highlight here that the PC mix had the lowest flowrate in comparison with the PFA and MF mixes. Most classical concrete technology books state that the use of SCMs can give a lower permeability, provided samples are cured under a suitable condition [2, 20]. This apparent anomaly might have been caused by the effect of different degrees of compaction of the three HPCs on their pore structure characteristics and transport properties, as highlighted by Banthia *et al.* [21]. However, the results in Figure 5 would suggest that, even though SCMs could be used to produce a potentially lower permeability concrete, this cannot be guaranteed for all manufacturing and exposure conditions. Therefore, direct on site measurements of concrete permeability are strongly recommended. The results also highlight that the proposed test method is able to identify the differences caused by these factors.

3.5 Requirements to obtain reliable water permeability results

It is necessary to specify the requirements to yield a reliable measurement for field applications. Provided the results after incremental immersion are considered as the 'reference' data, the moisture condition after other saturation techniques should be identical for achieving similar results. The saturation degree is the ideal parameter to describe this feature and the method proposed using the Archie's law [22] was used to estimate the degree of saturation in this study:

$$S = \left(\frac{R_O}{R_t}\right)^{\frac{1}{m}} \times 100\% \quad (1)$$

where S denotes the saturation degree of the pore system (%), R_o denotes the resistance at the saturation condition (Ω), R_t denotes the resistance at the time of measurement (Ω), m denotes the cementation factor reflecting the tortuous nature of the capillary pores.

To evaluate the value of saturation degree, R_o and m in Equation (1) need to be determined. A fully saturated condition is difficult to reach and, hence, the samples were immersed for 50 days (leaving one surface exposed to surface) and the resistance was measured to obtain R_o. For the value of 'm', it is generally in the range of 1.5 to 3 [23] and a middle value, 2, was used in this case.

Figure 6 shows the saturation degree for the different conditions, viz. after drying before immersion (AD), incremental immersion (IM, 6 days and 10 days respectively) and vacuum saturation (VS). As expected, the saturation degree after VS is generally higher than that of AD, especially at the surface region (around 20 mm). Meanwhile, the saturation degree after VS was lower than those after the two IMs. These results show the reason why the proposed vacuum saturation procedure was not as effective as the incremental immersion. Furthermore, the surface region, approximately 20 to 25 mm, achieved a higher degree of saturation (higher than 95%) after incremental immersion. This moisture condition can be considered as the requirement for the proposed steady state water permeability test, as there is a concentration of flow paths located within this region (confirmed by experiments [12, 24].

4. Conclusions

1) The vacuum saturation gives statistically similar results compared with results after incremental immersion when the moisture content inside the concrete is high. To remove the influence of moisture on permeability tests, only the low level of vacuum (40 mm Hg) and the long duration (6 hours) is effective.
2) Ponding for 48 hours before carrying out the new water permeability test is insufficient to remove the influence of moisture, as the flowrates of the new

water permeability test after ponding are approximately 3 times greater than after incremental immersion, which is mainly attributed to the fact that only the surface region (typically 1-2 mm) was saturated.

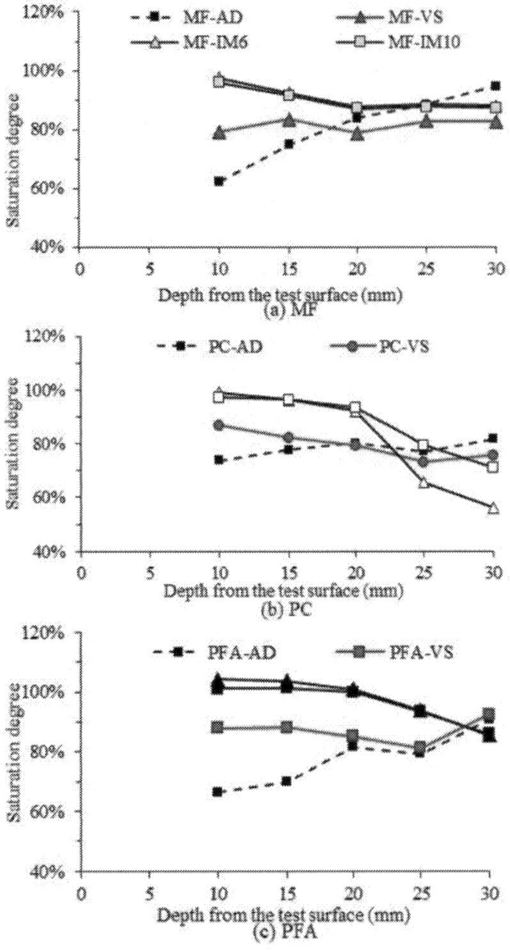

Fig. 6. Comparison of the saturation degree between vacuum saturation and incremental immersion

3) The results indicate that the test region up to a depth of 25 mm should be saturated for performing reliable site water permeability tests.
4) The permeability of HPCs is a complicated function of many factors, especially it is sensitive to the construction practice, curing regime and mix proportion. Therefore, direct measurements of permeability on site are highly recommended and the new test method could be a potential technique for this purpose.

References

1. S.A. Barbhuiya, P.A.M. Basheer, G.I.B. Rankin, M.W. Clark. *Concrete in aggressive aqueous environments - Performance, Testing, and Modelling*: RILEM Publications SARL (2009).
2. P.K. Mehta, P.J.M. Monteiro. *Concrete: Microstructure, Properties, and Materials*. McGraw Hill Professional, (2005).
3. S.C. Pattanaik, E. Gopalkrishnan, S.K. Patro. The Indian. Conc. J. (2015).
4. J.H. Bungey, Constr. and Bldg. Mater. **18**, (2004).
5. M.Q. Zhong, C. Wang, C.H. Yang, Y. Tang, R.Y. Wang, Technical report on the Properties of Concrete and Countermeasures for Bridge Engineering. Chongqing: Department of Transport; (2012)
6. K. Yang, P.A.M. Basheer, Y. Bai, B.J. Magee, A.E. Long, Mater. and Struct. **48**, (2015).
7. K. Yang, P.A.M. Basheer, Y. Bai, B.J. Magee, A.E. Long. NDT & E Int. **64**, (2014).
8. R.T. Torrent, M.G. Alexander, J. Kropp. *Non-Destructive Evaluation of the Penetrability and Thickness of the Concrete Cover RILEM TC 189-NEC: State-of-the-Art Report: RILEM* (2007).
9. P.A.M. Basheer. *Handbook of Analytical Techniques in Concrete Science and Technology: Principles, Techniques and Applications*, Noyes Publications, (2001).
10. P.A.M. Basheer, E.A. Nolan. Constr. and Bldg. Mater. **15**, (2001).
11. P.A.M. Basheer, R. Montgomery, A.E. Long. NDT & E Int. **12**, (1995).
12. D. Whiting, FHWA/RD-81/119: Federal Highway Administration, (1981).
13. K. Yang, P.A.M. Basheer, B. Magee, Y. Bai, A.E. Long. J Mater. Civ. Eng. **27**, (2015).
14. W.J. McCarter, B. Suryanto, H.M. Taha, S. Nanukuttan, P.A.M. Basheer. J Struct. Inte. Maint. **2**, (2017).
15. K. Schonlin, H.K. Hilsdorf. *Concrete Durability* ACI, (1987).
16. A. Elahi, P.A.M. Basheer, S. Nanukuttan, Q.U.Z. Khan. Constr. and Bldg. Mater. **24**, (2010).
17. P.A.M. Basheer, F.R. Montgomery, A.E. Long. Proc. ICE Struct. Bldg. **104**, (1994).
18. J. Kim, W.J. McCarter, B. Suryanto, S. Nanukuttan, P.A.M. Basheer, T.M. Chrisp. Cem. Concr. Comp. **72**, (2016).
19. D.C. Montgomery. *Statistical quality control: A modern introduction*. Oxford, England, Wiley, (2009).
20. A.M. Neville. *Properties of concrete*. Pearson, (1995).
21. N. Banthia, M. Pigeon, J. Marchand, J. Boisvert. J Mater. Civ. Eng. **4**, (1992).
22. M.R. Nokken, R.D. Hooton. Mater. Struct. **41**, (2007).
23. W.J. McCarter, T.M. Chrisp, G. Starrs, P.A.M. Basheer, S. Nanukuttan, S. Srinivasan. Int. J Struct. Eng. **6**, (2015).
24. D.L. Guth, P. Zia. ACI Mater. J. 98, (2001).

MATEC Web of Conferences **289**, 07001 (2019) https://doi.org/10.1051/matecconf/201928907001
Concrete Solutions 2019

Performance of patch repairs on historic concrete structures: a preliminary assessment

Ana Paula Arato Gonçalves[1,*], *Susan* Macdonald[1], *Élisabeth* Marie-Victoire[2], *Myriam* Bouichou[2], and *Chris* Wood[3]

[1]Getty Conservation Institute, 1200 Getty Center Drive, Suite 700, Los Angeles, CA, USA
[2]Laboratoire de Recherche des Monuments Historiques, 29 rue de Paris, 77420 - Champs-sur-Marne, France
[3]Historic England, 4th Floor, Cannon Bridge House, 25 Dowgate Hill, London, EC4R 2YA, UK

Abstract. Carrying out patch repairs to historic concrete buildings and structures needs to be done carefully if their cultural value and significance is to be maintained. This often means repairs using custom designed materials and mixes for compatibility with the original concrete, and with great care paid to good workmanship. But with most repairs, commercial mixes are used which are not compatible with the host concrete but are justified because they adhere well, cure quickly and require limited skills to implement. A research collaboration has been established to test the performance of both approaches. The Performance Evaluation of Patch Repairs on Historic Concrete Structures (PEPS) began in 2018 and is a collaboration between the Getty Conservation Institute, Historic England and the Laboratoire de Recherche des Monuments Historiques. Its purpose is also to better understand key design and specification parameters and application methods. The research is based on assessing case studies in USA, England and France within a variety of climatic and environmental conditions, typologies and repair materials. This paper will present the methodology adopted to evaluate the repairs in the first phase. This will also include historical research on the specification and application of the repairs, preliminary field assessment and some testing.

1 Introduction

As the building stock of the 20th century ages, more examples from that era are now being appreciated and identified as culturally significant. With this recognition comes a need for well understood repair strategies that better meet some of the typical principles applied to work on historic buildings and a body of practitioners who are able to specify and undertake this work. The importance and dominance of reinforced concrete as a construction material during the 20th century has left us with a diverse, creative and innovative concrete heritage. Today there is increasing interest and urgency in the development of sympathetic repair approaches and materials that can meet conservation demands.

The rapid development of reinforced concrete construction over the 20th century means that much of this concrete heritage, however carefully designed and built, did not benefit from the current level of knowledge about material behavior and durability that contemporary construction employs. Consequently, some of these structures were repaired at an early stage, bolstering the parallel development of the concrete repair industry. Nowadays, this multi-billion-dollar industry is supported by prolific research and professional organizations. However, though concrete conservation is a discrete sub-section of the much larger concrete repair industry, it is still in its infancy and there remains much work to be done to adequately equip this nascent field with the knowledge and skills to meet the growing demands.

2 Challenges in conserving historic concrete

The fundamental difference between concrete conservation and concrete repair is that any work targeting historic structures must prioritize the need to preserve its cultural significance [1]. Therefore, the development of a repair strategy for a historic concrete structure has to consider the cultural values supported by that material, such as aesthetic value in the case of an exposed concrete façade. Once the cultural significance of a structure is identified, typical conservation principles should be applied to ensure that its significance is transmitted to future generations. These principles include minimal intervention and minimizing risks by fully understanding the long-term effects of adopted repair materials and methods.

Despite the difference in principles, the methodologies used to develop appropriate repairs for ordinary reinforced concrete structures and for the conservation of culturally significant reinforced concrete share the same basic framework. Best-practice for both should base decision-making on a thorough analysis of the causes of damage, which includes the history of the building, past repairs, field and laboratory based scientific analysis. The

* Corresponding author: anaratogoncalves@getty.edu

design of the repair strategy has to take into consideration compatibility of the repair material to the substrate. In addition, the repair method of choice has to be appropriate for the type of repair and the available means, in terms of economy, availability of materials and workmanship. In all cases, the specified solution should be durable and satisfy all the performance criteria stablished for the structure.

The common roots shared by these two fields make it possible to adapt a lot of the materials and techniques developed for repair of ordinary structures to historic concrete. However, some of the biggest challenges relate to the use of standard repairs used on conservation projects, particularly the durability of patch repairs designed to preserve the aesthetic significance of the original concrete [2-4]. This knowledge gap is reflected in American [2, 5, 6] and European [7] concrete repair standards, which mention color and texture as properties to consider in specific cases, but fail to give specific information on how this can be achieved or measured successfully. It was also found that data on repair methods and materials contain little or no information about their effects on the appearance of the patch. In addition, almost no information was found on how these repairs age. Although durability of repairs is an ongoing concern in the concrete repair industry, long-term appearance, patina, and re-repair, are issues considered primarily by the conservation field.

Another challenge is that conservation principles promote the use of like-for-like materials to avoid the risks of introducing new materials into a system. Additionally, a conservation approach refrains from using materials and techniques that have not been through the test of time. However, materials such as polymer modified cement-based mortars and concretes, bonding agents, corrosion inhibitors, and various additives are widely used in the concrete repair industry to improve workability of the material, performance, and compatibility between the behavior of the new and the old materials. However useful, their long-term effects on behavior and aesthetic of the repair should be understood before being applied on historic structures.

Due to this preference for custom-designed repair materials, in addition to the inherent difficulty of matching final repair color and texture to original exposed concrete, the task of repairing historic exposed concrete requires an extremely high level of craftsmanship and quality control. Developing solutions is also highly specialized, as it requires professionals with experience in both concrete repair and conservation. Currently, the necessary skilled professionals are rare in most parts of the world.

An important contributing factor to this situation is the lack of literature and training opportunities focused on concrete conservation. The existing literature focuses on providing basic guidance on concrete conservation based on experience [4, 8, 9], and, more rarely, on scientific research [10, 11]. Practical challenges in developing repair solutions are found in articles and a few books that present isolated case studies [12]. Moreover, practitioners report different approaches being used to patch repair historic concrete with varying degrees of success.

However, there are no specific studies on the long-term performance of patch repairs designed to preserve the aesthetic significance of the original concrete. Research originated in the concrete repair field does not consider the same performance criteria as the conservation field, therefore the data is useful as a reference but results are not directly applicable.

An experts meeting held by the Getty Conservation Institute in 2014 concluded that one of the main research priorities for concrete conservation was patch repair materials and methods specifically focused on providing better guidance to practitioners [13]. However, this would have to be based on an evaluation of past repairs to understand current approaches and the conditions for successful application when related to their long-term performance.

3 Performance evaluation of patch repairs on historic concrete structures (PEPS)

As a response to this specific challenge, three institutions, the Getty Conservation Institute (GCI), Historic England (HE) and Laboratoire de Recherche des Monuments Historiques (LRMH) are collaborating in the development of a research project called the Performance Evaluation of Patch Repairs on Historic Concrete Structures (PEPS). The goal is obviously to evaluate the performance of patch repairs on exposed concrete in historic structures and buildings in order to better understand the key design and specification parameters and application methods required for a successful conservation approach. The aim is to produce practical guidance that will help those repairing historic concrete. As government agencies faced with providing advice to stewards of historic buildings, both LRMH and HE have been working for many years on many of the challenges of concrete conservation and now have a body of monitored research undertaken over many years that can help to inform this work.

The scope of the project includes the preliminary evaluation of thirty case studies, ten case studies in each country participating in this study: England, France, and the USA. Case studies were selected according to the following criteria:

- Concrete buildings or structures, where the surface is exposed or untreated;
- Repairs performed with the goal of conserving historic and aesthetic values;
- Access to documentation, and professionals who worked on the repair (architects, engineers, tradesmen and craftsmen);
- Easy access to the site and patches to be evaluated;

- Consent must be given from site owners/managers for access to site and data;
- Case studies should have diverse ages of parent concrete and ages of repair;
- Include repairs made with various levels of craftsmanship;
- Include repairs that used different materials (concrete, mortar, like-for-like, industrialized, etc.) and placement techniques (hand troweled, poured);
- Case studies can present combined use of patch repairs and other protection or treatment techniques;
- Complementary techniques applied should not obscure patches, such as coatings;
- Case studies should represent a good range of environmental conditions (coastal, urban…) and aspects.

The case studies will not be identified in articles and any other public dissemination unless otherwise agreed by site owners. A very wide range of case studies were selected in order to reflect the breadth of current approaches and the diversity of contexts for concrete repairs. The range of different materials, applications and sites, also means that it will not be possible to compare performance results between case studies. However, given the current lack of literature on the subject, there is significant value in developing benchmarking research that shows the current state of concrete conservation in relation to patch repairs. It is expected that results will indicate the need for future research in specific aspects and repair materials.

Patch repairs are defined as surface repairs with limited or no load bearing function. This focus was defined by the omnipresence of this type of intervention in concrete conservation. Complementary treatment techniques such as cathodic protection, realkalisation, penetrating water repellants, etc., will not be the focus of the assessment, but their influence on patches will be evaluated if they are present in any case studies.

The methodology developed consists of a preliminary evaluation and a later phase of in-depth study. The first phase comprises evaluation of all thirty case studies, which includes desk studies on building specifications and documentation undertaken during repair, complemented by conducting interviews with owners, asset managers or other relevant individuals to validate the history of the repair. This data will be used to understand how the repairs were developed and executed to support interpretation of current conditions observed on site. This phase also includes field assessments of each case study with the goal of documenting and evaluating the current state of the patch repairs through a visual survey, photography, condition mapping, and in-situ non-destructive evaluation techniques, such as sound and covermeter testing. The aim is to assess ten patches in each case study chosen as representative patches in various conditions and degrees of exposure. This phase is being conducted by the three teams, one in each country, and data is collected in forms before being transferred into a database. Using different assessment teams poses the challenge of maintaining a manageable degree of variance in data collection. This is achieved by ensuring frequent contact, constant sharing of results, and the collaboration of all in the development of the methodology, and trials.

The results of the first phase are expected to identify the different approaches currently being used in the repair of historic concrete and indicate issues that might appear over time. The analysis of these results will inform final planning and fine-tuning of the methodology for the second phase of evaluations. In addition, results will be used as base for the selection of the fifteen case studies for the in-depth study.

In order to assess the patch repairs beyond their surface appearance, the second phase of evaluation will include in-depth study and testing of five sites per country. Due to the historic and aesthetic values of the surfaces investigated, priority will be given to nondestructive in situ testing, but these results will be complemented with laboratory analysis to improve results' reliability. The goal of this phase of investigation is to detect underlying deterioration and identification of material characteristics that could help understand the conditions identified and the behavior of the repair. Testing will be performed on the same patches assessed on the previous phase and their adjacent areas as it is important to identify material characteristics to detect any significant differences in behavior between repair and concrete substrate. Samples from all three countries may be sent to the same laboratory to guarantee consistent results in the analysis. The testing program is still under development.

4 Planned outcomes from the research and next steps

Ultimately, this project intends to produce better understanding on:

- Current repair methods and practices;
- performance of patch repairs typically undertaken on historic concrete;
- efficacy and durability of patch repair materials currently used in historic concrete;
- efficacy of current repair methods.

A long-term goal of the work is to develop and disseminate best practice information to those involved in conserving historic concrete including:

- Guidance on selection of appropriate repair mortars;
- Practical guidance on repair methods;
- A methodology for recording, monitoring and evaluating repairs to historic reinforced concrete.

Additionally, these results will also help in the identification of specific future research needs.

The research project began in 2018 and the first phase of work will be undertaken in 2019, with later phases planned in 2020. The project will provide updates of the work at regular milestones.

References

1. H. A. Heinemann, H. Zijlstra, and R. P. J. van Hees. *Concrete Solutions 2011.* Grantham. Schneck, Mechtcherine (Eds), 55–67. CRC Press (2012)
2. American Concrete Institute Committee 546. ACI 546.3R-06 guide for the selection of materials for the repair of concrete. In *Concrete Repair Manual*, American Concrete Institute, Farmington Hills, MI; International Concrete Repair Institute, Des Plaines, IL, pp. 881-914 (2008)
3. S. Macdonald (ed). *Concrete: Building Pathology*. Oxford: Blackwell Science (2003)
4. D. Odgers (ed). *Concrete* Practical Building Conservation. London: Ashgate (2012)
5. American Concrete Institute Committee 562. *ACI 562-16 Code Requirements for Assessment, Repair, and Rehabilitation of Existing Concrete Structures and Commentary, Reported by ACI Committee 562.* Farmington Hills, MI: American Concrete Institute (2018)
6. American Concrete Institute Committee 563. *ACI 563-18: Specifications for Repair of Concrete in Buildings, Reported by ACI Committee 563*, Farmington Hills, MI: American Concrete Institute (2018)
7. British Standard Institution. BS EN 1504 Products and Systems for the Protection and Repair of Concrete Structures- Definitions- Requirements-Quality Control and Evaluation of Conformity- Parts 1 to 10. London, UK: British Standard Institution (2005-2017)
8. P. E. Gaudette, and D. Slaton. *Preservation of Historic Concrete*. Preservation Briefs 15. Washington, DC: National Park Service, Heritage Preservation Services (2007) https://www.nps.gov/tps/how-to-preserve/preservedocs/preservation-briefs/15Preserve-Brief-Concrete.pdf.
9. D. Farrell, and C. Wood. *Concrete Repairs: Traditional Methods and Like for Like Repairs.* In The Building Conservation Directory: Tisbury: Cathedral Communications (2015)
10. E. Marie-Victoire. *Les alterations visibles du béton: Définitions et aide au diagnostic*. Les cahiers techniques du Cercle des Partenaires du Patrimoine, no. 1. Champs-Sur-Marne: Cercle des Partenaires du Patrimoine (1996)
11. M. Bouichou, and E. Marie-Victoire. *Le nettoyage des bétons anciens: Guide des techniques et aide à la décision*. Les cahiers techniques du Cercle des Partenaires du Patrimoine, no. 4. Champs-Sur-Marne: Cercle des Partenaires du Patrimoine (2009)
12. C. Croft, S. Macdonald, and G. Ostergren (ed). *Concrete: Case Studies in Conservation Practice*. Los Angeles: Getty Publications (2019)
13. Custance-Baker, Alice and Susan Macdonald. *Conserving Concrete Heritage Experts Meeting: The Getty Center, Los Angeles, California, June 9-11, 2014*, Los Angeles, CA: Getty Conservation Institute, p 22 (2015) http://www.getty.edu/conservation/publications_resources/pdf_publications/pdf/Conserving_Concrete_Heritage_Experts_Meeting.pdf

A novel approach to in-depth façade assessments: Improved corrosion test methods for embedded steel framing in historic masonry clad buildings

Gina Crevello[1,*], *Irene* Matteini[1], and *Paul* Noyce[1]

[1]Echem Consultants LLC, 4 Jefferson Plaza, Ste 500, Poughkeepsie, NY, USA

Abstract. Corrosion of structural steel frames and associated steel assemblies within 'vintage' buildings circa 1880s to 1930s pose a health and safety risk to the public in major urban centers. The projecting masonry elements pose a particular concern when the underlying steel assemblies and anchorage begin to corrode. Failed masonry has fallen from buildings, leading to death in worst case scenarios. While some signs of masonry cracking or displacement are usually visible prior to failure, the level of degradation of the embedded steel is not. With the equipment available to test these unforeseen conditions, methodologies need to be shifted to understand unobservable conditions to assist in condition state ratings of embedded steel. In many cities, building owners are being faced with large expenditures to strip and replace terra cotta or stone elements where the underlying steel is in fair condition. This paper will discuss the field-testing programs where a building elevations' masonry clad, steel assemblies (outriggers, anchorage and cross bracing) were evaluated for corrosion. The testing program assessed various steel components which either projected from the structure or were embedded at great depth with a bespoke, in-depth advanced testing program geared towards the development of condition state ratings for the façade elements.

1 Introduction

Corrosion of embedded structural steel in masonry clad buildings can lead to extensive maintenance and repair costs to the owner, the loss of structural integrity of the corroding members, and can pose life safety issues to pedestrians should masonry fall from overhead. Often, the results of corrosion related damages to masonry cladding are evident on the façade, however the degree of internal degradation of the steel is unknown. Traditional methods of material analysis include visual inspections and sounding surveys to determine quantifiable damage to the cladding, non-destructive evaluations to understand cracking of the masonry, and water infiltration and laboratory testing to determine material performance properties of the masonry units. Stripping of masonry for probes to allow observations of the steel typically occurs when investigating steel frame conditions, but is often limited due to the time, cost and destructive nature of probing. In-depth corrosion testing programs can provide ample support to understand condition of the underlying steel.

A façade evaluation of masonry clad steel framing and associated embedded steel assemblies, can be an invaluable tool. Corrosion testing programs allow the owners and design team an understanding of complex conditions within the building envelope which are often unseen during a traditional inspection. In addition, condition states related to corrosion activity can be categorized, and corrosion related degradation can be modelled based on the building's microclimates and corrosion rate. The resulting analysis provides a correlation of the unobserved conditions with steel section loss, which minimizes destructive investigations and can provide categorizations of future corrosion conditions. The knowledge gathered allows the team to make proactive repair choices on established corrosion condition states and helps budget funding based on anticipated degradation.

This paper will focus on terra cotta clad steel frame assessments.

1.1 Steel Frame and Terra Cotta Construction

1.1.1 History

Steel frame construction was prolific from the 1880s to the 1930s, allowing designers and builders to achieve great heights in construction. This common construction method was chosen for fire resistance, durability and efficiency while the great selection of façade stones allowed for creative expression and architectural beauty to be achieved. While cladding types included granite, marble, limestone, porphyry, terra cotta became an ideal building material because of its fireproof, lightweight (lighter than cast or carved stone), and high compressive strength properties, and its ability to be easily molded and manufactured at a faster rate than stone could be carved. It could be used to mimic more expensive stone and the use use of glazing allowed for a wide color range

* Corresponding author: gcrevello@e2chem.com

for material selection, further enhancing terra cotta's versatility and waterproofing qualities.
In addition to structural steel framing Terra cotta cladding is typically four to eight inches in depth with some masonry units being greater in size. Brick is the typical infill material between the façade cladding and the steel frame. In most instances, a form of protection to the steel was utilized: lead paint, asbestos paper, or a cementitious parge coat. Corrosion of the steel had been observed and commented on as early as the 1894, by American architect George Post. During the demolition of the Tower Building in 1914, twenty-six years after construction, severe corrosion was noted, but it was dismissed and assumed to be caused by defective flashing.(1) Early engineering did not focus on material durability nor did construction details fully address moisture management, corrosion of the steel frame and associated steel assemblies . As the construction industry embraced these new systems, advances were made to improve building envelope performance.

Construction detailing became standardized by the National Terra Cotta Society first in 1914, and again in 1927. By 1927, further details were developed in relation to material performance, accounting for expansion of the terra cotta, structural failure, and water management within the building envelope. Shelf supports, provisions for movement, free standing construction, and flashing and drips were the most important principles upon which the standard was revised. It was also noted that proper care should be exercised to prevent corrosion of all steel supports and ties. Where such protection cannot be permanently secured through mortar or concrete, or with corrosion resistant metallic coatings, non-corrosive metals should be employed. (2)

1.1.2 Deterioration Mechanisms

The underlying causes of distress must be understood, so that a durable repair and appropriate maintenance program can be established.(3) Common modes of material failure for terra cotta include glaze and surface erosion, stress cracking from expansion, and deterioration caused by faulty craftsmanship.(4) Most deleterious to the material is corrosion of the anchorage and steel frame. Other than craftmanship defects, moisture ingress is a leading cause of the pervasive deterioration of terra cotta systems. Moisture penetrates the building envelope through failed joints, failed glazing, eroded surfaces caused by caustic cleaning or glaze (5) degradation from atmospheric contaminants, or other waterproofing failures of the building envelope.

The amount of water that penetrates a building increases as façade deterioration occurs. Moisture migration and movement through the building envelope of glazed terra cotta systems are prone to trapping moisture. This is due to the highly resistive and impermeable surface layer of the vitrified glaze, narrow mortar joints between units, and lack of weeps within the assembly. The material properties of the fired clay terra cotta body can retain moisture, keeping the infill and terra cotta unit damp. Retained moisture can lead to glaze spalling, cyclic freeze/thaw issues, and corrosion of the reinforcing steel.

Moisture and oxygen in the building envelope will incite corrosion of the steel frame and anchorage. This in turn cracks the masonry. As the cracks open, more oxygen and airborne particulates can reach the embedded steel. This increases rates of oxidation. Cracking caused by the accumulation of corrosion scale can destabilize terra cotta to the point of failure, leading to spalls and falling units. (6)

1.1.3 Understanding Time Frames for Failure

The greatest risk to façade deterioration is corrosion of the embedded steel frame and associated assemblies. A key advancement in understanding building envelope performance is the ability to forecast corrosion condition states through extensive, in-depth façade testing programs. This methodology has been successfully applied to steel frame buildings, in particular, those clad in terra cotta and glazed brick. To establish corrosion condition states adequate data is required, with a correlation and calibration of the data to the embedded steel. Observations are required though once visual conditions and corrosion data have been correlated, the testing programs become minimally invasive.

2.0 Case Study

To date this testing procedure has been performed on mass masonry wall systems with isolated, concrete encased steel members, independent armatures, full steel framing (electrically continuous), and on building systems with limestone, sandstone, granite and terra cladding. It is especially effective for glazed terra cotta and granite which are highly resistive materials. For this reason, a terra cotta façade is the subject of the case study.

2.1.1 Project Introduction

The subject building is located on the Upper West side of Manhattan in a historic district. While the building itself is not landmarked, it contributes to the historic character of the neighbourhood. The building was constructed in 1914, and designed by architects Gaetano Ajello and Rosario Candela, two of the most prominent pre-war luxury apartment building architects in NYC.

The building is thirteen stories in height. The exterior, street facing facades comprise a granite base, glazed brick cladding, decorative terra cotta balconies, balconettes, water tables, and a decorative terra cotta cornice. A character defining feature of the building is the Juliette balconettes which adorn the façade, yet also pose a high risk for failed masonry deficiencies if deteriorated.

During the local law inspection, by others, the Engineer of Record (EOR) for the inspection deemed that all the balconettes were unsafe. This would require stripping and replacement of all elements based on limited inspections. The associated costs with this work was approximately $2 million USD. Prior repair work had been carried out on some of the balconettes, where a cementitious parge was applied over damaged terra cotta. This and the noted ingress of moisture from leaking air conditioning units was causing deterioration of the steel outriggers which support the terra cotta balconette tiles.

Due to this widespread approach of full replacement, the owner of the building engaged the authors to establish the condition of the embedded steel, to determine efficacy of cathodic protection and to prioritize repairs based on steel condition states.

2.1.2 Testing Program

The testing program was developed to test the balconette outriggers and the cornice outriggers. The balconette outrigger testing program is the subject of the condition state modeling for the purposes of this paper and information on the visual inspection has not been presented for brevity.
The objectives were to:

- Understand the mechanisms of deterioration of the steel sub-component
- Determine sub-components' condition and remaining lifecycle through a condition state process
- Determine the feasibility of corrosion control systems (impressed current)
- Provide a budget and recommendations for the findings

The team utilized condition states to help understand where the materials were in their overall lifecycle and to provide a structured method of analyses for different materials, conditions or components. This is an excellent method for comparing large quantities of structural components to improve financial decision making based on in-depth understanding of long-term performance (8).

2.1.3 Methodology

For the testing program, a bespoke modified linear polarization resistance test method was utilized to test the corrosion activity at the cantilever position of the outrigger within the building façade. Downhole corrosion rate testing was performed at three depths within the building envelope, surrounding the outriggers or critical steel component. These depths were three inches (75mm), six inches (150mm) and 10 inches (ca 250mm). The purpose of testing within the building envelope was to gather meaningful data within two inches (50mm) of the embedded steel and at depths critical to the performance of the cantilevered steel. As this testing was "in-wall" it was deemed 3D versus traditional surface applied testing. Surface applied testing for glazed terra cotta and glazed brick is largely dependent on the surface ratio of the mortar joint to the probe head due to the highly resistive vitrified glaze. Testing within the wall cavity allows for readings that are closer to the steel and representative of the steel environment which impacts steel degradation.

The building has a total of twenty-four balconettes and four balconies equally split across the East and North elevation which are constructed from terracotta. In addition to this, the decorative cornice at the parapet stands 4'3" tall (ca 1.3m) and comprises five terra cotta masonry courses. There are sixty supportive and projecting modillions on the east elevation and fifty-two on the north elevation.

Fig. 1: Projecting Terra Cotta Façade Elements.

All of these subcomponents were included within the inspection and consist of equal sized terra cotta pieces for the balconettes (168 in total), different sizes for the balconies across both elevations (30 North, 42 East and 112 pieces for the cornice (Modillion).

A total of three hundred and twenty embedded steel subcomponents are supporting these terracotta pieces across the building as shown in Table 1.

2.1.4 Service Life Classification

The building was categorized as a Class II structure at medium risk of corrosion and material degradation based on function, environmental exposure and construction material. This addresses temperature, contaminants and the building location for assessment purposes.

Fig. 2. (Top) Terra Cotta Balconettes. **Fig. 3.** (Bottom) Exposed Corroding Outrigger.

Table 1. Test Components

Type	Quantity	Total
T-Section	6 per balconette	144
I-Section	14 per balcony on the North elevation	64
	18 per balcony on the East elevation	
Double L-Section	60 at the cornice on the East elevation	112
	52 at the cornice on the North elevation	
Total		320

The building falls under medium risk primarily due to its location and the vulnerability of contamination in the form of airborne chloride ions and carbon dioxide (carbonation) problems due to its age and exposure.

In order to understand the long-term durability of the structure, a condition state rating system was utilized which defines the current condition of the structure with regard to its degradation. This then provides a time frame for critical failures and the ability to demonstrate how repairs will provide a condition state extension, or when the structure components require replacement. Service life extension can be demonstrated based on what method of repair is utilized.

Technical Service Life, Functional Service Life and Economic Service Life are all means of assessing a structure as it ages and nears critical thresholds that reduce capacity or functionality. The technical service life for corrosion of the embedded steel can be defined on a five (5) state basis as follows:

Table 2. Interpretation of Condition States for Corrosion

Condition State	Section Loss (%)	Comments
5	<1	Very Low
4	1-9	Low to Medium
3	10-17	Medium
2	18-25	Medium to High
1	>25	Very High

The rationale for the condition states for corrosion of the embedded outriggers was the correlation of actual section loss observed and correlated with the measured corrosion rate values from the testing program. The Life 52® service life modelling system utilized provided the basis of the corrosion assessment by utilizing a five (5) state system based around section loss of the steel sub-component. As more steel section is lost the condition state changes from one to another. This method of approach allows for the development of models where the amount of section loss can be determined with time, ultimately changing states. The condition states are then correlated with limit states, as seen in Table 3.

When condition state 2 is met the sub-component is said to be at its' limit and any additional section loss will ultimately lead to obsolescence. In this instance, obsolescence is not necessarily defined as structurally deficient, but so far progressed in the deterioration curve that repair, and rehabilitation become financially impractical to implement. The limit states are also influenced by the risk factors affecting the concrete or masonry, and the financial implications of the cost of deterioration to the owner.

Although the best approach is to be able to measure section loss of the actual component this is often impossible to carry out due to the physical amount of work required to access the sub-component. As a result of this, the approach was to use a combination of inspection methods enabling the inspection team to create a site-specific condition state table utilizing the test results obtained. By knowing the condition state of each sub-component, repair work can be planned accordingly with the ultimate aim of being more cost effective.

The condition state classification of the embedded steel sub-component has been based around the corrosion rate data recorded under the investigative work on site. Table 3 is unique to each structure and developed based on test results and visual observations.

Table 3. Corrosion Condition States and Limit States

Condition State	Corrosion Rate (μm/yr)	Section Loss (%)	Limit State
5	<0.1	<1	0/3
4	0.1 – 1.16	1-9	1/3
3	1.16 – 3.34	10-17	2/3
2	3.34 – 11.6	18-25	3/3
1	>11.6	>25	4/3

3.0 RESULTS

3.1.1 Corrosion Testing Program - Linear Polarization Resistance

A total of four-hundred and seven (407) readings were taken across the structure. The following tables provide the analysis in accordance with the condition state rating for corrosion:

Table 4. Corrosion Rate Results Balconies and Balconettes

Condition State	Corrosion Rate (μm/yr)	West End Elevation		88th Street Elevation	
		# of readings (153)	(%)	# of readings (154)	(%)
5	<0.1	3	1.96	6	3.92
4	0.1-1.16	140	91.5	131	85.62
3	1.16-3.34	7	4.58	13	8.5
2	3.34-11.6	3	1.96	3	1.96
1	>11.6	0	0	0	0

Table 5. Corrosion Rate Results Modillions

Condition State	Corrosion Rate (μm/yr)	West End Elevation		88th Street Elevation	
		# of readings (50)	(%)	# of readings (50)	(%)
5	<0.1	5	10	4	8
4	0.1-1.16	44	88	42	84
3	1.16-3.34	1	2	2	4
2	3.34-11.6	0	0	2	4
1	>11.6	0	0	0	0

3.1.2 Building Envelope Results

A total of nine-hundred and eighty (980) readings were taken across the structure. The following tables provide the analysis in accordance with the condition state rating for corrosion:

Table 6. 3D Results Balconies and Balconettes

Condition State	Corrosion Rate (μm/yr)	West End Elevation		88th Street Elevation	
		# of readings (408)	(%)	# of readings (408)	(%)
5	<0.1	20	4.9	12	2.33
4	0.1-1.16	242	59.31	304	58.91
3	1.16-3.34	89	21.81	103	19.96
2	3.34-11.6	49	12.01	84	16.28
1	>11.6	8	1.96	13	2.52

Table 7. 3D Results Upper L-Section Parapet Wall

Condition State	Corrosion Rate (μm/yr)	West End Elevation		88th Street Elevation	
		# of readings (18)	(%)	# of readings (16)	(%)
5	<0.1	13	72.22	6	37.5
4	0.1-1.16	4	22.22	10	62.5
3	1.16-3.34	1	5.56	0	0
2	3.34-11.6	0	0	0	0
1	>11.6	0	0	0	0

Table 8. 3D Results Double L-Section Parapet Wall

Condition State	Corrosion Rate (μm/yr)	West End Elevation		88th Street Elevation	
		# of readings (14)	(%)	# of readings (8)	(%)
5	<0.1	3	21.43%	0	0
4	0.1-1.16	8	57.14%	8	100
3	1.16-3.34	2	14.29%	0	0
2	3.34-11.6	1	7.14%	0	0
1	>11.6	0	0.00%	0	0

3.1.2 Results Summary

Two methods were utilized in the in-depth inspection: the surface mounted testing on the façade of structure and the second deep within the building fabric. The surface measurements were inconclusive regarding the corrosion rate measurement, however the corrosion potential (E_{corr}) showed a realistic picture of corrosion activity occurring on the building. After successfully altering the test method to test within the building envelope an accurate assessment of corrosion was correlated between section loss, corrosion rates from which the team developed custom condition and limit states.

At the 88th St. balconies/balconettes, the potential provided an uncertain condition at 74% of the readings measured with an additional 5% with a 90% probability of corrosion. When comparing this with the 3D E_{corr} the data is fined tuned and the uncertainty values lie at 35% instead of 74% and the probability of corrosion risk has only increased by 1%. Overall the data with the 3D E_{corr} clarified the original surface position of 79% of corrosion activity down to 41%.

At the West End avenue balconies/balconettes, the potential provided an uncertain condition at 37% of the readings measured with an additional 16% with a 90% probability of corrosion. When comparing this with the 3D E_{corr} the data sets have shifted and there is uncertainty at 41% instead of 37% and the probability of corrosion risk has decreased to 9% from 16 %. Overall the data with the 3D E_{corr} clarified the original surface position of 53% of corrosion activity down to 50%.

When comparing both elevations, a more accurate picture was provided by the 3D E_{corr} at West End Avenue versus 88th St. The comparison provided an overall reduction of or risk of 38% at 88th St. and only 3% at West End Avenue.

When summarizing the 3D I_{corr} data, all data above Condition State 3 was deemed to indicate an obsolete component and required tile removal. These areas are regarded as corroded beyond acceptable section loss limits for the steel sub-component to function as designed. All data above Condition State 3 was assessed and considered as a contingency for repair rather than replacement. The contingency based on various condition states allowed the team to build in additional repair capacity by accounting for all corrosion activity, which will shift conditions.

Initially, the Local Law Façade Engineer had recommended that all projecting elements on the building be removed and replaced. After the testing program, damage was refined based on embedded steel condition and the overall recommendation for terracotta removal was based on corrosion activity rather than assumed deterioration. The values were obtained by taking an average of the three I_{corr} measurements from each side of the steel component. The evaluation of the values can be seen in Table 9, which provides the distribution of data for all areas.

In this table, the steel elements have been categorized by Condition States. State 5 and 4 require no action to be taken, Condition States 5 to 3 are areas where where the steel can be treated electrochemically (ICCP) as steel section is not beyond a structural limit state, and elements in Condition State 2 and 1 require intervention by physical repair or replacement.

Table 9. 3D Overall Summary

88th Street					
Min	**Max**	**Number of Readings**	**%**	**Condition State**	**Comments**
<	0.11	0	0	5	No Action
0.1	1.16	44	51.2	4	
1.16	3.34	22	25.6	3	50% used
3.34	11.6	17	19.8	2	100% Used
>	11.6	3	3.5	1	
West End					
Min	**Max**	**Number of Readings**	**%**	**Condition State**	**Comments**
<	0.11	0	0	5	No Action
0.1	1.16	37	54.41	4	
1.16	3.34	23	33.82	3	50% used
3.34	11.6	7	10.29	2	100% Used
>	11.6	1	1.47	1	

The terra cotta tile summary for removal based on the corrosion inspection is as follows:

Table 10. Tile Removal Summary

		Totals
88th Street	Balconies	2
	Balconettes	12
	Terracotta Tiles	114
	Previously Removed	12 (10%)
	Recommended Removal	14 (12%)
	Contingency Removal	7 (6%)
	Tiles Not Tested	3 (3%)
	Tiles Not Removed	81 (71%)
		Totals
West End Avenue	Balconies	2
	Balconettes	12
	Terracotta Tiles	122
	Previously Removed	4 (%)
	Recommended Removal	5 (12%)
	Contingency Removal	10 (8%)
	Tiles Not Tested	35 (29%)*
	Tiles Not Removed	103 (84%)

4.0 CONCLUSIONS

The testing program successfully identified various conditions of the steel without the removal of the terra cotta elements. With a statistically high level of reliability established by collecting data from all projecting elements, it was determined that 65% of the embedded steel was in fair condition and did not require replacement based on the 3D I_{corr} measurements The evaluation further provided information regarding degradation rates, modeling of future steel and masonry degradation, identified drivers of corrosion activity, and methods to reduce ongoing corrosion by limiting moisture ingress into the already compromised façade. In the end, the cost benefit to the client was a reduction of façade repairs, limiting the required amount of stripping necessary to treat the steel.

The overarching development of the condition states particular to the embedded steel component supporting the terra cotta tiles illustrates that an in-depth corrosion testing program, with a high level of calibration steel conditions, can reduce assumed repairs significantly.

The high level of detailed accuracy which can be gathered allows the team to determine rates of degradation, condition state modeling, and future performance while also fine-tuning repair programs that will minimize the impact of corrosion related degradation in the future.

ACKNOWLEDGEMENTS

The authors would like to thank the owners of the building, and the masonry contractor who provided swing stage support and assistance during the field program.

REFERENCES

1. P. Gibbs. *Technical Advice Notes 20: Corrosion in Masonry Clad Early Steel Framed Buildings.* (Edinburgh, Scotland. Historic Scotland: 2000). P.5
2. National Terra Cotta Society. *Terra Cotta Standard Construction. Revised Edition.* New York: NTCS: 1927).
3. Gerns, et. al. Understanding Terra Cotta Distress: Evaluation and Repair Approaches. Journal of Building Appraisal. Vol.2 No.4. (London, UK Palgrave MacMillan Ltd. 2007)
4. National Park Service. Preservation Briefs 7: The Preservation of Historic Glazed Architectural Terra-Cotta.
5. ASTM STP992. Diagnosis of Terra Cotta Glazed Spalling.
6. G. Crevello and M. Perez. "Terra Cotta Clad Steel Frame Building Repair Approach: Repair Options and Replacement Materials." Structure Magazine: May 2018.
7. C Gray. A Law and the Face of the City. New York Times. April 24, 2014.
8. Crevello, et al. Unpublished Client Reports.

9. ASTM C876-2015 (Latest Revision) "Standard Test Method for Standard Test Method for Half-Cell Potentials of Uncoated Reinforcing Steel in Concrete". (West Conshohocken, PA; ASTM)
10. J. P. Broomfield. *Corrosion of Steel in Concrete,* 2nd Ed. (London: UK. Spon Press: 2007).

Challenges of preserving modernist concrete

Ann Harrer[*1] and *Paul Gaudette*[2]

[1] Associate Principal, Wiss, Janney, Elstner Associates, Los Angeles, California, United States
[2] Principal, Wiss, Janney, Elstner Associates, Chicago, Illinois, United States

Abstract. Exposed architectural concrete is often a defining characteristic of modernist structures. Concrete, like other construction materials, eventually deteriorates over time and with exposure to environmental factors. With any historic structure, challenges exist in the development of an appropriately sensitive rehabilitation strategy to address existing distress, including structural repairs, use of compatible materials, development of repair procedures without unnecessary removal of historic material, and sustainable practices, as well as to meet the requirements of governing standards or jurisdictions. It is important to understand the structure's construction, including material constituents of the concrete, unique characteristics, and extent and causes of deterioration, as well as project requirements and goals. Following a condition assessment, the rehabilitation strategy can be developed, taking into account the sensitive nature of the structure. Appropriateness and compatibility of repairs, together with technical soundness, are key to the success of the repair design. In addition, poor or incomplete repairs will have a minimal service life and can result in additional distress or other problems with the structure. This paper will describe challenges encountered in the preservation of modern concrete structures, focusing on architectural concrete. Case studies of several iconic modernist concrete structures, including the Salk Institute for Biological Studies in La Jolla, California, for which the authors have developed that highlight approaches to address these challenges.

1 Introduction

Concrete has been a popular construction material since antiquity and has been utilized by the modern construction industry for well over a century. Concrete gained acceptance in modern construction because it offered the advantages of conforming to virtually any shape and was readily adaptable to various architectural finishes. As development of new admixtures and additives continues, the possibilities of concrete construction appear limited only by the creativity of the designer.

As a fireproof material, concrete's popularity increased as cities rapidly expanded in the early twentieth century. Later, as understanding of concrete's advantages and potential grew, it became extensively employed not only for the structural elements but also for exposed architectural surfaces at both the exterior and interior of buildings.

Concrete, like other construction materials, eventually deteriorates over time and with exposure to environmental factors. Primary causes of concrete deterioration are usually related to freezing and thawing of critically saturated concrete, corrosion of embedded metal, materials problems, or issues during original construction, such as insufficient concrete cover for reinforcement. The nature and severity of each type of deterioration depend upon the climate and other conditions of environmental exposure (such as pollutants and/or chlorides) [1]. Deterioration and distress can also be related to structural issues such as overloading, or natural disasters such as earthquakes.

While similar processes and procedures are utilized for the assessment and repair of any concrete structure, additional investigation may be necessary for historic structures.

Fig. 1. Exposed concrete finish at the Salk Institute for Biological Studies, designed by Louis Kahn. Photo by the authors.

* Corresponding author: aharrer@wje.com

In addition, the development of repairs may be complicated due to the original construction methods, exposed finish, conservation of historic fabric, or other factors. Additional challenges are presented by modernist concrete structures. Consideration of the significance of the concrete structure, preservation goals, and limitations on repairs and modifications need to be addressed as part of the repair design and should be reviewed with the owner of an architecturally significant and historic property during the initial stages of the project.

2 Modernist concrete characteristics

Widely used in Modernist architecture, concrete essentially defined the Brutalist architecture movement, named after the French term béton brut, meaning raw concrete. Architects such as Le Corbusier, Erno Goldfinger, Marcel Breuer, and others pioneered the Brutalist movement. Other modern architects such as Louis Kahn, Tadao Ando, Minoru Yamasaki, and Welton Becket continued the use of concrete as architectural elements in their designs.

Fig. 2. Exposed concrete finish at the Legislative Assembly in Chandigrah, India, designed by Le Corbusier. Photo by the authors, 2016.

One defining characteristic of modernist concrete structures is that the finish of the exterior of the building often is exposed concrete. The exterior is detailed and defined by the concrete material, placement method, and finishing procedures. Figures 1 through 3 are examples of exposed concrete finishes of modernist concrete structures.

In the design of concrete as a modernist architectural material, care and attention was paid to the material constituents utilized as well as to construction methods, as these factors would directly impact the overall aesthetic. By varying the amount and type of material constituents, including the aggregates and cements, the material properties and color of the concrete can be adjusted.

Construction methods and finishing techniques directly affect the final surface appearance and texture.

Fig. 3. Exposed interior concrete architectural elements at North Shore Congregation Israel in Glencoe, Illinois, designed by Minoru Yamasaki. Photo by the authors.

Some of the various finishing techniques used historically as well as today included exposing the aggregates, retaining the texture from the formwork for a rougher finish (board form), and rubbing to create a smooth finish (Figure 4).

Fig. 4. Examples of variations in the exposed concrete finish in modernist concrete structures. Photos by the authors.

3 Assessment

As an initial step in the development of a rehabilitation strategy for a modernist concrete structure, an assessment of the concrete structure should be performed to evaluate existing conditions and determine causes of distress and deterioration. The findings of the assessment provide the basis for development of a rehabilitation strategy that typically includes a repair and maintenance program. The assessment should also include identification and documentation of historically or architecturally significant components. An understanding of the original construction as well as of deterioration mechanisms provides information to support development of the repair and maintenance program.

The condition assessment should include, at a minimum, the following:

1. Significance - The significance of the concrete and defining features of the structure should be determined.
2. Research - Prior to the field investigation, information relevant to the original construction, techniques, materials, use, maintenance, and modification of the structure should be gathered and reviewed.
3. Field Investigation - The field investigation should include the observation and documentation of the structure, including verification of as-built construction and identification of architecturally or historically signifiant elements, and documentation of deterioration and distress (Figure 5).
4. Field and Laboratory Testing - Visual observations should be supplemented with nondestructive evaluation methods, exploratory openings, and laboratory analysis to determine deterioration mechanisms, material constituents, and as-built conditions.
5. Evaluation of Findings - Findings from the research, field investigation, and testing are compiled and analyzed to inform the appropriate scope of rehabilitation or repair.

The first step is to research the history and significance of the building, the concrete, and its character-defining features. For example, is the finish board-formed, exposed aggregate, or covered with an architectural coating? If exposed, what is the color and texture of the concrete surface? Research should include collecting and reviewing as much information as possible regarding the original construction, use, repair, previous assessments, and maintenance of the concrete.

The next step is to perform an on-site field investigation, and field and laboratory testing [2, 3]. The resulting data informs a report that serves as a summary of findings and, taking into consideration the owner's project requirements and significance of the concrete structure, provides the basis for the development of a conservation approach.

3 Conservation Approach Development

The conservation and repair approach for a modernist concrete structure is developed following the completion of a condition assessment. The findings from the condition assessment are studied to determine the underlying causes of observed deterioration and distress.

In some cases, structural assessment and analysis may be necessary, for example if as-built conditions are found to be different than anticipated (e.g., less reinforcement in the structure than shown in construction documents).

Fig. 5. Delamination and spalling at exposed concrete are examples of distress and deterioration that should be documented as part of a visual condition survey. Photo by the authors.

The causes of observed deterioration and distress should be understood in order to develop an appropriate scope and define the extent of repairs. The owner's intended use and maintenance of the structure as well as the significance and value of the structure should also be understood and considered. Through evaluation of the assessment findings, taking into consideration the owner's project goals, a conservation approach can be developed to address concerns while being sensitive to the preservation of the historic fabric.

The rehabilitation approach needs to consider technical concerns, such as deterioration mechanisms and durability, characteristics of the original concrete as well as the aesthetic considerations of matching the original concrete elements [4]. Most importantly, the rehabilitation approach should address the underlying causes of the distress or deterioration of the concrete to minimize continued deterioration or premature failure of the repairs. This may require removal of historic fabric, such as in cases where removal of concrete to access all sides of a corroded bar is necessary to mitigate the ongoing corrosion (Figure 6). Removal behind existing reinforcement bar also allows for repair material to be mechanically anchored.

Repairs to the concrete structure should be durable and extend the service life of the concrete structure. It is therefore important that the repair approach includes surface preparation of the substrate to achieve bond of the repair material to the original concrete substrate. The characteristics of the original concrete should be considered in the characteristics of the repair material and design of the repair itself [5]. For example, the compressive strength and modulus of elasticity of the repair material need to be compatible with the original or substrate concrete. In addition, the depth and shape of the repair should be selected to provide durable service life to both the repair and the substrate concrete.

As previously noted, in modernist concrete structures the exterior finish is often exposed concrete.

Therefore, the conservation approach should take into account matching of the existing overall aesthetic of the concrete including the color, texture, and finish, while maintaining the durability of the structure and addressing the underlying cause of distress.

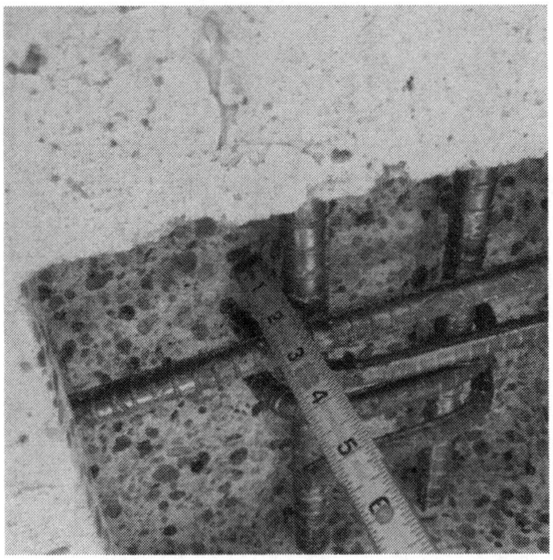

Fig. 6. Removal of concrete beyond the corroding reinforcement bar to allow removal of corrosion product and to address underlying distress mechanism. Photo by the authors.

Research and consideration of available cements, aggregates, and potentially admixtures to develop a compatible mix that is a color match to the original is necessary. Locally available materials may have been sourced during original construction that are available, however, it should be noted that material removed from the same quarry but from a different depth or location may not be similar to the original constituents. Available mineral pigments and admixtures should be considered to achieve a color match, however, the durability and color steadfastness of the materials needs to be understood. Matching the existing geometry and finish may affect the constructability of the repair, a factor that needs to be considered.

4 Repair Implementation

The results of architectural concrete repair work are dependent to a great extent on the experience of the concrete repair contractor and the level of craftsmanship of the person performing the work. Prior to the start of an architectural concrete repair project, it is important that a mock-up and trial repair program be performed. As the level of craftsmanship of anyone implementing the repairs is critical to the overall technical performance as well as the final aesthetic, it is imperative to establish the desired level of craftsmanship prior to the start of work. Mock-ups and trial repairs are necessary to confirm the developed rehabilitation approach (Figure 7).

The design of the concrete repair mix design needs to be developed to achieve technical specifications outlined in the rehabilitation approach as well as to match the adjacent original concrete. The mix development, review and approval process can require quite a bit of time and should be included in the construction schedule. It is also important that the concrete mix be designed for constructability to allow for placement. Typically formed concrete repairs are preferable over trowel-applied mortars for several reasons. For example, concrete repair materials will have more characteristics similar to the substrate concrete. The mix design developed for the repairs should be compatible with the characteristics of the original concrete. Prior to the start of work, off-building samples and mock-ups should be prepared to demonstrate constructability, performance, and aesthetic characteristics of the repair material.

The design of formwork and finishing procedures of the concrete need to be developed to match the existing concrete. In order to achieve compatible, durable, and aesthetically matching repairs, the creation of formwork, placement procedures, and finishing procedures need to be practiced and finalized through a series of off-building samples and mock-ups [6]. Only during the off-building samples and mock-ups can mixes, placement, and finishing procedures be fine-tuned and modified without resulting in unacceptable repairs on the building that will require removal.

Fig. 7. Completion of off-building trial samples and mock-ups prior to the start of repairs to confirm appropriate performance of the concrete mix design, placement procedures, and finishing techniques. Photo by the authors.

The successful completion of samples, mock-ups, and trial repairs will demonstrate the level of craftsmanship, the ability to achieve the necessary technical characteristics, and the match to the aesthetic details of the existing structure. Production repair work can then begin. Trial concrete repair work at the building should be completed to achieve the desired craftsmanship level established during the samples and mock-ups. During the implementation of the concrete repairs, craftsmanship needs to be maintained at a high level to achieve a final result that is acceptable and matches the existing concrete in color, finish, and texture. During the repair work quality control measures, including sampling and testing of plastic and hardened concrete is recommended. Field testing should include monitoring the batching and mixing of the concrete repair materials with slump, air content, unit weight, and other typical quality control testing. Cylinders should be cast and compressive strength testing conducted regularly. In addition, testing of the repairs, once adequately cured, to confirm adequate bond to the substrate should be completed. This testing should consist of hammer sounding to detect any areas of unsound or unbonded repair materials. Additional quality control testing, including tensile bond testing can also be performed. If any repairs deviate from the approved mock-ups, are unsound, or do not match the adjacent concrete, these can be removed and replaced.

5 Summary

The use of concrete as both an architectural and structural material has created challenges, particularly in modernist structures where the concrete is a defining feature of the building. Understanding the significance of a modernist structure and conserving the concrete defining features are vital to the preservation of these modernist structures. Concrete is susceptible to deterioration as a result of various distress mechanisms, from environmental exposure to problematic original design and construction. The first step in the preservation of a modernist concrete structure is the condition assessment to identify architecturally or historically significant characteristics, to understand as-built construction, and to determine the extent and cause of distress and deterioration. The preservation approach needs to address the underlying deterioration or distress mechanisms while matching the existing original color, finish, and texture of the concrete structure. The repair implementation requires a high level of contractor experience and craftsmanship to retain the overall design aesthetic of the structure. The completion of mock-ups and trial repairs is essential to establish standards for the overall repairs.

References

1. Gaudette, Paul, and Deborah Slaton. *Preservation Brief 15 : Preservation of Historic Concrete.* Washington D.C.: National Park Service, Heritage Preservation Services, 2007.
2. American Concrete Institute. *Guide to Evaluation of Concrete Structures before Rehabilitation.* ACI Committee 364, ACI 364.1R-19.
3. Gaudette, Paul, and Ann Harrer, "Assessment of Historic Concrete Structures," Practice Points Number 16, *APT Bulletin: The Journal for Preservation Technology* 48, no. 4, 2017.
4. American Concrete Institute. *Guide to Concrete Repair.* ACI Committee 546, ACI 546R-14.
5. Gaudette, Paul, Deborah Slaton, and David Patterson, "Morse and Ezra Stiles Colleges, Yale University, New Haven, Connecticut USA 1958–1963," *Concrete: Case Studies in Conservation Practice, Conserving Modern Heritage.* Los Angeles : Getty Conservation Institute, December 2018.
6. Caldwell, Jeffrey and Ann Harrer, "Repairs to Concrete at the Pilgrimage Theatre in Los Angeles, California," *APT Bulletin: The Journal for Preservation Technology* 48, no 1, 2017

Acknowledgements

Deborah Slaton, FAPT, a principal with Wiss, Janney, Elstner Associates, Inc., is gratefully acknowledged for her constructive and detailed reviews.

Influence of carbonation on the resistance of concrete structures to chloride penetration and corrosion

Abbas S. AL-Ameeri [1,2], *M. Imran Rafiq* [1] *and Ourania Tsioulou* [1]

[1] School of Environment and Technology, University of Brighton, Brighton, UK
[2] Engineering College, University of Babylon, Babylon, Iraq

Abstract. Chloride-induced corrosion of steel rebar embedded in concrete is one of the major concerns influencing the durability of reinforced concrete structures. It is widely recognized that the carbonation in concrete affects the chloride diffusivity and accelerates the chloride-induced reinforcement corrosion. However, only very limited studies have dealt with this issue in the literature. The presence of service load related cracks also affects the reinforcement corrosion. This study aims to investigate the potential impact of concrete carbonation on the chloride penetration resistance, and degree of corrosion, in RC structures subjected to service related microcracks.

The experimental programme involves casting of concrete prisms (100 x 100 x 500 mm) with different water-cement ratios (w/c) of 0.4, 0.5 and 0.6 and with four different crack widths (0, 0.05-0.15 mm, 0.15-0.25 mm and 0.25-0.35 mm). These samples were exposed initially to accelerated carbon dioxide (CO2) environment and then exposed to the accelerated chloride environment. Carbonation depth, chloride penetration, and the degree of corrosion (using half-cell) were experimentally measured. The results indicated that (i): The depth of carbonation increases with the increase in crack width and w/c ratio, (ii) chloride penetration depth in concrete structures increases significantly due to the influence of carbonation and (iii) half-cell corrosion potential increases significantly when carbonated concrete samples are exposed to the chloride environment relative to the uncarbonated concrete samples.

1 Introduction

Carbon dioxide and chloride ion attack seem to be significant factors which affect the sustainability of concrete structures. Furthermore, the temperature and relative humidity affect the penetration of chloride and sulfate in concrete structures. As a result, the diffusivity of carbon dioxide and chloride have a significant impact on corrosion in concrete structures [1, 2]. This study aims to investigate the influence of carbonation on the penetration of chloride in concrete and its degree of corrosion. Chloride ions may be transferred into concrete from external sources such as de-icing salt, seawater and groundwater, and internal sources through contaminants in concrete such as marine aggregate and chemical admixtures containing chloride ions [3].

The diffusivity of CO_2 in concrete leads to changes in the mineralogical composition (due to carbonation) of concrete and has the potential to reduce the pH of the pore water in concrete. It may also break down the chloride binding (calcium chloroaluminate), which free up more chloride leading to the deeper penetration of chloride in the concrete. It is a result of the reaction between one of the cement components, namely tricalcium aluminate (C3A) and chloride ions within the pore water in concrete [4]. Carbonation is a significant factor in reducing the pH and liberation of bound chlorides for cement mortar samples [5].

Cracks in concrete structures are expected due to its weak tensile strength. Cracks can be either non-structural or structural. Structural cracks tend to be wider (> 0.1 mm) while non-structural cracks are commonly finer. Structural cracks often result from one, or a combination, of the following: insufficient reinforcement, low strength of concrete, excessive loading magnitude or frequency, impact loading, permanent actions such as foundation settlement, creep, etc. While, the non-structural cracks may occur due to physical impact (shrinkable aggregate, drying shrinkage), chemical effect (carbonation shrinkage and alkali -silica reaction) and thermal impact (expansion/contraction due to external temperature variation and freeze and thaw cycle) [6,7]. The cracks may also be the result of deterioration and loss of durability of material or are worsened due to the degradation of material. The durability of concrete structures is controlled by their capacity to delay the transport of ions and harmful fluids inside the concrete. The transport of chloride within uncracked concrete is said to be a combination of the concentration gradient (diffusion), the pressure gradient (permeation) and capillary sorption [3, 8]. The cracks in concrete introduce an additional factor within the transport properties of concrete, namely the permeability of concrete, which is likely to significantly increase the transport of chloride ions inside concrete [9]. Park et al. [10] found the chloride diffusion coefficient in concrete increases 24 times for crack width of 0.1 mm and 145 times for crack width of 0.4 mm, when compared with the un-cracked concrete. These results are not fully in agreement with the results obtained in [9,11] due to natural formation of cracks in the sample (width, depth and tortuosity) or the method of assessing the diffusion coefficient of chloride.

The studies of durability assessment in literature, such as corrosion initiation, often excludes the effect of carbonation and cracks, voids and defects in concrete, thereby their effectiveness in predicting the durability of concrete structures is reduced [11]. Therefore, it is important to investigate the influence of carbonation and service cracks on the penetration of chloride and the degree of corrosion. Moreover, carbonation and chloride negatively affect the passive protective film (PPF) of steel reinforcement in concrete structures [1], which accelerates the corrosion of reinforced concrete structures [3, 13]. The focus of this study is to examine the effect of carbonation in reinforced concrete samples on the penetration of chloride and degree of corrosion in cracked and un-cracked concrete samples.

2 Experimental work

2.1 Materials and concrete mix design

Portland limestone cement (CEM II/A-LL 32,5R) with a specific gravity of 3.05 was used in this study. Chemical and physical properties of the cement comply with BS EN 197- 1: 2011 [14].

Natural sand was used as fine aggregate (particle size <5mm), and the coarse aggregate used was crushed gravel with the size ranging from 5-14 mm. The grain size distribution, and maximum chloride and sulfate content within the aggregates comply with BS 882:1983 [15]. Deformed or ribbed steel (B500A) (8mm) with mechanical properties compliant with BS 4449:2005+A2:2009 [16] was used to reinforce the concrete prisms to achieve the controlled crack widths in the samples, and to investigate the corrosion condition.

Three different w/c ratios were used to investigate their effects on the penetration of carbonation and chlorides. The Building Research Establishment (BRE) method [17] was used for concrete mix designs that are shown in Table 1.

Table1 Concrete mixes designs used in this study.

Mix symbol	Content per unit volume of concrete (kg / m^3)				
	w/c	Cement	Water	Sand	Gravel
M 0.4	0.4	513	205	653	980
M 0.5	0.5	410	205	711	1023
M 0.6	0.6	350	205	711	1041

2.2 Specimen Geometry

Three 100 mm cube specimens were used to establish the compressive strength of concrete, porosity and depth of carbonation measurements. Reinforced concrete prisms of 100 x 100 x 500 mm were used to measure the depth of carbonation (DoC), chloride penetration and corrosion condition (using half-cell corrosion potential) in both the cracked and un-cracked specimens. A deformed steel bar (B500A) of 8mm diameter was located at a cover depth of 2 cm to reinforce the concrete prisms to achieve the controlled crack sizes within the samples and to investigate the half-cell corrosion potential. The method of filling and preparation of moulds, compaction and levelling the surface of concrete, curing and transporting of samples were in accordance with BS EN 12390 - 2:2009 [18]. Samples were cast in two layers and each layer was vibrated to achieve homogenous concrete. The specimens were demoulded and cured using tap water until the time of testing, or exposure to CO_2 environment condition at the age of 28 days.

2.3 Specimen conditioning

The main objective of this study was to investigate the effect of carbonation depth on the chloride penetration and corrosion condition for different properties of concrete and crack width in concrete structures. The flexural method was used to induce the cracks in concrete samples as shown in Fig.1. In addition to the control samples (no cracks), three different crack width ranges, (0.05-0.15mm, 0.15-0.25mm and 0.25-0.35mm) were examined. The crack width was measured using a microscope meter with an accuracy of 0.01 mm (Fig. 2).

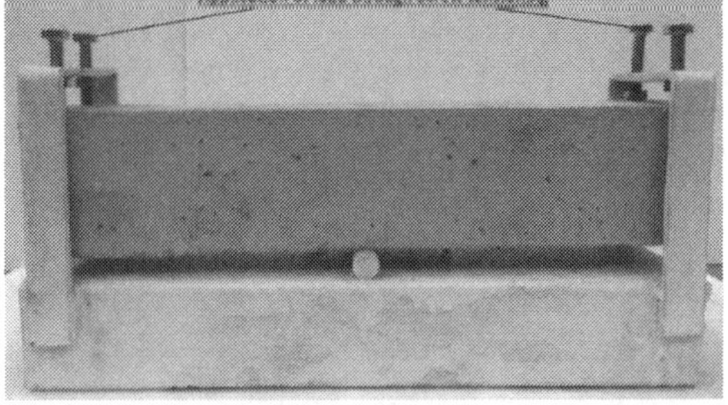

Fig. 1. Method for cracking in the prism samples

In this study, one face of the specimens was exposed to accelerated environment conditions whereas the other faces were sealed using coats of water-based alkyl polysiloxane resin. The control samples were exposed to a chloride fog environment (with 5% NaCl) within a Cyclic Chloride Test (CCT) chamber for 17 weeks in total.

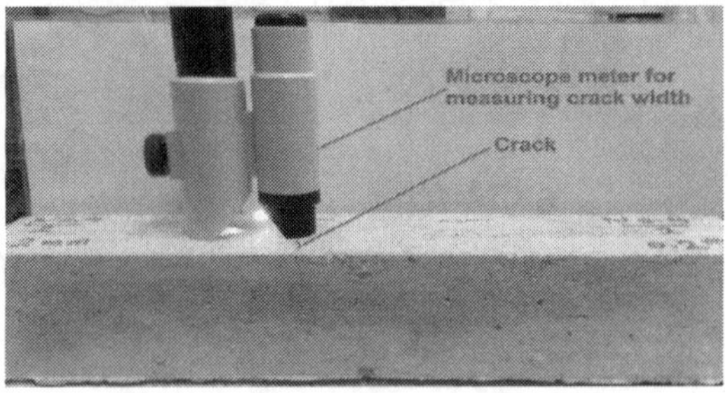

Fig.2. Microscope meter to measure the crack width

The rest of the samples were taken out of the CCT chamber after 10 weeks of chloride exposure and were exposed to CO_2 concentration of 5% within a CO_2 chamber (at a temperature of 25 °C and relative humidity of 65%) for 5 weeks before returning them back in the CCT chamber (chloride exposure) for the remaining 7 weeks.

2.4 Testing Methodologies

The compressive strength, ultra-sonic pulse velocity, porosity, depth of carbonation (DoC), chloride penetration, and half- cell potential were carried out for this study. Porosity and compressive strength test were conducted according to ASTM C642:2013 [20] and BS EN 12390-3:2000 [21] respectively. The chloride penetration depth was measured by splitting the conditioned samples into two parts (using a compression testing machine) and spraying 0.1% silver nitride solution in accordance with [22,23]. The silver nitrate solution reacts with chloride ions to form silver chloride, changing the colour of the concrete surface hence the depth of chloride penetration can be measured [22,23]. Three specimens were tested for every case and an average value is reported here.

In order to measure the depth of carbonation, CO_2 conditioned samples were split into two parts using a compression testing machine in accordance with Al-Amoudi et al. [24]. The first part was sprayed by a solution with one gram of phenolphthalein powder dissolved into a solution of 70 ml and 30 ml of ethanol and deionized water respectively [25,26]. The change of colour of the concrete surface was then used to determine the carbonation depth, DoC. The second face of the sample was drilled for control samples and for cracked samples (at the crack location) to collect concrete powder at different depths from the exposed face. Seven depth intervals were adopted in this study; 0-6 mm, 6-12 mm, 12-18 mm, 18-24 mm, 24-30 mm, 30-36 mm and 36-42 mm using dry drilling equipment. The collected concrete powder was then sieved using 150 micro-meter sieves to reduce the amount of coarse grains (obtained due to crushing of aggregates). The powder was then dried in an oven at 50°C for 24 hours and kept in sealed plastic bags till tested.

The apparent pH of the powder was measured by dissolving 1 g of concrete powder in 20 ml of deionized water. This mixture (powder and water) was stirred vigorously for 30 seconds and left to settle for 24 hours [27, 28]. A pH meter was used to measure the apparent pH of this solution in accordance with BS EN ISO 10523:2012 [29]. Three specimens were tested for every case and an average value is reported in the results.

Finally, the corrosion monitoring tests were carried out using non-destructive techniques. These techniques have been utilized to assess the corrosion activity and determine the corrosion condition of the reinforcement. One of the most used electrochemical techniques was used for monitoring the corrosion of steel in concrete, the half-cell potential. The half-cell potential test was performed in accordance with the ASTM C876:2015 [30] and the method proposed by Bouteiller et al., (2012) [31] as shown in Fig.3. Three specimens were tested for each case and the average results of these are reported in the results.

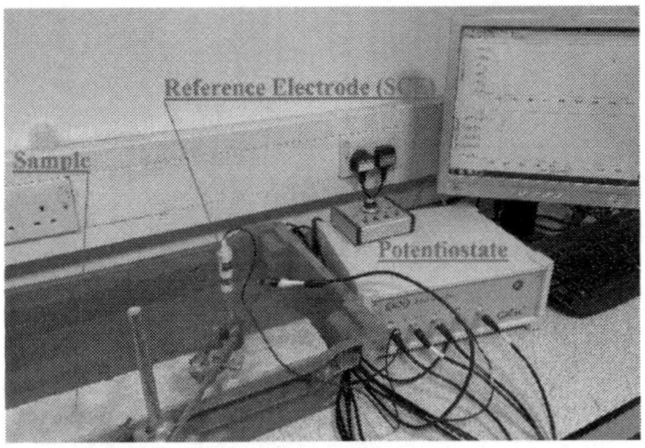

Fig. 3. Potentiostat for half-cell potential measurements.

3 Results and discussion

In this section, the results of the experimental work are presented and discussed. These are mainly focused on the impact of carbonation on chloride penetration and corrosion activity for different w/c ratio (representing the strength of concrete) and crack widths in concrete samples.

3.1 Carbonation Depth in Concrete

The depth of carbonation was established by measuring the depth of carbonation at different points along the sample width and computing the average of these measurements as shown in Fig.4.

The DoC as a function of w/c ratio for the mixes used in this study is shown in Table 2, where M 0.4, M 0.5 and M 0.6 refers to w/c ratio of 0.4, 0.5 and 0.6 respectively. The results illustrate that the carbonation depth increases significantly from the exposed surface with the increase in w/c ratio in concrete samples. The percentage increases in DoC for samples having w/c ratio 0.5 and 0.6 with respect to the concrete samples with w/c ratio of 0.4 are 25% and 113% respectively.

Fig.4. DoC in the concrete sample (M 0.6) exposed to CO_2, (5%) for 5 weeks

Table 2: Properties of concrete used in the study

Property	Concrete Mixes		
	M 0.4	M 0.5	M 0.6
DoC at 5% CO_2 (mm) *	8.0	10.0	17.0
Coefficient of variation (COV) of DoC	3.3	2.6	1.6
Porosity % *	10.1	11.1	11.5
Compressive strength (MPa)*	53.2	48.1	39.7

* These results are the average for three readings

The increase of DoC in concrete mixes as a function of the w/c ratio is a result of the increase in porosity of concrete (also shown in Table 2) and volume of permeable voids, i.e. the pore system of the hardened cement paste [1,32], that helps CO_2 to penetrate and react with water to form carbonic acid, the latter then reacts with $Ca(OH)_2$ to form the carbonation products ($CaCO_3$).

Fig. 5 presents the results of apparent pH measurements within the concrete depth, obtained using the dust powder collected at various depths within the concrete samples. The results clearly demonstrate a reduction in apparent pH within the surface region of concrete (affected by carbonation), which is more pronounced for higher w/c ratios. The increase in w/c ratio lead to an increase the penetration of CO_2, which in turn increases the consumption of the alkalinity sources in concrete (i.e. $Ca(OH)_2$ and C-S-H) [27, 28], hence the pH levels reduced within the depth of concrete are affected by carbonation.

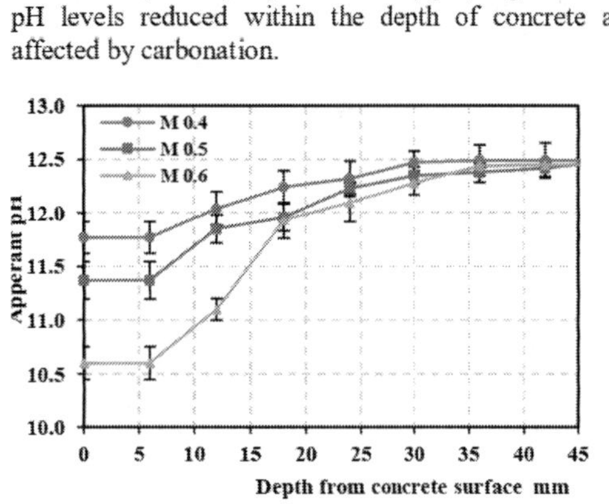

Fig. 5. Apparent pH profile in concrete due to carbonation

3.2 Chloride penetration

Two sets of concrete samples were tested for chloride penetration; the first set was exposed to a CO_2 environment in-between the exposure to the chloride fog environment, whilst the second set was just exposed to the chloride fog (control specimen). The difference between the series is the first was exposed to an accelerated CO_2 environment, while the second was exposed to a normal CO_2 environment and exposure time to CO_2 for both was 5 weeks. The silver nitrate ($AgNO_3$) colorimetric method was used to measure the chloride penetration depths. The $AgNO_3$ solution spray reacts with chloride ions to form a mixture of silver oxide and silver chloride that precipitates on the concrete surface and changes colour forming a boundary corresponding to the chloride penetration depth within the concrete specimen. Results for the two sets for different crack widths and w/c ratio are presented in Fig. 6. Key observations from these results are:

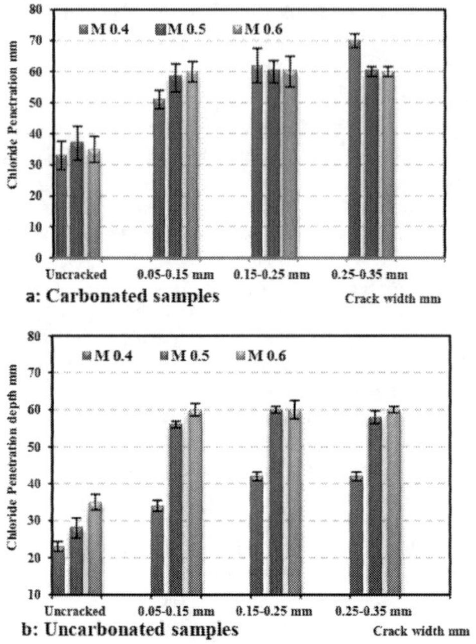

Fig. 6. Effect of w/c ratio and cracked width on chloride penetration(d_{cl-}) for un-carbonated and carbonated samples

(i) The results exhibit higher chloride penetration depths for samples exposed to CO_2 environment when compared with the control samples as shown in Fig.7 and Table 3, particularly for the uncracked concrete samples. This can be attributed to the fact that the carbonation reduces the pH of the pore water in concrete which is a significant factor in liberation of bound chlorides (calcium chloroaluminate) in cement mortar [5] hence increases free chloride concentration and ultimately the penetration depth.

(ii) The results depict an increase in chloride penetration with the increase in w/c ratio for both sets of concrete samples (i.e. with and without carbonation) particularly for the uncracked samples. This can be explained by the fact that the w/c ratio affects the volume of internal voids or porosity in the concrete that in turn affects the chloride ions transport mechanisms in concrete, i.e. diffusion, permeation, sorption and permeability [33]. The percentage increase in chloride penetration depth with the increase in w/c ratio are shown Table 4.

(iii) The chloride penetration depth increases with the crack widths for the all mixes till a threshold crack width is reached, which is estimated from the Fig. 7 to be around 0.1 mm. The crack width and depth affect the transport mechanism of chloride ion into the crack opening since wider cracks may allow the chloride solution to penetrate in the cracks [34,35]. The cracks of width less than 0.1 mm appear to have contributed only through an increase in the rate of transport of chloride ions (i.e. increase in permeability and diffusion rates of concrete). However, cracks of width greater than 0.1 mm

(iv) appears to have provided additional surface area (along the crack walls) for the transport of chloride ions into concrete. Hence, an increase in chloride penetration perpendicular to the crack walls can also be observed (Fig. 8), which is similar to the chloride penetration profile from the surface of an uncracked concrete. The percentage increase in chloride penetration depth for cracked samples in relation to the uncracked samples are shown in Table 5.

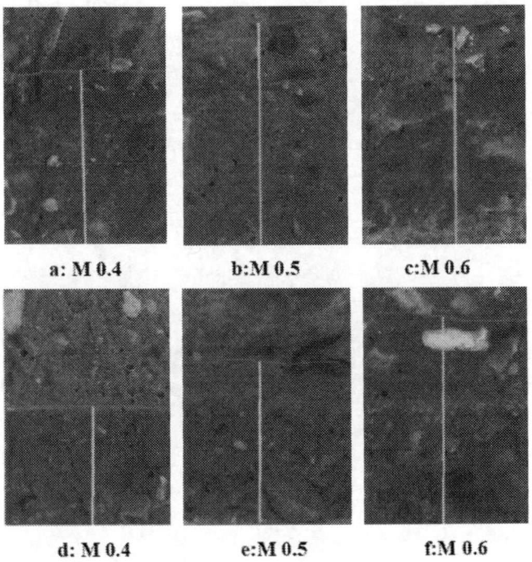

Fig. 7. Effect the w/c ratio and carbonation on d_{cl^-} for Un-cracked sample (a, b and c) carbonated samples and (d, e and f) Un-carbonated samples

Table 3 Percentage increase in chloride penetration depth in carbonated samples in comparison to un-carbonated samples

Sample	Percentage increase (%)			
	Un-cracked	0.05 - 0.15 mm	0.15-0.25 mm	0.25-0.35 mm
M 0.4	43	50	48	67
M 0.5	32	4	0	3
M 0.6	0	0	0	0

Table 4 Percentage increase in chloride penetration depth for M 0.5 and M 0.6 samples with respect to M 0.4

Series	Percentage increase (%)			
	Un-cracked	0.05 -0.15 mm	0.15-0.25 mm	0.25-0.35 mm
Series 1 carbonated	12	14	-3	-14
	6	18	-3	-14
Series 2 uncarbonated	22	65	43	38
	52	76	43	43

Table 5 Percentage increase in chloride penetration depth for cracked samples with respect to un-cracked sample

Series	Sample	Percentage increase (%)		
		0.05 - 0.15 mm	0.15-0.25 mm	0.25-0.35 mm
Series 1 carbonated	M 0.4	55	88	112
	M 0.5	57	62	62
	M 0.6	71	71	71
Series 2 uncarbonated	M 0.4	48	83	83
	M 0.5	100	114	107
	M 0.6	71	71	72

Fig. 8. Effect of w/c ratio on the penetration of chloride in the cracked sample

3.3 Half-cell potential (E_{corr}) results

The half-cell potential (electrode potential, E_{corr}) measured at various intervals during the exposure of samples to chloride fog, with the aim to detect the transition from passive to active corrosion condition, is presented in Fig. 9. ASTM C876-2015 [30] provides useful interpretation of these results. Half-cell potential values of higher than −126 mV measured with reference to the saturated calomel electrode (SCE) refers to more than 90% probability of inactive corrosion (passive state). The potential values in the range of −126 to −276 mV SCE represents intermediate corrosion risk. Whereas, the potential values more negative than −276 mV SCE represents a greater than 90% probability of active corrosion [31,35].

(i) Fig. 9-a, b and c on the next page presents the results of a control specimen (exposure to chloride fog without carbonation) for the three mix designs used in this study. An increase in exposure time of samples to chloride fog monotonically increases the more negative values of the electrode potential, i.e. the probability of corrosion increases with the duration of exposure. The electrode potential for many samples had exceeded the threshold potential (-126mV SCE) between the passive to intermediate corrosion risk condition after 40 days of exposure. The specimens M 0.5 and M 0.6 attained the potential values that are more negative than -276 mV SCE after 80 days of exposure for all samples.

(ii) Fig. 9-d, e and f presents the results of carbonated samples for the three mixes used in this study. The exposure of samples to the CO_2 environment has a considerable impact on the electrode potential of steel reinforcement. A step change in potential drop can be seen at around 80 days in all carbonated samples (during the period of CO_2 exposure), whereas the potential values for samples M 0.4 remained more positive than -126 mV SCE (> 90% probability of no corrosion) for the control samples. This suggests that the corrosion conditions have been altered (from passive to active corrosion state) for this concrete mix due to the presence of carbonation.

(iii) The negative values of half-cell potential generally increased with the increase in w/c of concrete (i.e. between M0.4, M0.5 and M0.6 concrete mixes) for both controlled and carbonated specimen. Table 6 details which samples had exceeded the threshold value of more negative than -276mV (>90% probability of active corrosion), and their respective exposure duration when the threshold was reached.

(vi) The cracks within the samples increases the transport of chloride, accelerating the transition of corrosion from passive to active state. This can be seen from Fig. 9 where the electrode potential value generally increased with the increase in crack width. Hence, the probability of corrosion is proportional to the crack width. For carbonated samples, the impact of crack opening on the potential drop was not clear since samples changed from passive to active corrosion condition during CO_2 exposure phase.

4 Conclusions

This study explores the impact of carbonation on the chloride penetration, and corrosion condition in concrete samples cracked by service loading. Chloride penetration, carbonation depth, and half-cell corrosion potentials were investigated by using an accelerated environment test programme for (CO_2 and Cl^-). The depth of carbonation, DoC was measured by phenolphthalein Indicator, whereas the chloride front (d_{cl^-}) was determined by $AgNO_3$ spraying. The half-cell corrosion potential was employed to establish the corrosion condition. DoC using phenolphthalein indicator and pH investigations made it possible to develop better understanding of the impact of carbonation on chloride penetration and half-cell corrosion potential (corrosion activity). The following conclusions can be drawn from the results;

1- The quality of concrete, i.e. w/c ratio, porosity, and compressive strength affects the depth of carbonation and chloride penetration depths.
2- The chloride penetration significantly increases within the concrete samples exposed to CO_2 environment for all mixes used in the study.
3- The half -cell corrosion potential is significantly higher in carbonated samples when exposed to the chloride environment and progressively increases with the exposure time.
4- For concrete specimens exposed to carbonation and chlorides, crack widths influence the corrosion activity and therefore should be considered in service life prediction models.

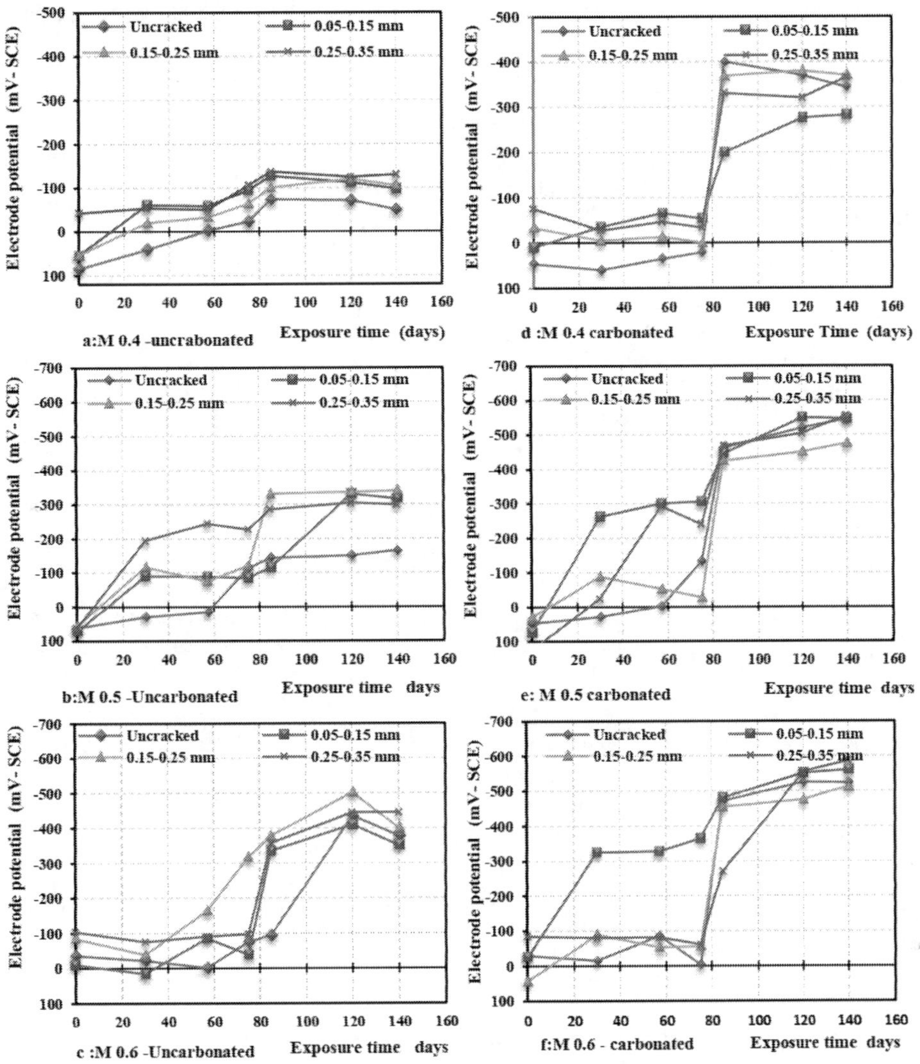

Fig. 9. Effect of w/c and carbonation on half-cell potential of reinforced concrete samples (a,b and c) uncarbonated samples and (d,e and f) carbonated samples

Table 6 Summary of half – cell potential less than -276 mV and the age

Series	Sample	Un-cracked		0.05-0.15 mm		0.15-0.25 mm		0.25-0.35 mm	
		E_{corr} ≤-276	Age* day	E_{corr} ≤-276	Age* day	E_{corr} ≤-276	Age* day	E_{corr} ≤-276	Age* day
Series1 carbonated	M 0.4	Y	85	Y	85	Y	85	Y	85
	M 0.5	Y	85	Y	85	Y	85	Y	85
	M 0.6	Y	85	Y	30	Y	85	Y	75
Series 2 Un-carbonated	M 0.4	-	-	-	-	-	-	-	-
	M 0.5	-	-	Y	120	Y	85	Y	85
	M 0.6	Y	110	Y	85	Y	75	Y	80

E_{corr}: Half-cell potential is more negative than -276 mV, high probability that corrosion is occurring.
* Age: Time of arrival half-cell potential-SCE is ≤ -276 mV.
Y: Yes, E_{corr} ≤ -276 mV.

References

1. Neville, A.M. *Properties of concrete*. (Five and Truly Final Edition) London:(2011) Pearson Education Limited.
2. Hussain, S., Rasheeduzzafar, Al-Mu&am, A. and Al-Gahtani, A., Factors affecting threshold chloride for reinforcement in concrete. Cem. and Con.Res. J, **25**(7) (1995) pp. 1543-1555.
3. Dyer, A., *Concrete durability*. New York: (2014). Taylor& Francis Group, LLC.
4. Broomfield, J.P. *Corrosion of steel in concrete*.2nd edition. London. Taylor &Francis (2007).
5. Wan, X., Wittmann, F., Zhao, T. and Fan, H., 2013. Chloride content and pH value in the pore solution of concrete under carbonation. Journal of Zhejiang University-Science A (Appl. Phy. & Eng.), **14**(1), (2013) pp71-78.
6. American Concrete Institute, ACI Committee 224. R, 01. Control of Cracking in Concrete Structures. (2001) Farmington Hills, Mich.: ACI.
7. Concrete Society. *Non-structural cracks in concrete*, TR **22** (2010) Concrete Society.
8. Morcous, G. and Louis, Z. Maintenance Optimization of Infrastructure Networks using Genetic Algorithms, J of Automat. in Const., **14** (1), (2005) pp 129 142.
9. Sheo-Feng, Z., Chun-Hua, L. and Rong-Gui, L, Experimental determination of chloride penetration in cracked concrete beams. International conference on advanced in engineering, proceeding **24** (2011) pp 380-384.
10. Park, S., Kwon, S., and Jung, S., Analysis technique for chloride penetration in cracked concrete using equivalent diffusion and permeation. Const.and Buil. Mat., **29**, (2012) pp. 183–192.
11. Kwon S., Na U., Park S.and Jung S., *Service life prediction of concrete wharves with early-aged crack: a Probabilistic approach for chloride diffusion*. Struct. Saf. J, **31**, (2009) pp 75–83.
12. Neville, A.M., Brooks, J.J., *Concrete Technology*. (Second Edition) London: (2010) Pearson Education Limited.
13. Pacheco, J. and Polder, R. Corrosion initiation and propagation in cracked concrete – a literature review. *Proceedings of the 4th International RILEM Ph.D. Workshop*, Madrid, Spain (2010).
14. BS EN 197-part 1 (2011). Cement-part 1: Composition, specifications and conformity criteria for common cement.
15. BS.882 (1983). Aggregate from natural sources for concrete. British Standard Institution.
16. BS 4449:2005+A3:(2016). Steel for the reinforcement of concrete. Weldable reinforcing steel. Bar, coil and decoiled product.
17. Teychenné, D.C., E Franklin, R.E. and Erntroy, H. C., *Design of normal concrete mixes*. Second edition. Building Research Establishment. Garston, Watford, (1988).
18. BS EN 12390 part- 2:(2009). Testing hardened concrete –; Part 2: Making and curing specimens for strength tests.
19. Al-Samaraai, M.and Raouf, Z. *Non –destructive tests for concrete*. University of Shariqia, UEA (1999).
20. ASTM C642:2013, Density, Absorption, and Voids in Hardened Concrete.
21. BS EN 12390, part 3:(2000). Testing hardened Concrete Part 3: Compressive strength of test specimens.
22. Yuan.Q. , Shi, C., He, F. Schutter, G. Audenaert, K., and Zheng, K., Effect of hydroxyl ions on chloride penetration depth measurement using the colorimetric method. Cem. and Con. Res. J ,**38** (2008) pp1177–1180.
23. He, F., Shi, C., Yuan, Q., An, X. and Tong, B. Calculation of chloride concentration at color change boundary of AgNO3 colorimetric measurement. Cem. and Con. Res. J **41** (2011) pp1095–1103.
24. AL-Amoudi, O., Rasheeduzzafar and Maslehuddin, M. Carbonation and corrosion of rebars in salt-contaminated OPC/PFA concretes. Cem. and Con. Res. J, **21**(1991), pp 38-50.
25. BS EN13295:(2004). Products and systems for the protection and repair of concrete structures — Test methods — Determination of resistance to carbonation.
26. DD CEN/TS 12390-10:(2007). Testing hardened concrete. Determination of the relative carbonation resistance of concrete, European Committee for Standardization.
27. Mc Polin, D., Basheer, M., Long, E., Grattan, V. and Sun, T. New Test Method to Obtain pH Profiles due to Carbonation of Concretes Containing supplementary Cementitious Materials, J of mat. in civil eng., ASCE, **19** November (2007), pp 936-946.
28. Wang, J., Basheer, M., Nanukuttan, S., Long, A., Bai, Y. Influence of service loading and the resulting micro-cracks on chloride resistance of concrete. Const.and Build. Mat.,Vol. **108** (2016) pp 56–66.
28. BS EN ISO 10523 (2012). Water quality — Determination of pH.
29. ASTM C 876(2015). Half-Cell Potentials of Uncoated Reinforcing Steel in Concrete.
30. Bouteiller, V., C. Cremona, V. Baroghel-Bouny.. Corrosion initiation of reinforced concretes based on Portland or GGBS cement: Chloride contents and electrochemical Cem. and Con. Res , **42**(11), (2012) pp1456-1467.
31. Chi, J. M., Huang, R.and Yanlg, C. C. Effect of Carbonation on Mechanical Properties and Durability of Concrete Using Accelerating Testing Method. ; J of Marine Sci. and Tech.y, **10**(1),(2002) pp 14-20.

32 Basheer, L., J. Kropp and D. J. Cleland. Assessment of the durability of concrete from its permeation properties: a review. Const. and buil. mat., **15**(2-3), (2001) pp 93-103.

33. Audenaert, K., G. De Schutter and L. Marsavina. Influence of cracks and crack width on penetration depth of chlorides in concrete. Eur. Journal of env. and civil eng., **13**(5), (2009) pp 561-572.

34. Suzuki, K., Ohno, Y., Praparntanatorn, S. and Tamura, H., . Mechanism of Steel Corrosion in Cracked Concrete, *Corrosion of Reinforcement in Concrete*, Ed. Page, C., Treadaway, K. and Bramforth, P., London: Society of Chemical Industry, (1990) pp. 19–28.

35. Song, H.W and Saraswathy, V. Corrosion Monitoring of Reinforced Concrete Structures - A Review, . Inte.J of Electro chem. Sci. , **2**(2007) pp 1-28,

Predicting residual service life of concrete infrastructure: a considerably controversial subject

Joost Gulikers*

*Ministry of Infrastructure and Water Management, Rijkswaterstaat-GPO, P.O. Box 2232, 3500 GE Utrecht, Netherlands

Abstract. There is an increasing demand from asset owners for service life prediction of existing reinforced concrete structures. This requires assessment of the current condition and modelling to allow for a prediction. This paper critically discusses a number of subjects relevant for service life prediction with respect to durability related to chloride-induced reinforcement corrosion. The subjects include the physical meaning and variability of the end-of-service-life criterion, the validity of the deterioration models, the availability, variability and reality level of input values for some model parameters, as well as some practical issues concerning site investigations. The findings are exemplified by calculation examples using both a deterministic as well as a full probabilistic approach. It is anticipated that in the future a full probabilistic approach will be adopted which makes service life predictions more prone to manipulation of input values, as literature provides a wide spectrum of values to choose from. Although a probabilistic approach seems very impressive to most asset owners it usually disguises the lack of knowledge, responsibility and liability of the consultant involved. It is concluded that asset owners will be easy prey for consultants to play a lucrative numbers game eventually providing a desirable and realistic outcome, mostly already known beforehand.

1 Introduction

The current stock of infrastructure contains a significant number of reinforced concrete structures reaching their originally intended design service life. In view of the economic importance of infrastructure assets in the national road and waterway network regular assessment of the current condition and prediction of the future condition is considered vital. This essential information on performance will allow timely measures to be taken on maintenance and repair, and allocate financial resources for eventual replacement.

At present the common approach is to perform visual inspections on a regular basis, which are incidentally complemented by non-destructive or minor-invasive testing. Based on these findings, maintenance activities are then planned for a time scale of 10 years. However, this approach has some drawbacks as it is a medium-term reactive strategy and mostly focusses on individual assets, or parts of them, rather than on the network (or part of it). In order to allow for a more cost-effective preventive maintenance strategy, prediction of the long-term performance until end of original design service life and beyond, including effects due to deterioration, are therefore advocated. In view of efficient and effective long-term asset management an increasing trend can be observed that infrastructure providers and asset owners explicitly request for the residual service life of individual infrastructure assets. In practice this request will usually be accomplished by assessment of the current condition taking the observed deterioration into account and performing predictions of future condition development based on numerical modelling.

This paper will address a selection of the major pitfalls and shortcomings involved in such predictions of residual service life with respect to chloride-induced reinforcement corrosion, as this is generally considered the largest threat to long-term durability of reinforced concrete structures.

2 Definition of end of (residual) service life

For the quantification of the residual service life, expressed in years, a substantiated sound end-of-service-life criterion should be established. With respect to the design of new concrete structures, according to EN 1990-1 (clause 1.5.2.8) [1], the design working life is defined as the "assumed period for which a structure or part of it is to be used for its intended use with anticipated maintenance but without major repair being necessary". In this definition a distinction is made between "maintenance" and "repair". In addition, it should be noted that maintenance is often subdivided into preventive and corrective maintenance. Repair is a typical reactive activity as it involves the elimination of physical damage, e.g. cracking and spalling of the concrete cover and loss of steel cross section in case of reinforcement corrosion. However, this definition of design working life is considered to be rather "loose" as it doesn't give any quantified information on the nature

* Corresponding author: joost.gulikers@rws.nl

and extent of the deterioration that is considered acceptable.

In the present standards, a distinction is made between ultimate limit states and serviceability limit states, the former related to load bearing capacity and collapse, whereas the latter "corresponds to conditions beyond which specified service requirements for a structure or a structural member are no longer met" [1]. However, both ultimate and serviceability limit states do not cover aspects related to durability. Therefore, it is suggested to introduce an additional series of limit states termed either as durability, maintenance or condition limit states. Whereas codes and standards provide clear values on the minimum value of the reliability index, β, regarding ultimate and serviceability limit states which have to be demonstrated through structural calculations, for durability no β-values have been specified, but in contrast so-called deemed-to-satisfy provisions are given.

For existing concrete structures it would be evident that a similar definition as for new structures should apply, with a clear quantifiable criterion for the durability (or maintenance or condition) limit state. In view of the predictions made by modelling and the stochastic nature of the exposure conditions, concrete properties etc., it is obvious that a probabilistic approach will be adopted. This implies that the eventual result of modelling will be expressed in terms of a probability that a specified "undesirable event" will occur.

With respect to reinforcement corrosion, a number of undesirable events can be distinguished e.g. depassivation, cracking of the concrete cover and spalling of the concrete cover. For the design of new structures, depassivation of steel reinforcement resulting from either carbonation or chloride ingress is considered the most appropriate durability limit state. However, for existing concrete structures both cracking and spalling of the concrete cover may be considered as viable alternative limit states.

Assuming an agreement has been achieved on the use in practice of some durability limit state, an acceptable probability of occurrence of the corresponding "undesirable event" should be introduced. Regarding depassivation, i.e. onset of corrosion, in the literature a wide range of values for P_{dep}, considered to be acceptable, can be found. According to [2] "A probability of the onset of corrosion of 10^{-2} (β=2.3) may be more appropriate.", whereas DuraCrete [3] makes a distinction made between 3 reliability levels depending on the "Cost of mitigation of risk related to the cost of repair" with β=3.72 (low cost), β=2.57 (normal cost), and β=1.28 (high cost), corresponding to $P_{dep}=0.01\cdot10^{-2}$, $0.5\cdot10^{-2}$ and $10\cdot10^{-2}$, respectively. It should be noted that the values proposed by DuraCrete [3] apply to "the acceptance criteria for a 50 year service life", and that "this set of acceptance criteria is only valid for analyses carried out for the specific probabilistic model used in DuraCrete". However, the first note suggests that the acceptable probability is dependent on service life (or age) of a structure (or structural member) but it remains unclear, why, and which value for a service life other than 50yr should be adopted. The second note raises serious doubt on which basis these acceptance criteria have been chosen or achieved since an acceptance criterion should be independent of the model, i.e. the mathematical expression, as well as being independent of the input values for the model parameters that have been used. According to fib [4] for depassivation, a fixed level of the reliability index β=1.30 ($P_{dep}=10\cdot10^{-2}$) is adopted independent of the consequence class, whereas in [5] a "probability less than approximately 10%" is considered acceptable for service life design with respect to chloride-induced corrosion. In [6] 2 reliability levels are distinguished: for exposure classes XC1, XC2, XC4, XD1, and XS1 β=0.5, corresponding to $P_{dep}= 30.9\cdot10^{-2}$, is advocated, whereas for exposure classes XC3, XS3, XD3, XS2, and XD2 β=1.5, corresponding to $P_{dep}=6.7\cdot10^{-2}$, is suggested. For XS3, XD3, XS2 and XD2 a note is made that in case of poor accessibility and if no inspections are possible the reliability level should be increased to β=1.5.

The wide range of β-values suggested for onset of corrosion in new structures indicates that there is hardly any agreement and consistency. Moreover, barely any reasoning is provided to support the chosen β-value. In practice, it is likely that the β-values suggested for new structures will simply be copied for condition assessment of existing structures as at present in the literature no values can be found explicitly related to durability of existing concrete structures. The durability consultant will be either thrilled or confused by the availability in (reviewed) literature of such a wide range of β-values as this will provide him/her with ample possibilities to pick the most appropriate value and bear hardly any risk himself/herself. Moreover, in documents it remains largely unclear what the basis is or has been for adopting or proposing these values and what their actual physical meaning is in terms of amount of steel affected by corrosion or in other relevant physical terms, which are understandable to the asset owner. It should be noted that the asset owner will be less interested in levels of probability, but will give more consideration to risk, i.e. that account should be given to the direct physical consequences of the occurrence of an undesirable event as well as to the indirect financial consequences. With respect to reinforcement corrosion, the event of depassivation merely induces a change of the electrochemical condition of the steel. If the corrosion process following depassivation is slow, e.g. due to a high concrete resistivity or a continuous lack of oxygen to fuel the cathode reaction, a high probability of depassivation may be considered acceptable.

In practice often a β-value is preferred, as a probability value for the onset of reinforcement corrosion may provoke questions by the client on what the actual physical meaning of P_{dep} is. However, there is a direct arithmetic relationship between the β-value and the probability, through the normal probability distribution, see Fig. 1.

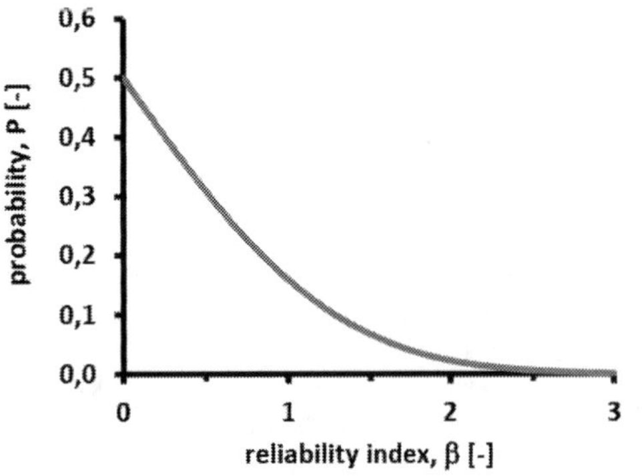

Fig. 1. Relationship between β-value and probability.

The effect of the β-value related to depassivation is illustrated in Fig. 2 regarding the (predicted) residual service life of an existing concrete structure for 3 ranges of the mean cover depth: μc=40, 50 and 60mm (all with a standard deviation σc=8mm). For a concrete structure or concrete element with a mean cover depth of 50mm this implies that the (residual) service life will be in excess of 120yr when β=0.50 (P_{dep}=0.309) is adopted as the end-of-service life criterion, whereas for β=1.3 (P_{dep}=0.097) and β=1.5 (P_{dep}=0.067), the service life is predicted to be ~50yr and ~40yr, respectively.

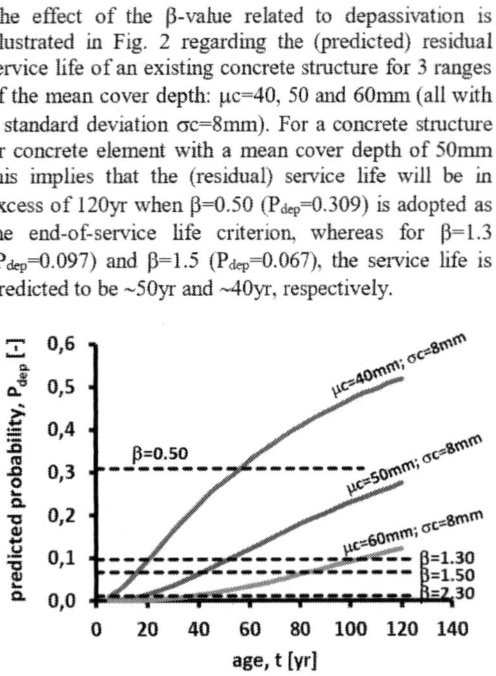

Fig. 2. Influence of the β-value on the predicted (residual) service life.

2 Deterioration models

In order to allow for a prediction of residual service life a suitable model for each relevant deterioration mechanism should be available. With respect to chloride-induced reinforcement corrosion a mathematical expression describing chloride transport into concrete as a function of time and depth will then be needed. In the literature, a wide variety of chloride ingress models can be found ranging from simple to highly complicated and complex taking into account a multitude of factors influencing transport of chlorides in cement-based concrete, e.g. binding, moisture content, temperature, other ionic species, time-dependent pore structure, amongst others. Most often these models are presented as either "validated" or "calibrated" models, and incidentally both qualifications are used. However, it should be appreciated that generally no further details are provided which specific requirements are fulfilled to deserve the qualification "validated" or "calibrated" and how it is demonstrated that the model proposed deserves either or both qualifications. In practice and in the literature the vagueness of these qualifications is mostly misused as they merely serve as a convincing "selling" argument.

With respect to chloride ingress in engineering practice the so-called error function solution to Fick's second law is most often used in view of its simplicity. In several case studies it has been concluded that the use of a (more) advanced model generally does not necessarily result in a better (or less worse) prediction of chloride ingress in the concrete cover. In [7] a benchmark has been performed based on measured chloride profiles from existing structures, however, the chloride profiles predicted at a later age demonstrate the significant variable outcome between the ingress models used by the respective developers.

The error function solution is given by:

$$C(x,t) = C_s - (C_s - C_i) \cdot \operatorname{erf}\left(\frac{x}{2\sqrt{D_a \cdot t}}\right) \qquad (1)$$

where C_s = chloride content at the exposed concrete surface (%m/m cement); C_i = initial chloride content, (%m/m cement); x = distance to the exposed concrete surface (mm); D_a = apparent chloride diffusion coefficient (mm^2/yr); t = exposure time or concrete age (t).

It should be noted that Eq. (1) represents the analytical solution to Fick's second law with 1-dimensional ingress for specific boundary and initial conditions. Most often these boundary conditions are forgotten when applying the error function solution, not only in engineering practice by consultants, but surprisingly also in publications by academics and scientists. The derivation of Eq.(1) is merely valid for a semi-infinite medium and in addition it is assumed that the surface chloride content, C_s, remains essentially constant over time. However, in real structures these boundary conditions are often compromised, e.g. in situations with continuous exposure to chlorides where it is frequently observed that C_s gradually increases over time, or at locations incidentally exposed to chlorides originating from de-icing activities during winter periods. Consequently, in most situations in practice Eq. (1) is invalid and should thus not be used. Moreover, Eq. (1) should not be applied on real concrete structures for exposure times less than 10yr, either for prediction or for calibration purposes. However, due to the lack of knowledge by users and the poor availability of suitable alternative models on an engineering level, Eq. (1) remains by far the most popular model for chloride ingress for application in both practice and in experimental research in the laboratory.

Based on field experience with chloride profiles measured on concrete cores drilled from existing structures, it is concluded that the apparent diffusion coefficient, D_a, is time-dependent according to an empirical power law relationship:

$$D_a(t) = D_a(t_{ref}) \cdot \left(\frac{t_{ref}}{t}\right)^n \quad (2)$$

where $D_a(t)$ = apparent chloride diffusion coefficient (mm^2/yr) at age t; $D_a(t_{ref})$ = apparent chloride diffusion coefficient (mm^2/yr) at reference age t; n = ageing factor (-).

However, it is widely known that this modified version employing a time-dependent chloride diffusion coefficient according to Eq. (2) does not satisfy the underlying differential equation, i.e. Fick's second law of diffusion, any more. Consequently, numerical models using consecutive time steps with the differential equation as the basis will result in a completely different outcome than the outcome obtained by the combination of Eq. (1) and Eq. (2). Therefore, it is generally strongly recommended not to apply software with the differential equation as the basis for use in practice by engineers.

A sensitivity analysis on the influence of individual model parameters has clearly demonstrated the dramatic impact of the ageing factor, n, on the chloride ingress rate and thus on time to depassivation of the embedded reinforcing steel. Therefore, behind the scenes in the absence of clients, the ageing factor n is frequently referred to as the "manipulation factor". Fig. 3 shows the relative influence of the ageing factor, n, on the predicted age at depassivation of the reinforcing steel for an existing concrete structure based on inspection results obtained at an age t_{ref} (=t_{insp})=20 yr with $D_a(t_{ref})$= $15mm^2/yr$ (equal to $0.475 \cdot 10^{-12} m^2/s$), C_s=3.0%, C_i=0.1%, and a critical chloride content, C_{crit}=0.5%, for 3 different cover depths, c = 40, 45 and 50mm, according to a deterministic calculation.

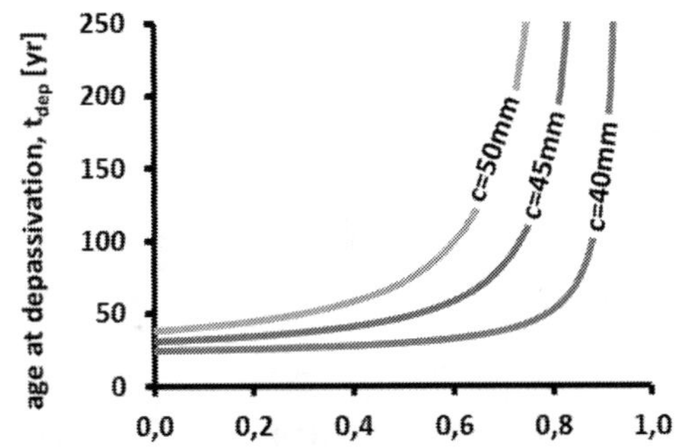

Fig. 3. Influence of the ageing factor, n, on the predicted age at depassivation, t_{dep}.

In most publications as well as in most calculations it is assumed that the ageing factor lies between $0.0 \leq n \leq 1.0$ [3,4]. An ageing factor n = 0 implies that the chloride transport properties as expressed by D_a will remain constant over time, e.g. $D_a(t) = D_a(t_{ref})$. Using a value n<0.0 implies that D_a will increase over time and thus will demonstrate gradually less resistance against chloride ingress whereas for n>0 this resistance will increase over time. At first glance such a time-dependent decrease of D_a seems reasonable considering the refinement of the pore structure and decrease of the porosity as a result of ongoing hydration of the binder, in combination with a decrease of the moisture content due to a gradual drying out of the concrete at greater depths. However, this reasoning becomes questionable for concretes having an age of more than 50yr as hydration will have come to an end, resulting in a static pore structure, whereas the moisture distribution in the cover zone may have come to a dynamic equilibrium with the exposure environment. Moreover, it could also be reasoned that, after some point in time, D_a may increase as a result of the occurrence of microcracking, e.g. in the tensile zone, or due to a change in the exposure conditions, e.g. through the occurrence of leakage. However, in most predictions of chloride ingress over time it is assumed that Eq. (2) is valid with a constant n-value until infinity.

For n = 1.0 the chloride profile will not change over time, and this is explained by a phenomenon termed "pore blocking" [8]. For n>1.0 the resistance to chloride ingress will increase over time at a rate that chlorides will withdraw and the chloride content will decrease. However, it is highly unrealistic that such a phenomena will occur in practice. Fig. 4 shows the predicted change over time of the chloride content at a fixed depth x = 40mm for n=-0.25, n=+0.00, n=+0.50, n=+1.0, and n=+1.25. The input used is based on a measured chloride profile at t_{ref} (or t_{insp}) =20yr using C_s=3.0%, C_i=0.1%, and $D_a(t_{ref})$ = $15mm^2/yr$ (this input is identical to that used for the previous example).

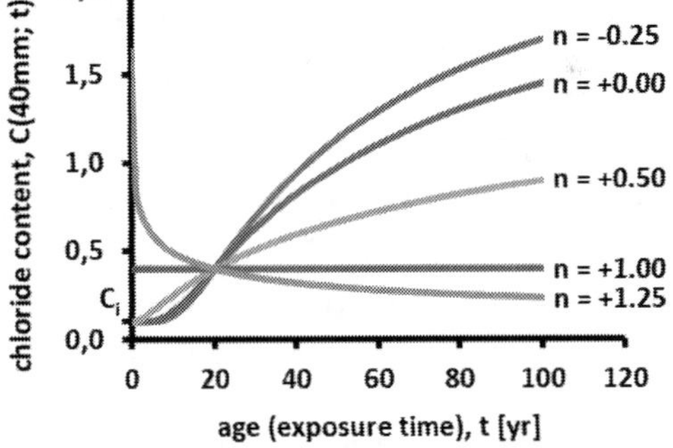

Fig. 4. Influence of the ageing factor, n, on the predicted development over time of the chloride content at a depth x=40mm.

Although values for the ageing factor n>1.0 are unrealistic, such values are frequently reported for binders with high amounts of supplementary cementitious materials, in particular blast furnace slag and fly ash [9, 10]. However, such findings may then raise considerable doubt either on the quality of the investigations, the validity of the empirical relationship or on the physical interpretation of ageing effects.

As a case study that has actually occurred in practice, the values reported in [10], 0.90 < n < 1.05 derived with results obtained from chloride migration tests, were proposed by a contractor in a project to demonstrate the superior resistance to chloride ingress of the concrete cover. The contractor expressed a clear preference for a numerical proof although the joint between precast concrete elements of a bridge deck showed serious leakage and shallow cover depths. For values n > 1.0 chloride contents were calculated merely at the depth of the reinforcing steel, x=30mm, at t= t_{design}=100yr, however this will surely not reveal that the predicted chloride contents will actually decrease over time, see Fig. 4. When confronted with this predicted highly unrealistic decrease of the chloride content over time, the contractor had not a word to say.

In the literature, a wide range of values for the ageing exponent, n, are reported depending on the type of cement and exposure conditions (humidity). However, generally no background is provided on the number and quality of the data that is available, nor on the type and age of the structures or on the local exposure conditions. It should be noted that the magnitude of the ageing factor cannot be measured but is derived by curve fitting of values for D_a obtained for a similar concrete composition and exposure conditions during a long period of time, preferably more than 50 years. In practice only a limited amount of data is often available coming from different structures and sources having apparently similar concrete composition (cement type) and exposed to apparently similar environmental conditions.

In practice, chloride ingress generally will have sufficiently advanced only after more than 15 years of frequent exposure. Consequently, there is a considerable lack of data on $D_a(t)$ covering the time period 0.1yr< t< 15yr. In [4] and [11] this is solved by introducing values obtained from laboratory tests, i.e. the chloride migration coefficient, D_{RCM}, as a substitute. This results in a mix of values for D_{RCM} (t) and $D_a(t)$ to quantify the ageing factor, n, by regression analysis, however it is well-known that the migration coefficient, $D_{RCM}(t)$ demonstrates a significantly different time-dependent behaviour than $D_a(t)$. Consequently, the magnitude of the ageing factor thus derived will generally be considerably less than the actual n-value based on D_a-values only, in particular for concretes made with binders containing SCM's [12].

3 Probabilistic approach

As there is significant uncertainty associated with most model parameters, in particular the ageing factor n and the critical chloride content, C_{crit}, it is considered more appropriate to use a probabilistic approach [13] This implies that all model parameters are subject to a statistical description, which are then treated as statistically independent random variables.

In the error-function solution to Fick's second law 5 stochastic parameters can be identified, i.c. the surface chloride content, C_s, the initial chloride content, C_i, the reference apparent diffusion coefficient, $D_a(t_{ref})$, and the ageing factor, n. The reference concrete age, t_{ref}, and the age for the prediction, t, are the only deterministic variables, having a predefined value. With respect to depassivation of the steel reinforcement, the thickness of the concrete cover, c, and the critical chloride content, C_{crit}, are also considered as stochastic variables. In principle, the statistical distribution adopted for each parameter, is reflected in a mean value, a standard deviation and a type of probability distribution, e.g. a lognormal distribution.

Frequently, there is a lack of data on a specific model parameter and in the absence of any other information it is common practice to disguise this lack of statistical data by giving the designation "based on expert opinion". In addition, due to this lack of enough data, a normal probability distribution is frequently chosen as this is the most obvious one. However, for certain parameters the pitfall is that a normal distribution may result in negative values, in particular when, probably based on arbitrary grounds, a large standard deviation is chosen. An illustrative example of the consequences is provided in [14], see Fig. 5, regarding the surface chloride content, C_s. For each exposure environment distinguished for marine structures, C_s is characterised by a mean value, μ, and a coefficient of variation, COV, with $\sigma = COV \cdot \mu$. The values adopted for COV are 0.80, 0.71, 0.50, and 1.33, which correspond to a probability that $C_s < 0$ of P = 0.106, 0.081, 0.023, and 0.227, respectively. It is obvious that negative chloride contents do not exist, and that a lognormal or a left truncated normal distribution would have been a more realistic choice to statistically describe C_s. On the other hand, the surface chloride content, C_s, has a relatively minor impact on the outcome, thus this unrealistic choice will probably not lead to significant changes of the outcome.

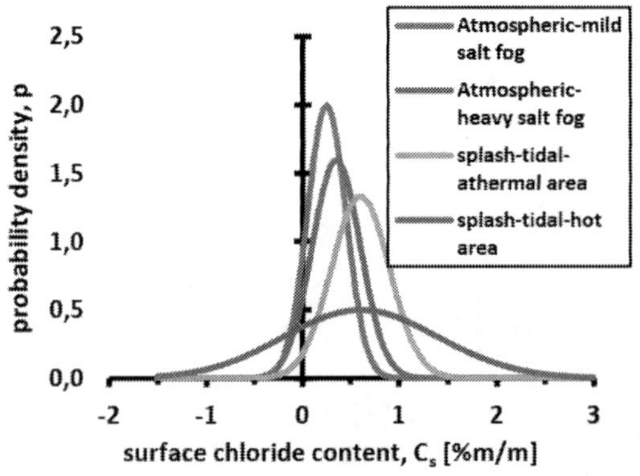

Fig. 5. Example of the use of a normal probability distribution for the surface chloride content. C_s, with a large standard deviation, resulting in negative values.

Regarding the critical chloride content, C_{crit}, literature provides a wide variety of input values [3, 4, 15, 16]. Some authors assume that C_{crit} is independent of exposure conditions or type of binder [4], whereas others indicate a strong dependency, especially from the exposure environment [3, 15]. Of particular interest for discussion is the detailed statistical distribution of C_{crit} provided in [15], as this characterization is based on results of long-term experiments in a natural exposure environment. For the atmospheric zone, a lognormal distribution was adopted whereas for both the splash as well as the submerged zone a beta-distribution was chosen, characterised by both a lower and an upper boundary. The probability density of C_{crit} for each environmental zone is presented in Fig. 6, whereas the corresponding (cumulative) probability is illustrated in Fig. 7.

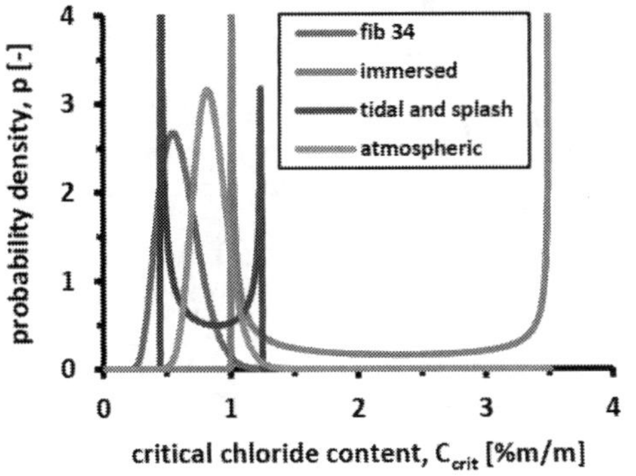

Fig. 6. Probability density distributions of C_{crit} according to [4] and [15]

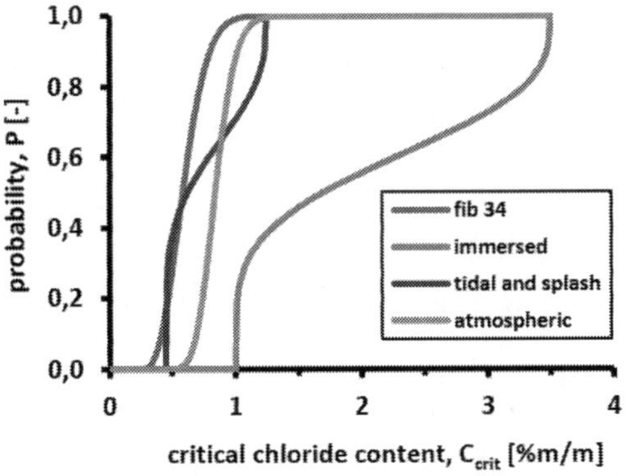

Fig. 7. Probability distributions of C_{crit} according to [4] and [15]

For both C_{crit} according to [4] and C_{crit} for the atmospheric zone in [15] the probability density distribution shows a familiar appearance with a peak value around the mean value. In contrast, according to [15], for both the immersed as well as the combined tidal and splash zone, the probability density distribution shows extreme density peaks at the lower and the upper boundary, whereas in the region between the lower and upper boundary, a nearly constant low probability density exists. Both probability density curves are also shown in [15], however no explanation is given, nor a discussion provided on their realistic nature. Without doubt such curves raise considerable doubt on their validity for use in practice on real concrete structures.

4 Practice-related issues

For assessment of existing reinforced concrete structures with respect to chloride-induced corrosion, it is common practice to drill a number of concrete cores from selected parts of the structure. Thereupon, the chloride profiles are composed giving the chloride content as a function of depth. For the assessment of the present condition, the chloride content measured at the level of the reinforcing bars is compared to the assumed critical chloride content. For prediction of the development of the condition over time, the measured chloride profiles are subjected to regression analysis using Eq. (1). This fitting procedure eventually results in values for the apparent chloride diffusion coefficient, $D_a(t_{insp})$, and the surface chloride content, C_s. Thereupon, by assumption, engineering judgement or expert opinion suitable values for the ageing factor, n, and the critical chloride content, C_{crit}, are introduced. Then, either a deterministic or a probabilistic approach is adopted to predict future chloride ingress and the time to depassivation of the reinforcing steel. Although this procedure seems straightforward, there are some critical issues involved.

Most often, measured chloride profiles are characterised by a limited number of relevant chloride contents: sometimes only 2 valid points are available, especially for structures, structural elements or parts of the structure with exposure times less than 15-20 years. This is due to a combination of shallow chloride penetration depths, the use of depth increments of 10mm or more, and the common neglect of the chloride content of the near-surface layer. Generally, chloride measurements executed by scientific institutes provide results within 10% of the real chloride content present, however, experience in practice with commercial laboratories indicate that a difference of more than 20% is more common. Without doubt, this inaccuracy will be reflected in the values obtained for $D_a(t_{insp})$ and $C_s(t_{insp})$, and eventually this will affect the validity of the predictions.

It is common practice that the chloride content of the near-surface layer is discarded as it does not fit in the chloride profile obtained by Eq. (1). This implies that in fact the surface chloride content, C_s, physically does not exist and instead its notional value is obtained by extrapolation of chloride contents measured at depths > 10mm to the exposed concrete surface, i.e. to x = 0. In practice, frequently chloride profiles are measured for which the second point also deviates from the chloride profile according to Eq. (1). In these cases either both the first and the second point are discarded, or merely the first point. Figure 8 exemplifies such a situation by 2 chloride profiles measured at the windward side of a box girder bridge exposed to an atmospheric marine environment. The initial rise in chloride going in is well known due to the effect of carbonation releasing bound chloride pushing a ramp of chloride ahead of the carbonation front. In addition, frequent exposure to rain can also influence the chloride profile in the near-surface layer.

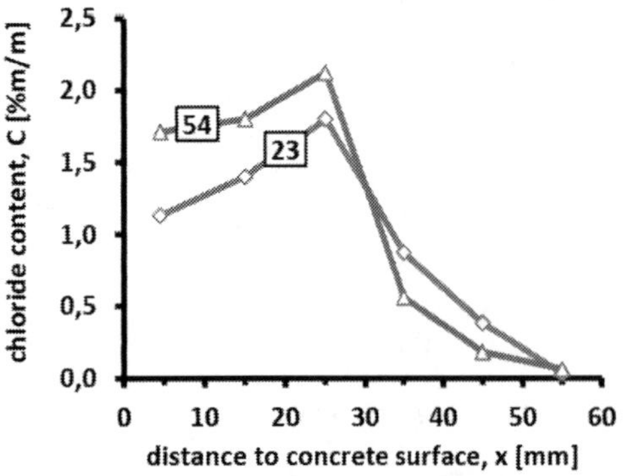

Fig. 8. Chloride profiles with contents measured at 5 and 15mm not complying with Eq.(1) (depth increments of 10mm).

Consequently, the results of the regression analysis regarding $D_a(t_{insp})$ and C_s will be seriously affected by the points chosen. For chloride profile 54, the use of points 2-6 will result in C_s=3.53% (m/m cement) together with $D_a(t_{insp})$=13.1mm^2/yr, whereas in case points 3-6 would have been used, regression analysis would have resulted into C_s=17.3% and $D_a(t_{insp})$ =4.0mm^2/yr, see Fig. 9. This significant difference in results obtained by regression analysis indicates that serious caution should be exercised in discarding measured chloride contents. Sometimes complete chloride profiles are discarded as too many points do not match with Eq. (1). However, it should be noted that also such "non-matching", undesirable profiles represent the facts and simply considering these profiles as outliers will be far too easy.

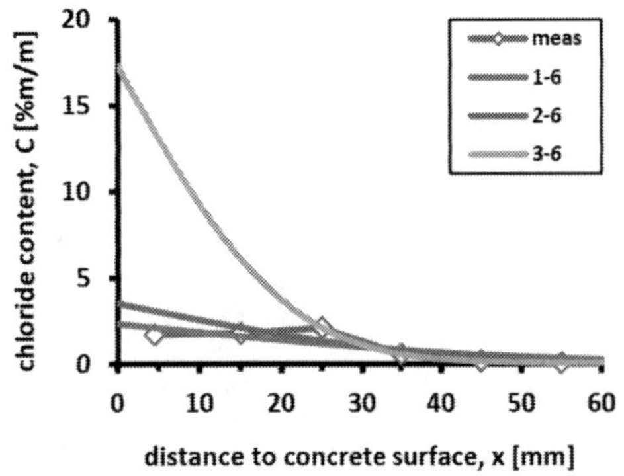

Fig. 9. Measured chloride profile no. 54 with corresponding chloride profiles fitted according to Eq. (1).

An alternative is to use smaller depth increments, e.g. 5mm, which provide more data points characterizing a profile and allows for a better identification of the depth at which depth the actual peak in the chloride profile occurs. However, this may have serious cost implications. In addition, the amount of material available for the chloride analysis will be less and may be such that it will impair the reliability of the quantification of both the amount of chloride and cement present in the concrete sample.

An additional serious issue regarding costs is the number of concrete cores required to obtain a reliable representative picture of chloride ingress over the concrete structure or part of it, e.g. a support. It should be noted that each concrete core retrieved from the structure corresponds to a hole, which should be repaired.

Spatial variability and accessibility are important aspects to be considered in choosing the number of cores and the appropriate locations for coring. Even for situations with apparently uniform exposure conditions and quality of the concrete cover, the differences between chloride profiles obtained from adjacent spots may be significant. Moreover, some locations in a structure, although known to be exposed to a more aggressive local environment than others, may not be easily accessible. Another area, at short distance from the heavily exposed area, may be easily accessible, but the chloride load in this alternative area may be far less, e.g. in situations with local leakage of bridge joints. Consequently, such chloride profiles will provide a far too optimistic picture on chloride ingress. In practice, chloride profiles are deliberately taken from areas in structures experiencing the most aggressive exposure conditions. However, these areas may be relatively small compared to the total surface area of the complete structure. It is obvious that a condition assessment and service life prediction of a concrete structure merely based on the results obtained from such areas will provide a far too pessimistic view of the overall state of the complete structure.

5 Concluding remarks

Prediction of residual service life of existing concrete structures is subject to serious doubt, not only with regard to the procedure adopted, the chloride profiling, the appropriateness of the deterioration models employed, the input values chosen, and the end-of-service-life criterion, but also on the usefulness of the outcome for the asset owner regarding maintenance and repair.

At present, a trend is observed towards the use of probabilistic approaches, explicitly taking into account the uncertainties in the model as well as the uncertainties in the quantification of the model parameters. However, there is a significant lack of quality data on the actual exposure conditions to which a concrete structure is subjected and its response, in particular in the long term. This implies that the statistical characterization of most model parameters is often assumed, invented or based on "expert opinion", however without providing any reasoning or evidence.

In practice there is a serious risk that statistical characterizations of model parameters published in papers authored by scientists will simply be copied by engineers, with a reference to the source, however without expressing any caution. It is obvious that adopting such a copy behaviour will have serious consequences for the reliability of residual service life predictions. As the majority of asset owners has hardly any feeling for modelling of deterioration processes, probabilistic approaches or reliability indices, he/she will be almost completely handed over to the professional integrity, expertise, objectivity, personal preference and randomness of the consultant, however without any control. Consequently, asset owners are an easy prey for consultants to play a lucrative numbers game eventually providing a desirable and realistic outcome, known beforehand.

References

1. EN1990, *Eurocode – Basis for structural design* (2002)
2. P. Bamforth, *Double standards in design*, Concrete, pp. 33-35. (2009)
3. DuraCrete, *General guidelines for durability design and redesign*, Final Technical Report, Document BE95-1347/R17, May 2000
4. fib *Model Code for Service Life Design*, bulletin 34, February 2006
5. CUR, *Durability of structural concrete regarding chloride-induced reinforcement corrosion, Guideline for the formulation of performance requirements– Background report* (in Dutch) (2009)
6. C. Gehlen, P. Schiessl, A. Schiessl-Pecka, *Position paper of DAfStB on implementation of the concept of performance-based design approaches considering DIN EN 206-1, Annex J.* Beton- und Stahlbetonbau, 103, Heft 12, pp. 837-839 (in German) (2008)
7. E.A.B. Koenders, *Modelling of chloride ingress in concrete based on benchmarking field results*, ICCRRR 2018, Cape Town
8. M. Maage, S. Helland, E. Poulsen, Ø. Vennesland, J.E. Carlsen, *Service life prediction of existing concrete structures exposed to marine environment*, ACI Materials Journal, November-December 1996, pp. 1-8
9. M.D.A. Thomas, P.B. Bamforth, *Modelling chloride diffusion in concrete – Effect of fly ash and slag*, Cement and Concrete Research 29 (1999) pp. 487-495.
10. G.J.L. van der Wegen, M.M.R. Boutz, A.J. Sarabèr, R.J. van Eijck, *Ageing coefficient of fly ash concrete and its impact on durability*, Proc. of the 1st Int. Conf. on Ageing of Materials & Structures, Delft, 26-28 May 2014, pp. 171-177
11. CUR, *Durability of marine concrete structures*, Publication 215 (in Dutch) (2005)
12. J.Gulikers, *A critical review of mathematical modelling of chloride ingress into concrete and the derivation of input data*, in Performance based Evaluation and Indicators for Concrete Durability, Proceedings of the International RILEM Workshop, Proceedings PRO 47, Madrid, 2006, pp. 165-175
13. D.V. Val, P.A. Trapper, *Probabilistic evaluation of initiation time of chloride-induced corrosion*, Reliability Engineering & System Safety, 93 (2008), pp. 364-372
14. X.P. Zhong, J. Zhu, C.B. Yuan, W.L. Jin, J. Xia *Durability design based on serviceability stage for corroded reinforced concrete structures*, Sixth International Conference on Durability of Concrete Structures, 18-20 July 2018, paper number DDS01, pp. 859-865
15. K. Li, Q. Li, X. Zhou, Z. Fang, *Durability design of the Hong Kong-Zhuhai-Macau sea-link project: Principle and procedure*, Journal of Bridge Engineering, 2015, 20(11), pp. 1-11
16. M. Thomas, *Chloride thresholds in marine concrete*, Cement and Concrete Research, Vol. 14, No. 4, 1996, pp. 513-519

Condition assessment and service life prediction of a marine concrete structure: an interesting case study

Joost Gulikers*

*Ministry of Infrastructure and Water Management, Rijkswaterstaat GPO, P.O. Box 2232, 3500 GE Utrecht, Netherlands

Abstract. On the windward side of an existing box girder bridge exposed to aerosol and incidentally spray chloride profiles were determined at an age of 18 and 32 years as to obtain data on the development of chloride ingress over time and to predict the time to onset of reinforcement corrosion. According to the chloride contents determined at the level of the reinforcing steel for an age of 18 years corrosion was likely to be present, however neither visual inspections nor large scale potential mapping revealed any spots suspect of corrosion. Chloride profiling performed at an age of 32 years indicated that chloride ingress was much slower than was anticipated. A full probabilistic approach was adopted to predict the development of chloride ingress over time and onset of reinforcement corrosion using updated values for the so-called ageing factor and the critical chloride content. However, the results obtained with these predictions proved to be far more pessimistic than is shown by the actual condition of the bridge. As a consequence it was decided not to rely anymore on predictions based on numerical models using probabilistic calculations but in contrast to adopt a pragmatic engineering approach using common sense as a basis.

1 Introduction

For marine concrete structures reinforcement corrosion induced by penetrating chlorides is considered the major threat for durability. Condition assessment by regular visual inspections is the common procedure adopted in practice. Generally, more in-depth investigations are considered after a reinforced concrete structure shows clear visual evidence of distress caused by corrosion of the embedded reinforcement. Such in-depth investigations may comprise non-destructive techniques e.g. potential mapping and cover depth measurements. However, the more common approach is to drill a number of cores, determine the chloride penetration profile and, based on the chloride content at the level of the reinforcing steel, judge if corrosion is likely to be present or will occur in the near future. It is frequently suggested that a pro-active strategy based on in-depth investigations performed on a regular basis, e.g. every 10 years, should be preferred as this in the long term will allow for a more cost-effective approach to maintenance and repair.

As part of a national program on condition assessment of existing concrete structures exposed to a marine environment, in 2002 a large number of concrete cores were drilled from a number of concrete structures along the Dutch coast [1]. The objectives of this research program were to collect data on chloride ingress and to validate the service life model developed in the European program DuraCrete [2]. At the time of inspection and measurements, none of the structures investigated, except one, showed any signs of damage resulting from corrosion.

A box girder bridge crossing the Eastern Scheldt was one of the structures investigated. Based on the findings obtained in 2002, during the time period 2005-2016 additional large-scale investigations were performed including potential mapping. Eventually it was decided to organise a large campaign in 2016 comprising a large series of drilled concrete cores to have an updated physical evidence of chloride ingress.

This paper presents and discusses the findings and interpretation of the chloride profile measurements obtained in 2002 and 2016 on the wall of a box girder bridge facing the North Sea (windward side).

2 Description of the box girder bridge

The box girder bridge is built over a storm surge barrier, see Fig. 1, and comprises 68 prestressed elements with 62 elements made from gravel concrete whereas the remaining 6 bridge elements (3×2abutments) have been constructed in lightweight concrete. All gravel concrete bridge elements have a span of 45m and the concrete composition used is based on 325kg/m^3 CEM III/B with a w/c = 0.44. Each bridge element consists of 2 precast sections which were combined by a closing joint which was cast in-situ. All precast sections were produced with steam-curing for 2 hours. The bridge elements were produced and placed in 1984.

* Corresponding author: joost.gulikers@rws.nl

Fig. 1. View of the location and exposure conditions of the walls of the box girder bridge.

The lowest part of the bridge is located at a level of +6m above MWL whereas the bridge deck is at a height of +11.75m above MWL. All elements of the box girder bridge are T-shaped which implies that the concrete faces on the underside are sheltered from rain and other precipitation. In addition, the presence of sliding doors located in front at the North Sea side will shield a significant part of the concrete surfaces. Consequently, the environment can be described as predominantly atmospheric with exposure to chlorides contained in aerosol, and occasionally seawater spray and incidentally splash (during periods of storm when the sliding doors are lowered).

2 Field investigations performed in 2002

2.1 Measured chloride ingress profiles

At the time of the first site investigations in 2002 the effective time of exposure amounted to 18 years. In order to collect representative data on chloride ingress over the complete length of the walls of the box girder bridge, a total of 80 concrete cores were retrieved. As on the entire surface of the bridge walls, no damage was observed; the locations for core drilling were more or less arbitrarily chosen from the perspective of relative ease of accessibility. In practice, this meant that cores were drilled approximately halfway up the height of the walls. With respect to the windward exposure side and gravel concrete bridge elements, 13 cores were drilled. Upon drilling, the cores were sent to a laboratory for chemical analysis yielding quantitative information on chloride ingress based on depth intervals of 10mm. These investigations result in a number of chloride penetration profiles providing information on the current electrochemical state of the reinforcing bars and the depth dependent profile may then be used for prediction of future ingress. Fig. 2 shows the results on chloride ingress profiles measured on 13 cores retrieved from 7 different bridge elements.

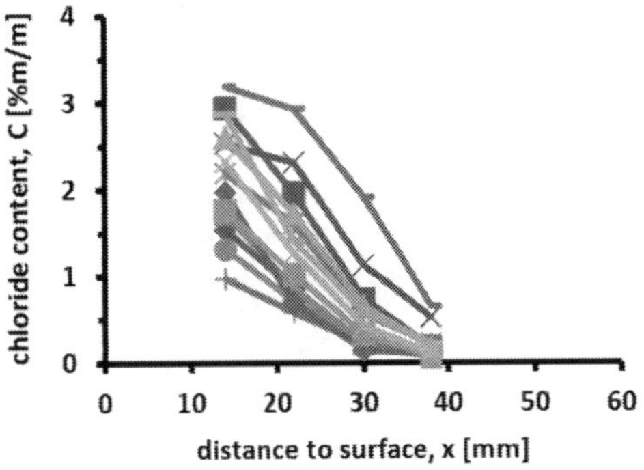

Fig. 2. Measured chloride profiles (2002) in the walls of the box girder bridge after 18 years of exposure (windward side).

It can clearly be seen that the scatter in chloride ingress is significant, despite the uniform concrete quality and exposure conditions, even at the level of the reinforcing steel (c =40mm), with chloride contents ranging from 0.05% to 0.80% (m/m cement). Generally, for chloride levels in excess of 0.7% the probability of depassivation of reinforcing steel is considered high, see Fig. 7 [3]. This conclusion resulted in great concern of the local asset owner on both the actual and future condition of the box girder bridge in view of timely and effective interventions to be considered.

2.2 Prediction of time to depassivation

In order to obtain quantitative information on the progress of chloride ingress and the resulting depassivation of the steel reinforcement over time, the measured chloride profiles were subjected to regression analysis to obtain best-fit values for relevant parameters, using an analytical solution to Fick's 2nd law of diffusion, commonly referred to as the error function solution, as the chloride ingress model:

$$C(x,t) = C_s - (C_s - C_i) \cdot \mathrm{erf}\left(\frac{x}{2\sqrt{D_a \cdot t}}\right) \quad (1)$$

where x is the distance to the exposed concrete surface, and t indicates either the age of the concrete or to the effective exposure time, whereas erf refers to the error function.

In fact, this expression contains 3 unknown parameters, i.e. the surface chloride content, C_s; the initial chloride content, C_i; and the apparent chloride diffusion coefficient, D_a. Fig. 3 shows the best-fit chloride profiles resulting from this regression exercise, using the least square method.

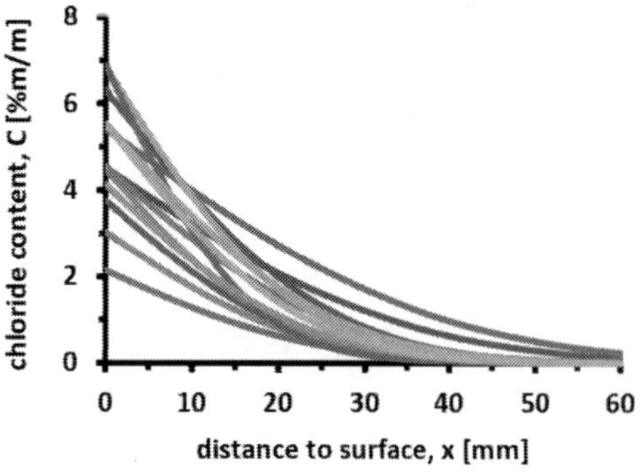

Fig. 3. Fitted chloride profiles (2002) in the walls of the box girder bridge after 18 years of exposure (windward side).

The values of the fitted surface chloride contents, $C_{s,fit}$, ranged from 2.1 to 6.9% (%m/m cement), whereas the values for the initial chloride content, $C_{i,fit}$, generally amounted to 0% and incidentally up to 0.06%. This level of C_i was as expected as no chlorides were allowed to be mixed in during concrete production. If significant levels for the chloride content at depths >50mm had been found then this could hint at a systematic bias in the chloride measurements. The apparent chloride diffusion coefficient, $D_{a,fit}$, was found to range between 4.7 to 24.7mm^2/yr, corresponding to $0.15 \cdot 10^{-12}$m^2/s and $0.78 \cdot 10^{-12}$m^2/s, respectively, with an average of 11.2mm^2/s ($0.35 \cdot 10^{-12}$m^2/s). These values for $D_{a,fit}$, as determined by regression analysis, for a concrete age of 18 years are indicative for a very dense concrete cover zone. This qualification is supported by other visual and microscopic observations on the concrete cores.

Adopting a conservative approach by assuming that the diffusion coefficient D_a remains constant over time and C_{crit} = 0.5% for individual concrete cores onset of reinforcement corrosion would occur within -7 to + 38 years.

However, for such predictions it should be borne mind that in practice it is generally observed that the chloride diffusion coefficient D_a gradually decreases over time, and the general trend is described by an empirical relationship given by:

$$D_a(t) = D_a(t_{ref}) \cdot \left(\frac{t_{ref}}{t}\right)^n \quad (2)$$

where n is the so-called ageing factor and t_{ref} is the reference age at which the reference value $D_a(t_{ref})$ is determined (from measured chloride profiles).

Moreover, there is significant uncertainty on the actual level of the critical chloride content, C_{crit}. According to [3] C_{crit} demonstrates a large variation which calls for a probabilistic approach.

An important point for the local asset manager is that at the time of inspection (2002), nowhere could damage be observed that could be directly or indirectly linked to the presence of reinforcement corrosion. This observation raised considerable doubt on the conclusions with respect to the calculated residual service life. As a result, additional large-scale investigations were proposed to either validate or invalidate the conclusions based on the 2002 inspection results.

3 Field investigations performed during 2005-2012

In view of the serious concern of the local asset manager on the actual and future condition of the bridge elements with respect to corrosion of the reinforcing steel, during the time period 2005-2014 a variety of non-destructive test methods were applied, including potential mapping, galvanostatic pulse and concrete resistivity. Regarding potential mapping, large scale investigations were executed in 2010 using a small grid size as to detect any spots suspected of active reinforcement corrosion. These essentially non-destructive measurements comprised the complete underside of several bridge elements, including the external surface of both cantilevers and walls, as well as the floor, see Fig. 4.

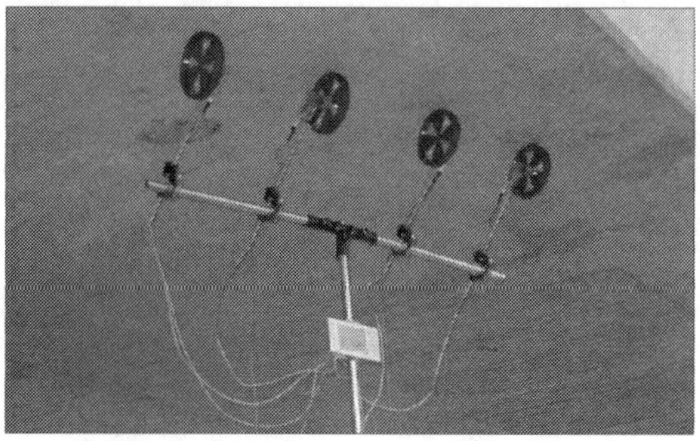

Fig. 4. Potential mapping executed on the floor of a bridge element using 4 potential wheels in parallel.

Generally, the potential values recorded were less negative than -200mV (vs $Cu/CuSO_4$) whereas resistivity measurements performed at the concrete surface generally demonstrated values well in excess of 500Ωm (equal to 50kΩcm). Consequently, none of these investigations indicated locations clearly suspect of reinforcement corrosion. This interpretation was also supported by the absence of any damage on the concrete surfaces investigated.

Eventually it was decided to perform a new series of chloride profiling in 2016 to validate the results of the chloride profiling in 2002. In addition, combining the results of 2002 and 2016 would allow a more reliable prediction of the future development of chloride ingress.

4 Field investigations performed in 2016

In view of the importance of the box girder bridge for traffic and the preference of the asset manager for preventive rather than corrective maintenance, in 2016 a new large campaign was organised to obtain quantitative information on chloride ingress. As the previous investigations were performed in 2002, i.e. at an age of 18 years, the information obtained at a significantly later age (32 years) would shed more light on possible ageing effects affecting chloride ingress over time and thus allow for a more reliable prediction of future ingress.

Prior to coring, visual inspections were performed to identify any possible locations suspected of reinforcement corrosion which could be used as preferential spots for coring, however, nowhere could damage related to corrosion be observed. In total more than 90 concrete cores were retrieved from several bridge elements, not only from the bridge walls but also from both cantilevers as well as from the bridge floor. In this paper only the data on chloride ingress in the wall on the windward side of the box girder bridge will be presented and discussed.

In contrast to the measurements executed in 2002, chloride profiling was performed at 3 different heights. In order to avoid any discussion on the method used for chloride analysis the cores were sent to the same laboratory that had performed the chloride analysis in 2002. As inspections executed prior to these investigation didn't reveal any damage on the concrete surface, the locations for coring were chosen according to pre-defined gridlines.

Chloride profiling was performed on 5 different bridge elements all made with traditional gravel concrete, comprising 18 concrete cores. The measured chloride profiles are shown in Fig. 5.

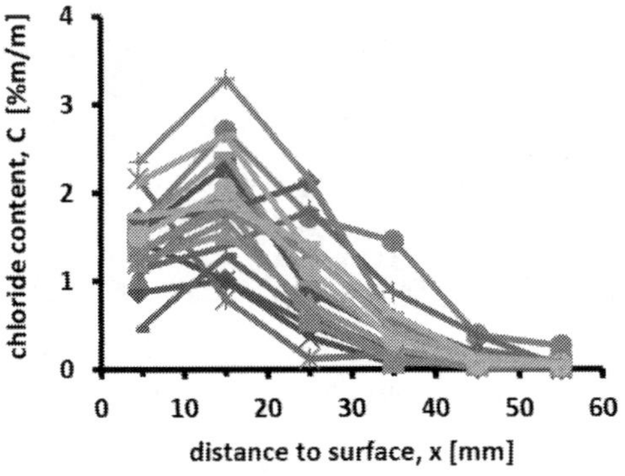

Fig. 5. Measured chloride profiles (2016) in the walls of the box girder bridge after 32 years of exposure (windward side).

With respect to onset of corrosion the chloride contents found at the level of the reinforcing steel, x = c, are relevant. According to the results of the regression analysis the chloride content at x = 40mm range from 0.05% to 0.84% with an average of 0.29%. However, these values achieved through regression analysis are generally higher than was actually measured at depths between 35 and 55mm.

5 Interpretation and prediction

Similar to the procedure adopted for the interpretation of the results from the previous inspection (2002) all measured chloride profiles obtained in 2016 were subjected to regression analysis yielding best-fit values for the model parameters, $C_{s,fit}$, $C_{i,fit}$ and $D_{a,fit}$, see Fig. 6.

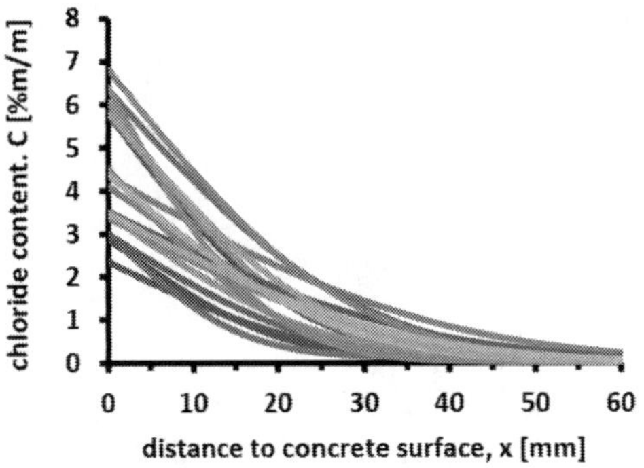

Fig. 6. Fitted chloride profiles (2016) in the walls of the box girder bridge after 18 years of exposure (windward side).

The values obtained through regression analysis for $C_{s,fit}$ ranged from 2.3 to 6.8% with an average of 4.4% (m/m cement), whereas the magnitude of $C_{i,fit}$ was almost 0. For $D_{a,fit}$ a minimum value of 2.3mm^2/yr (0.07·10^{-12}m^2/s) and a maximum value of 14.8mm^2/yr (0.47·10^{-12}m^2/s) were derived, whereas the average value amounted to 7.1mm^2/yr (0.23·10^{-12}m^2/s).

Comparing the average values obtained for $D_{a,fit}$ at t_{insp1} = 18yr and at t_{insp2} = 32yr the magnitude of the ageing factor was calculated to be n ≈ 0.78. This value is much higher than n = 0.45 which is reported in [3] as an average value for concrete produced with CEM III/B exposed to a marine environment.

In addition, in [4] it is advocated that a higher value for C_{crit} could be used for high quality concrete without showing cracks in the concrete cover. As the bridge elements are prestressed and the compressive strength exceeds 50MPa it seems reasonable to adopt a higher value for C_{crit}.

In view of the uncertainties involved for each chloride profile, a probabilistic approach was adopted assuming that for an individual chloride profile both $C_{s,fit}$ and $C_{i,fit}$ remain essentially constant over time, whereas the ageing factor n and the critical chloride content are regarded as stochastic variables.

For the ageing factor a beta distribution was used with a mean of μn = 0.75 and a standard deviation, σn = 0.05. For the lower and upper boundaries, a minimum a = 0.0 and a maximum b = 1.0 were chosen. For the critical chloride content, a beta distribution was applied with μC_{crit} = 0.75; σC_{crit} = 0.15; a = 0.20% and b = 2.00%, see Fig. 7.

The results of these probabilistic calculations for each chloride profile, using the modified statistical values of n and C_{crit}, identified as "realistic scenario", are shown in Fig. 8. For comparison Fig. 9 shows the results obtained according to the "fib 34 scenario" when the values according to [3] would have been adopted (n: μn = 0.45;σn = 0.20 and C_{crit}: μC_{crit} = 0.60%; σC_{crit} = 0.15%).

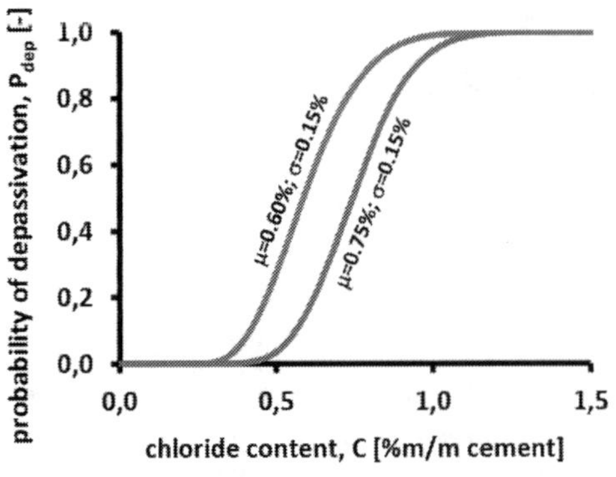

Fig. 7. Distribution of the probability of depassivation for the critical chloride content as adopted in the probabilistic calculations.

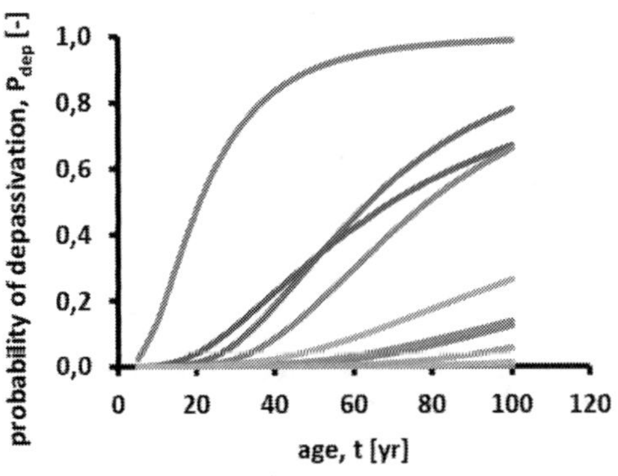

Fig. 8. Development of probability of depassivation for individual chloride profiles according to input values adopted for the "realistic scenario".

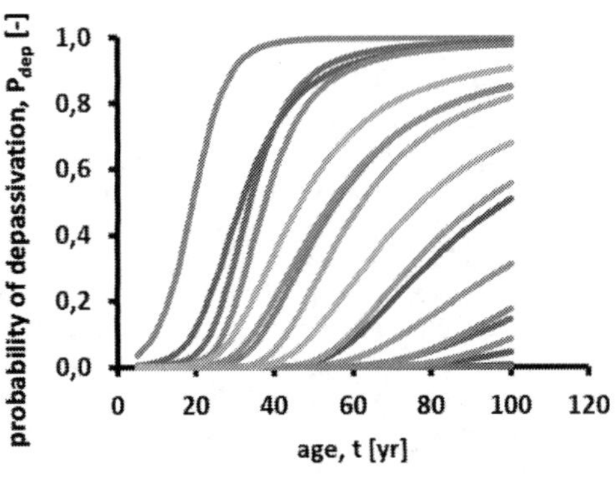

Fig. 9. Development of probability of depassivation for individual chloride profiles according to input values for the "fib 34 scenario".

However, even for the results according to the "realistic scenario", the conclusion of these predictive calculations is that localised onset of corrosion may start within less than a decade after the second inspection in 2016. On the other hand, in the area of the worst chloride profile found in the 2016 inspections no evidence of active reinforcement corrosion was found, although the probabilistic calculations resulted in a 78% probability of depassivation being present.

6 Discussion and outlook

During the time period 2002-2016 extensive investigations have been performed with respect to chloride ingress into the wall of a number of bridge elements. Although it has been established that the concrete cover is of high quality with very low permeability, as supported by the very low values for the apparent diffusion coefficient D_a, locally, chloride ingress has advanced quite deep. However, visual inspections and surface hammering have not identified any spots that are suspected of reinforcement corrosion.

As an explanation for the contrasting results of visual inspection on the one hand and chloride content present at the level of the reinforcing steel and the probabilistic predictions on the other, 2 major hypotheses can be established.

The first hypothesis is that the steel-concrete interface does not show any major defects which may act as preferential sites for depassivation. The complete circumference of all reinforcing steel bars is covered with a dense and adhering cement layer without weak areas. As the bridge elements are prestressed, cracking in the concrete cover is largely absent and under normal service conditions the tensile stresses in the reinforcing bars are low. Consequently, the bond between steel bars and surrounding concrete will be hardly impaired. According to this reasoning, the critical chloride content is likely to be much higher than would normally be assumed, e.g. according to the values given in [3]. Moreover, values for C_{crit} may be higher for concrete produced with blast furnace slag cement (slag content > 50%). It is well known that slag cements are able to bind more chlorides, thus the concentration of free chlorides dissolved in the pore solution to attack the passive layer will be less.

The second hypothesis is based on the high concrete quality which is reflected in a high concrete resistivity even in a continuously wet exposure environment. In the investigations performed in 2002, concrete resistivity was measured on a limited number of concrete cores retrieved from the walls. Generally, the resistivity values were in excess of 500Ωm. This was confirmed by measurements performed on site using the 4-point Wenner probe placed on the wetted concrete surface of the walls. As corrosion is an electrochemical process, in concrete demonstrating resistivities in excess of 500Ωm the corrosion rate is generally found to be very low [5]. This implies that the production of electrons released in the anodic reaction will be limited which explains why the potential values that have been measured are corresponding to a passive state.

Given the high uncertainties in the input values of the most dominant model parameters it was decided not to rely on predictions based on numerical modelling. Meanwhile in practice a probabilistic approach is often advocated to take into account these uncertainties, however, experience, with this paper as an example, has shown that such an approach eventually turns out to become a “numbers game” to disguise the lack of knowledge and responsibility.

The results obtained from the investigations executed in 2002 and 2016 have provided a better impression of the development of chloride ingress over time. However, the chloride profiles that have been measured show a significant scatter, despite the fact that both the quality of the concrete cover and the environmental load with respect to moisture and chlorides is considered quite uniform over the exposed concrete surface of the wall. Consequently, it is concluded that the spatial variability of chloride ingress is considerable, even for this relatively ideal situation. This implies that a considerable number of cores would be required to have a reliable impression of chloride ingress over the complete wall surface.

At present the local asset manager is tired of all these investigations resulting in more confusion on the actual condition of the box girder bridge with respect to reinforcement corrosion. Therefore, in the near future, no further investigations on site are envisaged but a pragmatic engineering approach based on common sense will be adopted.

References

1. CUR, Publication 215, *Durability of marine concrete structures*, 2005 (in Dutch)
2. DuraCrete, Final Technical Report, Document BE-95-1347/R17 (2000)
3. fib, Model code for service life design, bulletin 34, 2006
4. M. Breit, C. Dauberschmitt, C. Gehlen, C. Sodeikat, A. Taffe, U. Wiens, *Approach on the critical chloride content in reinforced concrete structures*, Beton- und Stahlbetonbau, 106 (2011), Heft 5 (in German)
5. P. Langford, J. Broomfield, *Monitoring the corrosion of reinforcing steel*, Construction Repair, 1987, 1(2)

ASR: Practical investigative techniques and field monitoring systems used to assess ASR for service life modeling.

Irene Matteini[1,*], *Paul Noyce*[2], *Gina Crevello*[3]

[1] Echem Consultants, Technical Operation Manager, New York, USA
[2] Echem Consultants, Technical Director, New York, USA
[3] Echem Consultants, Principal, New York, USA

Abstract. Alkali Silica Reaction (ASR) is a common deterioration mechanism affecting many concrete structures of any type and age. Initially identified more than 60 years ago (Stanton, 1930), this mechanism is based on the chemical reactions between certain siliceous minerals present in the aggregate and the alkalinity of the concrete in the presence of moisture (internal RH). While certain deterioration patterns are clearly associated with ASR, such as gel exudation, aggregate expansion, and characteristic cracking, the material degradation can often be misdiagnosed to the untrained eye. In addition, certain elements of a structure can be severely affected while neighboring elements of the same batch/ mix design do not bear signs of deterioration or impact. Thus far, in situ field monitoring of ASR affected structures is related to moisture measurements, electrical resistivity, expansion, service life models are based on fracture mechanics of the aggregate. The impact to the concrete is loss of integrity, decreased compressive strength, shear and tensile strength. Some observed structures have split, with such force, that the concrete structure had cracks greater than 25mm where steel retention bands have split. The authors of this paper were engaged in two instances to provide service life assessments for 'corrosion related degradation' on ASR affected structures. In all instances the elements which were assessed were structural, load bearing elements, which if failed could pose a significant risk to owner, user, or end recipient. The need to develop an assessment technique for monitoring and service life assessments which are practical and efficient is being developed. The paper will discuss the development of the approach, from visual indicators identifying condition hierarchies, to long term condition monitoring for various concrete parameters combined with laboratory testing (expansion and residual alkalis) and mathematical modeling. Three case studies will be presented to illustrate conditions and process.

1 Introduction

The expansion of concrete generated by the reaction between cement and aggregates is commonly referred to as Alkali Aggregate Reaction (AAR). The most common mechanism is referred to as Alkali Silica Reaction (ASR) which is defined as the reaction between the alkali ions present in concrete and reactive siliceous aggregates.

First studied by Thomas E. Stanton [1] of the California State Division of Highways in the 1930s, ASR is one of the most critical forms of deterioration that has occurred in concrete structures around the globe. In the United States, Parker Dam in Arizona was one of the first major structures to be diagnosed with ASR in the 1940's shortly after Stanton's study.[2] Built in 1934, this dam is still in use today after 86 years.

ASR is a chemical reaction that requires four main ingredients to be present for the reaction to occur: hydroxyl ions, alkali metal ions, reactive silica aggregates and sufficient moisture. In summary, the hydroxyl ions available in the pore structure will attack and provoke the destruction of the atomic bonds of the siliceous compounds causing the silica to dissolve.

Fig. 1 Aerial view of Parker Dam. Courtesy of Bureau of Reclamation

Then, the alkali ions react with the $Si(OH)_4$ complexes forming a (Na,K)-Si-OH Gel around the edges of the aggregate. In the presence of a sufficient level of moisture, the gel expands generating stresses which are responsible for the formation of cracks. [3] The reaction is illustrated in Fig. 2.

* Corresponding author: imatteini@e2chem.com

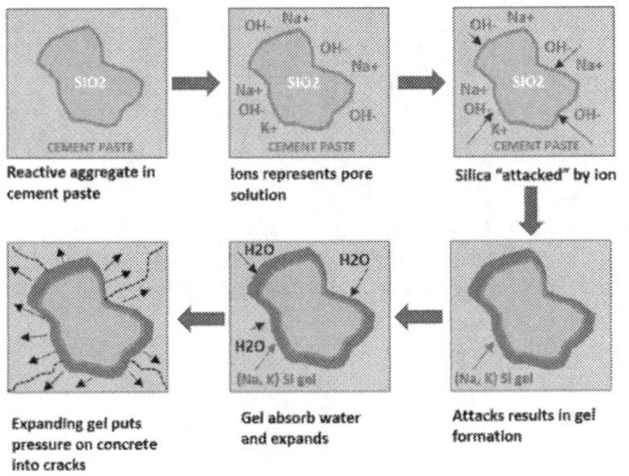

Fig. 2 Diagram illustrating the ASR Mechanism. Courtesy of Echem Consultants.

This swelling has the potential to induce tensile stresses within the concrete and can lead to the appearance of cracking with a pattern that depends on the geometry of the structure, and the layout of the reinforcement. [4] The typical cracking pattern is so called "map-cracking" which consists of a network of cracks randomly distributed from which a whitish gel will leach out of the concrete as illustrated in Fig. 3.

Fig. 3 Typical ASR Crack Mapping. Courtesy of Echem Consultants.

Besides the typical crack mapping of ASR, another typical ASR form of deterioration is referred to as "Pop-outs". This consists of a reactive aggregate undergoing expansion causing the concrete to detach leaving the reactive aggregate behind. [5] This is generally a localized phenomenon as shown in Fig. 4.

The main four ingredients available have an impact on the severity of the reaction that can occur in concrete structures. Each ingredient is briefly described below.

1.1 Reactivity of Aggregate

The quantity of reactive silica present in the aggregate mix determines the severity of the reaction. The reactivity of the silica minerals depends on their crystal structures.

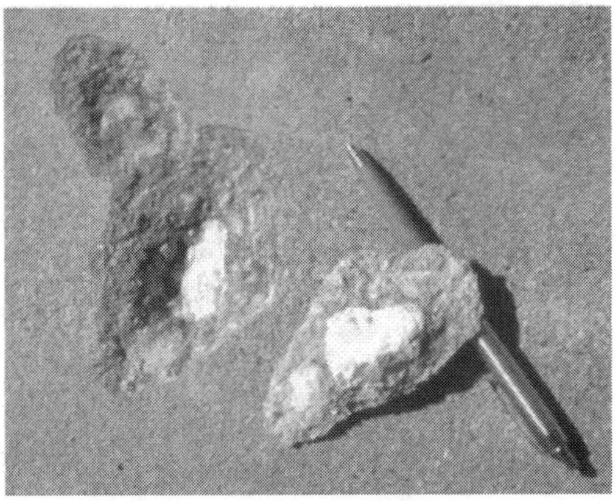

Fig. 4 "Pop-out" effect of ASR deterioration. [4]

For example, opal has a very disordered structure and is the most reactive form of silica. While, pure, undeformed quartz with an ordered structure is generally not reactive. [4]

1.2 Alkali Content

The alkali content present in the concrete plays an important role as well. Given the presence of reactive aggregate in the mix, using a low effective alkali content can prevent the reaction occurring. As shown in Fig. 5, if using a concrete with a Na_2O content less than $3kg/m^3$ in the presence of chalcedony, a typical aggregate with intermediate reactivity, it will result in a negligible reaction. [6]

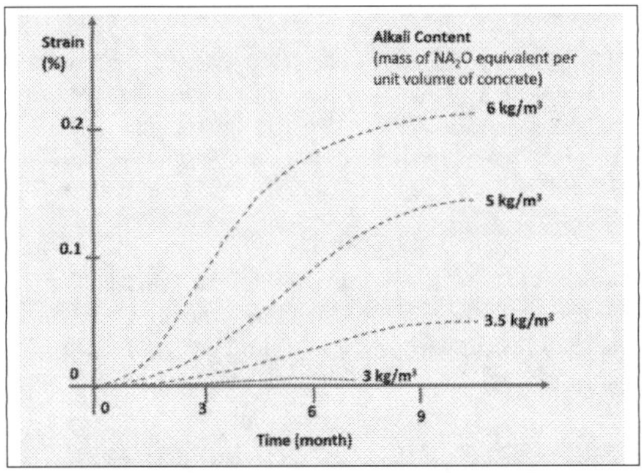

Fig. 5 Behaviour of concrete with reactive aggregate containing chalcedony as a function of Alkali content. [6]

External alkalis such as deicing salts or airborne salt in marine environments do not generally contribute significantly to the total alkali content and do not increase the risk of ASR given the slow rate of penetration into the concrete.[7]

1.3 Type and Quantity of Cement

Another ingredient that can be added as a form of prevention is the use of a pozzolan or blast furnace slag cement. Thus, this ingredient besides adding workability

 https://doi.org/10.1051/matecconf/201928908004

to the mix, it reduces the concentration of OH^- ions in the pore solution of the cement paste as the hydroxyl ions are being consumed by the pozzolanic reaction. Furthermore, the alkali movement is reduced as a result of the lower permeability of the pozzolan and blast furnace in the cement paste.

1.4 Environment

As previously discussed, the environment plays an important role as ASR can only occur in the presence of moisture. It has been observed that in an environment with a relative humidity below 80%, alkalis and reactive aggregate can coexist without causing damage. Temperature influences the reaction as well, warmer temperature facilitates the reaction to occur faster.

As engineering professionals, understanding the ingredients that influence the rate at which this reaction occurs is the first step to understand the durability and service life. ASR is responsible for a reduction in strength and stiffness of concrete components and in some instances of the entire structure.

2 Case Studies

In the following section, two (2) case studies with ASR related problems will be illustrated. Different field investigative techniques and monitoring programs used to study this mechanism will be discussed. The two structures presented are very different in size and use but both united by fundamentally the same problem.

2.1. Case Study One: Power Line Distribution Pole

The first case study is focused on power line distribution poles located on the northeast coast of the United States of America. Installed in the late 1970's, the poles are pre-cast and pre-stressed concrete structure used to provide bus and switch supports for electrical utility lines.

The poles, originally installed with a projected service life of 100 years, represent a critical support structure for the power plants. The structures, while still erected, have been suffering from several modes of concrete and material failure overtime as shown in Fig. 6.

In 1964, the poles were first built and utilized to rebuild and upgrade the power line from 34.5 to 115 kV. Each pole was a centrifugally-spun concrete cylinder of pre-determined length.

During these years, the use of pre-cast, pre-stressed poles was common for electrical and infrastructure with applications such as railway ties, and electrical distribution poles etc.

Between 1972 and 1974, M. Van Buren played an important role in the development of numerous patents in regard to the construction of pre-stressed reinforced concrete tapered poles *"which would provide uniform compressive units throughout the length of the utility pole"*.

Fig. 6 Image of one pole used for this investigation. Courtesy of Echem Consultants.

According to Van Buren, the high weight to strength ratio of concrete is desirable to taper elongated vertical support members made of concrete. This permits a more efficient structure, since each cross sectional area may be tailored to the total weight being supported by it. As a result a uniform unit stress may be maintained throughout the structure.[8]

Today, this type of pole is still being used for lighting, sub-station bus-supports, and electrical transmissions distribution power lines.

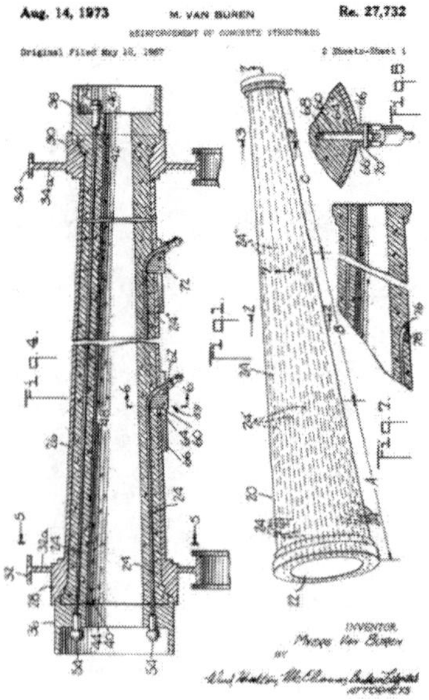

Fig. 7 Drawings extracted from M. V. Buren Patent, 1973

The work scope for this investigation included ten (10) structures of different size and shape which were no longer in service due to the advanced level of deterioration.

A testing program was developed that included the following:

Table 1 Summary Testing Program

Reference	Tests Carried Out						
	Corrosion Rate	Corrosion Rate/Half-Cell	Electrical Continuity	Resistivity	Compressive Strength	Ultrasonic Pulse Velocity	Concrete Cover
IR-Pole 1	☒	☒	☐	☒	☒	☒	☒
IR-Pole 2	☒	☒	☒	☒	☒	☐	☐
IR-Pole 3	☒	☐	☐	☒	☒	☐	☒
IR-Pole 4	☒	☐	☐	☒	☒	☐	☒
IR-Pole 5	☒	☐	☐	☒	☒	☒	☒
IR-Pole 6	☒	☐	☐	☒	☒	☐	☒
MF-Pole 1	☒	☐	☐	☒	☒	☒	☒
MF-Pole 4	☒	☐	☐	☒	☒	☒	☒
MF-Pole 5	☒	☒	☒	☒	☒	☒	☒
MF-Pole 6	☒	☐	☐	☒	☒	☒	☒

The structures appeared in poor condition showing parallel hairline cracks and map cracking along their entire elevation. In addition, gel exudation was visible on the surface as shown in Fig. 8.

Fig. 8 Image showing ASR mapping cracking and whitish gel leaching at the bottom of one of the pole. Courtesy of Echem Consultants.

At the base of each pole, a grounding wire was attached which was electrically discontinuous with the reinforcement of the structure.

Little to no corrosion was found in nearly every pole, even when the steel tendons or reinforcing steel was exposed. Dust samples were collected to determine chloride levels which were found to be of negligible quantity and below the threshold for corrosion. In addition, no reduction in alkalinity was detected from the phenolphthalein test.

As a result of the petrographic analysis, it was determined that Alkali-Silica Reaction was the root cause of the concrete deterioration of these structures. Multiple aggregate types utilized in the original mix design are known to be unstable in alkaline environments such as opal, and chert. Visible micro-cracking appeared in several of the aggregates showing gel exudation. In addition, a sufficient level of moisture was also detected due to a high water table on site.

Furthermore, given the pre-cast nature of the poles, Delayed Ettringite Formation (DEF) was also suspected as a potential deterioration mechanism that could have occurred in these structures over time. DEF is normally associated with steam cast cured members and when curing temperatures exceeds 158° F (70° C). DEF refers to the development of ettringite within the cement paste after the concrete has hardened. Like ASR, this mechanism will result in the formation of large cracks caused by the expansive reaction.

Beginning in the early 1980s, DEF was reported to have caused destructive cracking of concrete sleepers (railway ties) in Europe and other pre-cast products in the United States. [9]

Numerous studies have stated that ASR and DEF could possibly relate to one another. In some of these cases, it has been concluded that ASR is the primary cause of cracking and deterioration, and that the subsequent precipitation of ettringite into the cracks is a secondary effect that may or may not cause further damage. [10]

Furthermore, the presence of microcracks in the aggregates appears to provide sites where secondary ettringite can grow and produce damage. Most petrographic examinations have shown that if one wishes to find ettringite in a deteriorated concrete, the first place one should look is within the transition zone. [11] In the end, DEF was ultimately not found in these structures.

Another unanswered question in regard to this case study is related to the location of the ground wire and its role in the ASR Reaction. As described before, each pole has a ground wire for lightning protection and ground fault dissipation. This ground wire is bonded to the substation grounding system and runs for the entire height of the structure which is discontinuous to the main steel reinforcement. The ground wire has the potential to generate an eletromagnetic field which could have played a role in the degree of this reaction given the size of the cracks that appear on the pole. In many of the poles, it was noted that the largest cracks, which developed for the entire height of these structures, occured close to a ground wire as shown in Fig. 9.

In conclusion, a monitoring program is being developed as part of Phase II of this study. On a final note, it should be noted that the level of deterioration of the poles has progressed to the point that the original design capacity has been altered. These structures should be considered structurally deficient and should be monitored until they are completely taken out of service.

Fig. 9 Image showing a detail of one of the poles where the ground wire is located close to a large crack which extended for the entire height of the pole. Courtesy of Echem Consultants.

2.2. Case Study Two: American Football Stadium

A reinforced concrete football stadium located in a major city of the United States is our second case study. Built in 1903, the stadium was considered the first vertical concrete structure to employ reinforced structural concrete worldwide. Therefore, it was nominated as a National Historic Landmark in 1987.

Fig. 10 Football Stadium under construction. Courtesy of Wikipedia.

Designed in a classical revival style inspired by the Panathenaic Stadium in Athens Greece, the stadium was erected in just 4½ months for a total cost of $310.000.

This structure has been suffering from ASR for decades with numerous consultants and engineers working on it overtime. Multiple petrographic examinations have reported the presence of reactive aggregates in the cement paste, attributing the main cause of deterioration to ASR. Several components of the structure have been suffering from it including columns, beams, and the main exterior arches as shown in Fig. 11.

Fig. 11 ASR problem visible in structural component of the stadium. Echem Consultants.

Our scope of work was to develop a monitoring program to evaluate the existing structure. After an initial evaluation, the monitoring program included the use of the following techniques :

- Environmental Monitoring
- Ultrasonic Monitoring
- Digital Image Correlation (DIC)

Due to limited time and budget, one beam was selected for extensive monitoring. This beam representes a main component of the structure as shown on the drawing below.

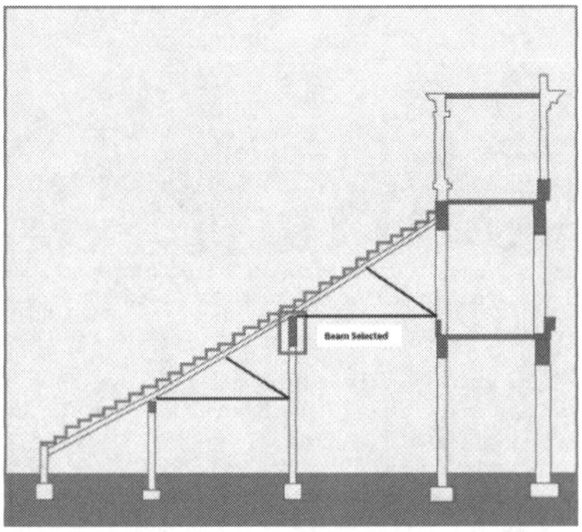

Fig. 12 Diagram representing the main cross section of the stadium. One beam was selected for this monitoring program. Courtesy of Echem Consultants.

2.2.1 Environmental Monitoring

A temperature/humidity sensor was installed at the selected location recording data every hour. The sensor was embedded in the concrete and sealed in order to obtain accurate information on the internal temperature and humidity of the material.

At this time, the data collected so far shows high constant level of moisture within the concrete as shown in Fig. 13. In addition, the humidity level recorded was not influenced by the atmospheric condition as it remained reasonably constant inside the concrete. This also supports the visible sign of ASR on the surface.
The majority of the structure is exposed to the elements with limited water-proofing. As moisture plays a fundamental role within the ASR, this demonstrates a long-term issue.

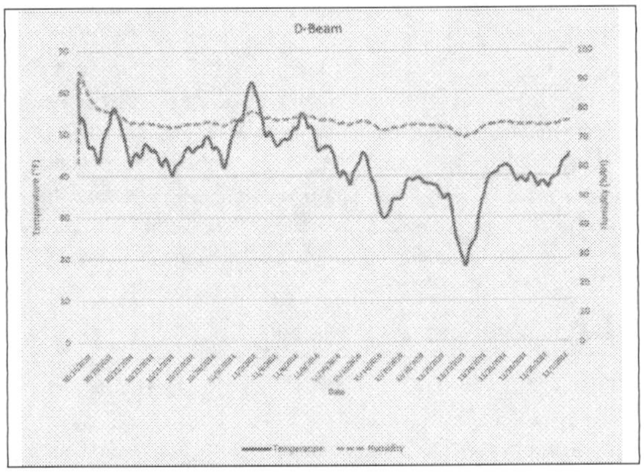

Fig. 13 Graph of Temperature/Humidity sensor at beam location. Courtesy of Echem Consultants.

2.2.2 Ultrasonic Wave Sensors

The main objective of this application, which is still in its development stage, is to monitor the progression of ASR.

This application is based on the use of the Coda Wave Comparison (CWC) technique [12]. Generally, ultrasonic signals can be divided into two portions: the coherent (early portion or p-wave) and the diffuse (late portion or coda wave). It is well known that the coherent portion of the wave is not sensitive to stresses until they reach a value that cause significant internal cracking. Hence, this approach is not suitable for service-life monitoring. In contrast, the coda wave field is influenced by any small changes in the material such as stress, temperature, and humidity variations, as well as aging/deterioration processes. The resulting internal changes are amplified by the multiple scattering properties observed in concrete affecting the stress wave field, which in turn alters the amplitudes and frequency content of the coda wave [12]. Fig. 14 illustrates the difference between conventional and coda wave ultrasonic monitoring, the main difference being that the latter is able to capture information from regions (highlighted in orange) that do not lie within the direct wave travel path. Over the last years, numerous researchers have started analyzing coda wave fields to monitor changes in concrete caused by cracking [14] and ASR [16]. A rigorous review of coda wave analysis techniques and their applications is provided in [15].

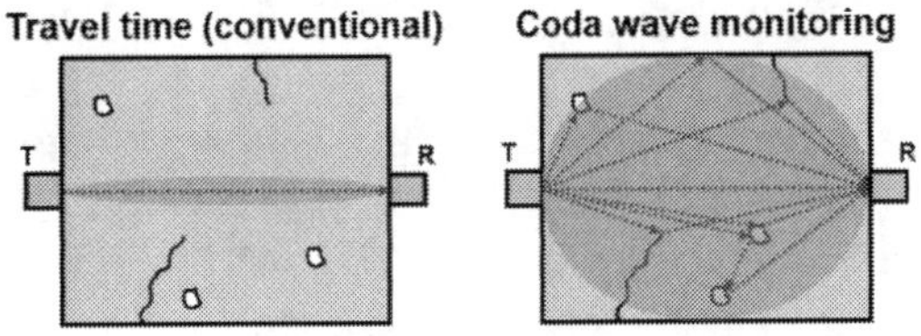

Fig. 14 Comparison of conventional (left) and coda wave (right) ultrasonic monitoring. Courtesy of Th. Schumacher.

A typical ultrasonic measurement setup is illustrated in Fig. 15: A constant and pre-defined pulse is transmitted into the material using an ultrasonic transducer and received by another one. This is done repeatedly at time intervals depending on the application. A high-speed data recorder captures and samples the ultrasonic waves arriving at the receiving transducer and stores them for later analysis.

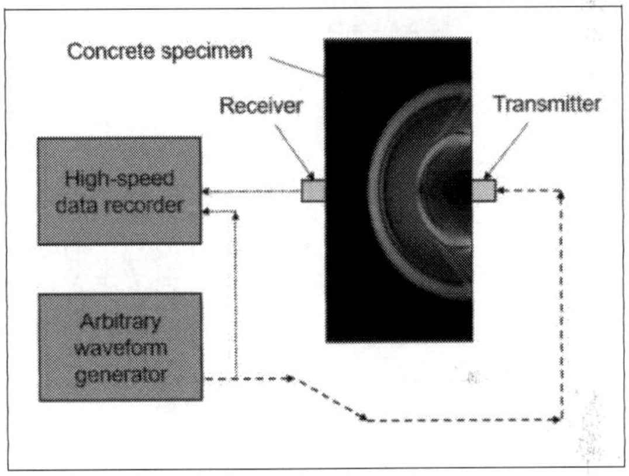

Fig. 15 Diagram illustrating the ultrasonic monitoring principle. Courtesy of Th. Schumacher.

In this application a commercial acoustic emission monitoring system (Vallen AMSY-5) is used to perform the measurements. The system both transmits and records ultrasonic signals at a sampling rate of 2 MHz. The transducers used were provided by Federal Institute for Materials Research and Testing (BAM) and are composed of a hollow piezoceramic cylinder of 20 mm diameter by 35 mm length. The electric connections are clamped to the piezoceramic part of the transducer. The total length of the transducer is 75 mm and 15 mm in diameter as shown in Fig. 16. [13]
A sample of a recorded ultrasonic signal is shown in Fig. 17. For each transducer, a 25 mm diameter hole was drilled to the middle of the beam and properly grouted, as shown in Fig. 18. Each transducer is connected to a pre-amplifier (Vallen AEP4) located close to the transducer. The transducers were embedded in one beam located 1.5 meter (5 ft) apart, acting both as transmitter and receivers, as shown in Fig. 19.

Fig. 16 Image of novel embeddable ultrasonic transducer. Source: [13].

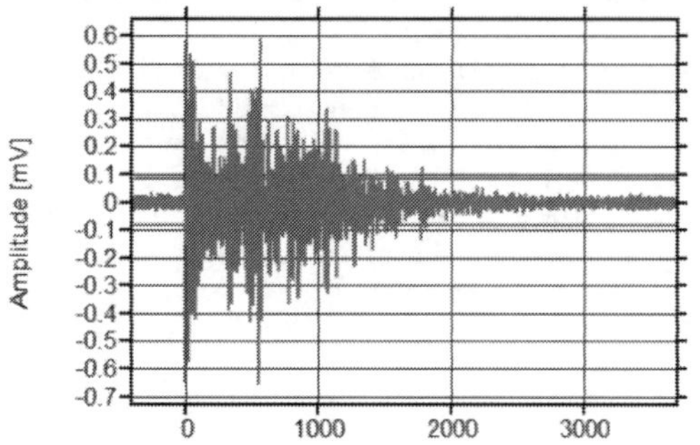

Fig. 17 Sample recorded ultrasonic waveform. Courtesy of Th. Schumacher

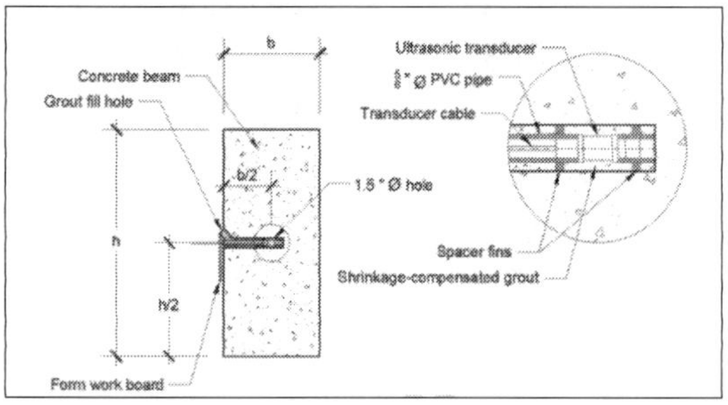

Fig. 18 Diagram showing the installation of the transducers at the selected beam. Courtesy of Th. Schumacher.

Fig. 19 Photo showing the installation locations of the three transducers at the beam. Courtesy of Echem Consultants.

While this project is still at an early stage, it is already showing promising results. One of the challenges is the fact that the stadium is actively used for sports and large events and is undergoing a large renovation project, which is adding noise to the system. Also, temperature and humidity changes affect the signals and this needs to be quantified.

On a final note, it is important to note that there are advantages of using this ultrasonic monitoring approach compared to the more traditional acoustic emission monitoring which has been employed at the stadium by others. With the proposed ultrasonic monitoring, which is an active technique, the source is controlled and repeatable and the recorded response can thus be directly associated with internal changes in the material. On the other hand, acoustic emission monitoring is a passive technique, recording active fracture processes in the material. Time, location, and nature of these sources are initially unknown, and occur spontaneously, i.e. they are not controlled by the operator and can thus not be repeated. Also, many of the other common sensors used, such as strain gauges and displacement sensors, are attached to the surface and thus represent a surface observation. In this application, the sensors are embedded and represent the internal condition of the material which is directly relevant for ASR.

2.2.3 Digital Image Correlation (DIC)

DIC (Digital Image Correlation) is an innovative non-destructive, non-contact optical technique used to measure strain and displacement. Originally developed by the National Physical Laboratory (NPL) in the UK, this technique works by comparing digital photographs at different stages of deformation. By tracking blocks of pixels, the system can measure surface displacement and build up full field 2D and 3D deformation vector fields and strain maps. [17] No special cameras or equipment is required which makes this technique very affordable.

At the stadium, a camera has been installed twenty-five feet (7.6m) from the beam in order to obtain an image representing the component in its entirely as shown in Fig. 21. A Fujifilm X-T2 mirrorless camera has been used to collect one picture every four hours. After collection, the pictures are processed by using Matlab. The software overlaps pictures and elaborates the pixels to determine changes in the component.

In conclusion, the aim of this study is to correlate data collected by the environmental monitoring, ultrasonic sensors and the DIC pictures. Although at an early stage, the goal is to develop a network of monitoring sensors installed in different components of the structure which will be able to provide a holistic picture of the on-going condition of the entire structure for service life prediction.

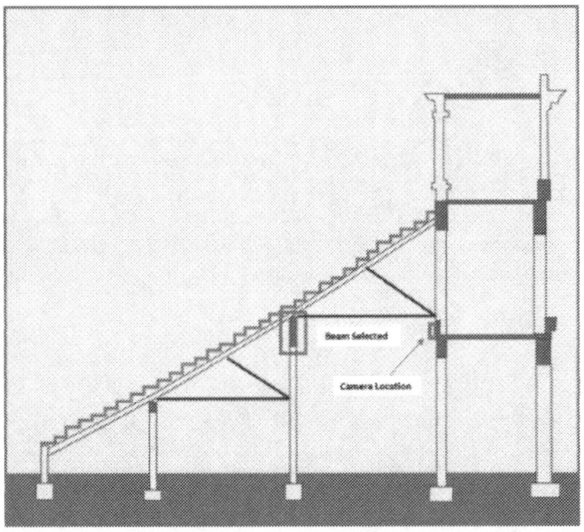

Fig. 20 Diagram showing the location of the camera in relation to the beam. Courtesy of Echem Consultants.

Fig. 21 Picture of the Camera Box installed at the stadium. Courtesy of Echem Consultants.

3 Conclusions

This paper attempts to illustrate the complexity of ASR and its potential impact in the service life of buildings and infrastructure. Using a diverse array of techniques, the aim of this study is to develop monitoring program in order to extend the service life of a component or of the entire building affected by ASR.

ASR is found in the cement paste coming from its reactive aggregate and the natural alkalinity of cement, therefore monitoring represents the best approach for any structure suffering from it.

Together with monitoring, several repair methods have been implemented overtime in order to help reduce the rate of deterioration. Unfortunately, in some cases, attempted repair methods have made the damage worse, for example by sealing the cracks. In addition, coating is another common method used to protect the concrete surface from excess moisture ingress. Different types of coatings have been used over time from silanes, acrylic rubber to epoxy etc. Unfortunately, in some instance, coatings have not been able to protect the concrete from moisture already present within the concrete and they do require to be reapplied over time. Gloablly, the use of lithium has shown some success, but the technique is less practicable for an entire structure.[18]

Thanks to new technology and advanced knowledge of this phenomenon, preventive measures have been implemented to prevent ASR in new construction in the United States and around the world by the use of non-reactive aggregates, limiting the alkali content of the concrete, use of supplementary cementing materials and ultimately the use of lithium-based admixtures.

Acknowledgement

A special thanks to Professor Thomas Schumacher of Portland State University and Ernst Niederleithinger of The Federal Institute for Materials Research and Training in Germany (BAM) for the on-going collaboration.

References

1. T.E. Stanton, "Expansion of concrete through reaction between cement and aggregate". *Engineering News-Record*, No. SP-249-1 (1940).
2. Meissner, H.S., Cracking in concrete due to expansive reaction between aggregate and high-alkali cement as evidenced in Parker Dam. *Proceedings, Am. Concrete Inst.*,**57**, pp.549-568, (1941)
3. Lima, Mehdi, CFRP Strengthening of ASR Affected Concrete Piers of Railway Bridges. *9th International Conference on Bridge Maintenance, Safety and Management*, (2018)
4. L. Bertolini, et al., Corrosion of Steel in Concrete, *Wiley-VCH*, pp.63-65 (2013)
5. FHWA-HIF-13-019, pp 43 (2013)
6. R.G. Sibbick, C.L. Page, Threshold alkali contents for expansion of concretes containing British aggregates, *Cement and Concrete Research*, **22**, pp.990-994 (1992)
7. J.H.M Visser, R.B. Polder, Alkali Silica Reaction in concrete penetration of Alkalis, *TNO Building and Construction Research Report 2000-BT-MK-R0207*, (2000)
8. M. V. Buren, Reinforcement of Concrete Structures United States Patent, Original Filed 1967, Reissued 1973 (1973)
9. W. G. Hime, Delayed Ettringite Formation – A concern for pre-cast concrete, *PCI Journal* **July-August**, (1996)
10. M.D.A. Thomas, Delayed Ettringite Formation in Concrete. Recent Developments and Future directions, *Report University of Toronto*, (1998).
11. FHWA/TX-06/0-4085-1, Alkali-Silica Reaction and Delayed Ettringite Formation in Concrete A Literature Review, (July 2001, Rev. February 2006)
12. A. Hafiz, T. Schumacher, Monitoring of Stresses in Concrete Using Ultrasonic Coda Wave Comparison

Technique, *Journal of Nondestructive Evaluation*, **37,** pp.73, (2018)

13. E. Niederleithinger , J. Wolf, F. Mielentz, H. Wiggenhauser S. Pirskawetz, Ultrasonic Transducers for Active and Passive Concrete Monitoring, *Sensors **2015**, 15(5), 9756-9772 (2015)*
14. J.B. Legland, Y. Zhang, O. Abraham, O. Durand, V. Tournat, Evaluation of crack status in a meter-size concrete structure using the ultrasonic nonlinear coda wave interferometry, *J. Acoust. Soc. Am.* ***142***, *pp. 2233–2241 (2017).*
15. T. Planes, E. Larose, A review of Ultrasonic Coda Wave Interferometry in concrete, *CEM Concrete Research* (2013)
16. D.P. Schurr, J.Y. Kim, K.G. Sabra, L.J. Jacobs, Monitoring damage in concrete using diffuse ultrasonic coda wave interferometry, *In: AIP Conference Proceedings*. pp. 1283–1290 (2011)
17. N. Mc Cormick, J. Lord, Digital Image Correlation, *Materials Today*, **10** (2013)
18. I. Sims, A. Poole, Alkali-aggregate Reaction in Concrete A World Review, *CRC Press*, pp.246-248 (2017)

Experimental study of the influence of crack width due to corrosion behaviour using different measurement methods and different exposures

Martin Schneider[1]*, and *Georg* Gardener[2]

[1]Carinthis University of Applied Sciences, 9800 Spittal/Drau, Austria
[2]Carinthia University of Applied Sciences, 9800 Spittal/Drau, Austria

Abstract. Corrosion of reinforcing steel has a great influence in reducing the lifetime of concrete structures; Carbonation of the concrete pore solution causes surface corrosion on the steel and diffusion of chloride ions through the capillary system of the concrete cover causes pitting corrosion on the steel surface. Corrosion of metals is highly dependent on the environmental conditions. Exposure to chloride ions can be critical to the service life of reinforced concrete structures. The durability of reinforced concrete structures exposed to deicing salt or marine environments can be affected by impact of chloride ions. Detection methods for the rate of corrosion of non-destructive and destructive procedures were analysed. The potential mapping applied on the concrete surface was discussed as a standard method for corrosion detection and will be explained in detail including the application boundaries of the method. It is assumed that the corrosion behaviour of reinforcing steel depends on crack widths. To analyse that, 8 coated and 8 uncoated test samples with different concrete strength classes were used. The concrete objects were exposed to a 3% sodium chloride solution. The corrosion behaviour of reinforcing steel is analysed by using potential mapping with different reference electrodes (Ag/AgCl and Cu/CuSO4). The results show a significant correlation between crack size and protection system on the surface. The maximum crack width with a low indication of corrosion was found to be 0.1 mm.

1 Cracks in concrete

The special feature of the composite material reinforced concrete is that the concrete protects the embedded reinforcing steel from corrosion. If this protection of the reinforcement steel is lost, it results in an influence on the durability. This can be caused by excessively large crack widths. The permissible crack width is specified in technical regulations and standards. Consequently, cracks with limited crack width are no longer considered a defect or damage. It should be noted, however, that cracks with small widths promote the penetration of harmful substances such as CO2 and chloride. For the assessment of the corrosion of reinforcement steel, electrochemical examination procedures are available. A relatively inexpensive non-destructive method is the half-cell potential measurement [1].

1.1 Exposure classes

To ensure the necessary durability, concrete elements must be sufficiently resistant to chemical and physical influences from environment or use. These effects on the building material are divided into exposure classes. The following exposure classes record effects on the reinforcement in the concrete: XC (Carbonation), XD (Deicing), XS (Seawater) [1].

1.2 Causes of Cracks

Crack formation in concrete occurs when the locally present stress exceeds the local tensile strength of the material. It is irrelevant at first whether the crack causing stress is generated from an external static or dynamic load or whether this results from an obstruction of the deformation effort of the concrete – residual stress or constrained stress. Of essential importance is the fact that concrete is not completely homogeneous and isotropic with regard to its properties, in particular its tensile strength [2].

• In fresh concrete cracks occur as a result of rapid volume reductions at the near-surface concrete layer - caused by unfavourable temperatures, solar radiation, wind and low humidity [3].
• In building practice, cracks may occur because of loss of the hydration heat or by shrinkage, in particular by early shrinkage [3].

In order to ensure sufficient protection of the reinforcement and the necessary durability, the crack widths are limited in standards and guidance. The limit values of the crack widths is dependent on the exposure class and the design class and are determined according to ÖNORM 1992 1-1.

* Corresponding author: m.schneider@fh-kaernten.at

https://doi.org/10.1051/matecconf/201928908005

Table 1. Exposure classes [4].

Exposure class	maximum crack width
X0, XC1	0,4 mm (appearance)
XD1, XD2, XD3	0,3 mm

1.3 Reinforcement corrosion caused by cracks

The type of crack plays an important role when it comes to assessment. For example, bending cracks usually have a lower risk potential for a component than separating cracks. CO_2, chloride and water can penetrate the component through separating cracks. In addition, separation cracks often cause a change in the static system. This can lead to extensive load redistribution and further crack formation. Surface cracks caused by inadequate post-treatment do not have the same serious consequences as large cracks caused by a lack of reinforcement steel. Cracks that occur along the reinforcement are particularly critical because they can lead to composite failure [5].

Fig. 1. Typical damage pattern of pitting corrosion 5].

A special characteristic of reinforced concrete is the corrosion protection of the installed reinforcing steel by the concrete. Due to the high pH value of the concrete pore solution ranging from pH 12.5 to pH 14, a passive film is formed on the surface of the reinforcing steel, a few atom layers thick, to protect the steel underneath from further degradation. If the concrete loses its alkalinity or if a critical chloride content is exceeded on the steel surface, the corrosion protection of the reinforcing steel may be lost. If pollutants such as chlorides or CO_2 penetrate the concrete via a crack, rapid damage processes can occur to both the reinforcement and the concrete itself. This leads to depassivation and corrosion of the reinforcement steel [8]. The danger of chloride-induced pitting corrosion (local corrosion) lies in the combination of high corrosion rates of the steel with simultaneous invisibility of the damage process from the outside. This local corrosion leads to high corrosion rates, because a small corrosion scar results in a large cathode surface which accelerates the corrosion process. The steel removal rates associated with these high corrosion rates can reduce the steel cross-section of the reinforcement within a short time. At the same time, the damage processes in the interior of the component go on, so that the reinforcement steel can already be significantly damaged, without any appearance of damage on the concrete surface [6].

2 Corrosion due to chloride ingress

The corrosion protection of the reinforcement steel can be lost if the surrounding concrete loses its alkalinity or if a critical chloride content on the steel surface is exceeded. Small amounts of chloride in the concrete can be contained in the raw materials: cement, water or concrete aggregates. They are defined as natural chloride content of the concrete. Chloride ions diffuse into the concrete or are transported by capillary absorbed water and interact with the passive film of the steel when they reach its surface. If a critical concentration of free chloride ions on the steel is exceeded, the passive film of the steel fails and corrosion can begin. This phase of chloride penetration before the depassivation of the steel is called the initiation phase. With the beginning of corrosion, the propagation phase starts. The chronological sequence of the accumulated damage to the reinforced concrete structures exposed to chlorides is shown in Figure 2 [7].

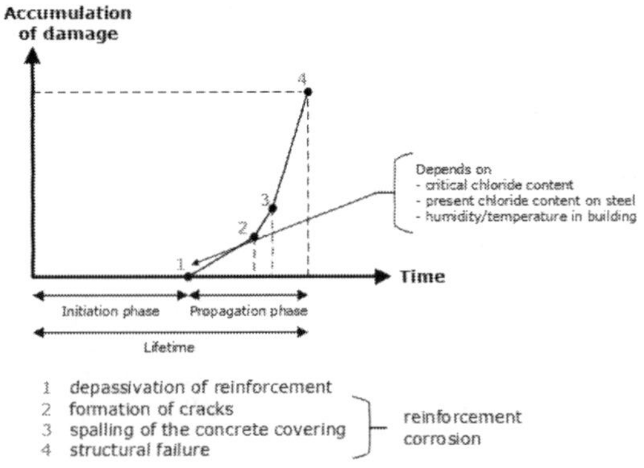

Fig. 2. Chronological sequence of accumulated damage to structures with chloride-induced corrosion [7].

2.1 Propagation Phase

A special type of corrosion is pitting corrosion. During this corrosion of steel in concrete, the positively charged iron ions (Fe^{2+}) go into solution, where the concrete pore fluid is the electrolyte. At the same time a reduction of the oxygen takes place. Under the formation of negatively charged hydroxyl ions (OH^-), the superfluous electrons at the steel/electrolyte interface react with water and oxygen, that is usually present and is dissolved in water in sufficient quantity. The formation of hydroxyl ions OH^- is called a cathodic sub-process. In reinforcing steel, but also in concrete (electrolyte), the charge balance is retained. Iron dissolution (formation of Fe^{2+}) is called an anodic sub-process. The processes described are simplified in Figure 3 below [8].

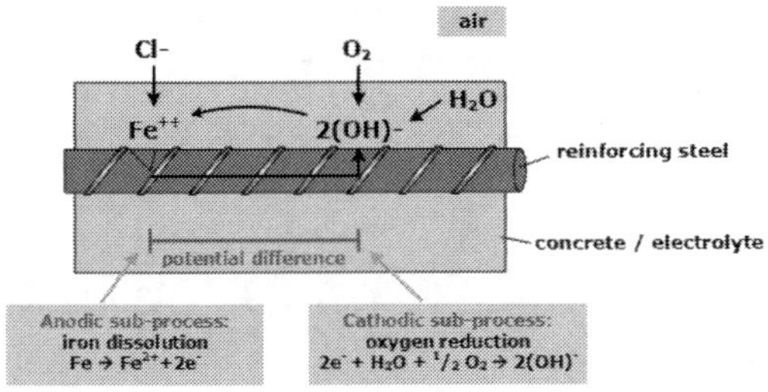

Fig. 3. Corrosion processes and elements involved [8].

Basically the corrosion process can be divided into two simultaneously running sub-processes. In the anodic sub-process, metal dissolution takes place. In this process the reinforcement steel loses two electrons to the environment and converts into Fe^{2+}.

$$Fe \rightarrow Fe^{2+} + 2\,e^- \qquad (1)$$

Oxygen is reduced in the cathodic sub-process. This absorbs the electrons emitted in the anodic reaction and reacts with the pore water to form OH^-.

$$H_2O + O_2 + 2\,e^- \rightarrow 2\,(OH)^- \qquad (2)$$

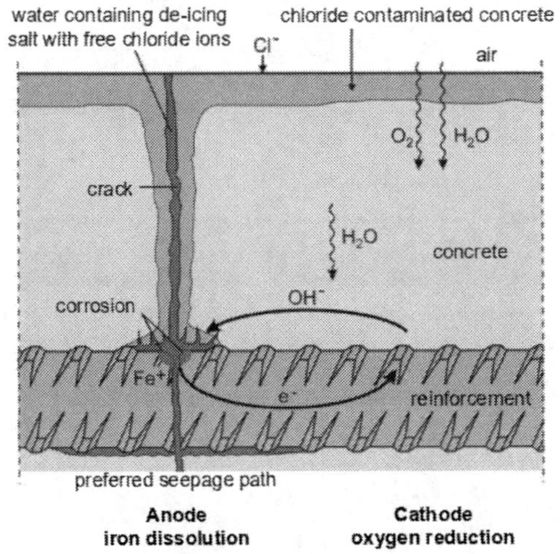

Fig. 4. Chloride-induced corrosion in the crack area [8].

2.2 Crack widths

According to ÖNORM B 1992 1-1, the crack width must be limited in order to ensure that the proper use of the structure, its appearance and durability are not affected. The standards show permissible characteristic crack widths (wk) in dependence on the exposure classes. For reinforced concrete elements exposed to exposure class XD, the crack width shall be limited to wk = 0.3 mm below the quasi-stable action combination. According to ÖNORM B 1992 1-1, if a crack width restriction is required, a minimum amount of reinforcement must be inserted in areas where tension is expected. The minimum reinforcement results from the balance of the tensile force before the incipient crack in the concrete and the tensile force after the incipient crack in the reinforcement. The actual load has no influence on the amount of steel, since the forces to be absorbed are purely a result of the tensile stresses in the concrete before the crack formation [4].

3 Experiment Setup

The aim of this laboratory experiment is to use the half-cell corrosion potential mapping to evaluate whether different crack widths have an effect on the corrosion behaviour of reinforcing steel. For this, 2 specimen series with two different exposure classes XD1 and XD3 as well as two different strength classes C30/37 and C35/45 were used. Half of the specimens were coated with epoxy resin. The cracks were artificially created by 3 point bending load. In order to determine the corrosion behaviour, the test specimens were coated with a sodium chloride solution and then observed during a period of one and a half months using half-cell corrosion potential measurement. Based on the resulting data, the behaviour of the reinforcement steel was evaluated.

3.1 Preparation of the test specimens

For the experiment, 8 coated and 8 uncoated test specimens with the dimensions 60x40x10 cm were produced and reinforced as shown in Figure 4.1 below.

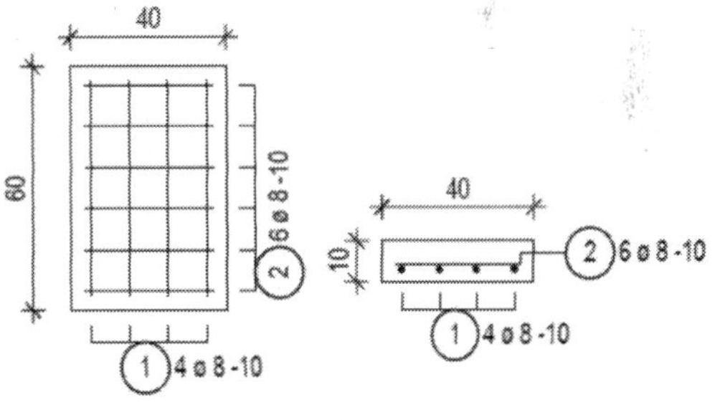

Fig. 5. Formwork and reinforcement plan of the test specimens.

The following concrete strength and exposure class was used for the production of the test specimens. Portland cement CEM I 42.5 N (270 kg) with an additive mixture consisting of slag sand, limestone and fly ash (40 kg) was used as binder.

Table 3. Concrete formulation.

class	exposure	w/c-ratio
C30/37	XD1	0.55
C35/45	XD3	0.45

Ribbed structural steel B500B was used as reinforcement for 14 test specimens and B550B for two others. The diameter was 8 mm and the bar-to-bar distance was 10cm. The concrete cover depth was 4cm. The coating used was a two-layer epoxy resin coating with a layer thickness of 0.5 mm.

3.2 Crack initiation

The cracks in the test specimen were created by a 3 point bending load. As shown in Figure 6, the concrete slabs were placed in a hydraulic testing machine and loaded in a controlled manner until the crack was created. The desired crack width was 0.1 - 0.3 mm. In order to prevent the crack from closing again, wooden wedges were attached to both ends of the crack.

Fig. 6. Crack initiation.

3.3 Salt exposure

All specimens were loaded by a central load according to the bending tensile test to initiate cracks with different crack widths. The test specimens were contaminated from 16.05.2017 to 22.06.2017. During this period, the concrete slabs were loaded with 3% sodium chloride solution. In the first 5 applications, the test specimen was loaded with sodium chloride for the rapid increase of the chloride. After that, the other treatments were only carried out with water.

3.3 Measurement setup

In the laboratory test, 16 plate-like test specimens with the dimensions 60 x 40 x 10 cm were used. The test specimens were stored under a roof for the entire duration of the measurement in an outdoor climate. To carry out the potential measurement, the reinforcement of the two test specimen groups was connected separately with a cable. As shown in Figure 8, the reference electrodes were attached to the top near the crack. The potential measurement was carried out with two different types of electrodes. On the one hand silver - silver chloride and on the other hand, copper - copper sulphate was used. This allows comparing several measurement results with each other.

Table 4. Cycle of application.

date	application
16.05.2017	100 ml NaCl solution, 3 % Cl
18.05.2017	100 ml NaCl solution, 3 % Cl
19.05.2017	100 ml NaCl solution, 3 % Cl
22.05.2017	100 ml NaCl solution, 3 % Cl
23.05.2017	100 ml NaCl solution, 3 % Cl
24.05.2017	100 ml water
29.05.2017	100 ml water
30.05.2017	100 ml water
31.05.2017	100 ml water
06.06.2017	100 ml water
13.06.2017	100 ml water
22.06.2017	100 ml water

Fig. 7. Experimental setup.

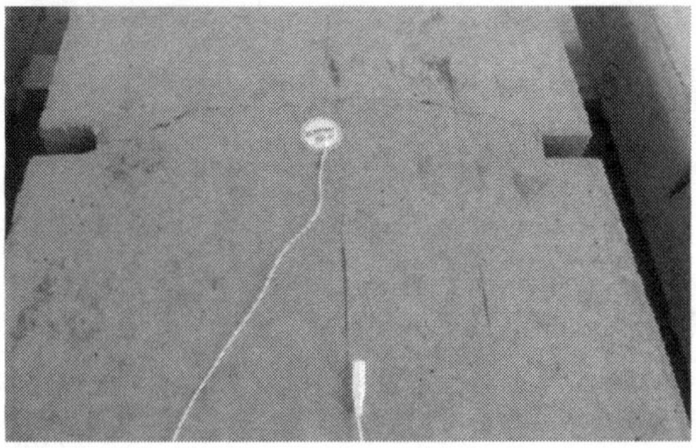

Fig. 8. Position of electrodes.

4 Results and evaluation

The electrochemical potential field measurement is a method for assessing the corrosion behaviour of the reinforcement of reinforced concrete components. By measuring the electrochemical potential difference, areas with a high corrosion probability can be detected. To determine the current potential at the steel/concrete interface of the reinforcement steel below the concrete surface, the potentials arriving at the surface are measured. The measurement is carried out by a high-impedance voltmeter connected to the reinforcement steel. If the electrode potential is determined at a standard electrode with the normal hydrogen electrode as reference, this is referred to as normal potential. The measured potential values can be interpreted as more negative potentials indicating a higher probability for corrosion processes. A difference is made here between two ranges. In the range above (more positive than) -200mV a low corrosion probability is given. In the range below (more negative than) -200mV a high corrosion probability applies [1].

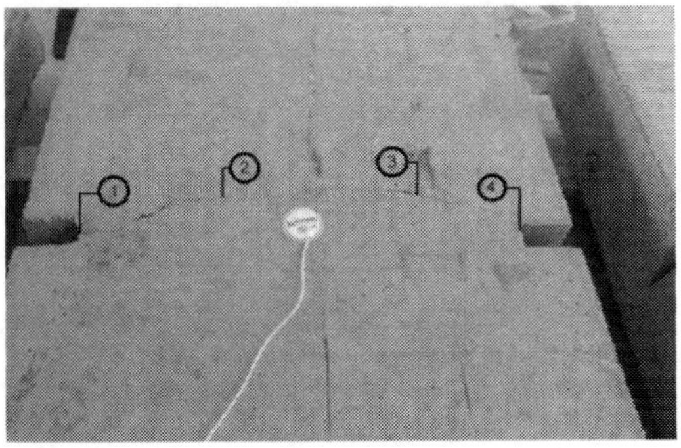

Fig. 9. Position of the measuring points

The measured crack widths are given in Table 5 below.

Table 5. Crack widths of uncoated and coated Specimens

Crack widths uncoated specimens [mm]					
	Measuring points				Mean value
Specimen No.	1	2	3	4	x
1	0,1	0,2	0,4	0,3	0,25
2	0,2	0,5	0,1	-	0,27
3	1	0,5	0,5	0,6	0,65
4	0,2	1	0,8	0,4	0,5
5	0,4	0,6	0,6	0,4	0,5
6	0,1	0,6	0,8	0,8	0,58
7	0,2	0,4	0,5	0,4	0,38
8	0,2	0,3	0,3	0,1	0,23

Crack widths coated specimens [mm]					
	Measuring points				Mean value
Specimen No.	1	2	3	4	x
1	-	0,3	0,4	-	0,35
2	0,6	1	1,2	0,6	0,85
3	0,4	0,5	0,4	0,2	0,38
4	0,5	1,4	1,2	0,6	0,93
5	1	1,4	1,4	0,7	1,13
6	0,1	0,2	0,3	-	0,20
7	0,3	0,4	0,4	0,2	0,33
8	0,1	0,2	0,2	0,1	0,15

4.1 Detection of cracks

In order to detect the crack widths as accurately as possible, these were measured at four points on the specimens, as shown in Figure 9. The results are summarized in the following diagrams.

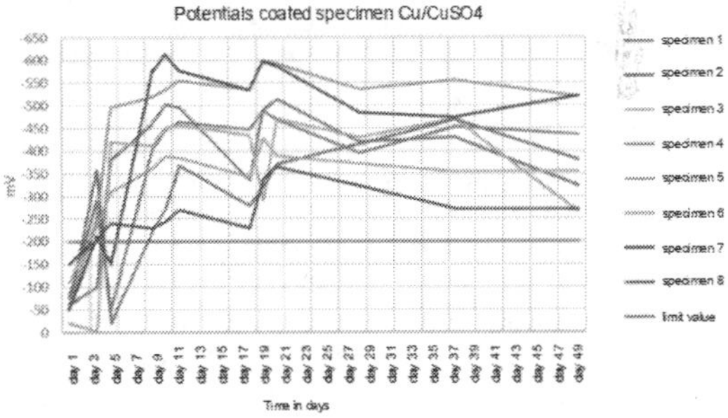

Figure 10. Potential mapping of coated samples $Cu/CuSO_4$.

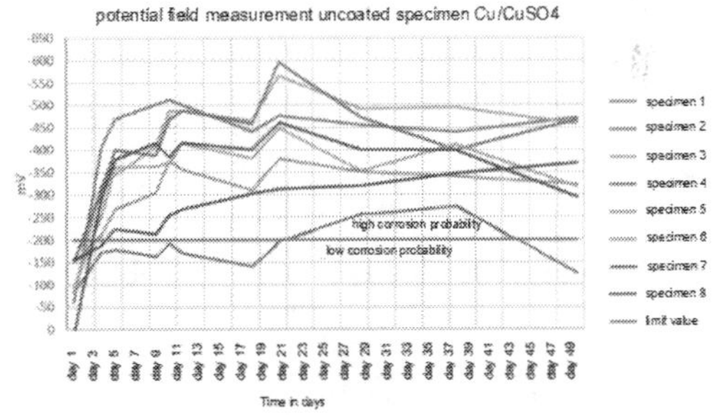

Figure 11. Potential mapping of uncoated samples $Cu/CuSO_4$.

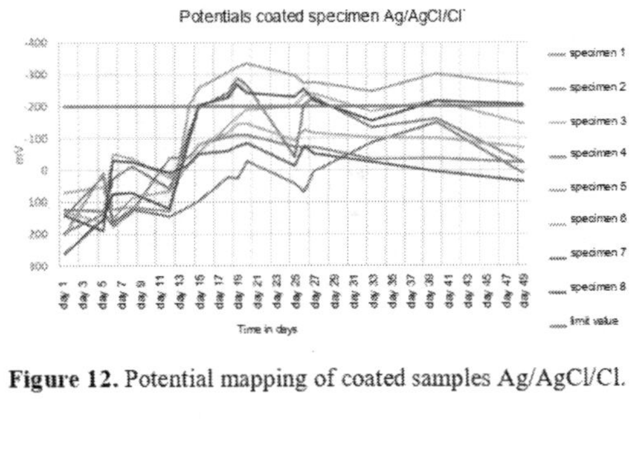

Figure 12. Potential mapping of coated samples Ag/AgCl/Cl.

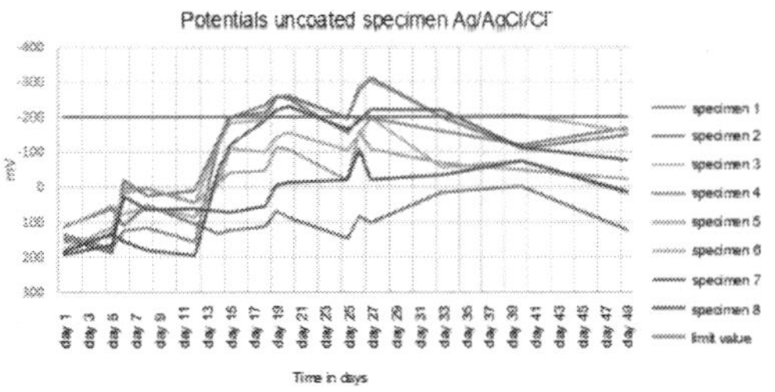

Figure 13. Potential mapping of uncoated samples Ag/AgCl/Cl.

4.3 Evaluation of the potential measurement with $Cu/CuSO_4$ electrode system

As can be seen in Figures 10 and 11, there are increased fluctuations in the corrosion potential during early application. These variations are due to the salt solution application, as the first five applications were carried out using sodium chloride. At the beginning of the measurements the coated specimens show greater variations than the uncoated specimens. However, the amplitudes of the corrosion graphs drop again right after the application. After further application, some peaks are still visible.

4.3.1 Coated Specimen

After the second application, 2 samples with a measured potential of -550 to -600 mV stand out: Test specimen 5 with a mean crack width of 1.13 mm and test specimen 8 with a mean crack width of 0.15 mm. After the fifth application, all 8 specimens were in a range below the -200mV limit. The fluctuation range of the measured potentials of the test specimens was -230mV after 28 days.

4.3.2 Uncoated Specimen

During the early applications using sodium chloride, the test specimens reacted quickly with few fluctuations. This effect is most clearly visible in sample 3, which also shows the largest average crack width of 0.65 mm. After 15 days, all samples exceeded the critical -200mV range. After 28 days the fluctuation range of the determined potentials was -300mV.

4.3.3 Comparison of the test series

In the comparison of the two test series, there should be a difference between the coated and uncoated series, as the moisture in the crack area is rather higher and cannot leak through the surface. However, no such effect could be observed. After being exposed to sodium chloride, 15 of the 16 samples of the two groups of test specimens reached the critical -200mV limit in which corrosion is very probable. After 28 days, the different crack widths did not show any significant similarity. No correlation was found between the crack widths and the corrosion potential of the individual specimens.

4.4 Evaluation of potential measurement with the Ag/AgCl/Cl electrode system

The measurement with the silver-silver chloride electrode system began 10 days before the first application of sodium chloride in order to check the functional capability of the electrode system by corresponding reference measurements. After the first application of sodium chloride a clear reaction was observed. As can be seen in Figures 12 and 13, the further applications are only weakly visible on the diagrams. This effect can be observed in both groups of the test specimens. This could be due to a contact problem of the reference electrodes. In comparison with the copper - copper sulfate electrode clear differences were determined within the potential measurement.

5 Summary

The aim of this study was to determine whether different crack widths may have an influence on the corrosion behaviour of reinforcing steel. Two different electrode systems were used: Silver - silver chloride - electrodes and copper - copper sulphate - electrodes. For the specimens. two different strength classes and two exposure classes were used. The measured potentials, Figs. 12 and 13, show that the silver -silver chloride – electrodes in this case were not suitable for this type of measurement because a contact problem of the electrodes led to irregular measurement results. Therefore, it was not possible to make a representative evaluation of the measurement results. For this reason, the results of the measured potentials of the copper - copper sulfate - electrodes were used. These measurement results show that no noticeable improvement can be achieved using an epoxy resin coating. Neither with the strength class nor with the exposure class could clear trends be observed. A correlation between the different crack widths and the corrosion rate could not be established. As a result, it is clear that different crack widths have no influence on the corrosion behaviour of the reinforcing steel.

References

1. DGZfP (Deutsche Gesellschaft für Zerstörungsfreie Prüfung), *Merkblatt B 03, Elektrochemische Potentialmessung zur Detektion von Bewehrungsstahlkorrossion*, Berlin (2014)
2. S.H. Müller, U. Nolting, M. Haist, *Symposium Beherrschung von Rissen in Beton*, Universität Karlsruhe (2010)
3. D. Bosold, A. Grünewald, *Zement-Merkblatt Betontechnik B18 "Risse in Beton"* (2014)
4. ÖNORM B 1992-1-1, Austrian Standard Institute (2018)
5. H.S. Müller, U. Nolting, M. Haist, *Symposium Dauerhafter Beton, Grundlagen, Planung und*

Ausführung bei Frost- und Taumittel-Beanspruchung, Universität Karlsruhe (2009)

6. M. Raupach, M. Kosalla, L. Wolff, L. Meyer, *Einfluss von Rissen auf die Korrosionsgefahr der Bewehrung in Parkdecks und deren Instandsetzung – Neue Erkenntnisse aus der Forschung und Praxis*, Technische Akademie Esslingen (2014)
7. M. Beck, *Zur Entwicklung der Eigenkorrosion von Stahl in Beton*, RWTH (2010)
8. C. Dauberschmidt, *Bauwerksdiagonose bei chloridbelastetem Beton- immer ein Entsorgungsfall?*, DBV-Tagung (2010)

Comparison between the predicted performance curve and the Markov Chain models for structural performance of infrastructure components

Nabil Semaan[1,*], and *Youssef* Dib[2]

[1]University Of Balamand, Civil and Environmental Engineering Department, Kelhat, Koura, Lebanon
[2]University Of Balamand, Math Department, Kelhat, Koura, Lebanon

Abstract. This paper compares the PPC model to a Markov Chain (MC) stochastic deterioration model. First, inspection data from the *Société de Transport de Montréal* (STM) is gathered and analyzed. Then Transition Probability Matrices (TPM) are developed, and, using Matlab, MC deterioration curves are developed. Comparison between MC and the PPC deterioration curves is performed for subway station walls and slabs. The comparison has shown that the useful service life can be as low as 2 years for components having many inspection history records, and very high as 30 years for components having very few inspection history records. The PPC model has always a higher useful service life estimate. Also, the MC has a ten times higher deterioration rate (0.2 per year) compared to the PPC model (0.02 per year). It can be concluded that the MC deterioration model requires a high amount of inspection data, and it is mathematically difficult to generate since most practicing managers and engineers have no background in Markov Chain modeling.

1 Introduction

Public transit contributes substantially to the quality of life enjoyed by citizens. Although continuously growing, symptoms of aging transit infrastructure and thus overdue maintenance are becoming a burden to municipalities and governments. The *American Society of Civil Engineers* (ASCE) Report Card assigned a grade D- (i.e., very poor) for the public transit system in the United States (US) in 2017. The US aging infrastructure has a 90 billion USD rehabilitation backlog in 2017, expected to grow to 122 billion USD in 2032 [1]. Hence, an urgent need of structural infrastructure management planning is required. Both assessing the structural performance of infrastructure components, and predicting the future performance of structures form the core of any infrastructure asset management.

A "Structural Performance Index" (SPI) model, developed in 2014 by Semaan and Zayed [2], was developed. The SPI model assesses both the physical condition (based on condition defects) and the functional performance (based on structural cracks) of a structural component. A "Predicted Performance Curve" (PPC), developed in 2016 by Semaan and Zayed [3], has also been developed. The PPC model evaluates the structural performance curve of a concrete component based on the SPI. The PPC utilizes a reliability-based cumulative Weibull function in order to construct the structural performance curve. The PPC can be easily developed using scarce inspection.

On the other hand, most existing infrastructure deterioration models use the Markov Chain concept. Engineers, managers and researchers are faced with difficulties when modeling subway structures components deterioration using Markov Chain (MC. These difficulties are listed as follows: (i) difficulty of data collection, (ii) difficulty in translation of physical conditions to mathematical problems, and (iii) the Markov Chain is a memory less model, i.e. the Markov Chain model cannot show the states before the studied period and hence the current state is independent of its history. Thus PPC challenges the mostly used MC model.

The objective of this research project is to study the structural performance curve of a subway component using a Markov Chain model and compare it to an existing Predicted Performance Curve (PPC) developed by Semaan [4].

2 Background

Deterioration models are very important for asset management since they can predict the future performance of an asset based on both previous and current performance. Performance, by definition, is the inverse of deterioration. The performance measured as a function of time is evaluated in order to predict the service life of the component.

Service life is the predicted life of a structure before demolishing [5], [6]. The useful service life of a project is the time where all the structures' components physical, functional and safety performance are at a minimum acceptable threshold according to international standards.

* Corresponding author: nabil.semaan@balamand.edu.lb

1.1 Deterioration modelling

Previous researchers identified three main categories of deterioration models: deterministic models, stochastic models and artificial intelligence models. Each category uses different techniques as shown in Table 1.

Table 1. Existing deterioration models and techniques.

Deterioration models	Deterioration techniques
Deterministic	Straight-line extrapolation
	Regression
	Curve fitting
Stochastic	Markov Chain
	Reliability
	Failure rate function
Artificial intelligence	Artificial neural networks
	Case-based reasoning

1.1.1 Markov chain models

The discrete time Markov Chain (MC) technique studies the long-term behavior of two sets (space-time) that are event-independent as in the case of deterioration. The Markov Chain model is the most common stochastic model used to model infrastructures' deterioration or performance. Some widely known research that has used MC for different infrastructure application is given in Table 2.

Table 2. Existing research using Markov Chain models

Research	Research topic
Butt *et al.*, 1987 [7]	Sewer deterioration
Wirahadikusumah, 1999 [8]	Sewer deterioration
Yang *et al.*, 2005 [9]	Pavement deterioration
Baik *et al.*, 2006 [10]	Pavement deterioration
Hong and Protzi, 2006 [11]	Pavement deterioration
Morcous, 2006 [12]	Bridge deterioration
Ortiz-Garcia *et al.*, 2006 [13]	Pavement deterioration

The Markov Chain model translates the deterioration of an infrastructure component to a stochastic probabilistic matrix that involves time. This model collects the deterioration data from the inspection of the components over time.

In the process of understanding the MC model, S is defined as the set of states, $S = \{S_1, S_2, \ldots\}$. Generally, researchers consider five (5) states, $S = \{S_1, S_2, S_3, S_4, S_5\}$.

The MC starts by transiting from one state S_i to another state S_j in a discrete manner. The Markov Chain defines '$p_{j,i}$' as the probability of transition from state S_i to state S_j. While, for a component that remains in the same state after a certain time interval 't', this probability is '$p_{i,i}$'. The probabilities for all possible transitions are gathered in one matrix called the stochastic Transition Probability Matrix (TPM).

Although the Markov Chain model is commonly used to study the performance and predict the behavior of structures with time to determine their service life, some drawbacks shall be considered. Table 3 presents the remarks on the Markov Chain principle by some researchers.

Table 3. Researchers' remarks on Markov Chain models

Research	Remarks
Sianipar and Adams, 1997 [14] Van Noortwijk and Frangopol, 2004 [15]	It is difficult to consider the interactions between components
Madanat *et al.*, 1997 [16] Destephano and Grivas, 1998 [17]	It assumes that each state is independent
Madanat and Ibrahim, 1995 [18]	It does not predict the condition improvement after repair
Tokdemir *et al.*, 2000 [19]	The transition probabilities are estimated in terms of subjective engineering judgment and require updating when new data is obtained

Hence, Markov Chain models, although highly used, are data hungry, requiring a very large amount of data in order to develop the TPM. In the case of subway network infrastructure components, collecting inspection data is to be done only when the station is closed after midnight. Thus, collecting a very large sample for subway stations requires more inspection time on one hand, and frequent inspections on the other, which incurs a high cost.

Also, most practicing managers and engineers have no background in Markov chain mathematical modeling, making it difficult to implement it, especially, again to subway networks.

Finally, the Markov Chain method cannot show the states before the studied period, so that the current state is independent of its history. Thus it is a memory-less technique.

1.1.2 The Predicted Performance Curve (PPC) model

Semaan and Zayed (2016) [3] developed the Predicted Performance Curve (PPC). The PPC develops a performance (or deterioration) curve of the structural components of concrete infrastructure. The PPC uses the Weibull distribution cumulative function in order to develop a structural performance curve. In order to construct the PPC, three steps are needed: i) develop an Ideal Performance Curve (IPC) using a Weibull distribution cumulative function; ii) introduce to the Ideal Performance Curve (IPC) at least one set of inspection data in order to develop an Updated Performance Curve (UPC); iii) after the last inspection data is introduced the PPC is obtained.

The main objective of a Predicted Performance Curve (PPC) model is that it predicts the structural deterioration of concrete components and their useful estimated service life.

The Predicted Performance Curve (PPC) model is based on a structural component assessment index model which was developed by Semaan and Zayed (2014) [2]. Their model combines a physical condition (based on condition defects) performance index, and a functional index. Then, this model combines these two indices into one - the "integrated performance index" P_I. Semaan and Zayed (2014) [2] stressed that although P_I is applied to subway components, it can generically be used for any infrastructure concrete component. The model assesses the components' performance based on the integration of the physical performance with the functional one. Hence, an integrated performance index could be used for the assessment of any infrastructure system.

The Ideal Performance Curve is defined in Equation 1.

$$P_I(t)^{IPC} = 1 \cdot e^{\left[\frac{\ln(0.2)t^3}{100^3}\right]} \quad (1)$$

Where: *P_I is Performance index*
IPC is Ideal Performance Curve
t is time

When t equals zero, the equation gives a P_I value of 1.0 which is the maximum, and it is usually achieved when the element is first constructed. When 't' equals 100 years, P_I is ideally estimated at 0.2 which means that the components inspected are at their minimum performance. The model also assumes that at the time where P_I has a value of 0.4, the useful service life or the performance threshold has been reached.

The curve representing the IPC is illustrated in Figure 1.

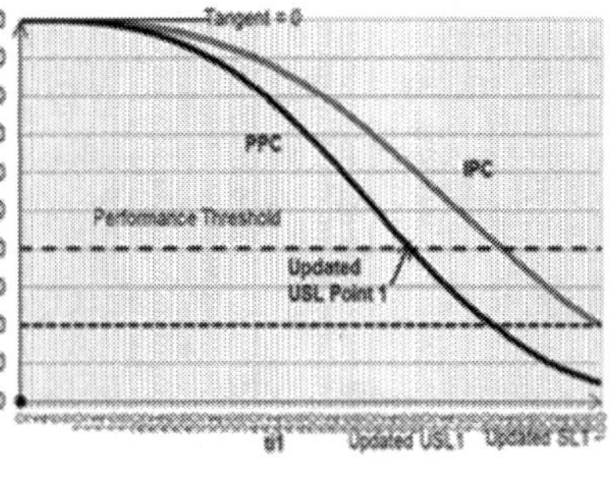

Fig. 1. The Ideal Performance Curve (IPC).

The Predicted Performance Curve uses Equation 2, as follows:

$$P_I(t)^{PPC} = 1 \cdot e^{\ln(P_{I1})\left(\frac{t}{t_1}\right)^3} \quad (2)$$

This curve is an update of the ideal performance curve after each inspection has been made. When each value of a P_I is calculated, it is plotted in the graph and then a new curve is drawn passing through that point. Figure 2 shows an example of such a curve.

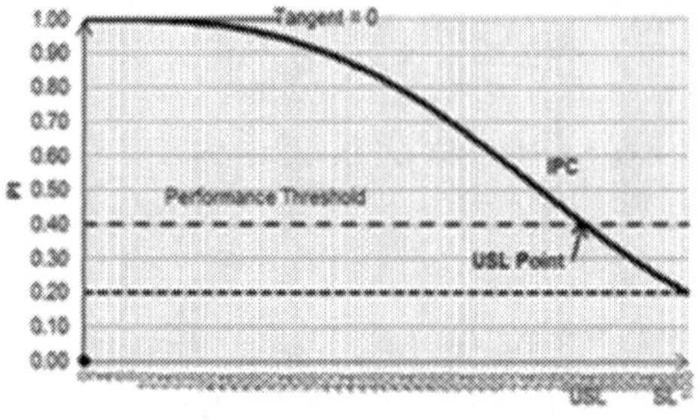

Fig. 2. The Predicted Performance Curve (PPC).

3 Markov Chain deterioration curves of subway structure components

A deterioration curve for subway structural components has also been developed using a stochastic Markov Chain model entitled the 'Subway Structures Markov Chain Deterioration Model' (SSMCD). Figure 3 illustrates the SSMCD Model methodology.

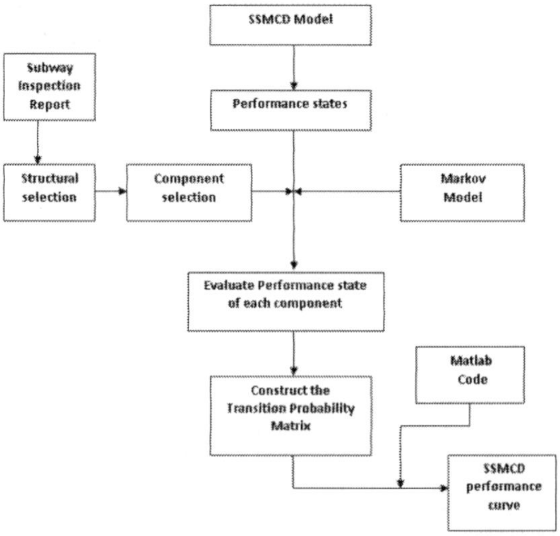

Fig. 3. The Subway Structure Markov Chain Deterioration (SSMCD) model flowchart

A Markov Chain considers discrete deterioration states. A deterioration state can be identified only in the following states and cannot have a condition in between states. Table 3 illustrates the SSMCD deterioration states in relation to the Integrated Performance Index (P_I) and its respective performance description.

Table 3. Definitions of the state's deterioration

Deterioration states	Integrated Performance Index (P_I)	Description of deterioration states
State 1 (S_1)	1.0	Very good condition Very low deterioration
State 2 (S_2)	0.8	Good condition Low deterioration
State 3 (S_3)	0.6	Average condition Average deterioration
State 4 (S_4)	0.4	Low condition High deterioration
State 5 (S_5)	0.2	Very low condition Very high deterioration

Table 3 shows that a P_I of 0.4 means a high deterioration, meaning public security is at risk. Hence structural components having a P_I less than 0.4 are not meeting their functional and physical performance; i.e. are in urgent need of rehabilitation. Moreover, if maintenance and rehabilitation were not done at these structural components, deterioration could be as low as 0.2. In this case the component is to be replaced completely and/or reconstructed. Managers should not let this case happen at all.

The first step in the Transition Probability Matrix (TPM) development is to divide the data into discrete time intervals: from year of construction to the first inspection period, and then from first inspection to second inspection, and so on. Next, in each time interval, analyze the transition of states from state (S_i) to state (S_j), where state (S_i) is an initial state at the time of previous inspection, or at the year of construction (S_1). While, state (S_j) is the condition state an element reached after a certain time interval '*t*'. The possible state transition is illustrated in Figure 4. The arrow represents a one direction transition from a state to another, since no structural improvement is considered.

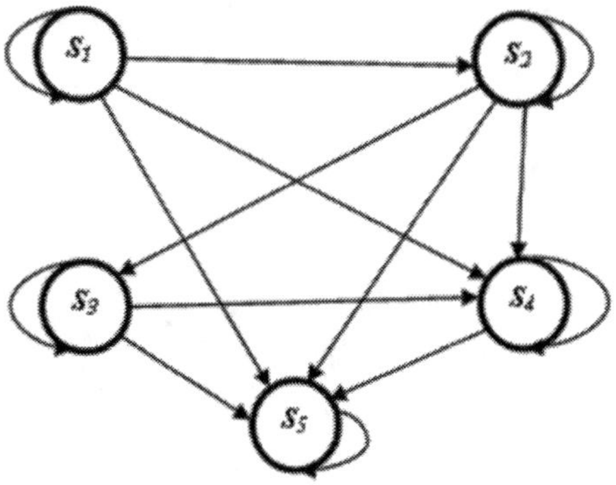

Fig. 4. Possible Deterioration State Transition

A probability of occurrence is evaluated for each transition from state *Si* to state *Sj*, as per Equation 3.

$$p_{j \cdot i} = \frac{NE_{j \cdot i}}{\sum NE} \quad (3)$$

Where: $p_{j,i}$ *is the probability of transition from state* S_i *to state* S_j,
$NE_{j,i}$ *is the number of structural components moving from state* S_i *to state* S_j *in the time interval*
NE is the total number of structural components under consideration.,

Referring to the possible state transitions in Figure 4, and Equation 3, a typical TPM is evaluated as shown in Table 4.

Table 4. The SSMCD Typical Transition Probability Matrix

	S_1	S_2	S_3	S_4	S_5
S_1	$p_{1,1}$	0	0	0	0
S_2	$p_{2,1}$	$p_{2,2}$	0	0	0
S_3	$p_{3,1}$	$p_{3,2}$	$p_{3,3}$	0	0
S_4	$p_{4,1}$	$p_{4,2}$	$p_{4,3}$	$p_{4,4}$	0
S_5	$p_{5,1}$	$p_{5,2}$	$p_{5,3}$	$p_{5,4}$	1

Finally, using the TPM, the SSMCD uses Matlab in order to develop the element deterioration curves.

4 The SSMCD application to the *Société Transport de Montréal (STM)*

The SSMCD model has been applied to the *Société de Transport de Montréal* (STM) stations. Three stations were selected, and structural slabs and walls components were chosen. Deterioration curves using Markov Chain were developed for the both the structural slabs and wall components. Table 5 shows information of each station, mainly the name, construction year, inspection periods, and number of structural slabs and wall components.

Table 5. STM Stations Information

Station	Construction Year	Inspection Years	Number of Slabs	Number of Walls
Beaudry	1966	1992 1997 2005	44	34
Sherbrooke	1966	1997 2005	16	14
Champs-de-Mars	1966	1997 2005	8	14

Applying the SSMCD model, the TPM for the components under consideration are developed. Tables 6 to 11 show the TPM of each station components.

Table 6. TPM of Beaudry Station Slab Components

	S_1	S_2	S_3	S_4	S_5
S_1	20/44	0	0	0	0
S_2	22/44	20/44	0	0	0
S_3	1/44	22/44	20/44	0	0
S_4	1/44	1/44	22/44	22/44	0
S_5	0	1/44	2/44	22/44	1

Table 7. TPM of Beaudry Station Wall Components

	S_1	S_2	S_3	S_4	S_5
S_1	16/34	0	0	0	0
S_2	18/34	16/34	0	0	0
S_3	0	18/34	16/34	0	0
S_4	0	0	18/34	16/34	0
S_5	0	0	0	18/34	1

Table 8. TPM of Sherbrooke Station Slab Components

	S_1	S_2	S_3	S_4	S_5
S_1	8/16	0	0	0	0
S_2	1/16	8/16	0	0	0
S_3	7/16	1/16	8/16	0	0
S_4	0	7/16	1/16	8/16	0
S_5	0	0	7/16	8/16	1

Table 9. TPM of Sherbrooke Station Wall Components

	S_1	S_2	S_3	S_4	S_5
S_1	12/14	0	0	0	0
S_2	0	14/14	0	0	0
S_3	2/14	0	14/14	0	0
S_4	0	2/14	0	12/14	0
S_5	0	0	2/14	2/14	1

Table 10. TPM of Champs-de-Mars Station Slab Components

	S_1	S_2	S_3	S_4	S_5
S_1	3/8	0	0	0	0
S_2	3/8	3/8	0	0	0
S_3	2/8	3/8	3/8	0	0
S_4	0	2/8	3/8	3/8	0
S_5	0	0	2/8	5/8	1

Table 11. TPM of Champs-de-Mars Station Wall Components

	S_1	S_2	S_3	S_4	S_5
S_1	8/14	0	0	0	0
S_2	3/14	8/14	0	0	0
S_3	3/14	3/14	8/14	0	0
S_4	0	3/14	3/14	8/14	0
S_5	0	0	3/14	6/14	1

5 SSMCD versus PPC deterioration curves

The deterioration curves using SSMCD developed for the structural slab and wall components for each station, are compared to the deterioration curves developed using PPC. Both deterioration curves are plotted on the same Figure for comparison purposes. It should be noted that the developed PPC deterioration curves are developed for each structural component separately. Thus, the figure shows PPC curves for external slabs (SE), internal slabs (SI), external walls (WE), and internal walls (WI) of each level (i.e. SE0, SE1, SE2, etc…).

Figures 5 to 10 show the MC and PPC deterioration curves of the subway stations structural components of the three stations.

The dotted lines illustrate the PPC curve, and the bold full line the SSMCD curve.

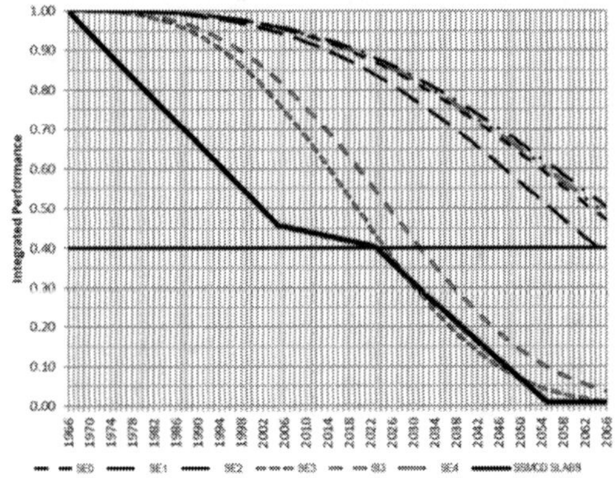

Fig. 5. Beaudry Station Slabs SSMCD vs PPC Deterioration Curves

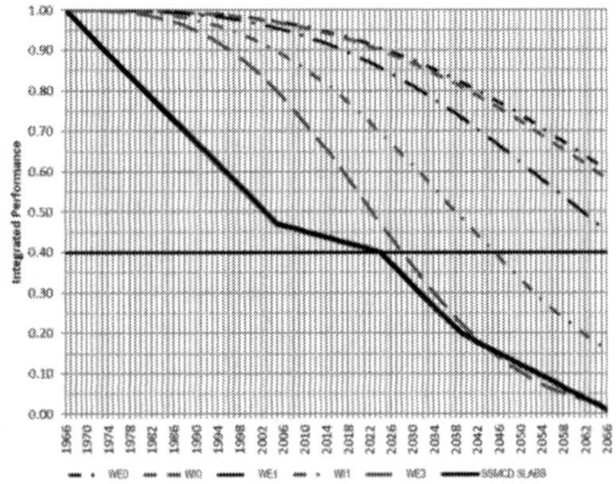

Fig. 6. Beaudry Station Walls SSMCD vs PPC Deterioration Curves

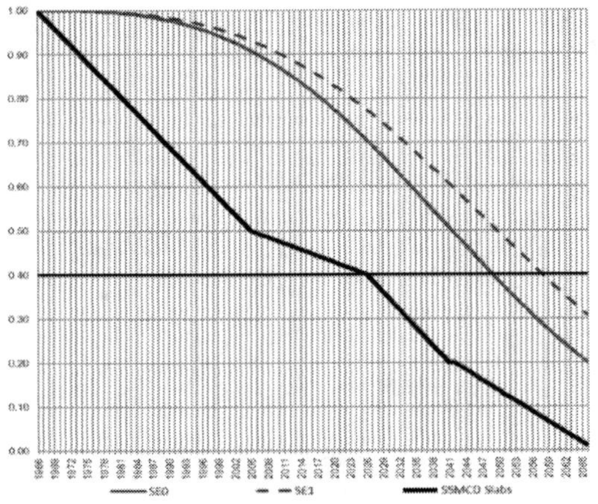

Fig. 7. Sherbrooke Station Slabs SSMCD vs PPC Deterioration Curves

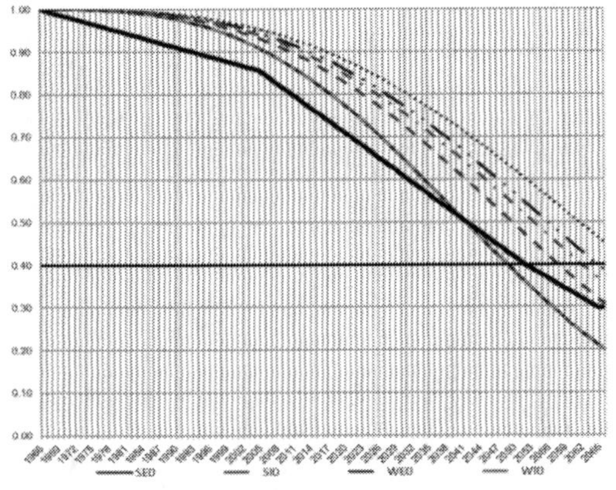

Fig. 8. Sherbrooke Station Walls SSMCD vs PPC Deterioration Curves

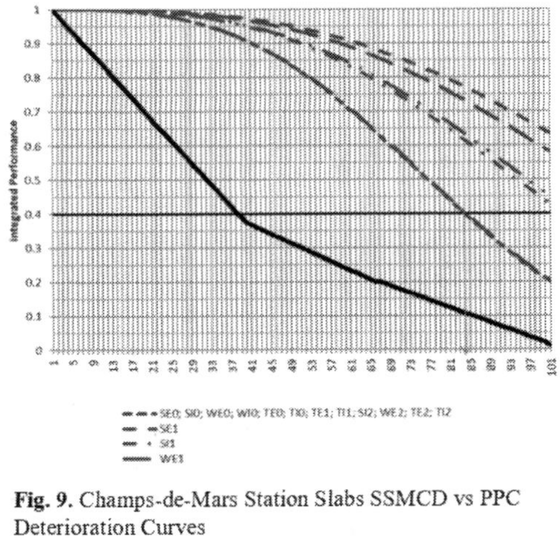

Fig. 9. Champs-de-Mars Station Slabs SSMCD vs PPC Deterioration Curves

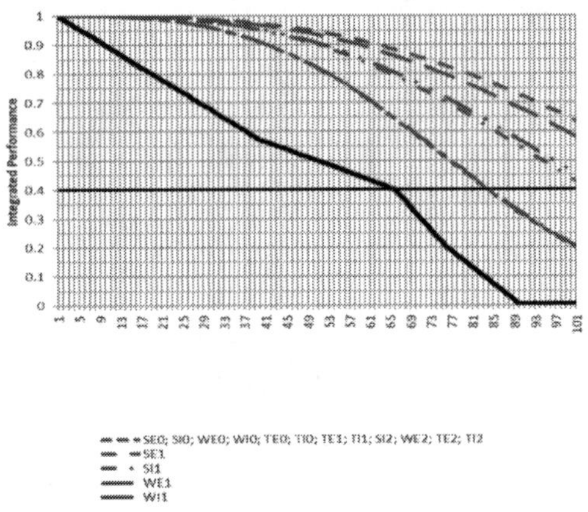

Fig. 10. Champs-de-Mars Station Walls SSMCD vs PPC Deterioration Curves

The comparison between the SSMCD and PPC deterioration curves is analyzed for each station and type of structural component, and summarized in Tables 12 to 17. Three parameters are compared and analyzed: (i) the Service Life which is the year when the element reaches a performance index of 0.2, i.e. highly deteriorated, (ii) the Useful Service Life which is the year when the element reaches a performance index of 0.4, i.e. not highly deteriorated but not safe to public, and (iii) the deterioration rate which is the change of the performance index per unit of time.

Table 12. Beaudry Station Slabs SSMCD vs PPC Comparison

Model	Service Life	Useful Service Life	Deterioration Rate
SSMCD	2039	2023	0.2 per year
PPC	2037	2025	0.03 per year

Table 13. Beaudry Station Walls SSMCD vs PPC Comparison

Model	Service Life	Useful Service Life	Deterioration Rate
SSMCD	2039	2024	0.2 per year
PPC	2040	2028	0.03 per year

Table 14. Sherbrooke Station Slabs SSMCD vs PPC Comparison

Model	Service Life	Useful Service Life	Deterioration Rate
SSMCD	2042	2026	0.2 per year
PPC	2066	2049	0.02 per year

Table 15. Sherbrooke Station Walls SSMCD vs PPC Comparison

Model	Service Life	Useful Service Life	Deterioration Rate
SSMCD	-	2052	0.04 per year
PPC	-	2049	0.02 per year

Table 16. Champs-de-Mars Station Slabs SSMCD vs PPC Comparison

Model	Service Life	Useful Service Life	Deterioration Rate
SSMCD	2031	2020	0.28 per year
PPC	-	2050	0.02 per year

Table 17. Champs-de-Mars Station Walls SSMCD vs PPC Comparison

Model	Service Life	Useful Service Life	Deterioration Rate
SSMCD	2051	2031	0.16 per year
PPC	2066	2049	0.02 per year

Analyzing the Useful Service Life prediction, it can be noticed than a difference of 2 years (between SSMCD and PPC) exist for components having many historical inspection records. While, the USL can reach a difference of 30 years for components having very few (to no) historical inspection records. This fact proves that the SSMCD deterioration curve does not accurately predict the USL when few data are available. It can be also observed that the PPC predicts a further USL than SSMCD.

On the other hand, the deterioration rate of PPC is around ten (10) less than that of SSMCD. This observation confirms the fact that the discrete MC model for few inspections data is more conservative.

Also, although the SSMCD in Figures 5 to 10 connects the line to the construction time ($t=0$), in reality, this line does not exist, due to the independency of past states to the current states.

6 Conclusions and recommendations

A Markov Chain deterioration model for the structural components of infrastructure has been developed, entitled the Subway Station Markov Chain Model (SSMCD). The SSMCD model has been applied to the *Société de Transport de Montréal* (STM) stations. Then the SSMCD is compared to the developed PPC deterioration model for subway networks.

Comparing both deterioration models, it is observed that the PPC is more accurate, requires less historical inspection records, is easier to develop, relates historical deteriorated states to current ones and predicts service life easily.

The Markov Chain model was found mathematically complicated to apply, and less accurate for small historical inspection records.

Thus, the PPC is highly recommended as a new deterioration model for infrastructure components.

References

1. ASCE, www.asce.org/reportcard/2017, (2017)
2. N. Semaan, T. Zayed, Int. Jour. Sci. Env. And Tech., **3**, 6, pp. 1943–1968, (2014)
3. N. Semaan, T. Zayed, *Proceedings of Concrete Solutions, 6th Int. Conf. on Conc. Rep.,* Taylor & Francis, pp. 429-436, (2016)
4. N. Semaan, *Structural Performance Model for Subway Networks*, Concordia University Ph.D. Thesis, (2011)
5. A. Sarja, E. Vesikari, *Durability Design of Concrete Structures*, RILEM report 14, Technical Research Center of Finland, E&FNSpon, (1996)
6. W. R. Hudson, R. Hass, W. Uddin, *Infrastructure Management*, McGraw Hill, NY, USA, (1997)
7. A. A. Butt, M. Y. Shahin, K. J. Feighan, S. H. Carpenter, Jour. Trans. Res. Boa., 1123, pp. 12-19, (1987)
8. R. Wirahadikusumah, Jour. Per. Cons. Fac., **23**, 1, pp. 47-54, (2009)
9. J. Yang, M. Gunaratne, J. J. Lu, B. Dietrich, Jour. Trans. Eng., **131**, 11, 99. 861-872, (2005)
10. H. Baik, H. S. Jeong, D. Abraham, Jour. Wat. Res. Plan. Man., **23**, 1, pp. 15-24, (2006)
11. F. Hong, J. Prozzi, Jour. Inf. Man., **12**, 2, pp. 77-86, (2006)
12. G. Mourcos, Jour. Per. Cons. Fac., **20**, 2, pp. 146-155, (2006)
13. J. Ortiz-Garcia, S. B. Costello, M. S. Snaith, Jour. Trans. Eng., **132**, 2, pp. 141-161, (2006)
14. P. M. M. Sianipar, T. M. Adams, Jour. Inf. Sys., **3**, 3, pp. 103-110, (1997)
15. J. M. Van Noortwijk, D. M. Frangopol, Prob. Eng. Mech., **19**, 4, pp. 345-359, (2004)
16. S. Madanat, M. Karlaftis, P. S. McCarthy, Jour. Inf. Sys., **1**, 3, pp.4-9, (1997)
17. P. D. DeStephano, D. A. Grivas, Jour. Inf. Sys., **4**, 2, pp. 56-62, (1998)
18. S. Madanat, W. H. W. Ibrahim, Jour. Transp. Eng., **121**, 3, pp. 267-272, (1995)
19. O. B. Tokdemir, C. Ayvalik, J. Mahammadi, *Poc. 17th Int. Symp. on Auto and Rob. in Cons.*, pp. 1091-1098.

The influence of SAPs on chloride ingress in cracked concrete

Tim Van Mullem[1], Robby Caspeele[1], and Nele De Belie[1,*]

[1]Magnel Laboratory for Concrete Research, Department of Structural Engineering, Faculty of Engineering and Architecture, Ghent University, Technologiepark Zwijnaarde 60, B-9052 Gent, Belgium

Abstract. Super Absorbent Polymers (SAPs) have proven to be effective as a self-healing agent for regaining the liquid tightness of cracked concrete. This is due to their large swelling capacity which allows them to (partially) block cracks which are in contact with water or moisture. Additionally, they are able to release this water when the climate becomes drier, thereby promoting the autogenous healing capacity of the concrete matrix. The effect SAPs have on chloride migration into cracked concrete is still unknown. The swelling capacity of the SAPs might partially block the crack, but this does not necessarily mean that the chloride ingress into the crack is lower. Especially, since the porosity of concrete with SAPs is slightly higher when additional water is added to compensate for the loss in workability. This paper compares the chloride ingress in cracked mortar with and without SAPs. The specimens were saturated in a chloride solution during 1 or 5 weeks after which the chloride ingress could be visualised using silver nitrate. The specimens which healed prior to chloride saturation had a significantly lower chloride ingress. The SAPs were able to delay the chloride ingress, as well as limit the influence of the crack on the chloride ingress.

1 Introduction

Concrete is a brittle material with a low tensile strength. As a result, it is prone to cracking. Cracks can be the result of the structure being subjected to shear, bending, tension or torsion resulting from either direct loading or restrained or imposed deformations. Additionally, cracks can also be formed due to e.g. plastic shrinkage and expansive action of corroding steel.

The formed cracks are a preferential pathway for the ingress of harmful substances, especially chloride ions. Cracks often cross the structural tensile reinforcement, and in consequence chloride ions can reach the steel reinforcement directly, immediately starting chloride induced corrosion.

In order to prevent an untimely failure of the structure, the cracks have to be repaired. Yet, manual repair of cracks is costly as a result of: specific repair products, skilled labour and economic loss due to downtime of the structure. Additionally, the locations where repair is needed are often not easily accessible. Self-healing concrete is a possible solution which can save money over the lifetime of a structure. When the concrete cracks, the concrete will repair itself without the need for manual repair.

A lot of research has already been performed on the influence of cracks on chloride ingress [1, 2]. However, the influence of self-healing on chloride ingress is scarcely reported in the literature. Most data can be found for encapsulated polymers: Maes et al. [3], Van den Heede et al. [4], and Van Belleghem et al. [5] investigated the effect of adding glass capsules filled with polyurethane on chloride resistance. Erşan et al. [6] studied the influence of nitrite producing bacteria on chloride ingress indirectly by investigating reinforcement bar corrosion.

Superabsorbent polymers (SAPs) are another type of healing agent used in self-healing concrete. [7, 8]. As they come into contact with water they swell and immediately seal the crack. The water that the SAPs absorb can be released over time during dry conditions, thereby stimulating the intrinsic autogenous healing capacity [9]. Up until now, their influence on chloride resistance has only been studied for uncracked concrete [10, 11]. If no additional mixing water is added, addition of SAPs can result in a lower chloride ingress.

This research investigates the influences of SAPs on the transport of chloride ions in cracked mortar. Prior to the transport tests, the ability of the SAPs to seal and heal cracks was investigated by a water permeability test. The variability on these results is often large, partly due to the variability of the crack width between different specimens. To reduce the variability of the crack width, cracked specimens were manufactured using an active crack width control technique [12]. For chloride ingress, a distinction is made between specimens which were cracked and healed prior to chloride diffusion and specimens which were cracked and were not healed. After a saturation period of 1 or 5 weeks, cracked and uncracked specimens were sawn and the chloride ingress was visualised by spraying silver nitrate $AgNO_3$ on the sawn surface.

* Corresponding author: Nele.DeBelie@UGent.be

2 Materials

2.1 Super Absorbent Polymers

The type of SAP which was used in this study was FlosetTM 27 CC. It is commercially available from the company Floerger SNF. This type of SAP is a cross-linked copolymer of acrylamide and acrylate. The particle size was below 600 µm.

2.2 Mortar composition

For each test two different mortars were made: a reference mortar (REF) and a mortar containing SAPs (SAP).

2.2.1 Water permeability

To test the sealing efficiency, standard mortar prisms with dimensions of 40 mm by 40 mm by 160 mm were made. The mortar was made with a cement type CEM I 42.5 N and a standardised sand 0/2. The sand to cement ratio was equal to 3 and the water to cement ratio was equal to 0.5. The mixing procedure of EN 196-1 was followed.

For the mortar containing SAP, 1% of SAP was added with respect to the weight of cement. To obtain a comparable workability between the REF and SAP series, 20 g of water was added per gram of SAP. The mixing procedure was identical to the one of the REF series, except for 30s of dry mixing the cement, the sand, and the SAPs.

2.2.2 Chloride ingress

For testing the chloride ingress, cylinders with a diameter of 100 mm and a height of 200 mm were cast. The mortar mix composition for the specimens used to test the chloride ingress was similar to the mortar mix composition for the specimens used to test the sealing efficiency. The sand to cement ratio was equal to 3 and the water to cement ratio was equal to 0.5. The cement which was used was a CEM I 52.5 N. Due to the larger dimensions of the specimens sand 0/4 was used. The dry components were mixed for 1 min, then water was added and the mixture was mixed for another 2 min.

For the mortar containing SAP, 1% of SAP was added with respect to the weight of cement. 24 g of water was added per gram of SAP to obtain a comparable workability between the REF and SAP series. The SAP particles were mixed together with the dry components. The rest of the mixing procedure was identical to the REF mixture.

3 Methods

3.1 Water permeability

The water permeability of cracked specimens is usually measured before and after healing. The accuracy of permeability tests is strongly influenced by variations of the crack width. In order to have the highest accuracy possible, an active crack width control technique was used to manufacture cracked specimens exhibiting a low variability of their crack width. Subsequently, a water flow test was used to test the permeability. This method of controlling the crack width actively before performing a water flow test has already been used previously [12, 13]. From the results of the water flow test it was possible to calculate different sealing efficiencies.

3.1.1 Active crack width control technique

Unreinforced mortar specimens (40x40x160 mm^3) were made with a cast-in hole (ø 5 mm) positioned with its centre 1.5 cm from the bottom side. After curing, one side of the hole of the specimens was slightly enlarged and provided with a tube in order to connect the specimen to the water flow setup. The other side of the hole was sealed using silicone. Additionally, a Carbon Fibre Reinforced Polymer (CFRP) laminate (PC CARBOCOMP UNI from the company TRADECC) was glued on the top side of the specimens using epoxy resin. At an age of 28 days the specimens were cracked in a three-point bending setup. The CFRP laminate at the top did not act as reinforcement but guaranteed that the two halves of the specimens remained connected upon crack formation. However, due to the lack of tensile reinforcement, the crack between the two halves of the specimens was too big. To reduce the crack width a screw jack was mounted on the bottom of the specimens which allowed the application of a force to push both halves of the specimens back together. By iteratively applying more force and measuring the crack width, the desired crack width of 150 µm at the bottom side could be reached with a good accuracy in each specimen. Fig. 1 shows a specimen for which the crack width was restrained using the described active crack width control technique.

Fig. 1. Specimen for which the crack width was restrained using an active crack width control technique.

3.1.2 Water flow test

After cracking and subsequent restraining of the crack widths, the crack at the sides of the specimens was sealed, thus only leaving the crack at the bottom side unsealed. The specimens were then saturated for 2 days in demineralised water. After the saturation period, the tube of the specimens was connected to the test equipment, see Fig. 2. Via this tube, a constant water head of 50±2 cm was applied on the specimens. Water leaking out of the specimens through their crack was captured in a container which was mounted on an electronic balance which recorded the weight in function of time. The weight was recorded for a minimum of 5 min out of which an average water flow Q could be determined in g/min.

After the initial water flow test the specimens were placed for 14 days in wet/dry cycles (12 h wet/ 12 h dry). They were then saturated for 2 days and tested again in the water flow setup. Finally, they were placed back in the wet-dry cycles for 14 days and after 2 days of saturation they were assessed in another water flow test.

After this water flow test, the specimens were tested at higher water pressures. In order to do this the connection to the open reservoir at a height of 50 cm was replaced by a connection to a pressure vessel. For 2 min the specimens were tested at a pressure of 1 bar, after which they were tested for 2 min at 2 bar. Finally, the specimens were again tested at a pressure of 50 cm for 5 min.

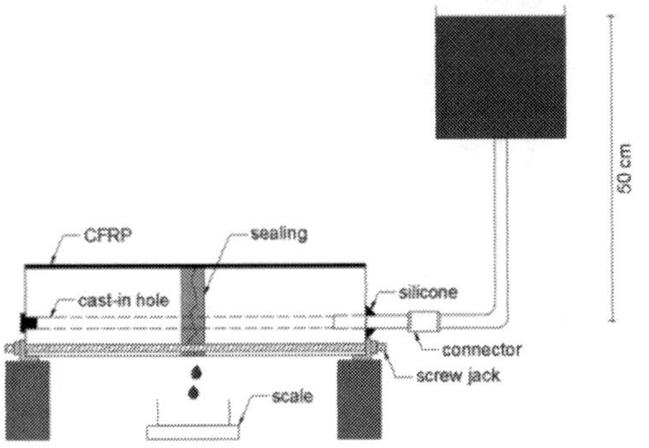

Fig. 2. Water flow setup (adapted from [12]).

3.2 Chloride ingress

The chloride ingress was tested on both REF and SAP mixtures. Half of them underwent prior healing before coming into contact with the chloride solution. So, in total, 4 series were tested: REF (not healed), SAP (not healed), REF (healed), and SAP (healed).

3.2.1 Brazilian splitting

Mortar specimens with a diameter of 100 mm and a height of 200 mm were sawn so that 3 disc specimens with a diameter of 100 mm and a height of 50 mm were obtained (the top and bottom part of the cylinders were thrown away). At an age of 28 days, the discs were completely cracked in a Brazilian splitting test. Subsequently, the two halves of a specimen were tied back together using metal straps. To create a specific crack width silicon sheets with a thickness of 100 µm were placed at the outer ends. These silicon sheets acted as spacers. In this way the crack length was reduced from a nominal length of 100 mm to a nominal length of 70 mm.

After this, the lateral surface and one of the circular surfaces was coated. This way, only one circular surface remained free of coating.

Fig. 3. Disc specimen (diameter of 100 mm and a height of 50 mm) tied back together using metal straps with silicon sheets in the crack as spacers. The silicon sheets protruding out of the crack are highlighted by the red ellipses.

3.2.2 Chloride saturation

The 2 series which underwent healing prior to chloride saturation (REF (healed), and SAP (healed)) were placed for 28 days in wet-dry cycles (12 h wet/ 12 h dry), similar to the total healing period of the specimens used to test the sealing efficiency, see section 3.1.2. To obtain a similar saturation degree between the healed specimens which were placed in wet-dry cycles and the unhealed specimens, all specimens were saturated for 2 days in water prior to being saturated in the chloride solution.

The chloride solution had a NaCl concentration of 150 g/l. Each series was placed in a separate container. The ratio between the total exposed area of all specimens in a container in cm^2 and the volume of exposure liquid in dm^3 was equal to 60. Every week the solution was stirred. The temperature throughout the test was equal to 20°C.

After 7 days (8 days for SAP (healed)) and 35 days (36 days for REF (not healed)) 3 cracked and 3 uncracked specimens were removed from the chloride solution to determine the chloride ingress. For each series also 3 uncracked specimens were removed after 14 days (15 days for REF (not healed)).

3.2.3 Chloride ingress by colour change boundary

To determine the chloride ingress, the disc specimens were sawn perpendicular to their crack. The sawing was executed dry in order not to flush out any of the chlorides. After sawing, one sawn half was sprayed with $AgNO_3$ (concentration of 0.1 mol/l). A line was drawn at the colour change. The chloride ingress perpendicular to the exposed surface was measured every 5 mm starting from the crack.

3.3 Optical microscopy

An optical crack microscope (Leica S8APO mounted with a DFC295 camera) was used to measure the crack width of both the specimens used for testing the sealing efficiency as well as the specimens used for testing the chloride ingress. For each crack three locations were chosen which were representative for the studied crack. For each location, 4 to 5 measurements of the crack width were performed. The crack widths reported in this paper are the average of all the measurements done at the three locations.

4 Results and discussion

4.1 Sealing efficiency

4.1.1 Crack width measurement

After cracking and subsequent active crack width control, the crack width was recorded for all prismatic specimens. Table 1 gives the mean crack width, as well as the standard deviation, of each specimen. The mean crack width μ of the REF series is nearly identical to the mean crack width of the SAP series (154 μm for REF and 153 μm for SAP). This similarity is a result of the active crack width control technique. This also explains why the difference between the maximum and the minimum crack width is only 10 μm for the REF series and 23 μm for the SAP series. In the literature, far wider acceptable crack width ranges have been reported e.g. 40 mm [14] and 50 mm [15] (both for crack-width-controlled 3-point bending). While still acceptable, it should be noted that the standard deviation σ, and subsequently the coefficient of variance COV of the SAP series is approximately 2 times larger than for the REF series.

Table 1. Mean crack width w (μm) and standard deviation σ (μm) of 5 REF specimens and 6 SAP specimens used to investigate the sealing efficiency.

	REF		SAP	
	w	σ	w	σ
1	151	12.8	159	10.5
2	159	21.4	141	8.2
3	157	17.7	152	11.4
4	149	15.6	164	15.0
5	153	10.4	147	22.5
6			156	15.7
μ	154		153	
σ	4.2		8.3	
COV	2.8%		5.4%	

4.1.2 Sealing efficiency

After the crack width was measured, the specimens were saturated for 2 days in demineralized water, after which an initial water flow test was performed on them. After 14 and 28 days of healing in wet-dry conditions, the water flow test was executed again. Fig. 4 shows the recorded flow Q at these 3 different moments in time for both the REF and the SAP series. The water flow of the SAP specimens is clearly lower than the water flow of the REF specimens for all three considered times.

SAPs immediately absorb water the moment they come in contact with it. This allows them to swell and (partially) close the crack. This immediate sealing SE_{imm} with reference to the REF specimens can be evaluated using equation 1 [14]:

$$SE_{imm} = 100\% * (Q_{REF,\ initial} - Q_{SAP,\ initial})/Q_{REF,\ initial} \quad (1)$$

in which $Q_{REF,\ initial}$ is the average initial flow of the REF specimens in g/min and $Q_{SAP,\ initial}$ is the average initial flow of the SAP specimens in g/min. The use of the averages is required since results from different specimens are used in this equation. Gruyaert et al. [14] state that the averages can only be used if the crack widths of all specimens are comparable. For an intended crack width of 150 μm they propose a crack width range from 130μm until 170 μm. All studied specimens are in this range, see Table 1. The immediate sealing SE_{imm} is equal to 31.2% (with $Q_{REF,\ initial}$ equal to 11.6 g/min and $Q_{SAP,\ initial}$ equal to 7.9 g/min).

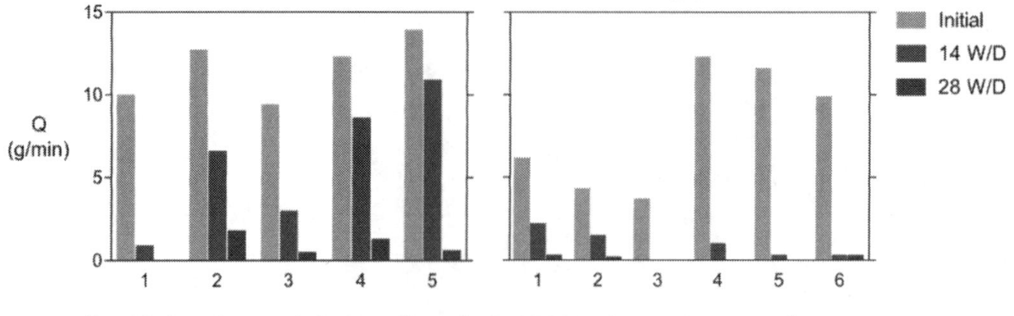

Fig. 4. Water flow Q before, after 14 and after 28 wet/dry cycles for 5 REF specimens and 6 SAP specimens.

By placing the specimens in wet-dry cycles they undergo a continuous process of hydration and dehydration. During dry periods, the SAP particles release the water that they have absorbed during the wet period. As a result of this the SAP specimens are wet much longer than the REF specimens.

It is possible to calculate the sealing efficiency as a consequence of the wet-dry cycles $SE_{W/D}$ by using equation 2 [14]:

$$SE_{W/D} = 100\% * (Q_{initial} - Q_{healed}) / Q_{initial} \quad (2)$$

in which $Q_{initial}$ is the average initial flow in g/min and Q_{healed} is the average flow in g/min after 14 or 28 wet-dry cycles. To calculate $SE_{W/D}$ for the REF series, respectively SAP series, both $Q_{initial}$ and Q_{healed} of the REF series, respectively SAP series, are used. $SE_{W/D}$ calculated out of the flow after 14 or 28 wet-dry cycles is given in Table 2. $SE_{28\ W/D}$ of the SAP specimens is very high (99.0%), although it should be noted that $SE_{28\ W/D}$ for the REF specimens is also quite high (93.6%). The effect of the SAPs is more clearly visible when comparing the sealing efficiency after 14 wet-dry cycles $SE_{14\ W/D}$: 89.9% for SAP against 48.9% for REF. These results show that the mortar containing SAPs can seal its cracks which are subjected to wet-dry cycles quicker than the REF mortar.

Table 2. Sealing efficiency due to wet-dry cycles after 14 days $SE_{14\ W/D}$, respectively 28 days $SE_{28\ W/D}$, for both the REF and the SAP series.

	REF	SAP
$SE_{14\ W/D}$	48.9%	89.9%
$SE_{28\ W/D}$	93.6%	99.0%

After the completion of the wet-dry healing cycles, it is possible to also calculate the total sealing coefficient SE_{tot} of the SAP specimens, which is a combination of immediate sealing and sealing due to healing:

$$SE_{tot} = 100\% * (Q_{REF,\ initial} - Q_{SAP,\ healed}) / Q_{REF,\ initial} \quad (3)$$

in which $Q_{REF,\ initial}$ is the average initial flow of the REF specimens in g/min and $Q_{SAP,\ healed}$ is the average flow of the SAP specimens in g/min after 28 wet-dry cycles. Note that $Q_{SAP,\ healed}$ can be replaced by an equivalent $Q_{REF\ healed}$ of the REF series to calculate the total sealing coefficient SE_{tot} of the REF series, but that equation (3) then becomes identical to equation (2). After all, REF series do not have an immediate sealing behaviour and as a consequence their total sealing only arises from the healing in the wet-dry cycles. In other words, SE_{tot} is equal to $SE_{28\ W/D}$ for the REF series.

The total sealing SE_{tot} of the SAP series is equal to 99.4%, as opposed to 93.6% for the REF series. Half of the SAP specimens had a flow Q equal to 0 g/min after the 28 wet-dry cycles, and the other half had a flow which was lower than 0.2 g/min, see Table 3. In comparison, only one REF specimen had sealed perfectly, while the other specimens still had a flow Q between 0.35 and 1.70 g/min after healing.

It can be concluded that, aside from promoting a faster sealing of cracks, SAPs also result in a more complete sealing of cracks. They stimulate the intrinsic autogenous healing capacity of mortar by increasing the availability of water in dry conditions allowing for further hydration and calcium carbonate precipitation [9].

After healing of the cracks, the strength of the healed products sealing the cracks was tested under higher water pressures. These tests at higher water pressures were executed immediately after the normal water flow test after 28 wet-dry cycles. After the tests at higher water pressures, the flow was measured again in a normal water flow setup at a water head of 50 cm. The results of these 4 water flow tests is given in Table 3. Out of the 3 SAP specimens that had a perfect sealing after 28 wet-dry cycles, 2 were able to resist a water pressure of 1 bar. The other specimen exhibited a very small flow (0.05 g/min). However, the sealing of all these specimens failed at 2 bar. For the REF series there was only 1 specimen that was perfectly sealed after 28 wet-dry cycles. This specimen was not able to resist a water pressure of 1 bar.

Table 3. Water flow Q in g/min after 28 wet-dry cycles measured at a water pressure of 0.05 bar, 1 bar, 2 bar and finally 0.05 bar again.

	Q (g/min)			
	0.05 bar	1 bar	2 bar	0.05 bar
REF 1	0.00	2.79	4.91	0.51
REF 2	1.70	4.21	8.50	0.05
REF 3	0.35	12.40	19.58	0.26
REF 4	1.17	25.09	50.44	2.52
REF 5	0.48	5.75	8.61	0.40
SAP 1	0.17	25.73	91.18	4.67
SAP 2	0.12	14.62	122.94	4.08
SAP 3	0.00	0.00	1.53	0.02
SAP 4	0.00	0.05	79.00	3.95
SAP 5	0.00	0.00	59.74	3.80
SAP 6	0.16	1.18	2.32	0.22

Comparing the results of both series, it is evident that the SAP series performs the best under a water pressure of 1 bar as a result of the 3 specimens which still had a (nearly) perfect sealing at this water pressure. At a pressure of 2 bar the REF specimens performed better; 4 out of 6 SAP specimens had a higher flow than any of the REF specimens. It could be seen with microscopic images made after testing that some of the SAP particles had been pushed out near the crack. At other locations it was the healing product which was pushed out by the high pressures. It is possible that the formed crystals in the SAP specimens were less strong as a result of their faster formation; as the cracks heal quicker, less carbon dioxide is able to enter the crack, hence reducing the impact of the carbonation reaction.

Although the sealing was broken in many specimens due to the high water pressures, it appears that the water which leaked out of the crack at the end of the tests (when the specimen was subjected again to a water head of 50 cm for 5 min after testing at 1 and 2 bar) was at least not higher than the water that leaked out during the initial water flow before healing, see Fig. 4. Thus, the healing was only damaged partially due to the high pressure.

4.2 Chloride ingress

4.2.1 Uncracked specimens

At approximately 1 week, 2 weeks, and 5 weeks, three uncracked specimens of each series were removed from the chloride bath. Fig. 5 shows the chloride ingress for each series. When comparing specimens which underwent the same healing cycle (healed or not healed), the chloride ingress in the SAP specimens is clearly larger than in the REF specimens. It has been reported in

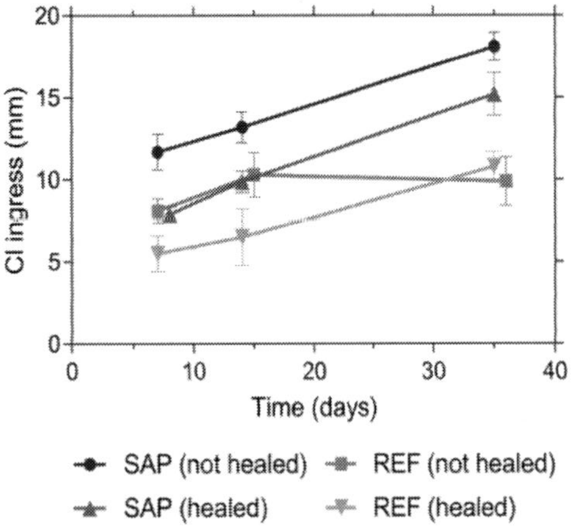

Fig. 5. Chloride ingress in uncracked REF and SAP specimens, respectively subjected and not subjected to wet-dry cycles. Error bars represent 2 times the standard deviation.

literature that the amount of air voids in mortars containing SAPs (where additional water was added to compensate for a loss of workability) is higher than in reference mortars [16] thereby allowing the chloride ions to migrate quicker through the matrix. When comparing specimens with the same mortar composition (REF or SAP), the specimens subjected to healing cycles exhibit a lower chloride ingress (except for the REF specimens at 5 weeks for which the chloride ingress is comparable). The lower chloride ingress for the healed samples can be explained by a densification of the mortar matrix as a result of the healing regime.

4.2.2 Unhealed specimens

The average crack width of the REF (not healed) series, respectively SAP (not healed) series, was equal to 117 µm (COV 9.9%), respectively 153 µm (COV 8.1%). The crack width of the SAP series is clearly larger than the crack width of the REF series. This difference might influence the chloride migration, although it should be reported that Jin et al. [17] assumed in the development of a simplified calculation model that the chloride ion concentration in the crack is equal to the solution for crack widths larger than 100 µm.

Fig. 6 shows the chloride ingress after 1 week and 5 weeks of contact with the chloride solution for the REF, respectively SAP, series. The chloride ingress after 1 week of exposure is quite similar. The chloride ingress away from the crack is slightly bigger for the SAP series, as could be expected based on section 4.2.1. The fact that the chloride ingress at the location of the crack is similar means that the immediate sealing effect of the SAPs is not sufficient to prevent an ingress of chlorides. The immediate sealing reported in section 4.1.2 was 31.2%, meaning that the crack was not completely sealed. Additionally, it should be pointed out that the cracks in the water permeability specimens have a triangular shape. The crack width will thus be smaller than the intended crack width of 150 µm (comparable to the crack width of the uniform cracks in the SAP specimens used for chloride ingress) at the location of

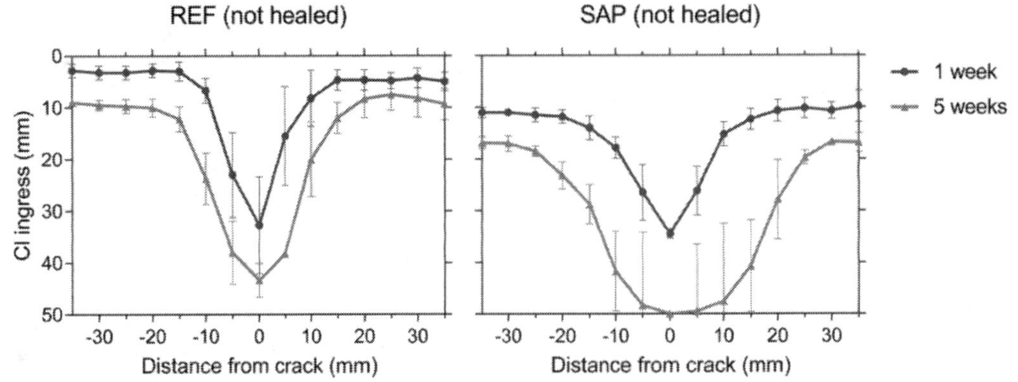

Fig. 6. Chloride ingress after 1 week and 5 weeks of saturation in a chloride solution of REF and SAP specimens which were not healed prior to contact with the chloride solution.

the hole where the water is introduced into the specimen during water flow testing. As a result of the incomplete sealing of the cracks, the chloride ions were able to penetrate into the crack.

The chloride ingress after 5 weeks is more severe for the SAP series than for the REF series. For the SAP series the chloride ions have migrated through the entire height of all the specimens, while this was not the case for any of the REF specimens. Additionally, the chloride front is much wider around the crack for the SAP specimens than for the REF specimens. Similar as for the uncracked specimens this can be caused by a higher content of air voids.

4.2.3 Healed specimens

Immediately after cracking and before healing, the average crack width of the REF (healed) series, respectively SAP (healed) series, was equal to 107 µm (COV 16.3%), respectively 134 µm (COV 8.1%). These two averages are more or less equal to one another (effective significance level of 1.1%).

Fig. 7 shows the chloride ingress after 1 week and 5 weeks of contact with the chloride solution for the REF, respectively SAP, series which were healed in wet-dry cycles prior to contact with the chloride solution. The beneficial influence of the healing regime is clearly evident by comparing these results to the results of the unhealed specimens.

After only 1 week of contact with the chloride solution, the chloride ingress in the healed SAP specimens is distinctly lower than in the healed REF specimens. After 5 weeks, the chloride ingress in the REF specimens is nearly unchanged. At the location of the crack the chloride ingress for both the REF and the SAP series is comparable. However, it should be noted that for the SAP series there was one specimen which had a much larger chloride ingress at the crack than the other two specimens which were analysed (39 mm of chloride ingress from the exposure surface at the location of the crack versus 29 mm and 24 mm). Notwithstanding the equal average chloride ingress at the crack, the influence of the crack seemed to be smaller for the SAP series than for the REF series; a constant chloride ingress was obtained much closer to the crack for the SAP series than for the REF series.

It can thus be concluded that adding SAPs to mortar which is allowed to heal prior to coming in contact with chloride ions results in a lower chloride ingress. At early ages the chloride penetration reduces and at later ages the influence of a crack on the chloride ingress reduces. This beneficial effect is a result of the stimulating effect SAPs have on the autogenous healing which results in a sealing of the cracks, as was also explained in section 4.1.2.

5 Conclusion

In this study the effect of SAPs on the chloride ingress in cracked mortar was studied. The following conclusions can be drawn:

- SAPs promote a faster and more complete sealing of cracks;
- Up until a water pressure of 1 bar the sealing of SAP specimens is better than the sealing of REF specimens, but at a water pressure of 2 bar healing products and SAP particles are pushed out;
- When additional water is added to compensate for a loss in workability (with a REF water to cement ratio = 0.5), the chloride ingress in uncracked SAP mortar is higher than in REF mortar;
- The chloride ingress in unhealed cracked specimens is higher in SAP specimens than in REF specimens as a possible consequence of the incomplete immediate sealing of the studied SAP particles and the larger air void content;
- The chloride ingress in cracked and subsequently healed specimens is lower than in unhealed specimens;
- At early ages the chloride penetration in cracked and healed SAP specimens is lower and at later ages the influence of a crack on the chloride ingress is reduced compared to cracked and healed REF specimens.

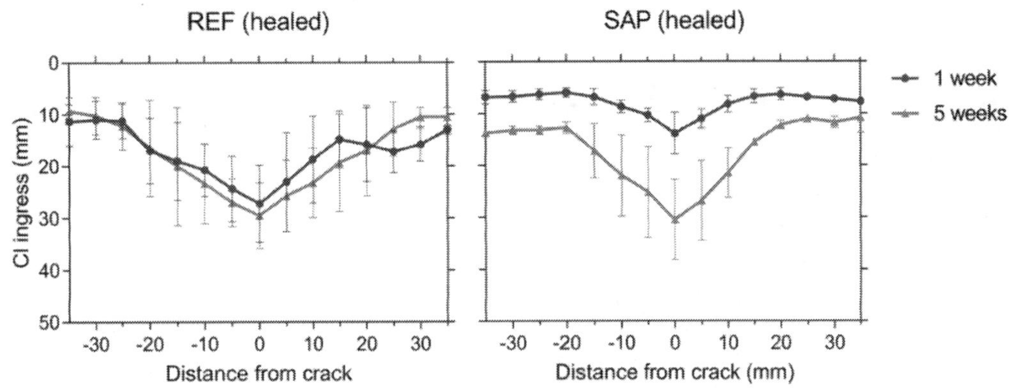

Fig. 7. Chloride ingress after 1 week and 5 weeks of saturation in a chloride solution of REF and SAP specimens which were not healed prior to contact with the chloride solution.

This research was supported by a grant (19SCIP-B103706-05) from the Construction Technology Research Program funded by the Ministry of Land, Infrastructure and Transport of the Korean government. N. Lampens and T. Stulemeijer from the Magnel Laboratory of Concrete Research are thanked for their technical support during the execution of the tests. TRADECC and ENCI are thanked for their generous donation of CFRP, respectively cement.

References

1. C.-p. Gu, G. Ye, W. Sun. A review of the chloride transport properties of cracked concrete: experiments and simulations. *Journal of Zhejiang University-SCIENCE A*. (2015);**16**(2):81-92.
2. B. Šavija, E. Schlangen. Chloride ingress in cracked concrete-a literature review. *Advances in modeling concrete service life*: Springer; 2012. p. 133-42.
3. M. Maes, K. Van Tittelboom, N. De Belie. The efficiency of self-healing cementitious materials by means of encapsulated polyurethane in chloride containing environments. *Construction and Building Materials*. (2014);**71**:528-37.
4. P. Van den Heede, B. Van Belleghem, M. A. Pereira Gomes de Araújo, J. L. Garcia Feiteira, N. De Belie. Screening of different encapsulated polymer-based healing agents for chloride exposed self-healing concrete using chloride migration tests. *(2018) KEY ENGINEERING MATERIALS*. (2018).
5. B. Van Belleghem, P. Van den Heede, K. Van Tittelboom, N. De Belie. Quantification of the service life extension and environmental benefit of chloride exposed self-healing concrete. *(2017) MATERIALS*. (2017).
6. Y. Ç. Erşan, K. Van Tittelboom, N. Boon, N. De Belie. Nitrite producing bacteria inhibit reinforcement bar corrosion in cementitious materials. *Scientific reports*. (2018);**8**(1):14092.
7. D. Snoeck, P. Dubruel, N. De Belie. Superabsorbent polymers to prevent water movement in cementitious materials. *INTERNATIONAL JOURNAL OF 3 R'S*. (2012);**3**(3):432-40.
8. D. Snoeck, S. Steuperaert, K. Van Tittelboom, P. Dubruel, N. De Belie. Visualization of water penetration in cementitious materials with superabsorbent polymers by means of neutron radiography. *Cement and Concrete Research*. (2012);**42**(8):1113-21.
9. D. Snoeck, J. Dewanckele, V. Cnudde, N. De Belie. X-ray computed microtomography to study autogenous healing of cementitious materials promoted by superabsorbent polymers. *Cement and Concrete Composites*. (2016);**65**:83-93.
10. M. T. Hasholt, O. M. Jensen. Chloride migration in concrete with superabsorbent polymers. *Cement and Concrete Composites*. (2015);**55**:290-7.
11. H. Beushausen, M. Gillmer, M. Alexander. The influence of superabsorbent polymers on strength and durability properties of blended cement mortars. *Cement and Concrete Composites*. (2014);**52**:73-80.
12. T. Van Mullem, E. Gruyaert, B. Debbaut, R. Caspeele, N. De Belie. Novel active crack width control technique to reduce the variation on water permeability results for self-healing concrete. *Construction and Building Materials*. (2019);**203**:541-51.
13. T. Van Mullem, K. Van Tittelboom, E. Gruyaert, R. Caspeele, N. De Belie, editors. Development of an improved cracking method to reduce the variability in testing the healing efficiency of self-healing mortar containing encapsulated polymers. Concrete Repair, Rehabilitation and Retrofitting (ICCRRR2018); 2018: EDP Sciences.
14. E. Gruyaert, B. Debbaut, D. Snoeck, P. Díaz, A. Arizo, E. Tziviloglou, et al. Self-healing mortar with pH-sensitive superabsorbent polymers: testing of the sealing efficiency by water flow tests. *Smart Materials and Structures*. (2016);**25**(8):084007.
15. E. Tziviloglou, V. Wiktor, J. Wang, K. Paine, M. Alazhari, A. Richardson, et al., editors. Evaluation of experimental methodology to assess the sealing efficiency of bacteria-based self-healing mortar: round robin test. RILEM Conference on Microorganisms-Cementitious Materials Interactions; 2016: RILEM.
16. A. Mignon, D. Snoeck, D. Schaubroeck, N. Luickx, P. Dubruel, S. Van Vlierberghe, et al. pH-responsive superabsorbent polymers: A pathway to self-healing of mortar. *Reactive and Functional Polymers*. (2015);**93**:68-76.
17. W. Jin, Y. Yan, H. Wang. Chloride diffusion in the cracked concrete. *Fracture Mechanics of Concrete and Concrete Structures-Assessment, Durability, Monitoring and Retrofitting*. (2010):880-6.

Pretensioned BFRP reinforced concrete beams: Flexural behaviour and estimation of initial prestress losses

Ana Pavlović[1*], *Ted* Donchev[1*], *Diana* Petkova[1], *Mukesh* Limbachiya[1], and *Refad* Almuhaisen[1]

[1] *Kingston University London, Department of Civil Engineering, London, UK*

Abstract. Fibre Reinforced Polymers (FRP) are becoming a popular reinforcement option for RC elements mainly due to good strength to weight ratio and resistance to corrosion. The main limitation for their wider application is their relatively low Young's modulus, which results in unfavourable serviceability performance, in terms of early development of deflections and cracks. Among others, prestressing has been suggested as one of the possible approaches to addressing this issue, with encouraging results from research conducted so far. This experimental study aimed to explore prestress losses of basalt fibre reinforced polymer (BFRP) reinforced pretensioned concrete beams. Five beams were produced, three of them internally reinforced with 6mm diameter BFRP bars, pretensioned to 20%, 30% and 40% of the ultimate load level of prestress. Additionally, two beams, acting as control samples, were reinforced with unprestressed BFRP and steel bars of same cross-sectional area, respectively. The dimensions of all samples were 125x200x1900 mm. Prestress losses were monitored with the aid of strain gauges attached to the reinforcing bars, as well as load cells. The strain readings were continuously taken during the pretensioning process, from initial application of the prestressing force, during casting and curing of concrete, until releasing of the beams from prestressing devices after curing. Ultimately, all samples were subjected to a quasi-static, load-controlled, four-point bending test until destruction. The results provide the information about the flexural behaviour of pretensioned BFRP reinforced beams, along with insight into some of the initial prestress losses of these elements.

1 Introduction

Until the last decades of the 20th century, steel reinforcement was exclusively used for internally reinforced concrete structures. With the increasing popularity of Fibre Reinforced Polymers (FRP), they have been considered a viable alternative for steel in certain applications, such as marine structures and bridges, where exposure to aggressive environments can cause corrosion of the steel.

In addition to excellent chloride corrosion resistance, FRPs also have high strength-to-weight ratio, non-ferrous and non-conductive nature, which make them suitable for various other structural applications. Nonetheless, they are often characterised as costly, especially in case of Carbon FRPs (CFRP). Hence other fibres, such as glass and, more recently, basalt have been utilised as a more economic substitute for carbon. Glass and Basalt FRPs (BFRP), on the other hand, have a much lower Young's modulus in comparison with CFRP, which results in poor performance as per serviceability limit state (SLS) criteria, with larger deflections at lower levels of loading and intensive development of cracking. To overcome this issue, prestressing of the FRP reinforcement has been proposed as a possible solution. This paper discusses experimental results of a study of pretensioned BFRP reinforced concrete beams.

2 Background

Prestressing of FRP materials has been a topic of research since the turn of the last century. A notable contribution was made in 1998 through the construction of the Taylor bridge in Canada. One tenth of the 40 support girders were produced with prestressed CFRP, and the deck has also been reinforced with GFRP bars [1].

An early contribution was made by Pearson and Donchev [2], who conducted experiments with post-tensioned BFRP reinforced concrete beams with and without grouting of the prestressing bars. The results indicated that the effect of the prestressing is much higher in the case of grouting of the bars and with the highest level of prestress (75%), resulting in an increase of the applied external load at the SLS limit by 70%.

Thorhallson and Jonsson [3] conducted a similar study with BFRP prestressed members, with no shear reinforcement. This study did not observe increase in capacity with prestressing of the member; however, it did agree with the conclusion that the deformation performance of pretensioned (PT) samples was improved in terms of lower deflections.

Furthermore, Crossett *et al.* [4] conducted research on self-compacting concrete (SCC) beams prestressed with 12mm diameter BFRP bars. The experimental investigation showed that prestressing of BFRP tendons resulted in increase in capacity over non-prestressed samples, as well as improved serviceability performance, both in terms of more limited deflections and reduced cracking. The increase of ultimate capacity was attributed to the higher reinforcing ratio and concrete crushing failure mode noted for all BFRP reinforced beams.

Encouraging results were also published by Mirshekari *et al.* [5]. The study was conducted on four beams of the same dimensions, with different levels of prestressing; namely, 0%, 20%, 30% and 40% of the ultimate tensile capacity of the BFRP bars. Additionally, a steel reinforced unprestressed beam was also produced as a control sample. The beams were otherwise equally reinforced with steel top reinforcement and shear links and submitted to a four-point bending test until failure. The analysis of the results confirmed that prestressing BFRP over 30% f_{fu} improves the serviceability performance of the elements to a level comparable to that of a steel reinforced element. In addition, the authors pointed out that all BFRP reinforced beams demonstrated a higher capacity than the steel reinforced control sample, where increase of the level of prestress attributed for an increase in the capacity as well.

Lago *et al.* [6] constructed a large scale BFRP prestressed bridge deck and tested it by investigating its flexural behaviour and presenting numerical analysis of the structural member. The results of the study were also positive with regards to structural performance; although, problems with the mechanical anchorage of the bars were identified.

Despite the considerable research conducted on BFRP prestressed elements, the long-term behaviour and prestress losses remain insufficiently investigated. A recent contribution was made by Pavlović *et al.* [7] who experimentally investigated short and long-term prestress losses of six samples of pretensioned BFRP reinforced beams over a period of three months.

This paper presents results of a parallel experimental study conducted at Kingston University London to give a contribution to the knowledge of flexural behaviour and prestress losses of BFRP pretensioned RC beams.

3 Methodology

Five samples were produced and tested in the Structures Laboratories at Kingston University London. The samples were concrete beams with a rectangular 125x200mm cross section, 1900 mm long.

The reinforcement cages were constructed with two 6mm diameter high-yield steel top reinforcement bars and 6mm diameter mild steel shear links with 100 mm spacing. The pure bending zone, mid 600 mm of the span, was not reinforced with any shear links. The main (tensile) reinforcement is given in Table 1.

Table 1 Reinforcement specification of samples

Sample ID	Main (tensile) reinforcement	Prestress force
S0	N6 high yield steel	0% f_{fu}
BFRP0	Ø6mm BFRP RockBar	0% f_{fu}
BFRP20	Ø6mm BFRP RockBar	20% f_{fu}
BFRP30	Ø6mm BFRP RockBar	30% f_{fu}
BFRP40	Ø6mm BFRP RockBar	40% f_{fu}

Additionally, for the three prestressed samples, stainless steel sleeve bonded anchorage was installed on either end of the bars, in accordance with the recommendations given in ACI440R [8], using Webertec EP Epoxy Structural Adhesive. The sleeve was used as a connection between the BFRP bars and steel threaded bars, which were utilised during the prestressing process. The initial prestress force was applied using a manual hydraulic jack on both bottom bars simultaneously and monitored with NovaTech F313 Low Profile Donut 100kN load cells.

After prestressing was completed, all 5 beams were cast using the same batch of C30/35 ready-mix concrete; to verify the strength and class of concrete, standard cubes and cylinders were also prepared. The samples after casting are shown in Fig. 1. The concrete was cured for 28 days in laboratory conditions; ambient temperature, as well as internal concrete temperature were also monitored using thermocouples.

Fig. 1 Samples immediately after casting of concrete

After 28 days curing, the prestress was gradually released from the external anchorage and transferred to fully cured concrete by manually unscrewing the anchor nuts, whilst clamping the anchor sleeves to prevent any torsional effects.

Throughout the application of prestress, casting and curing of concrete, until the release and transfer of prestress, the load in the bars was monitored using NovaTech 100kN load cells, and strains of the BFRP bars in the midspan and close to supports were monitored using Vishay 120±0.3% Ω foil strain gauges.

Results were continuously recorded using VPG System 8000 data acquisition system at a 1Hz rate.

Following the release of prestress and cutting off sleeve anchors, the beams were subjected to destructive four-point bending tests, over a 1700mm span and 600mm spacing between point loads. The test was load-controlled, at a rate of 2kN/5min, applied using a hydraulic jack mounted on a reaction frame. During the testing, load was monitored using a load cell, deflections using LVDTs, and strains of reinforcement using strain gauges, and the data was recorded using the VPG System 8000 data acquisition system at 60Hz rate. Crack development was manually traced and noted.

4 Experimental results

4.1 Initial strain level monitoring

The presented results were obtained from strain gauges installed on the BFRP bars at 50mm from the edge of the beam. All the gauges showed decrease of strain throughout monitoring. In the initial 24h from the application of the external prestress, the decrease was noticeably faster, at an approximate rate from 0.14% to 0.28% per hour. This period accounted for the largest total loss of from 3.3% to 6.7% of the initially applied strain.

Following the casting of concrete and the described initial period of intense losses, the rate of decrease became much slower, with strain decreasing at approximately less than 0.1% per day. This rate was maintained at an approximately constant level throughout the curing period, up until the release of the bars from the external anchorage and transfer of the prestressing force to the cured concrete elements.

4.2 Final Testing

As described, after transfer of the prestressing force to the cured concrete beams, the samples were subjected to four-point bending test until failure, and the results are described in this section. Fig. 2 shows the load versus deflection graphs of all five tested samples.

As seen on the graph, the steel control sample exhibited a near-linear behaviour until a load of approximately 8kN was applied, corresponding to the appearance of the first crack. After this, the deflections were developing more rapidly with increase of the load, followed by typical yielding plateau with reaching of the ultimate load of 22.5kN, and the decrease of load after failure.

The behaviour of the BFRP reinforced unprestressed sample was similar until the appearance of the first crack, which also occurred at approximately 8kN of load. After this point, the development of deflections was much more rapid than for the steel reinforced sample, reaching the ultimate deflection of approximately 60mm, at the ultimate load of over 40kN.

All the prestressed samples exhibited an increase in stiffness in comparison with the unprestressed BFRP sample. In addition to that, each increase of the level of prestressing resulted in a delay of the point of change of stiffness, which occurred at approximately 14kN, 22kN and 26kN for BFRP20, BFRP30 and BFRP40 samples respectively. There was no increase in the ultimate load in comparison with BFRP0, with all samples failing after reaching a load of over 40kN. However, the midspan deflection of each prestressed sample was significantly lower compared to BFRP0; reaching maximum values of 42.2mm, 34.4mm and 21.4mm for BFRP20, BFRP30 and BFRP40 respectively.

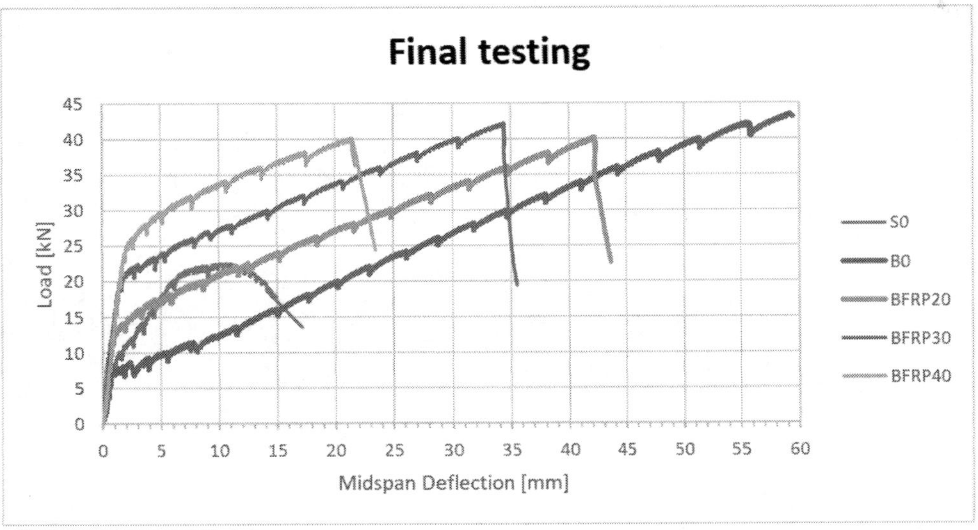

Fig. 2 Four-point bending test results: Load Vs Deflection

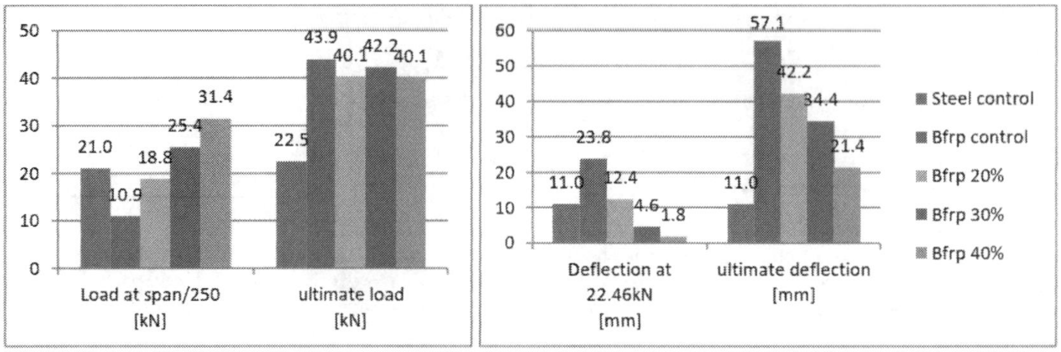

Fig. 3 Summary of ULS and SLS load and deflection results

4 Discussion

The summary of the results of the final four-point bending test is given in Fig. 3. The analysis of ultimate load reached during four-point bending test of the beams shows that the increase in load-bearing capacity of BFRP reinforced samples in comparison with the steel reinforced control sample is significant, from around 78% to 95% of the ultimate load of S0. It is also worth noting that the prestressed samples did not reach a higher ultimate load than the unprestressed BFRP sample.

On the other hand, the increase of the ultimate deflection of B0 compared to S0 was significant, over 400%. Prestressing of the beams evidently had a beneficial effect on the development of deflections, i.e. with increase of the level of prestressing, the ultimate deflection measured at the midspan decreased.

Moreover, the comparison of load reached for each sample at the serviceability limit state (SLS) criteria of span/250 (6.8mm) [9] also reflects the positive effect of prestressing regarding limiting the deflections. Samples which were prestressed to 30% and 40% of the ultimate tensile capacity of the BFRP bars exceeded the performance of the steel reinforced beam, by achieving 21% and 50% higher load than S0 at the specified limit, respectively.

In addition, analysis of modes of failure of each sample was also conducted. As shown in Fig. 4 the crack pattern of all beams was that of a flexural failure, with cracks extending at a near-vertical angle, with some approaching a more diagonal angle near the load application points.

The destruction of S0 was characteristic of an under-reinforced beam; the sample failed by yielding of tensile reinforcement, which is also evident on the load vs deflection graph (Fig. 2). On the other hand, the destruction of BFRP0 was brittle, with deflections and cracks progressing continually until failure, which occurred by rupture of the BFRP bars. Similarly, the failure of all prestressed samples was distinctively brittle, with little to no prior warning before the failure, which also occurred by rupture of the bars. This was especially pronounced in the case of BFRP30 and BFRP40. For practical applications, this issue could be addressed in several ways, such as the usage of hybrid prestressed/unprestressed reinforcement, structural health monitoring etc. The prestressed samples developed a smaller number of cracks in comparison with BFRP0, with the least number of cracks observed for BFRP40.

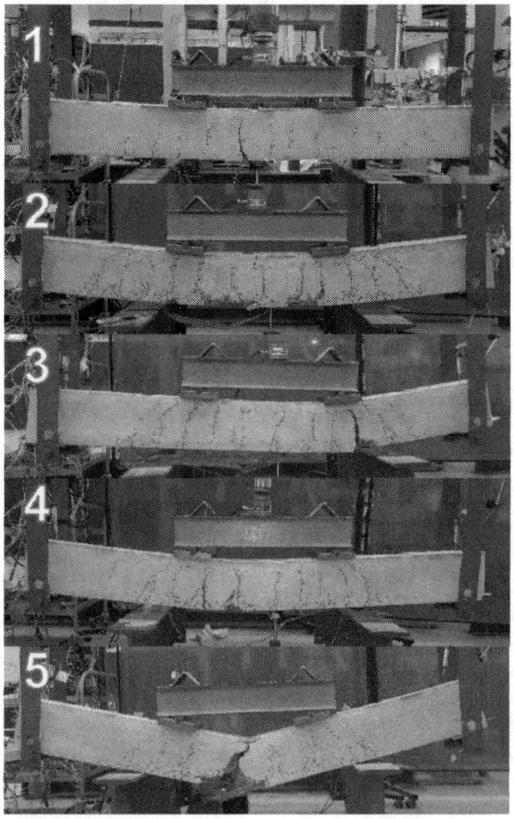

Fig. 4 Photographs of all samples at the end of the four-point bending test (1 – S0; 2 – BFRP0; 3 – BFRP20; 4 – BFRP30; 5 – BFRP40)

Additionally, an overview of initial crack load increments is given in Fig. 5. The results show that both the steel and BFRP reinforced unprestressed samples had a first visible crack at a load of approximately 8kN. The opening of cracks is related to exceeding of the tensile strength of concrete on the tensile face of the flexural member; therefore, by introduction of a prestressing force, the appearance of cracks was delayed for prestressed samples. Furthermore, the increase of the level of prestress resulted in a further delay of the appearance of the first crack on the bottom of the beams. It should be also noted that all the initial cracks appeared in close proximity to the midspan of the loaded samples.

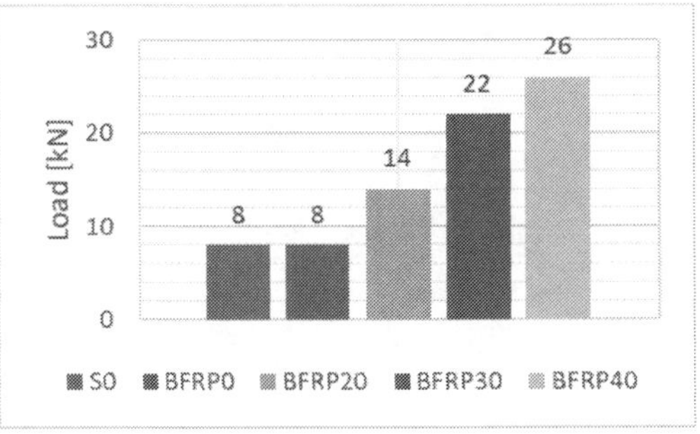

Fig. 5 Load corresponding to initial crack appearance

5 Conclusions

Based on continuous monitoring of the initial strain level and subsequent analysis, the following conclusions can be made:

- The fastest rate of decrease of strain, of approximately 0.20% per hour on average, was observed for the initial period of around 24h from the application of the external prestressing force.
- The subsequent period was characterised by a much more gradual continuous decrease of strain, at a rate of around 0.1% per day.

Additionally, based on the flexural behaviour of five samples tested under four-point bending until destruction, the following can be concluded:

- Prestressing of BFRP reinforced samples with over 30% of the ultimate tensile capacity of the bars improved the serviceability performance of the beams to a level higher than that of the steel reinforced sample.
- Ultimate deflections of all prestressed samples were reduced in comparison with the unprestressed BFRP sample.
- The initial appearance of cracks on the tensile face of the samples was delayed for the prestressed samples, correspondingly with the level of prestressing applied.

6 Acknowledgements

The authors would like to express their gratitude to MagmaTech UK Ltd. for sponsoring the project with RockBar BFRP bars, as well as to the technical staff of the Structures Laboratories at Kingston University London.

7 References

[1] Shehata, E. et al. (1999) Taylor Bridge - *A Bridge For the 21 st Century*. Winnipeg Manitoba, World Wise 99

[2] Pearson, M., Donchev, T. (2013) Experimental Study on the Behaviour of Prestressed Concrete Beams with Internal BFRP Reinforcement. *Proceedings of the 11th International Symposium FRPRCS11, Guimaraes, Portugal*, June 26-28, 2013

[3] Thorhallsson E.R., and Jonasson, B. (2012) Test of Prestressed Concrete Beams with BFRP tendons. *The 6th International Conference on Fibre-Reinforced Polymer (FRP) Composites in Civil Engineering (CICE 2012)*, 13-15 June 2012, Rome, Italy

[4] Crossett, P., Taylor, S., Robinson, D., Sonebi, M., Garcia-Taengua, E., Deegan, P., and Ferrara, L. (2015) The Flexural Behaviour of SCC Beams Pre-stressed with BFRP. *Proceedings of the 7th Beinnial Conference on Advanced Composites in Construction (ACIC 2015)*, 09-11 September 2015, Cambridge, UK

[5] Mirshekari, M., Donchev, T., Petkova, D., and Limbachiya, M. (2016) Deformability of Pretensioned PC Beams with BFRP Reinforcement. *The 8th International Conference on Fibre-Reinforced Polymer (FRP) Composites in Civil Engineering (CICE 2016)*, 14-16 December 2016, Hong Kong, China

[6] Lago, B., Taylor, S., Deegan, P., Ferrara, L., Sonebi, M., Crosset, P., and Pattarini, A. (2017) Full-scale testing and numerical analysis of a precast fibre reinforced self-compacting concrete slab pre-stressed with basalt fibre reinforced polymer bars. *Composites Part B*, **128**, pp. 120-133

[7] Pavlović, A., Donchev, T., Petkova, D. and Limbachiya, M. (2019). Initial estimate of short and long-term prestress losses of BFRP pretensioned RC beams. *Proceedings of the 14th International Symposium FRPRCS14*, Belfast, UK, June 4-7, 2019

[8] ACI Committee 440 (2004) 440.3R-04: *Guide Test Methods for Fiber-Reinforced Polymers (FRPs) for Reinforcing or Strengthening Concrete Structures*. American Concrete Institute, USA

[9] The European Committee for Standardisation (1992) EN 1992-1-1: Eurocode 2: *Design of concrete structures - Part 1: General rules and rules for buildings*, Section 4.4.3.1 (5)

Modelling the behaviour of concrete shear walls with BFRP reinforcement

Hamid Rahman[1*], *Ted* Donchev[1*], and *Diana* Petkova [1]

[1] Faculty of Science, Engineering and Computing, Kingston University London, United Kingdom

Abstract. Corrosion of steel reinforcement in concrete structures is one of the main challenges for the construction industry. FRP reinforcement could be used as alternative to steel reinforcement providing several advantages, such as high resistance to corrosion, high tensile strength and opportunity for developing a more ductile mode of destruction. The last characteristic is extremely beneficial in the aspect of lateral load resisting systems including RC shear walls. The presented modelling results consist of preparing finite element models of reinforced concrete shear walls utilizing Ansys 19.2 - Solid65 element which is capable of both cracking and crushing. The results were verified with experimental medium-scale concrete shear walls reinforced with steel and BFRP bars. The models were loaded under cyclic lateral loading following a modified ATC (Applied Technology Council guidelines for seismic testing) 24 protocol. BFRP reinforced models developed similar ultimate capacity and significantly higher energy dissipation in comparison with steel reinforced models. The promising results could provide a momentum toward construction of shear walls using FRP reinforcement with the aims of improving durability and energy dissipation.

1 Introduction

The problem of steel corrosion in concrete structures can be overcome by alternative reinforcement of Fibre Reinforced Polymers (FRP) bars. The expectation is that the usage of FRP reinforcement for RC shear walls will result in higher energy dissipation due to their lower Young's Modulus and higher deformability of the FRP. Standards and formal design deficiency is a significant barrier for extensive use of FRP in (Reinforced Concrete) RC [1].

Change in the traditional design philosophy of concrete structures is needed for FRP reinforcement as mechanical behaviour of FRP reinforcement differs from the behaviour of conventional steel reinforcement. FRP materials are anisotropic and are characterized by high tensile strength and only in the direction of the reinforcing fibres, which affects shear behaviour and dowel action of the FRP bars as well as bond performance. Furthermore, FRP materials do not yield and they are elastic until failure and the design procedure must account for a lack of ductility in structural concrete members reinforced with FRP bars [2].

Currently there are areas where there is limited knowledge of the performance of FRP reinforcement including fire-resistance, durability in outdoor or severe exposure conditions, bond fatigue, and bond lengths for lap splices. These are the areas which need further research and additional information according to the ACI 440 committee report on FRP reinforcement (ACI, 2006).

Mohamed et al., [1] evaluated GFRP reinforced shear walls by conducting experiments on concrete shear walls reinforced with GFRP bars. The outcome of those studies shows that GFRP reinforced shear walls are capable of higher energy dissipation in comparison with steel reinforced shear walls [1].

2 Finite Element Modelling

2.1 Model Geometry

One of the essential components of defining a mathematical model is to define its geometry in a finite element method. The geometry of the model will define how the elements are compiled. The geometry of the shear wall which is the subject of this study was modelled using Ansys 19. Since the shear wall was made of concrete with reinforcement layout in two layers and Ansys "Solid 65" element was used, therefore appropriate modelling is via application of 3D geometry [3].

In modelling the geometry of the shear wall some features that do not contribute significantly in design of the actual model were removed; this is also known as "defeaturing". Therefore the connection details to the strong frame and loading cell were not modelled in Finite Element Analysis (FEA). Connection details are elaborated in Section 3.

*Corresponding authors: Ted Donchev (T.Donchev@kingston.ac.uk) and Hamid Rahman (Hamid.Rahman@kingston.ac.uk)

2.2 Element Selection

To analyse the crack and crush properties of the shear wall under quasi-static loading, the wall was modelled using solid element and the reinforcement was modelled in link element. Selecting appropriate elements is the most important and fundamental decision that has to be taken before starting the analysis which depends on the geometry of the model and type of loading applied. The geometry of this model was volumetric hence 3D solid was used for concrete. Solid65 in Ansys is a three dimensional element used to model solids, especially concrete. Using a Link element option allows the analysis of the element in tension, compression and bending of the element. The element will allow definition of the section properties to model the cross sectional shape and accurately predict bending behaviour.

The Solid65 3D option can represent the concrete with and without reinforcement. The solid element is capable of cracking in tension and crushing in compression. The element is defined by eight nodes, each node having three degrees of freedom in the x, y and z directions. The element is capable of analysing the nonlinear properties of the concrete and cracking in three orthogonal directions, crushing, plastic deformation and creep. The geometry, node locations and coordinate system are shown in Figure 1 for Solid65 element. This element is defined by isotropic material properties.

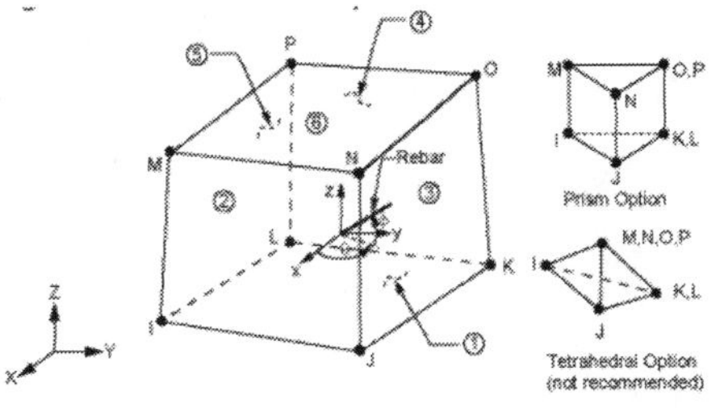

Figure 1: 3D Ansys Solid65 element geometry

Link180 was used to represent the reinforcement in the wall which was an uniaxial tension-compression element with three degree of freedom at each node. The three degrees of free include translation in the nodal x, y and z directions. The element capacities include plasticity, creep, rotation, large deflection, and large strain capabilities. The link element has an option to represent tension-and-compression, tension-only or compression only states as required. The geometry, node locations and the coordinate system for this element are shown in Figure 2.

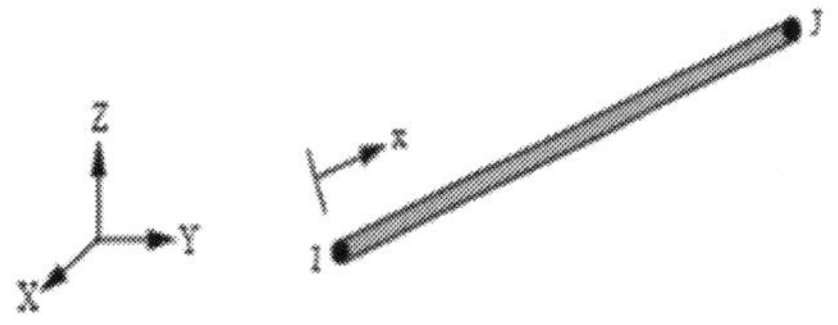

Figure 2: Ansys Link180 element geometry

The element (Link180) is defined by nodes I and J in (Figure 2), cross sectional area, mass per unit length and material properties. The element supports elasticity, isotropic hardening plasticity, kinematic hardening plasticity, Hill anisotropic plasticity, Chaboche nonlinear hardening plasticity and creep formations. The element can allow change in the cross-sectional area as a function of axial elongation while preserving the volume of the element after deformation which is quite suitable for elastoplastic application. The element also offers the ability to keep the cross-sectional area constant or rigid by facilitating its key options. Other features of the element include support element birth and death, initial state, large deflection, large strain, nonlinear stabilization and stress stiffening.

In summary the elements chosen for analysis of this problem are listed below:

- Solid65 to represent concrete
- Link180 to represent reinforcement.

The basic geometry of the shear wall model is indicated in Figure 3. The concrete volume is modelled using Solid65 element and the reinforcement is modelled using Link180 element by connecting the coincident element nodes. The connection of element nodes by Link180 element generates a perfect bond between the reinforcement and the concrete. The mesh allows for closely spaced nodes in both elements.

Figure 3: Volume generated by ANSYS for the shear wall model.

The shear wall model in Figure 3 is a modelling representation of the concrete volume and reinforcement cage as well as the boundary condition and the application force. The cyclic load is applied at the top right corner of the wall representing the cyclic lateral load generated in the case of a heavy earthquake on the building in this case consisting of the modelled reinforced concrete shear wall. The two variation of the shear wall models are listed in Table 1.

For both the concrete and reinforcement elements the real constants and section properties were assigned accordingly. Solid65 was represented by real constant 1 and Link180 was configured with section 1 Rebar, 2 Rebar and 3 Link; user defined. The section 1 represents 10mm diameter bars, section 2 represents 6mm bars and section 3 represents the link reinforcement. Since the reinforcement is defined as link element (Link180); therefore, the real constants for Solid65 are defined as zero value for models 1, 2, 3 and 4. The zero value of real constant waives the smeared capacity of the elements in the model.

Table 1: Basic configuration of the RC wall models

Model no	Rebar type	Rebar size (mm)	Stirrups (mm)	Breath (mm)	Height (mm)	Depth (mm)
1	BFRP	6	6	1000	1000	100
2	Steel	6	6	1000	1000	100

2.3 Material Properties of the Model

The material properties for each of the three components that are used to make the model in this study were different. The components included the steel, Basalt Fibre Reinforced Polymer (BFRP) and the concrete. The material models for each of the components are described below.

2.3.1 Steel and BFRP

The material model for the reinforcements including BFRP bars, steel bars and the stirrups was defined as linear elastic and isotropic material. The following two properties were defined for the steel reinforcement;

- Elastic modulus (E_s)
- Poisson's ratio (v)
- Yield Stress (f_y)
- Tangent modulus (E_s')

Poisson' ratio was taken as 0.3 and elastic modulus was taken as 210 GPa for the steel incorporated in the reinforcement of the models [4]. Two wall models were generated using steel rebars. One wall model was reinforced with 6mm steel bars and the second model was reinforced with the 6mm BFRP bars. All stirrups were 6mm steel bars.

The linear and bilinear properties defined for the BFRP reinforcement are outlined below.

- Elastic modulus (E_s)
- Poisson's ratio (v)
- Yield Stress (f_y)
- Tangent modulus (E_s')

The above property values were extracted from the manufacturers' brochure for the bars. The information provided by the manufacturer showed that the Poisson's ratio for the BFRP was 0.23 and the elastic modulus was 50 GPa [5]. Two wall models were generated. One wall model was reinforced with 6mm BFRP bars and the second model was reinforced with 10mm BFRP bars. The bar properties are summarized in Table 2 below. All stirrups incorporated were 6mm steel bars.

Table 2: Material Properties of Steel and BFRP

No	Name	Linear Isotropic		Nonlinear Isotropic	
		E_s (MPa)	v	f_y (MPa)	E_s' (MPa)
1	Steel	210x10^3	0.3	500	2100
2	BRFP	50x10^3	0.23	1000	500

2.3.2 Concrete

The material properties for the concrete were defined as nonlinear to reflect the actual inelastic nature of the concrete. So in addition to the elastic properties as defined for the reinforcement, nonlinear properties are also defined. The following linear properties were defined for the concrete:

- Poison's ratio (v)
- Elastic modulus (E_c)

The Poisson's ratio for the concrete material was assumed to 0.2 for all four models [6]. In addition to the above, the following properties were defined to demonstrate the crack behaviour when incorporating Solid65 element for the analysis.

- Shear transfer coefficient for an open crack (β_t)
- Shear transfer coefficient for a closed crack (β_c)
- Ultimate uniaxial tensile stress (f_t)
- Ultimate uniaxial compressive stress (f_c)

The mean compressive cylinder strength and modulus of elasticity were calculated according to EN 1992-1-1:2004 [8]. The compressive strength values were obtained from conducting compressive tests. The tension calculation can be calculated as per equations below:

$$f_{cm} = f_{ck} + 8 \text{ (MPa)} \quad (1)$$

$$f_{ctm} = 0.3 \text{ x} f_{ck}^{(2/3)} < \text{C50/60} \quad (2)$$

Where f_{ctm} is the mean tensile strength and f_{cm} is the mean compressive strength.

The shear transfer coefficient value ranges from 0 to 1. This coefficient characterizes the crack face conditions. A zero coefficient value generates a smooth crack i.e. full loss of shear transfer. A coefficient value of 1 produces a rough crack i.e. no loss of shear transfer. It is recommended that the shear transfer coefficient for an open crack shall be less than 0.2 to avoid convergence problems [7]. The concrete property values are summarized in Table 3 below. The concrete properties were kept the same for both samples to make sure the types of reinforcement were the only variables under study.

Table 3: Summary of concrete material properties and constants as defined in ANSYS

Model	E_c (GPa)	f_c (MPa)	f_t (MPa)	β_t	β_c	ν
1 & 2	33	38	2.9	0.2	1	0.2

Default properties of zero were chosen for the following ANSYS configurations which can also be assigned other than the zero values:

- Biaxial crushing stress (f_{cb})
- Ambient hydrostatic stress, for use with constant f_1 and f_2 (σ_h^a)
- Biaxial crushing stress under the ambient hydrostatic stress state (f_1)
- Uniaxial crushing stress under the ambient hydrostatic stress state (f_2)
- Stiffness multiplier for crack tensile condition (T_c)

The ambient hydrostatic stress is required only when f_1 and f_2 are not defined as the default value of zero. The above stress state is defined as a function of the principal stress in all three principal directions (Ansys, 2019).

The Solid65 element also requires multilinear isotropic hardening material properties to model the concrete in compression. The multilinear isotropic properties of concrete are calculated utilising following formulae.

- $f = E_c\varepsilon/1{+}(\varepsilon/\varepsilon_0)$ (3)
- $\varepsilon_0 = 2f_c/E_c$ (4)
- $E_c = f/\varepsilon$ (5)

Where:

- f = stress at any strain ε
- ε = strain at stress f
- ε_0 = strain at the ultimate compressive strength f_c.

The numerical relationships above including Equations 1, 2 and 3 were used to generate a stress-stress curve (Figure 4) for the concrete under compression in this study.

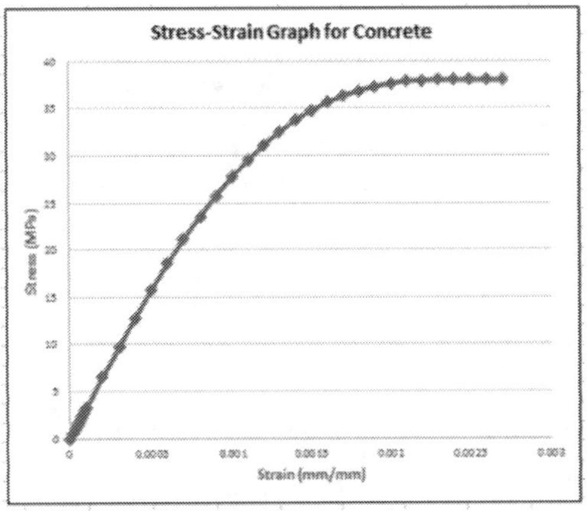

Figure 4: Concrete stress-strain relationship curve for C30/37

The isotropic material properties for the concrete per the above formulae and curve are tabulated below (Table 4);

Table 4: Linear and Multilinear Isotropic properties of concrete and Poisson's Ratio

Linear Isotropic (C30/37)		
E_c (MPa)	$33\text{x}10^3$	
ν	0.2	
Multilinear Isotropic (C30/37)		
	Stress f, [MPa]	Stain ε, [m/m]
0	0	0
1	3.294	0.0001
2	6.551	0.0002
3	9.735	0.0003
4	12.813	0.0004
5	15.757	0.0005
6	27.765	0.001
7	34.756	0.0015
8	37.625	0.002
9	37.872	0.0025

2.4 Meshing

The volumes were meshed using the standard meshing option in the Ansys Workbench platform as shown in Figure 5 below.

There are multiple options for defining the meshing size in the Ansys Workbench platform.

Choosing different sizes of the mesh will provide different results, converging-ability and calculation-time.

The above means that an ideal type of mesh is required to be able to calibrate the model with experimental results and have a converged solution with a reasonable calculation time.

Different meshing strategies were selected to be able to obtain a converged solution in a reasonable time calibrating with the experimental data.

The appropriate size of the mesh that would provide fairly close results to the experimental data is shown in Figure-5 below. This is a "medium" size mesh approximately 3.333 cm. A Medium size mesh is the "coarse and fine" mesh as defined by the Ansys software.

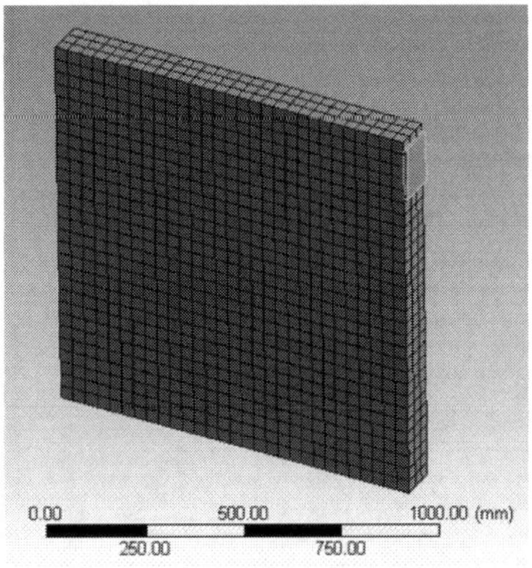

Figure 5: Meshing of the RC shear wall model.

2.5 Loads and boundary conditions.

When the basic model was made and the geometry, element, material and meshing were set, then the boundary condition and loads were defined.

Boundary conditions to restrain displacement are required to constrain the model and get a unique solution per condition. It is essential to set the boundary conditions to stop large displacements invalidating the results. The boundary condition was set as per the physical model that was to be devėloped in the KU civil engineering lab. The boundary conditions were used in all models.

The bottom surface of the model's volume was constrained in all direction to stop rotation and displacement of the bottom plate in all directions. The basic summary of boundary conditions is depicted in the model (Figure 6).

The load was applied to the shear wall as in-plane displacement at top of the wall. The load was applied on the top surface. The load was designed to push and pull the wall in-plane, until an eventual failure.

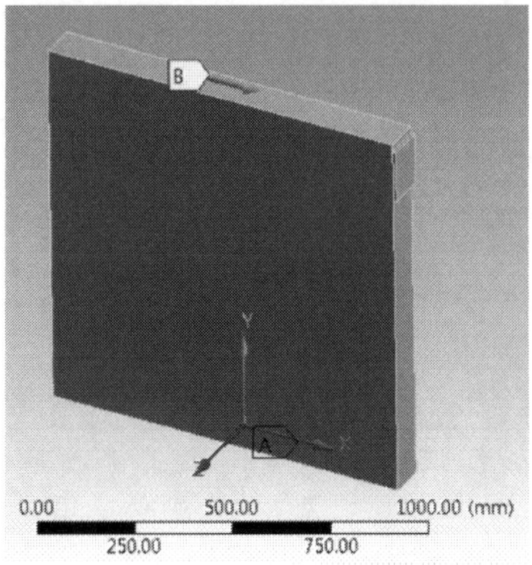

Figure 6: Basic boundary conditions for the wall models fully fixed at bottom and lateral load applied at the top.

The loading cycles applied were the ATC 24 [9] protocol assuming 1 cycle per amplitude and the loading regime used in the modelling session. Fourteen cycles of displacement load amplitudes were sat up to be applied to the test specimen ranging from 0.2 mm to 40 mm as indicated in Table-6 below. No more displacement loads were applied once the specimen failed to sustain more loads.

Table 6 - Load cycle – displacement amplitude table

Loading Cycles	Displacement Amplitude (mm)
1	0.2
1	0.4
1	0.8
1	1.2
1	2.5
1	3.5
1	5.0
1	10
1	15
1	20
1	25
1	30
1	35
1	40

3. Experimental tests

The experimental programme was intended to verify and calibrate the modelling results comprised of testing two 1000x1000x100mm reinforced concrete shear walls under quasi-static loading until failure [10]. One sample was reinforced with steel and the second sample was reinforced with BFRP. The reinforcement details and general dimensions are shown in Figure 7 below.

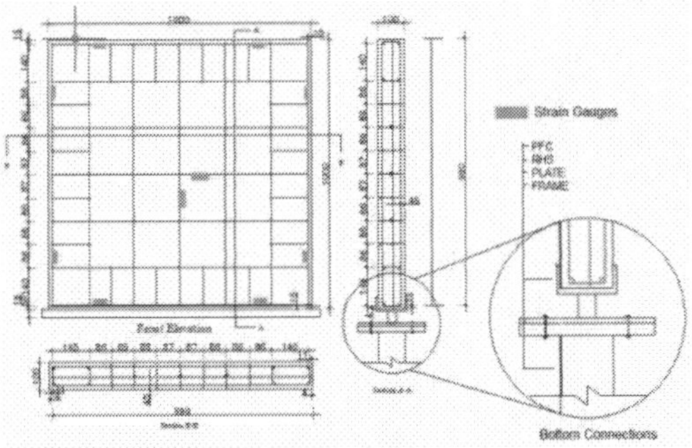

Figure 7: Reinforcement detail and bottom connection and strain gauges.

The strength and properties of the concrete used and the size and properties of the steel and BFRP bars used were the same as the strength, properties and size used to create the FEA model. The loading and boundary condition used were the same as the loading and boundary condition used to create the model.

The samples were cast incorporating PFC (Parallel Flange Channel) steel section at the bottom of the wall with anchoring steel bars. Additional RHS (Rectangular Hollow Sections) was welded to the PFC section on one side and to the perforated steel plate attached to the strong testing frame on the other side (Figure 7).

Each of the shear walls was fixed to the strong frame through the RHS welded to the steel anchor plates at the bottom surface of the wall. This made the base of the wall fixed to the strong frame. The top end of the wall was laterally supported in the out-of-plane direction. Rollers were attached to the strong frame to will allow movement of wall in in-plane direction.

The lateral load was applied through a screw jack model BD Benziler Screw Jack 500 kN to apply the quasi-static loading. The jack was fixed to the reaction frame. The system was installed to insert in-plane horizontal quasi-static load to the loading plate of the specimen. The loading plate was connected to the opposite end of the wall through 10 mm diameter screw rods bolted to another loading plate installed opposite to the load cell end of the wall to counteract the tensile force applied to the wall by the screw jack when pulling and translate it to compression lateral compression force at the end of the wall opposite to the loading cell. The load was controlled through the data from displacement gauges.

Rosette gauges were added to the concrete surface of the shear wall at the middle of the wall on both sides. The gauges where applied in central surface locations of the wall. The Rosette had a combination of three gauges in two orthogonal directions and one in a diagonal direction to provide the information about the direction and magnitude of the principal strain on the wall surface.

In addition to the above CDS (Cable-extension displacement sensors), LVDT (Linear Variable Displacement Transducers) and dial gauges were also used in many critical positions to capture deformation of the wall due to the cyclic loading and displacement. The CDS' and LVDT's were attached between the strong frame and the specimen. LVDT measured displacements of up to 5 mm and larger displacements where captured by corresponding CDSs. Dial gauges were used to compare the digitally measured displacements with non-digital readings for double checking.

4. Results

The results of the force displacement amplitude are shown in figures 8a and 8b below. The graph is the result of experiments and finite element analysis on the BFRP and Steel reinforced samples.

The comparison of results from the experimental data show that BFRP reinforced model dissipate more energy than the steel reinforced model. The comparison of the FEA results on both types of reinforcement confirms the same. Energy dissipation is the area under the curves in figures 8a and 8b.

The results of the experiments show that the BFRP reinforced sample graph continues till 40 mm displacement amplitude almost constantly after 20 mm displacement amplitude; however, the steel reinforcement curve shows no result after 15 mm displacement amplitude because the reinforcement bars broke at this stage.

The result of the FEA shows continuation of the curve after 15 mm displacement amplitude for the steel reinforced sample. This is because bilinear curve properties were used for the steel reinforcement. The tangent modulus of the bilinear curve will always be more than zero so there will be no breaking of steel bars in a theoretical analysis.

Both BFRP and steel reinforced shear wall FEA models were intensively calibrated to represent the experimental test results obtained.

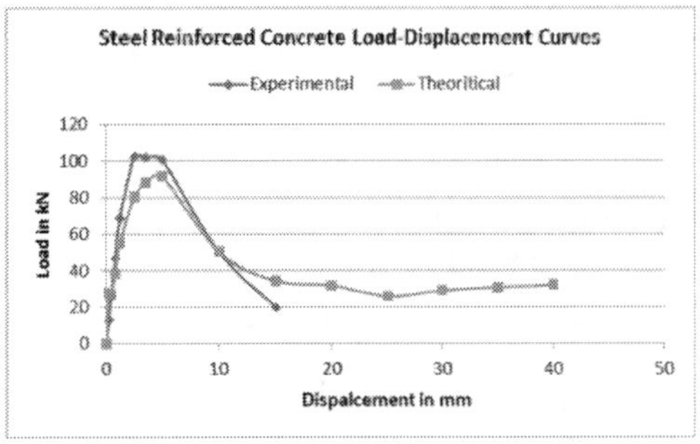

Figure 8a: Experimental and FEA results comparison curves for steel reinforced concrete

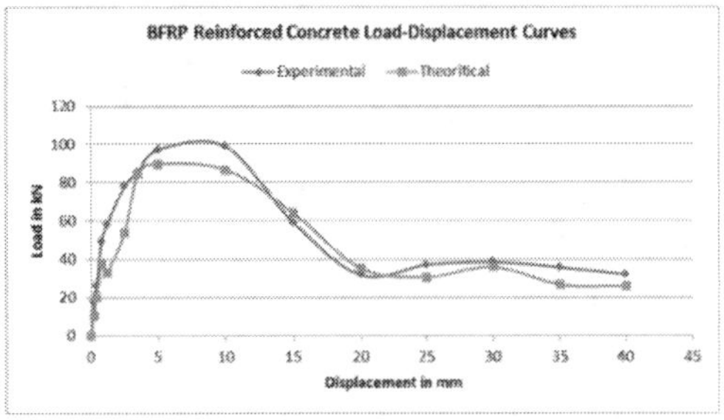

Figure 8b: Experimental and FEA results comparison curves for the BFRP reinforced concrete

5. Conclusions

The experiments conducted and the finite element analysis employed confirm the followings:

1. BFRP reinforced shear wall samples develop higher cumulative energy dissipation than the steel reinforced wall samples.
2. The experiment with steel reinforced sample was terminated when reaching the 20 kN residual resistance load at 15 mm displacement due to developing of a significant crack. The FEA results indicate potential remaining resistance above this level of displacement.
3. Obtained initial similarity between experimental and theoretical results is encouraging. Further refining of the FEA mesh will allow achieving even better correspondence between the obtained results.

6. References

1. N. Mohamed, A. Farghaly, B. Benmokrane and K. Neale, *Evaluation of GFRP-Reinforced Shear Walls*, CSCE General Conference, Quebec Canada, May (2013)
2. ACI-440.1R-06, *Guide for the Design and Construction of Structural Concrete Reinforced with FRP Bars*, American Concrete Institute (ACI), USA (2006)
3. Ansys help documentations, Ansys help systems, Ansys (2019)
4. J.M. Gere, *Mechanics of Material (6th edition)*, Thomson Engineering, Belmont, CA, USA (2003)
5. MagmaTech, *RockBar Brochure*, link http://magmatech.co.uk/downloads/RockBar.pdf (2019)
6. FIB-Bulletin No. 43, *Structural Connections for Precast Concrete Buildings*, Guide to Good Practice, Federation Internationale du Beton (FIB), Lausanne, Switzerland, (2008)
7. D. Kachlakev, T. Miller and S. Yim, *Finite element modelling of reinforced concrete structure strengthened with FRP laminates*, Final report No. SPR 316, USA, (2001).
8. EN-1992-1-1:2004, Eurocode 2: Design of concrete structures - Part 1-1: General rules and rules for buildings, (2004)
9. Applied Technology Council (ATC) 24: *Guidelines for Cyclic Seismic Testing of Components of Steel Structures*, Redwood City, CA (1992)
10. H. Rahman, T. Donchev, T. Petkova, D. Jurgelans and G. Hopartean, *Behaviour of Concrete Shear Wall with BFRP Reinforcement for Concrete Structures,* Proceedings of FRPRCS-14 Conference, Belfast, UK (2019)

Durability of concretes prepared with crystalline admixtures

Kosmas K. Sideris*, A. Chatzopoulos, Ch. Tassos, P. Manita.

Democritus University of Thrace, Department of Civil Engineers, Building B', University Campus of Xanthi, Kimmeria, Greece

Abstract. The objective of this work was to study the influence of crystalline admixtures on the durability of concrete. Four concrete mixtures – two reference concretes and two alternative mixtures-were produced during the first phase of the research. Influence of curing on the activation of the crystals was investigated on concrete slab specimens. The properties measured were the compressive strength and different durability indicators. The results revealed that crystalline admixtures enhanced the strength and the durability of the alternative mixtures.

1 INTRODUCTION

Concrete is the most commonly used construction material worldwide and the material with the second greatest consumption globally, just after water. One can say that it is concrete that has been used for the major infrastructures built during the last century all over the world and therefore is the material strongly connected with the development of our society and the improvement of our way of living. However concrete structures do not last forever; deleterious substances penetrate from the surrounding environment into concrete and react with its components resulting in its deterioration and/or corrosion of the embedded steel. In almost all cases, the reactants responsible for concrete deterioration penetrate into the concrete matrix with water. Since durability of concrete is of major concern, both for financial as well as for sustainability issues, reducing or even better minimizing the penetration of water should be of major priority. Water can enter concrete though the network of pores and capillaries that forms during cement hydration, or through cracks and other voids in the concrete. Therefore, almost all concrete structures require protection from water.

It is well known that a carefully designed concrete mixture which is properly produced and executed (transferred on site, cast, vibrated and cured) using a low w/cm ratio will result in a final product with low permeability and enhanced durability. However, no concrete is absolutely waterproof [2]. Concrete is a porous material and water can penetrate into its mass through pores and microcracks (formed initially or during the service life of the structure) due to capillary absorption, or due to hydrostatic pressure. Capillary absorption is the movement of water through the small pores in concrete in the absence of an externally applied hydraulic pressure. This is the result of surface tension between the water and the pore walls [3]. Permeability of concrete is the movement of water due to a pressure gradient, such as water in contact with a concrete structure installed underground [3]. A well-known and widely accepted method for reducing the permeability of concrete mixtures and therefore increase their durability is the addition of supplementary cementing materials (SCMs) into concrete mixtures [4,5]. As stated in ACI212.3R-16, a class of materials referred to as permeability-reducing admixtures (PRAs) have been developed to improve concrete durability through controlling water and moisture movement [6] as well as by reducing chloride ion ingress [7-8] and permeability [4]. According to ACI212.3R-16 [3], admixtures intended to reduce water ingress split into two subcategories: PRAs for concrete exposed to nonhydrostatic conditions (PRAN) and PRAs for concrete exposed to hydrostatic conditions (PRAH) [3].

2 EXPERIMENTAL PROGRAM

The research was divided into two phases. During the first phase of this research, four concrete mixtures were produced [7] - two conventional and two alternatives with added crystalline admixtures. The concrete mixtures were of C25/30 and C30/37 strength class, according to EN206-1. CEM I 42.5R and CEM II 32.5 (B-M) N cements were used to produce the mixtures. Aggregates were of limestone origin with a maximum grain diameter of 31.5 mm. Appropriate doses of retarder and polycarboxylate superplasticizer were used to achieve the desired workability and setting time.

The mixing ratio as well as the properties of the fresh concrete mixtures are shown in Table 2. The C30/37 PRAH mixture was produced with CEM II 32.5N cement instead of CEMI 42.5R in order to reduce the production cost of the alternative mixture.

Cubic specimens of 150X150X150mm and cylindrical specimens of 100x200mm were prepared. All samples were maintained in a curing chamber (T = 20°C, RH> 98%) until the test age. Cylindrical specimens with dimensions of 60X100mm were also used for measuring the carbonation depth.

* Corresponding author: kksider@civil.duth.gr

The 150mm cubes were used to measure the compressive strength at the ages of 2 and 28 days as well as to determine the water absorption. Resistance to chloride penetration of concrete was measured in cylindrical specimens with a diameter of 100mm and a height of 50mm cut off from the 100x200mm specimens. These specimens were maintained as above until the age of 28 days. The chloride diffusion coefficient Dnssm was then evaluated according to the procedure described in NT Building 492.

Table 1. Mixing ratio of concrete mixtures.

kg/m3	C 25/30	C 25/30-PRAH	C 30/37	C 30/37-PRAH
CEM I 42,5R	0	0	330	0
CEM II/(B-M) 32.5N	400	320	0	350
Sand	1042	1105	1133	1133
Gravel	180	190	195	195
Coarse	575	610	625	625
Water	200	160	165	160
superplasticizer	1.40	1.12	1.16	1.16
retarder	5.60	4.80	5	3.5
PRAH	--	2.56	--	2.8
W/C	0.50	0.50	0.50	0.46
S0 (cm)	25	23	24	21
S30 (cm)	20	22	22	19

Carbonation resistance was measured in cylindrical specimens 60x100mm. These specimens were maintained in the above-mentioned wet chamber for 3 days and then remained in a laboratory environment until the age of 28 days. After this age they were inserted into the accelerated carbonation chamber (RH = 55-60%, T = 20°C, CO2 = 1%) where they remained for 60 days.

During the second phase of this research we wanted to examine the effect of curing on activating the crystals. For this reason we produced five concrete slabs (5.0X5.0X0.15m) (Figure 1). The reference concrete used was a C20/25 strength class concrete according to EN206-1. CEM II 32.5 (B-M) N cement was used. Aggregates were of limestone origin with a maximum grain diameter of 31.5 mm. This concrete was used for casting of two slabs. The remaining three slabs were cast using the same concrete with the addition of crystalline admixtures (P-c20/25) at a percentage of 0.8% per weight of cement. The mix design characteristics of both concretes are presented in Table 2.

Fig. 1. Concrete slab specimens produced

Table 2. Mixing ratio of C20/25 concrete mixtures.

kg/m3	Reference-R-C 20/25	PRAH C20/25 P-C 20/25
CEM II/(B-M) 32.5N	280	280
Sand	1150	1150
Gravel	180	180
Coarse	655	655
Water	180	180
superplasticizer	1.10	1.10
retarder	0.56	0.56
PRAH	--	2.24
W/C	0.64	0.64
S0 (cm)	19	19
S30 (cm)	15	16

Cubic specimens of 150X150X150mm were prepared and used for measuring the compressive strength at different ages. All samples were maintained in a curing chamber (T = 20°C, RH> 98%) until the test age.

After casting, the concrete slabs were cured using different curing scenarios. Slabs 1 and 2- cast with reference concrete- were cured under wet burlap for one and seven days respectively. Slab 3–cast with the concrete containing crystalline admixtures- was cured under wet burlap for seven days. Slabs 4 and 5 were cured using a water jet machine (Figure 2). In this case a water jet was applied four seven and fourteen days respectively. The water jet was applied three times per day, 10 minutes every time.

Fig. 2. Curing PRAH containing concrete slabs with water under pressure.

At the age of 28 days, core specimens were taken from all slab specimens. The core dimensions were 100X100mm and 60X100 mm and were used for measuring the durability indicators of the elements.

3 EXPERIMENTAL RESULTS AND DISCUSSION

3.1 Compressive strength.

The compressive strength was measured for all concrete mixtures at the ages of 2 and 28 days. These values are shown in Table 3.

Table 3. Compressive strength of concrete mixtures (MPa)

(MPa)		C25/30	C25/30-PRAH	C30/37	C30/37-PRAH	REF-C20/25	C20/25-PRAH
2 days	fc_2	26.9	21.2	31.4	31.6	20.5	20.8
	S_{dev}	0.35	0.49	0.1	0.28	0.42	0.58
28 days	fc_{28}	39.3	41.1	54.2	64.0	32.8	34.6
	S_{dev}	0.3	0.1	0.7	0.5	0.51	0.28

The values of the above table show the positive effect of crystalline admixture on the compressive strength of the mixtures. Both C30/37 mixtures showed the same compressive strength at the age of 2 days. However, the mixture prepared with addition of crystalline admixtures exhibited an increased strength of 9.8 MPa (18.15%) relative to the reference mixture at the age of 28 days.

In the case of C25/30 mixtures, the 2-day strength of the alternative concrete is reduced by 5.65 MPa (21.04% reduction) relative to the reference mixture, due to the 20% reduced cement content in this mixture. However, the gradual growth of the crystals resulted in an increase in the later strength of the alternative mixture, which at 28 days was measured at 41.10 MPa, i.e. 1.8 MPa higher than the reference mixture (4.6 percent increase). In the case of C/20/25 concretes, both the 2 and 28 days compressive strengths were increased.

3.2 Water absorption

The water absorption of the concrete mixtures was estimated according to EN 13057 on 150mm cube specimens. The specimens remained in the laboratory curing chamber until the age of 28 days. Then they were dried at 65 °C until their weight was stabilized. The specimens were then properly prepared (wrapped around the insulating tape to ensure absorption only from the bottom of the specimen) and after weighing they were placed in basins of clean water in such a way that the water covered 3-5 mm of height of the specimen.

Measurements of weight change of the specimens were performed at 10 minutes, 30 minutes, 60 minutes, 90 minutes, 120 minutes, 180 minutes, 240 minutes, 480 minutes and 24 hours. These weights were divided by the exposed surface of the specimen into the water to calculate the water absorbance in gr/m^2.

Table 4. Water absorption of concrete mixtures

$kg/(m2.h0,5)$	C25/30	C25/30-PRAH	C30/37	C30/37-PRAH
$S_{coefficient}$	1.21	0.76	1.09	0.81

The rate of water absorption was measured on slab specimens using the Karsten method also described by EN1323. This method estimates the surface water absorption under atmospheric pressure. Plastic tubes were moulded on the surface of the slabs and filled with water

The rate of water absorption is presented for all slab specimens in Fig. 3.

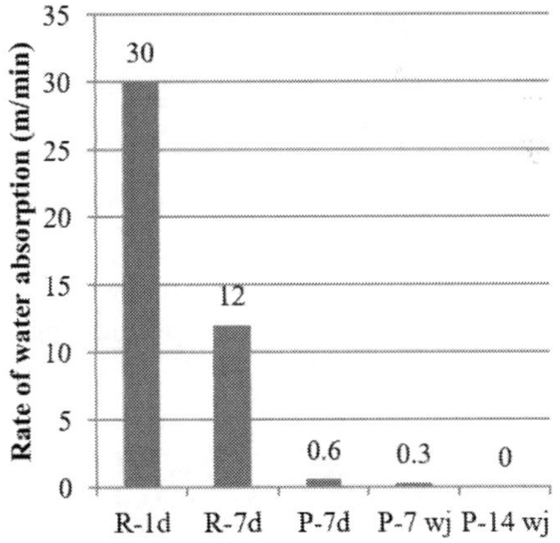

Fig. 3. Water absorption of concrete slabs cured with different curing scenarios.

It is well known that extension of the wet curing period facilitates the hydration of cement and therefore the more proper closure of porosity. Thus the extension of wet curing decreased the rate of water absorption of R-7d concrete slab by 60%. In the case of concrete slabs produced with crystalline admixture concrete, water absorption against atmospheric pressure was almost completely diminished.

3.2 Chloride penetration coefficient

The Dnssm (non-steady-state migration) chloride diffusion coefficient values were measured according to NT Building 492 Standard [8] and are presented for all mixtures of the first phase in Table 5.

Table 5. Dnssm chloride penetration coefficient (x10^{-12}m^2/s) of concrete mixtures.

x10^{-12}m^2/s	C25/30	C25/30-PRAH	C30/37	C30/37-PRAH
Dnssm	15.27	12.54	11.6	8.61

The Dnssm chloride diffusion coefficient values were measured according to NT Building 492 on standard diameter concrete cores taken from the slab specimens and are presented for these mixtures in Table 6.

Table 6. Dnssm chloride penetration coefficient (x10-12m2/s) of concrete slabs cured with different curing scenarios.

x10^{-12}m^2/s	R-1d	R-7d	P-7d	P-7dwj	P-14dwj
Dnssm	34.30	22.51	17.73	10.38	8.98

The above coefficients were used to calculate the lifetime of reinforced concrete constructions cast with these mixtures. For this purpose the Life 365 program was used. The following assumptions were made: Exposure class XS1, time for the propagation of corrosion 10 years, 50 mm reinforcement cover. The total service life time is shown in the following diagram.

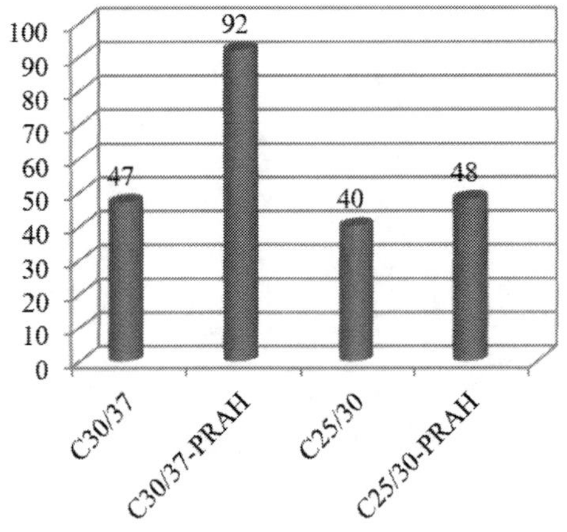

Fig. 4. Service life of concrete mixtures.

The figure above shows the particularly positive effect of the crystalline admixtures on the service life of the concrete. The C30/37-PRAH mixture develops twice the lifetime of penetration of chlorides compared to the reference C30/37, while the C25/30-PRAH mixture has an increased lifetime of 20% over the reference C25/30. Crystal activation in both cases led to a drastic reduction in porosity, thus preventing penetration of chlorides into the mixtures.

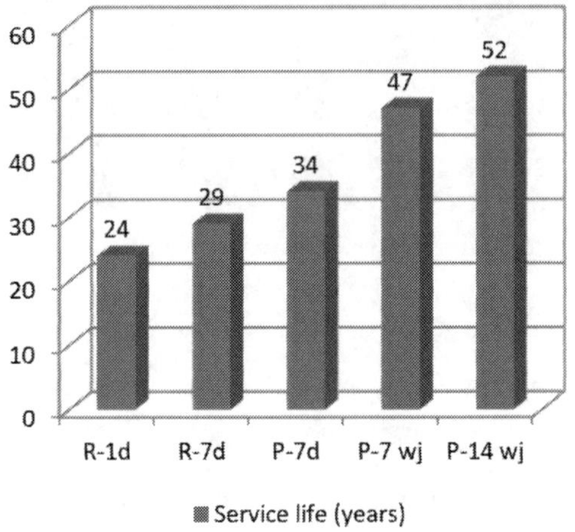

Fig. 5. Service life of concrete slabs cured with different curing scenarios.

The figure above shows the positive effect of the combination of using crystalline admixtures and extending the curing period on the service life of the concrete elements. Wet burlap curing increased the service life of the slab produced with crystalline admixture (P-7d) by 15% due to partial activation of the crystals. Since crystalline admixtures are hydrophilic at the beginning of their action, more water is needed for their activation. This additional water was provided by water jet. As indicated above, this method resulted in the formation of greater amounts of crystals that filled the pores and positively affected the resistance of concrete mixtures against ingress of chlorides. Thus the service life of P-7d wj and P-14d wj slabs was increased by 69% and 116% as compared with the reference slab (R-1d). What is more impressive is that even this poor concrete with the high w/c ratio fulfilled the requirements of Eurocodes for a design service life of 50 years when crystalline admixtures were added in the mixture and the concrete elements were properly cured.

3.3 Carbonation depth

The carbonation depth was measured according to the procedure described in EN 14630 Standard [9]. The specimens remained for 60 days in the Accelerating Carbonation Chamber. The carbonation depth of all mixtures of the first phase is shown in Table 7.

Table 7. Carbonation depth of concrete mixtures.

	C25/30	C25/30-PRAH	C30/37	C30/37-PRAH
Carbonation depth (mm)	8,7	9,1	2,9	4,5

The carbonation depth of the concrete slabs was evaluated on 60X100 mm core specimens drilled from the slabs at the age of 28 days. After this age, core specimens were inserted in the accelerated carbonation chamber and remained there for 60 days. The carbonation depth, measured according to the EN 14630 Standard, is shown in Table 8.

Table 8. Carbonation depth of concrete slabs cured with different curing scenarios.

	R-1d	R-7d	P-7d	P-7d wj	P-14d wj
Carbonation depth (mm)	19,8	18,8	16,5	14,2	12,5

Based on the values of Table 8, it appears that the carbonation depth of the prepared mixtures is affected by the addition of the crystalline admixtures in a different way. In particular, the C25/30-PRAH mixture exhibits a slightly increased 5% carbonation depth relative to the reference mixture while the C30/37-PRAH mixture exhibits a carbonation depth increased by 55% relative to the C30/37 reference mixture. This may be due to the non-activation of the crystals; the specimens remained in a wet curing chamber for only three days, unlike the specimens used in the previous tests and which were maintained in the laboratory's wet curing chamber for 28 days. In addition, it should be noticed that the alternative C25/30-PRAH mixture was prepared with a 20% lower cement, while the C30/37-PRAH alternative was prepared using CEM II (B-M) 32.5N cement more prone to carbonation due to its high pozzolan content.

The significance of curing slabs containing crystalline admixture with water under pressure for extended curing periods is clearly demonstrated in Table 7. Slabs produced with PRAH concrete (P-7d) showed a 5.3% lower carbonation depth as compared with slabs produced with the reference concrete (R-7d). In the case of PRAH concrete slabs cured under water pressure, the carbonation depth was decreased by 21.3% (P-7dwj) and 31.91% (P-a4dwj) respectively.

Long-term outdoor exposure experiments are underway in the Building Materials Laboratory in Xanthi in order to calculate the carbonation coefficient k and to assess more accurately the service life against carbonation of reinforced concrete structures exposed to environmental class XC4.

4 Conclusions

The crystalline PRAH additives used in the present study appear to be an effective way of sealing the porosity of the concrete prepared therewith. The presence of moisture is a catalytic factor for the activation of crystals which thus clog the pores of the cement and prevent the penetration of harmful fluids and gases. The alternative mixtures with crystalline PRAH additives that have been tested in the present study showed increased compressive strength, lower water absorption coefficient, and significantly lower permeability to chlorides than reference mixtures, thus enhancing the lifetime against seawater originated chlorides of reinforced concrete structures. These mixtures also exhibited a higher carbonation depth, due to the use of a different dosage and type of cement as well as to the low water curing period, which did not help to activate the crystals.

It was also found that curing concrete specimens with water under pressure is essential for the formation of crystals and the increase of durability of concrete elements.

Crystal admixtures and chemical additives used in the above experiments were provided by Penetron Hellas SA and NORDIA SA. Their support to the research work of the Building Materials Laboratory is greatly acknowledged.

References

1. *EN 206-13, Concrete - Specification, performance, production and compliance*, CEN, (2016)
2. P. Perkins, *Repair, Protection and Waterproofing of Concrete Structures,* Elsevier Applied Science Publishers, (1986), pp. 301
3. ACI Committee 212, *Report on Chemical Admixtures for Concrete (ACI 212.3R-16)*, American Concrete Institute, Farmington Hills, MI, (2016), pp. 43-48
4. R.L. Munn, G. Kao, and Z.T. Chang, Durability Performance of Australian Commercial Concrete Modiied with Permeability Reducing Admixture, *The 22nd Biennial of the Concrete Institute of Australia*, Melbourne, Australia, (2005)
5. N. De Belie, M. Soutsos, E. Gruyaert, *Properties of Fresh and Hardened Concrete Containing Supplementary Cementitious Materials, State-of-the-Art Report of the RILEM Technical Committee 238-SCM,* Working Group 4, Springer Publishing, (2018), ISBN 978-3-319-70605-4
6. S.K. Roy, and D.O. Northwood, Admixtures to Reduce the Permeability of Concrete, *Durability of Concrete: Proceedings of the Fourth CANMET/ACI Inter-national Conference*, SP-170, V. M. Malhotra, ed., American Concrete Institute, Farmington Hills, MI, (1999), pp. 267-284
7. K.K. Sideris, C. Tassos, A. Chatzopoulos, P. Manita, Influence of PRAH crystalline admixtures on the durability of concrete. *Proceedings of the 18th Greek Concrete Conference*, (2018), pp. 83
8. *NT-Build 492, Chloride migration coefficient from non-steady-state migration experiments*, Nordtest methods, (1999)
9. EN 14630. Products and systems for the protection and repair of concrete structures - Test methods - *Determination of carbonation depth in hardened concrete by the phenolphthalein method,* CEN, (2006)

Field investigation of slab deterioration due to carbonation in a concrete building

Eleni B. Araya[1,*], *Stephen* Ekolu[1] and *Fitsum* Solomon[1]

[1]Department of Civil Engineering Science, University of Johannesburg, Auckland Park 2006, Johannesburg, South Africa

Abstract. This paper presents an investigation on a deteriorated slab of a two-story building located in Johannesburg, South Africa. The structure has been in use as a residential building for more than 60 years. The top slab of the first floor was delaminated and reinforcement bars were severely corroded. Condition assessment of the building was done using the Schmidt hammer test, cover-meter survey and half-cell potential measurements. In addition to the non-destructive tests done, X-ray diffraction and scanning electron microscopy were also conducted on the deteriorated concrete.

Key words: Carbonation, Cover, Corrosion, Non-destructive tests, Reinforced concrete, Spalling, Site inspection.

1. Introduction

Reinforced concrete is the most widely used construction material due to its great versatility. However, the steel reinforcement is thermodynamically unstable under normal atmospheric conditions and converts into its natural state which is iron oxide or rust. This phenomenon of steel conversion into rust or iron oxide is the corrosion process [1]. Carbonation and chloride attack are the two main causes of corrosion of steel in concrete. The interaction of carbon dioxide in the atmosphere with the alkaline hydroxides in the concrete causes a reduction in pH which can lead to carbonation-induced corrosion. When carbon dioxide dissolves in pore water it forms carbonic acid that neutralizes the alkalis and forms calcium carbonates. Gradually as this reaction continues, it reduces the pH of concrete from 12.5 to a level below 9.0 at which steel corrosion can take place with available sufficient air and moisture. Chloride ions can penetrate into concrete from sea water, salty groundwater or de-icing salts. Once they come into contact with the steel in concrete, chlorides facilitate the breakdown of the passive layer of oxide on the steel and lead to corrosion [2-6]. Corrosion in reinforced concrete structures results in the loss of steel cross-sectional area and loss of bond between steel and the surrounding concrete. Furthermore, the products of corrosion expand and induce tensile stresses in the surrounding concrete. This stresses lead to internal cracking, cover spalling and delamination [7-8].

Before commencing any repair strategy, an appropriate diagnosis of the deterioration process is necessary. Diagnosis of distress can be conducted using condition assessment methods that commonly involve non-destructive (NDT) field testing and laboratory analyses on core or dust samples [9-11]. One of the most widely used NDT corrosion assessment methods is the half-cell potential procedure standardized in ASTM C876 [12].

In this study, various in-situ tests and laboratory analyses were done including visual inspection, Schmidt hammer test, cover-meter survey and half-cell potential measurements. In addition, X-ray diffraction (XRD) and Scanning Electron Microscopy (SEM) were also conducted on the concrete samples taken from the deteriorated slab.

2. Background

An investigation was conducted to evaluate corrosion attack in a reinforced concrete building that had been in service for more than 60 years in Johannesburg, South Africa. This was done by conducting condition assessment. From the visual inspection conducted, it was observed that there was extensive spalling of concrete cover and severe corrosion of exposed reinforcement at the soffit of the roof slab shown in Fig. 1. To extend the service life of the building, repair and maintenance of the structure would be required. Therefore, conducting a detailed investigation was necessary to evaluate the condition of the building for continuation of its service to the residents.

Johannesburg is an inland city with tropical climate conditions. As such, no chloride exposure conditions arise. However, it is a major urban city with substantial industrial pollution responsible for carbonation as the main mechanism of steel corrosion, as already established in past studies [4-5, 9-10].

* Corresponding author: arayaeleni@yahoo.com

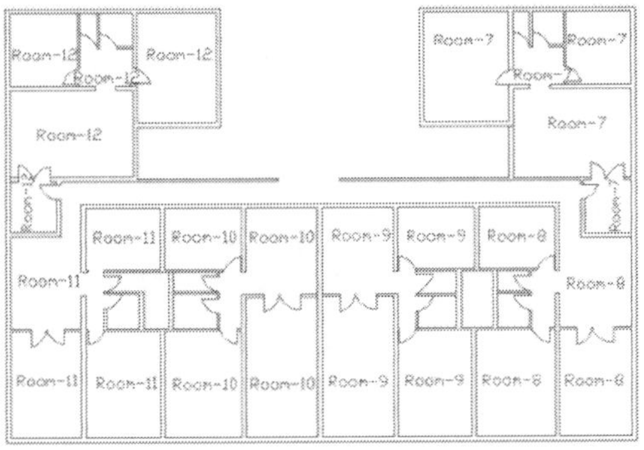

Fig. 1. First floor plan of the reinforced concrete building labeled from Room-7 to Room-12.

During this study, the structure was in use as a community residential building. Accordingly, the regime of tests were restricted to only non-destructive measurements. The main objectives of this research were:

- To assess and identify the type and mechanism of corrosion occurring at the two-story reinforced concrete building by conducting condition assessment.
- To investigate the corrosion attack by conducting an analytical study of the deteriorated concrete.

3. Condition assessment of the building

3.1 Field measurements

3.1.1 Visual inspection

Visual inspection is typically the first step in condition assessment. Hence, the assessment commences visual inspection with photographic recording and documenting the visible deteriorations such as spalling, loss of concrete cover and corroded reinforcement bars. Fig. 2 gives a distressed area showing total loss of the concrete cover and severe corrosion of bars at the soffit of the roof slab.

Fig. 2. Exposed rebars and delamination at Room-9 of Fig. 1.

3.1.2 Schmidt hammer test

The Schmidt hammer test was done to determine the in-situ compressive strength of the concrete. The strength measurements were conducted on all 26 rooms (from Room-7 to Room-12) in 1m x 1m grid space of the slab in Fig. 1. Altogether 1294 data points of rebound values were collected. Using the manufacturer's data, the rebound values have been converted into corresponding compressive strengths. Fig. 3 shows the Schmidt hammer test in progress.

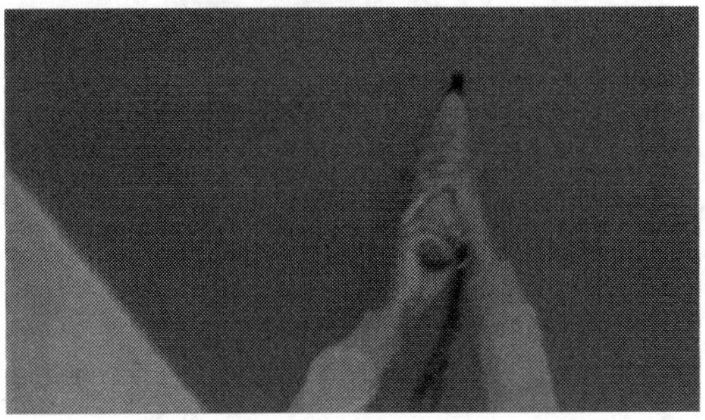

Fig. 3. Schmidt hammer measurement at the soffit of the roof slab.

The indicative concrete strength data obtained have been plotted in a histogram shown in Fig. 4. The mean compressive strength was found to be 17.1 Mpa with a standard deviation of 4.9.

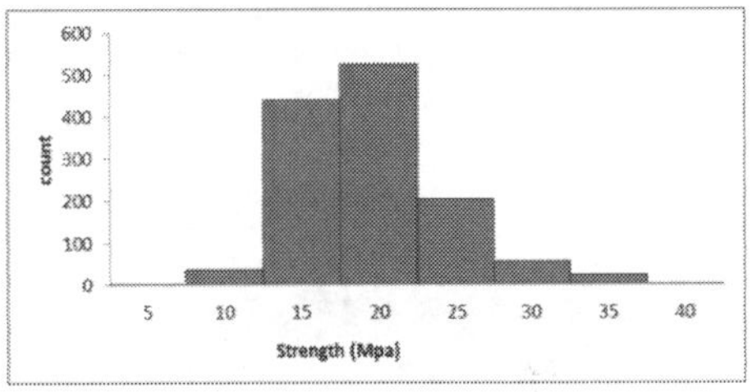

Fig. 4. Histogram of Schmidt hammer test data.

3.1.3 Cover-meter survey

A cover-meter survey was conducted to determine the concrete cover depth and to check bar location. The same grid of 1m x 1m as mentioned in Section 3.1.2 was used for the cover-meter survey as well. Fig. 5 shows the actual cover meter measurements being conducted. Fig. 6 gives a plot of the concrete cover results in a histogram. The 2580 data measurements collected, gave a mean concrete cover of 19.5 mm with a standard deviation of 5.2.

Fig. 5. Cover-meter measurement at the soffit of the roof slab.

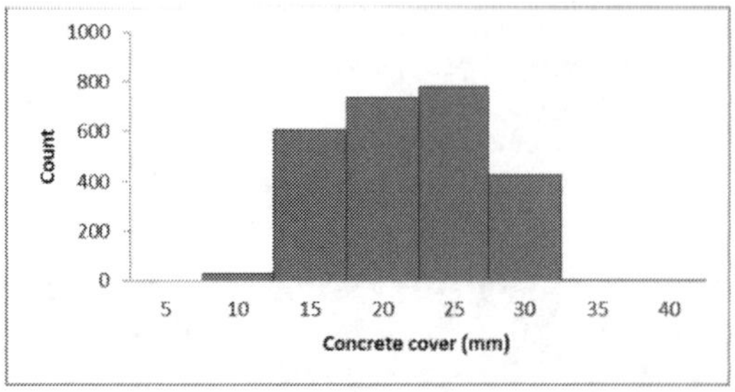

Fig. 6. Histogram of the cover-meter results.

3.1.4 Half-cell potential measurement

The half-cell potential method was used to asses the corrosion level of steel reinforcement. A total of 155 data sets were collected. Fig. 7 shows the half-cell potential measurement in progress. The measurements were taken at random spots across the slab soffit. Also given in Table 1 are data showing the minimum and maximum values obtained at the various locations. It can be seen that the readings ranged from +85 mV to -687 mV. All the data obtained has also been plotted in a histogram shown in Fig. 8. The mean half-cell potential reading was -202.9 mV with a standard deviation of 138.0. Most highly negative half-cell potential readings were recorded at points where there was roof leakage and delamination. These highly negative readings were likely due to oxygen starvation.

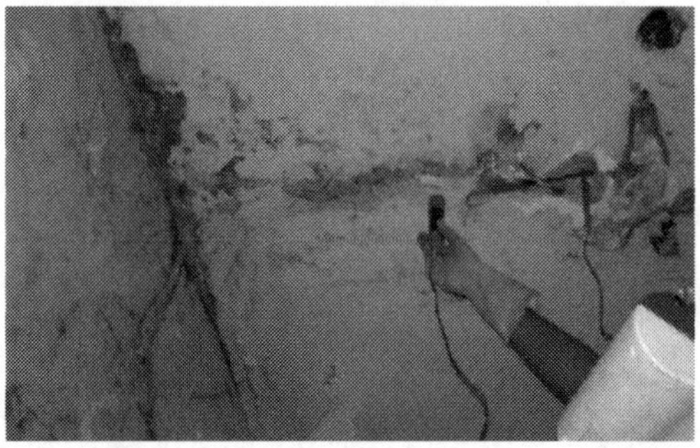

Fig. 7. Ongoing half-cell potential measurement at the soffit of the roof slab.

Table 1. Maximum and minimum values of the half-cell potential readings.

Flat	Half-cell potential reading (mV)		Number of data collected
	Maximum	Minimum	
Room-7	85	-247	58
Room-8	-246	-481	18
Room-9	-15	-687	29
Room-11	-114	-420	50
Total			155

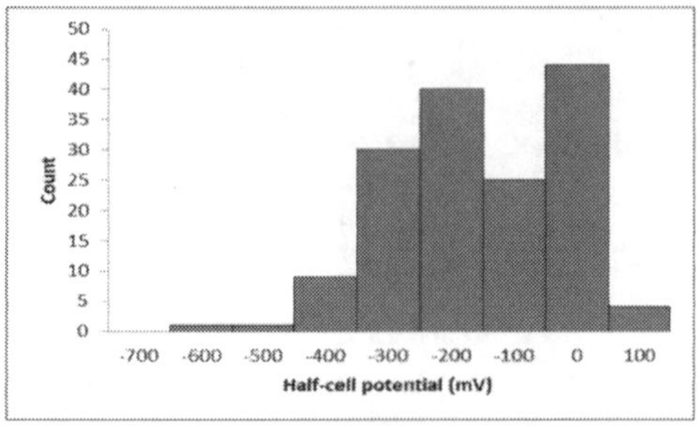

Fig. 8. Histogram of the half-cell potential measurement data.

3.2 Analytical study

After the field condition assessment was done, an analytical study was conducted on concrete samples collected from the RC building.

3.2.1 X-ray diffraction

XRD analysis was conducted at 40 KeV, 15 mA using a Rigaku MiniFlex 600 Benchtop diffractometer with Cu-Kα radiation (λ = 1.5418 Å) over 2θ angle of 0° - 90° at room temperature. Fig. 9 shows the presence of calcium carbonate, which may be attributed to carbonation as per Equation (1).

$$Ca(OH)_2 + CO_2 \longrightarrow CaCO_3 + H_2O \quad (1)$$

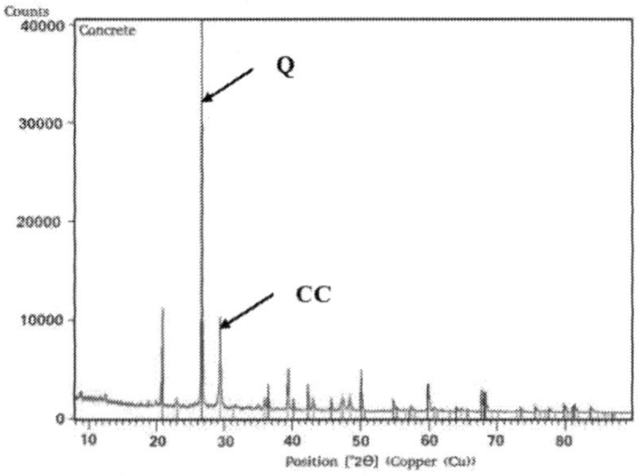

Fig. 9. XRD pattern of the concrete sample, showing calcium carbonate (**CC**) and quartz (**Q**).

3.2.2 Scanning Electron Microscopy

SEM coupled with Energy Dispersive Spectroscopy (EDS) were used to examine the morphology (see Fig. 10) and the presence of metal ions (see Fig. 11) in the concrete sample, respectively. The presence of calcium (Ca), silicon (Si), carbon (C), and oxygen (O_2) in the concrete sample was identified using EDS analysis (see Fig. 11) which supported the XRD results.

Fig. 10. SEM image of the concrete sample. Symbol C indicates the location where the EDS analysis was performed.

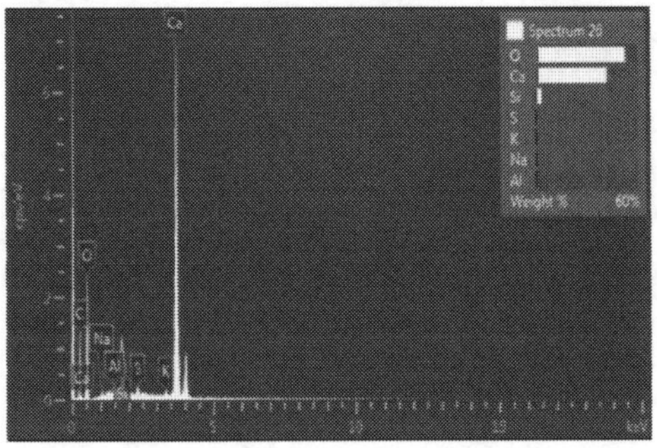

Fig. 11. The result of EDS analysis taken at the point marked with symbol C in Fig. 11.

4. Conclusion

Field investigation involving condition assessment was conducted to evaluate deterioration of a community residential structure. In addition, an analytical study of the concrete samples taken from site was done which indicated carbonation attack to be the mechanism responsible for the observed corrosion of steel reinforcement in the reinforced concrete building. Concrete in the soffit of the roof slab was delaminated and steel reinforcement was severely corroded due to carbonation-induced corrosion. Further research is ongoing to determine appropriate repairs that may be employed.

Acknowledgement

The paper is based on the BERTRAMS Community Upliftment Project conducted under the NRF community engagement fund through the collaboration of Tshwane University of Technology (TUT) Architecture Department, CARINBE and Civil Engineering Science Department of the University of Johannesburg (UJ). The author would also like to acknowledge and thank the Global Excellence Stature (GES) of UJ, for the scholarship funding of Eleni Araya's masters study program.

References

1. PCA., *Types and causes of concrete deterioration*, Portland Cement Association (PCA): Skokie, IL, USA (2002)
2. J.P. Broomfield, *Corrosion of steel in concrete: understanding, investigation and repair*, CRC Press, 17-21 (2003)
3. S.O. Ekolu, M.D.A. Thomas and R.D. Hooton, Pessimum effect of externally applied chlorides on expansion due to DEF - proposed mechanism, Cement and Concrete Research, **36**, 688-696 (2006)
4. S.O. Ekolu, A review on effects of curing, sheltering, and CO_2 concentration upon natural carbonation of concrete, Construction and Building Materials, **127**, 306-320 (2016)
5. S.O. Ekolu, Model for practical prediction of natural carbonation in reinforced concrete: Part 1-formulation, Cement and Concrete Composites, **86**, 40-56 (2018)
6. S.O. Ekolu and M. Allan, *Treatment of a stochastic service life prediction model to an evaluation of a distressed two-story RC building*, Proc. 3rd Intl. Conf. on Concrete Repair, Rehabilitation and Retrofitting (ICCRRR), Cape Town, South Africa, 451-457 (2012)
7. S. Guzmán and J.C. Gálvez, Modelling of concrete cover cracking due to non-uniform corrosion of reinforcing steel, Construction and Building Materials, **155**, 1063-1071 (2017)
8. V. Talakokula, S. Bhalla, R.J Ball, C.R. Bowen, G.L. Pesce, R. Kurchania, B. Bhattacharjee, A. Gupta and K. Paine, Diagnosis of carbonation induced corrosion initiation and progression in reinforced concrete structures using piezo-impedance transducers. Sensors and Actuators A: Physical, **242**, 79-91 (2016)
9. W.F van Ede and S.O. Ekolu, *Condition assessment of a Johannesburg skyscraper*, Proceedings of the International Conference on Construction Materials and Structures (ICCMATS), Johannesburg, South Africa, 1052-1059 (2014)
10. S.O. Ekolu, *Model verification, refinement and testing on independent 10-year carbonation field data*, Proc. 3rd Intl. Conf. on Concrete Repair, Rehabilitation and Retrofitting (ICCRRR), Cape Town, South Africa, 445-450 (2012)
11. P. Gu & J. Beaudoin, Obtaining effective half-Cell potential measurements in reinforced concrete structures, Corros. Sci., 97 (1998)
12. ASTM International C876-15, *Standard test method for corrosion potentials of uncoated reinforcing steel in concrete*, West Conshohocken, PA, (2015)

Structural condition assessment of a reinforced concrete driveway

Piotr Berkowski[1,*], *Grzegorz* Dmochowski[2]

[1]Wroclaw University of Science and Technology, Faculty of Civil Engineering, Wyspiańskiego Blvd., 27, Wroclaw, Poland
[2]Wroclaw University of Science and Technology, Faculty of Civil Engineering, Wyspiańskiego Blvd., 27, Wroclaw, Poland

Abstract. The reinforced concrete structure of a driveway built in the eighties of the 20th century is under consideration. It is a single-storey structure with a length of approx. 66.00 m, made in the form of a reinforced concrete skeletal structure, consisting of 7 frames with cantilevers, spaced at 9.00 m: 3 frames in the horizontal part of the driveway are three-span ones, and 4 in the sloping part are single-span frames. The driveway frame main beams have a rectangular cross-section and are based on pillars that have been placed on the spread footing - these elements form a monolithic structure. Prefabricated reinforced concrete hollow floor plates were laid on them in the central part of the driveway, and in the outer parts the driveway floor was made of hollow clay slab blocks (Ackerman type). At the time of investigation, the upper layers of the driveway consisted of a surface waterproofing layer, a concrete slab of approx. 10 cm thickness made on a 1 cm sand layer. However, no internal waterproofing layer was found as was in the original design. The paper describes the process of determining the current technical condition of the reinforced concrete driveway structure, paying attention to the design and construction of the object, its operation (including the issues of durability of concrete under the conditions of possible impacts of the marine environment). Static and strength calculations were also performed (supported by concrete and reinforcement tests) and the conditions for further operation of the driveway were analyzed as well as concepts of rehabilitation.

1 Description of the driveway structure

The reinforced concrete structure of a driveway was built in the early eighties of the 20th century in the immediate vicinity of the sea coast. Its structural layout results from the terrain on which it is located, and it is functionally used as an access ramp to the level of the third floor of the adjacent building (Fig. 1).

Fig. 1. Location of the driveway.

The driveway, in addition to the access function, is used as a parking place for private cars. Besides, parking spaces are also located under its horizontal part.

From the structural point of view the driveway is a single-storey structure made in the form of a reinforced concrete skeletal structure, and consists of 7 frames (Fig. 2) with cantilevers: 3 frames in the horizontal part of the driveway are three-span ones, and 4, in the sloping part, are a single-span ones. At the level of the upper girders, frames are monolithically connected by main beams, situated along the length of the driveway.

Fig. 2. Bearing frames and main beams of the driveway.

The slab of the structure is made of prefabricated, hollow plates in the central part of the driveway, and in its outer

*Corresponding author: piotr.berkowski@pwr.edu.pl

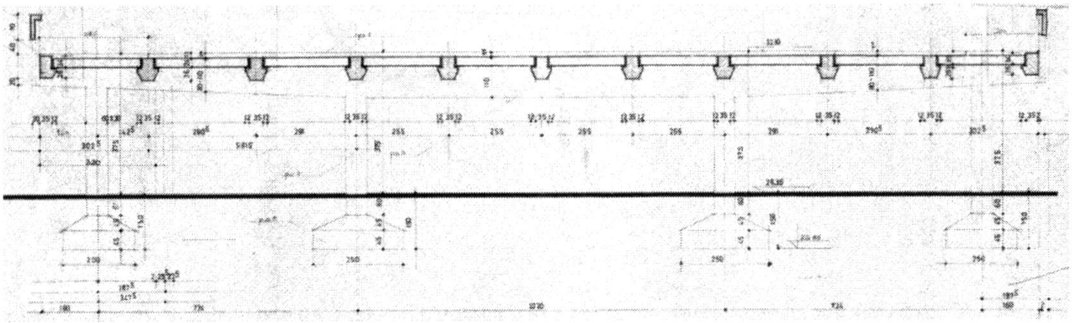

Fig. 3. Cross-section of the driveway bearing structure [2].

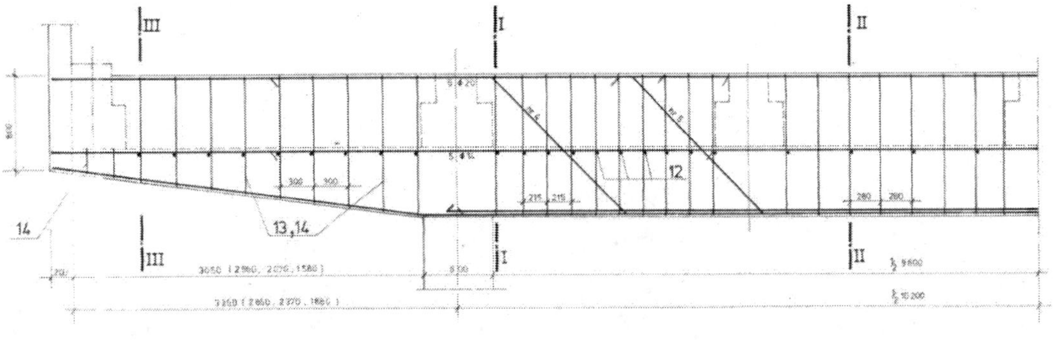

Fig. 4. Reinforcement of the transverse frame [3].

parts the plate is constructed in the form of the ribbed floor slab made of the Ackerman hollow clay blocks [1]. The pavement of the driveway is made in the form of a concrete slab.

1.1 Analysis of the existing old technical documentation

The original design documentation from 1973 consists of two parts: architectural design [2] and constructional one [3].

The first of them contains a technical description and architectural drawings of the driveway with all functional, geometrical and material data. In this design the platform and passage were designed as a monolithic reinforced concrete structure with transverse frames and longitudinal ribs (Fig. 3). The platform deck was designed from prefabricated hollow plates, and its trapezoidal parts – from Ackerman blocks. The surface of the pedestrian part was designed from flagstones and the driveway with an asphalt pavement.

The second part of the documentation consists of technical description, structural calculations and constructional drawings (Fig. 4) of the platform and driveway based on the architectural design. The driveway was calculated for a service load of 5 kN/m^2 or from a vehicle of 20 kN. The concrete strength of R_w = 200 kg/cm^2 (20 MPa) for prefabricated elements and R_w = 170 kg/cm^2 (17 MPa) for in-situ elements were assumed. Ribbed reinforcement steel with a Q_r = 4.200 kg/cm^2 (420 MPa) and a smooth reinforcement with Q_r = 2.500 kg/cm^2 (250 MPa) were adopted.

The following structural models were applied in the static calculations:

- for Ackerman floor – a single-span beam, freely supported;
- for longitudinal beams – 4- or 5-span continuous beams;
- for transverse frames – for girder of a 1-span frame: a single-span beam with cantilevers; for girder of a 3-span frame: a 3-span continuous beam with cantilevers;
- for transverse frames columns – axially compressed columns;
- for foundations – axially loaded footings.

A material assessment regarding the quality of concrete in the construction of the driveway was conducted in 2005 [4]. The documentation contains results of concrete tensile strength tests (adhesion) using the "pull-off" method, chemical analysis for SO_4^{2-} and Cl^- ion content as well as determination of the pH value. The results were as follows:

- tensile strength of concrete ranging from 0.51 to 1.78 MPa in monolithic elements that proved the diversity of concrete made at the construction site; for a prefabricated slab, the tensile strength of concrete was 2.56 MPa which proved the better quality of concrete in the prefabricated panels;

- the content of SO_4^{2-} ions (sulfates) was determined from 0.021 to 0.112%, which is the quantity normally found in concrete;
- the content of Cl^- ions in the concrete of the bottom parts of the structure, despite pillars, was at the level that did not cause corrosion of reinforcement and concrete; the pH value in all samples tested, except for the columns, was greater than the limit value of 11.4 pH (or 12.0 pH for the occurrence of Cl^- ions); only in one of the pillars was the pH value a little lower than the limit that could start the corrosion process.

1.2 Current structural system of the driveway

The driveway consists of 3 main functional elements: a slanted part with a slope of approx. 6° and a trapezoidal plan of approx. 13.0-18.0 per 45.0 m, a horizontal parking platform with dimensions of approx. 21.0-28.0 m and the remnants of the pedestrian communication platform (terrace). The total length of the construction of the driveway is approx. 66.0 m (Fig. 5).

Fig. 5. General view of the driveway in 2018.

The supporting structure of the driveway is a system of 7 monolithic reinforced concrete frames with cantilevers: 4 frames are single-span and 3 are three-span ones. The length of the central span is approx. 10.20 m and of the extreme spans – approx. 7.24 m. The cantilevers have variable outreaches of 2.08, 2.57, 3.06 and 3.55 m. The frame girders have a cross-section of 60x120 cm and at the end of the cantilever part – 38x60 cm. The main columns of the frames have a cross-section of 40x60 cm with 10 bars of 14 mm diameter, while the end columns of two-span frames and a pedestrian bridge – 30x60 cm with 12 bars of 14 mm diameter. The frame girders are bases for longitudinal ribs in the shape of an inverted T-letter sections with dimensions 60x60 cm, spacing 2.55 to 2.90 m, and on the cantilevers – side ribs with a cross-section of 35x60 cm, running along the driveway. The side rib along the pedestrian bridge has a width changing from 47 to 35 cm and a height of 60 cm. The transverse frame system (transoms with brackets and posts) and longitudinal ribs form a monolithic system.

Fig. 6. View of the driveway main bearing structure.

On the shelves of the ribs, prefabricated hollow plates with a thickness of 24 cm, width 120 cm and length 217 cm (reinforced with 10 mm bars) are mounted, and in the trapezoidal end areas a ceiling is made of Ackerman airbricks (reinforced in ribs with 18 mm bars) with a 4 cm concrete upper layer (Fig. 7).

Fig. 7. View of the Ackerman slab.

The foundation of the columns was designed as the footing 250x250 cm with slants and height 90 cm, on lean concrete. Foundations were placed 1.50 m below ground level.

The driveway is equipped with reinforced concrete balustrades with a height of 1.10 m, consisting of posts 25x30 cm, mounted at the ends of cross-frame brackets, and topped with inverted L beams, 70 cm high and 15 cm thick (Fig. 8).

The following layers of driveway floor were found on the basis of the core drillings made in the places where the prefabricated hollow plates constitute the slab - from the top of the slab (Fig. 9):

- coating of waterproofing material on the leveling layer – 1.0-1.5 cm;
- main concrete layer – 10 cm;
- sand and cement bedding – 1 cm;
- prefabricated hollow slab – 24 cm;
- plasterboard – 0.5 cm.

https://doi.org/10.1051/matecconf/201928910002

Fig. 8. Driveway concrete balustrades.

Fig. 9. Core drilling through the driveway slab.

In the outer, trapezoidal, bays instead of prefabricated hollow plates an Ackerman hollow clay blocks ceiling was used with 20 cm thickness and an upper concrete layer of 4 cm thickness.

At the time of conducting the assessment, the upper pavement layer at the whole surface of the driveway was protected with different kinds of waterproofing materials as the owner tried to found out the best solution to prevent the continuous leakage of rainwater and melted snow in order to stop concrete destruction (Fig. 10).

Fig. 10. Different protection layers on driveway surface.

2 Assessment of the actual driveway technical condition

A lot of problems connected with leakage of rainwater and melted snow (with de-icing salts) [5, 6] through the platform slab occurred during the previous exploitation of the driveway. At the same time, several technical assessments were prepared and the main causes of the degradation of the driveway structural elements were defined as well as solutions of repair or protection proposed.

As the life of the driveway structure was close to the serviceability limit, a decision about comprehensive technical assessment [7, 8] and development of the repair concept was taken by the owner. Below, elements of the conducted evaluation process are described.

2.1 Description of driveway damage

The following characteristic types of damage of structural and finishing elements [9] were found:

1. Spalling on the surfaces of reinforced concrete elements (Fig. 11).
2. Local surface destruction and material losses due tocorrosion of bars and reinforcement due to insufficient thickness of concrete cover, especially for stirrups (Fig. 11).
3. Local corrosion damage (defects in material) of the exposed main reinforcement bars connected with the exposure of the reinforcement cover (Fig 12).
4. Local losses of the material of the Ackerman blocks (dropping of the lower parts of blocks) and corrosion damage (destruction and loss) of the exposed reinforcement (delamination and cracking of the ceramic as a result of freezing water penetrating the leaky surface) (Fig. 13).
5. Local damage of hollow-core slabs (dropping of lower plate fragments) and corrosion damage of exposed reinforcement (delamination and loosening of concrete lagging as a result of freezing water penetrating through the leaky surface) (Fig. 14).

Fig. 11. Spalling and lack of proper rebar cover.

https://doi.org/10.1051/matecconf/201928910002

Fig. 12. Reinforcement corrosion.

Fig. 13. Destruction of Ackerman hollow clay blocks slab.

6. Local cavities and delamination of concrete, corrosive damage (destruction and defects of material) of reinforcement bars in balustrades.
7. Delamination (de-bonding) at the edge contact of the end ribs and the curb of the driveway plate.
8. Organic contamination (overgrowing with mosses), resulting in the destruction of plaster and surface concrete (Fig. 15).
9. Freeze-thaw destruction of the driveway surface and material losses and delamination of the pavement surface (Fig. 16).
10. Prolonged leaching of concrete material components as a result of water penetration – water stains and stalactites on the lower surfaces of plates as well on frames and ribs surfaces (Fig. 17).

Fig. 14. Destruction of prefabricated hollow slabs.

Fig. 15. Destruction due to organic contamination.

Fig. 16. Destruction of driveway pavement surface.

Fig. 17. Salt infiltrations on the surface of structural elements.

2.2 Description of measurements and tests carried out

In the procedure of assessing the technical condition of the driveway [7, 8, 9] the following material and structural examinations were made:

- reinforcement openings in different elements of the driveway structure;
- testing of concrete strength with Schmidt hammer;
- an open-pit of the foundation footing of the transverse frame column.

In order to verify the structural reinforcement, its identification was carried out at 16 points: in 5 new pits and in 11 places with damaged concrete cover where the reinforcement was exposed. A PMD10 Bosch metal detector was used to determine the reinforcement situation and a YATO YT-7201 electronic caliper was used to measure reinforcement diameters. Generally, in all the examination points, the reinforcement was as designed. Moreover, in new reinforcement openings no corrosion of the reinforcement was revealed (Fig. 18). However, in previously made openings and in the places where the reinforcement was exposed due to damage of concrete cover, corrosion of the reinforcement was found, with varying intensity (Fig. 19).

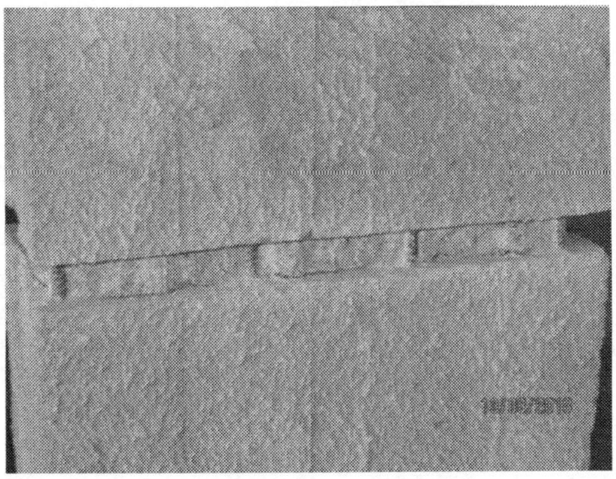

Fig. 18. Reinforcement checking opening in column.

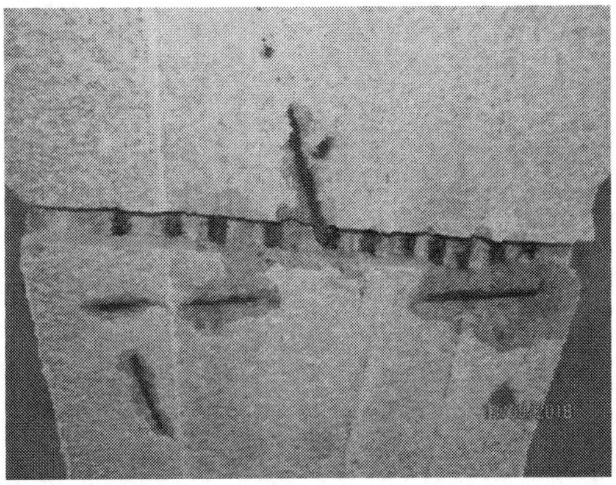

Fig. 19. Reinforcement checking opening in frame girder.

A Proceq Schmidt type N-34 hammer was used to verify the strength of the concrete. On the basis of performed control tests, the following values of concrete strength were calculated statistically:

- in one-span frame girder – B20 (C16/20);
- in three-span frame girder – B15 (C12/15);
- in frame column – B15 (C12/15);
- in main beam – B17,5 (C12/15-C16/20);
- in side beam – B17,5 (C12/15-C16/20).

Concrete with R_w = 170 kg/cm^2 (17 MPa), according to the standards in force in 1973, was assumed in the original design [2].

In order to verify the geometry of the foundations of the main columns of the supporting frames, a trench was made (Fig. 20). Measurements showed that the foundation had different dimensions than those assumed in the project [3]. In view of the above, the dimensions from the inventory were used to calculate the bearing capacity of the foundations.

Fig. 20. Foundation inventory.

2.3 Control static calculations

Analyses of calculation assumptions adopted for the original design [2, 3] were made in order to stay coherent with them during verification of the structural capacity for current loads, considering its technical state as well as for predicted rehabilitation solutions.

Based on the results of the conducted strength tests, class B17,5 concrete was adopted for static calculations.

For the control static and strength calculations, the number and layout of the reinforcement bars were taken according to the documentation after verification on the basis of the actual measurements made. The strength of the reinforcement was assumed as for A-III class.

Due to the simple structural layout of the driveway structure and the lack of linking of the driveway slab with ribs and frame girders, a bar model was adopted for the calculation, reflecting the arrangement of existing ribs and transverse frames. Analogous static schemes were assumed by the authors of the driveway project [3].

The following structural elements were checked both for the current state and loads and for predicted rehabilitation designs and loads, including temperature:

- bearing capacity of prefabricated hollow slabs;
- bearing capacity of longitudinal ribs;
- bearing capacity of single-span lateral frames (girders, cantilevers, columns);
- bearing capacity of three-span lateral frames (girders, cantilevers, columns);
- bearing capacity of foundations.

2.4 Some remarks on the driveway technical state

Conclusions from the analysis of the design documentation and identification of structure:

1. The most important constructional error was the lack of a waterproofing layer that was recommended in the project [2, 3].
2. The design error was not taking into account the environmental conditions influencing the structural material and lack of adoption of sufficient thickness of reinforcement cover.
3. Another design error was lack of execution of an effective rainwater drainage system from the driveway surface.
4. The last design and execution error was also the lack of a proper system of expansion joints.
5. The occurrence of damage to the construction and finishing elements of the driveway results from the coupling of internal factors (design and execution errors) and external (environmental influences) that have resulted in a reduction of the durability of the structure.
6. The direct external cause of damage was the destructive environmental impact on concrete and reinforcement of the structure, in the form of a combined effect: water (rain, snow melts), salts used for defrosting during the winter and temperature changes (especially freezing and thawing cycles, mainly in relation to the cover zone of elements directly exposed to atmospheric and maritime region influences).
7. The direct execution cause of damage of the driveway structure was not constructing it in accordance with the design, i.e. the lack of any waterproofing layer in the driveway surface. Another factor contributing to the destruction process was the destruction of the surface pavement, also made contrary to the original design. The combination of these two elements enabled easy and continuous penetration of water through the driveway slab, long-term destruction of the concrete, and, consequently, local strong corrosion damage of concrete and reinforcement. However, it should be emphasized that only in a few places was damage of this type so significant that in the future, in the absence of repair interventions, there may be a risk of loss of load-bearing capacity of these construction elements.
8. The protecting layers made on the surface of the driveway only partially protect the structural elements from further degradation. However, no damage was found in the driveway structure, such as cracks, deformations, which could result from improper work of the foundations (e.g. settlement, rotation, slippage).

3 Driveway rehabilitation concepts

The owner of the driveway was presented with 4 possible rehabilitation concepts depending on the estimated life cycle cost [10] and on the amount of costs necessary to pay. The following rehabilitation ideas were considered:

I. Demolition of the existing structure and constructing a new one with the designed durability of the new structure for 60-80 years.

II. Demolition of the concrete pavement and the entire slab of the driveway, as well as stairs and balustrades, and leaving only the transverse supporting frames and longitudinal beams (ribs). Then, execution of a new platform slab, in a composite version or without an assembly with the remaining elements of reinforced concrete structure, with effective waterproofing - surface, expansion, drainage system, balustrades and stairs, followed by material reconstruction and surface protection of reinforced concrete elements left behind - life service expectancy of the structure 30-40 years (Fig. 21).

III. Demolition of the concrete pavement and the slab made of the Ackerman hollow clay blocks, as well as stairs and balustrades, and leaving the transverse support frames, longitudinal beams (ribs) and floor slab made of prefabricated hollow plates. Then, execution of a new, reinforced concrete top slab,

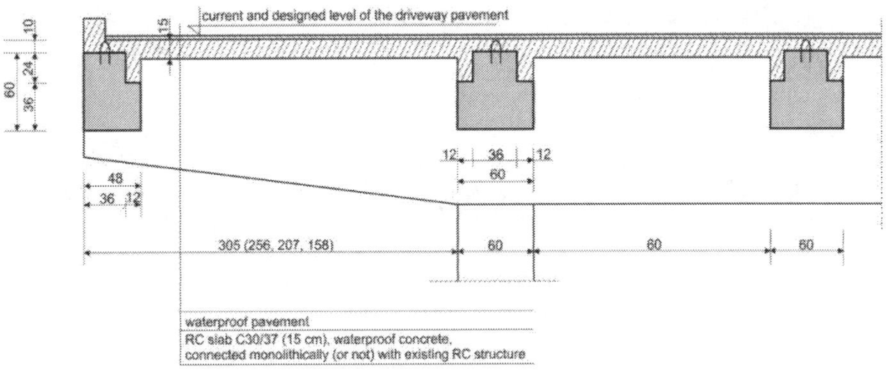

Fig. 21. Driveway renovation concept no. II.

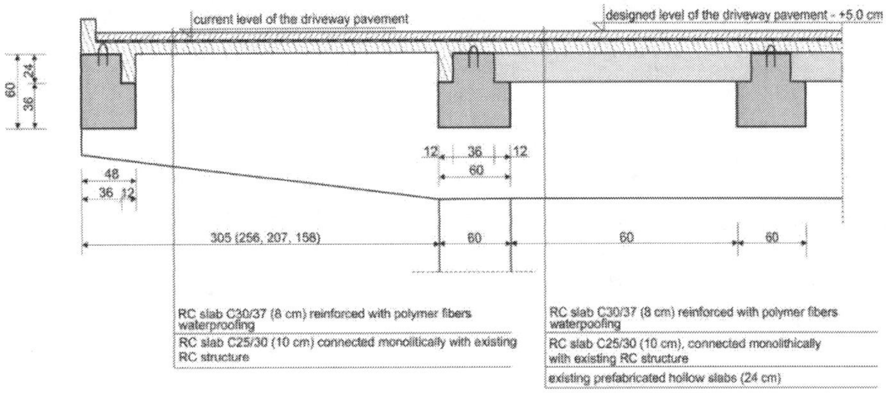

Fig. 22. Driveway renovation concept no. III.

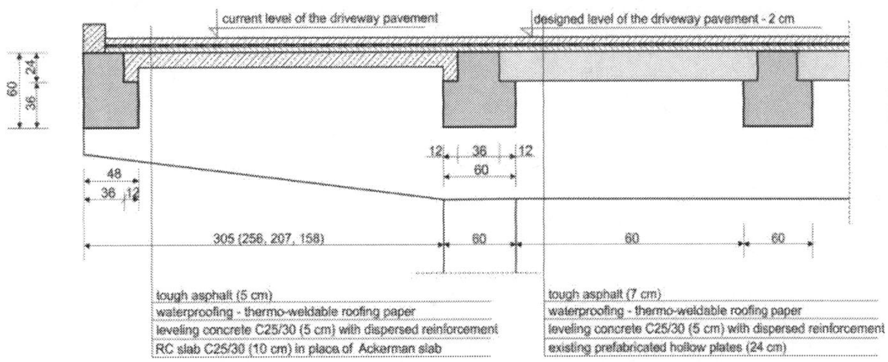

Fig. 23. Driveway renovation concept no. IV.

combined with the remaining construction elements, with a concrete surface reinforced with polymer fibers, effective waterproofing, drainage system, expansion joints, balustrades and stairs; material reconstruction and surface protection of reinforced concrete elements left behind - life expectancy of 15-20 years (Fig. 22).

IV. Demolition of the concrete pavement, as well as stairs and balustrades and a slab of hollow clay Ackerman blocks and replacing it with a reinforced concrete slab; leaving transverse supporting frames, longitudinal girders (ribs) and floor slabs of prefabricated hollow RC plates. Then, execution of an additional concrete leveling layer on the entire ceiling, with effective waterproofing; making a new surface made of solid asphalt, with a drainage system, proper expansion joint system, new balustrades and stairs; material reconstruction and surface protection of left reinforced concrete elements - life expectancy of the structure 15-20 years (Fig. 23).

In the proposed alternative solutions no. II, III and IV there was also defined a range of other works necessary to be implemented:

- Removal of the existing reinforced railings and mounting new steel road type balustrades.
- Design a drainage system due to the large area of rainwater collection as well as introducing an effective waterproofing system.
- Constructing a new and appropriate expansion joint system.
- Cleaning off salt, spalled plaster and concrete from the surfaces of all the elements of the driveway structure to be used in the future.
- Reparation of corrosion damage of reinforcement and concrete by proper cleaning, restoration of reinforcement and placing reprofiling layers providing proper cover of rebar.
- Applying a new cover layer on all the surfaces by spraying on the mesh or protecting the surfaces of all concrete elements with concrete repair system materials resistant to the seaside environmental influences.

4 Final remarks

The main scope of the conducted technical state assessment and structural analysis of the driveway was development of possible concepts of its rehabilitation.

To reach the aims of the conducted research project analyses of original design as well as of later studies on technical condition of the driveway through its exploitation life were performed. Furthermore, material investigation, structural inventory and structural modelling and calculations were made. Conclusions regarding current technical state of the driveway, identification of the reasons of their occurrence and the possibility of using the driveway until the repairs were defined.

Finally, four rehabilitation concepts were presented with emphasis on the relationship between the possible period of extending the service life of the driveway and the range of rehabilitation.

References

1. M. Fischer, *Steineisendecken im Deutschen Reich 1892-1925*. Dissertation. Brandenburgische Technische Universität Cottbus (2008) (in German)
2. Technical project. Driveway to the main building. Architectural design. Design Office of Maritime Construction, Gdańsk (1973)
3. Technical project. A platform and a driveway to the main building. Structural design. Design Office of Maritime Construction, Gdańsk (1973)
4. E. Grzes, Technical expertise regarding the quality of concrete in the construction of the driveway at the building M.I.R. in Gdynia (2005)
5. P. Berkowski, M. Kosior-Kazberuk, Proceedia Engineering, **172** (2017)
6. P. Berkowski, G. Dmochowski, M. Kosior-Kazberuk, Proceedia Engineering, **57** (2013)
7. Ł. Dorbiec, R. Jasiński, A. Piekarczyk, *Diagnostics of Reinforced Concrete Structures. Methodology, field tests, laboratory tests of concrete and steel* (PWN, Warsaw, 2010) (in Polish)
8. A. Zybura, M. Jaśniok, T. Jaśniok, *Diagnostics of Reinforced Concrete Structures. Tests of reinforcement corrosion and protective properties of concrete* (PWN, Warsaw, 2011) (in Polish)
9. M. Raupach, T. Büttner, *Concrete Repair to EN 1504. Diagnosis, Design, Principles and Practice* (CRC Press, Boca Raton, 2014)
10. A.P. Johnson, G.J. Klein, J.S. Lawler, SEI, vol. 29 (2019).

https://doi.org/10.1051/matecconf/201928910003

Concrete repair challenges in the Middle East

*Rene Brueckner**

Mott MacDonald, 10 Temple Back, Bristol, BS1 6FL, UK

Abstract. The root cause analysis, determination of the extent of concrete defects and identification of an appropriate repair strategy can be straightforward, but it also often provides significant challenges to both contractors and engineers. The challenges can be due to a lack of QA/QC documentation, locally available investigation equipment, repair material and techniques and accepting actually feasible solutions.

The paper presents a case study of a bridge in the Middle East where the root cause analysis was found to be straightforward but the determination of the extent of non-visible defects and the implementation of a feasible repair strategy proved to be much more complicated. The root cause analysis of the defects was carried out by visual inspection, representative intrusive investigations and a comprehensive document review. The determination of the extent of the defects using GPR techniques proved to be challenging under the local conditions. Even more difficulties were encountered by implementing the developed repair strategies to address the various defects to ensure that the required 120 years design life in a very aggressive environment can be achieved. Several revisions to the repair strategy were necessary to identify the most appropriate solution and to accelerate the programme.

1 Introduction

The design of a new 56km highway in the Middle East comprised the construction of 5 main interchanges with approximately 7km of bridges and elevated ramps. The majority of the structures consisted of multiple-span post-tensioned concrete box girders with span lengths varying from 30m to 45m that were constructed span by span using precast segments. The precast design was beneficial for a rapid and efficient construction with a high-quality durable product. The contractor, who was experienced in the construction of these types of structures, started with the precast bridges and elevated ramps and decided to undertake the widening of a bridge at an existing junction using in-situ construction methods towards the end of the project.

The existing main bridge to be widened consisted of two symmetrical bridge decks with a 1.6m deep post-tensioned multi-celled box girder with fours spans each and a length of 28 to 30m per span. The demand for the new highway and upgrade of the existing bridge required the bridge to be widened by 9.1m to increase the total width of each bridge deck to 27.15m. The new widening bridges were designed to be a four span post-tensioned cast in-situ voided slab deck with the same span configuration as the existing bridges. Furthermore, the design required the widening to be fully integrated with the existing structure along its cantilevered slabs of the existing structures. A general cross-section of the new widening section is shown in Figure 1.

Due to the inexperience of the contractor in the construction of in-situ bridges and the decision to undertake the works towards the end of the project it was apparent at an early stage that problems were likely to occur. Upon the completion of the first two spans of the bridge widening, several areas affected by honeycombing were observed on the soffit of the widening decks. Following initial breakouts and repairs it was found that the defects extended over a wider area and were not only limited to the visually identified honeycombed areas.

2 Defects

Honeycombing on the soffits of two spans was observed upon the removal of the formwork which initiated a more detailed investigation and assessment of the extent of the defects. This resulted in the identification of the following defects:

- Honeycombing resulting in non-embedment / lack of bond of the lowest reinforcement in localised areas;
- Voids below lapping bars – lack of bond;
- Cold joints within the lowest reinforcement layer;
- Gap between void formers and internal concrete cover to the lowest reinforcement layer which reduces the internal concrete cover;

* Corresponding author: rene.brueckner@mottmac.com

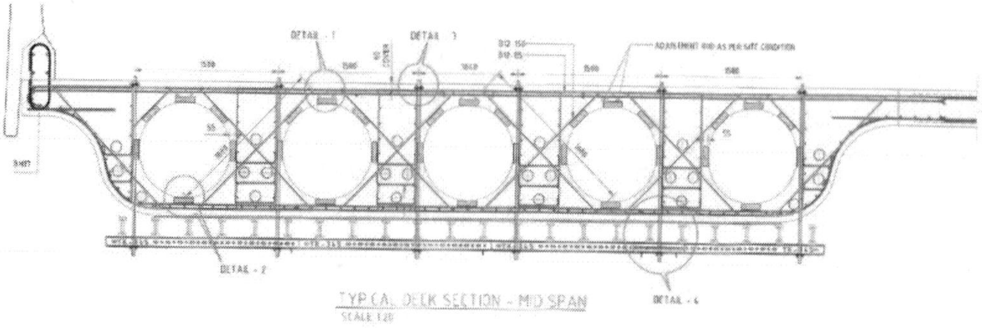

Fig.1. Cross-section of bridge widening

- Reduced level of compaction of the concrete between void formers and lowest reinforcement;
- Reduced localised cover;
- Cracking on top of slab;
- Excessive debris, such as reinforcement caps, tie-wires, reinforcement pieces in soffit reducing the effective cover;
- Polyethylene sheets above bottom reinforcement layer preventing encapsulation of the reinforcement.

The defects were mainly located below the five void formers of the voided post-tensioned widening slabs. Figure 2 shows a general view of the most-affected area of one span where the lines of the five void formers are marked on the soffit (yellow line).

Fig.2. General condition of soffit

Honeycombing was identified over a large area but the extent of the honeycombing into the slab varied significantly. Near-surface honeycombing defects were observed on the majority of the soffit whereas honeycombing extending beyond the lowest reinforcement layer was mainly observed below the void formers (Figure 2 to Figure 4). The maximum extent of honeycombing into the slab was found to be approx. 180mm, i.e. up to the bottom of the void formers. The honeycombing resulted locally in a lack of bond around the reinforcement but there was no honeycombing below the post-tensioning ducts away from the void formers. The soffit also suffered from an excessive amount of debris that was not cleaned before the concrete pour and included reinforcement pieces, caps and large polyethylene sheets.

Cold joints were found between the 1st and 2nd pour layer which is located approximately between the two lower layers of transverse reinforcement or between bottom reinforcement and void former. A gap was observed between the void formers and the lowest reinforcement in several locations, reducing the internal concrete cover to the lowest reinforcement. The gap was mainly encountered where the bottom spacers for the void former (concrete blocks: 200x200x55mm^3) were located at a distance of 100mm from the joints of void formers.

It was observed that the size of the gap varied and that there was a small gap potentially along the entire length of the void formers. The latter does not reduce the protection by the internal concrete cover as it was mainly a result of the void former floating up during concreting and the specified internal cover was obtained.

Cracking was identified on top of the slab and appeared to be a result of early thermal cracking and drying shrinkage. No cracking was observed on the soffit due to the post-tensioning.

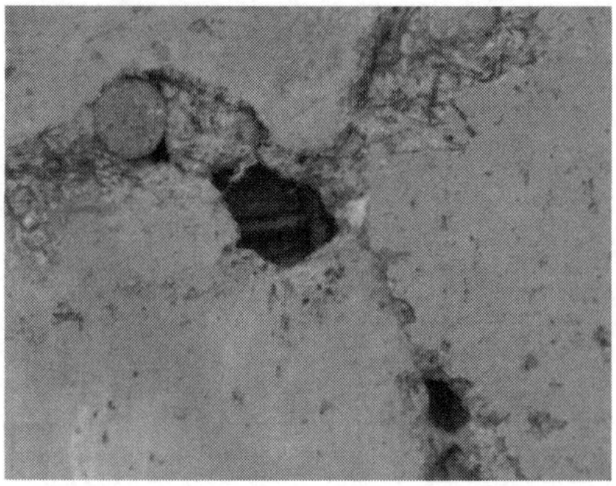

Fig. 3. Honeycombing with visible reinforcement

Fig.4. Void below void former and non-embedded rebar

3 Root cause analysis

The root cause of the defects was identified to be a combination of varying concrete consistency and concreting / compaction operations. The consistency of the C45/55-10mm concrete (OPC+35% PFA+7.5% MS), as delivered to site, ranged from 160mm to 210mm with periods of lower slumps (130-140m^3 of 490m^3), as was observed through compaction issues at the time of the pour and can be seen in Figure 5. Such consistency may be considered suitable provided that the concrete is fully compactable in all locations and the pour layers are interlocked during compaction without the risk of cold joint formation.

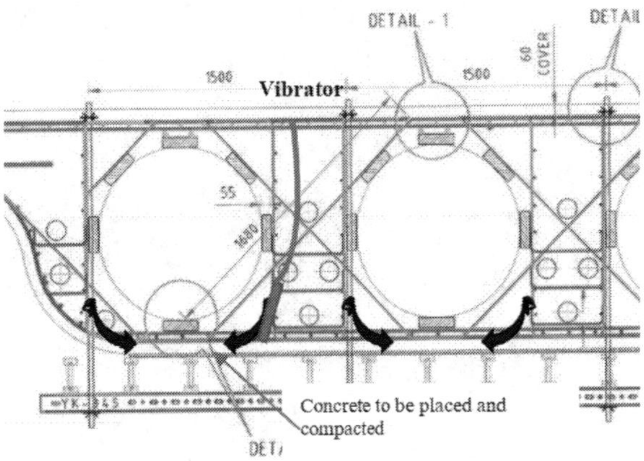

Fig.6. Cross-sectional arrangement

It was also reported that the concreting sequence consisted of pouring an initial layer of concrete that only covered the bottom transverse reinforcement before proceeding to build up the full height of the section that included the bottom longitudinal and second transverse reinforcement layer. This pour method was implemented to reduce the uplift of the void formers. The time between finishing the bottom layer and continuation of the full height concrete pour was reported to have resulted in total loss of workability of the 1st layer of concrete so that the 2nd layer could not be poured 'fresh on fresh'. Compaction of the 1st layer was also inadequate.

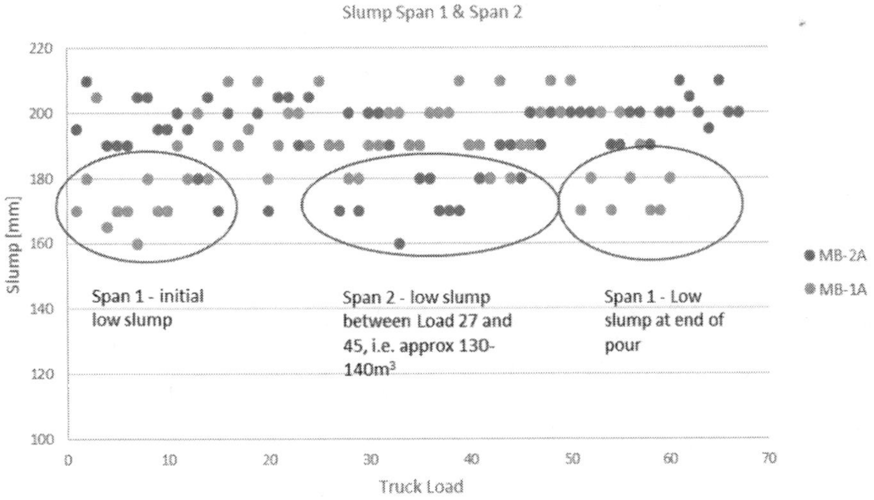

Fig.5. Slump of Slab 1 and Slab 2 pours

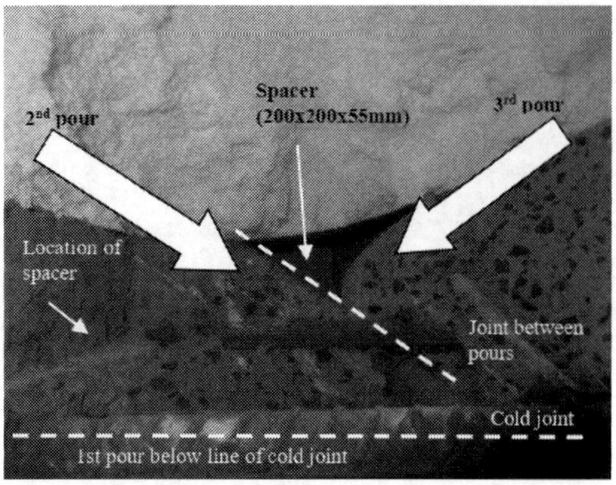

Fig. 7. Root cause schematic

The formation of a cold joint between the two layers was prevented below the ducts; however, cold joints were identified under the void formers. This was a result of the arrangement of the void formers, spacers, post-tensioning ducts, reinforcement bars that were spaced at 125mm and 150mm and the delayed time between the two layers in addition to the partially reduced consistency of the concrete. The arrangement made it difficult to insert vibrating pokers and compact the concrete below the void formers whereas compaction around the ducts was achievable, as indicated in Figure 6 and shown in Figure 7, and also discussed by Grantham [1].

There was also a high degree of debris contained within the concrete cover zone which was attributed to poor housekeeping, cleaning and workmanship.

4 Determination of the extent of defects

The extent of the defects on the two slabs was determined using both intrusive and non-intrusive testing methods. The latter comprised Impact Echo and GPR scanning to determine sub-surface defects but both techniques proved unsuitable due to interferences from vibrations from the traffic on the existing neighbouring bridge and the extent of reinforcement, respectively. The verification of negative anomalies in the GPR scan data with intrusive break-outs provided unreliable results. Based on this the extent of repairs was determined by a combination of visual inspection, investigation holes and representative break-outs.

Investigation break-out holes of approx. 100mm diameter at a spacing of 3m along the line below each void former and representative break-outs were used to determine the extent of gaps, sub-surface honeycombing and loss of bond between concrete and reinforcement that could have effects on the durability and structural integrity of the deck.

5 Remediation strategy

The environmental exposure conditions in the Middle East are very aggressive in particular with regards to chloride induced reinforcement corrosion when located in proximity to the Gulf. In addition to the exposure conditions, the design life requirements of the new highway structures for this project was specified to be 120 years.

The effects of the various defects identified on the soffit of the decks were assessed to be loss of resistance to chloride ingress from external sources due to lack and loss of cover as a result of sub-surface and near surface honeycombing in combination with localised lack of reinforcement embedment below the void formers. As the specification did not require a waterproofing system to be installed on the deck of the bridge and existence of cracks there was a high risk that moisture and chlorides may enter the voids resulting in chloride ingress from the internal face. The resulting reinforcement corrosion of the bottom layer and associated spalling was considered to have a high potential to develop into a hidden defect.

The lack of reinforcement embedment below the void formers and its effects on the structural integrity were not significant but had to be repaired to ensure the as-design conditions.

The effects on the specified design life of 120 years were required to be mitigated by reinstating the resistance against chloride ingress and carbonation of external and internal faces. As the original durability design only relied upon the resistance of an uncracked concrete cover zone around the bottom reinforcement layer by post-tensioning a protective coating system was not specified. At the time the defects were observed and repairs were planned to be carried out, the structure had already been post-tensioned and therefore the removal of the entire soffit was not desirable as the status of the cover zone would change from compression into tension following repair and the time required to complete the demolition and reinstatement was unacceptable. However, initial discussions included the removal of the entire soffit of the worst affected deck in combination with the application of a protective coating system.

Taking into account the various defects and project durability requirements a repair methodology was developed and proposed to the contractor which comprised:

- Repair of the gaps below the void former and reinstatement of the internal concrete cover by injection of cementitious grout which also addressed the lack of reinforcement embedment in interconnected voids.
- Repair of small-scale honeycombing and inspection break-outs by using a cementitious shrinkage compensated concrete repair mortar with a minimum equivalent compressive strength of a C45/55 concrete.
- Repair of large defect areas by adopting one of three specific reinstatement methods as in the following paragraph.
- Application of a protective coating system to mitigate the risk of deterioration from any unidentified and hidden defects.

https://doi.org/10.1051/matecconf/201928910003

Method 1 of the large area repairs comprised gravity pouring of high flow mortar from the top of the bridge whereas Method 2 was suggested to repair the defects on the soffit by pumping a repair mortar from the soffit, see Figure 8. The third and considered most suitable repair method (Method 3) was sprayed concrete application.

In addition to the three repair methods a preplaced aggregate method in combination with self-compacting mortar pumped from the bottom was also considered but excluded due to the increased labour cost, the required skills and levels of control. The most likely outcome would have been incomplete grouting due to inexperience and dry and hot aggregate in addition to areas with lack of aggregate as a result of insufficient aggregate placement.

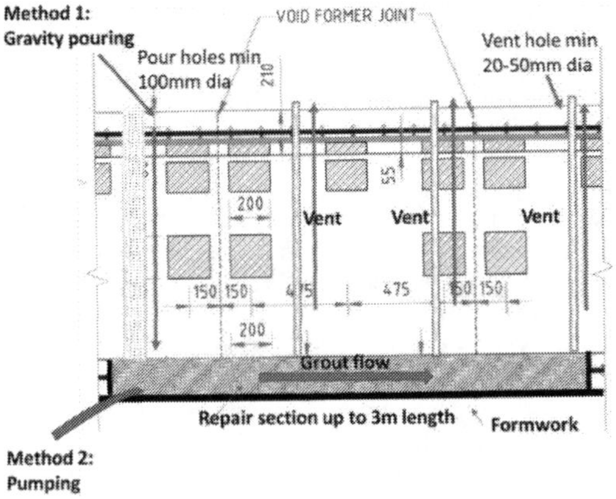

Fig.8. Repair schematic of Method 1 and 2

6 Implementation

6.1 Large defects

Removal of defective areas and repairs commenced at the worst affected span and comprised the removal of the entire length below two of five void former lines as part of the initially considered replacement of the full deck soffit. The methodology consisted of two stages, Stage A comprised the removal of the concrete below void formers 2 and 4 including reinstatement and Stage B was planned to cover void former lines 1, 3 and 5, see Figure 9. The specified demolition method was hydro-demolition but this could not be sourced and the concrete was broken out using hand-held pneumatic breakers.

The reinstatement of the large strips proved to be challenging due to various reasons:

- A competent and experienced sprayed concrete subcontractor could not be identified and Method C had to be discarded;
- A suitable pump for Method B could not be sourced.
- Method A, gravity pouring from the top of the deck, created further difficulties and challenges.

The challenges comprised the opening up of the pour and vent holes and removal of all polystyrene debris from the void formers, sealing and support of the formwork and manpower for mixing and pouring of the concrete repair mortar. The repair of the first 3m long by 1.25m wide section resulted in severe leakage from the formwork, missing of pour holes and insufficient manpower for mixing to fill the repair within the workability time of the mortar. This resulted in opening up of an additional pour hole during the pour causing polystyrene debris to fall into the formwork. Hence, a section of the trial repair had to be repaired.

The repair mortar was a shrinkage compensated, high flow and high strength pre-packaged concrete repair mortar that may be used with additional aggregate for larger thicknesses. The incorporation of aggregate was assessed and found to cause additional problems as there was a high risk that the aggregate may not be stored in an air-conditioned environment. Pre-bagged aggregate stored in the shade with ambient temperatures of 35 to 45°C, as at the time of the repairs, significantly reduced the workability of the mortar.

The difficulties encountered during the repairs of the first two strips and observations that the concrete was acceptable below the post-tensioning ducts and outside the areas of the worst visible surface defects led to an optimisation of the repair procedure. Further verifications using only smaller strip break-outs of 1.5m x 0.3m at representative locations along void former lines 1, 3 and 5 were carried out. The 1.5m x 0.3m strip break-outs comprised a hit and miss strategy of three break-outs covering a length of 7.5m along one line. Signs of honeycombing around the reinforcement and up to the void former required continuous hit and miss break-outs including removal of all honeycombed concrete under all void former lines of the respective span. The three hit and miss trials, that were undertaken at the worst visually affected areas without obvious full depth honeycombing, confirmed that full removal of the deck soffits and large strip break-outs below each void former could be avoided. However, this method identified the presence of a large cast-in PE sheet above the bottom reinforcement layer, see Figure 10.

Nevertheless, any areas showing visible surface defects and cast-in debris were removed and extended to sound substrate if required. This resulted in both shallow and deep break-outs requiring repair using Method A or hand applied overhead application.

Small-scale deep repairs were carried out by using a hand applied high-strength concrete repair mortar that was suitable for overhead applications.

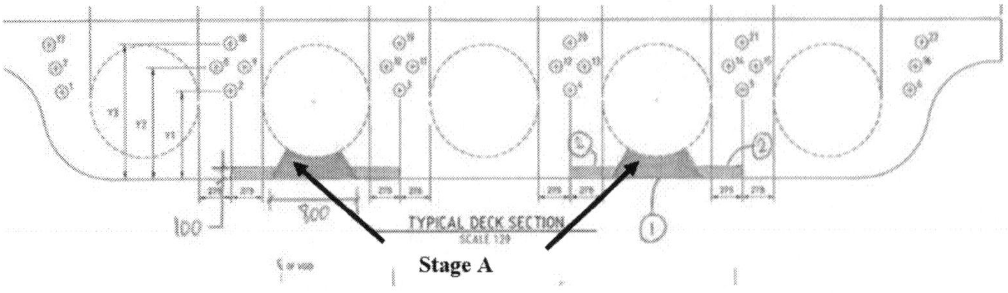

Fig.9. Stages of full soffit demolition and reinstatement

Fig.10. Debris below void former

6.2 Gap below void former

The gaps below the void formers and any interconnected voids within the bottom rebar layer were repaired by injection of a low viscosity cementitious grout that was able to flow below and around the void former.

The injection methodology included an initial investigation to establish a gap and connectivity below the void former by injecting water as initial repairs showed that curing compound egressed at various locations indication sub-surface interconnected voids and honeycombing. Following trials, injection packers were installed at a spacing of 300mm below the centre line of the void formers, sealed and grout was injected by utilising an injection pressure up to 1 bar at the injection port. The effectiveness and pressure limits of the injection were determined prior to commencement of the works using mock-ups to prevent the deformation / compression of the void former. Injection pressures in excess of 2 bar were assessed to compress the polystyrene void formers by more than 10% causing an increase of dead load and affecting the design load assumptions.

The effectiveness of the procedure during injection was verified by visual confirmation to ensure that the grout flowed from the injection port to the adjacent port and the packer was sealed once grout flowed. The challenges encountered during injection were pressure control and appropriate seals at the injection ports to mainly avoid health and safety issues.

The grout injection was also used to mitigate the risk of any honeycombed areas below the void former that could not be visually identified from the soffit. Interconnected voids within the lower reinforcement layers were sealed as well.

6.3 Hidden defects

Any hidden defects that were not identified during the inspection and repair works and the lack of external cover at the curved sections was addressed by the application of a polymer-modified cementitious coating despite initial reluctance of the contractor.

The coating provides an effective barrier to waterborne salts such as chlorides and sulfates and is able to compensate for the lack of external cover and any hidden defects to achieve the required design life of 120 years. Additionally, it provided a mitigation measure in the case of cracking of the shrinkage compensated repair material due to both loss of compressive stresses in the repair areas and inadequate curing and workmanship as had been observed.

7 Conclusions

A remediation strategy was identified with the objective to mitigate the risk of deterioration of the structure caused by reinforcement corrosion due to ingress of aggressive media within the 120 years design life as a result of the various defects observed following construction. Furthermore, it was verified that the bond between concrete and reinforcement was sufficient. Lack of significant bond would have affected the structural integrity of the reinforced concrete below the void formers and the post-tensioning ducts.

The repair methodology comprised the following:

- Removal of all visually identified honeycombed areas and surface-near debris up to sound substrate;
- Reinstatement of the broken-out defective concrete;
- Filling of the gap below the void formers and any hidden interconnected voids;
- Application of a protective coating providing a permanent barrier to airborne chlorides and sulfates to address the lack of cover at the curved sides of the deck and at other unidentified locations at the soffit.

The development of the repair strategy and methodology and its implementation on site provided various significant challenges. Despite the expected high standards in construction and materials availability it became a major challenge to identify experienced specialist contractors to undertake the works and therefore, a simple repair method such as sprayed concrete application could not be undertaken. The methodology had to acknowledge the capability of the main contractor and further simplify the repairs until it was achievable to the expected quality for the available resources. The environmental conditions during the repairs, i.e. summer in the Middle East, also played an important role and required consideration during the design stage.

The main lesson learnt during the repair process was to adjust the repair methods to the capabilities of the contractor and not to expect that novel and common standard repair techniques, that are developed in an office environment, can be applied in all projects. The time constraints imposed by the construction programme and leaving the only in-situ structure to be built after completion of all other structures in combination with the inexperience of the contractor in in-situ works played another main role in cause and rectification. Additionally, the availability of materials and simple equipment also affected the repairs.

A major lesson learnt by the contractor during the construction of the bridges was to not rely on the workmanship to ensure adequate compaction and instead to use self-compacting concrete. The use of self-compacting concrete was initially recommended but dismissed by the contractor at the beginning due to the associated higher costs. The final costs to the main contractor to complete the widening as a result of the the repairs and delays to the project can be considered significantly higher than investing a comparative insignificant amount at the initial stages to prevent costly repairs, as highlighted by Sfikas [2].

Despite the various challenges experienced during the repair, all identified defects were satisfactorily repaired and the risk of any potentially hidden defects affecting the durability within the specified design life were mitigated.

References

1. Grantham, MG.: Sustainable, durable concrete – Are specifications always fit for purpose – A case study ICCRRR2018, Cape Town. MATEC Web of Conferences.
2. Sfikas, Ioannis. (2017). Self-Compacting Concrete: History & Current Trends. Concrete (London). 51. pp. 12-16.

Structural failure of old reinforced concrete river wharf

Grzegorz Dmochowski[1], *Piotr* Berkowski[2,*]

[1]Wroclaw University of Science and Technology, Faculty of Civil Engineering, Wyspianskiego Blvd., 27, Wroclaw, Poland
[2]Wroclaw University of Science and Technology, Faculty of Civil Engineering, Wyspianskiego Blvd., 27, Wroclaw, Poland

Abstract. According to the archival documentation the river harbour wharf was constructed around 1912. The quay was divided into 6 separated segments with a length of 32.0 m each. The structure of the wharf consists of the reinforced concrete frames situated perpendicularly to the riverbed. At ground level, the frames are connected by a reinforced concrete deck and longitudinal beams, located along the side of the river and the quay. The frames are also connected along the riverside by the concrete cap situated at the top of the retaining wall made of prefabricated panels and embedded in the bottom of the river. That wall constitutes a direct strengthening and protection of the embankment against the influence of the flowing river water. Basing on the performed inventory inspections, material examination, structural calculations and hydro-geotechnical analyses, the main reasons for the failure of the reinforced concrete wharf of the Odra River were defined. They were: very poor technical condition of the structure, lack of stabilizing anchors at the level of the wharf structure frames, and sudden fluctuations in water level in the river and their impact on the structure of the wharf. Moreover, from a static point of view, the embankment under the wharf was in an unstable equilibrium and at any time it could be damaged if additional horizontal forces were applied to it. Additionally, the possible impact of anchoring to the piers of the wharf of two river barges on wharf structure capacity was also analyzed.

1 Description of the wharf structure

The discussed wharf which, according to the archival documentation, was constructed around 1912 (Fig. 1), is situated on the left bank of the Odra's southern channel in Wroclaw. The wharf was designed and erected by the Schlesichse Betonbau Gesellshaft for the Schlesichse Dampfer Compagnie founded in 1887. In the following years, the company took on a special position in German inland shipping. It was a shipping and warehouse company and owned two shipyards. Its shipping was operated on the Oder, Elbe and the connecting channels.

Fig. 1. Historical view of the wharf.

Fig. 2. General view of the wharf before failure.

In the past, the wharf served, among other, as a dockside quay for a brewery located nearby. In recent years, the wharf has been sporadically used for mooring barges. Even though the Odra embankments have been assessed [1] and renovated in recent years over a considerable length. the considered part was left without any technical intervention, repair or protection (Fig. 2).

1.1 Analysis of the archival documentation

The archival documentation consists of several drawings of the whole dockside quay with industrial buildings and

*Corresponding author: piotr.berkowski@pwr.edu.pl

https://doi.org/10.1051/matecconf/201928910004

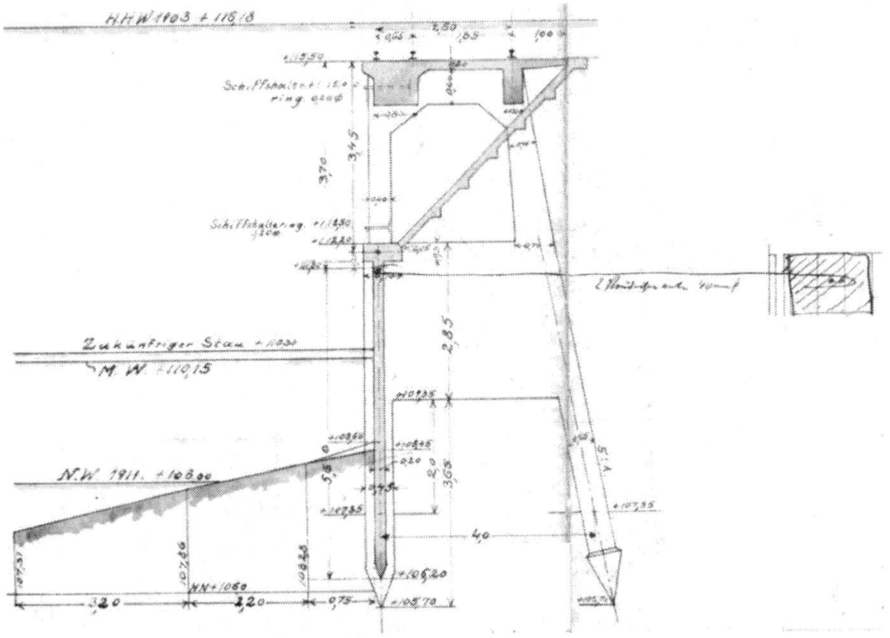

Fig. 3. Archival drawing of the wharf bearing structure.

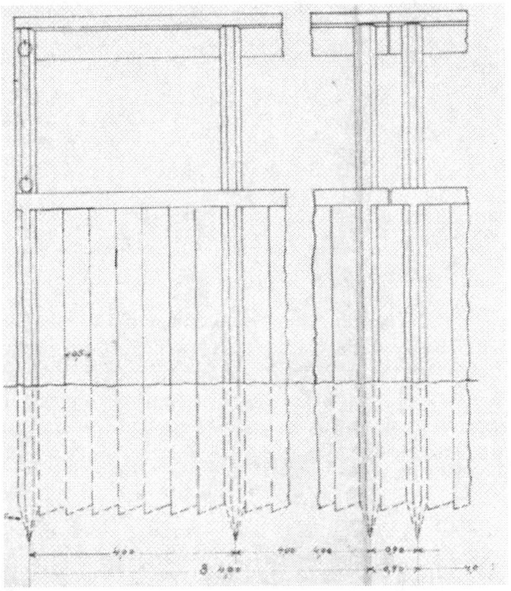

Fig. 4. View of the wharf with retaining wall.

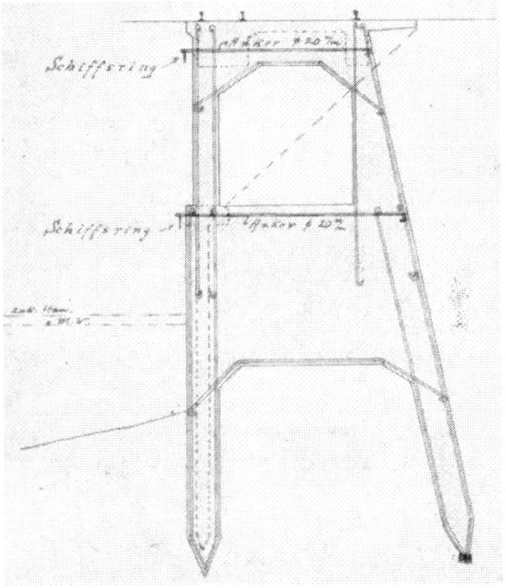

Fig. 5. Archival drawing of the wharf bearing structure rebar.

selected structural drawings of the wharf frames and the retaining wall.

In the archival design, the total length of the wharf was 192.0 m and it was divided into six segments with a length of 32.0 m each. The structure of the wharf consists of the reinforced concrete "H" frames situated perpendicularly to the riverbed (Fig. 3). At ground level, the frames were connected by a reinforced concrete deck and two longitudinal beams located along the side of the river and the quay. The frames were also connected along the riverside by the concrete cap situated at the top of the retaining wall (Fig. 4) made of prefabricated panels and embedded in the bottom of the river. That wall constituted a direct strengthening and protection of the embankment (with an inclination of about 45°) against the influence of flowing river water. The total height of the bottom columns of the frame was 6.50 m and the upper part columns was about 3.30 m in height. The axial spacing of the lower pillars was 4.0 m and the width of the top deck was 3.80 m (Fig. 5).

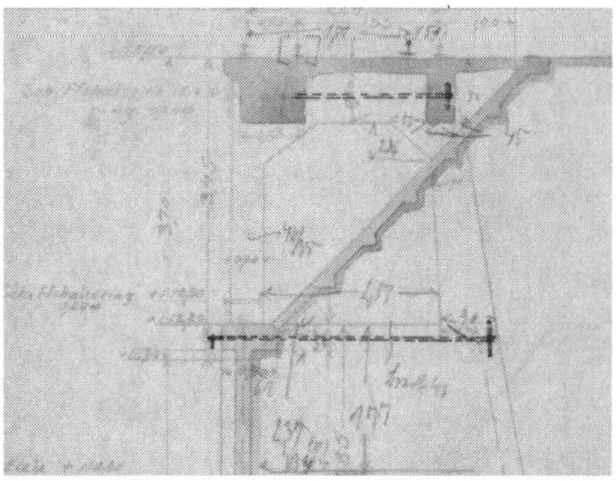

Fig. 6. Archival drawing of the frame with freehand sketches.

In two of the archival drawings it can be seen with the freehand sketches, that the reinforced concrete frames were to be additionally stabilized with steel anchors. In Fig. 3 it is designed by use of a 40 mm diameter rod at the level of the top of the cap covering the retaining walls (Fig. 3) and fixed in blocks located out of the embankment. The second proposal (Fig. 6) consists of inserting additional steel rods at two levels: between upper longitudinal beams and between columns at the level of cap over the retaining wall.

That fact will have some influence on the behavior of the wharf during failure.

1.2 Description of the current structure of the wharf

Currently, the whole wharf at this part of the river embankment is divided into 6 segments separated from each other by expansion joints). The first - no. I - consists of 3 fields of 4.10 m, no. II of 2 fields, and no. III, IV, V and VI have 10 fields each. The upper slab, based on the frames. girders and longitudinal beams, is approx. 3.3 m above the top of the lower cap constructed on the vertical retaining wall. Under the wharf board, a slope of about 45° was made, connecting the retaining wall cap with the upper edge of the deck. The slope surface is strengthened by use of octagonal concrete blocks, while on the horizontal part of the embankment a stone pavement was made.

At present, the following elements of the reinforced concrete wharf structure can be distinguished (Fig. 7):

- A reinforced concrete retaining wall (sheet piling) made of prefabricated panels with a rectangular section 0.20x0.50 m and 5.60 m long, combined with transverse frames and topped with a reinforced concrete cap 0.30x0.70 m;
- reinforced concrete transverse frames, consisting of the lower part (6.50 m high) in the form of 2 piles with a cross section of 0.40x0.45 m, from which the pile introduced towards the wharf is an inclined pile, and the second one in the retaining wall line, is a vertical pile; these piles are connected with a reinforced concrete girder having height from 1.75 to 2.85 m below the upper level of the retaining wall, with a thickness of 0.40 m; in the upper part (3.30 m high), from the river side, the frame is made of a vertical pillar with an octagonal cross section of 0.35x0.40 m and an oblique pillar with a width of 0.35 m and variable height from 0.50 to 0.90 m; both pillars are connected at the top by a beam with a cross-section of 0.35 x 0.55 m with slants;
- longitudinal beams, based on pillars of the upper girder of the supporting frames, with a cross section of 0.40x0.55 m; in the segment no. V beams from the river side have a cross-section of 0.30x0.75 m, and from the land side - 0.80x0.75 m;
- the upper slab (deck) of the wharf with a thickness of 0.20 m and width 3.80 m, based on the longitudinal beams, situated along the river and with cantilever of 0.90 m along the line of the land.

Fig. 7. Current view of the wharf structure.

2 Assessment of the technical state of the destroyed wharf

In May 2017 a structural and constructional failure of the reinforced concrete of the wharf occurred over a length of over 100 m, mainly in the area of segments IV, V and VI. At those parts the embankment sloped down towards the river bed and its construction was locally completely destroyed (Fig. 8).

Fig. 8. General view of the destroyed embankment structure.

Fig. 9. View of the displaced wharf structure.

During the inspection made immediately after the failure it was also observed that the upper part of the wharf structure experienced about 15 cm displacement towards the river as well as along the shoreline, which was evident in the displacements experienced by the pillars and the reinforcement of columns and beams (Fig. 9).

At the time of the accident at the quay, two barges were moored in parallel to the embankment. Finally, it was found that the barges broke away from the wharf at one end and stood perpendicular to the current of the river channel, virtually blocking its total width for some time.

The assessment of the technical condition of the quay before the catastrophe was made on the basis of its post-catastrophic review, including some damaged and undamaged quays. A very bad, practically pre-emergency technical condition of the entire wharf structure was found. In many places there had been very large corrosion losses of concrete and reinforcement. At some parts practically the entire reinforcement cover had been completely destroyed and had lost its protective properties. The vast majority of reinforcement had corroded up to 30-40%, some of the reinforcement was virtually completely destroyed. The total reinforcement loss was estimated to be at least 50%.

Fig. 10. Corrosion of concrete cover and reinforcement.

Fig. 11. Corrosion of concrete cover and reinforcement.

Already in the study from 2003 it had been indicated that the technical condition of the wharf was bad and further postponement of its repair might lead to a failure. Since then, no significant repairs had been carried out in the waterfront area and its technical condition had significantly deteriorated (Figs 10 & 11).

In the procedure of assessing the technical condition of the driveway [2, 3, 4] the following material and structural examinations were made:

- analysis of archival documentation;
- geometrical and geodesic inventory of the structure;
- damage inventory (including underwater filming);
- drillings and geotechnical tests;
- concrete and steel strength examination;
- structural modelling and static calculations.

2.1 Structural inventory

The geometrical dimensions and reinforcement inventory of the upper part of the wharf structure were determined on the basis of archival documentation and authors' own measurements and on-site examination holes. While the bottom part was under water and covered with ground and was unavailable for measurement, so its inventory was made only based on archival documentation.

Verified with measurements, the geometrical dimensions of the wharf structure are described in the paragraph 1.2. In the case of dimensions and disposition of the reinforcement (smooth steel was used) it was conducted in places where the reinforcement was exposed as a result of the destruction of the quay structure. The following results were obtained and then used in static calculations:

- top deck slab of the wharf:
  - bottom 16 mm bars every 12 cm, top 16 mm bars every 18 cm, distribution rebar 10 mm every 30 cm;
- longitudinal beams:
  - bottom 6 rods 18 mm , with 3 slanted rods bent up to the top of the span; stirrups 6 mm every 17-30 cm, in a denser spacing at supports;
- transverse girders of the upper frames:
  - bottom 7 rods 18 mm;

- columns of the upper frame from the river side:
  - 3 rods 18 mm distributed on each surface parallel to the river;
- columns of the upper frame from the land side:
  - 3 rods 18 mm from the water side, 4 rods 18 from the land side.

The reinforcement of the bottom frame columns was assumed according to the archival documentation as 10 rods 25 mm on the perimeter of the cross-section. The archival documentation also shows that reinforced concrete frames had been additionally stabilized with steel anchors with a diameter of 40 mm at the level of the top of the bottom girders connected the piles. After the failure of the wharf, a lot of extreme upper edges of those girders were exposed. In most cases, those anchors were not found. Only two anchors were found for the 2 frames in segment no. VI. They were rods with a diameter of 25 mm fixed to the reinforced concrete wall of one of the buildings located at the wharf. Those rods were completely corroded. In the calculations of the quay stability, it was assumed that originally stabilizing anchors with a diameter of 25 mm were present, but at the time of the failure they were not present or did not fulfill their function.

2.2 Damage inventory

During the failure a great part of the embankment sloped into the river. Geodesic measurements of its shape showed that it moved at the depth of about 3 to 4 meters. Simultaneously, also the reinforced concrete structure of the wharf moved towards the river causing massive cracking of the structural elements (Figs 12 & 13) along with pulling out the reinforcement at the connection points of the posts and the frame girders (Figs 14).

Fig. 12. Damaged wharf strucutre.

A few days after the disaster, the technical condition of the reinforced concrete piers in the segments V and VI was also checked below the then water table using a submarine camera. The level of the water table coincided with the top of the reinforced concrete cap. During the inspection, surface corrosion of the prefabricated concrete retaining wall panels was found, numerous defects of concrete on their surface and lack of tightness of contacts between them. However, taking into account the scale of displacements of the wall, in the upper part it has been preserved in the aforementioned review in a surprisingly good technical condition.

Fig. 13. Damaged wharf strucutre.

Fig. 14. Column rebars torn out of the frame..

2.3 Geotechnical examination

Due to covering of the disaster area with fragments of the destroyed construction, the test could only be carried out in two available places (Fig. 15). Soil research showed that under the 2 m thick layer of loose embankment there are fine compacted sands with a degree of density $I_D = 0.46$ to 0.59. Results of those tests were used to define embankment soil model for structural calculations.

During the geotechnical tests, the depth of the river bed was also measured in three places near the retaining wall. The bottom of the river bed (or the surface of the ingoing layers) was found at a level of approx. 4.5 m below the top of the cap connecting the panel wall. As the measurements were made after the landslide, the bottom of the river at the time of the failure could have been much lower. Therefore, it was at least 1.0 m below the level of the bottom assumed in the archival documentation, which meant a much shorter anchorage of the retaining wall in the bottom of the river.

https://doi.org/10.1051/matecconf/201928910004

Fig. 15. Soil examination.

2.4 Concrete and steel strength examination

In order to determine the strength of reinforcing steel and concrete, several concrete blocks were removed and a number of reinforcing bars cut from the construction (Figs 16 & 18). Results of those tests (Figs 17 & 19) were used to define the wharf structure material models for structural calculations.

Fig. 16. Concrete probes.

Fig. 17. Concrete testing.

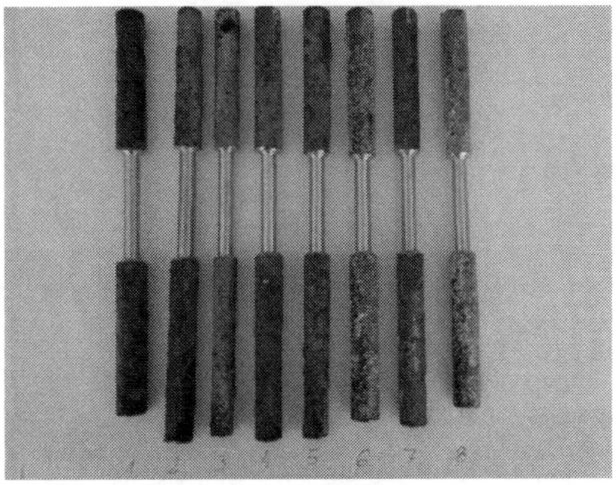

Fig. 18. Rebar probes.

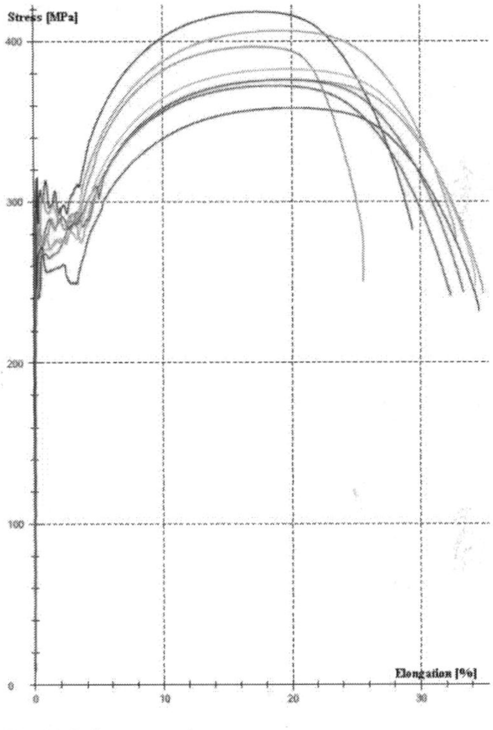

Fig. 19. Rebar test results.

A number of cylindrical samples were prepared and destroyed in accordance with the applicable standard (PN-EN 12390-3:2011) in order to determine the concrete class. The final test results and the strength value determined on the basis of a simple statistical analysis are presented below:

f_{csr} = 29.7 MPa, σ = 2.0 MPa, f_G = 29.7-1.64x2.0 = 26.4 MPa, f_{ck} = 0.8x26.4 = 21.1 MPa, f_{cd} = 21.1/1.5 = 14.0 MPa,

where: f_{csr} - mean value of strength from *n* samples; σ – standard deviation; f_G – guaranteed compressive strength; f_{ck} – characteristic compressive strength; f_{cd} – compressive strength.

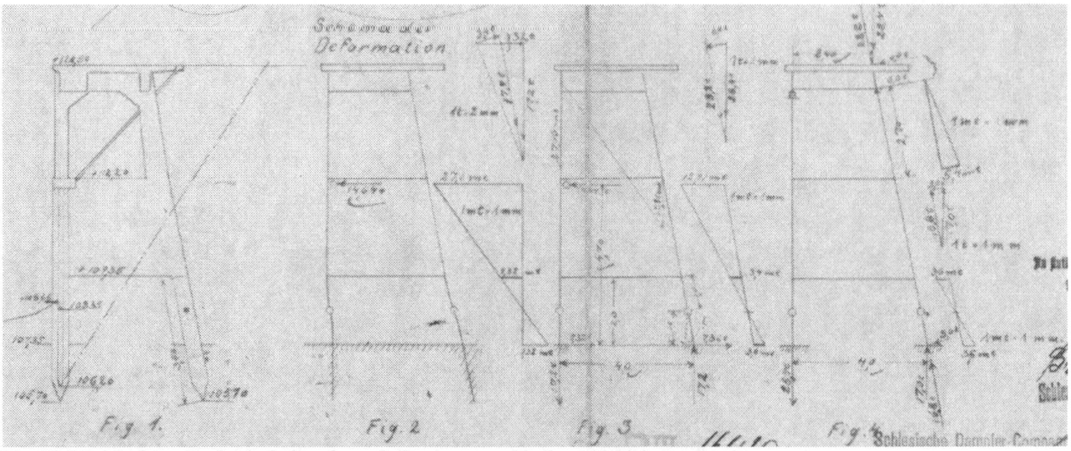

Fig. 20. Model of structural calculations for main frame assumed in the achival documentation.

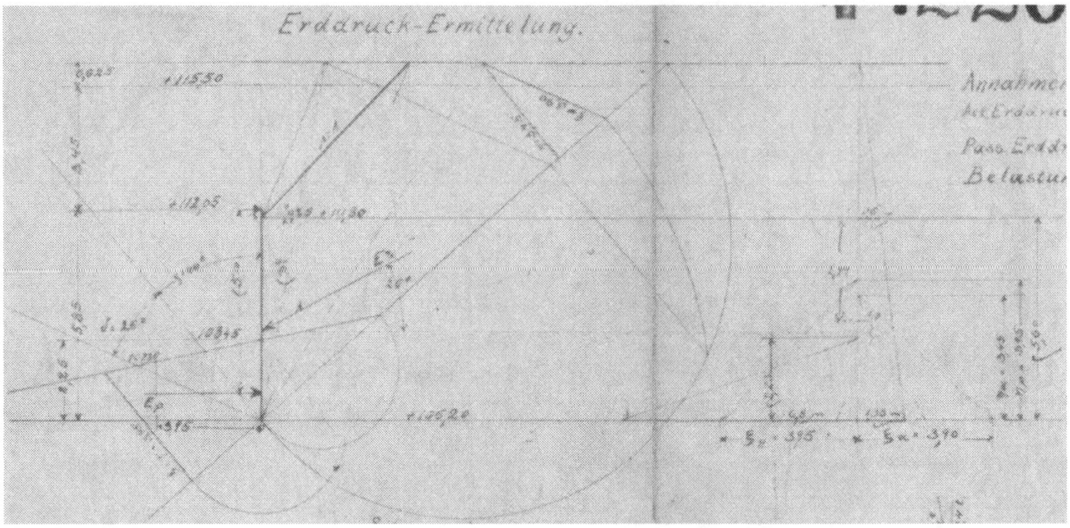

Fig. 21. Model of structural calculations assumed in the achival documentation for embankment stability.

Due to the very diverse state of concrete in the wharf structure, an average of class B15 (C12/C15) was assumed, although the results of the study indicates higher class.

To determine the strength of the reinforcing steel 8 samples of reinforcing bars were taken and broken in a strength machine. As a result of the tests, the following yield values were obtained:

R_{sr} = 386 MPa, σ = 20 MPa, R_G = 386-1.64x20 = 353 MPa, f_{yk} = 0.8x353 = 282 MPa, f_{yd} = 282/1.15 = 245 MPa,

where: R_{sr} - mean value of strength from n samples; σ – standard deviation; R_G – guaranteed tensile strength; f_{yk} – characteristic tensile strength; f_{yd} – tensile strength.

The yield limit f_{yd} = 210 MPa for steel was taken for the structural calculations.

2.5 Structural modelling and calculations

In the archival documentation, some drawings with graphical calculations for the main frame design (Fig. 20) and for verification of embankment stability (Fig. 21) were found.

In the calculations related to the assessment of the technical condition of the wharf structure, in order to determine the causes of the catastrophe, simplified static and strength calculations of the upper part of the wharf structure were performed as well as verification of its stability for various calculation situations.

The results of calculations of the bearing capacity of the frame elements allowed us to conclude that even with the assumption of 50% reinforcement loss, the wharf structure should have been able to transfer service vertical loads of 5 kN/m^2, and the forces generated from the barges should not cause direct damage to the wharf structure.

The scope of analyzes regarding the stability test of the quay included 2D numerical simulations in a plane deformation state for five wharf calculation and work models:

- state immediately after its construction;
- state after lowering the bottom of the river;
- state after loosening the ground behind the piling wall;
- state at the moment of the loss of load-bearing capacity of the anchor rods;
- state after applying external load from barges mooring forces.

Based on the calculations carried out, which also included the history of the construction of the wharf structure, it was found that assuming a structure without anchoring or with severe corrosion of the anchors (insufficient load capacity), the embankment was practically in a state of unstable equilibrium and could have been damaged due to the existence of some additional factor. The calculations of the wharf were conducted with the assumptions which were generally favorable for its stability, namely the adoption of high water level and the lowering of the river bed only about 1 m below archival design. In the considered case, the additional factor that could have caused the waterfront failure might be forces from mooring barges and greater than the assumed depth of the river. Assuming loading the wharf with the forces from the barges (when the barges break off at one end from the wharf), the model was unstable with the possibility of large displacements of the structure.

3 Final remarks

Basing on the performed assessment of the damaged river reinforced concrete wharf main reasons of its failure were defined. The most important factor was a very poor technical condition of the structure due to more than 100 year of service without proper conservation and repair. Additionally, from a static point of view, the embankment under the wharf was in an unstable equilibrium and at any time it could be damaged if additional horizontal forces or phenomena related to the flow of the river appeared. The analyzes carried out showed that the factors that could initiate the destruction process of the wharf included: lack or destruction of the stabilizing anchors, sudden fluctuations in river water level [5] and horizontal forces originating from barges moored to the piers of the quay bearing frames. Coupling all of the above factors caused the destruction of the part of the wharf and the necessity of its reconstruction.

Currently, the undestroyed part of the wharf has been protected using a sheet pile retaining wall with Larsen profiles (Fig. 22), while the destroyed part is going to be completely rebuilt.

Fig. 22. Sheet piling retaining wall.

Fig. 23. Reconstruction works on the destroyed wharf.

References

1. J. Bień, T. Kamiński, P. Rawa, *Emergency conditions of masonry embankments of the Odra river in Wrocław*, Proc. of 27th Scientific and Technical Conference Construction Failures, Międzydroje 2015 (in Polish)
2. Ł. Dorbiec, R. Jasiński, A. Piekarczyk, *Diagnostics of Reinforced Concrete Structures. Methodology, field tests, laboratory tests of concrete and steel* (PWN, Warsaw, 2010) (in Polish)
3. A. Zybura, M. Jaśniok, T. Jaśniok, *Diagnostics of Reinforced Concrete Structures. Tests of reinforcement corrosion and protective properties of concrete* (PWN, Warsaw, 2011) (in Polish)
4. M. Raupach, T. Büttner, *Concrete Repair to EN 1504. Diagnosis, Design, Principles and Practice* (CRC Press, Boca Raton, 2014)
5. K. Parylak, K. Witek, *Complexity of conditions of the causes of the disaster of the Odra reloading wharf*, Acta Sci. Pol. Architectura, 17 (3) 2018

Life cycle assessment of refurbishment scenarios to confined tank structures at a waste water treatment plant

Michel Donadio[1], *Catrina* Heider[2], *Zeno* Dan[3]

[1]*Sika Services AG, Switzerland*
[2]*Sika Technology AG, Switzerland*
[3]*Sika Europe Management AG, Switzerland*

ABSTRACT: The paper will show a Life Cycle Assessment (LCA) of three different refurbishment scenarios for confined tank structures at a waste water treatment plant. – One using a traditional Portland cement based repair mortar, one combining this cement repair mortar with a reactive protective coating, and one using 100% calcium aluminate based repair mortar technology. In order to evaluate the sustainability and potential environmental impact of each of the three scenarios, the Life Cycle Assessments (LCA) were performed, which included Cumulative Energy Demand (CED), Global Warning Potential (GWP) and Photochemical Ozone Creation Potential (POCP). The results are compared and analyzed in this paper, particularly in relation to extending the structural service life, which shows how innovative systems can give advantages to designers and contractors, which in turn are also of benefit to their customers.

1 Introduction

Sewers and confined tanks in waste water treatment plants often suffer from biogenic corrosion in their areal zones. This biogenic corrosion [1] occurs because of the transformation of the sulfates present in domestic waste water into hydrogen sulphide gas and then subsequently into sulphuric acid by different strains of bacteria.

The sulphuric acid attacks all types of mineral substrates severely. Traditionally, concrete repair systems with reactive resins are used, but these require defect-free conditions and application, which are often impossible to achieve in such difficult environments.

In the last 20 years, 100% calcium aluminate based mortars (cement & aggregates) have been used to successfully renovate these structures. One of the main features of this technology is its ability to block the transformation of hydrogen sulphide into sulphuric acid, due to the bacterio-static effect of calcium aluminate mortar. This has already been explored in various published papers [1], [2], [3].

Therefore, choosing the right repair strategy and product quality not only significantly reduces the frequency of refurbishment cycles, but will also have a positive impact on the structure's longevity, and on the overall economic performance of the structure.

The main sustainability aspects for refurbishment are thus related to the normal three pillars of sustainability (social, economic and environmental), through the main drivers:

Materials input, time efficiency, economic costs and environmental impacts.

To examine these, three different refurbishment scenarios for the refurbishment of a confined tank structure (1,000 m^2), are compared in this paper for their environmental impact, durability and cost effectiveness.

2 Method

Life Cycle Assessment (LCA) [4] provides a method to quantify and evaluate potential environmental impacts throughout a product's life cycle. It starts with the raw materials, through the production phase, on to the site; then through the products' use, service life and end of life treatment, either to waste incineration, or its recycling can be included. This is commonly referred to as cradle to grave in the standard ISO 14044 (ISO 2006). In this study, the production module (cradle to gate) and the surface pre-treatment were considered in the assessment.

An LCA for a construction material manufacturer can produce a quantitative environmental profile within a defined framework, which can help to improve or differentiate between products and solutions in terms of their potential environmental benefits. This LCA of alternative refurbishment systems and products was performed by the Sika Global Product Sustainability Department according to ISO 14044 (ISO 2006) and EN 15804 (CEN/TC 350 2012).

3 Scope of the analysis

The functional unit taken for the LCA is the refurbishment of an arbitrary confined reinforced concrete tank in a waste water treatment with a surface area of 1,000 m^2 over a 50 year life span.

It is assumed that this internal tank surface is subjected to biogenic corrosion on the areal part of the structure. To perform the LCA, an inventory analysis was necessary, i.e., data collection and calculation procedures to quantify all of the relevant inputs and outputs of the procedures and products in each system.

The product data collection was carried out by the respective manufacturer and corresponds to the measured or qualified data, including the product formulations, transportation, packaging and production. The systems were then modelled in the LCA software GaBi 8.7.0 [5], based on this data and datasets from commercial databases such as ELCD (European Reference Life Cycle Database), Thinkstep and Ecoinvent. The materials' transport to the site, their installation (excepting the surface pre-treatment) and their end of life treatment were excluded from this LCA.

The impact assessment phase associates inventory data with environmental impact categories (calculated with the CML 2001 method). Three impact categories are considered to be the most relevant for this example, as these three depict the sustainability drivers in refurbishment: the carbon footprint, resource (and energy) efficiency, and pollutant emissions (also related to health and safety of the workers).

3.1 Cumulative energy demand

Cumulative Energy Demand (CED) reflects the total input of primary energy from renewable and non-renewable resources. It is measured in MJ.

3.2 Global warming potential

The global warming potential (GWP) is the potential contribution to climate change due to greenhouse gases emissions. It is measured in kg (or tons) of carbon dioxide (CO_2) equivalent for a 100-year time horizon.

3.3 Photochemical ozone creation potential

Photochemical ozone creation (POCP) is the potential contribution to summer smog, related to ozone induced by sunlight on volatile organic compounds (VOC) and nitrous oxides (NOx). It is measured in kg of ethene equivalents (also known as ethylene).

4. Retrofitting scenarios analysed

The different systems are calculated over a 50 year period based on an arbitrary surface area of 1,000 m^2. The scenarios and products are described below.

3.1. Description of the scenarios

There were three different scenarios selected for comparison. The first scenario assumes the refurbishment is carried out with a CEM III based concrete repair mortar complying with EN 1504-3 and achieving the highest performance classification of R4 (Cement used being sulfate resistant). Based on site returns and lab assessment, it is assumed that the full cover of 50 mm would need to be replaced every 7 years to provide corrosion protection to the reinforcing bars.

The second scenario uses the a polymer modified R4 class CEM III mortar complying to EN 1504-3 followed by an epoxy cement combination coating that provides a temporary moisture barrier and a smooth surface suitable for the use of a 100% solid epoxy top coating. Based on site returns, we have assumed this would need to be replaced every 10 years.

The third scenario uses a repair mortar based on calcium aluminate technology (both the cement matrix and aggregates being calcium aluminate based). As the aluminate matrix inhibits the transformation of the H_2S into sulphuric acid [1], longer term durability [6] can be achieved, so for this LCA, we assumed this would need to be replaced only once over the 50 year design period.

3.2. Detailed application information

3.2.1 Scenario 1

Retrofitting every **7 years** with repair mortar; no protective coating after the mortar application:

- Surface Preparation by high pressure water
- Reprofiling with concrete repair mortar over 50 mm with wet spray application (100 tonnes per retrofitting).

3.2.2 Scenario 2

Retrofitting every **10 years** with repair mortar; epoxy protective coating after the mortar application:

- Surface Preparation by high pressure water
- Reprofiling with concrete repair mortar over 25 mm by wet spray application (50 tonnes per retrofitting)
- Resurfacing and temporary moisture barrier over 3 mm thick by wet spray application (1030 kg per retrofitting)
- Concrete protection with 2 coats of epoxy coating @500g/m^2 per coat, by airless application (16 tonnes per retrofitting)

3.2.2 Scenario 3

Retrofitting **once** over the designed period:

- Surface Preparation by high pressure water
- Reprofiling with Calcium Aluminate concrete repair mortar over 25 mm by wet spray application (60 tonnes).

https://doi.org/10.1051/matecconf/201928910005

3.1. Product description

Table 1: The refurbishment materials

Product Type	**Description**
Scenario 1	
Repair mortar	1-component, non-polymer modified (CC), R4 repair mortar complying with EN 1504-3
Scenario 2	
Repair mortar	1-component, polymer modified (PCC), R4 repair mortar complying with EN 1504-3
Resurfacing mortar	3-component resurfacing mortar complying with EN 1504-3 and -2, not requiring any curing and allowing fast over-coating with an epoxy resin
Epoxy coating	Protective coating complying with EN 1504-2 – 100% solids content
Scenario 3	
Repair mortar	1-component, 100% Calcium Aluminate mortar, R4 repair mortar complying with EN 1504-3

4. LCA Results

The actual LCA analysis is a complex analysis. For simplicity the illustrated graphs below present a summary from the study, showing the overall impact on the three categories (refer to Fig. 1, 2, 3, 4 & 5).

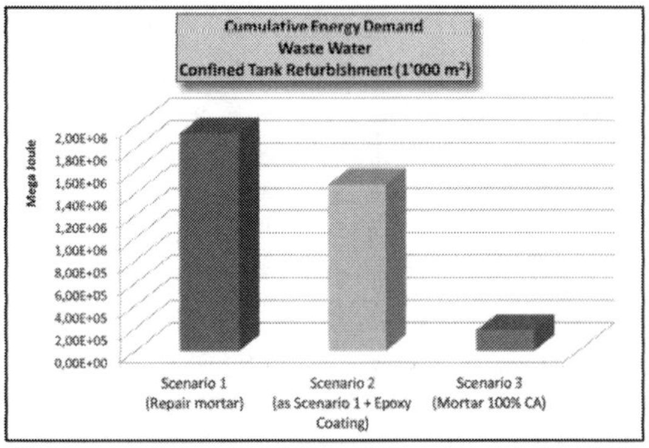

Fig. 1: Cumulative Energy Demand for a lifespan of 50 years.

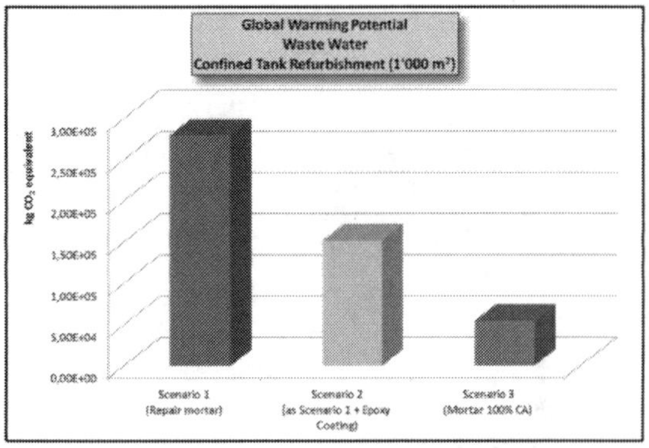

Fig. 2: Global Warming Potential, for a lifespan of 50 years.

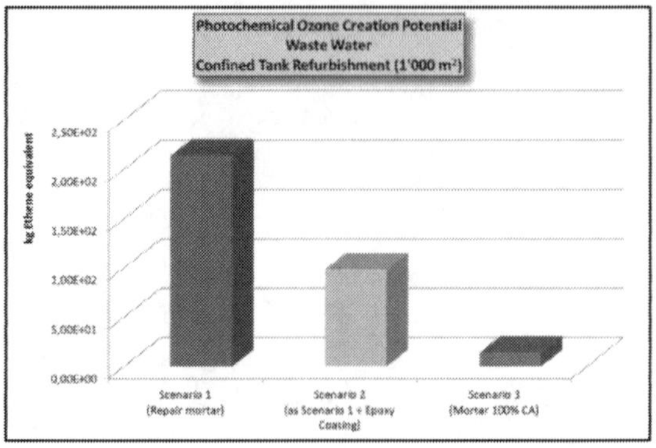

Fig. 3: Photochemical Ozone Creation Potential for a lifespan of 50 years.

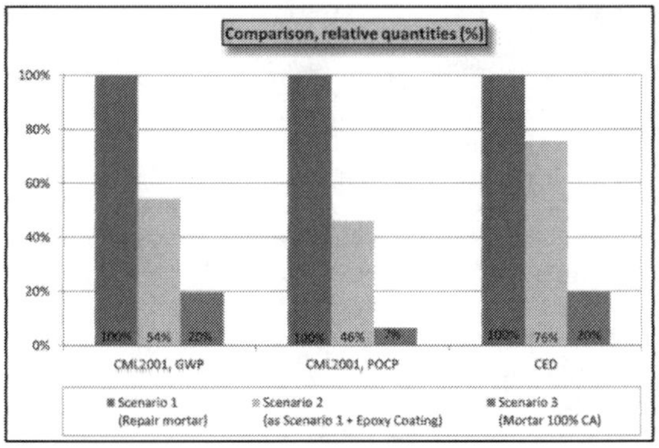

Fig. 4: Comparison of the three categories – **Scenario 1** being used as reference.

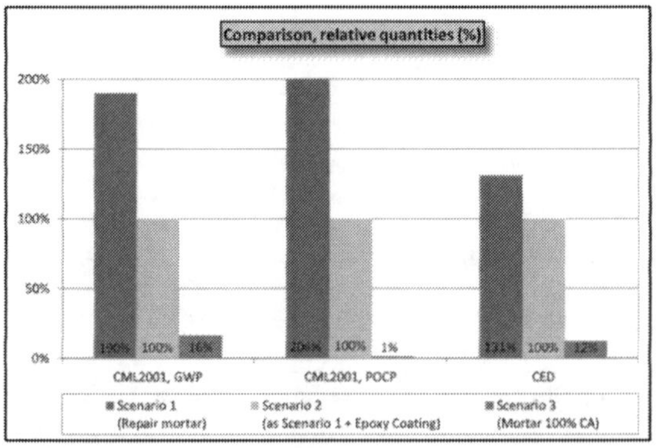

Fig. 5: Comparison of the three categories – Scenario 2 being used as reference.

5. Discussion

5.1 Scenario-1

Scenario-1 entails higher environmental impact in the three categories compared to the others. This is due to the necessary repetition of the retrofitting, as all types of Ordinary Portland Cement based system are very sensitive to acid attack generated by biogenic activities.

 https://doi.org/10.1051/matecconf/201928910005

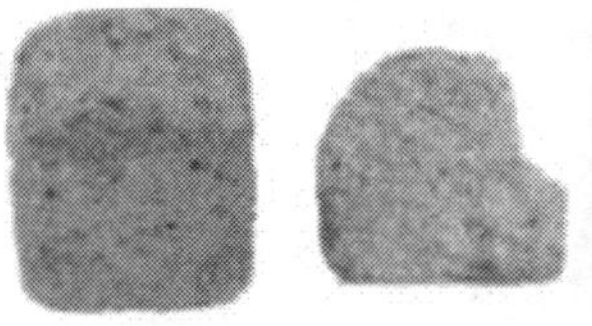

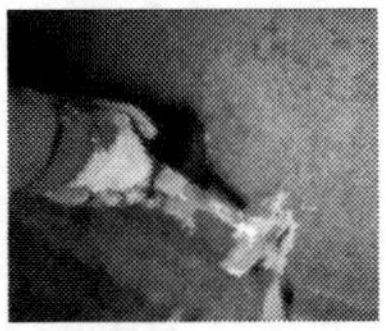

Fig. 6 & 7: OPC mortar containing fly ash, silica fume and silica sand after 1 year in an incubation chamber simulating the biogenic corrosion – the picture on the left being the cube at the start of the test and the one in the middle at 1 year later (3).

Fig. 8: The picture on the right is in a large waste collector near Paris, where a repair mortar is being scratched off easily, due to its deterioration by biogenic corrosion, just a few years after the installation.

5.2 Scenario-2

Scenario-2 also has significant impact in each of the three categories; though lower than Scenario-1.
This is because the system comprises epoxy resin based materials that have some environmental impact.

Additionally, the relatively frequent repetition of the retrofitting being required, also increases this impact.

5.3 Scenario-3

Scenario-3 has the lowest environmental impact in all of the three categories.

This is due to the long service life and durability of the calcium aluminate based repair mortar in this type of environment. Investigations in incubation chambers [6] and in real-life similar environments have shown that these mortars have a very low degradation rate.

Fig. 9: Picture taken in Israel where a system similar to Scenario-2 has shown premature failure after just 5 years in service.

5.4 Cumulative energy demand

Scenario-3 has the lowest impact to primary energy from renewable and non-renewable resources.

When compared to the Scenario-1, the energy saving per year using the 100% calcium aluminate mortar is equivalent to the energy required to produce ~40 tons of oil (and to ~28 tons of oil when compared to the system used in Scenario-2).

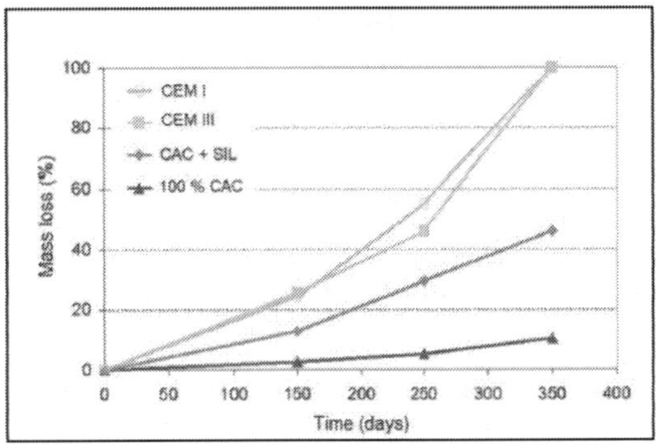

Fig. 10: Weight loss over time from four cementitious materials exposed in the Hamburg Simulation Chamber [6] for one year.

Fig. 11: Cube of 100% CAC mortar exposed in the Hamburg Simulation [6] for 1 year. Left-hand picture was taken at the start of the test, with the right-hand picture taken 1 year later.

5.5 Global warming potential

Using the 100% Calcium Aluminate technology will minimise the potential contribution to climate change by significantly reducing the emissions of carbon dioxide.

When compared to Scenario-1, the CO_2 saving per year using the 100% calcium aluminate mortar is equivalent to the CO_2 issued during the transport of 24.7 Tonnes (typical full truck capacity) of product by truck for over ~150'000 km, which is equivalent to more than three times the worlds circumference (the figure is still approx. half this amount for Scenario-2 at ~67'000 km).

5.6 Photochemical ozone creation potential

Scenario-3 also has the lowest impact in the category of ozone creation potential.

When compared to the Scenario-1, the VOC saving per year using the 100% calcium aluminate mortar is equivalent to saving ~115 tons of solvent based coating (the figure is ~49 tons when compared to Scenario-2).

6. Conclusions

Biogenic corrosion is a growing problem in sewers and waste treatment plants due to various different factors (e.g. the reduction of water consumption, and the reduction of odour leading to the enclosing of many open tanks, etc.).

Field returns have shown repeated failures of traditional repair and retrofitting systems such as those described in Scenario-1 & Scenario-2, leading to repeated retrofitting sequences.

Repetition of these retrofitting procedures not only has an economic impact (materials cost, loss of service, temporary by-pass, etc.), but this imposes a significant impact on the environment as well.

When a suitable 100% calcium aluminate technology is used to retrofit surfaces affected in this way, or to protect new structures, significant savings can be achieved. This is because, once the application is done, the protected concrete surfaces in the areal zone of the structure will not be affected by biogenic corrosion in future as sulfuric acid is not being produced any longer.

In addition to providing substantial cost & time savings, the use of this type of mortar significantly decreases the impact to the environment (reductions of 90 to 99% when compared to Scenario-1 - refer to Fig. 4 above; and reductions of 98 to 83% when compared to Scenario-2 – refer to Fig. 5 above).

In conclusion, when compared to traditional refurbishment solutions, the achievements of such a durability orientated system for these confined tanks subject to potential biogenic corrosion over a service life of 50 years can be summarised as follows:

- Higher performance = lower refurbishment frequency and intensity = environmental benefits, plus materials, time and cost savings
- Energy savings (equivalent to ~40 tonnes of oil)
- Carbon savings (equivalent to transporting a full truck ~150’000 km)
- VOC emissions savings (equivalent to saving the use of ~115 tonnes of solvent based paints).

References

1. J. Herisson, Thesis: *Bio-degradation of cement material in sewer refurbishment – comparative study of aluminate cement and Portland cement.* (2012)
2. F. Saucier et all; *Use of Calcium Aluminate Technology;* The Institute of Concrete technology, (Year book 2015-2016)
3. V.A. Geoffroy et all, *Calcium Aluminate Cements: Proceeding of the Centenary Conference,* Fentiman CH, Mangabhai RJ and Scrivener KL. (Editors), (2008)
4. CEN/TC 350 2012, *Sustainability of construction works - Environmental product declarations - core rules for the product category of construction products,* EN 15804:2012, (2011).
5. GabBi 8.7.0; *Software-System and Database for Life Cycle Engineering*. Thinkstep AG (Leinfelden-Echterdingen), (1992-2018)
6. S. Ehrich & all, Journal of Materials in Civil Engineering, *Biogenic and Corrosion Sulfuric Acid Corrosion of Mortars*, (1999)

Hybrid sandwich panels for building uses: focus on glass fibre reinforced polymer and mineral matrix interface.

Khaled Djama*, *Laurent* Michel, *Emmanuel* Ferrier, and *Aron* Gabor

Université de Lyon, Université Lyon 1, Laboratory of Composite Materials for Construction (LMC2), 82 bd Niels Bohr, 69622 Villeurbanne Cedex, France

Abstract. Sandwich structures are being used increasingly in civil engineering because of their high strength, stiffness, and stiffness-to-density ratio. The studied sandwich structure was made of glass fibre reinforced polymer (GFRP) skins, and GFRP reinforced core. Two case studies are presented in this paper: a use as lightweight floor in building rehabilitation, and an application as façade panels. In both cases, the GFRP sandwich structure can be associated with a mineral matrix because of conventional construction methods and acoustic floor insulation in the first case, and because of architectural issues in the second. To design the hybrid sandwich panel and ensure monolithic mechanical behaviour, a finite element method (FEM) that predicts the interface failure was introduced. To implement the FE model by mechanical interface properties, pull-off and push-out tests were performed to assess the mode I and mode II stress limits. Four GFRP surface roughnesses and two configurations with chemical additions were tested. The three configurations that performed the best were tested by submitting the hybrid sandwich structure to three-point bending loading. The prediction of the interface failure by the FEM was assessed by comparison to the experimental data. Finally, full-scale panels were experimentally tested and designed for the named two uses cases thanks to the FE model.

1 Introduction

Sandwich panels were introduced during World War II [1] to ensure high mechanical performance using low-weight structures. The demand for low-weight structures was high in the warplane sector. Nowadays, this high stiffness-to-weight ratio requirement remains valid in several industrial sectors [2] like civil aerospace, automobiles, rail cars, etc. In civil engineering, numerous applications can be found. Chróścielewski et al. [3] presented a sandwich footbridge in 2017. The full-scale structure was subjected to static load under real climatic conditions. Cementitious materials are omnipresent in civil engineering sandwich panels, for their mechanical properties, fire resistance, and construction patterns. Hopkins et al. [4] investigated the creep behaviour of concrete roof and floor sandwich panels. Other applications as façade panels can be cited. Sharaf and Fam [5] analysed the bending behaviour of large-scale (9145 × 2440 × 78 mm) cladding sandwich panels, simply supported at six points. Pascual et al. [6] emphasised the physical feasibility of fabricating composite sandwich structures by bonding glass face sheets to GFRP core profiles.

Conventional sandwich panels are composed of thin stiff skins and a low-density core. Generally, the cores are made of polyurethane, polystyrene, aluminium, or concrete foams. The skins could be metallic, mineral, or composite. Structural issues of conventional sandwich panels were well studied in the last decades. Mainly, the low mechanical properties of foams lead to high shear deflection and shear failure, and small skin thicknesses caused buckling under compression load. To address these issues, the introduction of metallic and composite connectors and ribs in conventional cores was investigated and presented in several articles [7-9]. Numerous enhancements, which are centred on the idea of increasing the core stiffness while maintaining a low density, have been proposed in the so-called unconventional core. Currently, the widely used unconventional core is known as a cell-designed core. It includes honeycomb, re-entrant honeycomb, auxetic-strut, and auxetic honeycomb core structures [2]. The use of truss core sandwich structures significantly increased during the last decade because of major advances in cutting and three-dimensional printing processes [10-12]. Five types of truss cores can be outlined [2]: corrugated, pyramidal truss, tetrahedral truss, Xtype truss, and Kagome truss cores.

Be that as it may, sandwich panels are multi-layered structures. Because of the different mechanical properties between adjacent layers, stress concentrations at the interface appear, making the panel subject to interface failure. Many authors have addressed this issue: Mitra et al. [13] assessed the interfacial delamination of a polymer foam-cored sandwich composites through peeling tests; Pietrek and Horst [15] proposed a finite element model that correctly reproduced the skin delamination and the buckling behaviour of an asymmetric sandwich panel. In brief, implementation and validation of models that can

* Corresponding author: khaled.djama@univ-lyon1.fr.

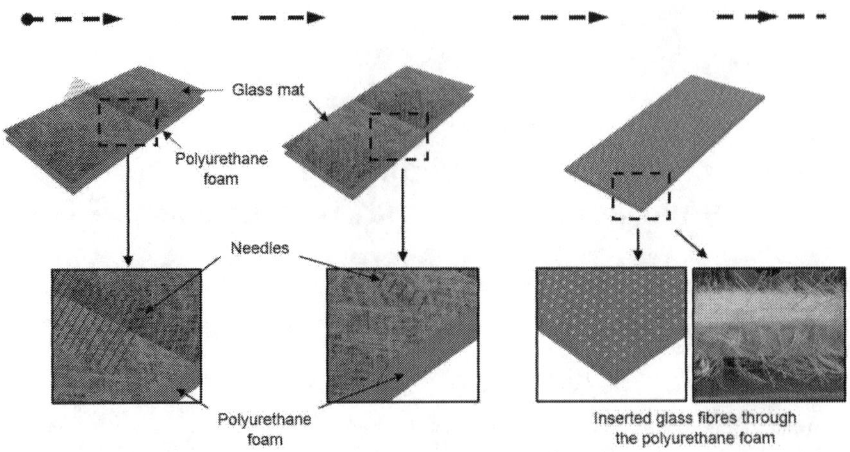

Fig. 1. First stage of the manufacturing process: Glass fibres are inserted through polyurethane foam [14].

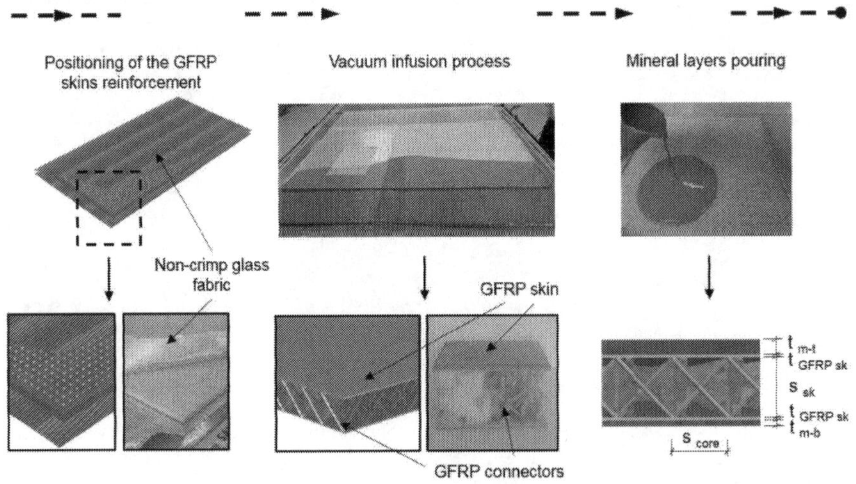

Fig. 2. Second stage of the manufacturing process: Sandwich panel fabrication [14].

well predict the interface delamination is a vital issue in the design procedure.

This study investigated the mechanical properties of the mineral-GFRP interface of hybrid skins, which were connected by GFRP-strengthened polyurethane foam. The truss core was fabricated by a novel patented manufacturing method [16] detailed in the following section. In this work, the mechanical properties of the interface were first assessed through an experimental campaign. Six different interfaces were tested. Then, a finite element model was implemented to predict the interface failure thanks to surface-based cohesive analysis [17]. The simulation responses were compared to experimental responses of three-point bending tests. The use of the presented panel in the field of building refurbishment was assessed through two applications: cladding panels and lightweight floor panels.

2 Manufacturing and materials properties

The studied sandwich structure was made of hybrid mineral-GFRP skins and polyurethane foam. The skins were connected by GFRP connectors that crossed the polyurethane foam. In the manufacturing process, 20 mm spaced needles crossed, numerous times, a stretched glass mat fabric and the polyurethane foam, in order to insert connecting glass fibre reinforcement (cf. Fig. 1). Then, the GFRP skin's reinforcement was positioned (non-crimp glass fabric [0°/90°]), and a vacuum infusion of an epoxy matrix was performed. Finally, the top and bottom mineral layers were poured (cf. Fig. 2). Fig. 3 presents the truss core geometry. The section parameters are presented in Table 1, and the geometric parameters of the truss core are given in Table 2.

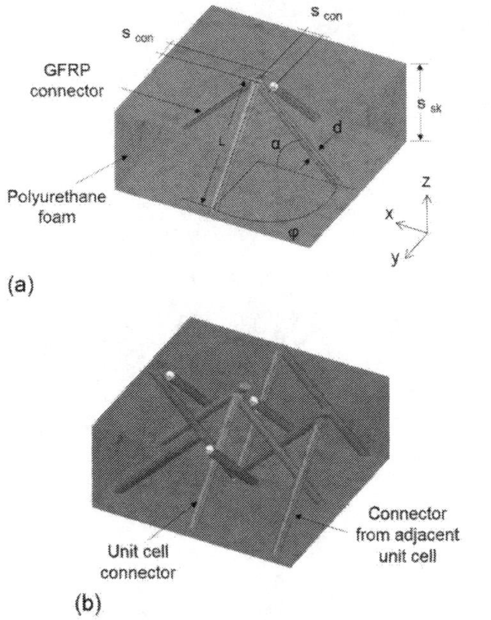

Fig. 3. Truss core geometry: (a) relevant parameters of the truss core unit cell; (b) complete unit cell: core unit cell with connectors from adjacent truss core unit cells [14].

Table 1. Cross-section parameters.

Parameter	Value (mm)
$t_{m\text{-}t}$: thickness of the top mineral layer	5
$t_{GFRP\ sk}$: thickness of the GFRP skins	1
s_{sk}: spacing between skins	20
$t_{b\text{-}t}$: thickness of the bottom mineral layer	2
s_{core}: spacing between truss core unit cells (between connector's nodes)	20

Table 2. Truss core geometric parameters.

Parameter	Value
s_{sk}: skins spacing	20 mm
s_{con}: connector spacing	5 mm
L: connector length	28 mm
d: connector diameter	2 mm
α: connector inclination	45°
φ: connector orientation	90°

The mechanical properties of the panel's components were determined through various tensile and compression tests. The assessed components were the mineral layer, the GFRP skins, and the GFRP connectors. Table 3 summarized the extracted mechanical parameters.

Table 3. Mechanical properties of panel's components.

Material	Test	Parameter	Value (MPa)
Mineral layer	Compression [18]	E	13 269 (4.7%)
		σ	42.5 (1.77%)
		0.4 σ	17.3 (0.69%)
	Tension [18]	E	15 700 (1.62%)
		σ	4.5 (6.19%)
GFRP skins	NF EN 527	E	9 597 (9.15%)
		σ	137.8 (19.04%)
GFRP connectors	Compression	E	1 593 (5.6%)
	Tension	E	2 331 (11.6%)
Polyurethane foam	NF EN 826	E	1.8*

* manufacturer value

3 Interface characterisation

The mechanical properties of six interfaces were investigated through tensile and push-out tests. Fig. 4 presents the tested interfaces. Interface (a) is the reference that has been obtained simply by removing peel-ply fabric after the vacuum infusion ($R_a^{(a)} = 23.26\,\mu m$; $R_q^{(a)} = 29.79\,\mu m$). Interfaces (b), (c) and (d) had different roughnesses. Interface (b) and (c) were obtained by using different peel-ply fabrics ($R_a^{(b)} = 94.63\,\mu m$; $R_q^{(b)} = 117.61\,\mu m$; $R_a^{(c)} = 111.66\,\mu m$; $R_q^{(c)} = 138.63\,\mu m$). Interface (d) was composed of interleaf sand layer between the reference interface and the mineral layer ($R_a^{(d)} = 148.84\,\mu m$; $R_q^{(d)} = 189.38\,\mu m$). The manufacturing steps that lead to this interface are detailed in Fig. 5. The sand sieving properties and its chemical composition are presented in Table 4. In interface (e), an adhesive primer was applied on the reference interface before the mineral layer pouring. In interface (f), an adhesive powder was mixed to the mineral one.

Table 4. Sand characteristics.

Sieve (μm)	Passing (%)	Composition	
630	99.9	SiO_2	98.4%
500	98.7	Fe_2O_3	0.021%
400	90.7	Al_2O_3	0.848%
315	64.2	TiO_2	0.024%
250	30.6	CaO	0.008%
200	11.8	K_2O	0.493%
160	3.8		
125	1.1		
100	0.4		
0	0		

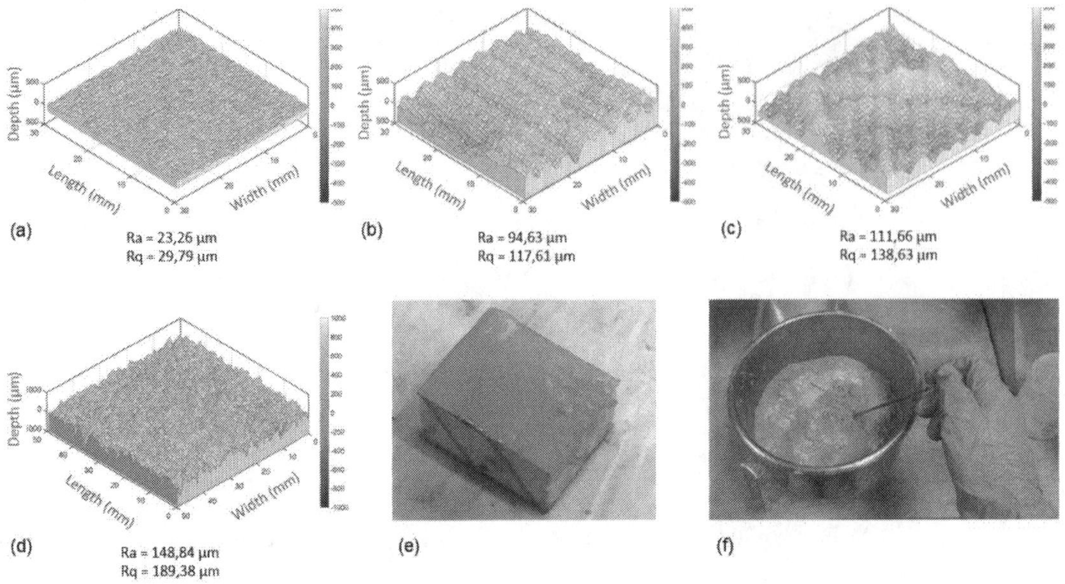

Fig. 4. Tested interfaces: (a) reference interface; (b) and (c) use of different peel-ply to increase the roughness; (d) interleaf sand layer (cf. Fig. 5); (e) adhesive primer; (d) addition of 0.05% of polyvinyl alcohol.

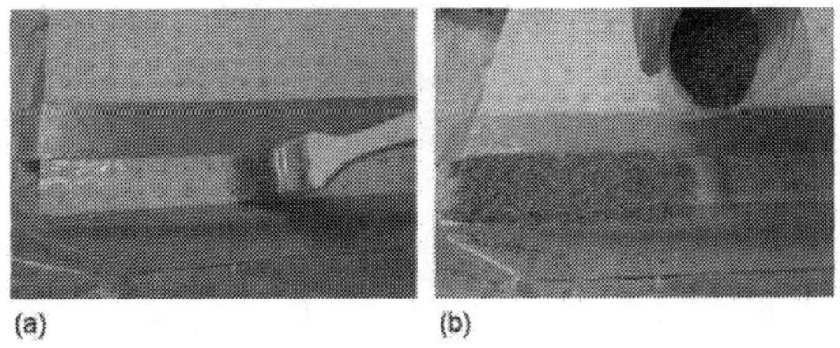

Fig. 5. Manufacturing steps of interface (d): (a) epoxy resin application. (b) sand pouring.

The additive was 0.05% of polyvinyl alcohol. This has a good interfacial bonding strength with silicon dioxide SiO_2. The interfaces' roughness was assessed by laser scanning.

3.1 Mode I

The interface strength of mode I was assessed by tensile tests (cf. Fig. 6). The sizes of the mineral layer and the GFRP sandwich were designed to enforce failure at the interface. For each interface configuration, five specimens were tested.

Fig. 7 presents the obtained results by means of average values and standard deviation. Even though the failure modes were validated, the errors in the mean values can be explained by planarity issues and other geometry imperfections. However, the strand deviation values were acceptable. The experimental results showed that interfaces (d) and (e) performed the best. As expected, comparison between interfaces (a), (b), (c) and (d) highlighted that the substrate's roughness improved the interface strength of mode I. Indeed, in addition to mechanical anchorage, the increase of the roughness increased the contact area between the substrate and the poured mineral layer. The chemical bond was therefore enhanced. The good strength of interface (e) was due to the improvement of the substrate wettability by the adhesive primer. In comparison with the reference interface, the addition of polyvinyl alcohol in interface (f) increased the strength of mode I, but not significantly in comparison with the other configurations.

3.1 Mode II

Mode II interface strength was assessed thanks to push-out tests (cf. Fig. 8). The specimens' sizes were designed to lead to interface failure. The free length of the mineral layer was 10 mm. For each interface configuration, seven specimens were tested. Because of geometry imperfections, we registered inadequate failures as mineral layer fracture or GFRP skins buckling, numerous times. In this case, the results were cancelled.

The results of the push-out test presented in Fig. 9 were calculated from four specimen responses, at least. Although there were stress concentrations at the endpoints of the interface, the mode II interface strength was calculated as a uniform distribution of the applied force over the contact surfaces.

As for mode I, interfaces (d) and (e) performed the best. The substrate's roughness and its wettability are the key parameters to understand the results of push-out tests. In brief, the remarks made in the previous section for mode I are still valid to explain the experimental data of mode II.

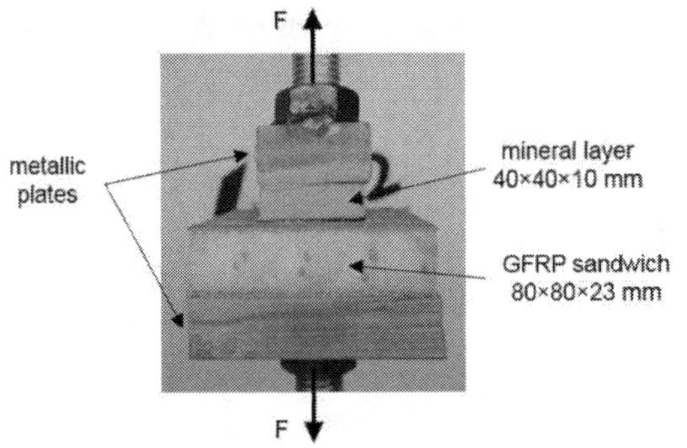

Fig. 6. Tensile test: interface strength of mode I.

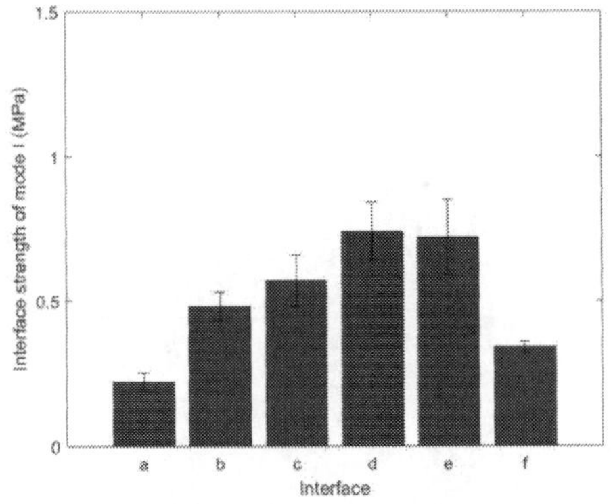

Fig. 7. Interface strength of mode I.

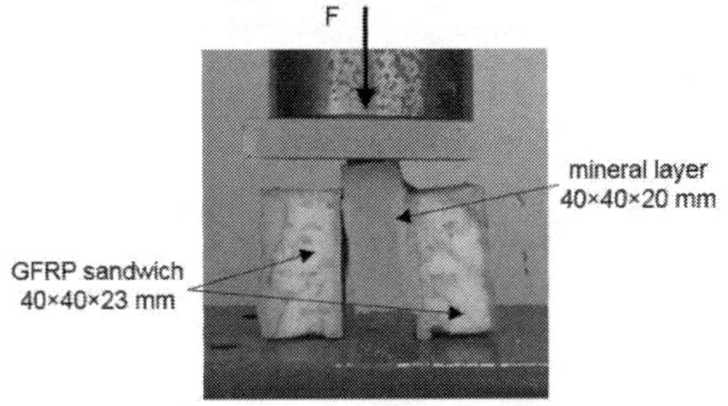

Fig. 8. Push-out test: interface strength of mode II.

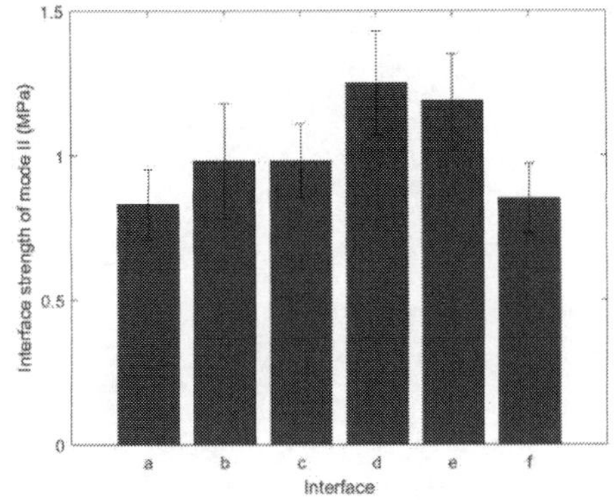

Fig. 9. Interface strength of mode II.

4 Three-point bending test and FE modelling of interface behaviour

The interface behaviour is modelled in this section. The purpose was to predict the interface failure using a finite element model implemented with the interfaces' strength that was previously assessed. To validate the model by comparison with experimental data, sandwich beams were tested under three-point bending load. The beam section fitted the one presented in Fig. 2 (the bottom layer of 2 mm thick was added to prevent fire issues). Two configurations were tested: configuration with interface (a) as a reference, and configuration with interface (d) that performed the best under tensile and push-out tests. Three specimens were tested for each configuration. The specimens' sizes were 90 mm width and 300 mm length. The test configuration was a short span one (250 mm), to enforce the interface failure. It was conducted at a displacement rate of 1 mm/min. The data acquisition of the applied force and the upper deflection was at a frequency of 2 Hz.

The experimental results of the three-point bending test are presented in Fig. 11. The initial stiffness and the pre-peak behaviour are similar for the three tested configurations. The average initial stiffness, calculated between the deflection points of 0.4 mm and 0.8 mm, was 912.7 N/mm (6.14 %). The forces at which the failure initiated are given in Table 5. The difference and similarity of the values are explained by the observed failure modes. As shown in Fig. 11, the failure mode of configuration (a) was mineral layer delamination, while the failure mode of configuration (d) was mineral layer multi-cracking and crushing. Therefore, the improvement of the interface mechanical properties changed the failure mode.

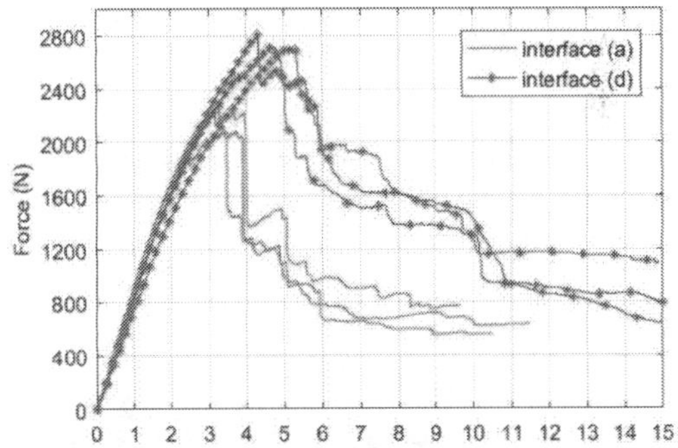

Fig. 10. Experimental results of the three-point bending test.

Table 5. Force at which the failure initiated.

Configuration	Interface (a)	Interface (d)
Force (N)	2016	2733
CoV (%)	6.10	1.97

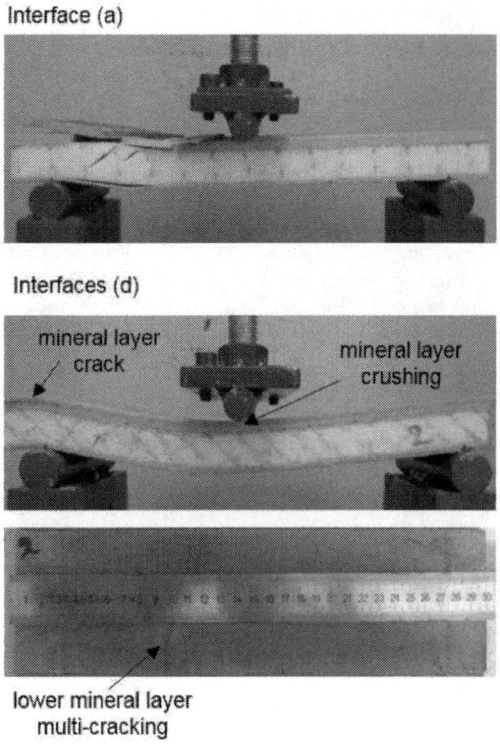

Fig. 11. Illustration of specimens' failure modes.

The FE model was implemented in Abaqus software [17]. The mineral layers were modelled with the Concrete Damage Plasticity (CDP) model [17, 19]. The retained material parameters are presented in Table 3 and Table 6. Hexahedral elements were retained for this material. To predict potential damage of GFRP skins and GFRP connectors, specifically in the post-peak phase, the Hashin model [17, 20] was used to model the GFRP failure. The GFRP material parameters are presented in Table 3 and Table 7. The retained values were those of material characterisation and epoxy matrix strength. The GFRP materials were modelled with continuum shell elements. The mechanical polyurethane foam contribution was neglected and the foam was not modelled. A convergence study permitted to retain elements' size of 5×5×1 mm for the mineral layers, 5×5 mm for the GFRP skins and approximatively 0.4×0.4×0.7 mm for the GFRP connectors. An implicit scheme was chosen, and the Newton-Raphson method was used for the equations' resolution. An automatic time incrementation was adopted with an initial increment size of 0.02 and a maximum one of 0.1. To reduce the number of nodes and calculation tin, a half specimen was modelled and symmetric conditions were applied at mid-span.

The mineral-GFRP interface was modelled using surface-based cohesive behaviour [17]. Contrary to cohesive elements, the named model doesn't assign material properties but interaction properties. Therefore, the interface thickness can be defined as nil, as is the case in the presented simulation. The linear behaviour is governed by equation 1, where vectors t and δ are respectively the nominal traction vector and the separation vector. K is the interaction stiffness matrix. The traction-separation behaviour was setting uncoupled. Therefore, the stiffness variables K_{ns}, K_{nt} and K_{st} were nil. The normal stiffness K_{nn} was equal to the mineral material elastic modulus (15700 MPa), and the shear stiffness K_{ss} and K_{tt} were equal to the mineral material shear modulus (5947 MPa).

Table 6. CDP model parameters.

Dilation angle	Eccentricity	f_{b0}/f_{c0}	K
34	0.1	1.16	0.667

Table 7. Hashin model parameters.

GFRP connectors				
Longitudinal tensile	Longitudinal compression	Transversal tensile	Transversal compression	Shear
$X^T = 18.4$ MPa	$X^C = 70$ MPa	$Y^T = 70$ MPa	$Y^C = 70$ MPa	$S = 14$ MPa
$G^c_{ft} = 0.14$ N/mm	$G^c_{fc} = 1.18$ N/mm	$G^c_{mt} = 1.24$ N/mm	$G^c_{mc} = 1.24$ N/mm	/
GFRP skins				
Longitudinal tensile	Longitudinal compression	Transversal tensile	Transversal compression	Shear
$X^T = 137.8$ MPa	$X^C = 70$ MPa	$X^T = 70$ MPa	$Y^C = 70$ MPa	$S = 14$ MPa
$G^c_{ft} = 4.70$ N/mm	$G^c_{fc} = 1.20$ N/mm	$G^c_{mt} = 1.43$ N/mm	$G^c_{mc} = 1.43$ N/mm	/

$$\begin{Bmatrix} t_n \\ t_s \\ t_t \end{Bmatrix} = \begin{bmatrix} K_{nn} & K_{ns} & K_{nt} \\ K_{ns} & K_{ss} & K_{st} \\ K_{nt} & K_{st} & K_{tt} \end{bmatrix} \begin{Bmatrix} \delta_n \\ \delta_s \\ \delta_t \end{Bmatrix} \quad (1)$$

The damage initiates when a stress value of t^0 is reached. The implemented values were those presented in Fig. 7 and Fig. 9. The chosen post-peak behaviour was linear (cf. Fig. 12). The final separation δ^f was setting as $1.1 \times \delta^0$. Mixed mode parameters were calculated according to equation 2. G^c_T is the total dissipated energy defined as the addition of G^c_n, G^c_s and G^c_t.

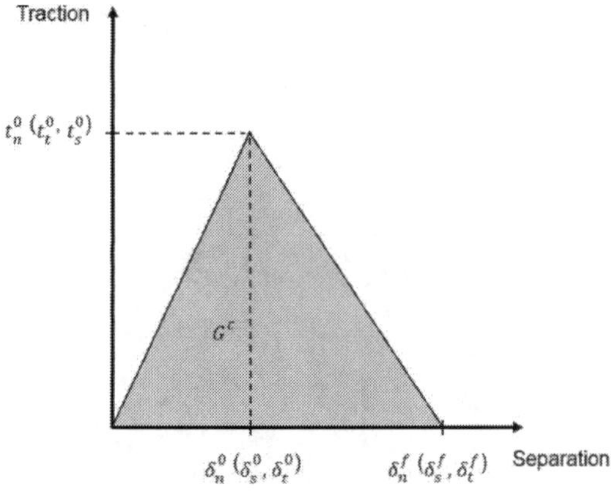

Fig. 12. Interface traction-separation low.

https://doi.org/10.1051/matecconf/201928910006

$$m_1 = \frac{G_n^c}{G_T^c} \quad ; \quad m_2 = \frac{G_s^c}{G_T^c} \tag{2}$$

The FE responses are presented with experimental data in Fig. 13. The results exhibited good agreement in terms of stiffness, maximum force, and post-peak behaviour. The failure modes were reproduced. Fig. 14 illustrated the prediction of the interface failure by the FE model, in configuration (a).

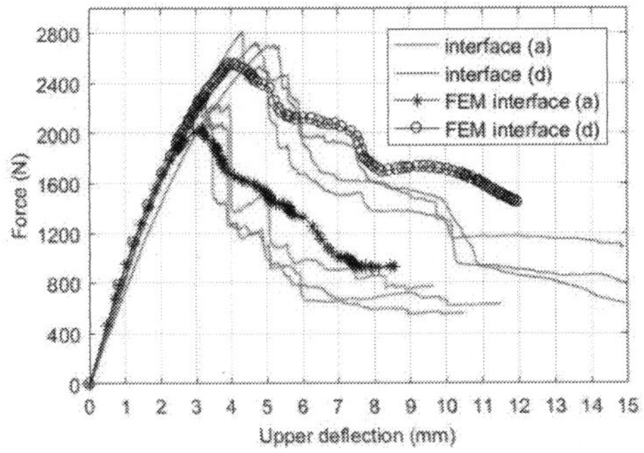

Fig. 13. Comparison of FE responses with experimental data.

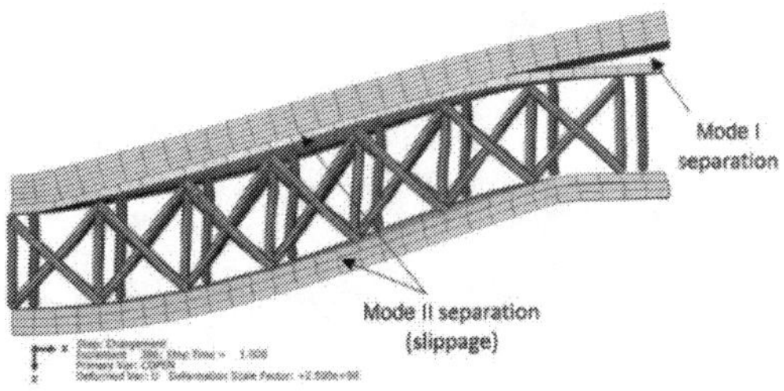

Fig. 14. FE interface failure of interface (a).

5 Building applications

In this section, the use of the presented sandwich panel in building refurbishment was assessed through a façade panel application and a lightweight floor panel use. On one side, several new thermal regulations in Europe (RT 2012 in France for instance) enforce the refurbishment of old buildings and the application of external thermal insulation to fit the new thermal requirements. In the case of cladding, façade panels with high mechanical properties can reduce the number of stud frame, decrease the application time and the solution's cost. On another note, sandwich panels can be very useful for damaged wood floor refurbishment. In numerous cases of buildings rehabilitation, heavy equipment can face access issues and lightweight pre-slab panels may be the appropriate solution. Furthermore, such panels avoid the addition of significant load to the existing structure. In what follows, the presented sandwich panel was designed for these two applications.

5.1 Façade panels

Two panels of 120 cm length and 60 cm were tested under distributed load (cf. Fig. 15). The section presented in Table 1 with interface (d) was retained for this use. The panel's mass density was 22.9 kg/m². The test was conducted by load increment and a 2 Hz data acquisition frequency was selected.

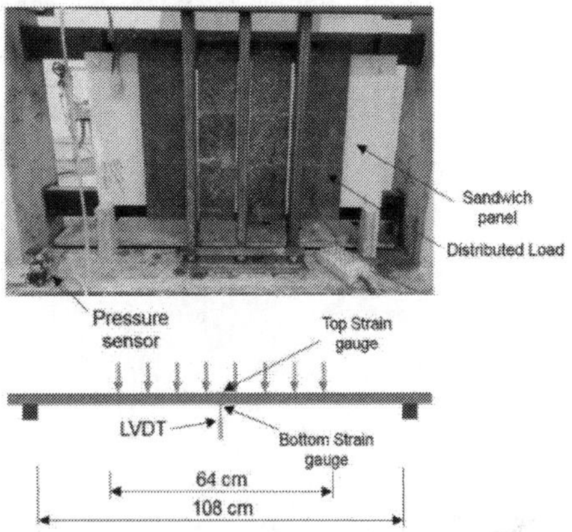

Fig. 15. Full-scale panel test for cladding use.

Thanks to the previously presented FE model, this test was simulated. The comparison between the model and the experimental responses is presented in Fig. 16 and Fig. 17. The FE responses fitted the experimental in terms of global and local behaviour. Regarding the French reVETIR standard [21], the requirements of the V_3 class were respected since no damage was observed before 1600 Pa. A multi-cracking phase that led to the presented bilinear curve (cf. Fig. 16) initiated around a pressure of 2600 Pa. No mineral layer delamination was experimentally observed. The FE simulation permitted shear stress assessment. Fig. 18 presented the shear distribution at a span of 8 cm for pressure of 1600 Pa. The interface stress in the bottom skin was 0.12 MPa (calculated as the difference of stress values in the mineral and GFRP layers). In accordance with the experimental observations, this value was lower than the mode II interface strength (cf. Fig. 9).

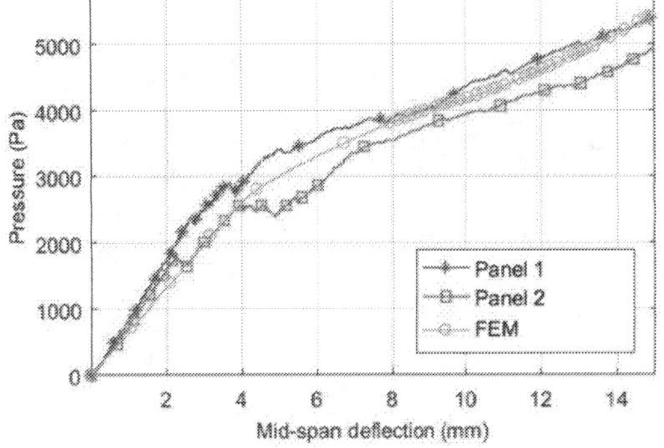

Fig. 16. Load-deflection curve of full-scale panel test: comparison of experimental and FE responses.

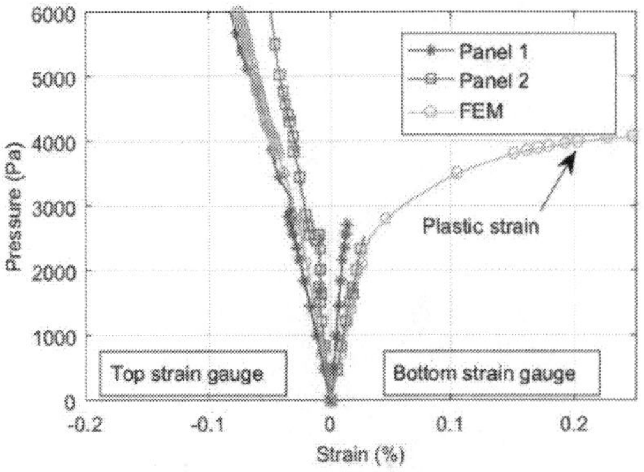

Fig. 17. Experimental and FE strain responses of full-scale panel test.

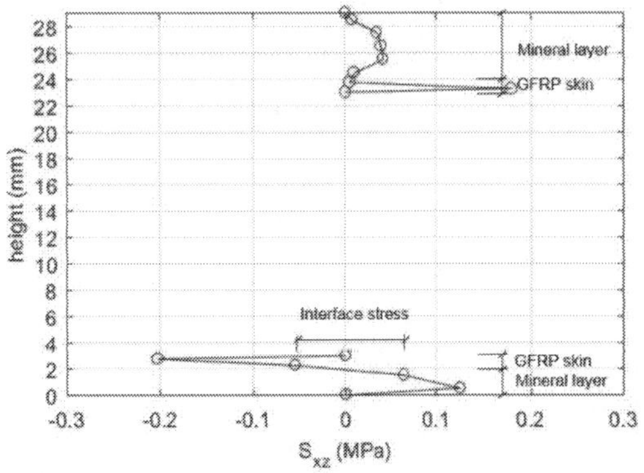

Fig. 18. Shear stress distribution at $x = 8\ cm$ and interface stress estimation.

5.1 Pre-slab panels

Panels of 250 cm length and 100 cm width were designed for a pre-slab panel application. The case study is presented in Fig. 19. An operating load of 150 kg/m² and a permanent load of 100 kg/m² were retained for the design. The permanent load was calculated as the weight addition of dead load (25 kg/m²), floor covering (20 kg/m²), plaster partitions (40 kg/m²), and ceiling panels (15 kg/m²).

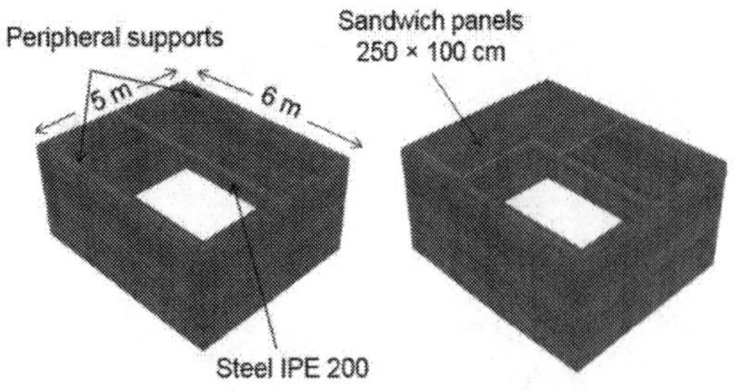

Fig. 19. Concept mapping for pre-slab use.

The FE design led to a section with a skin spacing (S_{sk}) of 80 mm, and a top mineral layer thickness (t_{m-t}) of 8 mm. No bottom mineral layer was retained to avoid premature cracks (fire issues should be fixed with appropriate insulation).

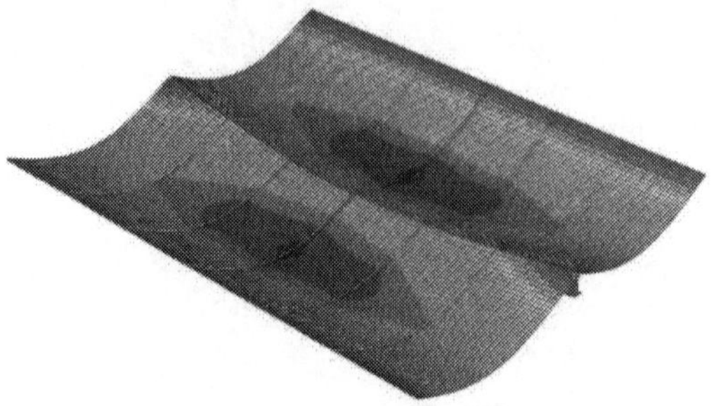

Fig. 20. Deflection shape of the presented pre-slab example.

The deflection shape presented in Fig. 20 highlighted the critical section in terms of deflection criterion. The deflection response was then assessed at this point and compared to a deflection criterion L/250. Fig. 21 indicated that the criterion was verified regarding the limit state of serviceability. Stress analysis showed that the stress distribution among the section and the mineral-GFRP interface was not detrimental.

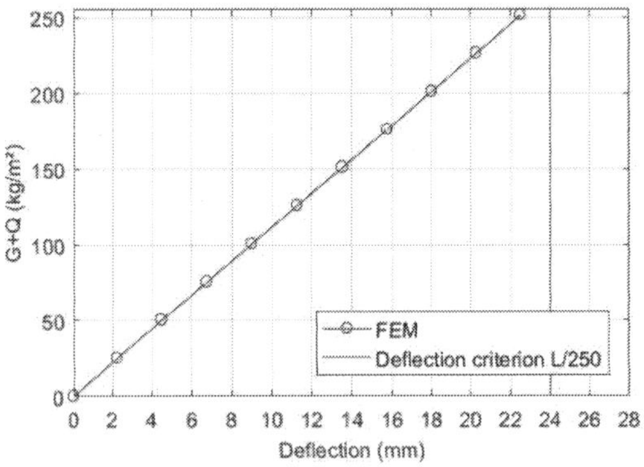

Fig. 21. Deflection response of the critical section of the presented pre-slab example.

6 Conclusions

The presented manufacturing process led to high-performance sandwich panels by resolving the shear deflection issues thanks to GFRP connectors. The interface strength was assessed through tensile and push-out test. The results of the different tested configurations indicated that the addition of an interleaf sand layer at the mineral-GFRP interface increased 3.8 times the mode I strength and 1.7 times the mode II strength. The presented FE model reproduced the sandwich panel mechanical behaviour, including interface failures thanks to surface-based cohesive analysis. Designs for façade cladding and pre-slab uses were presented. The analysis showed that the responses fulfilled criteria in terms of stresses and deflection. The mass density of the panels (less than 25 kg/m²) allows easy and quick fastening. Since the pre-slabs are intended for structural use, additional analysis should be performed regarding dynamic loads and fatigue.

References

[1] J. R. Vinson, Springer Dordrecht, *Sandwich Structures: Past, Present, and Future*, 3–12, (2005)

[2] V. Birman and G. A. Kardomateas, Compos B Eng, *Review of current trends in research and applications of sandwich structures*, **142**, 221–240, (2018).

[3] J. Chróścielewski, M. Miśkiewicz, Ł. Pyrzowski, B. Sobczyk, and K. Wilde, Compos B Eng, *A novel sandwich footbridge - Practical application of laminated composites in bridge design and in situ measurements of static response*, **126**, 153–161, (2017).

[4] P. M. Hopkins, T. Norris, and A. Chen, Compos. Struct., *Creep behavior of insulated concrete sandwich panels with fiber-reinforced polymer shear connectors*, **172**, 137–146, (2017)

[5] T. Sharaf and A. Fam, Thin-Walled Struct., *Analysis of large scale cladding sandwich panels composed of GFRP skins and ribs and polyurethane foam core*, **71**, 91–101, (2013)

[6] C. Pascual, J. Montali, and M. Overend, Compos. Struct, *Adhesively-bonded GFRP-glass sandwich components for structurally efficient glazing applications*, **160**, 560–573, (2017).

[7] N. Williams Portal, M. Flansbjer, K. Zandi, L. Wlasak, and K. Malaga, Compos. Struct, *Bending behaviour of novel Textile Reinforced Concrete-foamed concrete (TRC-FC) sandwich elements*, **177**, 104–118, (2017)

[8] R. Lameiras, J. Barros, I. B. Valente, and M. Azenha, Compos. Struct, *Development of sandwich panels combining fibre reinforced concrete layers and fibre reinforced polymer connectors. Part I: Conception and pull-out tests*, **105**, 446–459, (2013)

[9] M. Mastali, I. B. Valente, and J. A. O. Barros, Compos. Struct, *Flexural performance of innovative hybrid sandwich panels with special focus on the shear connection behavior*, **160**, 100–117, (2017)

[10] S. Daynes, S. Feih, W. F. Lu, and J. Wei, Mater Design, *Optimisation of functionally graded lattice structures using isostatic lines*, **127**, 215–223, (2017)

[11] I. Ullah, J. Elambasseril, M. Brandt, and S. Feih, Compos. Struct, *Performance of bio-inspired Kagome truss core structures under compression and shear loading*, **118**, 294–302, (2014)

[12] J. Mei, J. Liu, and J. Liu, Compos A Appl Sci Manuf, *A novel fabrication method and mechanical behavior of all-composite tetrahedral truss core sandwich panel*, **102**, 28–39, (2017)

[13] N. Mitra, A. Patra, S. Mondal, and P. K. Datta, Eng. Struct, *Interfacial delamination crack profile estimation in polymer foam-cored sandwich composites*, **189**, 635–643, (2019)

[14] K. Djama, L. Michel, A. Gabor, and E. Ferrier, Compos. Struct, *Mechanical behaviour of a sandwich panel composed of hybrid skins and novel glass fibre reinforced polymer truss core*, **215**, 35–48, (2019)

[15] M. Pietrek and P. Horst, Compos. Struct, *Analysis and numerical prediction of the delamination behavior of debonded asymmetric sandwich shells with a thin-walled skin considering plastic deformation*, **188**, 220–232, (2018)

[16] *Core material for composites: SAERfoam®*, [Online], Available: http://www.saertex.com/en/products/saerfoam, (Accessed: 09-Mar-2018)

[17] *Abaqus Analysis User's Guide (6.14)*, [Online], Available: https://www.sharcnet.ca/Software/Abaqus/6.14.2/v6.14/books/usb/default.htm, (Accessed: 25-Jan-2019)

[18] T. Brockmann, *Mechanical and fracture mechanical properties of fine-grained concrete for textile reinforced composites*, Mainz, Aachen, (2006)

[19] J. LUBLINER, J. OLIVER, S. OLLER, and E. ONATE, Int J Solids Struct, *A plastic-damage model for concrete*, **25**, 299–329, (1989)

[20] Z. Hashin, J. Appl. Mech, *Failure Criteria for Unidirectional Fiber Composites*, **47**, 329–334, (1980)

[21] CSTB, *Classement reVETIR des systèmes d'isolation thermique des façades par l'extérieur*, (1996)

Corrosion at low moisture content in both carbonated and chloride polluted concrete – Villa E-1027, a case study

Elisabeth Marie-Victoire[1,*], *Myriam* Bouichou[2]

[1] Laboratoire de Recherche des Monuments Historiques, Centre de Recherche sur la Conservation (CRC-USR 3224), Muséum national d'Histoire naturelle, CNRS, Sorbonne Universités, 29 Rue de Paris, 77420, Champs-sur-Marne, France

Abstract. Corrosion in reinforced concrete is generally attributed to either carbonation or chloride presence in the vicinity of the bars. But in the field of cultural heritage, especially for the most ancient monuments, it is not rare to encounter both carbonated and chloride polluted concrete, inducing heavy corrosion, as was the case in the Villa E-1027 in Roquebrune-Cap-Martin, on the French Mediterranean seashore. The villa was designed by Eileen Gray and Jean Badovici between 1926 and 1929. Due to aggressive environmental conditions and a period of dereliction, the concrete of the villa was quite heavily decayed and a deep restoration was led between 2000 and 2006. But after a little more than 10 years, and despite active maintenance, the villa is again facing corrosion induced decay. Prior to the definition of a new restoration protocol, to better evaluate the corrosion activity, in 2017 a permanent monitoring of moisture and temperature both in the air and in the concrete was installed. In the meantime, a series of instant electrochemical measurements was performed from 2017 to 2018. A first analysis of the results of the monitoring and the non-destructive tests clearly evidences that probably due to the conjunction of the carbonation of the concrete and external active chloride pollution, corrosion can happen at quite low moisture content..

1 Introduction

If carbonation induced corrosion and chloride induced corrosion were extensively studied as separate mechanisms [1] [2] [3] [4], research on their combined pollution, which is often encountered in the field of historic concrete, is rare [5]. For the latter, carbonation is usual, and can affect the concrete cover up to several centimetres [6].When chlorides, either linked to sea proximity or to de-icing salts, reach the reinforcing bars, then heavy corrosion can occur. In France, among the 850 historic monuments made of concrete, 27 % are concerned by chloride pollution and most of them are affected by carbonation [7]. The Villa E-1027 is one of those monuments. The purpose of the present research was to study the corrosion mechanisms affecting this villa and to determine the most suitable restoration technique.

2 Context

Built between 1926 and 1929 by Eileen Gray and Jean Badovici in Roquebrune-Cap-Martin near Nice, the villa E-1027 faces the Mediterranean sea (Fig.1). Its structural elements and floors are made of concrete, while the non-structural components are more heterogeneous (bricks…). It was partially rendered and totally covered with a lime wash.

After being abandoned for a long time, in 2000, the villa was decayed in correlation with water infiltration, ground movement and rebar corrosion. A first deep restoration operation was conducted between 2000 and 2006, including the use of a mineral migrating corrosion inhibitor for the concrete reinforcement and foundation treatments. Nevertheless, eleven years after this restoration and despite periodic maintenance, the concrete was again decaying due to rebar corrosion.

Fig. 1. General view of the Villa E-1027, facing the sea.

The first step of diagnosis performed in 2000 [8], evidenced, for the concrete design, a mix of CEM I cement (300kg/m^3) and calcareous aggregates with siliceous sand. The performance of this concrete was quite low with an open porosity of 17% and compressive strengths ranging between 9.6MPa (patio beam) and 13.3 MPa (patio pillar). In terms of decay, if carbonation had

* Corresponding author: elisabeth.marie-victoire@culture.gouv.fr

exceeded the rebar level in most of the areas examined, moderate external sulfate and chloride pollution were encountered, leading to the formation of Friedel Salts and Ettringite in the cement matrix. Potential mapping evidenced significant gradients in the beams and the underside of the floor of the patio.

A second step of diagnosis was carried out in 2016 [9], showing significant chloride (0.05 up to 2.14% by weight of cement) (Table 1) and alkali pollution (1.19 up to 22.5 eqNa2O kg/m^3) on both the patio and the balconies of the villa, and a moderate sulfate pollution (1.18% up to 5.36% by weight of cement). Carbonation had also progressed (20 to 70mm), and corrosion was still active as confirmed by instant electrochemical measurements performed with a Galvapulse© (Table 2).

Table 1. Chloride profiles (% m/m of cement) on the pillars of the patio and on the balconies.

	0-10 (mm)	10-20 (mm)	20-30 (mm)	30-40 (mm)
Pillar n°2	0.43%	1.85%	0.45%	0.31%
Pillar n°4	1.12%	1.34%	1.83%	/
Pillar n°9	0.99%	0.44%	0.45%	0.31%
South balcony (Sea side)	1.86%	0.26%	0.23%	0.66%
East balcony	0.15%	0.05%	0.15%	/
Underside of the floor	0.31%	2.14%	1.45%	/

Table 2. Electrochemical measurements performed with a Galvapulse© equipped with a copper/copper sulfate reference electrode on 3 testing areas of the underside of the southern balcony (Z3) and of the floor of the patio (Z1 and Z2).

Testing area	Potential min (mV)	Potential max (mV)	Max potential gradient (mV)	Corrosion current (µA/cm²)
Z1	78	224	146	11
Z2	-82	136	218	5.5
Z3	-14	365	379	6

Taking into account the results of these 2 steps of diagnosis, the restoration options were limited. The durability of a standard patching operation, combined with a migrating corrosion inhibitor had shown its limits (lower than 10 years). As the chloride pollution was inevitably active, the villa facing the sea, electrochemical chloride extraction could only be a temporary solution. Finally a waterproof paint couldn't be envisaged due to recurrent infiltration problematic in a context of a concrete loaded with different salts (sulfates, chlorides, alkali). The last durable option was cathodic protection. But its invasiveness represented an issue for an historic concrete which has to be as preserved as possible.

Considering this problematic preservation, prior to the final choice of restoration strategy and taking into account that complementary maintenance work had just been performed to eradicate some infiltration problems on the balconies that had also an impact on the concrete of the patio, a last phase of diagnosis was scheduled with the aim of evaluating the residual corrosion activity on the balconies and the patio (Fig. 2), after this maintenance. Two types of analysis were considered: instant electrochemical measurements and permanent monitoring over one year.

Fig. 2. Detailed view of the patio.

3 Analytical protocol

3.1 Non-destructive testing

Three types of non-destructive testing were carried out: resistivity mapping, potential mapping, and polarization resistance measurements, using a Proceq© corrosimeter and a Gecor10©, both equipped with copper-copper sulfate reference electrodes. Tests were performed 4 times between 2017 and 2018, respectively in April 2017 and March, June and October 2018 (Table 3). Three zones selected on step 2 of the diagnosis were considered (Z1, Z2, Z3, plus a beam of the patio), but only the results obtained in zone 3 (southern balcony) and on a beam of the patio (partially cracked due to corrosion), which are representative of all the measurements, will be presented in this paper. In each zone, polarization resistance was measured on two points corresponding to potential and resistivity gradients.

Table 3. Environmental conditions at the testing dates.

Date	25/04/2017	01/03/2018	26/06/2018	15/10/2018
T (°C)	16	6.8	24.3	22.6
RH (%)	76	78	67	59

3.2 Permanent monitoring

For the permanent monitoring, 2 active corrosion spots previously identified by non-destructive testing were selected respectively on a beam of the patio and on the underside of the southern balcony (Z3). Three types of measurements were considered:

- Temperature and relative humidity in the air (weather station, Fig. 4),
- Temperature and relative humidity in the concrete (Vaisala© sensors, Fig. 3),
- Resistance, potential and polarization resistance (dedicated sensor designed by the Spanish company Sensecorr, Fig. 3) equipped with manganese/ manganese oxide reference electrodes.

All the sensors were connected to a data acquisition system remotely interrogated by cell phone (Fig. 4).

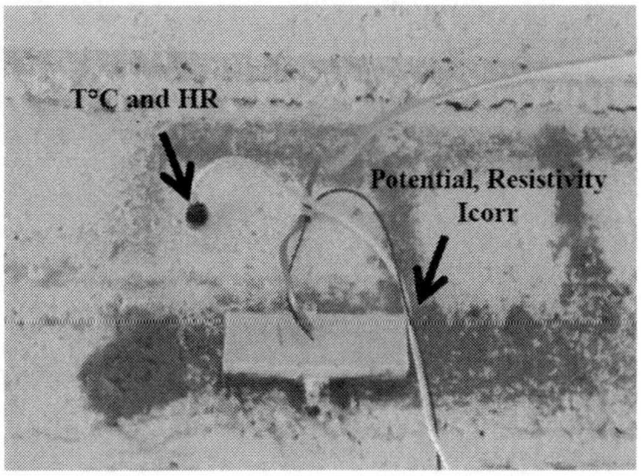

Fig. 3. Temperature, relative humidity and corrosion sensors.

Fig. 4. Weather station (left) and data acquisition system remotely accessible (right).

4 Results

3.1 Non-destructive testing

3.1.1 Balcony

The first map of resistivity performed in April 2017 on the underside of the southern balcony (Fig. 5), was indicative of an overall moderate risk of corrosion according to the RILEM recommendation [10], with values ranging between 9 and 124 kΩ.cm and an average of 43 kΩ.cm.

On the potential map (Fig. 6) significant gradients [11] (average value: -196mV, maximum gradient: 160mV) could be observed on the lower zone of the map and especially on the point C6.

Polarization resistance measurements were performed on two points of the lower part of the zone and were indicative of a high level of corrosion according to the RILEM Recommendation [12] (C3: 1.550µA/cm^2 ; C4: 1.213µA/cm^2).

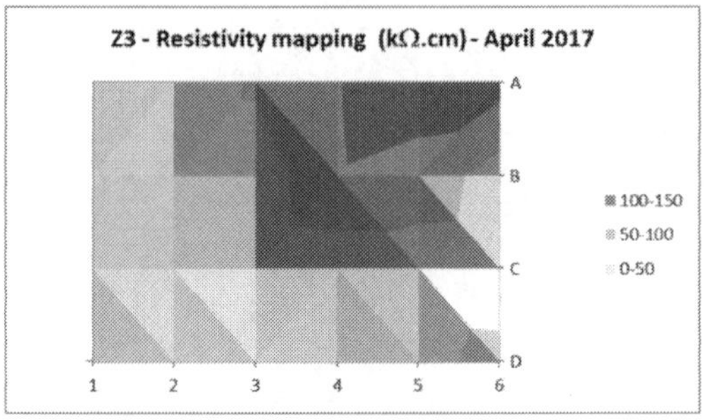

Fig. 5. Resistivity mapping obtained on zone 3 in April 2017.

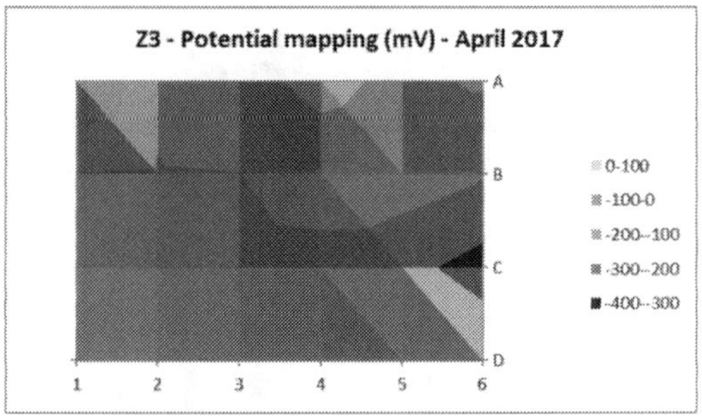

Fig. 6. Potential mapping obtained on zone 3 in April 2017.

On the following three series of measurements performed in March, June and October 2018, almost similar resistivity and potential maps were observed, with nevertheless an increase of corrosion risk on the lower part of the beam (Fig. 7). A clear correlation could be observed between the evolution of the relative humidity in the air and the resistivity and potential values. Thus, when the relative humidity decreases, the resistivity increases and the potentials tend to less electronegative values (Fig. 8 and 9).

The same tendencies were observed for the corrosion currents except in June 2018, where the higher values were obtained (Fig. 10). This could be linked to connection difficulties encountered during the tests that lead to a repetition of the measurements and to an increased moistening of the concrete in points C3 and C4. But during the period of measurement, that lasted one year and a half, the corrosion current on the balcony was indicative of a corrosion level varying between moderate and high.

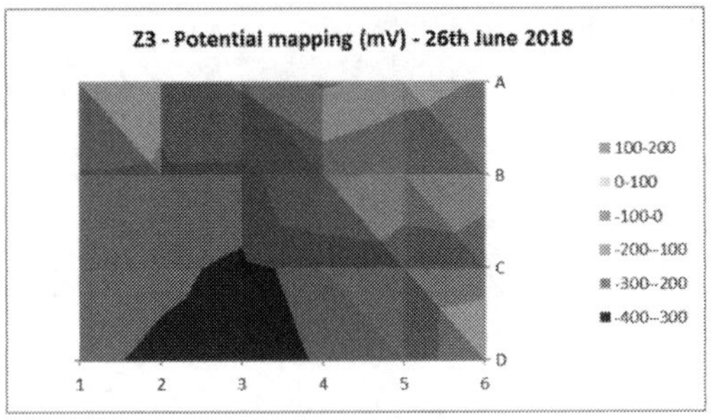

Fig. 7. Potential mapping obtained on zone 3 on June 2018.

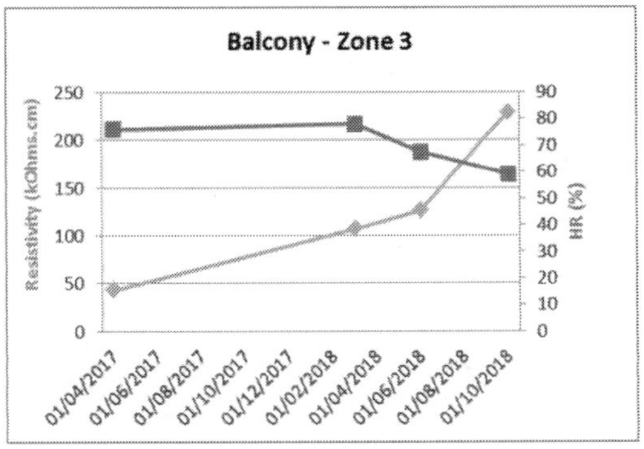

Fig. 8. Evolution of the average resistivity in zone 3 and of the relative humidity in the air.

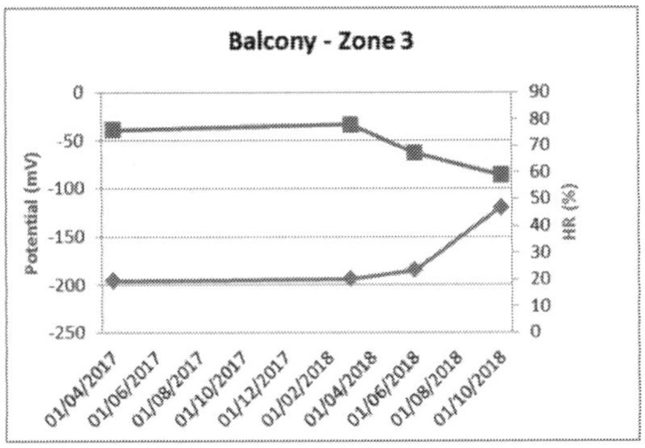

Fig. 9. Evolution of the average potential in zone 3 and of the relative humidity in the air.

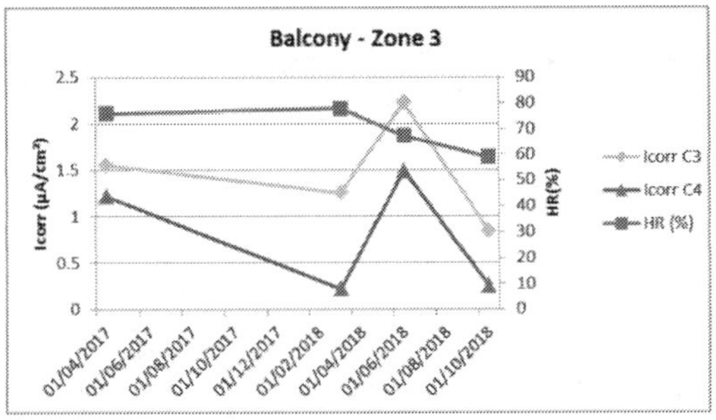

Fig. 10. Evolution of the average corrosion current in zone 3 and of the relative humidity in the air.

3.1.2 Beam

On the beam, the resistivity measured was a little higher than in zone 3 and indicative of a risk of corrosion varying from negligible [10] on the upper part of the beam, to moderate on the lower part (Fig. 11). The values ranged between 28 and 176 kΩ.cm and an average of 109 kΩ.cm.

The potential map (Fig. 12) was consistent with the resitivity mapping, indicating a probable risk of corrosion [11] on the lower part of the beam (average value: -230mV, maximum gradient: 95mV). Nevertheless, as for zone 3, it is to be noticed that the potential values are not very electronegative for a concrete both carbonated and chloride polluted.

Polarization resistance measurements were performed at two locations identified on the potential map, on the lower part of the beam. They were indicative of a low level of corrosion [12] (B23: 0.140µA/cm²; C4: 0.204µA/cm²).

As for the balcony area, similar results were observed on the three following three series of measurements, with again a clear correlation between the evolution of the relative humidity in the air and that of the resistivity and potential (Fig. 13 and 14). If the corrosion currents measured at point B12 were also well correlated (Fig. 15) and indicative of a negligible level of corrosion, at point B23 a continuous increase of the corrosion current could be noticed, in correlation with the opening of a crack (Fig. 16).

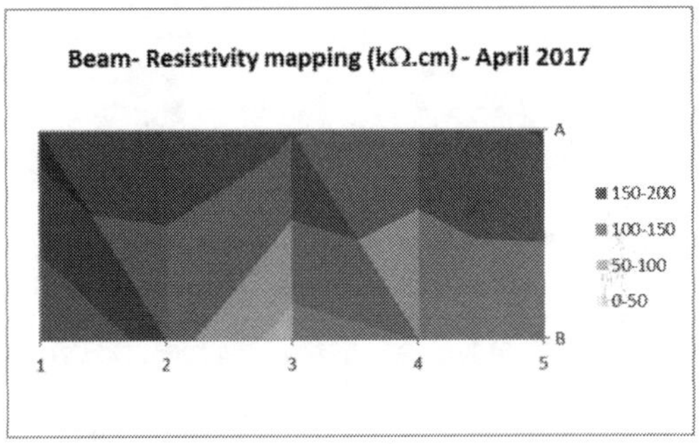

Fig. 11. Resistivity mapping obtained on the beam on April 2017.

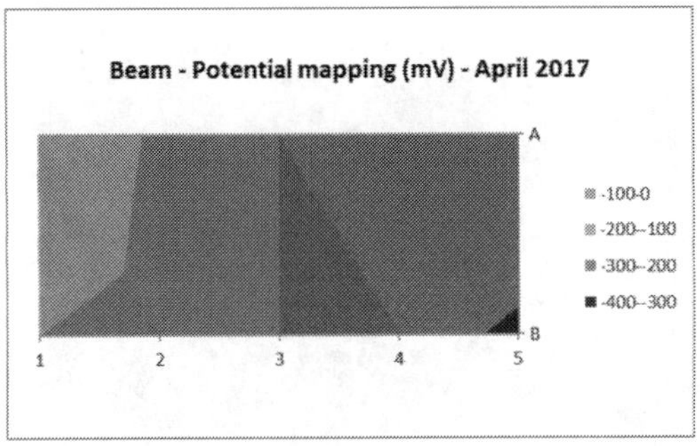

Fig. 12. Potential mapping obtained on the beam in April 2017.

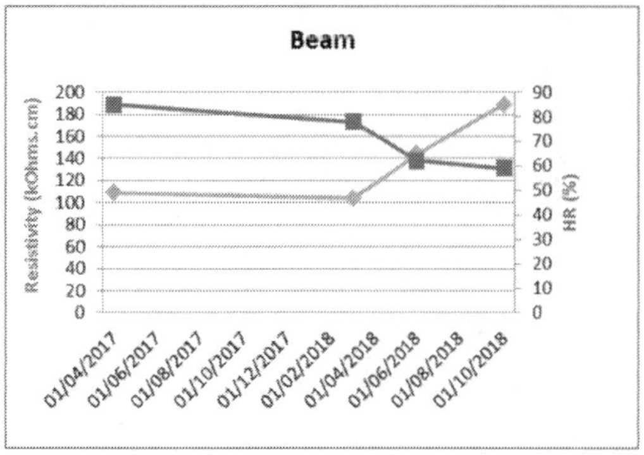

Fig. 13. Evolution of the average resistivity in the beam and of the relative humidity in the air.

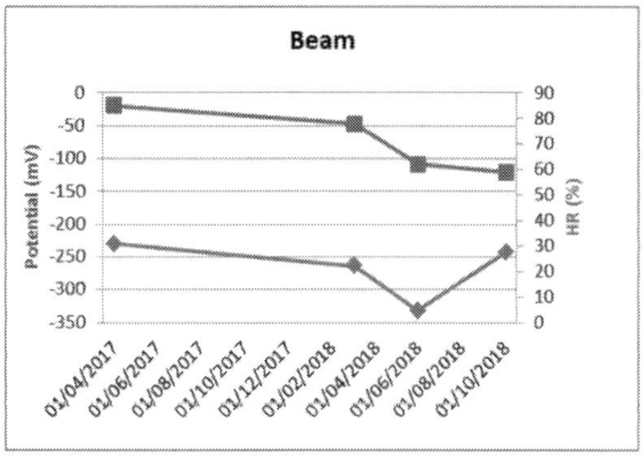

Fig. 14. Evolution of the average potential on the beam and of the relative humidity in the air.

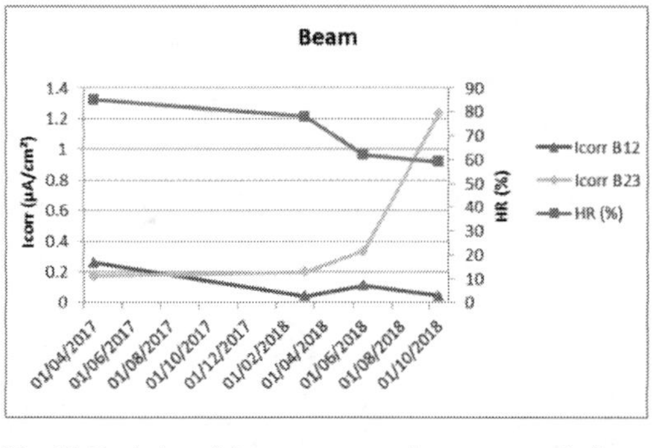

Fig. 15. Evolution of the average corrosion current on the beam and of the relative humidity in the air.

Fig. 16. Opening of a crack on the beam in the vicinity of the measurement point B23.

3.2 Monitoring

Data were recorded between April 2017 and September 2018.

3.2.1 Temperature

As can be observed on Fig. 17, the temperature measured in the concrete either under the balcony or in the beam, are very similar but also very close to that measured in the air. The three curves are overlapping. During the monitoring period, temperatures in the concrete ranged between 0.5 and 33°C, with a maximum daily amplitude of 5°C.

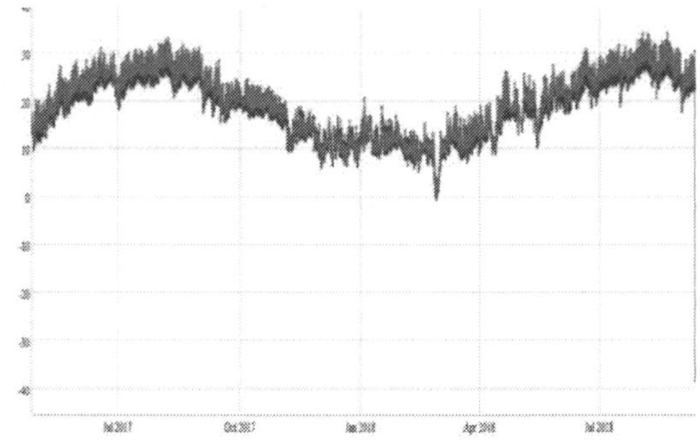

Fig. 17. Temperature (°C) in the air (brown curve) and in the concrete in the balcony and in the beam (green curve).

3.2.2 Relative humidity

Contrary to the temperature, relative humidity in the air and in the concrete was clearly distinct. Thus it fluctuated a lot in the air, while it was much more stable in the concrete (Fig. 18), which was already observed by several authors [13]. When comparing the results obtained in the 2 zones monitored, it appears that the relative humidity was slightly higher but also less fluctuating in the beam (average: 68%) than in the balcony (average: 63%). Globally, the relative humidity in the concrete was lower in winter and higher in spring. Thus, in the beam, the maximum was reached the 13rd of June 2018 (78%) and the minimum in December 2017 (60%). On the Balcony, the maximum was observed on the 21st of June 2018 (77%) and the minimum in December 2017 (53%).

3.2.3 Electrochemical monitoring

As the two zones monitored were protected from the rain, electrochemical monitoring was an issue. Effectively, the contact mortar used between the sensors and the concrete gradually dried and the measurements became impossible, quite rapidly for the beam and soon after for the balcony (Fig. 19). Nevertheless, the first data were consistent with the non-destructive testing, with corrosion current values lower in the beam than in the balcony (Fig. 20).

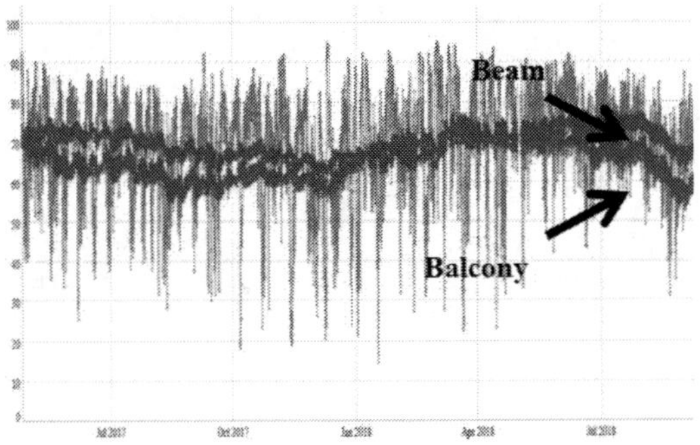

Fig. 18. Relative humidity (%) in the air (blue curve) and in the concrete in the balcony and in the beam (purple curve).

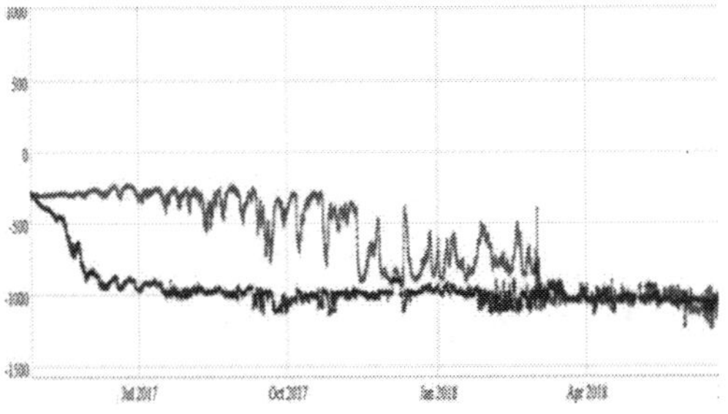

Fig. 19. Potential (mV) on the balcony (ochre curve) and on the beam (blue curve).

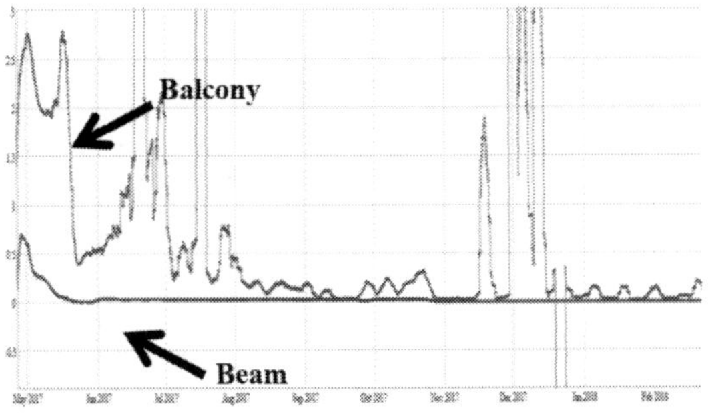

Fig. 20. Corrosion current (μA/cm²) on the balcony and on the beam.

5 Discussion

Due to contact mortar issues in the testing zones which were protected from the rain, electrochemical monitoring couldn't be performed during the year and a half initially scheduled. Nevertheless, the temperature and relative humidity monitoring evidenced a certain stability in the concrete on the two zones examined (Fig. 21) both on a daily and a yearly basis, so that the non-destructive instant electrochemical measurements could be extrapolated to the whole year, and were indicative of active corrosion in both the balcony and the beam examined.

As the testing zones were sheltered from the rain, the relative humidity in the concrete was quite low (yearly average between 63 and 68%). Several studies evidenced a critical threshold of relative humidity between 60% and 70%, below which corrosion is supposed to be negligible [14] [15]. On this basis, the corrosion currents observed on the 2 tested zones of the villa could be surprising. But other authors evidenced that this threshold can vary depending on the exposure of the concrete (rain and temperature…) [13], and the carbonation and/or chloride pollution of the concrete [3]. Thus for a carbonated concrete, depending on the chloride content the relative humidity below which corrosion is negligible, can decrease to 40% [3]. This could be linked to both the hygroscopicity of the chloride salts and to the aggressiveness of the combined carbonation and chloride pollution for the corrosion of reinforcing bars.

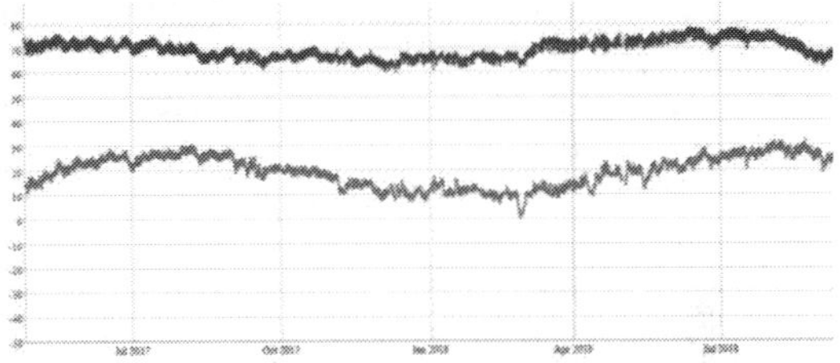

Fig. 21 Temperature (°C, green curve) and relative humidity (%) in the beam (purple curve).

6 Conclusions

Eleven years after a massive restoration of the Villa E1027 built in the late 1920's on the French Mediterranean seashore, due to corrosion of the rebars a new repair operation was planned in 2016. Prior to the elaboration of the restoration strategy, a series of analyses were set up in order to evaluate the levels of carbonation and chloride pollution and to try to quantify the corrosion activity. The diagnosis combined chemical analysis, environmental parameters follow-up and electrochemical measurements. The latest included instant non-destructive testing and monitoring for 18 months.

The results of this diagnosis evidenced active corrosion induced by combined carbonation and chloride pollution, despite a low yearly average relative humidity in the concrete.

As a consequence, the possible durable restoration strategies were limited to one option: cathodic protection. But as this technique can be quite invasive and in addition difficult to implement in historic concrete due to an erratic and non-connected rebar network, a pilot impressed current cathodic protection was installed on a beam, a pillar and 1m² of the southern balcony underside. The purpose was to try to limit the cuts and holes to respect the conservation principle of historic monuments and to follow-up the efficiency of such a treatment during 6 months, prior to the installation of the system on the whole patio.

Acknowledgements

The authors would like to thank the French ministry of Culture for their financial support and their interest in the project (specially Mr. Christophe Curial from DRAC-PACA), but also the Sensecorr and Freyssinet companies for their technical support, Mrs. Carmen Andrade for her expertise and her help, the architect in charge of the project (Mrs. Claudia Devaux), and the association Capmoderne in charge of the management of the villa.

References

1. M. Thiéry, "Modélisation de la carbonatation atmosphérique des matériaux cimentaires", Ouvrage d'art OA52, Etudes et recherches des Laboratoires des Ponts et Chaussées, Laboratoire des Ponts et Chaussées, Paris, 337p, (2006)
2. M. Hergenroeder, R. Rackwitz, "Investigations on the statistics of carbonation depths in concrete", 2nd International RILEM/CEB symposium on quality control of concrete structures, Ghent, pp. 451-460, (1991)
3. L. Bertolini, B. Elsner, P. Pedeferri, R. Polder, Corrosion of steel in concrete, Wiley-VCH, Weinheim, Germany, 392p (2004)
4. U. Angst, B. Elsener, C.K. Larsen, O. Vennesland, Critical chloride content in reinforced concrete – A review. Cement and Concrete Research, 39, 1122-1138 (2009)
5. J. Mai-Nhu, « Corrosion des armatures du béton : couplage carbonatation/chlorures en présence de cycles hydriques », thèse de doctorat de l'université de Toulouse, Toulouse, France, 244p (2013)
6. E. Marie-Victoire, E. Cailleux, A. Texier, Carbonation and historical buildings made of concrete, , In Journal de physique IV, Proceedings of NUCPERF 2006: Corrosion and long term performance of concrete in NPP and waste facilities, 27-30 March 2006, Cadarache, France, EFC Event 284, vol. 136, p. 305-318 (2006)
7. E. Marie-Victoire, M. Bouichou, T. Congar, R. Blanchard, Concrete cultural heritage in France: inventory and state of conservation, In Concrete repair, rehabilitation and retrofitting IV: Proceedings of the 4th international conference on concrete repair, Rehabilitation and Retrofitting (ICCRRR-4), Leipzig, Germany (2015)
8. A. Boule, Rapport LERM n°00.5048.001.01.A, Arles, France, 34p (2007)
9. M. Renouard-Opigez, Rapport CEBTP N°CNI7.G.229, Nice, France, 104p (2016)
10. R. Polder with contributions from C. Andrade, B. Elsener, O. Vennesland, J. Gulikers, R. Weidert and M. Raupach, RILEM TC 154-EMC, Recommendations: Test methods for on-site measurement of resistivity of concrete, Materials and Structures, **33**, pp 303-611 (2000)
11. B. Elsener with contributions from C. Andrade, J. Gulikers, R. Polder and M. Raupach , RILEM TC 154-EMC, Recommendations: Half-cell potential measurements – Potential mapping on reinforced concrete structures, Materials and Structures, **36**, pp 461-471 (2003)
12. C. Andrade and C. Alonso with contributions from J. Gulikers, R. Polder, R. Cigna, O. Vennesland, M. Salta, A. Raharinaivo and B. Elsener, RILEM TC 154-EMC, Recommendations: Test methods for on-site corrosion rate measurement of steel reinforcement in concrete by means of the polarization resistance method, Materials and Structures, **37**, pp 623-643 (2004)
13. C. Andrade, C. Alonso, J. Sarría, Corrosion rate evolution in concrete structures exposed to the atmosphere, Cement & Concrete Composites, **24**, p. 55–64, (2002)
14. K. Tuutti, Corrosion of steel in concrete, Swedish cement and concrete institute, Stockholm, Sweden, 473p (1982)
15. J.N. Enevoldsen, C.M. Hansson, The influence of internal relative humidity on the rate of corrosion of steel embedded in concrete and mortar, Cement and Concrete Research, **24**, pp. 1373-1382, (1994)

Investigation of the prestressed beams and glass-window panels at a shopping centre

Andreea-Terezia Mircea[1,*]

[1]Technical University of Cluj-Napoca, Department of Civil Engineering and Management, 400114 Cluj-Napoca, Romania

Abstract. In today's rapidly changing demands of commercial buildings, designers, manufacturers, contractors, and installers face the necessity to advance their methods, to improve techniques, and generally to stay competitive under all constraints of time, budget, codes and client desires. Building with precast concrete elements provides quick and cost-efficient production, along with flexible, independent interior design. The show-window of the investigated commercial building is divided into several glass panels. Having only upper and lower grip, the glass window panels with very tall heights, are subject to deformation. In order to resist, panels are provided with U-shaped profiles or/ and perimeter rails. The critical points are the glass panels' support, so the ceiling and floor fastenings have to be very solid, as well as qualitatively superior designed and executed. This paper aims to highlight some relevant aspects regarding the investigation of the metal fixing system designed to connect show-window panels to prestressed concrete ceiling beams at a commercial building.

1 Introduction

In recent years, the construction sector, as one of the most dynamic and flexible economy sectors, faces rapidly changing demands. Especially for commercial buildings like shopping centres (Figure 1), the designers, manufacturers, contractors, and plumbers go along with the necessity to advance and integrate their methods, to improve techniques, and generally to stay competitive under all constraints of time, budget, codes and client desires.

Fig. 1. The investigated commercial building.

In this regard the main advantage of building with precast concrete elements is a considerable shortening of construction time without climatic influences. This building method provides quick and cost-efficient production, along with flexible, independent interior design, also with significant advantages concerning fire prevention, sound protection and thermal insulation.

The show-window of the investigated commercial building is divided into several glass panels. Having only upper and lower grip, the glass window panels with very tall heights, are subject to deformation. In order to resist, fixed panels are positioned with U-shaped profiles or/and perimeter rails. The critical points are the glass panels' support, so the ceiling and floor fastenings have to be very solid, as well as qualitatively superior designed and executed.

2 Detailing Analysis

Fastening systems are used to create non-permanent connections that can be removed or dismantled without damaging the joining components.

The investigated fastening system placed between the show-window panels and the ceiling beams consists of U-shaped metallic profiles mounted to the prestressed concrete beams through on-site drilled channels onto the bottom flange. These were made by penetrating the concrete core between the reinforcement bars, without damaging them. The perforated channels have a slightly larger diameter than the screws that were inserted by hammering. The weakening of a section through channels and perforations drilled between reinforcement (active or passive bars) results in generating loose zones where stress concentrations get at higher levels (in this case above the level of service) and have, as a result, cracks directed to these areas.

However, the problem that can arise in the analyzed situation is the redistribution [1] of the bending moments along the ceiling beams (i.e. secondary beams), due to the change of the static scheme over time, due to the

* Corresponding author: andreea.mircea@ccm.utcluj.ro

change of the static scheme over time consequent to the beams' gravitational time-dependent movement.

The fixing method of the fastening system is presented in Figure 2.

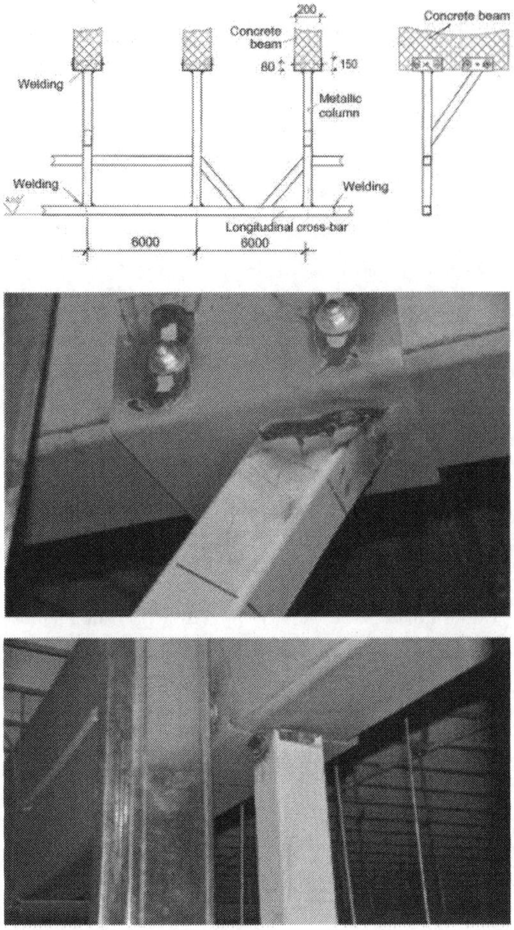

Fig. 2. Fixing method of the fastening system on the beams.

The rail for elastic fastening of the glass panels is shown in Figure 3.

Fig. 3. The rail for elastic fastening of the glass panels

Analyzing the overlapping layout of the glass window panels with the structure, it can be considered that the ceiling beam, as a simply supported beam, has to resist, beside its self-weight, also the following loads, presented in Table 1:

Table 1. Loads at the roof level.

Load	Type of load	Characteristic [kN/m^2]	Long term [kN/m^2]
1. Self weight plated sheet	Permanent	0.20	0.20
2. Mineral wool	Permanent	0.25	0.25
3. PVC membrane	Permanent	0.05	0.05
4. Equipment	Permanent	0.25	0.25
5. False ceiling	Permanent	0.25	0.25
6. Snow	Variable	1.60	0.64

3 Results and Discussion

Structural analysis of the secondary prestressed concrete beams reveals the time dependency of the deflections as shown in the diagrams presented in Figures 4-6.

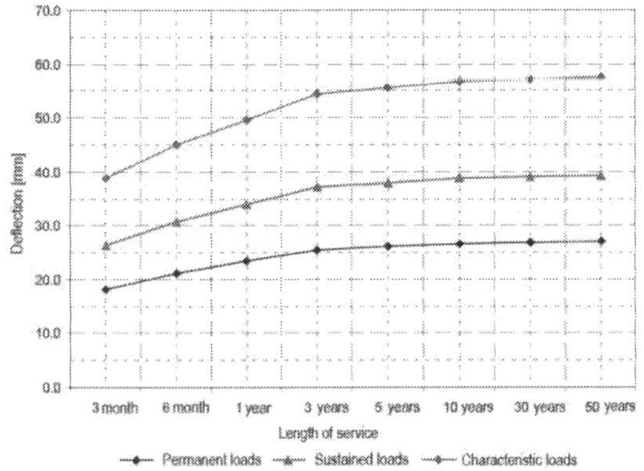

Fig. 4. Deflection variation of the simply supported secondary beam with service time.

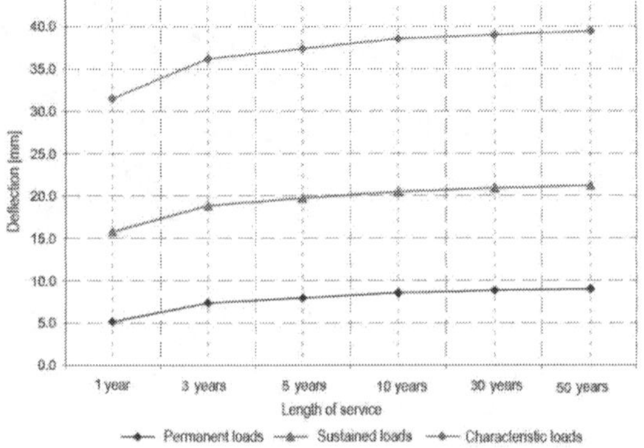

Fig. 5. Deflection increase under service conditions within 3 month, with load level corresponding to permanent loads.

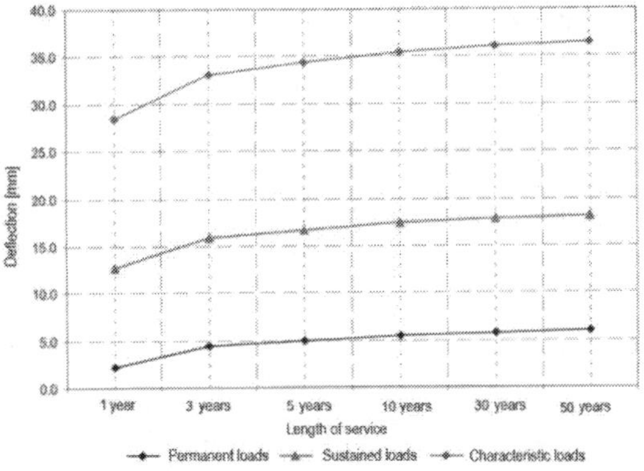

Fig. 6. Deflection increase under service conditions within 6 month, with load level corresponding to permanent loads.

The deformation of a structure under sustained load, can cause it to change shape. Creep does not necessarily cause concrete to fail or break apart; the deformation increases at a decreasing rate during the period of loading [2]. Thus, it is observed that deflection increased to 15.7 mm at 3 months and to 12.7 mm at 6 months. Approximately, 25-30 % of the total creep takes place in the first month under sustained loading, and about 50-75 % of the total creep occurs during the first half year of sustained loading in concrete sections of moderate size [3]. Consequently, it can be presumed that at about 1 year, the static scheme will be changed, middle intermediary supports will occur.

A vertical movement of the glass panel's rail support up to a maximum of 10 mm was estimated. In order to determine its safety and suitability, as well as to find out if the glass panel has enough stiffness to ensure effort redistribution, a second order Finite Element Analysis was performed using the RISA 3D Program.

Glass has a standard Young's modulus of 70 GPA, and a compressive strength of 100 MPa, in accordance with the technical files of the manufacturer. For a thickness of 10 mm and 20 mm, Figure 7 reveals that a show-window panel has a high stiffness, and can be considered as a rigid support.

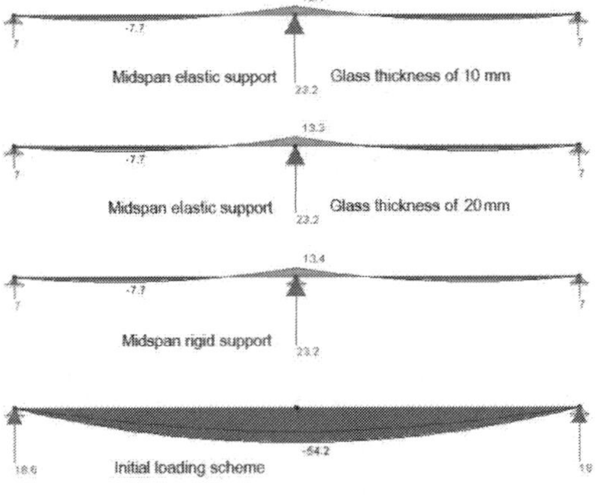

Fig. 7. Redistribution of bending moments under sustained loads.

Stress redistribution takes place because statically indeterminacy generates secondary bending moments due to prestressing. This modifies the time behaviour of the secondary beam, as the time dependent deformations increase, the cracking pattern will change and the safety level will decrease.

Related to the fastening system placed between the show-window panels and the ceiling beams, another aspect is that in most of the joints, and even in the areas of contact with the prestressed concrete elements (see Figure 2), the anticorrosive film was destroyed by welding. These areas are vulnerable to accelerated corrosion, and the contact of corroded elements with the surface of the concrete will lead to electro-chemical corrosion of steel rebar reinforcement, thus reducing the durability of the concrete prestressed elements.

As presented in Figures 8, 9 and 10, the redistribution of the bending moments will take place in every scenario. The stress concentrations under the contact zone may reach 17 MPa, much below the compressive strength value of 100 MPa, without local buckling. Also the generally buckling of the secondary beam is excluded, even for a deflection of 10 mm (Span/400).

Any future agglomeration of snow that leads to overtaking the level of long-lasting loads is likely to accelerate the phenomenon, and might lead to an earlier failure of the secondary beams.

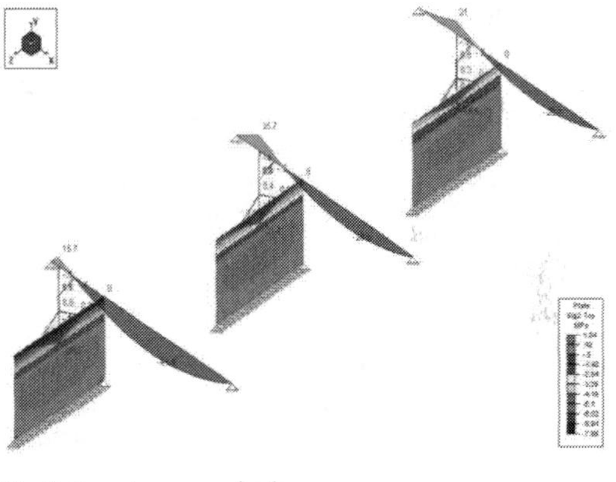

Fig. 8. Long term snow loads

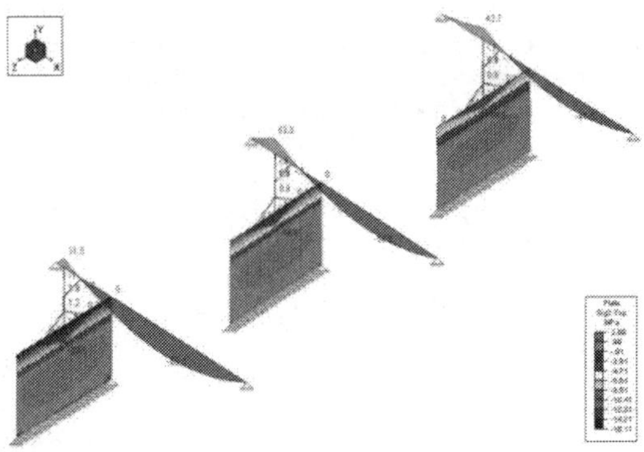

Fig. 9. Total sustained loads

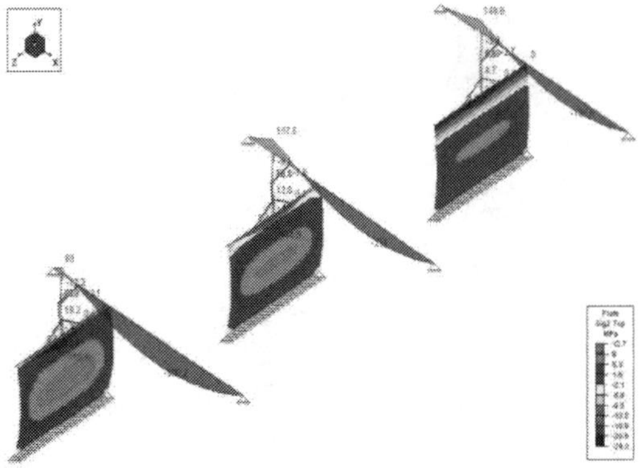

Fig. 10. Efforts at ultimate limit states.

4 Conclusions and Recommendations

The full life-cycle approach of any construction has to be considered with all its interdependencies, and from the most early design phases [4]. Identifying opportunities, precise formulation, setting goals, priorities, and systematic seek for pragmatic results [5] have to be part of the building companies' management objectives for any construction project. Innovative methods have to be applied, so as structural performance to be improved, and the architectural characteristics to be preserved [6].

The conclusions of this study can be drawn for all prestressed concrete beams from the investigated commercial building having this type of fastening system, placed between the show-window panels and the ceiling beams (secondary beams) bottom flange. So, the investigations carried out lead to the following conclusions:

- Given the fact that drilling of channels through the prestressed concrete beam didn't damage the reinforcement bars, the safety level of the element remains virtually unaffected;

- The drilling of channels through dense reinforced areas of prestressed concrete beams is a risk that may not be covered by documents made on site (e.g. possible loss of steel area, vulnerability to corrosion);

- The fixing method of the show-windows is defective because it doesn't ensure the time dependent control of the long time deformation and secondary (parasite) stresses can occur in the prestress beams consequently to the interaction with the show-window panels (i.e. changing of the statically scheme).

To avoid these situations it is recommended:

- To restore the anticorrosive film protection of the fastening system placed between the glass window panels and the ceiling beams;

- To monitor the fastening system behaviour;

- If monitoring results state the compensation of the displacement tolerance, the structural designer shall be immediately consulted, in order to implement the necessary measures (e.g. fastening the show-windows by areas of minimum bending stresses – in the middle of the beams' height with adequate time dependent tolerances).

References

1. T.Y. Lin, N.H. Burns, Design of Prestressed Concrete Structures, ISBN 10: 0471018988 / ISBN 13: 9780471018988, Published by Wiley (1981)
2. A.K. Gambali, N.K. Shanagam, Creep of Concrete, IJEDR | **2**, Issue 4 | ISSN: 2321-9939 (2014)
3. EN 1992-1-1:2004 Eurocod 2: Design of concrete structures - Part 1-1: General rules and rules for buildings (2004)
4. A.T. Mircea, Selecting Environmentally Compatible Materials for Priority Integration into the Design Process, *17th International Multidisciplinary Scientific GeoConference SGEM 2017*, Vol. 17-62, Section Green Buildings, Technologies and Materials, Albena, Bulgaria, pp. 409-416. (2017) *doi: 10.5593/sgem2017/62/S26.052*
5. S.A. Ciplea, Costs Reduction in Civil Engineering Through Optimization Methods, *The 2nd Review of Management and Economic Engineering Management Conference (RMEE)*, Management of Crisis or Crisis of Management, Cluj-Napoca, Romania, pp. 71-76. (2011)
6. D. Sucala, I.N. Sucala, S.A. Ciplea, Technologically Consolidation Fundations Using Insulation Method Based on the Construction, *17th International Multidisciplinary Scientific GeoConference SGEM 2017*, Vol.17-14, Applied and Environmental Geophysics, Albena, Bulgaria, pp. 437-444. (2017) *doi: 10.5593/sgem2017/14/S05.055*

Flexure Behaviour of Reinforced High Strength Concrete Elements Affected by Corrosion

Camelia Negrutiu[1,*], *Ioan* Sosa[1], and *Bogdan* Heghes[1]

[1]Technical University of Cluj-Napoca, Faculty of Civil Engineering, Cluj-Napoca, 28 Memorandumului St, 400114, Romania

Abstract. Corrosion of the reinforcement is a constant vulnerability for reinforced concrete structures exposed to aggressive environments. High strength concrete is known to prevent corrosion of the reinforcement, in a non-cracked state, when exposed to aggressive environments. The purpose of this study is to assess the opportunity of using high strength concrete in cracked elements exposed to corrosion and compare them with non-exposed elements. A series of simply supported reinforced high strength concrete beams with concrete cover of 25 and 50 mm were pre-cracked, up to a service life crack of 0.1 mm, further exposed to accelerated corrosion through a process of electrolysis and finally tested to failure. A series of non-exposed witness specimens were also tested to failure. All elements were designed with the same bending capacity. The flexure behaviour was assessed by plotting experimental and theoretical ultimate limit state position of the neutral axis at midspan and the results show no significant differences in the overall behaviour, despite the affected reinforcement, between the corroded and non-corroded elements. Moreover, the design bending moments were approximately 40% lower than the experimental ones, even for corroded beams, which can be a significant strength reserve of the beams, useful in aggressive environments.

1 Introduction and scope

The corrosion of the reinforcement poses a serious concern in the service life and durability in reinforced concrete structures, due to the high costs of maintenance and repair. Degradations such as the reduction of the structural loading capacity, accelerated fatigue and increased deformation in service life are common consequences in structures exposed to aggressive environments. Although worldwide researchers have concentrated their efforts in developing effective procedures to evaluate the effects of corrosion, very few structural engineers actually implement durability safety measures. These are the evaluation of the concrete cover as a function of the exposure class and basic concrete composition recommendations expressed in current standards [1].

Reinforced concrete structures present structural cracks in their service life, which are an open path for environmental aggressive agents to reach the reinforcement and initiate corrosion. The initiation and the propagation of corrosion are influenced by the crack width: less than 0.1-0.2 mm as maximum crack width under static loads is reported not to have a significant impact on the corrosion process; the wider the cracks, the faster the corrosion and the time to failure [2, 3]. However, micro-cracks developed under alternating loads are increasing the corrosion rate [4]. The thickness of the rust layer on the reinforcement is also correlated to the crack width [5]. Furthermore, the intensity of the current influences the reinforcement section loss. A moderate corrosion rises up to 0.5 $\mu A/cm^2$ which can produce a loss of section of 5.7 μm/year and an oxide layering up to 17.3 μm/year, whereas stronger corrosion is produced by an intensity current of 1 $\mu A/cm^2$, which corresponds to a maximum of 11.5 μm/year and an oxide layering of maximum 34 μm/year [6]. Bond deterioration between steel and concrete was found to be a major contributor to the reduction of bending capacity [7]. Ultimate flexural strength decreases proportional to the decrease in the rebar cross-sectional area [8] which is exacerbated by the loading levels [9]. The deformability of the elements is not significantly influenced by the short-term corrosion, but the presence and the distribution of the cracks can facilitate the initiation and progression of the reinforcement corrosion [10]. Corrosion in longitudinal tension reinforcement has a significant negative influence on seismic behaviour [11] and it changes the failure mode to brittle failure and loss of ductility [11, 12]. Corrosion of the stirrups also leads to loss of strength and ductility [13, 14], but stirrup presence can be beneficial on the overall performance [15]. Other reports mention the beneficial influence of the increased concrete cover and low water/cement ratio of the surrounding concrete [15].

High strength concrete is an emerging, innovative, high performance construction material with improved durability resistance. Thick and impermeable concrete cover can stop the initiation of the corrosion and the subsequent deterioration [16], which could recommend high strength concrete as the material choice for sustainable concrete structures. Moreover, the self-

* Corresponding author: camelia.negrutiu@dst.utcluj.ro

healing capacity of the high strength concrete due to the subsequent hydration of the binder particles can delay the propagation of the corrosion or even partially regenerate concrete integrity [17, 18, 19, 20].

The current paper studies the performance of structural, reinforced high strength concrete elements exposed to artificial accelerated corrosion compared to non-exposed specimens in terms of flexural capacity and neutral axis depth. The aim is to demonstrate the beneficial uses of high strength concrete in aggressive environments, even when the corrosion process is initiated in the embedded reinforcement.

2 Experimental program

2.1 Concrete and reinforcement

The testing specimens and the reinforced concrete elements were made with high strength concrete of concrete grade C70 according to [21], with a water/binder ratio of 0.245. The type of the longitudinal tensile reinforcement was European steel S500 C, whereas the compressed longitudinal and the transversal reinforcement was mild OB37 of local Romanian production. The concrete composition is detailed in Table 1.

Table 1. Concrete composition C70.

Components	Quantity (Kg/m^3); *(l/m^3)
Portland Cement CEM I 52.5R	520
Silica fume	52
River sand (0-4mm)	530
Crushed gravel (4-8mm)	530
Crushed gravel (8-16mm)	706
* Superplasticizer	11.44
* Water	140

The concrete's compressive strength on cubes $f_{cm,cube}$, the flexural and splitting tensile strength $f_{ct,fl}$ and $f_{ct,sp}$, the secant modulus of elasticity E_{cm} were determined at the age of 28 days and are presented in Table 2. The yield strength of the reinforcement f_{yk} is guaranteed by the manufacturer.

Table 2. Test characteristics of concrete and reinforcement.

Concrete C70				Reinforcement	
				S500 C	OB37
Mechanical properties [MPa]				Yield strength [MPa]	
$f_{cm,cube}$	$f_{ct,fl}$	$f_{ct,sp}$	E_{cm}	f_{yk}	
93.9	9.60	6.95	46905	500	255

2.2 Reinforced concrete elements

The experimental study consisted in testing two types of beams:

- Type a - Non exposed elements, tested to failure
- Type b – Pre-cracked to a service limit state, exposed to accelerated corrosion, and then tested to failure

The characteristics of the reinforced concrete elements, such as element ID, cross section b×h, concrete cover "c" and the type of element are detailed in Table 3.

Table 3. Characteristics of the reinforced concrete elements.

ID	b×h [mm]	c [mm]	Type
GDA 1-1	125×125	25	a
GDA 1-2	125×125	25	
GDA 2-2	125×150	50	
GDS 1-1	125×125	25	b
GDS 1-2	125×125	25	
GDS 2-2	125×150	50	

The beams were different in concrete cover and cross-section, but equal at design bending capacity (same reinforcement and effective depth). In order to accelerate the artificial corrosion effects, type b elements were loaded and cracked to a maximum crack width of 0.1 mm before the corrosion test. The pre-cracking of the elements is designed to facilitate the corrosion process and extensive crack width analysis is performed in previous scientific articles. All elements were tested to failure.

2.3 Loading test

The specimens used in this study were one-point loading, simply supported reinforced concrete beams, as seen in Figure 1. The strains were measured using analogue strain gauges at mid span.

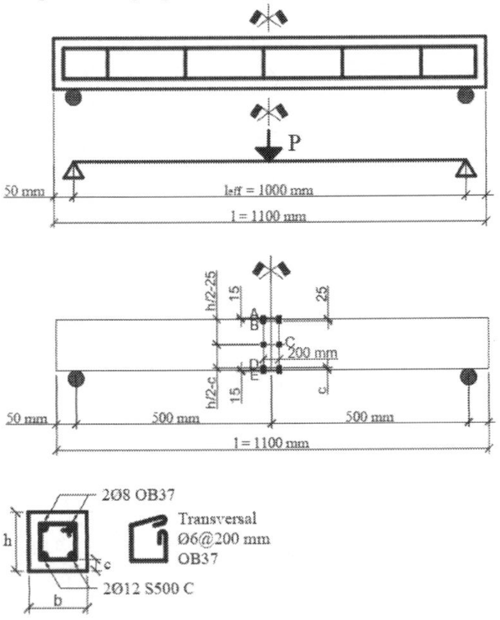

Fig. 1. Reinforcement detailing and set-up test.

2.4 Accelerated corrosion test

Simulation of reinforcement corrosion due to environmental actions is a long-term procedure. Therefore, a simplified, more aggressive action was set in place through a process of electrolysis. The set-up for the accelerated corrosion is similar to other research [22]. A schematic view and a picture of the corrosion equipment are presented in Figures 2 and 3.a. The circuit of the electric current generated by an external source consisted of an anode (the embedded reinforcement), the cathode (a stainless-steel plate covering 3 faces of the element) and the electrolyte, a 5% NaCl solution. The intensity of the current was set to 3 mA/cm^2 and the elements were exposed for 7 days.

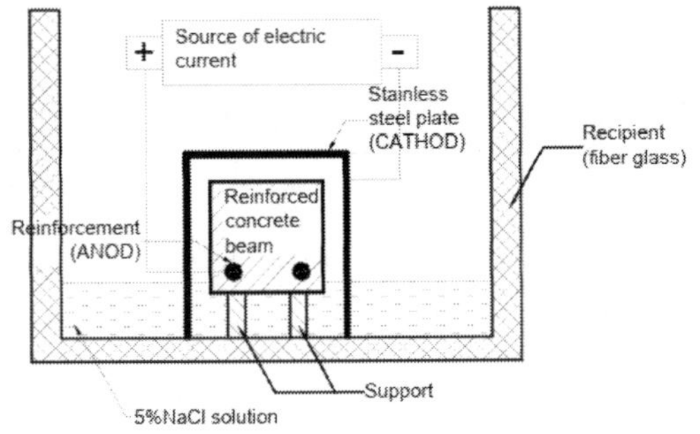

Fig. 2. Schematic view of the corrosion process

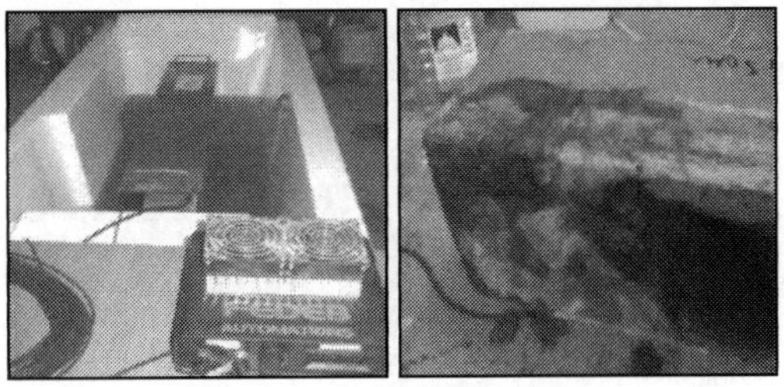

Fig. 3. (a) Corrosion equipment; (b) Visual aspects of a corroded specimen.

Type b elements displayed service limit crack widths of 0.1 mm when the process of electrolysis started. The cracking state of the elements favours the ingress of the aggressive media in real structures and for the current study, it constituted the path of the electrolyte to the reinforcement. After only 1 day of exposure, the intensity of the electric current dropped under 0.5 mA/cm^2, an indication that the salt bridge of the electrolysis could no longer close the circuit and reach the embedded steel. Moreover, an inspection of the initial cracks immediately after the exposure indicated a light-coloured filling material, which was very likely a mixed cement paste. Although not tested at microscopic level, a process of self-healing due to the un-hydrated cementitious particles in low water/binder compositions might be the source of the filling material. Nevertheless, corrosion took place, as demonstrated by the specific red spots on the surface of the beams (Figure 3b), but only the transversal reinforcement located in the cracking zone was mainly affected.

3 Results and discussions

3.1 Ultimate versus design bending moments

Table 4 presents the experimental values recorded during the loading tests of the elements with concrete cover c=25 mm and c=50 mm. The bending design capacity M_{Rd} was also calculated, using common design equilibrium equations, with high strength concrete coefficients η=0.895 and λ=0.748, as in relations (3.20) and (3.22), EC2 [21]. Safety coefficients for concrete (γ_c=1.5) and steel (γ_s=1.15) are also taken into consideration. The characteristic compressive strength was f_{ck}=71 MPa.

The corrosion of the reinforcement has little effect on the bending capacity of reinforced high strength concrete beams. The average ultimate bending moment M_u for beams with concrete cover of 25 mm is 2.47% higher for beams type a than for beams type b. However, less than 5% in experimental testing could usually be allowed, due to execution procedures or minor differences in the material properties and not solely to the influence of the corrosion process. This hypothesis is also confirmed on 50 mm concrete cover beams, which present an inversed percentage: type b ultimate moment M_u is higher than type a with 2.13%.

Table 4. Bending moment - design and ultimate.

Property	c= 25 mm				c= 50 mm	
	Type (a)		Type (b)		Type (a)	Type (b)
	GDA 1-1	GDA 1-2	GDS 1-1	GDS 1-2	GDA 2-2	GDS 2-2
M_{Rd} (kNm)	8.32	8.32	8.32	8.32	8.32	8.32
M_u (kNm)	16.50	14.55	15.37	14.92	13.80	14.10

Figure 4 presents the difference between the design resistant bending moment M_{Rd} and ultimate experimental bending moment of both 25 mm and 50 mm concrete cover beams. It can be seen that all the beams, even the corroded ones, display a strength reserve of 39 to 45% up to failure.

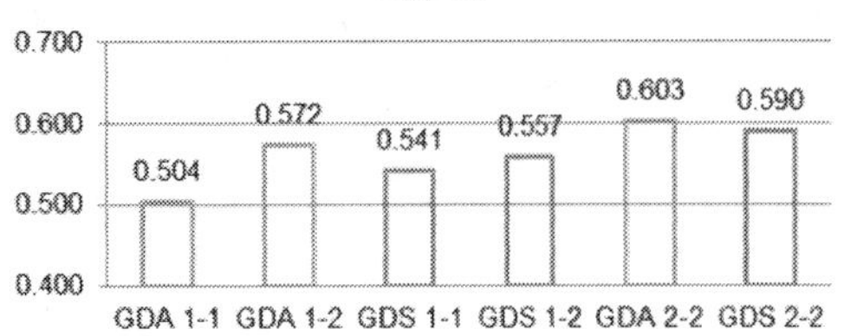

Fig. 4. Ultimate versus design bending moments of non-exposed and exposed beams.

The inverse ratio M_u/M_{Rd} gives a significant safety factor for the beams: for 25 mm concrete cover, an average of 1.86 for non-exposed elements, type a, between the ultimate experimental and the design bending moments and an average of 1.82 for exposed elements, type b. For 50 mm concrete cover, the differences are lower than for 25 mm concrete cover, due to the brittleness of the concrete cover when cracked: 1.66 for type a and 1.69 for type b.

3.2 Neutral axis depth analysis

The neutral axis position was constructed based on the strain gauges readings at all loading stages for each beam. The behaviour of the beams in terms of x/d ratio (neutral axis depth / effective depth) of the cross section, was assessed at the load level that initiated the yielding of the reinforcement.

Therefore, the longitudinal tensile reinforcement was extracted from type b beams, inspected for signs of corrosion and weight loss and tested for tensile force. No weight loss was recorded, although occasional red, corrosion like spots appeared, mainly at the point of contact with the stirrups. As seen in Figure 5, the yielding started at a stress of about σ=580 MPa and based on the conventional modulus of elasticity of the steel E_s = 200000 MPa, the yielding strain ε_y was determined 2.9‰.

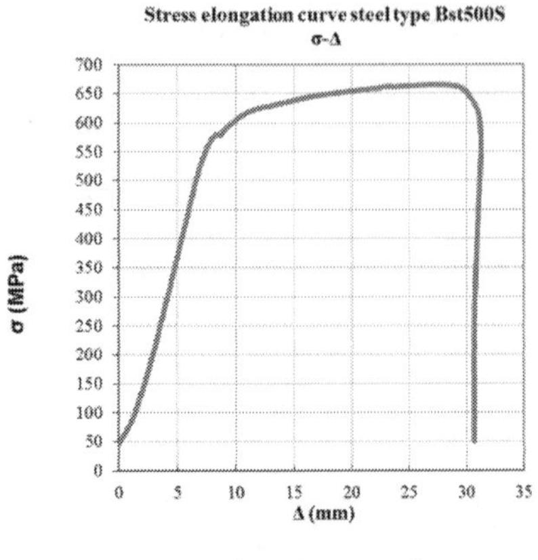

Fig. 5. Steel stress – elongation curve for S 500 C

Furthermore, ductility requirements in the ultimate limit state design are that the reinforcement should reach the yielding strain ε_{yd} in the tension zone and the ultimate compression strain ε_{cu2} in the compression zone at the same time, as seen in Figure 6. The ductility equations for high strength concrete and S500 C tensile reinforcement considering Figure 6, Eurocode 2 [23] and fib Model Code 2010 [21] values are:

$$\left(\frac{x}{d}\right)_{design} = \frac{\varepsilon_{cu2}}{\varepsilon_{cu2}+\varepsilon_{yd}}=0.554 \quad (1)$$

$$\varepsilon_{yd} = \frac{f_{yk}/\gamma_s}{E_s}=0.00217 \quad (2)$$

$$\varepsilon_{cu2} = 0.0027 \quad (3)$$

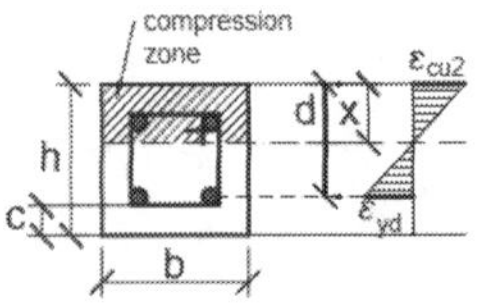

Fig. 6. Ultimate limit state strain diagram.

Figures 7 and 8 present the strain diagrams for a yielding point of the tensile reinforcement when it reaches the yielding point of 2.9‰, or in very close proximity. Displacements of the strain gauges occurred for beam type (a) GDA1-1 and thus invalid measurements were recorded. GDA1-1 was therefore dismissed from further comparisons.

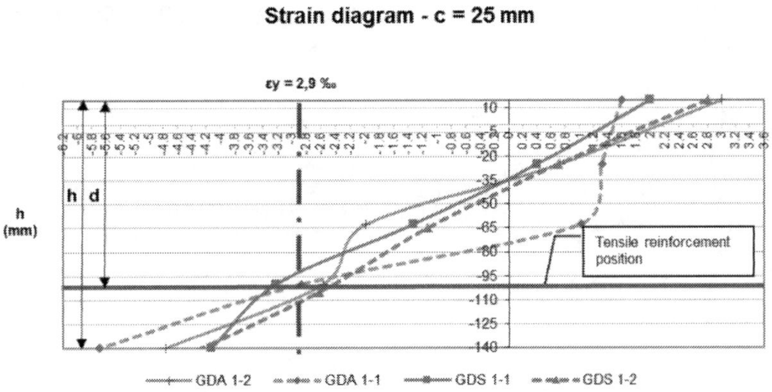

Fig. 7. Neutral axis depth in strain diagrams for beams with 25 mm concrete cover.

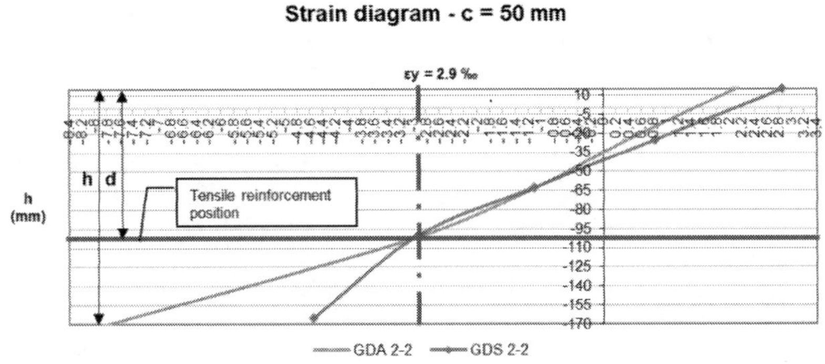

Fig. 8. Neutral axis depth for beams with 50 mm concrete cover.

Figure 9 displays the experimental value of the (x/d) ratio for each beam along with the design value. Figure 10 shows a general graphic representation of the experimental and design strain deformations. It can be seen that the beams are in the range of 0.67-0.79 of the design limitations, confirming the safety percentages are met for the resistant bending moment. Furthermore, the corroded beams presented slightly higher (x/d) ratios, with approximately 13% in the case of 25 mm concrete cover and 15% in the case of the 50 mm. Although the longitudinal tensile reinforcement was not affected by corrosion, the transversal reinforcement presented corrosion pitting, acting like a sacrificial anode. The resulted products filled the existing cracks, putting an additional stress on the crack walls, thus increasing their width and the deformation of the reinforced concrete elements, as expressed by higher (x/d) ratios. Even so, the corroded elements still possess significant residual flexural strength and deformability. These results can be very important in designing reinforced high strength concrete structures in harsh environments.

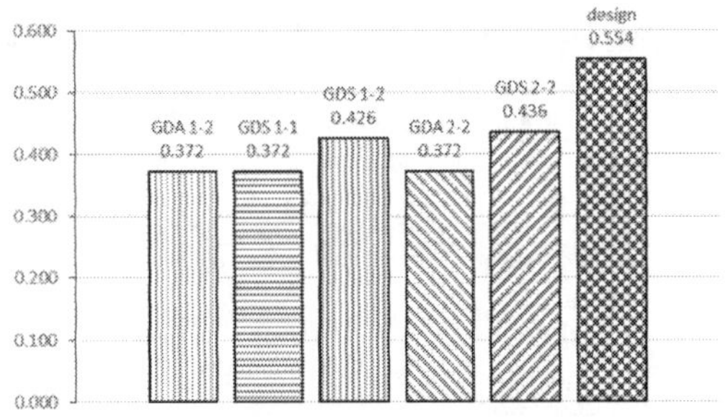

Fig. 9. Type a and b beams (GDA1-1 dismissed): $(x/d)_{exp}$ vs. $(x/d)_{design}$

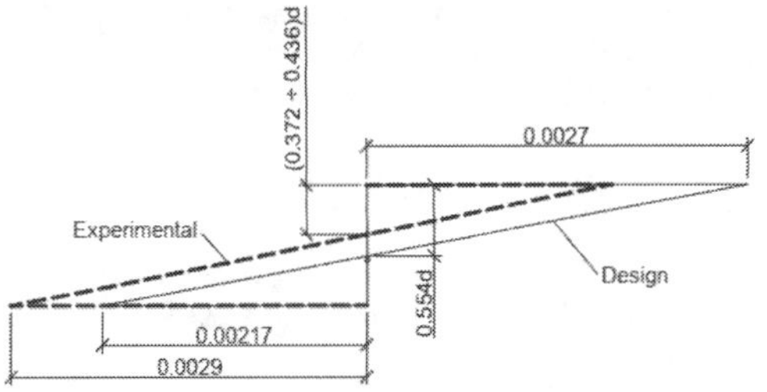

Fig. 10. Experimental and design strain diagram

4 Conclusions

Corrosion of the embedded reinforcement leads to serious deterioration of reinforced concrete elements, such as flexural and shear degradation, with a significant financial impact due to maintenance and repair measurements.

This paper studied the bending moment and neutral axis depth of reinforced high strength concrete elements, with 25 mm and 50 mm concrete cover, pre-cracked to a service limit state and then exposed to accelerated corrosion, in comparison to non-exposed elements. These are the main findings of the study:

- The intensity of the applied electrical current significantly decreased after one day of exposure due to the obstruction of the electrolyte's pathway through the structural cracks with a concrete self-generated filling material. However, a prolonged exposure could clear up the cracks and re-establish the initial current intensity, which will be the focus of further studies.
- The longitudinal reinforcement was not affected by corrosion;
- The transversal reinforcement intersected by bending cracks was affected by pitting corrosion. Occasional red spots on the surface of the elements also appeared in other areas;
- Ultimate bending capacity was above the design load, with an average of 39% for corroded beams and 45% for non-exposed beams;
- The ultimate bending moments of the exposed and non-exposed beams were within a range of 3% for both 25 mm and 50 mm concrete covers;
- The experimental (x/d) ratios of all the beams were in the range of 0.67÷0.79 of the design ratio;
- The corroded beams presented slightly higher (x/d) ratios, with approximately 13% in the case of 25 mm concrete cover and 15% in the case of the 50 mm than non-exposed beams;

Based on the research results, the corroded reinforced high strength concrete elements possessed significant residual flexural strength and deformability. Using high strength concrete in aggressive environments can be beneficial for extended structural service life. However, a continuation of the study is in order for the evaluation of the self-healing capacity of the high strength concrete, a complete mapping of the stirrup corrosion and the influence of the service cracks or concrete defects on the structural behaviour in harsh environments.

References

1. European Standard "Concrete —Part 1: Specification, performance, production and conformity", EN 206-1
2. S. Yoon, K. Wang, W. J. Weiss, et al., ACI Mat. J. **97(6)** (2000)
3. W. Li, W. Liu, S. Wang, Adv. in Mat. Sc.& Eng. (2017), 11 pg, ID 3968578
4. A. Blagojevic, D.A. Koleva, J.C. Walraven, XIII Conf. On Durab. of Build. Mat. And Comp., 834-841, (2014)
5. F. Chuanqing, J. Nanguo, et al., Corrosion Science 117, **11-23** (2017)
6. K. A. Mokhtar, J. M. Loche, et al, Report no. 2-2. Concrete in marine environment, MEDACHS, Interreg IIIB Atlantic Space – **Project no. 197** (2006)
7. L. Chung, H. Najm, et al., Cement & Concrete Composites **30**, 184–193 (2008)
8. F. Jnaid, R. S. Aboutaha, Eng Struct. **119**, 198–216 (2016)
9. Y. Liu, N. Jiang, et al., Constr. and Build. Mat. **117**, 302–310 (2016)
10. A. Phetkaysone, T. Ishida, et al., Concrete Durability and Service Life Planning – ConcreteLife'09, RILEM Pub. SARL, 55 – 62 (2009)
11. Y-C. Ou, N. Dang, Eng. Struct. **126**, 210–223 (2016)
12. W. Zhu, R. François, Constr. and Build. Mat. **51**, 461–472 (2014)
13. Z-H. Lu, Y-B. Oua, et al., Constr. and Build. Mat. **127**, 293–305 (2016)
14. A. K. El-Sayed, Constr. and Build. Mat. **134**, 176–184 (2017)
15. A. S. Al – Harthy, J. Mullard, et al.,, CONSEC'07 Tours France, Vol. **1**, 383 – 390 (2007)
16. A. N. Sanchez, A. A. Sagues, et al., RILEM Int. workshop on performance-based specification and control of concrete durability, 533 – 540
17. A. Castel, T. Vidal, et al., CONSEC'07 Tours France, Vol. **1** (2007)
18. M. Sahmaran, M. Li, et al., ACI Mat. J. **104 (6)**, (2007)
19. V. Baroghel – Bouny, P. Rougeau, et al., Bulletin des Laboratoires des Points et Chaussées **217**, 75 – 84 (1998)
20. P. C. Aïtcin, E&FN SPON, ISBN: 0419192700, (1998)
21. *fib* Bulletin 55, Model Code 2010, ISSN 1562-3610, ISBN 978-2-88394-095-6
22. A. K. Azad, S. Ahmad, et al., ACI Mat. J., **104(1)**, (2007)
23. European Standard, "Eurocode 2: Design of concrete structures - Part 1-1 : General rules and rules for buildings", EN 1992-1-1

Study on estimation of creep behaviour of concrete at early age considering temperature effect

Kayo Ohashi[1*], *Jun-ichi* Arai[2], and *Toshiaki* Mizobuchi[3]

[1]Student of master course, Department of Civil and Environmental Engineering, Hosei University Graduate School, Japan
[2]Student of doctoral course, Department of Civil and Environmental Engineering, Hosei University Graduate School, Japan
[3]Professor, Department of Civil and Environmental Engineering, Hosei University, Japan

Abstract. Clarifying the creep behaviour of concrete at early age not only improves the accuracy of temperature stress analysis but also contributes to prediction accuracy and control measures in cracks caused by thermal stress. However, most past researches on creep behaviour were investigated after 28 days. Currently, it is difficult to accurately perceive the creep behaviour of concrete at an early age in the test method of creep which is generally carried out. Therefore, it is necessary to evaluate the creep behaviour of concrete at early age and to establish a convenient test method to estimate the creep behaviour. Therefore, in this study, experiments were carried out for concrete at early age within one week. As the result of the experiments, it was shown that the creep strain is proportional to the load stress of concrete at an early age and the strain of specific creep decreases as the loaded age increases. In addition, based on the experimental results, an estimation equation for creep strain at early age was proposed. Within the scope of this experimental result, it was confirmed that the estimation equation proposed in this study accurately represented the creep behaviour of concrete at early age.

1 Introduction

The purpose of this study was to clarify the creep behaviour of concrete at early age.

Thermal cracking caused by volume change during heat hydration of cement is not only an initial defect but also greatly affects durability of concrete structures. The thermal stress analysis is carried out in order to predict and control thermal cracking. However, the result of prediction of the thermal stress analysis may be different from results of stress behaviour in the existing structure. One of the cause is that the creep behaviour of concrete at early age has been clarified.

Most past research on creep behaviour has been investigated after 28 days. There are few studies on the creep behaviour of concrete at early age during heat hydration of cement. Among them, the estimation equation that can be used for the thermal stress analysis have been proposed for Morimoto's research [1], Yoshitake's research [2], Sakata's research [3], Ozawa's research [4], and Iriya's research [5]. However, the estimation equation of creep strain at early age that comprehensively evaluated the influence factor including stress/strength ratio, water to cement ratio, and curing temperature has not been proposed so far.

Therefore, in this study, experiments were carried out for concrete at early age within one week to acquire the data for the creep behaviour of concrete at early age. In addition, based on the experimental results, the estimation equation of creep strain at early age was proposed to clarify the creep behaviour of concrete.

2 Outline of Experiment

2.1 Case studied

In this study, the subject of investigation was creep of concrete made with Portland blast-furnace slag cement type B (hereinafter referred to as BB) in JIS (Japanese Industrial Standards). The curing temperature was constant at 20°C, 30°C, 40 °C, and 50°C. The stress/strength ratio was set to three levels of 0.2, 0.3, and 0.4. For the 20°C curing temperature, the water to cement ratio was set to three levels of 0.45, 0.5 and 0.6. For the other curing temperatures, the water to cement ratio was set to 0.5. In this experiment, the target slump was 12±1.5cm in consideration of workability, and the target air content was 4.5±1.0%. Table 1 shows the cases studied and Table 2 shows the mix proportion of concrete.

Table 1. Mix proportion of concrete.

Kinds of cement	Curing tem-perature (°C)	W/C	Unit quantity (kg/m^3)			
			W	C	S	G
BB	20	0.45	157	349	775	1017
		0.5	154	308	798	1043
		0.6	154	257	815	1072
	30, 40, 50	0.5	157	314	771	1053

Table 2. Study cases

* Corresponding author: kayo.ohashi.7f@stu.hosei.ac.jp

Kinds of cement	Curing temperature (°C)	W/C	Age of testing (days)	Stress/strength ratio
BB	20	45 50 60	1 3 5 7	0.2 0.3 0.4
	30	50	3 5 7	
	40			
	50			

2.2 Experimental method

In this study, experiments using the convenient creep test equipment developed in our laboratory were carried out for concrete at early age. Fig. 1 shows details of the creep test equipment used for loading. As the method of experiment, the steel rod is pulled by the jack with a hole in the center, and the nut is tightened when the predetermined load is reached.

The concrete is stressed using the force of the steel rod as it tries to return to its original length. In the experiment, the cases shown in Table 2 were carried out using one specimen for each age and stress/strength ratio. The creep strain was measured by the strain gauge attached the steel rod. Fig. 2 shows a schematic of the test object to be loaded. The details of specimen used in the experiment is shown in Fig. 2.

Also, the compressive test was carried out using oil pressure type universal testing machine to find the stress/strength ratio. The compressive test was carried out using three specimens. The specimen is a cylinder with a diameter of 100 mm and a height of 200 mm. Table 3 shows loading stress.

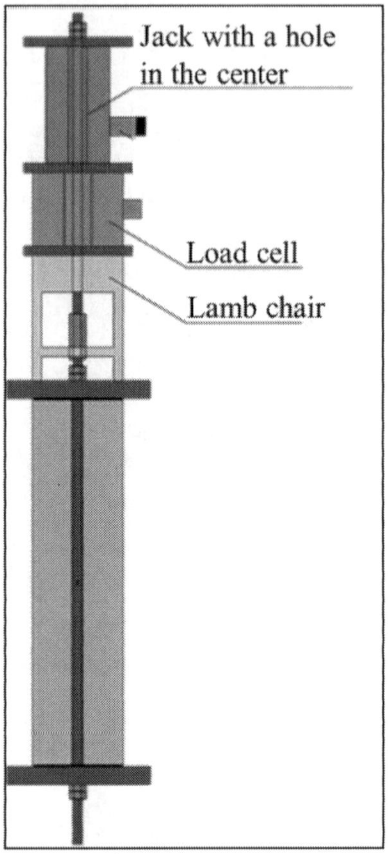

Fig. 1. Detail of the creep test equipment.

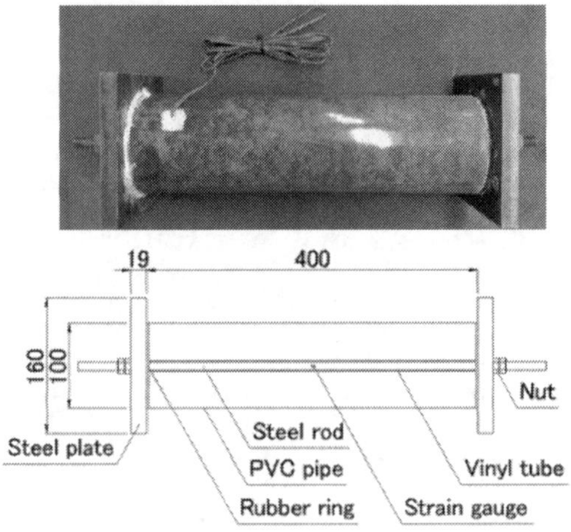

Fig. 2. Details of specimen.

3 Experimental results

3.1 Relationship between water to cement ratio and strain of specific creep

Table 4 shows all experimental results. In Table 4, the case displayed as "-" didn't have a result because the experiment failed.

Table 3. Loading stress

Curing tempera-ture (℃)	W/C	Age of testing (days)	Loading stress (kN)		
			Stress/strength ratio		
			0.2	0.3	0.4
20	0.45	1	8.2	15.2	16.2
		3	26.2	40.2	53.6
		5	35.5	52.2	70.9
		7	37.0	54.7	73.9
	0.5	1	5.7	9.1	12.5
		3	20.7	32.0	42.8
		5	27.9	41.9	55.9
		7	44.7	62.0	83.2
	0.6	1	4.4	6.5	10.0
		3	15.7	22.7	31.2
		5	23.1	35.0	46.2
		7	25.7	42.0	54.8
30		3	21.9	33.8	43.7
		5	27.5	41.2	55.0
		7	33.4	50.0	66.7
40	0.5	3	22.3	33.4	44.5
		5	30.1	45.2	60.2
		7	37.3	55.9	74.6
50		3	18.0	27.0	36.0
		5	29.4	44.1	58.8
		7	34.3	51.5	68.7

Table 4. All experimental results

Curing tempera-ture (℃)	W/C	Age of testing (days)	Specific creep strain (10^{-6}/Mpa)		
			Stress/strength ratio		
			0.2	0.3	0.4
20	0.45	1	106.7	79.4	53.1
		3	23.4	20.8	21.1
		5	16.0	15.7	14.3
		7	12.7	12.7	13.2
	0.5	1	88.6	65.2	34.8
		3	18.4	20.5	20.6
		5	14.5	14.7	15.7
		7	14.7	16.0	15.9
	0.6	1	148.9	111.4	85.1
		3	37.3	32.8	33.1
		5	21.7	25.1	22.0
		7	14.5	13.1	15.6
30		3	-	1.4	9.5
		5	13.1	10.6	8.0
		7	-	7.1	7.9
40	0.5	3	25.4	14.6	35.4
		5	29.0	28.2	22.6
		7	14.1	15.7	20.2
50		3	-	139.3	133.4
		5	97.2	-	-
		7	105.2	97.7	96.5

Fig. 3 shows one case of the relationship between water to cement ratio and strain of specific creep (curing temperature: 20°C, the age of testing: 3days, stress/strength ratio: 0.4). For water to cement ratios between 0.45 and 0.5, the strain of specific creep was nearly constant. However, for water to cement ratios between 0.5 and 0.6, the strain of specific creep increased by approximately 15 $\times10^{-6}$/MPa. This result shows, under the conditions of this experiment, when the water to cement ratio is 50% or more, the strain of specific creep increases.

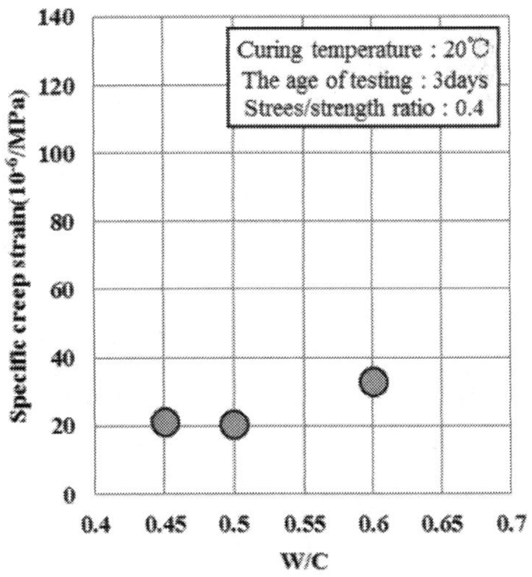

Fig. 3. Relationship between water to cement ratio and strain of specific creep (curing temperature: 20°C, the age of testing: 3days, stress/strength ratio: 0.4).

3.2 Relationship between stress/strength ratio and strain of specific creep

Fig. 4 shows relationship between stress/strength ratio and strain of specific creep. Hiramoto's research [6] shows that when the ratio of stress/strength is less than 0.5, the strain of specific creep is constant since the creep strain is proportional to the load stress. Fig. 3 shows that the strain of specific creep is constant when water to cement ratios is 50%. Therefore, the creep strain is also proportional to the load stress in this study too.

Also, Fig. 4 shows that the specific creep increases when curing temperature is 50°C.The creep strain is caused by the water movement in the concrete [7]. Therefore, the reason why the specific creep increases when curing temperature is 50°C is the water movement in the concrete becomes active at high temperature. The same goes for Fig. 5.

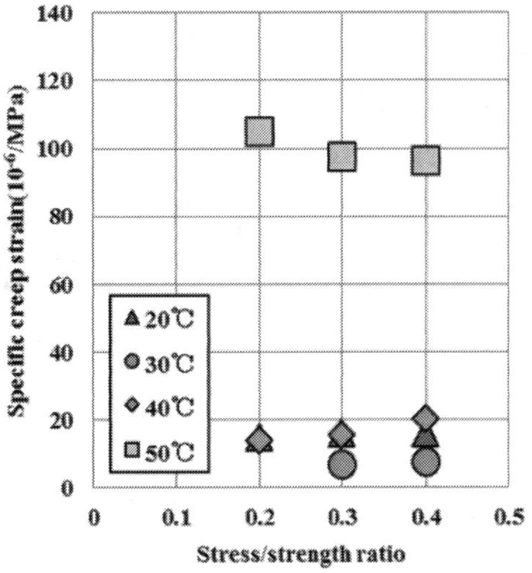

Fig. 4. Relationship between stress/strength ratio and strain of specific creep.

3.3 Relationship between loaded age and strain of specific creep

Fig. 5 shows relationship between loaded age and strain of specific creep. The strain of specific creep decreases as the loaded age increases. This is because the strength development of concrete progresses as the age of concrete increases. And the strain of specific creep decreases faster as the curing temperature is higher, since the higher the curing temperature, the faster the strength development of the concrete. The same tendency was shown in Hiramoto's research. [6]

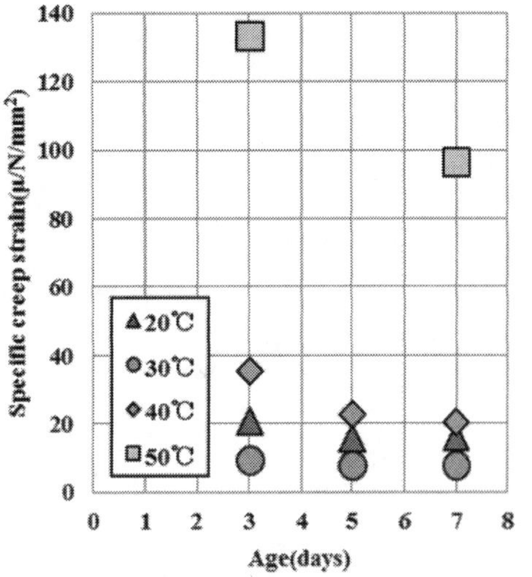

Fig. 5. Relationship between loaded age and strain of specific creep.

3.4 Relationship between curing temperature and strain of specific creep

Fig. 6 shows the relationship between curing temperature and strain of specific creep. The strain of specific creep increases as the curing temperature increases. In Hattori's research [8], the creep behaviour of concrete in early age has the tendency that the water movement in the concrete becomes active at high temperature and creep strain increases. This results of the experiments shows the same tendency as Hattori's research. [8]

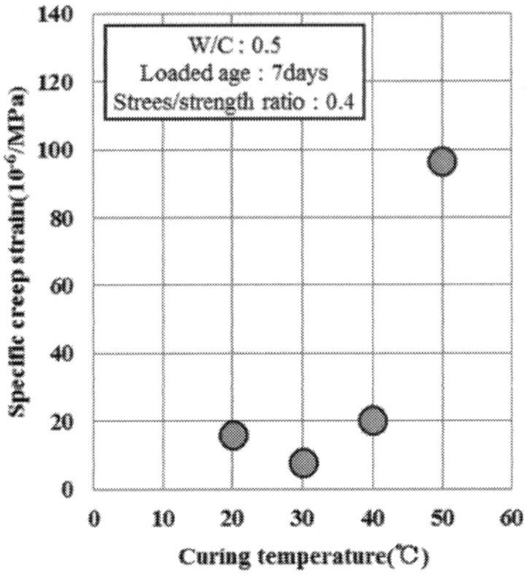

Fig. 6. Relationship between curing temperature and strain of specific creep.

3.5 The estimation equation of creep strain in early age concrete

Since the experimental results are same tendency shown in past research, the experimental results are assumed to be correct, and the estimation equation was created to match the experimental results. The estimation equation was created based on the CEP-FIP model code 90 [9].The estimation equation of creep strain at early age is shown below.

$$\varepsilon = \alpha * [\, 1 / (1 + 0.01 * exp\{-0.01 * (t - t_0)\}\,)\,] * \varepsilon^{\infty} * [\, 1 - exp\{-\beta * (t - t_0)^{0.5}\}\,] \quad (1)$$

$$\alpha = exp\,(T^{0.7}) * 9.51 * 10^{-7} + 0.654 \quad (2)$$

$$\varepsilon_{t14} = \{0.1 * exp\,(W/C * 10) + 10\} * \{4.66 * exp\,(-t_0) + 0.828\} \quad (3)$$

$$\beta = 0.01 * exp\,(0.457 * t_0) + 0.661 \quad (4)$$

here

ε : the strain of specific creeep (10^{-6}/MPa)

t_0 : loaded age (days)
t : age (days)
T : curing temperature (℃)
ε_{t14} : the strain of specific creep at management materials age 14 days (10^{-6}/MPa)
W/C : water to cement ratio
α, β : coefficient

One case of the comparison of experimental results and the estimation values calculated by equation (1) to (4) are shown in Fig. 7. In addition, the estimation values calculated by the design equation from the design standards [10] are also shown in Fig. 7 too. From Fig. 7, within the scope of this study, it was confirmed that the estimation equation proposed in this study is able to estimate the creep behaviour of concrete at early age with relatively high accuracy than the design equation from the design standards [10].

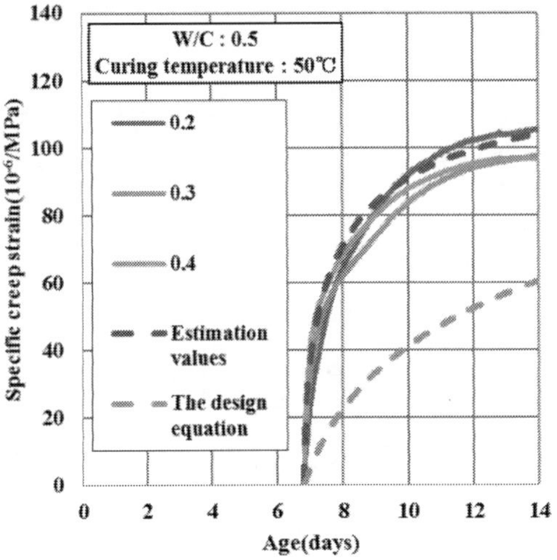

Fig. 7. Comparison of experimental results, estimation values calculated by equation (1) to (4) and the design equation.

Fig. 8 shows comparison of the experimental results and the estimation values calculated by equation (1) to (4) at an age of 10 days after loading. Within the scope of this study, it was also confirmed that the estimation equation proposed in this study accurately represented the creep behaviour of concrete at early age, even when the curing temperature and the stress/strength ratio are changed.

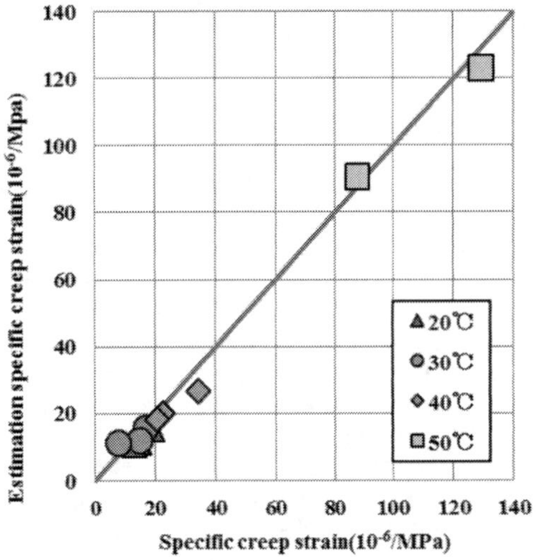

Fig. 8. Comparison of experimental results and estimation values calculated by equation (1) to (4) at age of 10 days.

4 Conclusions

In this study, experiments using the convenient creep test equipment developed in our laboratory were carried out for concrete at early age within one week and we acquired the data for the creep behaviour of concrete at early age. In addition, based on the experimental results, the estimation equation of creep strain at early age was proposed to clarify the behaviour of young concrete. The results are shown below.

1. When water to cement ratio is 60% or more, the strain of specific creep increases.
2. The creep strain is proportional to the load stress.
3. The strain of specific creep decreases as the loaded age increases. And the strain of specific creep decreases faster as the curing temperature is higher.
4. The strain of specific creep increases as the curing temperature increases.
5. The result of experiments using the convenient creep test equipment developed in our laboratory shows the same tendency as past research.
6. The estimation equation of creep strain at early age we propose makes it possible to estimate creep strain at early age.

References

1. H. Morimoto, T. Iwamoto, T. Kurihara, K. Koyanagi, Compressive and tensile creep properties of concrete at early age, *Cement Science and Concrete Technology*, **No.47**, pp356-359, (1993)
2. I. Yoshitake, H. Nakamura, S. Nagai, S. Hamada, A Study on the Tension Creep Equation During Hydration in the Early Age, *Journal of Japan Society of Civil Engineers*, **No.634/V-45**, pp43-53, (1999)

3. K. Sakata, K. Ayano, Study on non-linear creep prediction equation of concrete, *Journal of japan society of civil engineers,* **No.451/V-17**, pp179-188, (1992)
4. M. Ozawa, R. Kunimori, K. kurumada, H. morimoto, On the Creep Properties of High Strength Concrete at Early Ages, *Proceedings of the Japan Concrete Institute,* **Vol. 22**, No.2, (2000)
5. T. Goto, T. Uehara, H. Umehara, Studies on Creep Behavior of Early Age Concrete, *Proceedings of the Japan Concrete Institute,* **vol.17**, No.1, (1995)
6. M. Hiramoto, K. Iriya, S. Gupta, H. Umehara, Age and Stress Dependence of Creep in Concrete at Early Ages, *Proceedings of the Japan Concrete Institute,* **vol.19**, No.1, (1997)
7. T. Tanabe, Nonlinear analysis method and program of RC structure considering initial stress, *Tokyo: Gihodo shuppan,* (2004)
8. T. Hattori, K. Iriya, T. Uehara, H. Umehara, Study on Temperature Effects in Tensile Creep of Early Aged Concrete, *Proceedings of the Japan Concrete Institute,* **vol.20**, No.2, (1998)
9. CEB:CEB-FIP Model Code 1990, *ComiteEuro-International Du Beton,* pp27-38, (1990)
10. JSCE Concrete Committee, Standard Specifications for Concrete Structures [Design] -2012, (2012)

The influence of defects at the steel/concrete interface for pitting corrosion initiation studied through X-ray Computed Tomography and image analysis

Emanuele Rossi[1,*], *Timo* Nijland[2], *Oğuzhan* Çopuroğlu[1], *Rob* Polder[1,3] and *Branko* Šavija[1]

[1]Delft University of Technology, Faculty of Civil Engineering & Geosciences - Department of Materials & Environment,Stevinweg 1, 2628 CN Delft, The Netherlands
[2]TNO Buildings, Infrastructures and Maritime, PO Box 155, 2600 AD Delft, The Netherlands
[3] RPCP, Fluwelensingel 106, 2806 CH Gouda, The Netherland

Abstract. Although corrosion of reinforcement is a well-known issue for the construction industry, there are still open questions about some fundamentals of corrosion in reinforced concrete. These points include, among others, which are the most sensitive locations of the steel/concrete interface for pitting corrosion to initiate and to propagate. In this study, X-ray computed tomography (CT-scan) is used to characterize eight 20-years-old reinforced concrete cores naturally deteriorated due to chloride-induced corrosion. The volume loss due to corrosion of the reinforcement was quantified through image analysis of CT-scans. The volume loss of the steel was found to be higher for steel rebars embedded in Portland cement specimens rather than in blended cement specimens. Furthermore, CT-scans revealed that the deepest and most frequent corrosion pits, as well as the consequent highest volume loss of steel, were present at the portion of the reinforcement closer to the outdoor environment and in proximity to air voids at the steel/concrete interface. As a consequence, the highest decrease of structural performance of the rebars would be likely localized at those locations. Therefore, the presence of interfacial air voids should be considered as relevant factor when assessing the risk of corrosion of reinforced concrete structures.

1 Introduction

Chloride-induced corrosion of steel reinforcement is a common cause of deterioration for reinforced concrete structures [1]. When assessing the risk of corrosion of structures, the critical chloride content (C_{crit}) that causes corrosion initiation is one of the main parameters considered [2]. However, the tolerable chloride content for corrosion initiation is influenced by characteristics of the steel/concrete interface and of the concrete environment, such as by macro-pores, interfacial air voids and cracks [3-4].

Some studies reported that defects at the steel/concrete interface (SCI) are required for initiation of corrosion of the reinforcement [5-8]. When no interfacial imperfections are present, strong and dense contact between steel and concrete is reported to provide physical and chemical protection for corrosion of the reinforcement [5, 9]. Therefore, the higher the number of interfacial defects, the lower the corrosion resistance [9, 10]. However, recent studies reported that defects at the SCI did not have any influence on corrosion initiation under certain environmental conditions [4, 11]. Furthermore, Angst et al. [4] observed by visual inspections that the location of corrosion onset did not coincide with the location of interfacial air voids. Although many studies have been conducted on the influence of interfacial defects for chloride-induced corrosion [4-12], on-going investigations and discussions on the topic highlights the importance to clarify if, and how, interfacial voids may affect initiation and propagation of corrosion of steel in concrete.

In this study, characterization of eight 20-year-old reinforced concrete specimens was conducted through X-ray Computed Tomography (CT-scan) and image analysis. The main aim of this study was to observe and to quantify the depth and the amount of corrosion pits with regard to the presence of defects at the SCI and to the inside/outside exposed portion of the reinforcement. Also, the volume loss due to corrosion of reinforcement embedded in different concrete mixes after 20 years of natural exposure was quantified through image analysis.

2 Materials and methodology

2.1. Specimen preparation and X-ray Computed Tomography

Eight reinforced concrete cores were analyzed through X-ray Computed Tomography (CT-scan). The cores were drilled out of 20-years-old prisms cast in 1998 by Polder et al. [13] with dimensions equal to 300 mm x 100 mm x 100 mm (Figure 1). The prisms were cast with four different cement types (CEM I, CEM II/B-V, CEM III/B, CEM V/A), identical water/binder ratio (0.55) and

* Corresponding author: e.rossi@tudelft.nl

they contained siliceous aggregates with a maximum diameter (D_{max}) of 8 mm. The steel rebars were cast parallel to the concrete casting direction.

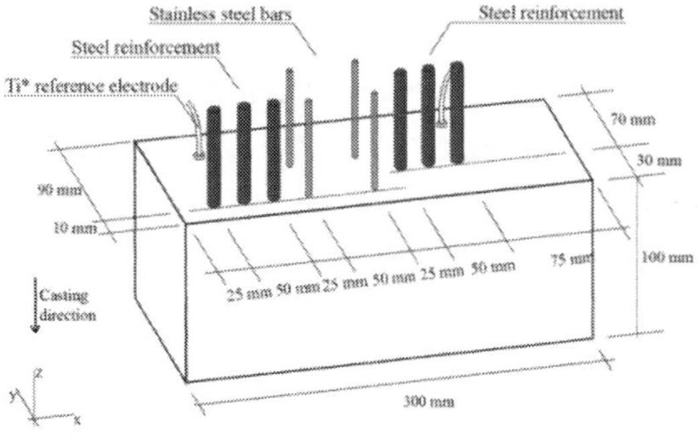

Fig. 1. Schematic representation of reinforced concrete prisms

The concrete compositions are summarized in Table 1.

Table 1. Mix proportions for each mix design (w/c of 0,55; D_{max} of 8 mm) [13]

Mix code	Cement content (kg/m^3)	Aggregate content (kg/m^3)	Admixtures (% by mass of cement)
I	287	1827	3.0
II/B-V	280	1830	2.0
III/B	286	1864	2.0
V/A	279	1839	2.0

The prisms were exposed to 3% NaCl solution wet/dry cycles for the first 6 months after casting. Each cycle consisted of applying the solution of the front side of the specimens for 24 hours and let the specimens dry for 6 days at 20°C and 50% relative humidity (RH). According to previous research, at 26 weeks after casting the chloride content at the locations of the rebars (10 mm depth) was higher than 2.5% by mass of cement for all the specimens analyzed in this research. Also, according to steel potential and corrosion rate measurements, corrosion initiation for the specimens analyzed in this study very likely occurred within the first year after casting [14]. After the exposure to wet/dry cycles with NaCl solution, the specimens were exposed unsheltered to the outdoor environment for around 18 years. From each prism, a 20-mm diameter concrete core was drilled embedding one single steel rebar which was cast with 10 mm of cover depth. Prior to drilling, the bottom side of the prism was marked with an arrow indicating the direction of the mould face, to keep track of which side was exposed to the outside environment (labeled as N-direction). Cores were then labelled depending on their cement type, as reported in Table 2, which also includes the condition at which cores were observed before being tested. Cores cast with identical cement type (i.e. with same mix code) have been drilled out from different prisms with equal composition.

Table 2. Labels and condition of reinforced concrete cores before testing

Label	Cement type	State of the specimen
I-1	CEM I 32.5 R Portland cement	Surface cracks and rust stains, heavily corroded
I-2	CEM I 32.5 R Portland cement	No signs of corrosion
I-3	CEM I 32.5 R Portland cement	Many surface rust stains, modestly corroded
II/B-V-1	CEM II/B-V 32.5 R (27% fly ash)	Many surface rust stains, modestly corroded
III/B-1	CEM III/B LH HS 42.5 (75% slag)	A few surface rust stains, lightly corroded
III/B-2	CEM III/B LH HS 42.5 (75% slag)	No signs of corrosion
V/A-1	CEM V/A 42.5 (25% fly ash, 25% slag)	Surface cracks and rust stains, heavily corroded
V/A-2	CEM V/A 42.5 (25% fly ash, 25% slag)	Surface cracks, no observed rust

After drilling, cores were kept drying for 72 hours in a room at 20°C and 45% RH, then CT-scans were performed using a Phoenix Nanotom. Images were acquired at 150 kV of transmission acceleration voltage with a spatial resolution of 20 μm. The total scanned height was around 45 mm starting at 10 mm from the top side of the core. Different components of each specimen (i.e. steel, corrosion product, air voids, cracks and concrete) were segmented based on their different grey scale values using the image analysis software Fiji. A typical CT-scan showing different phases of the material is shown in Figure 2.

2.2 Volume loss quantification and corrosion pits distribution

After segmentation, the volume of each steel rebar at the time of testing was quantified through pixel and voxel statistics according to Dong et al. [15]. The volume loss due to corrosion was determined as the percentage difference between the initial volume and the volume at the time of testing according to (1):

$$V_{loss} = [(V_0 - V_{20}) / V_0] * 100 \qquad (1)$$

where V_0 is the initial volume of the reinforcement (mm^3) and V_{20} is the volume of the reinforcement 20 years after casting (mm^3).

 https://doi.org/10.1051/matecconf/201928910011

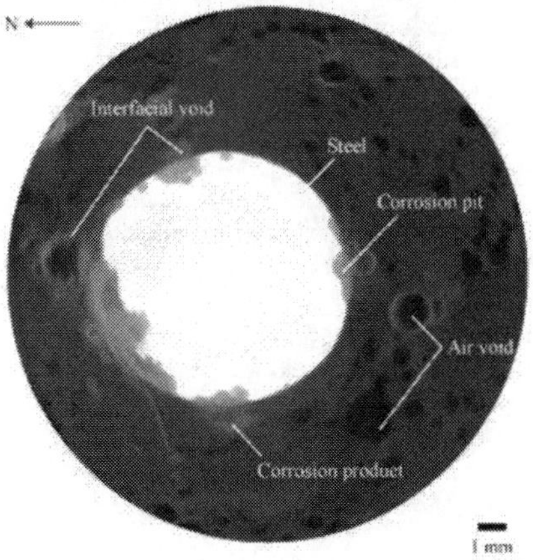

Fig. 2. CT-scan showing different phases (from I-1)

The initial volume of the reinforcement, V_0, was calculated through measuring the diameter of non-corroded portions of the specimens.

A selection of 2D CT-images was used to analyze the relation between corrosion pits, exposure side of the steel and presence of interfacial defects. For each CT-image, pit depths were clockwise measured at different angles from the center of the bar around 360° with 30° interval as conducted by Sun et al. [16], as shown in Figure 3. The depth of the pits, d, is measured (2):

$$d = R - R_0 \quad (2)$$

where R_0 is the initial radius of the rebar and R is the radius of the rebar 20 years after casting. The range between 180° and 0° clock-wise, *N-direction*, indicates the direction to the mould face (i.e. the portion closer to the outside environment). After measuring pit depths, they were divided in four groups depending on their orientation (N- and S-direction) and if they were in proximity to air voids at the SCI or not: type 1 (T1, in orange) and type 3 (T3, in green) were representative for corrosion pits formed close to air voids at N- and S-orientation respectively. Type 2 (T2, in yellow) and type 4 (T4, in blue) were representative for pits formed where the cement paste is dense at the SCI at both the orientations (N- and S-direction respectively). An example of this distinction is shown in Figure 3.

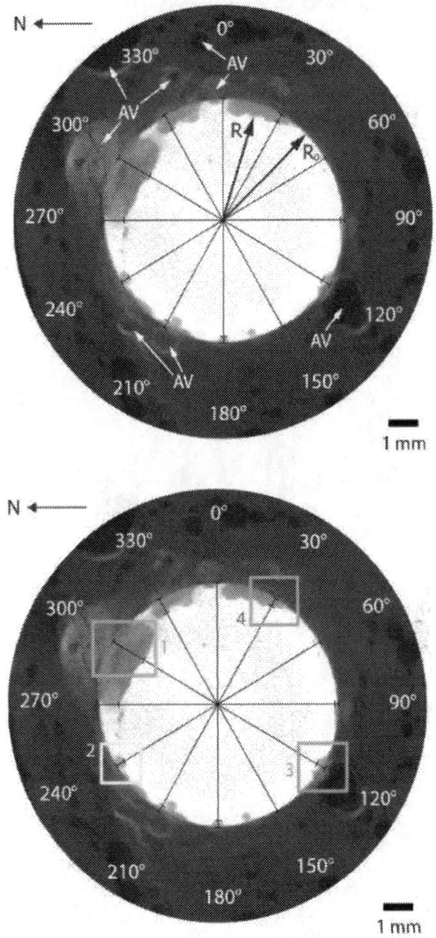

Fig. 3. (top) Depth and orientation of corrosion pits (AV=air voids). (Bottom) corrosion pits distinction: N-oriented pit in proximity of an air void (type-1, orange); N-oriented pit coincident to dense cement paste at the SCI (type-2, yellow); S-oriented pit in proximity of an air void (type-3, green); S-oriented pit coincident to dense cement paste at the SCI (type-4, blue).

https://doi.org/10.1051/matecconf/201928910011

3 Results and discussion

3.1 Steel volume loss due to corrosion

The average volume loss of reinforcement due to corrosion is reported in Figure 4.

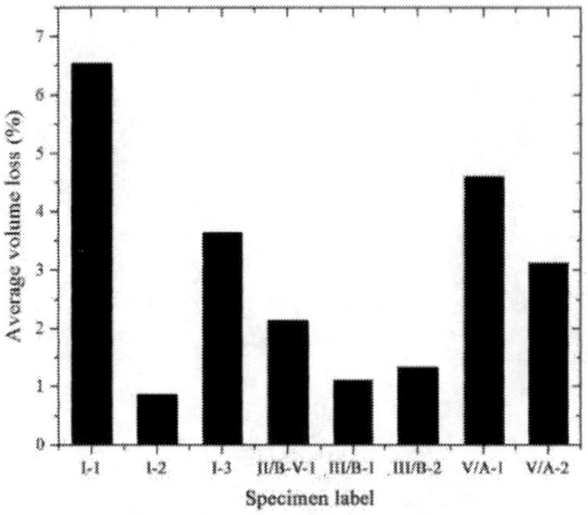

Fig. 4. Average volume loss (%) of steel reinforcement due to corrosion after 20 years of exposure

On average, rebars embedded in blended cement-concrete had lower volume loss due to corrosion than those cast with OPC (Ordinary Portland Cement). Negligible volume loss was measured for CEM III/B specimens, equal to around 1% for both III/B-1 and III/B-2. Slightly higher average volume loss of around 2% was found also for II/B-V-1. The highest average volume loss for blended cement specimens was measured for V/A-1. However, surface cracks were noticed upon inspecting the specimen. Cracks probably allowed for more aggressive ingress of electrolyte and harmful agents, resulting in higher average volume loss than rebars embedded in blended cement-based concrete and especially than V/A-2, which was cast with identical cement type (i.e. CEM V/A) but lost 3% of steel volume because it was less cracked than V/A-2. Higher variability was observed for rebars embedded in CEM I concrete since volume loss of around 6.5%, 0.9% and 3,5% was measured for I-1, I-2 and I-3, respectively.

Concrete cast with blended cement had higher corrosion resistance than that cast with OPC, probably thanks to the denser microstructure of the former ones, which is known to improve both the chloride penetration resistance and the resistivity of the mix [14]. Similar improvements can be achieved by lowering the water/binder ratio of the concrete [4]. Interestingly, I-2 had the lowest average volume loss of all the specimens, equal to around 0.9%. Since this volume loss is significantly lower than that of other rebars embedded in specimens cast with OPC, and they were exposed to the same environmental conditions, doubts were raised about the composition of I-2. Indeed, during the 20 years after casting, the specimens were re-located and some of them might be mis-labeled. Specimens were then re-identified non-destructively, as described more in detail elsewhere [17]. It might be the case that the composition of I-2 was not identified accurately, and that its water/binder ratio would be lower than 0,55. Since lower water/binder ratio improves the corrosion resistance of concrete [3], it would explain why I-2 had, unexpectedly, the lowest volume loss of reinforcement.

It must be specified that volume losses of steel quantified through CT-scan analysis could be misestimated due to segmentation inaccuracies and resolution limitation, as reported by others [18]. Nevertheless, since specimens were subjected to equal image acquisition and processing, potential drawbacks related to compare one test to each other are considered negligible.

3.2 Corrosion pit depths, orientation and relation with interfacial defects

The relation between corrosion pit depths, orientation and presence of interfacial defects is shown in the boxplot of Figure 5.

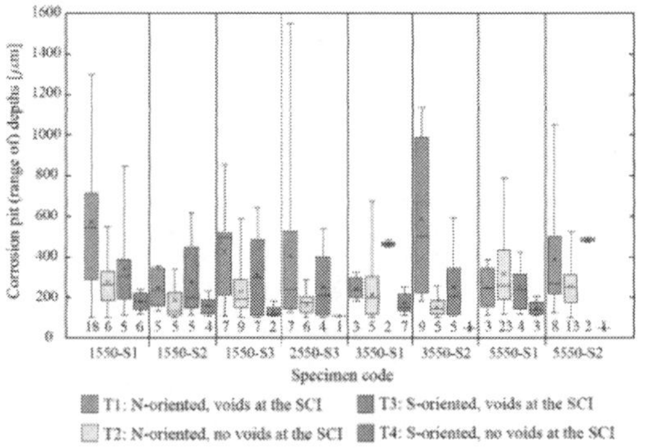

Fig. 5. Range of depth of types of corrosion pits (µm) with outside/inside orientation and proximity to concrete defects at the SCI. The top, middle and bottom line of the boxes correspond to the 75-, 50- and 25-percentile value (x_{75}, x_{50} and x_{25}) respectively. The whiskers show the minimum and maximum values. The number below each series represents the number of pits of each type per specimen.

Comparing T1 to T2 and T3 to T4, the influence of air voids at the SCI on corrosion pit depth can be analyzed. T1 and T3 are generally deeper than T2 and T4 respectively. This statement is not valid for III/B-1 and V/A-1 only. However, in the former case only a few air voids were present at the SCI, hence compromising the comparison with other specimens, while V/A-1 was heavily cracked, which probably led to more generalized corrosion along the steel portion oriented towards the outside environment (i.e. N-oriented side).

Comparing T1 to T3 and T2 to T4 the influence of the steel orientation can be analyzed. For most of the specimens, N-oriented pits are more and deeper than S-oriented pits, likely due to more frequent exposure to wet and dry cycles of the electrolyte and higher ingress of harmful agents (i.e. chlorides and carbon dioxide) due to the closer proximity of this side to the outside environment. On average, the combination of exposure to the outside environment and presence of interfacial air voids (i.e. T1 pits) revealed to be the most aggressive conditions for corrosion pits to firstly form and then to grow; on the contrary, the S-oriented side of the reinforcement revealed to be the least corroded side, particularly when no air voids at the SCI were found (T4). To give a better understanding of the occurrence of pits in proximity to interfacial voids, 3D renders of the front side of the corroding rebars with and without highlighted corrosion product around the reinforcement are shown in Figure 6. Corrosion pits in proximity to interfacial voids are indicated by black arrows. From Figure 6 it is qualitatively visible that the deepest corrosion pits occur where interfacial voids are present. This influence might not be clearly marked for the most corroding rebars (i.e. I-1 and V/A-1), for which the whole front side is heavily corroding. On the contrary, deep corrosion pits occurring at those locations seem obvious for the other specimens, especially for I-3, II/B-V-1, III/B-2 and V/A-2. Interestingly, these specimens seem to be subjected to relatively deep corrosion pits occurring in proximity to interfacial voids that are 5-15 mm apart from each other. Around the deep pits, reinforcement looks slightly corroding.

In this study it was observed that defects at the steel/concrete interface (i.e. macro-voids) appear to be

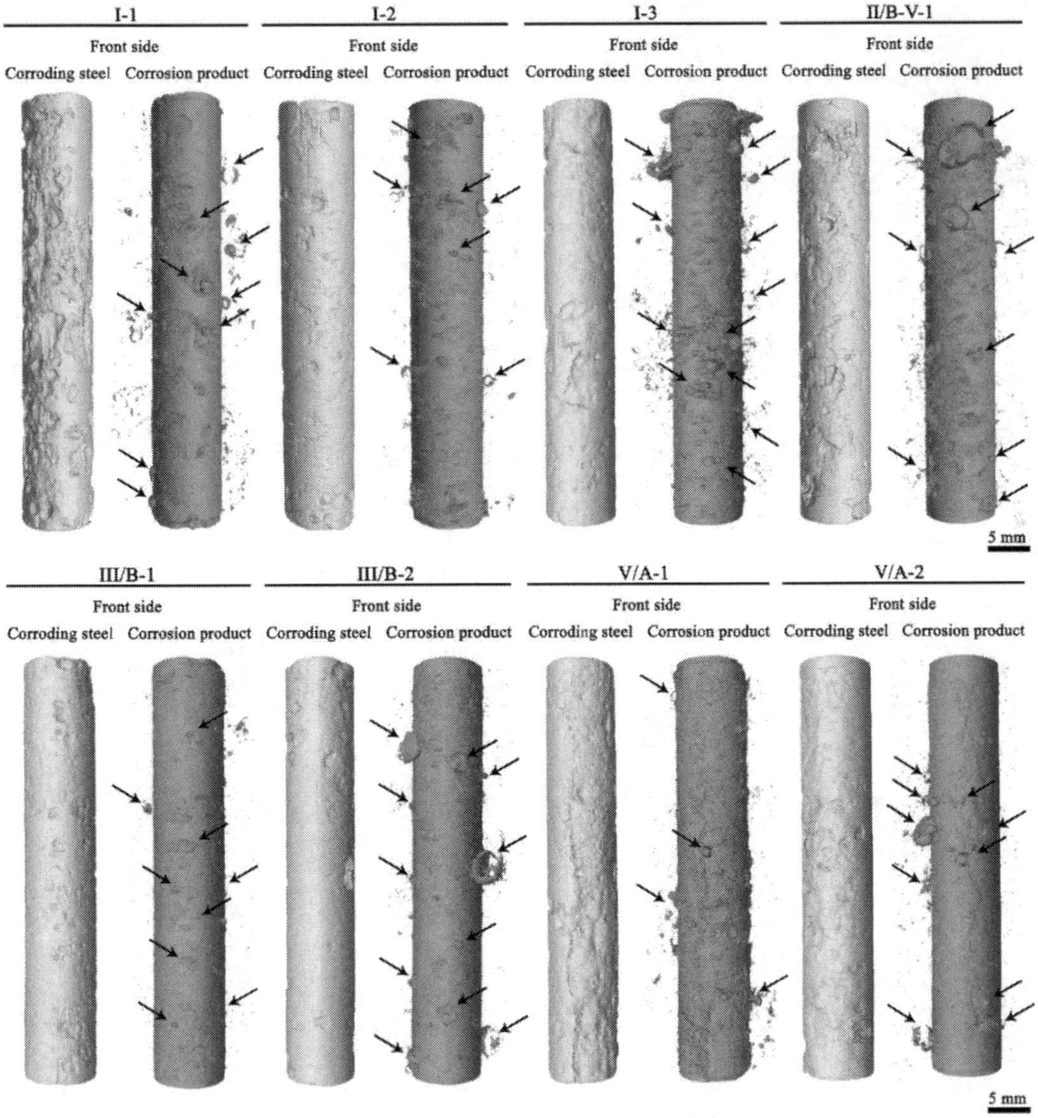

Fig. 6. Render of corroding reinforcement (grey) and corrosion product around the steel rebar (brown). Black arrow indicates pits in proximity to interfacial air voids. Only the front side of the rebar is reported.

the most preferential locations for corrosion pits to be found. Moreover, pits formed in proximity to interfacial voids are on average deeper than those formed where the cement paste is dense and compact around the reinforcement. Occurrence of pits in proximity to interfacial voids is also confirmed by Environmental Scanning Electron Microscope (ESEM) images of polished sections of the SCI, of which an example is reported below (Figure 7).

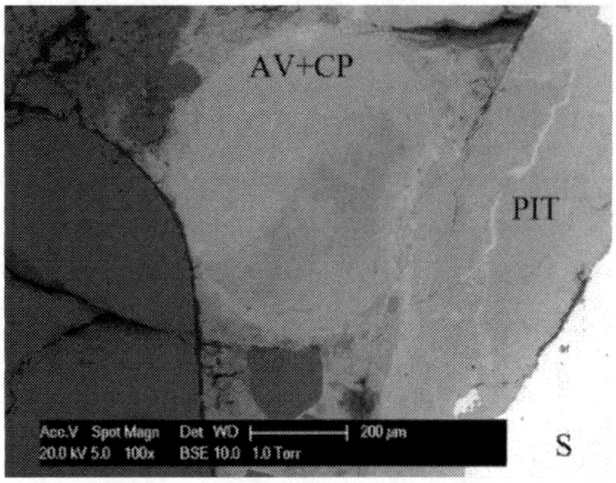

Fig. 7. Corrosion pit (PIT) in proximity of an interfacial air void filled by corrosion product (AV+CP). (S=steel rebar)

The results of this study are in agreement with what was pointed out by Glass et al. [10] about the inversely proportional relation between corrosion resistance and percentage of interfacial voids. Corrosion initiation in proximity to interfacial voids was recently observed by Van Steen et al. [18], which localized anodic-induced corrosion initiation sites through CT-scan and acoustic emission. Once initiated, corrosion pits tent to propagate at those locations. This fact may probably be caused by the higher availability of oxygen and electrolyte present at the void locations as well as by their accommodating open space, which results in no constraint from the surrounding concrete that might block oxide formation [19]. A similar mechanism was observed by Savija et al. [20] when monitoring the progress of anodic-accelerated corrosion of concrete cores through CT-scan, during which non-uniformities of the corrosion layer and consequent higher localized volume loss occurred in proximity to interfacial air voids. However, the rebars of the studies of Van steen et al. [18] and Savija et al. [20] were subjected to anodic-accelerated corrosion, which might influence where corrosion would initiate.

Based on the results of this study, interfacial air voids are the locations where corrosion propagates the most, resulting in higher localized steel volume loss in proximity to interfacial defects. This finding might be related to the influence that interfacial voids might have on either corrosion initiation or corrosion propagation. On the one hand, it might be assumed that the deepest pits are those that initiated the earliest, hence the results of this study would suggest that interfacial voids would be the most sensitive locations for corrosion initiation. This mechanism would presuppose that the corrosion rate does not depend on the local characteristics of the steel/concrete interface, and that would be constant for any corrosion pit. However, it is likely that the corrosion rate at pits would be locally influenced by, for instance, the availability of oxygen, the electrolyte and the local chloride content. Hence, concluding that interfacial voids are the most sensitive locations for corrosion initiation might be inaccurate. More continuous observations of corrosion initiation are needed to better understand this mechanism. On the other hand, this study showed unequivocally that corrosion propagated the most in proximity to interfacial voids. This mechanism might be completely unrelated to the influence that interfacial voids might have on corrosion initiation. In other words, corrosion could initiate anywhere the chloride content is higher than the critical threshold, regardless of the presence of interfacial defects. Interfacial voids might then be the locations where corrosion pits would propagate only because, for instance, no physico-chemical constraint would be provided by the surrounding matrix. Propagation of corrosion in proximity to interfacial voids might also mitigate the growth of neighbouring pits due to, for example, the cathodic flow occurring in proximity to the growing pit. This mechanism would result in deep pits localized where interfacial defects are present, and no corrosion pits in the surroundings, as qualitatively suggested by Figure 6. To clarify the influence that interfacial air voids have on both corrosion initiation and propagation, more investigations are encouraged. Nevertheless, the results of this study suggest that interfacial air voids should be considered detrimental factors for corrosion resistance of reinforced concrete.

In this study, the most sensitive rebar side to corrosion was that exposed to the outdoor environment, contrary to what has been reported by other authors [4, 21]. Angst et al. [4] observed that in rebars cast perpendicularly to the concrete casting direction, the bottom side of the steel was the most corroding due to the higher amount of localized defects caused by plastic settlement and bleeding. However, the rebars analyzed in the present study were placed parallel to the casting direction, not causing localization of defects at one specific side. Since the quality of the two steel sides could be considered ideally the same, it is reasonable that the outer portion of the steel would be more sensitive to corrosion on average, due to, for instance, higher availability of the electrolyte, more frequent wet/dry cycles and higher content of harmful agents coming from the outside environment. It must be specified that the rebars analyzed in this study had 10 mm of cover depth, which is not very representative for good practice. It is very likely that for thicker cover depth (i.e. 30 mm), the influence of the exposure side would not be so marked because of the higher physical protection provided. In that case, the preferential locations for corrosion to propagate would probably be in proximity to interfacial defects regardless of the exposure side of the steel reinforcement.

4 Conclusions

From the present study, the following major conclusions can be drawn:

- After 20 years of natural exposure, the volume loss of reinforcement embedded in OPC concrete was on average higher than that of rebars embedded in blended cements-concrete,
- The deepest corrosion pits were observed at the portion of the steel exposed to the outside environment and in proximity to interfacial defects (i.e. air voids), which appeared to be the most preferential locations for corrosion propagation,
- In cracked concrete (i.e. 5550-S1) whole side of the steel oriented towards the crack was corroding, suggesting a transition from pitting to more distributed corrosion,

In the authors' opinion, the use of X-ray Computed Tomography could very much help to better understand some fundamentals of corrosion in reinforced concrete, such as monitoring the corrosion initiation sites during time of more recently cast specimens. For future research, understanding where corrosion sites initiate would be valuable for both the scientific community and the industry since it would help to be aware of what are the potentially most sensitive locations for corrosion to happen, with consequent more specific inspections and effective maintenance.

The authors would also like to thank Mr. Arjan Thijssen for his assistance with the X-ray Computed Tomography experiments.

References

1. L. Bertolini, B. Elsener, P. Pedeferri, R.B. Polder, Wiley-VCH Verlag GMBH & Co., Weinheim (2004)
2. U. Angst, B. Elsener, C.K. Larsen, Ø. Vennesland, Cem Concr Res, **39**, 1122–1138, (2009)
3. U. Angst, M. Geiker, A. Michel et al., Mater Struct, **50**, 143, (2017)
4. U. Angst, B. Elsener, C.K. Larsen, Ø. Vennesland, Corros Sci, **53**, 1451–1464, (2011)
5. T. Soylev, R. Francois, J Mater Civ Eng, **17**, 447-455, (2005)
6. T. Yonezawa, V. Ashworth, R. Procter, Corros Eng **44**, 489-499, (1988)
7. T. Mohammed, N. Otsuki, M. Hisada, ACI Mater J, **96**, 154-159, (1999)
8. T. Mohammed, N. Otsuki, H. Hamada, T. Yamaji, ACI Mater J, **99**, 149-156, (2002)
9. C. Christodoulou, G. Glass, Australasian Corrosion Association 2012 Annual Meeting, 159, (2012)
10. G. Glass, B. Reddy, L. Clark, Const. Mat, **160**, 155-164, (2007)
11. U. Angst, PhD thesis, NTNU, Trondheim, (2011)
12. T. Mohammed, N. Otsuki, H. Hamada, J Mater Civil Eng **15**, 460–469, (2003)
13. R.B. Polder, P. Russo, 98-BT-R1664, TNO Building and Construction Research, Rijswijk, TNO, (1998)
14. R.B. Polder, W.H.A. Peelen, Cem Concr Compos, **24**, 427–435, (2002)
15. B. Dong, W. Ding, S. Qin, G. Fang, Y. Liu, P. Dong, S. Han, F. Xing, S. Hong, Constr Build Mater **168**, 11-20, (2018)
16. H. Sun, D. Zhao, Y. Gu, S. Memon, Z. Ren, B. Liu, X. Zhang, F. Xing, D. Li, Constr Build Mater, **143**, 24-32, (2017)
17. J. Pacheco, MSc thesis, Materials&Environment, TU Delft, (2010)
18. C. Van Steen, L. Pahlavan, M. Wevers, E. Verstrynge, Constr Build Mater, **197**, 21-29, (2019)
19. T.A. Soylev, R. Francois, Cem Concr Res, **33**, 1407–1415, (2003)
20. B. Savija, M. Lukovic, S.A.S. Hosseini, J. Pacheco, E. Schlangen, Mater Struct, **48**, 2043–2062, (2015)
21. U. Angst, C.K. Larsen, Ø. Vennesland, B. Elsener, in: P. Castro-Borges, E.I. Moreno, K. Sakai, O. Gjørv, N. Banthia (Eds.), CONSEC'10 6th Int. Conf. on Concrete under Severe Conditions, Environment and Loading, **1**, Taylor & Francis, London, 359–366, (2010)

Research and practices of large composite external wall panels for energy saving prefabricated buildings

Yunxing Shi*, *Yangang* Zhang, *Kun* Ni, *Wei* Liu, and *Ye* Luo

Technical Center, China State Construction Engineering Corporation, Beijing, 101300, China

Abstract. The production process and application of large composite external wall panels (composite panels for short) are introduced in this paper. Composite panels with both load bearing and thermal insulation were formed by pouring normal concrete (NC) and ceramsite foamed concrete (CFC) continuously according to particular technological requirements, which made two layers into a seamless whole. The layers of NC and CFC are for load bearing and thermal insulation respectively. The composite panels were manufactured in the scale of industrial production, and applied to several energy saving prefabricated buildings successively, instead of polystyrene sandwich composite panels (sandwich panel for short) as external wall panels. There are several obvious advantages of the composite panel over the sandwich panel or outer benzoic board. Firstly, it solved the problems of durability of polystyrene and the complex production process of the sandwich pane, the production process of the external wall was thus greatly simplified. In addition, the fire risk was much reduced.

1 Introduction

Nowadays efficient construction and green building are regarded as the development goals of the building industry in China, and a perfect combination of prefabricated construction with an energy saving system is considered an effective way to achieve the goal. However, the external wall thermal insulation in use popularly with outer benzoic board shows obvious shortcomings, such as flammable, poor durability and easy to become detached. Although a concrete sandwich slab with a core of EPS or XPS as insulation layer prefabricated in the concrete product plant gets over some defects of outer benzoic board system, it is still a complex process and durability problem as well. However, a large composite external wall panel, developed by the China State Construction Technical Centre, is a fundamental solution for the problems above. The composite panels were made of normal concrete (NC) and ceramsite foamed concrete (CFC) with a continuous pouring process to achieve a seamless interface of two layers. The former is for load bearing, and the latter for thermal insulation respectively. An illustration of a full size composite panel in practical application is shown in Fig.1.

The composite panel has good thermal insulation due to a network of closed fine pores of the CFC layer, and its outer surface has architectural decoration as well as a water proof effect. Other decorative facade to imitate stone texture, wood grain and brickwork and so on are also available.

2 Preparation of concrete and production of composite panel

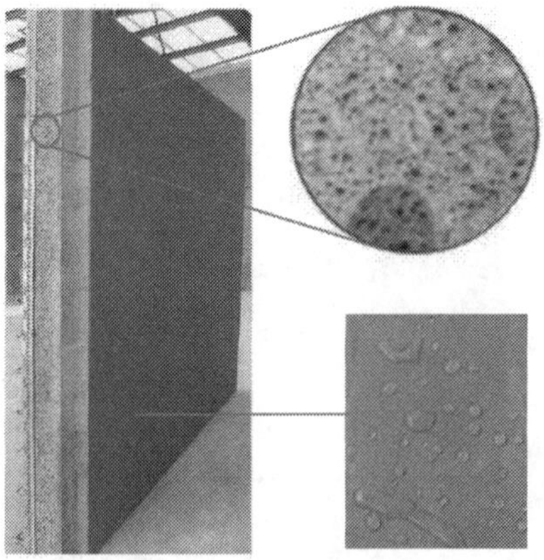

Fig.1 Appearance of the composite panel

2.1 Preparation of concrete

The NC was used for the load bearing layer, the strength grade of which should meet requirement of design, C30~C40 in general, according to Chinese standards. The fresh state of concrete should show good workability to avoid segregation and bleeding, so as to ensure a firm combination between NC and CFC layers.

* Corresponding author: yunxing_shi@sina.com

https://doi.org/10.1051/matecconf/201928910012

The basic mix proportions of the CFC are given in Table1, in some cases, the proportions may need to be modified in a small range considering the situations of raw materials. As lightweight coarse aggregate of CFC, a kind of sintered expanded clay ceramsite with average size 20mm was used (as shown in Fig.2), and preformed foam from a large capacity foaming machine, developed by the Technical Center of CSCEC, was added to the mixing chamber directly (as shown in Fig.3), then mixed with the concrete mixture until a uniform state was reached in 30~60sec. Some losses of foam volume may occur during the mixing. The fresh CFC with stability foam showed a slump flow 450~500mm (as shown in Fig.4).

Table 1. Mix proportions of CFC (kg/m^3)

Cement (P.O 42.5)	280~ 330
Fly ash (II)	50~90
Ceramsite bulk volume (m^3)	0.4~0.55
Foam volume (m^3)	0.45~0.6
Water	105~125
Superplasticizer	(polycarboxylic liquid) 2~3.5

Fig.2 Appearance of expanded clay ceramsite

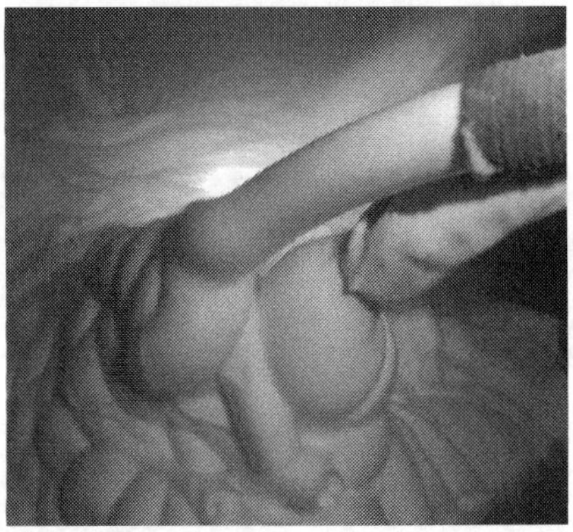

Fig.3 Preformed foam to the mixing chamber

Fig.4 Slump flow of fresh CFC

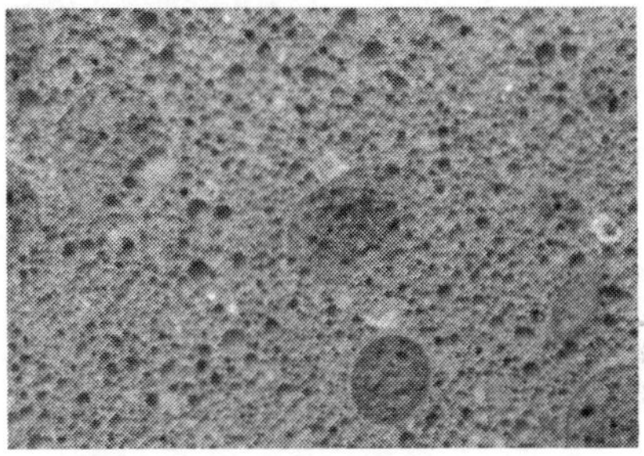

Fig.5 Morphology of internal pores of CFC

Fig.5 is an observation of internal morphology of the hardened CFC; from the image of section, it could be seen that enormous numbers of closed pores were present in uniform distribution inside, whose most probable value of size was about 1mm. This was the reason why the composite panel showed sound performance of thermal insulation.

2.2 Production of the composite panel and its thermal insulation

In automatic production lines, the composite panel series have realized industrial scale production. The production flowchart is shown in Fig.6, among all of the operations mentioned in the flowchart, formwork, reinforcement work and concrete work were the major steps, moreover, concrete work was done after the reinforcement work, which was arranged according to the structural load and the codes concerning structural design. The concrete work consists of two parts, namely, pouring of NC and CFC in turn. Besides, for avoiding crack appearance along the connection of the two layers in the longitudinal direction, the time interval between the two pours is as small as possible, less than 40 minutes usually. A prolonged time interval may cause a negative effect for the combination between the NC and CFC layers. Of course, the workability of fresh concretes of NC and CFC need to be in a good phase, without segregation and bleeding, specially, the distributed foam in the fresh CFC ought to be kept in a uniform and stable state until final setting.

https://doi.org/10.1051/matecconf/201928910012

Fig.7 is the appearance of the reinforcement cage with lifting lug after completion of the NC pour, one of the steps of process operation in the production line. Fig.8 and Fig.9 show pouring the CFC and the composite panel in the demoulding operation respectively.

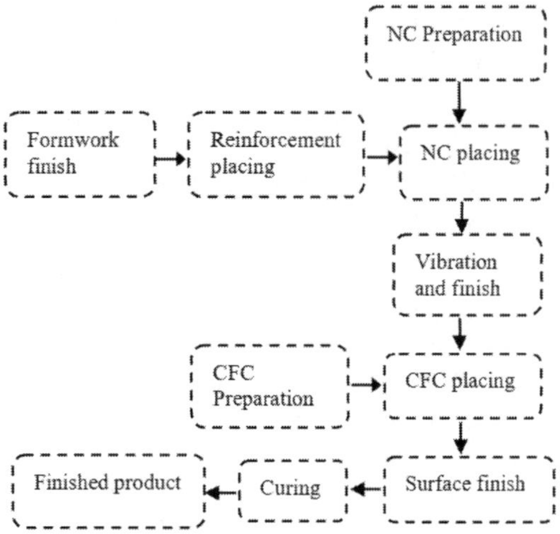

Fig.6 Production flowchart of the composite panel

The measured values of heat transfer coefficient of composite panels in different thickness are shown in Table 2. The performance of the thermal insulation of the composite panels could be changed by adjusting density and thickness of the CFC layer to satisfy the requirements of practical engineering.

Fig.7 Reinforcement cage after completion of the NC layer.

Fig.8 Operation of pouring CFC

Fig.9 Composite panel in demoulding operation

Table 2. Heat transfer coefficient of the composite panel

Thickness of RC (mm)	Thickness of CFC (mm)	Heat transfer coefficient ($W/m^2 \cdot k$)
60	100	1.0
60	150	0.737
60	180	0.7

3 Examples of engineering application

The composite panels have been successfully applied in four demonstration projects, which are all energy saving prefabricated buildings. The construction efficiency was greatly raised, and the energy saving effect was effectively improved as well due to the advantages of composite panels prefabricated in an industrialized scale.

3.1 Demonstration project of an office building with green construction in Chengdu

The project was located in the capital city of Sichuan province, Chengdu, Southwest China. The production date of the composite panels was from Nov.2016 to Feb.2017, and the assembling construction date was from Mar. to May of 2017. The sum of area of the composite panels applied in the project was over 3000m^2.

Erection construction was the next important stage of construction after completion of production of the composite panels. Fig.10 is the appearance inside of composite panels fixed to the RC frame structure on its facade by bolted connection.

Fig.10 Composite panel on frame structure inside

 https://doi.org/10.1051/matecconf/201928910012

The building was finished after 3 months and Fig. 11 is a view of the outside of the building.

Fig.11 The outdoor scene of the building completed

3.2 The other three demonstration projects

After the first demonstration project in Chengdu, three other projects were also finished successively. They were located in Fuzhou city in East China, Changsha city and Wuhan city in Central China respectively. Among them, the appearances of two projects, in Fuzhou as well as in Wuhan, are shown in Fig.12 and Fig.13 separately.

Fig.12 The building in Fuzhou after major structure completed

Fig.13 The building in Wuhan

3.3 Another project under construction

Besides the four projects completed above, another project under construction is a luxurious hotel for VIP guests (Fig.14), which is one of a collection of projects for the Winter Olympic Games to be held in Beijing and Zhangjiakou city in the winter of 2022 simultaneously. The project is located in the Chongli playing area in Zhangjiakou city. The thermal insulation external wall together with high fidelity stone facade was required by the design.

Additionally, agreeably to the design and green construction code, the sand and stone excavated in the construction site would be used as raw materials of the external wall as much as possible. The mass of mixture of sand and stone excavated out in the construction site is shown in Fig.15. The stone blocks accumulated are a kind of coarse-grained granite containing a part of sand gravel caused by weathering of rock. The sand and stone blocks after the treatment would be used as raw materials in the production of external walls. Fig.16 is a demonstrated panel of the prefabricated external wall for the project following the requirement of design. The external wall comprises a CFC main body for thermal insulation and a surface layer for decoration, moreover, the surface layer was made with a high-fidelity texture of the stone from the site, as shown in Fig.16.

Fig.14 The appearance of the hotel in the design drawing

Fig.15 Sand and stone excavated out in the site

Fig.16 Demonstrated panel for the external wall

At the present time, large composite panels consistent with the demonstrated panel are on their way to production on an industrialized scale. It may be expected that such an enclosure structure with integration of thermal insulation and decoration will give the project plenty of advantages in green features and building energy saving.

4 Concluding remarks

1. The large composite external wall panels with integration of thermal insulation and decoration were prefabricated in a process of continuous pouring of NC and CFC in turn on a scale of industrial production. Moreover, large numbers of composite panels have been applied to several energy saving projects successfully.

2. The composite panels and their application in engineering showed the remarkable advantages of the production process, assembling construction, durability and building energy saving.

3. Sand, stone and raw soil excavated out in the site could be used as constituent materials of the composite panels. It has shown obvious benefit of resource saving and ecosystem protection.

References

1. Hubert Bachmann, Alfred Steinle, *Precast Concrete Structures*, Ernst & Sohn, 2011,Berlin, Germany
2. Yunxing Shi, Lihong Jiang, Zhongnan Song, *Ceramsite foamed concrete and its application in energy saving buildings*, China Building Materials Press, Beijing, China (in print)
3. Yunxing Shi, Yangang Zhang, et al, *An engineering example of energy saving renovation of external wall of original building with lightweight insulation composite panel,* Proceedings of Concrete Solutions 6th International Conference on Concrete Repair, Taylor and Francis, June 2016, Thessaloniki, Greece
4. Yunxing Shi, Zhongnan Song,Yangang Zhang, et al, *A new type of integrative composite panel with energy saving and decoration*, China Invention Patent (Number CHINA: ZL201210489490.7) 2012 (in Chinese).

Mechanical properties of alkali activated geopolymer paste using different Romanian fly ash sources – experimental results

Adrian Lăzărescu[1,2,*], *Călin* Mircea[2], *Henriette* Szilagyi[1], and *Cornelia* Baeră[3]

[1]NIRD URBAN-INCERC, Cluj-Napoca Branch, 400524 Calea Florești 117, Cluj-Napoca, Romania
[2]Technical University of Cluj-Napoca, Civil Engineering Faculty, 400020 Daicoviciu 15, Cluj-Napoca, Romania
[3]NIRD URBAN-INCERC, Timișoara Branch, 300223 Traian Lalescu 2, Timișoara, Romania

Abstract. As concrete demand is constantly increasing in recent years and also considering that cement production is both a consumer of natural resources and a source of carbon dioxide release into the atmosphere, there have been worldwide investigations into green alternatives for making concrete environmentally friendlier and simultaneously to satisfy the development of infrastructure facilities. The use of fly ash as a component of cementitious binders is not new but when considering the specific case of alkaline activation and fly ash representing the only source for the binder formation, it necessitates a more complete understanding of its specific reactions during the alkaline activation process. Since the fly ash varies dramatically, not only from one source to another, but also from one batch to another even when provided by the same power plant, its chemistry in obtaining alkali-activated materials during the geopolymerisation process and the final mechanical properties are considered crucial for the performance of geopolymer concrete. This paper will provide a review of the experimental results concerning the physical and mechanical evaluation of the alkali-activated fly ash-based geopolymer materials, developed with different types of fly ash, for a better understanding of geopolymer concrete production control.

1 Introduction

Concrete is the second most widely used material in the world, being considered the most versatile, durable and reliable building material - therefore the need to produce large quantities of Portland cement (PC) as binder for concrete are constantly increasing. The cement industry is considered to be the second highest carbon dioxide generator, being preceded only by the automotive industry. Each tonne of produced cement requires between 60 and 130 kg of liquid fuel or its equivalent, depending on the type of cement and the manufacturing process used, and approximately 110 KWh of electricity [1]. One tonne of produced cement releases between 0.8 and 1.1 tonnes of CO_2 into the atmosphere, as a subsequent consequence of combustion and calcareous calcination [2]. Only in 2014, the world cement production represented approximately 4.2 billion tonnes [3], indicating an obvious increase related to the previous years.

Environmental problems associated with ordinary Portland cement production are extremely well known and they represent a problem which is carefully monitored, in terms of the amount of carbon dioxide released into the atmosphere during its production. Constant efforts are therefore made in order to counteract this effect in future years. As the industry is constantly growing, the involvement of modern building materials in the economic world is more than necessary. Alkaline activation of the fly ash represents a special procedure, able to generate the solidification of fly ash powder when mixing it with a certain type of alkaline activator and creating a new binding material. Consequently, it has opened the perspective of obtaining a new building material when incorporating aggregates, as an alternative to traditional concrete and cement-based composites [4-7]. Developing these types of composites, with similar or even superior, properties to cementitious composites, without using cement in the composition, creates an extraordinary opportunity for the environment and for the construction industry, constituting itself as an alternative to traditional technology, both due to mechanical properties and high resistance in aggressive environments [8].

The large amount of fly ash resulting from the energy industry in Romania can create new opportunities to use this waste as a substitute for Portland cement in the production of new materials. In 2017 only, one power plant in Romania produced nearly 650,000 tonnes of fly ash, 50,000 tonnes of slag and 50,000 tonnes of gypsum. Despite the slag recovery by recycling means its reuse into the coal supply of boilers (43,000 t/year) and the fly ash delivery to the construction industry (cement, concrete, mortar for stabilizing soils, roads etc.), representing approximately 162,000 t/year, the unused

* Corresponding author: adrian.lazarescu@incerc-cluj.ro

https://doi.org/10.1051/matecconf/201928911001

quantity of fly ash from the power plant is still an important environmental problem, with almost 565,000 t/year stored in waste dumps [9].
When fly ash is used as a main binding material in the production of alkali activated materials, a careful consideration is needed when choosing the primary materials used for the mix. Chemical properties of the components have to be taken into consideration so that the geopolymerisation reaction can properly occur [10]. The solid part (fly ash) and the liquid part (alkaline activator) in the system should not be considered as additions, as they represent the main constituents of the binder [11].

The physical, chemical and mechanical properties of the alkali-activated geopolymer materials are strongly influenced by the properties of raw material such as fly ash [12, 13]. Since the properties of fly ash vary, depending on several parameters (the quality of coal, temperature and process of combustion etc.) [12-15], understanding the mechanisms that generate the geopolymerisation process have to be well known.

The aim of this paper is to present the preliminary results obtained on alkali-activated fly ash-based geopolymer paste using different Romanian power-plant source materials for producing this type of material, based on efficient source materials, available at national level and also presenting the effect of their physical and chemical properties on the compressive strength of the binder.

2 Experimental method – materials and methods

This chapter summarizes both the state-of-the art information about the raw components used for the development of alkali-activated fly ash-based geopolymer pastes (FAGP) and also about the applied particular experimental methodology (mix design, moulding methods, curing and also testing methods).

2.1 Materials

2.1.1 Fly ash

Fly ash (FA) is a by-product obtained by electrostatic or mechanical precipitation of the pulverized particles resulting from the coal-fired combustion gases of the furnaces in power plants. It is a fine powder, consisting mainly of spherical shaped glass particles [16, 17]. Although some certain characteristics proved to be common for all fly ash types, due to factors like coal type as raw material for FA production and the burning process as well, fly ash powders show significant differences when coming to their chemical composition. The raised variation of the chemical composition, leading to different performance of FA as binding material, when used as primary raw material for producing alkali-activated fly ash-based binding paste (FAGPP), represents an essential aspect for the general development of the concept [18].

For the current research, Class F fly ash was used for preparing the FAGPP specimens and was characterized as follows [19]: $SiO_2 + Al_2O_3 + Fe_2O_3 > 70\%$, $SO_3 < 5\%$ and L.O.I. <6%.

Two different Romanian power plant, low-calcium, fly ashes (FA.S1 and FA.S2) were used in this study (Fig. 1a and Fig. 1b), focused on the development and implicitly, the comparative analyses of FAGPP mechanical properties [20]. The chemical composition of the FA samples was established by X-ray fluorescence analysis (XRF analysis), presented in Table 1. Also their relevant physical properties are presented in Table 2.

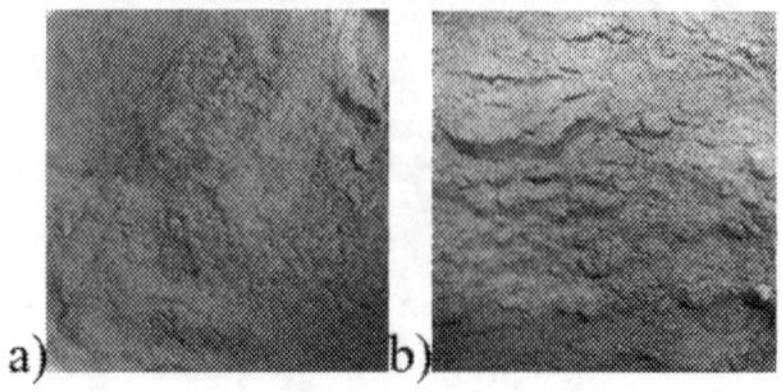

Fig. 1. a) FA.S1 sample b) FA.S2 sample

Table 1. XRF analysis data for the fly ash samples.

Oxides	FA.S1 %	FA.S2 %
SiO_2	53.61	54.32
Al_2O_3	26.16	22.04
Fe_2O_3	7.58	9.02
CaO	2.42	5.85
MgO	1,49	2,48
SO_3	0.26	0.20
Na_2O	0.59	0.54
P_2O_5	0,12	0,16
TiO_2	1.04	0.86
Mn_2O_3	0.08	0.06
L.O.I.	3.57	3.05

Table 2. Physical properties of fly ash samples

Properties	FA.S1	FA.S2
Colour	Light grey	Greyish white
Residue retained on 45 μm, %	37.6	59.4

2.1.2 Alkaline Activator

When in contact, the reactive solids from the fly ash and the alkaline solution react and a hard, water-resistant material is generated due to the formation of an alumino-silicate network [21] and geopolymerisation occurs, resulting from the dissolution of the Si and Al compounds of the fly ash [22].

By using an alkaline hydroxide activating solution and a silicate solution, the SiO_2 / Al_2O_3 and Na_2O / SiO_2 ratios in the mixture are optimized to obtain higher mechanical performances [23, 24].

The most commonly used combination for the production of the alkaline activator is sodium hydroxide (NaOH) or potassium hydroxide (KOH), together with a sodium silicate solution ($nSiO_2Na_2O$) or a potassium silicate solution ($nSiO_2K_2O$) [25]. Waterglass adds silicon atoms and mixed liquid, both of which play an important role in producing the geopolymerisation process.

It is recommended to use sodium hydroxide of purity ranging between 94 and 99%. Regarding the sodium hydroxide solution, it should be prepared 24 hours before use, but not earlier than 36 hours because it loses its properties as an activator [26]. Sodium hydroxide concentration has an important influence on the final mechanical properties of the alkali-activated fly ash-based geopolymer material [27, 28].

The alkaline activator used for this experimental study was a combination of sodium hydroxide solution (NaOH) and sodium silicate solution (Na_2SiO_3), as shown in Fig. 2a and Fig. 2b. The sodium silicate solution was purchased from the local market and its chemical composition was SiO_2=30%, Na_2O=14% and H_2O=56%. The sodium hydroxide solution was prepared by dissolving the NaOH flakes into water until the desired concentration of the solution was achieved. Two distinct NaOH solutions, of 10M and 12M respectively, were prepared for FAGPP mixes. The 10M NaOH solution was prepared by dissolving 400g of NaOH flakes into water for 1 litre solution (40g x 10 = 400g NaOH/litre, where 40g is the molecular weight of NaOH) [20].

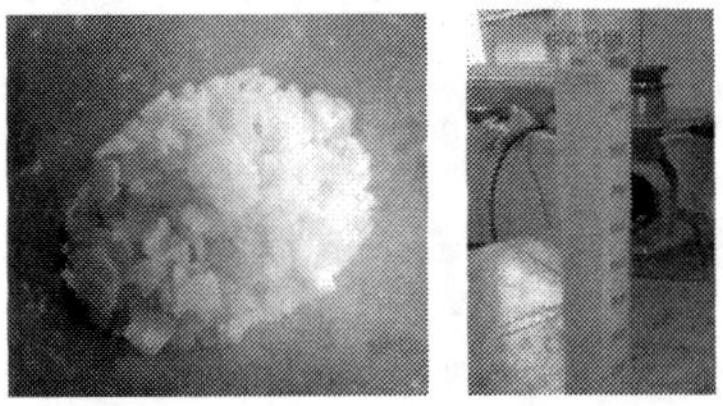

Fig. 2. a) NaOH flakes b) Na_2SiO_3 solution

2.2 Mix design of FAGPP

Producing FAGPP does not imply a different technology than in the case of conventional mortars, the use of a pallet mixer is convenient for a proper mix. Preliminary results obtained on FAGPP, when using Romanian local raw materials, indicated that an alkaline activator/fly ash ratio (AA/FA) of 0.5 for FA.S1 and 1.0 for FA.S2 of fly ash provided good workability [29]. The Na_2SiO_3/NaOH solution ratios ranged between 0.5 and 2.5, for a comparative analysis regarding the variation effect on the material behaviour. Table 3 summarizes the fly ash-based geopolymer paste mixes developed in the current experimental work.

The FA and AA were mixed together for three minutes, until a homogeneous paste was obtained. The mixtures were then placed in 40 mm x 40 mm x 160 mm moulds and heat cured at 70°C for 24 hours. A glass film was placed on top of every mould in order to prevent excessive water release from the mixtures. After demoulding, the FAGPP specimens were stored in the climatic chamber at the temperature T (20±1)°C and relative humidity RH (60±5)% until the age of 7 days, when their mechanical performance was tested. Fig. 3 shows aspects related to the preparation of the FAGPP samples.

Table 3. Mix design of FAGPP.

FA type	AA / FA ratio	Na_2SiO_3 / NaOH ratio	NaOH molarity
FA.S1	0.5	0.5	10M and 12M
		1.0	
		1.5	
FA.S2	1.0	2.0	
		2.5	

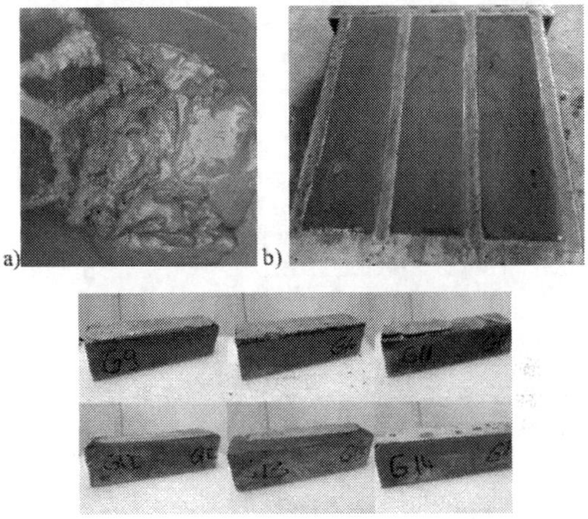

Fig. 3. a) Final mixing sequence b) Moulding c) FAGPP specimens after de-moulding

2.3 Testing method of FAGPP samples

The FAGPP compressive strength testing (as shown in Fig. 4) was performed at the age of 7 days, using three prismatic specimens for each type of alkali-activated fly ash-based geopolymer paste; the mean value of the results was considered relevant for the data interpretation. The testing method was in accordance to EN 196-1:2006, namely the standard testing methodology for ordinary Cement mortars. Early age testing at 7 days was considered relevant for the comparative evaluation, as previously experiments proved that generally the FAGPP reach most of their compressive potential by this age [20].

Fig. 4. FAGPP samples after compressive test

3 Results and discussions

Alkali-activated fly ash-based geopolymer materials properties can be notably influenced by several important factors such as the source and quality of the fly ash used as the precursor material, the NaOH solution concentration and the Na_2SiO_3/NaOH solution ratio. The physical and chemical properties of the fly ash samples, as well as the different ratios of alkaline liquid ratios, at the age of 7 days, and FAGPP compressive strength were consider relevant for the comparative evaluation of the above nominated parameters.

3.1 Fly ash parameters

The oxide composition of the two samples of fly ash were analysed by XRF analysis and are listed in Table 1. Based on literature [30], both samples of fly ash are categorized as Class F fly ash, because the amount of SiO_2, Al_2O_3 and Fe_2O_3 is more than 70% (Table 4). A high difference occurs in CaO content, where FA.S2 has a higher percentage than FA.S1. The literature suggests [31] that a low content of CaO in the chemical composition of the fly ash leads to better compressive strength of the material.

Table 4. Class F compliance of the FA samples.

FA	$SiO_2 + Al_2O_3 + Fe_2O_3$ %
FA.S1	87.35
FA.S2	85.38

The reactivity of fly ash samples highly depends on the LOI, which represents the unburned material [31, 32]. Both samples of fly ash have LOI less than 6%, as recommended by standards [30] and also less than 5%, as recommended by literature [31]. The fineness of the fly ash is defined as the percentage of particles retained on the 45μm mesh sieve. As shown in Table 2, FA.S1 has a lower percentage of retained particles on the mesh sieve (37.6%), therefore it has a higher fineness.

Although the chemical composition and the L.O.I. parameters of the fly ash samples used (FA.S1 and FA.S2) are only slightly different, as shown in Table 1, it can be easily noticed (Table 3) that the FAGPP mix design for each FA sample is significantly different. In order to achieve the desired workability and to obtain a proper geopolymerisation process, the alkaline activator/fly ash ratio for FA.S1 was 0.5 and for FA.S2 was 1.0. The FAGPP samples prepared with FA.S2 showed lower compressive strength values for all the Na_2SiO_3/NaOH ratios and for both, 10M and 12M NaOH solution concentration. the fineness of the two fly ash samples being different (Table 2), shows that this parameter plays an important role in the workability of the mix and further on, in the compressive strength development of the alkali activated material (Fig. 5 and Fig. 6): – the finer the fly ash, the better the workability of the mix, therefore, a lower alkaline activator/fly ash ratio.

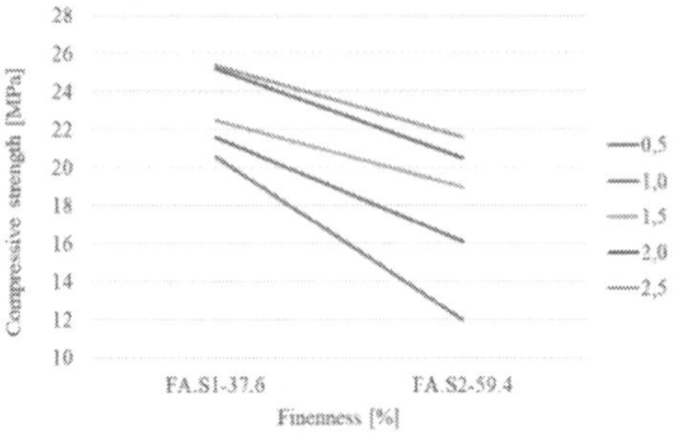

Fig. 5. Effect of fineness to compressive strength (10M NaOH solution – 0.5 to 2,5 alkaline activator ratio)

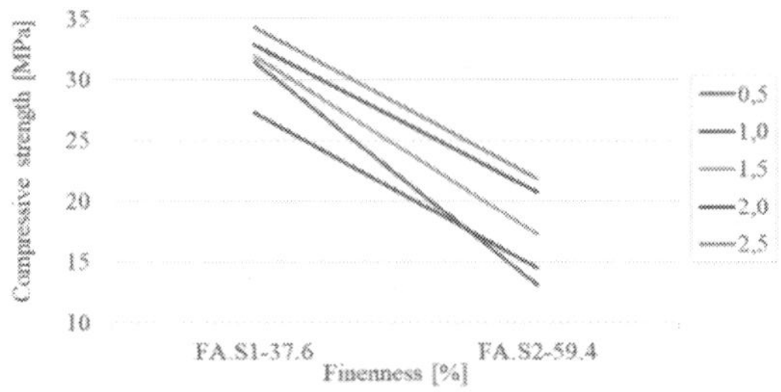

Fig. 6. Effect of fineness to compressive strength (12M NaOH solution – 0.5 to 2.5 alkaline activator ratio)

The influence of fly ash fineness (in terms of particles retained on the 45μm sieve) on the compressive strength is graphically emphasised in the figures above. The graphs indicate that the compressive strength decreases as the percentage of particles retained on the 45 μm sieve increases, therefore the fly ash sample with a lower fineness generates lower compressive strength values. A low percentage of particle fly ash retained on the sieve shows higher specific surface area, therefore a higher reactivity of the fly ash and subsequently, an increase in the compressive strength of the FAGPP.

The decrease in the amount of fly ash retained on the 45 μm sieve causes a decrease of the final pore size of the binder and also of the total porosity, with an increase of the density of the material and, therefore, a higher final compressive strength. This happens because fly ash is more reactive as smaller particles, compared to bigger ones, in its physical composition [33,34,35].

3.2 Influence of the Na_2SiO_3/NaOH ratio and the NaOH solution concentration on the compressive strength

The effect of sodium silicate solution to sodium hydroxide solution (Na_2SiO_3/NaOH ratio), by mass, on the compressive strength of the FAGPPs specimens has been observed by analysing it for each of the molar concentrations of sodium hydroxide solution and for each fly ash sample, FA.S1 and FA.S2. The results are graphically presented in Fig. 7 and Fig. 8, both proving that the best compressive values were obtained for the Na_2SiO_3/NaOH ratio of 2.5 (for both samples of fly ash). For the mixtures prepared using FA.S1, the ratios that induced lower compressive strength were generally generated by a lower geopolymerisation reaction during the mixing of the components. The compressive strength evolution strongly depends the complexity of the geopolymerisation reaction and the concentration of the alkaline activator constituents.

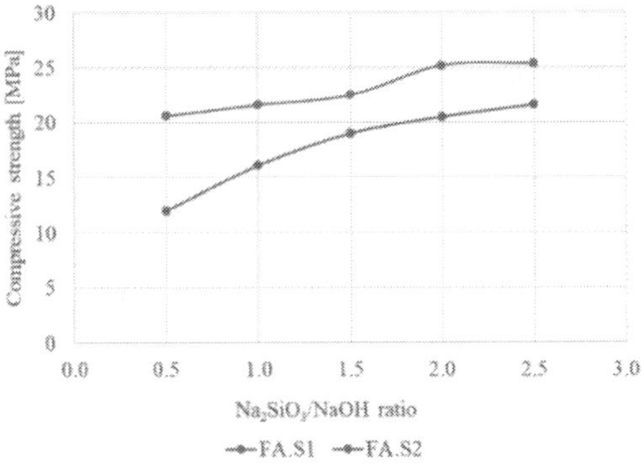

Fig. 7. Effect of Na_2SiO_3/NaOH ratio on the compressive strength (10M NaOH solution)

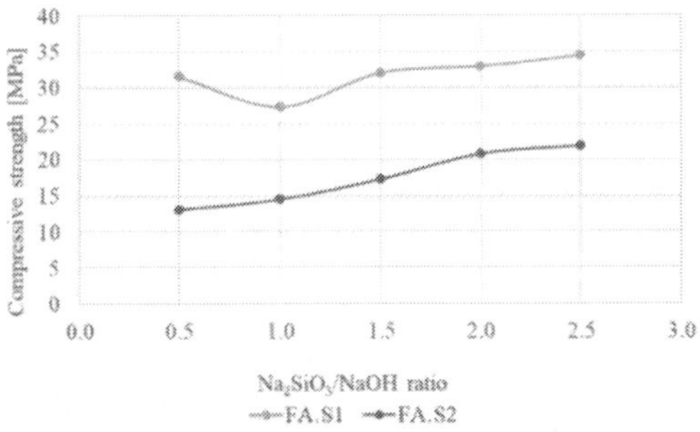

Fig. 8. Effect of Na_2SiO_3/NaOH ratio on the compressive strength (12M NaOH solution)

Fig. 7 and Fig. 8 also illustrate the effect of NaOH solution concentration on the compressive strength of the FAGPP specimens. The compressive strength of the geopolymer binder increases with the increase in the concentration of the NaOH solution and is also significantly affecting the microstructure of the alkali-activated geopolymer binder in terms of geopolymer synthesis [20]. A higher concentration of sodium hydroxide solution leads to a greater dissolution of precursor (fly ash), hence a higher compressive strength is achieved.

4 Summary and conclusions

The experimental procedures for obtaining FAGPP, using Romanian local raw materials, were performed in controlled laboratory conditions, regarding the temperature, the relative humidity, the pressure and the air flow velocity, in order to collect proper data that could be used for future studies in the topic. Careful investigation of specific ratio variations within the paste when using fly ash from different sources provided important conclusions regarding further approaches of the mix design of the geopolymer material. Optimal dosage of the geopolymer paste for superior compressive performance will be used in geopolymer mortar and concrete design.

The best results were provided by the alkali-activated FAGPP when using FA.S1, on mixes developed by using the 2.5 Na_2SiO_3/NaOH solution ratio and 10M NaOH concentration: 25.4 MPa, respectively 12M NaOH concentration: 34.4 MPa.

Fly ash FA.S1 provided both proper workability of the mixes and better compressive strength of the material. The compressive strength of the FAGPP increased with the decrease of the percentage of particles of the fly ash retained on the 45 μm sieve. The compressive strength of the geopolymer paste was influenced by both the physical and chemical characteristics of fly ash, however, the compressive strength was more affected by the fineness of the fly ash. Also, the compressive strength of the FAGPP samples increased with the increase of the molar concentration of the NaOH solution molarity.

The results obtained on FAGPP specimens contribute to the necessity of ongoing research on the identification and exploitation of the various Romanian sources of by-products, resulting both from secondary materials and wastes, from technological processes in various industries. Besides the economic advantages, an important aspect is the ecological one, namely the re-use of industrial wastes and by-products (sources of environmental pollution) into the economic circuit. Likewise, total replacement of cement in innovative, smart materials strongly contributes to the reduction of carbon dioxide emissions.

Further similar studies will be performed using different types of fly ash, from different Romanian power-plants, for a stronger confirmation of the present conclusion. The influence of fly ash chemical composition on geopolymer paste matrix, with respect to the evaluated NaOH concentration and the Na_2SiO_3/NaOH solution ratio and the optimization of the mix design, represent the near future challenges of the ongoing research.

Acknowledgements

The current experimental program was carried out in the framework of the PhD. Research Program of the first author at the Technical University of Cluj-Napoca. All the experimental work has been carried out in the laboratory of NIRD. URBAN-INCERC, Cluj-Napoca Branch, Romania and was supported by the Research Programmes for Smart Specialization, Sustainable Territorial Development, Environment Preservation and Resilience of Building Heritage – "CONCRET", PN 18 35 04 03: "Research for valorising the inert, hydraulic late or pozzolanic mineral additions in innovative cementitious materials for resilient structures, in the context of implementing the "Circular Economy" concepts in Romania and PN 19 33 04 02: "Sustainable solutions to ensure health and safety of the population in the open innovation context and environmental protection", financed by the Romanian Government.

References

1. V.M. Malhotra, *Role of supplementary cementing materials in reducing greenhouse gas emissions. Infrastructure regeneration and rehabilitation improving the quality of life through better construction: a vision for the next millennium*, Sheffield (1999)
2. International Energy Agency, *Carbon emissions reductions up to 2050*, Cement Technology Roadmap (2009)
3. CEMBUREAU: World Statistical Report Edition (2017)
4. J. Davidovits, *Synthesis of new high-temperature Geopolymers for reinforced plastics and composites*, SPE PACTE'79, Costa Mesa, California, Society of Plastics Engineers, USA (1979), pp. 151-154
5. J. Davidovits, *Chemistry of Geopolymeric Systems. Terminology*, International Conference on Geopolymer Concrete, France (1999)
6. J. Davidovits, *High-Alkali Cements for 21st Century Concretes*, in Concrete Technology, Past, Present and Future, Proceedings of V. Mohan Malhotra Symposium, Ed.: P. Kumar Mehta, ACI **144** (2003), pp. 383-387
7. A. Palomo, M.W. Grutzeck, M.T. Blanco, *Alkali-Activated Fly Ashes: A Cement for the Future*, Cem. Con. Res. **29** (1999) 8, pp. 1323-1329
8. P. Duxon, J.L. Provis, G.C. Luley, J.S.J. van Deventer, *The role of inorganic polymer technology in the development of 'green concrete'*, Cem. Con. Res. **37** (2007), pp. 1590-1597
9. NIRD URBAN-INCERC, *Research for valorising the inert, hydraulic late or pozzolanic mineral additions in innovative cementitious materials for resilient structures, in the context of implementing the 'circular economy' concepts in Romania*, National Programme 18 35 04 03, Ministry of Research and Innovation, Romania (2018)
10. A. Bilodeau, V.M. Malhotra, *High-volume fly ash system: Concrete solution for sustainable development*, ACI Mater. J. **97** (2000) 1, pp. 41-48
11. J.L. Provis, J.S.J. van Deventer, *Alkali Activated Materials*, State-of-the-Art Report, RILEM 224-AAN (2001)
12. J. van Jaarsveld, J. van Deventer, G. Lukey, *The characterisation of source materials in fly ash-based geopolymers*, Mater. Lett. **57** (2003), pp. 1272-1280
13. E.I. Diaz, E.N. Allouche, S. Eklund, *Factors affecting the suitability of fly ash as source material for geopolymers*, Fuel **89** (2010), pp 992-996
14. P. Chindaprasirt, U. Rattanasak, *Utilisation of blended fluidized bed combustion (FBC) ash and pulverized coal combustion (PCC) fly ash in geopolymer*, Waste Management **30** (2011), pp. 667-672
15. P. Chindaprasirt, U. Rattanasak, C. Jaturapitakkul, *Utilization of fly ash blends from pulverized coal and fluidized bed combustions in geopolymeric materials*, Cem. Con. Comp. **33** (2011), pp 55-60
16. ASRO, *SR EN 450-1:2012 Fly ash for concrete. Definition, specifications and conformity criteria*, Romanian Standards Association, Bucharest, Romania (2012)
17. ASRO, *SR EN 197-1:2011 Cement. Composition, specification and conformity criteria for common cements*, Romanian Standards Association, Bucharest, Romania (2011)
18. A. Lăzărescu, *Mix design and testing of the preliminary fly ash-based geopolymer mixes*, PhD. Report, Technical Univeristy of Cluj-Napoca, Romania (2017)
19. J. Davidovits, *Properties of geopolymer cements, Alkaline Cements and Concretes*, Kiew, Ukraine (1994), pp. 131-149
20. A. Lăzărescu, H. Szilagyi, C. Baeră, A. Ioani, *The Effect of Alkaline Activators Ratio on the Compressive Strength of Fly Ash-Based Geopolymer Paste*, IOP Conf. Series: Materials Science and Engineering (2017)
21. F. Pacheco-Torgal, J. Castro-Gomez, S. Jalali, *Alkali-activated binders: A review. Part 1. Historical background terminology, reaction mechanisms and hydration products*, J. Constr. Build. Mater. **22** (2008), pp. 1305-1314
22. K. Somna, C. Jaturapitakkul, P. Kajitvichyanukul, P. Chindaprasirt, *NaOH-activated ground fly ash geopolymer cured at ambient temperature*, Fuel **90** (2011), pp. 2118-2124
23. B.V. Rangan, *Mix design and production of fly ash based geopolymer concrete*, Ind. Conc. J. **82** (2008), pp 7-15
24. V.M. Malhotra, *Introduction: Sustainable Structures in the 21st Century*, ACI Concr. Int. **23** (2001) 3, pp. 57-63
25. V.F. Barbosa, K.J. Mackenzie, C. Thumaturgo, *Synthesis and characterisation of materials based on inorganic polymers of alumina and silica sodium polysialate polymers*, Int. J. **2** (2000), pp. 309-317

26. R. Anuradha, V. Sreevidya, R. Venekatasubramani, B.V. Rangan, *Modified guidelines for geopolymer concrete mix design using Indian standard,* AJCE **13** (2012) 3, pp. 353-364
27. B.V. Rangan, *Geopolymer Concrete for Environmental Protection,* Indian Concr. J., Special Issue – Future Concrete (2014), pp. 41-59
28. M. Sofi, J.S.J. van Deventer, P.A. Mendis, G.C. Lukey, *Engineering Properties of Inorganic Polymer Concretes (IPCs),* Cem. Concr. Res. **37** (2007) 2, pp. 251-257
29. H. Xu, J.S.J. van Deventer, *The Geopolymerisation of Alumino-Silicate Minerals,* Int. J. Miner. Process. **59** (2000) 3, pp. 247-266
30. ASTM, *Standard Test Methods for Sampling and Testing Fly Ash or Natural Pozzolans for Use in Portland-Cement Concrete, ASTM C618*, in Annual Book of ASTM Standard, West Conshohcken, USA (2004)
31. A.M. Fernandez-Jimenez, A. Palomo, *Characterisation of Fly Ash. Potential Reactivity as Alkaline Cements,* Fuel **82** (2003), pp. 2259-2265
32. C.-H. Huang, S.-K. Lin, C.-S. Chang, H.-J. Chen, *Mix proportions and mechanical properties of concrete containing very high-volume of Class F fly ash,* Constr. Build. Mater. **46** (2013), pp. 71-78
33. P. Chindaprasrt, C. Jaturapitakkul, T. Sinsiri, *Effect of Fly Ash Fineness on Compressive Strength and Pore Size of Blended Cement Paste,* Cement Concrete Comp. **27** (2005), pp. 425-428
34. A.-T. Mircea, *Study upon ApplyingvMulti-Criteria Analysis to the Selection of Building Materials*, 28th International Business-Information-Management-Association Conference, Seville, Spain **1-4** (2016), pp. 1354-1360
35. O. Corbu, A.M. Ioani, M.M. Al Bakri Abdullah, V. Meiţă, H. Szilagyi, A.V. Sandu, *The Pozzolanic Activity Level of Powder Waste Glass in Coomparisons With Other Powders,* International Conference on Innovative Research ICIR – EUROINVENT 2015 **660** (2015), pp. 237-243

MATEC Web of Conferences **289**, 11002 (2019) https://doi.org/10.1051/matecconf/201928911002
Concrete Solutions 2019

Development of sustainable concrete repair materials via microencapsulated agents

Chrysoula Litina[1*] and *Abir* Al-Tabbaa[1]

[1]Engineering Department, University of Cambridge, Trumpington Street, Cambridge CB2 1PZ, UK

Abstract Identification of non-structural damage in concrete infrastructure and actuation of preventive repair solutions is an established approach to avoid further structural damages and more expensive repair regimes. However the repair of concrete itself is not infallible with 55% of reported repairs in the EU failing within 5 years of service. Thus the already once repaired concrete structure is then subject to a constant cycle of repeated repair and a cumulative associated life cycle cost. The development of external repair material with self-healing capabilities, can affect a real step-change on the life-cycle costs and maintenance of existing and new infrastructure. Developed polymeric microcapsules containing liquid sodium silicate were used to impart autonomic self-healing to readily available commercial repair mortars for the first time. These materials cover a range of potential real time repair applications. Initially the compatibility between the developed self-healing agents and commercial products was established and the self-healing performance of the novel composite system was then evaluated. The results underlined the huge potential for the proposed composite systems as a stepping stone toward commercial uptake of self-healing microcapsule-based cementititious materials.

1 Introduction

Identification and correct diagnosis of the underlying cause of damage in concrete is the first step towards a reliable and durable repair strategy [1]. Although adequate design and selection of appropriate materials is a prerequisite, it is not the only limiting factor for robust repairs. To tackle this, most previous efforts have focused on improving the repair material performance and reducing or controlling the expected damage. These attempts can only result in marginal improvements as they do not address the underlying problem, the inherently brittle behaviour of concrete and its natural propensity to crack. Self-healing design can be a promising solution to prolong the functionality of a repaired section by imbuing repair materials with a built-in capability to repair inflicted damage.

Self-healing in concrete is an acknowledged intrinsic behaviour occurring autogenously through a combination of physical, chemical and mechanical processes [2]. Yet the phenomenon can be erratic and its efficiency limited since a number of factors need to be met for its occurrence. The phenomenon has been investigated extensively and different healing systems to enhance and benefit from this behavior have been put forward over the last twenty years; including expansive mineral additives, super absorbent polymers, microcapsules, bacterial carbonate precipitation and microfibers [3–7]. Amongst these technologies the encapsulation approach, namely the embedment of repair agents in vessels within the matrix to be released at the location and time of damage and actuate repair on demand is most promising.

A particular category of encapsulation is microencapsulation. The principle of self-healing through microencapsulation is that microcapsules are embedded in the bulk of the host matrix during manufacture and trigger the release of their healing compound upon rupture of their shell [8,9]. The latter is initiated by stress built-up during crack formation. Chemical interactions between the released core material(s) and the surrounding matrix heal the crack restoring partially or fully the bulk material properties. Such a system that does not require external trigger, poses no restriction to normal casting and can address multiple simultaneous cracking offers the greatest flexibility for field applications and direct incorporation in ubiquitous repair materials.

More recently collaborative research at the University of Cambridge [10] led to the development of polymeric microcapsules that contain a liquid sodium silicate core. The healing potential of the microcapsules under different cracking regimes was confirmed [11] and the effect of these healing additives was investigated on both fresh (viscosity - setting time) and hardened properties (modulus of elasticity, compressive and flexural strengths) [12,13]. The potential of the developed microcapsule-based concrete material was demonstrated in new concrete elements with full-scale field trials [14,15] but also assessed for use in existing infrastructure as a repair material.

This paper focuses on the latter reporting on the development of microcapsule based self-healing concrete

* Corresponding author: cl519@cam.ac.uk

materials for reliable concrete repairs. Different commercially available products covering a range of potential real time repair applications were targeted; including materials for cementitious hand-placed, poured and sprayed repairs, structural and non-structural. An initial objective was to establish compatibility between the developed healing-promoting microcapsules and the commercial products. Subsequently the potential of the novel composite system for self-healing applications was assessed and established.

2 Experimental Methodology

1. Materials and mixture proportions

Five repair mortars were selected for this project provided by Fosroc Ltd; a high build repair mortar (HBRM), a dry spray repair mortar (DSRM), a low alkali repair mortar (LARM), a patching repair mortar (PRM) and a self-levelling underlayer repair mortar (SLRM). All materials were provided as single component, packed products that only required the addition of a predetermined amount of water. For comparison, an overview of the compositions, in terms of their Portland cement blends, is presented in Fig. 1.

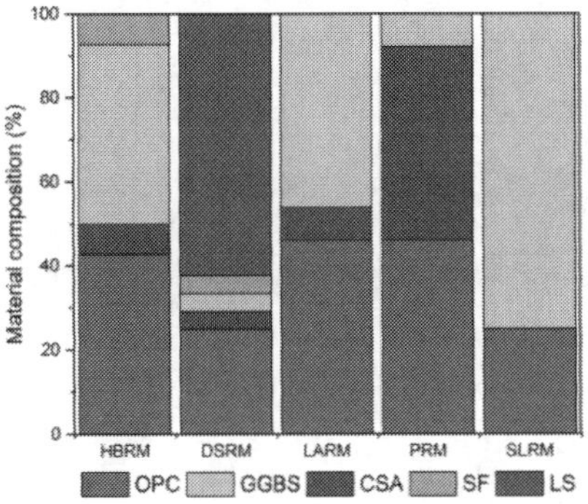

Fig. 1. Relative range of cementitious material composition of commercial repair mortars considered in this investigation as provided by the manufacturer. LS: limestone, SF: silica fume, GGBS: ground granulated blast furnace slag, CSA: calcium sulphoaluminate cement, OPC: ordinary Portland cement.

The self-healing additives used in this study were microcapsules developed previously by Lambson in collaboration with the University of Cambridge [10]. The microcapsules have a polymeric shell made of cross-linked gelatine and gum arabic and their cargo material consists of a solution of mineral oil and liquid sodium silicate (40% weight in H_2O) and they have an average size of 290 μm and a density of ~1.0 g/cm^3. Images of the microcapsules are shown in Fig. 2. Sodium silicate was selected, as the healing compound, because it can form products of similar characteristics to the host cementitious matrix. The potential for aqueous sodium silicate as a healing agent in cement-based materials has previously been illustrated both in terms of crack closure and durability [13,16–20].

Fig. 2. Microscope images of microcapsules used in this study.

Considering previous results of microcapsule content on fresh and hardened material properties and in terms of healing potential [11,13], three different microcapsule contents were proposed within the so far investigated range as part of the compatibility testing; 8%, 16% and 32% by volume of cement.

2. Experimental programme

Fresh properties such as setting time and workability were tested to ensure compatibility and suitability for practical applications, while mechanical properties were tested in terms of compressive and flexural strengths. These were determined to be the more critical elements for compliance with BS EN 1504. Tests were conducted at 1, 7 and 28 days for all mixes.

Optimum self-healing additive fractions were determined and further investigation on the in situ self-repair potential of these novel materials was carried out. These tests included the evaluation of both mechanical integrity and durability recovery potential through three-point bending and sorptivity testing. Seven and 28 days of initial cracking were selected to establish the healing efficiency of the composite material. Cracked samples were placed in high relative humidity (~68%) containers to allow healing to take place for 28 days. The latter condition was selected as both the most representative for real-time field applications and for conservative self-healing estimation. To monitor crack initiation, opening and closure of the crack faces in all the specimens were monitored over time using a stereo microscope. A Leica S6D digital stereoscope was used. Digital images were captured at three different positions on the crack face. The cracks were photographed on the day of the cracking and at the end of the healing period at the exact same locations. Image analysis software (Image J) was then used to analyse the acquired captions and the average crack opening was calculated for each case. The values obtained from the image analysis were used to quantify the crack mouth healing percentage (CMH) in each case as shown in Equation 2:

$$CMH(\%) = \frac{w_i - w_h}{w_i} \text{x} 100\% \qquad (1)$$

Where w_i is the initial crack width and w_h is the healed crack width. To complement the macroscopic observations of the crack healing an Proceq ultrasonic pulse velocity test instrument was used for the determination of the evolution of the crack depth with healing time for each prismatic sample following the guidelines in BS-1881: Part 203. Probes were both placed on either side of the crack at a certain distance x from the crack for the first measurement and 2x from the crack for the second measurement. The time required for the ultrasonic wave to travel between these two points was recorded and the crack depth *d* was calculated as follows:

$$d = x\sqrt{\frac{4t_1^2 - t_2^2}{t_2^2 - t_1^2}} \qquad (2)$$

Where x is the distance between the probe location and the crack (30 mm in this case) and t_1 and t_2 is the time taken for the ultrasonic wave to travel through the concrete for the first and second measurement respectively. Tables 1 and 2 provides an overview of the investigated properties during both preliminary compatibility and self-healing performance stages.

Table 1: The investigated compatibility properties and testing protocols using microcapsule dosages of 0, 8, 16 and 32% by volume.

Properties		**Testing protocols**	**Standard**
Mechanical properties	Compressive strength	40 x 40 x 40 mm at 1, 7 and 28 days	EN 196-1
	Flexural strength	40 x 40 x 160 mm at 1, 7 and 28 days	EN 196-1
Workability	Flowtable		EN 1015-3

*In all experimental procedures triplicate samples were used

Table 2: The investigated self-healing properties and testing protocols.

Properties		**Testing protocols**
Mechanical properties	Flexural strength recovery	Controlled crack width at maximum flexural loading capacity, @7 and 28 days (28 days healing)
Sealing	Liquid capillary absorption coefficient	Controlled crack width at maximum flexural loading capacity, @7 and 28 days (28 days healing)
	Ultrasonic wave transmission (crack depth)	@7,14 and 28 days (28 days healing)
Macroscopic observations	Crack area measurements	Optical microscopy @7,14 and 28 days (28 days healing)
Microscopic observations	Scanning electron microscopy	

*In all experimental procedures triplicate samples were used

3. Results and Discussion

3.1 Compatibility with commercial repair mortars

The inclusion of microcapsules into the mix did have an effect on the viscosity of the mix, in some cases substantial (Fig. 3). PRM and LARM appeared more sensitive to the addition of the healing agent to their composition. That effect became more pronounced as the microcapsule dosage was increased. Then the maximum increase in viscosity when compared to the control mix was more than 20%. This affected the level and quality of compaction and hence the performance of the hardened material as confirmed by compressive strength results. However, the compressive strength did not show a consistent decrease with increasing concentration of microcapsules (Fig. 4). Although there was an observed loss of strength at 1 day, the 28 days' values for compressive strength for samples with microcapsules were for some repair materials higher than the control mixes prepared without self-healing additives. HBRM, DSRM and SLRM seemed to benefit from the presence of the microcapsules with higher strength achieved – compared to the control- at 28 days for 8% microcapsule addition. In fact DSRM exhibited comparable strength to the reference specimen even with 32% by volume fraction addition. For these three materials the late development of strength compensates for the lower early age strength.

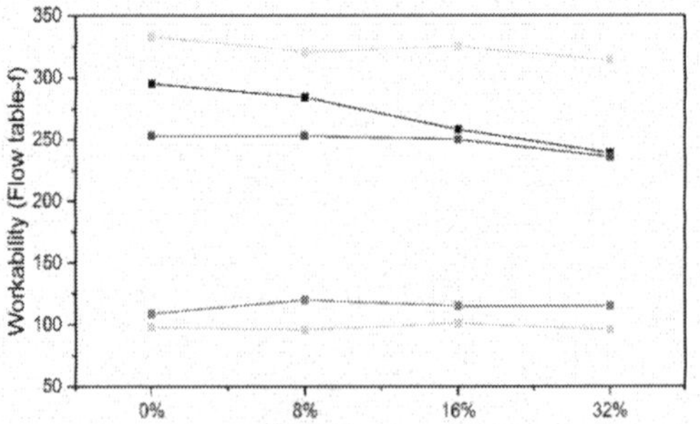

Fig. 3: Workability (flowtable results) variation with increasing loading percentage of microcapsules with respect to cement volume.

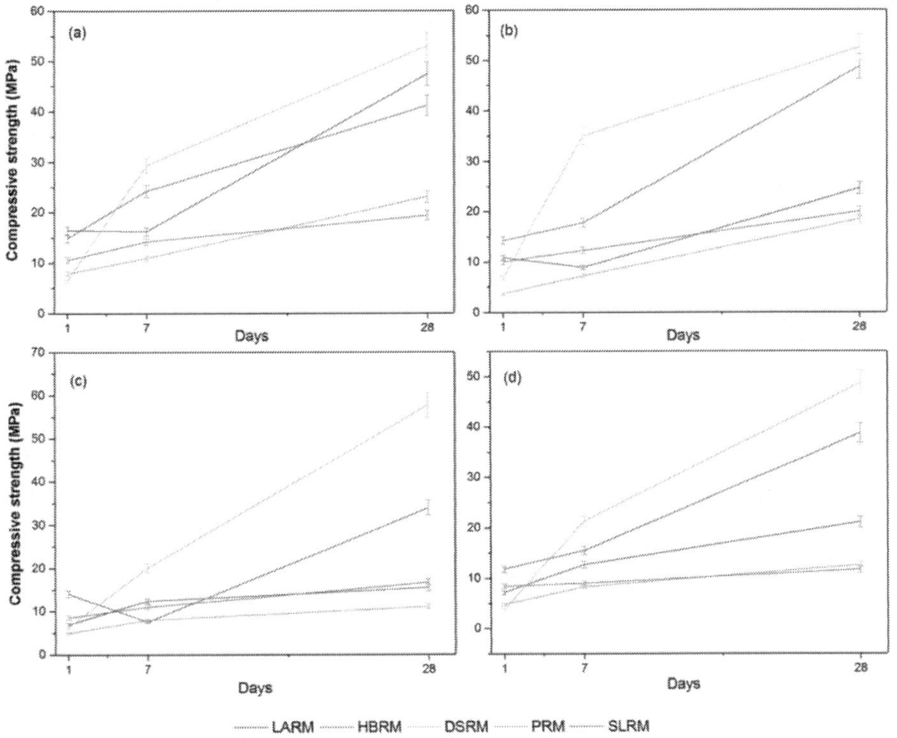

Fig. 4: Compressive strength variation at 1,7 and 28 days with increasing volume fraction (Vf) of microcapsules; (a) Control, (b) 8%, (c) 16% and (d) 32% for LARM, HBRM, DSRM, PRM and SLRM.

3.2 Base-line healing ability

These initial compatibility results suggested three materials; HBRM, DSRM and SLRM and two microcapsule contents 8% and 16% by volume as most suitable for further exploration of the self-healing/self-repairing ability of the composite materials. The intrinsic self-healing performance of the repair materials as measured by crack mouth healing and strength recovery helped further establish these repair materials as the ones with the greatest potential for the microcapsule enhancement studies. Indicative results for strength recovery are presented in Fig. 5. The results at 7 and 28 days showed HBRM > DSRM > SLRM > PRM > LA in order of the highest recovery potential. Macroscopic results of the crack mouth healing confirmed the mechanical recovery results as presented in Fig. 6.

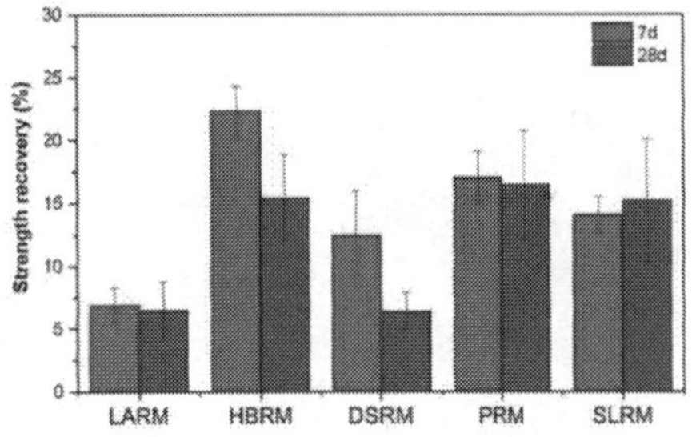

Fig. 5: Measured strength regain for different investigated materials according to initial age of cracking.

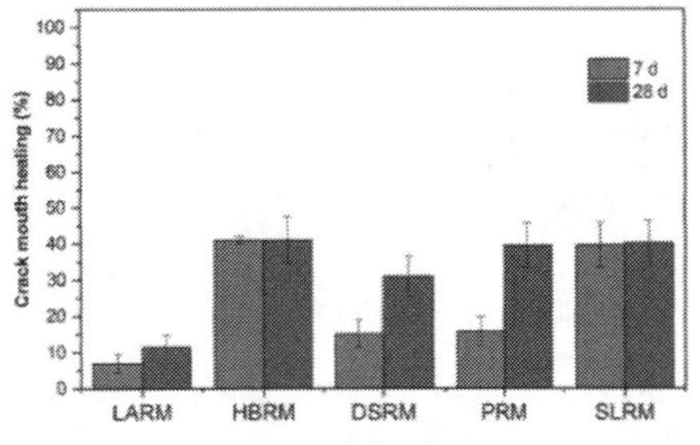

Fig. 6: Crack mouth healing for different investigated materials according to initial age of cracking.

3.3 Improved healing ability and durability restoration

The inclusion of microcapsules improved the crack closure and reduced the water absorption significantly compared to the cracked specimen. However larger volume fractions (16%) were found to have slightly increased sorptivity coefficients with relatively large deviation in the results (Fig. 7). Yet a significant improvement in flexural strength was achieved for 2 of the investigated materials (Fig. 8a), DSRM and SLRM even at the lowest investigated microcapsule content (8% Vf). This improvement outweighed the initial loss of flexural performance due to the addition of the microcapsules. It is implied that the intrinsic self-healing mechanism of the repair material itself is enhanced by the further interaction of the sodium silicate core released at the crack plane. The aforementioned results correlate with the macroscopic observation for crack mouth healing (Fig. 8b).

HBRM sees a total crack mouth healing value fluctuating around 45%, marginally impaired by the higher microcapsule fraction as opposed to DSRM and SLRM. These show an improved crack closing potential linked to an increase in microcapsule concentration, approaching 75% and 65% for DSRM and SLRM, respectively, for 16% by volume microcapsule addition. The most pronounced effect can be observed on DSRM; where the intrinsic self-healing is improved almost four times as the microcapsule content doubles. It is interesting to see whether this macroscopic external crack mouth reduction links to a reduction in depth of the crack (Fig. 9). These measurements were performed at regular time intervals to monitor the progress of healing as indicated in Table 2. Results indicated that the addition of microcapsules increased the crack depth recovery; However, that improvement was only marginal. The greatest benefit seems to be in the reduction of crack opening and thus durability improvement rather than mechanical recovery.

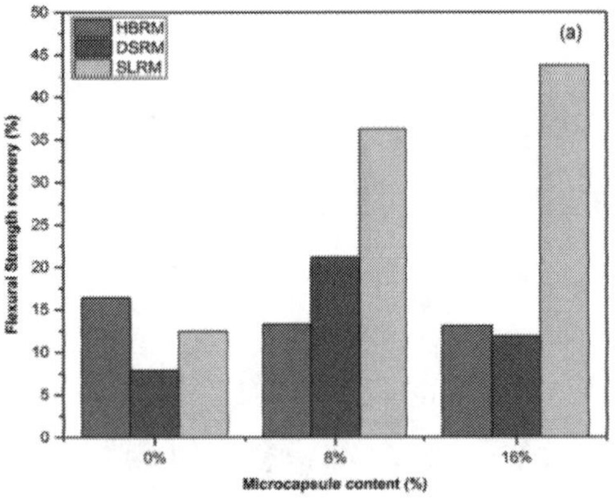

Fig 8 (a)

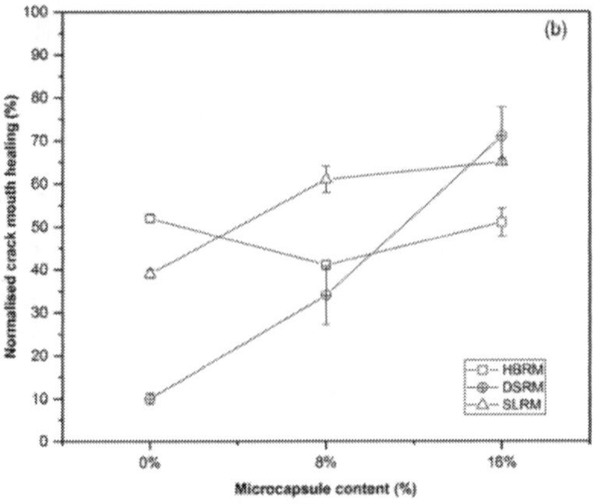

(b)

Fig. 8: (a) Measured strength regain for different investigated materials according microcapsule content; (b) Normalised crack healing achievable after 28 days healing period for samples with microcapsules cracked at 7 days.

4 Conclusions

The implementation of self-healing techniques as part of repair methodologies in extending the service life of concrete structures has great relevance as it can guarantee direct and timely ratification in the construction industry. The experimental results proved the compatibility of the proposed composite cement-based repair materials in terms of fresh properties and mechanical strength. Optimum self-healing additive fractions were selected for three end applications and further investigation on the in situ self-repair potential of these novel composite materials was carried through. The results showed improved crack-width and crack-depth reduction and recovery in permeability. The technical and practical considerations encountered during the project reported herein will assist the future development and accelerated commercialisation of targeted self-healing concrete applications using microencapsulated healing agents.

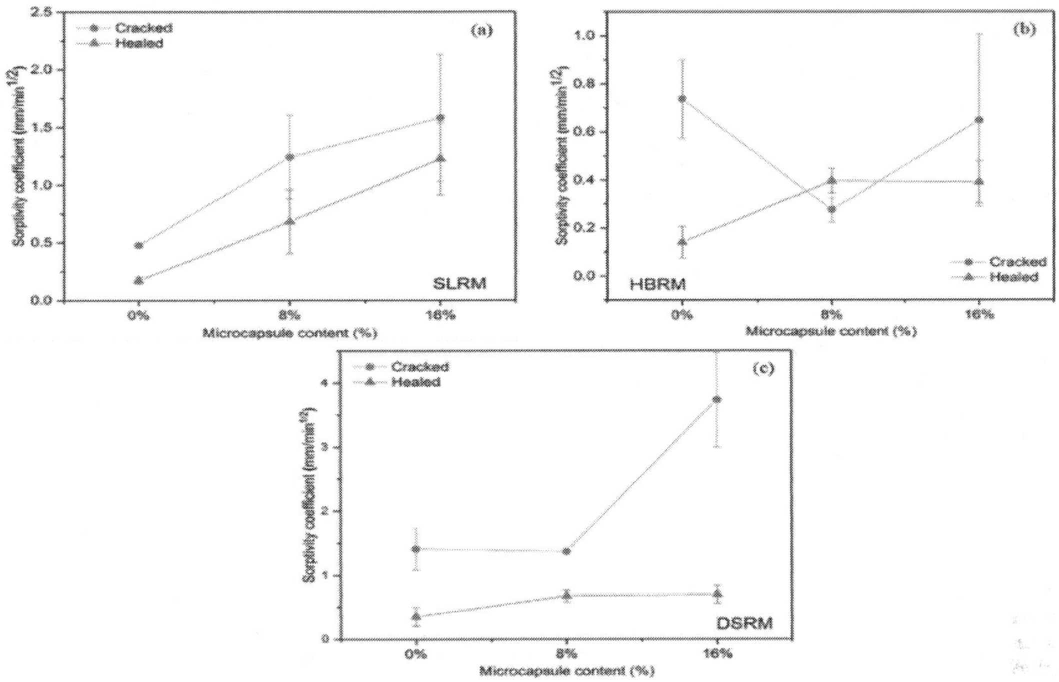

Fig. 7: Measured sorptivity coefficients for cracked, and healed samples; (a) SLRM, (b) HBRM and (c) DSRM.

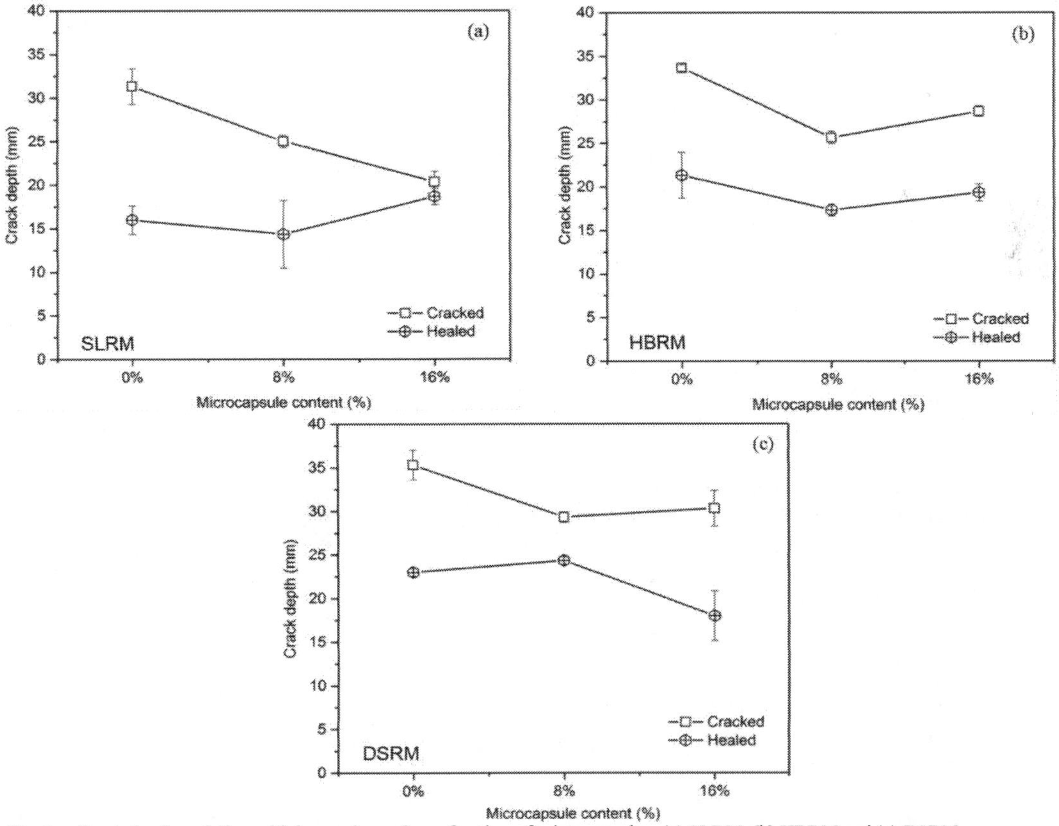

Fig. 9: : Crack depth variation with increasing volume fraction of microcapsules; (a) SLRM, (b) HBRM and (c) DSRM.

Acknowledgements

Financial support from the UK Engineering and Physical Sciences Research Council (EPSRC-UK) for this study (Grant Ref. NMZJ/137, RG81609) is gratefully acknowledged.

References

1. A. Garbacz, L. Courard, and B. Bissonnette, Bull. Polish Acad. Sci. Tech. Sci. **61**, (2013)
2. M. de Rooij, K. Van Tittelboom, N. De Belie, and E. Schlangen, *Self-Healing Phenomena in Cement-Based Materials State-of-the-Art Report of RILEM Technical Committee 221-SHC: Self-Healing Phenomena in Cement-Based Materials* (RILEM 2013, 2013)
3. D. Snoeck, K. Van Tittelboom, S. Steuperaert, P. Dubruel, and N. De Belie, J. Intell. Mater. Syst. Struct. **25**, 13 (2012)
4. K. Van Tittelboom, N. De Belie, W. De Muynck, and W. Verstraete, Cem. Concr. Res. **40**, 157 (2010)
5. K. Van Tittelboom and N. De Belie, Materials (Basel). **6**, 2182 (2013)
6. T.-H. Ahn and T. Kishi, J. Adv. Concr. Technol. **8**, 171 (2010)
7. K. Sisomphon, O. Copuroglu, and E. A. B. Koenders, Cem. Concr. Compos. **34**, 566 (2012)
8. S. K. Ghosh, in *Self-Healing Mater. Fundam. Des. Strateg. Appl.*, edited by S. K. Ghosh (WILEY-VCH Verlag GmbH & Co. KGaA, 2009), pp. 1–25
9. J. P. Youngblood and N. R. Sottos, MRS Bull. **33**, 732 (2008)
10. A. Kanellopoulos, P. Giannaros, D. Palmer, A. Kerr, and A. Al-Tabbaa, Smart Mater. Struct. **26**, 045025 (2017)
11. P. Giannaros, A. Kanellopoulos, and A. Al-Tabbaa, Smart Mater. Struct. **25**, 084005 (2016)
12. C. Litina and A. Al-Tabbaa, in *5th Int. Conf. Self-Healing Mater.* (Durham, USA, 2015)
13. A. Kanellopoulos, P. Giannaros, and A. Al-Tabbaa, Constr. Build. Mater. **122**, 577 (2016)
14. A. Al-Tabbaa, C. Litina, P. Giannaros, A. Kanellopoulos, and L. Souza, Constr. Build. Mater. **208**, 669 (2019)
15. O. Teall, R. Davies, M. Pilegis, A. Kanellopoulos, T. Sharma, K. Paine, A. Jefferson, R. Lark, D. Gardner, and A. Al-Tabbaa, in *Proc. 11th Fib Int. PhD Symp. Civ. Eng.*, edited by K. Maekawa, A. Kasuga, and J. Yamazaki (Tokyo, Japan, 2016), pp. 639–646
16. A. Kanellopoulos, T. S. Qureshi, and A. Al-Tabbaa, Constr. Build. Mater. **98**, 780 (2015)
17. M. Pelletier, R. Brown, A. Shukla, and A. Bose, *Self-Healing Concrete with a Microencapsulated Healing Agent* (Kingston, USA, 2011)
18. J. Gilford, M. M. Hassan, T. Rupnow, M. Barbato, A. Okeil, and S. Asadi, J. Mater. Civ. Eng. 886 (2013)
19. E. Mostavi, S. Asadi, M. M. Hassan, and M. Alansari, J. Mater. Civ. Eng. **27**, 04015035 (2015)
20. R. Alghamri, A. Kanellopoulos, and A. Al-Tabbaa, Constr. Build. Mater. **124**, 910 (2016)

AUTHOR INDEX

Akhundzada Hikmatullah 186
AL-Ameeri Abbas S. 301
Albkeirat Mohammad 186
Almuhaisen Refad 355
Alt Philip Henes 63
Al-Tabbaa Abir 456
Aminu Alhassan Yunusa 43
Annede-Villeau Camille 107
Apeh Sunday 43
Arai Jun-ichi 431
Arato Gonçalves Ana Paula 285
Araya Eleni B. 372
Asan Muhammet 261
Assmann Alessander 57
Baerã Cornelia 1, 215, 245, 449
Bai Yun 279
Bakalli Mirdash 238
Basheer Muhammed 279
Berkowski Piotr 376, 392
Bhuiyan Shamir 89, 113
Blaga Sergiu 255
Boinski Frédéric 93
Bouichou Myriam 267, 285, 414
Bouteiller Véronique 93
Brueckner Rene 385
Caspeele Robby 347
Chatzopoulos A. 367
Cheaitani Atef 100
Cheytani Samir 100
Christodoulou Christian 168
Copuroðlu Oðuzhan 437
Corbett David 274
Crémona Christian 267
Crevello Gina 289, 324
Dãbãcan Ovidiu 255
Damien Guillaume 93
Dan Zeno 238, 400
Davies Robert 25
Davison Nigel 161, 168
De Belie Nele 17, 84, 347
Debraekeleer Mats 84
Dejager Elien 84
Dib Youssef 340
Djama Khaled 405
Dmochowski Grzegorz 376, 392
Donadio Michel 168, 238, 400
Donchev Ted 186, 210, 355, 360
Ebell G. 148
Eissmann Diana 57
Ekolu Stephen 372
Ersan Yusuf Cagatay 9
Erþan Yusuf Caðatay 17
Faur Andrei 191
Ferrara Liberato 35
Ferrier Emmanuel 405
Fisher Gregor 202
Freeman Brubeck 25
Gabor Aron 405
Gardener Georg 77, 333
Gaudette Paul 296
Georgopoulos Costas 210
Ghafoori Nader 49, 53
Giorgini Roberto 107
Glass Gareth 161
Goodwin Fred 232
Grantham Michael 274
Grebeni°an Elvira 245
Gruyaert Elke 84
Guades Ernesto 202
Gulikers Joost 310, 318
Gunes Oguz 261
Hallopeau Xavier 107
Harrer Ann 296
Hasnat Ariful 49, 53
Heghes Bogdan 425
Hegyi Andreea 1, 215, 245
Heider Catrina 400
Hopartean George 210
Huang I-Wen (Evan) 232
Imran Rafiq M. 301
Jacint Virág 191
Jefferson Anthony 25
Karatosun Serkan 261
Kawaai Keiyu 29
Kesikidou Fotini 63
Khadra Mahdi 267
Law David 89, 113
Lãzãrescu Adrian 1, 449
Lesieutre Olivier 107
Limbachiya Mukesh 210, 355
Litina Chrysoula 456
Liu Wei 444
Long Adrian 279
Luo Ye 444
Macdonald Susan 285
Maler Matthew O. 49, 53
Manita P. 367
Marie-Victoire Elisabeth 267, 414
Marie-Victoire Élisabeth 285
Matei Claudiu 1
Matteini Irene 289, 324
Mazanec Oliver 57
Meillier Arnaud 107
Mezquida-Alcaraz Eduardo J. 35
Michaux Christophe 93
Michel Laurent 405
Mircea Anamaria 215
Mircea Anamaria Cãtãlina 1

AUTHOR INDEX

Mircea Andreea-Terezia....421
Mircea Călin....215, 245, 449
Mircea Călin GR....255
Mircea Diana-Maria....251
Misak Lubos....274
Mizobuchi Toshiaki....431
Morales Cruz Cynthia....222
Moucadeau Eugène....107
Najimi Meysam....49, 53
Nanukuttan Sreejith....279
Negrini Alberto....35
Negrutiu Camelia....425
Ni Kun....444
Nicholls Peter....113
Nijland Timo....437
Noyce Paul....289, 324
Ohashi Kayo....431
Papayianni Ioanna....63
Parnada Noel....210
Paulini Peter....69
Pavloviæ Ana....355
Peelen Willy....120
Petkova Diana....186, 210, 355, 360
Pichlhöfer A....148
Polder Rob....120, 437
Rahman Hamid....360
Rathod Nikita....153
Raupach Michael....222
Roberts Adrian....161
Roig-Flores Marta....35
Rossi Emanuele....437
Saliba Justin....89
Šavija Branko....437
Schmidt Jacob....202
Schneck Ulrich....130, 138
Schneider Julien....93
Schneider Martin....77, 333
Schwarz W....148
Selvarajoo Tharmesh....25
Semaan Nabil....340
Sergi George....153
Serna Pedro....35
Seveviratne Gamini....153
Seyhan Engin Cueneyt....232
Shi Yunxing....444
Sideris Kosmas K....367
Simpson David....153
Slater Peter....153
Solomon Fitsum....372
Sonmez Merve....9
Sosa Ioan....425
Srebrenkoska Vineta....186
Stang Henrik....202
Stone Christian....161, 168
Sugitani Saaya....29
Szilagyi Henriette....1, 245, 449
Szilágyi Henriette....215
Tang Luping....126
Tassos Ch....367
Tsioulou Ourania....301
Ujike Isao....29
Van Belleghem Bjorn....17
van den Hondel A.J. (Hans) MSc....177
Van Mullem Tim....347
Van Tittelboom Kim....17, 84
Vildaer Stéphanie....267
Völker T....148
Ward Liam....89
Wilsch Gerd....148
Wood Chris....285
Yang Kai....279
Yazdi Mohammad Ali....84
Zhang Emma Qingnan....126
Zhang Yangang....444